HANDBOOK OF RESEARCH IN MARINE PHARMACEUTICS

Exploring Oceanic Microbial Diversity for Human Health and Wellness

HANDBOOK OF RESEARCH IN MARINE PHARMACEUTICS

Exploring Oceanic Microbial Diversity for Human Health and Wellness

Edited by

Md. Faiyazuddin, PhD

Khalid Rehman Hakeem, PhD

AAP | APPLE ACADEMIC PRESS

First edition published 2025

Apple Academic Press Inc.
1265 Goldenrod Circle, NE,
Palm Bay, FL 32905 USA

760 Laurentian Drive, Unit 19,
Burlington, ON L7N 0A4, CANADA

CRC Press
2385 NW Executive Center Drive,
Suite 320, Boca Raton FL 33431

4 Park Square, Milton Park,
Abingdon, Oxon, OX14 4RN UK

Library and Archives Canada Cataloguing in Publication

Title: Handbook of research in marine pharmaceutics : exploring oceanic microbial diversity for human health and wellness / edited by Md. Faiyazuddin, PhD, Khalid Rehman Hakeem, PhD.
Names: Faiyazuddin, Md., editor. | Hakeem, Khalid Rehman, editor
Description: First edition. | Includes bibliographical references and index.
Identifiers: Canadiana (print) 20240454928 | Canadiana (ebook) 20240454936 | ISBN 9781774917428 (hardcover) | ISBN 9781774917435 (softcover) | ISBN 9781003537403 (ebook)
Subjects: LCSH: Marine pharmacology.
Classification: LCC RS160.7 .H36 2025 | DDC 615.3/29—dc23

Library of Congress Cataloging-in-Publication Data

CIP data on file with US Library of Congress

ISBN: 978-1-77491-742-8 (hbk)
ISBN: 978-1-77491-743-5 (pbk)
ISBN: 978-1-00353-740-3 (ebk)

DEDICATION

I dedicate this book with immense love and admiration to my living legendary beloved parents, Sabiha Khatoon and Md. Moinuddin Ahmad. Your unwavering support, wisdom, and unconditional love have been the bedrock of my journey. Your guidance has shaped my character, and your sacrifices have fuelled my aspirations. This book stands as a testament to the values and lessons you've instilled in me. Thank you for being my constant inspiration and for believing in me even during the toughest times.

—Prof. (Dr.) Md. Faiyazuddin, PhD

This book is dedicated to my beloved mother, Hajrah Begum, who left this world so early. To me, she was the whole world. She was the most loving, humble, compassionate, understanding, family-oriented woman. I am going to miss you so much, Mom. I pray to Allah (SWT) to bless your soul and grant you Jannatul-Firdous (the highest level of heaven).

"O Allah, forgive and have mercy upon her, excuse her and pardon her, and make honorable her reception. Expand her entry, cleanse her with water, snow, and ice, and purify her of sin as a white robe is purified of filth. Exchange her home for a better home. Admit her into the Garden; protect her from the punishment of the grave and the torment of the Fire."

—Prof. (Dr.) Khalid Rehman Hakeem, PhD

About the Editors

Md. Faiyazuddin, PhD
Professor of Pharmaceutics, and Founder Dean R&D, Al-Karim University, School of Pharmacy, Katihar, Bihar, India

Md. Faiyazuddin, PhD, is a distinguished academician and researcher with extensive experience in working across government, public, and private sectors. He has 17 years of experience in pharmacy teaching, drug development, research consultation, and social pharmacy outreach activities in academia and industry. Renowned for his prolific contributions to the field, Professor Faiyazuddin is an avid researcher, delving into various facets of formulation science. Currently, he is a Professor of Pharmaceutics and Founder Dean Research & Development at Al-Karim University, School of Pharmacy, Katihar, Bihar, India. Before this, he was the Dean and Professor of Pharmaceutics at The R. C. University, India (2019–2021); Principal and Professor of Pharmaceutics/ Drug Delivery at Bihar College of Pharmacy (2016–2019); Assistant Professor and Quality coordinator at the University of Hail, KSA (2014–2016); and Assistant Professor and Chief Investigator Nano Drug Delivery Research Laboratory at the Integral University, India (2007–2014). He is a visiting scientist and adjunct professor at reputed universities at the national as well as international level, sharing outstanding ideas and cutting-edge technologies with laureates in health and the pharmaceutical sectors. He was awarded a prestigious DST Young Scientist Award by the Govt. of India to pursue research in cancer nanotechnology. Holding a PhD in Pharmaceutics from Jamia Hamdard, New Delhi, India, Dr. Faiyazuddin specializes in pharmaceutical nanotechnology and earned an executive strategic leadership management degree at the Jacobs-Abbey Global Institute for Leadership Studies, Woodbridge, Virginia, USA. He has extensive research experience in pharmaceutical nanotechnology, advanced drug delivery, marine bioactive, and recently pharmaceutical AI and probiotics medicines. Moreover, he lends his expertise to the editorial boards of more than 10 esteemed international pharmacy journals, including *Journal of Interdisciplinary Nanomedicine* (Wiley), *Current Indian Science* (Bentham), *International Journal of*

Pharmacy Teaching & Practices (Hindawi), *Pharmaceuticals* (MDPI), and *Pharmacologia*. This is reflected by his authorship of more than 115 published papers (in ACS, Elsevier, Wiley, Bentham, RSC, and Frontier Q1 journals), nine issued patents and patent applications, 12 books, 48 book chapters, eight awards, seven fellowships, member of editorial boards conference chair, and more than 50 invited conference presentations. Dr. Faiyazuddin is a member of many prominent professional organizations, including the Institute of Nanotechnology, Sterling University, Scotland; the British Society of Nanomedicine; the University of Liverpool UK; the International Society of Aerosol in Medicine; and the American Lung Association, USA. Dr Faiyazuddin has secured multiple major federal and non-federal research grants and completed research grants from various R&D agencies, including SERB-DST and foreign agencies. Outside his day job, Dr. Faiyazuddin has a keen interest in connecting worldwide researchers to practice safe drugs under the flagship of the International Pharmaceutical Forum.

Khalid Rehman Hakeem, PhD

Professor, King Abdulaziz University, Jeddah, Saudi Arabia

Khalid Rehman Hakeem, PhD, is a Professor at King Abdulaziz University in Jeddah, Saudi Arabia. After completing his doctorate in Botany with a specialization in Plant Eco-physiology and Molecular Biology from Jamia Hamdard in New Delhi, India, in 2011, he briefly worked as a lecturer at the University of Kashmir in Srinagar, India. He later joined the Universiti Putra Malaysia in Selangor, Malaysia, working as a Doctorate Fellow in 2012 and as a Fellow Researcher (Associate Professor) from 2013 to 2016. Dr. Hakeem has over 10 years of teaching and research experience in plant eco-physiology, biotechnology, molecular biology, medicinal plant research, plant-microbe-soil interactions, and environmental studies. He has received several fellowships at both national and international levels and has served as a visiting scientist at Jinan University in Guangzhou, China. Prof. Hakeem is a Fellow of the Royal Society of Biology in the UK. Currently, he is involved in various international research projects with different government organizations. Dr. Hakeem has authored and edited more than 80 books with international publishers, including Springer Nature, Academic Press (Elsevier), CRC Press, and Apple Academic Press. He has also published over 185 research articles in peer-reviewed international journals and 80 book chapters in edited volumes with international publishers. Dr. Hakeem serves as an editorial board member and reviewer for several high-impact international scientific journals published by Elsevier, Springer Nature, Taylor and Francis, Cambridge, and John Wiley. He is on the advisory board of Cambridge Scholars Publishing in the UK. Additionally, he is a fellow of the Plantae Group of the American Society of Plant Biologists, a member of the World Academy of Sciences, a member of the International Society for Development and Sustainability in Japan, and a member of the Asian Federation of Biotechnology in Korea. Dr. Hakeem has been listed in *Marquis Who's Who in the World* from 2014 to 2020. Currently, he is focused on studying plant processes at eco-physiological and molecular levels.

Contents

Contributors

Mohit Agrawal
School of Medical and Allied Sciences, K.R. Mangalam University, Gurugram, Haryana, India

Usama Ahmad
Associate Professor, Department of Pharmaceutics, Faculty of Pharmacy, Integral University, Lucknow, Uttar Pradesh, India

Wasim Ahmad
Department of Pharmacy, Mohammed Al-Mana College for Medical Sciences, Dammam, Saudi Arabia

Bushra Ahmed
Department of Pharmaceutics, School of Pharmaceutical Education and Research (SPER), Jamia Hamdard, New Delhi, India

Mohammed Gulzar Ahmed
Department of Pharmaceutics, Yenepoya Pharmacy College and Research Center, Yenepoya (Deemed to be University), Mangalore, Karnataka, India

Asad Ali
Faculty of Pharmacy, Integral University, Lucknow, Uttar Pradesh, India

Shaukat Ali
School of IT and Management, Al-Karim University, Katihar, Bihar, India

Sumel Ashique
Assistant Professor, School of Pharmaceutical Sciences, Lovely Professional University, Phagwara, Punjab, India

Kamran Ashraf
Faculty of Pharmacy, Universiti Teknologi MARA (UiTM), Cawangan Selangor, Kampus Puncak Alam, Bandar Puncak Alam, Selangor Darul Ehsan, Malaysia

Ahsan Ali
Department of Pharmaceutics, School of Pharmaceutical Education and Research (SPER), Jamia Hamdard, New Delhi, India

Amelia M. Avachat
Sinhgad College of Pharmacy, Vadgaon (Bk.), Pune, Maharashtra, India

Uzma Azeem
Government College Malerkotla, Department of Botany, Malerkotla, Punjab, India

Bela Rani Bhagat
Central Tasar Research and Training Institute, Central Silk Board, Ministry of Textiles, Ranchi, Jharkhand, India

Adrija Bhunia
Department of Pharmaceutics, Yenepoya Pharmacy College and Research Center, Yenepoya (Deemed to be University), Mangalore, Karnataka, India

Adwiza Chakraborty Bishakha
Drug Delivery and Therapeutics Lab, London IG2 6TT, United Kingdom

Iffat Ara Brishti
Drug Delivery and Therapeutics Lab, London IG2 6TT, United Kingdom

Shilpa S. Chapadgaonkar
School of Biology, Dr. Vishwanath Karad MIT World Peace University, Kothrud, Pune, Maharashtra, India

Hema Chaudhary
School of Medical and Allied Sciences, K.R. Mangalam University, Gurugram, Haryana, India

J. C. Darshan
Department of Pharmacy Practice, Yenepoya Pharmacy College and Research Center, Yenepoya (Deemed to be University), Mangalore, Karnataka, India

Kamal Dua
Faculty of Health, Australian Research Center in Complementary and Integrative Medicine, University of Technology Sydney, Ultimo, NSW, Australia; Discipline of Pharmacy, Graduate School of Health, University of Technology, Sydney, Ultimo, NSW, Australia

Ashvin V. Dudhrejiya
B. K. Mody Government Pharmacy College, Gujarat Technological University, Rajkot, Gujarat, India

Zolkapli Eshak
Faculty of Pharmacy, Universiti Teknologi MARA (UiTM), Cawangan Selangor, Kampus Puncak Alam, Bandar Puncak Alam, Selangor Darul Ehsan, Malaysia

Fredrick Nwude Eze
Faculty of Pharmaceutical Sciences, Prince of Songkla University, Hat Yai, Songkhla, Thailand

Md. Faiyazuddin
School of Pharmacy, Al-Karim University, Katihar, Bihar, India; Nano Drug Delivery®, Raleigh-Durham, NC, USA

Huda Ghazanfar
Institute of Soil and Environmental Sciences, University of Agriculture, Faisalabad, Pakistan

Gaurav Gupta
School of Pharmacy, Suresh Gyan Vihar University, Jagatpura, Jaipur, Rajasthan, India; Uttaranchal Institute of Pharmaceutical Sciences, Uttaranchal University, Dehradun, Uttarakhand, India; Department of Pharmacology, Saveetha Dental College, Saveetha Institute of Medical and Technical Sciences, Saveetha University, Chennai, Tamil Nadu, India

Khalid Rehman Hakeem
Department of Biological Sciences, Faculty of Science, King Abdulaziz University, Jeddah, Saudi Arabia

Jasmin Hassan
Drug Delivery and Therapeutics Lab, London IG2 6TT, United Kingdom

Afzal Hussain
Department of Pharmaceutics, College of Pharmacy, King Saud University, Riyadh, Saudi Arabia

Musawar Hussain
Institute of Soil and Environmental Sciences, University of Agriculture, Faisalabad, Pakistan

Zeenat Iqbal
Department of Pharmaceutics, School of Pharmaceutical Education and Research (SPER), Jamia Hamdard, New Delhi, India

Anas Islam
Faculty of Pharmacy, Integral University, Lucknow, Uttar Pradesh, India

Pooja Jain
Department of Pharmaceutics, School of Pharmaceutical Education and Research (SPER), Jamia Hamdard, New Delhi, India

Sougata Jana
Department of Health and Family Welfare, Directorate of Health Services, Kolkata, West Bengal, India

Yash Jasoria
School of Medical and Allied Sciences, K.R. Mangalam University, Gurugram, Haryana, India

Titilope John Jayeoye
Department of Chemistry, Faculty of Science, Chulalongkorn University, Bangkok, Thailand

Sejal Kabra
School of Pharmacy, Dr. Vishwanath Karad MIT World Peace University, Kothrud, Pune, Maharashtra, India

Dayanand Kannur
SCES's, Indira College of Pharmacy, Pune, Maharashtra, India

Mohd. Muazzam Khan
Faculty of Pharmacy, Integral University, Lucknow, Uttar Pradesh, India

Rahmuddin Khan
Department of Pharmaceutics, School of Pharmaceutical Education and Research (SPER), Jamia Hamdard, New Delhi, India

Sumaiya Islam Khan
Drug Delivery and Therapeutics Lab, London IG2 6TT, United Kingdom

Prashant Kumar
SRM Modinagar College of Pharmacy, SRMIST Delhi-NCR Campus, Modinagar, Ghaziabad, Uttar Pradesh, India

Sreejan Manna
Department of Pharmaceutical Technology, Brainware University, Barasat, Kolkata, West Bengal, India

Mohd. Aamir Mirza
Department of Pharmaceutics, School of Pharmaceutical Education and Research (SPER), Jamia Hamdard, New Delhi, India

Neeraj Mishra
Amity Institute of Pharmacy, Amity University Madhya Pradesh (AUMP), Gwalior, Madhya Pradesh, India

Sourav Mohanto
Department of Pharmaceutics, Yenepoya Pharmacy College and Research Center, Yenepoya (Deemed to be University), Mangalore, Karnataka, India

Hannis Fadzillah Mohsin
Faculty of Pharmacy, Universiti Teknologi MARA (UiTM), Cawangan Selangor, Kampus Puncak Alam, Bandar Puncak Alam, Selangor Darul Ehsan, Malaysia

Nongnuj Muangsin
Department of Chemistry, Faculty of Science, Chulalongkorn University, Bangkok, Thailand

Sugandha V. Mulgund
Bharati Vidyapeeth (Deemed to be University), Poona College of Pharmacy, Pune, Maharashtra, India

Vikas Musale
Department of Pharmaceutics, R. C. Patel Institute of Pharmaceutical Education and Research, Shirpur, Maharashtra, 425404, India

Madhavi Patel
Parul Institute of Pharmacy, Faculty of Pharmacy, Parul University, Vadodara, Gujarat, India

Prashant L. Pingale
GES's Sir Dr. M. S. Gosavi College of Pharmaceutical Education and Research, Nashik, Maharashtra, India

Satish Polshettiwar
School of Pharmacy, Dr. Vishwanath Karad MIT World Peace University, Kothrud, Pune, Maharashtra, India

Taslima Sultana Poly
Drug Delivery and Therapeutics Lab, London IG2 6TT, United Kingdom

Bhupendra Prajapati
Shree S. K. Patel College of Pharmaceutical Education and Research, Ganpat University, Kherva, Gujarat, India

Bijal Prajapati
Parul Institute of Pharmacy, Faculty of Pharmacy, Parul University, Vadodara, Gujarat, India

Syed Jalal Q. Rahman
Novo Nordisk Inc., NJ, USA & University of Cumberlands, KY, USA

Safiya Mohammed Saeed
Drug Delivery and Therapeutics Lab, London IG2 6TT, United Kingdom

Olivia Sen
Department of Pharmaceutical Technology, Brainware University, Barasat, Kolkata, West Bengal, India

Sunny R. Shah
B. K. Mody Government Pharmacy College, Gujarat Technological University, Rajkot, Gujarat, India

Muhammad Tahir Shehzad
Institute of Soil and Environmental Sciences, University of Agriculture, Faisalabad, Pakistan

Shilpa N. Shrotriya
Bharati Vidyapeeth (Deemed to be University), Poona College of Pharmacy, Pune, Maharashtra, India

Ayesha Siddiqui
Department of Pharmaceutics, School of Pharmaceutical Education and Research (SPER), Jamia Hamdard, New Delhi, India

Sudarshan Singh
Faculty of Pharmacy, Chiang Mai University, Chiang Mai, Thailand

Jenny Sivakumar
Department of Microbiology, Cauvery College for Women (Autonomous), Tiruchirappalli, Tamil Nadu, India

Rokeya Sultana
Department of Pharmacognosy, Yenepoya Pharmacy College and Research Center, Yenepoya (Deemed to be University), Mangalore, Karnataka, India

Gowri Sundaram
PG & Research Department of Physics, Cauvery College for Women (Autonomous), Tiruchirappalli, Tamil Nadu, India

Fouzia Tasnim
Drug Delivery and Therapeutics Lab, London IG2 6TT, United Kingdom

Md. Jasim Uddin
Department of Pharmaceutical Technology, Faculty of Pharmacy, Universiti Malaya, Kuala Lumpur, Malaysia

Shahid Nazir Wani
Chitkara College of Pharmacy, Chitkara University, Rajpura–140401, Punjab, India;
Aman Pharmacy College, Dholakhera, Udaipurwati Jhunjhunu, Rajasthan, India

Ishrat Zahoor
Department of Pharmaceutics, MM College of Pharmacy, Maharishi Markandeshwar (Deemed to be University), Mullana–Ambala, 133207, Haryana, India

Preface

Researchers are looking at novel bioactive natural products as a result of the pharmaceutical industry's growing need for new lead chemicals. Oceans are abundant in new bioactive secondary metabolites, and they have drawn the attention of many scientists across the globe. Natural substances discovered in the ocean have been found to be significant biological modulators and possible sources for new drugs. These compounds exhibit distinct structural variations, a wide range of biological functions, and innovative modes of action. We have covered some of these novel antibacterial, antiviral, and anti-cancer chemicals identified from marine sources in the current book, along with their potential applications in the eradication of disease and commercial exploitation of these compounds for potential medication development.

Our book is a compilation of the latest information available on marine drugs. This book comprises 20 chapters on various related topics. This book can serve as a manual for researchers involved in the fields of pharmaceutical sciences, microbiology, biotechnology, and molecular biosciences. We hope the academicians and research students at postgraduate and doctoral levels in colleges, research laboratories, industries, and universities could use this book as a reference book.

This is our chance to express our gratitude to the authors who selflessly gave their time to meet the deadlines for each chapter. We hope that the readers, especially the students, will benefit from reading this overview of marine bioresources as much as we have benefited from reading, evaluating, and editing the chapters.

Last but not least, we are thankful to Apple Academic Press, particularly Mr. Rakesh Kumar and Mr. Ashish Kumar, for helping us during the process of compiling this volume.

—*Editors*

CHAPTER 1

Global Marine-Derived Drug Industry and Market Analysis

MD. FAIYAZUDDIN,[1] SYED JALAL Q. RAHMAN,[2] KAMRAN ASHRAF,[3] and SHAUKAT ALI[4]

[1]*School of Pharmacy, Al-Karim University, Katihar, Bihar, India*

[2]*Novo Nordisk Inc., NJ, USA & University of Cumberlands, KY, USA*

[3]*Faculty of Pharmacy, Universiti Teknologi MARA (UiTM), Cawangan Selangor, Kampus Puncak Alam, Bandar Puncak Alam, Selangor Darul Ehsan, Malaysia*

[4]*School of IT and Management, Al-Karim University, Katihar, Bihar, India*

ABSTRACT

The marine environment, which is abundant in ecological and biological diversity, plays a crucial role in the discovery of new drugs. Marine natural products, known for their unique structural and biological properties, have proven to be effective in the treatment of various diseases. The global market for drugs derived from marine sources was valued at $28.26 billion in 2021 and is expected to reach $52.14 billion by 2028. This growth is being fueled by the interest of numerous pharmaceutical and biotechnology companies in marine-derived drugs for serious illnesses such as cancer. Key players in this market include Abbott Lab, BASF, DSM, and Pfizer. Some of the clinically approved drugs derived from marine sources are cytarabine, trabectedin, and omega-3-acid ethyl esters. North America is the leading market, followed by Europe and China. Challenges in this industry include high investment costs and the necessity for interdisciplinary collaboration. However, ongoing

Handbook of Research in Marine Pharmaceutics: Exploring Oceanic Microbial Diversity for Human Health and Wellness. Md. Faiyazuddin & Khalid Rehman Hakeem (Eds.)

research and development investments offer the promise of advancements in marine pharmacology and new opportunities for the treatment of diseases.

1.1 INTRODUCTION

Evidence suggests that life originated from the oceans, which occupy 70% of the earth's surface and represent the ecological, chemical, and biological diversity of the earth's biosphere (Globe Newswire Industry Research, 2021).

With evolving life forms, the patterns of diseases are changing and emerging due to the ever-changing environments. Marine natural products have been found to be effective and proven biological agents and are one of the key sources in the drug discovery process. These products offer unique structural modifications, diverse biological actions, and novel mechanisms of action. Aquatic life forms have been analyzed by researchers for various medicinal properties such as analgesic, antibacterial, anticancer, antifungal, anti-inflammatory, antimalarial, antimicrobial, immunomodulatory, and neuroprotective effects (Malve et al., 2016; Ghareeb et al., 2020).

1.2 MARKET OVERVIEW OF MARINE-DERIVED DRUGS

A classic example of a marine-derived substance that has been used for centuries is fish oil. There has been growing awareness and focus on understanding the diverse medicinal properties and effects derived from the aquatic habitat in the last couple of years (Business Research Insights, 2022). It is also gaining attention because of the rising concerns of severe diseases like cancer.

The global market for marine-derived drugs was approximately $28.26 billion in 2021 and is estimated to reach around $52.14 billion by 2028 with a Compound Annual Growth Rate (CAGR) of 8–10%. The growth could be attributed to the increasing interest and demand for marine-derived drugs (Business Research Insights, 2022).

1.3 MAJOR PLAYERS

Some of the key players in the marine drugs market are Abbott Lab (US), Amway (US), BASF (Germany), Croda (UK), DSM (Netherlands), Eisai

(Japan), GSK (UK), Nordic Naturals (US), Pelagia (Norway), Pfizer (US), Pharma Mar (Spain), Seattle Genetics (US), Takeda (Japan), TerSera Therapeutics (US), and others (GlobeNewswire Industry Research 2021 and Business Research Insights 2022).

The above companies are involved in potential drug discoveries and growth strategies to achieve a competitive edge across the different stages of the value chain (MarketWatch, 2022).

1.4 LIST OF SOME OF THE APPROVED DRUGS

Some drugs derived from marine life have been clinically approved (for cancer, pain, Alzheimer's disease, etc.), such as cytarabine (Cytosar-U®), vidarabine (Vira-A®; discontinued in the US), trabectedin (Yondelis®), ziconotide (Prialt®), eribulin mesylate (Halaven®), plitidepsin (Aplidin®), brentuximab vedotin (Adcetris®), and omega-3-acid ethyl esters (Lovaza®) (Jimenez et al., 2019). Currently, there are around 20 marine compounds in various phases of clinical trials in the drug discovery pathway.

1.5 MARKETED DRUGS BY THE FDA AND/OR THE EMA

Drugs approved and marketed by the European Medicines Agency (EMA) and/or the Food and Drug Administration (FDA) are listed in Table 1.1.

1.6 REGIONAL MARKET SHARE

North America has the majority market share of 35%, followed by Europe and China with a combined share of around 35%, and the remaining 30% by the rest of the world (GlobeNewswire Industry Research, 2021).

North America is projected to drive the market with higher growth rates compared to other regions owing to the region's well-developed medical infrastructure and significant investments and funding in research and development pertaining to marine pharmaceuticals (Business Research Insights, 2022).

TABLE 1.1 List of Marketed Drugs

Brand Name(s)	Generic Name	Date of Marketing Authorization	Natural Source	Chemical Class	Clinical Use
Cytosar-U Aracytin C. Hospira	*Cytarabine*	1969 (FDA)	Sponge	Nucleoside	Leukemia
Vira-A	*Vidarabine*	1976 (FDA)	Sponge	Nucleoside	Antiviral
Fludara	*Fludarabine*	1992 (FDA) 1994 (EMA)	Sponge	Nucleoside	Leukemia
Prialt	*Ziconotide*	2004 (FDA) 2005 (EMA)	Mollusk	Peptide	Chronic pain
Lovaza (US) Eskim (EU) and others	Omega-3 acid ethyl esters	2004 (FDA) 2005 (EMA)	Fish	PUFA*	Hypertriglyceridemia
Arranon (US) Atriance (EU)	*Nelarabine*	2005 (FDA) 2007 (EMA)	Sponge	Nucleoside	Leukemia
Yondelis	*Trabectedin*	2007 (EMA) 2015 (FDA)	Tunicate	Alkaloid	Ovarian cancer, soft tissue sarcoma
Halaven	*Eribulin*	2010 (FDA) 2011 (EMA)	Sponge	Macrolide	Breast cancer
Adcetris	*Brentuximab vedotin*	2011 (FDA) 2012 (EMA)	Mollusk/cyanobacterium	ADC*	Lymphomas
Zepzelca	*Lurbinectedin*	2020 (FDA)	Tunicate	Alkaloid	Ovarian cancer
Polivy	*Polatuzumab vedotin*	2019 (FDA) 2020 (EMA)	Mollusk/cyanobacterium	ADC*	Breast cancer
Padcev	*Enfortumab vedotin*	2019 (FDA) 2021 (EMA)	Mollusk/cyanobacterium	ADC*	Urothelial cancer
Blenrep	*Belantamab mafodotin*	2020 (FDA) 2020 (EMA)	Mollusk/cyanobacterium	ADC*	Multiple myeloma

**ADC*: Antibody drug conjugate; PUFA: Polyunsaturated fatty acids.

Source: Reprinted with permission from Cappello & Nieri, 2021.

1.7 DRIVING AND RESTRAINING FACTORS FOR THE MARINE-DERIVED DRUGS MARKET

1.7.1 DRIVING FACTORS

1.7.1.1 EMERGENCE OF THE ONCOLOGY SEGMENT TO DRIVE THE MARKET DEVELOPMENT

Cancer is one of the leading causes of death in the world. The market for oncology drugs will expand significantly during the forecasted period. Although there has been substantial progress in the area of cancer during the last decade, the need for urgent development of newer medications still exists. The high incidence and prevalence of diverse cancer types around the globe will drive the sales of marine pharmaceuticals in the years to come (Business Research Insights, 2022; Lindequist et al., 2021).

1.7.1.2 RISE IN INTEREST FOR REMEDIES TO PROMOTE INDUSTRY PROGRESS

The interest in the marine ecosystem has increased to find a cure for wide-spread ailments and is currently being tested for various properties like antimicrobial, antifungal, analgesic, etc. This eagerness will also propel the development of future marine drugs and ultimately drive growth (Business Research Insights, 2022; Lindequist et al., 2021).

1.7.2 RESTRAINING FACTORS

1.7.2.1 HIGH INVESTMENT AND LOW SUCCESS RATE COUPLED WITH LONGER DURATION OF DRUG DEVELOPMENT

The approval for any drug would ideally require a long time (around a 10–15-year period) and a huge investment (average of $900 million) from drug discovery to launch in the market. Additionally, this comes with a very high risk of failure as numerous promising marine drugs have failed in the past during the development processes due to toxicity, challenges in supply deliverability in a cost-effective and sustainable manner, and other facts (Business Research Insights, 2022; Lindequist et al., 2021).

1.7.2.2 INTERDISCIPLINARY COLLABORATION

For the marine pharmaceuticals drug development to be successful, it would highly rely on the interdisciplinary collaboration between biologists, biotechnologists, chemists, doctors, pharmacists, scientists, etc., and between organizations, hospitals, and universities, which could be challenging and difficult in certain situations (Business Research Insights, 2022; Lindequist et al., 2021).

1.7.2.3 LACK OF EFFECTIVE TECHNIQUES FOR HARVESTING MARINE COMPOUNDS

Compounds typically found in deep seas instead of ocean floors and other water bodies can prove to be a barrier to the continuous availability of raw materials for marine drugs (Business Research Insights, 2022; Lindequist et al., 2021).

1.8 FUTURE OF MARINE-DERIVED DRUGS

With the growing interest in marine drugs, there are a lot of drugs in different stages of various ongoing clinical investigations. Some of the ongoing trial drugs are listed in Table 1.2.

1.9 CONCLUSION

The recent advances in marine-derived drugs open future avenues and opportunities for more potential discoveries and cures for different diseases and ailments. Moreover, with the continued investments in research and developments in marine-derived drugs and awareness of marine compounds, the drug manufacturing of marine compounds is expected to increase in the next couple of years (Medgadget, 2019).

In the last decade, the number of marine drugs launched in the EU and/or USA market has nearly doubled, and there are a large number of clinical trials registered, the majority of which are in the first two phases. Most of the marine drugs available for clinical use and/or under trial are either for anticancer therapy or conjugated to an antibody (ADCs). Though in the terrestrial environment, the main source of pharmacological compounds is

TABLE 1.2 Marine Drugs in Various Stages of Clinical Trials Subdivided into Three Phases of Clinical Investigation

Generic Name	Natural Source	Chemical Class	Clinical Use
	Phase III		
Tetrodotoxin	Pufferfish	Alkaloid	Chronic pain
Plinabulin	Fungus	Diketopiperazine	Cancer
Marizomib (salinosporamide A)	Actinomycetes	γ-lactam-β-lactone	Cancer
Plitidepsin	Tunicate	Depsipeptide	COVID-19
	Phase II		
Bryostatin	Bryozoan	Macrolide lactone	Alzheimer
Plocabulin (PM060184)	Sponge	Polyketide	Cancer
Tisotumab vedotin	Mollusk/cyanobacterium	ADC	Cancer
Ladiratuzumab vedotin (SGN-LIV1A)	Mollusk/cyanobacterium	ADC	Cancer
Telisotuzumab vedotin	Mollusk/cyanobacterium	ADC	Cancer
CAB-ROR2 (BA-3021)	Mollusk/cyanobacterium	ADC	Cancer
CX-2029	Mollusk/cyanobacterium	PDC	Cancer
W0101	Mollusk/cyanobacterium	ADC	Cancer
	Phase I		
GTS-21 (DMXBA)	Worm	Alkaloid	Obesity
Samrotamab vedotin	Mollusk/cyanobacterium	ADC	Cancer

**ADC*: Antibody Drug Conjugate; PDC: Probody Drug Conjugate.

Source: Reprinted with Permission from Cappello & Nieri, 2021. Copyright © 2021 by the authors. Licensee MDPI, Basel, Switzerland. https://creativecommons.org/licenses/by/4.0/

the plant kingdom, in the evolving marine environment, the key source of marketed drugs is animals (Capello et al., 2021).

In conclusion, with the advancements in marine pharmacology and pharmacognosy with the use of nanotechnological techniques, many marine drugs are expected with improved pharmaceutical efficacy and/or safety (Capello et al., 2021).

KEYWORDS

- **compound annual growth rate**
- **European Medicines Agency**
- **Food and Drug Administration**
- **marine-derived drugs market**
- **nanotechnological techniques**
- **terrestrial environment**

REFERENCES

Capello, E., et al. (2021). From life in the sea to the clinic: The marine drugs approved and under clinical trial. *Life (Basel)*, *11*(12), 1390. https://www.ncbi.nlm.nih.gov/pmc/articles/PMC8704254/ (accessed on 7 June 2024).

Ghareeb, M., et al. (2020). Insights about clinically approved and preclinically investigated marine natural products. *Elsevier*, *2*, 88–102. https://doi.org/10.1016/j.crbiot.2020.09.001.

Jimenez, P. C., Wilke, D. V., Branco, P. C., Bauermeister, A., Rezende-Teixeira, P., Gaudêncio, S. P., Costa-Lotufo, L. V., (2020). Enriching cancer pharmacology with drugs of marine origin. *Br J Pharmacol.*, *177*(1), 3–27. doi: 10.1111/bph.14876. Epub 2019 Dec 23. PMID: 31621891; PMCID: PMC6976878.

Lindequist, U., et al. (2016). Marine-derived pharmaceuticals – challenges and opportunities. *Biomolecules & Therapeutics (Seoul)*, *24*(6), 561–571. https://doi.org/10.4062/biomolther.2016.18.

Malve, H. (2016). Exploring the ocean for new drug developments: Marine pharmacology. *Journal of Pharmacy & Bio allied Sciences*, *8*(2), 83–91. https://doi.org/10.4103/0975-7406.171700.

Marine pharmaceuticals and marine-derived drugs market size and share 2021 global industry analysis by trends, key findings, future demands, growth factors, growth strategy, emerging technologies, leading players updates, and forecast 2027. (2021). *GlobeNewswire Industry Research.* https://www.globenewswire.com/news-release/2021/10/10/2311473/0/

en/Marine-Pharmaceuticals-and-Marine-Derived-Drugs-Market-Size-and-Share-2021-Global-Industry-Analysis-By-Trends-Key-Findings-Future-Demands-Growth-Factors-Growth-Strategy-Emerging-Te.html (accessed on 7 June 2024).

Marine pharmaceuticals market size, share, and industry analysis, by type (drugs and health supplement), by application (hospital & clinic, pharmacy and other) regional forecast (2022–2028). (2022). *Business Research Insights*. https://www.businessresearchinsights.com/market-reports/marine-pharmaceuticals-market-100954 (accessed on 7 June 2024).

Marine-derived drugs market global investor, latest trends, industry growth, size, segmentation, future demands, by regional forecast to 2028. (2022). *Market Watch*. https://www.marketwatch.com/press-release/marine-derived-drugs-market-global-investor-latest-trends-industry-growth-size-segmentation-future-demands-by-regional-forecast-to-2028–2022–12–01?tesla=y.

Marine-derived drugs market is growing at a CAGR of 11.20% and is expected to reach $21,955.6 million by 2025 – exclusive report by Infinium Global Research. (2019). *Medgadget*. https://www.medgadget.com/2019/07/marine-derived-drugs-market-growing-at-a-cagr-of-11-20-and-expected-to-reach-21955-6-million-by-2025-exclusive-report-by-infinium-globalresearch.html.

CHAPTER 2

Marine Natural Product Database for Exploring New Drugs

MOHIT AGRAWAL,[1] YASH JASORIA,[1] HEMA CHAUDHARY,[1] BHUPENDRA PRAJAPATI,[2] and MD. FAIYAZUDDIN[3]

[1]*School of Medical and Allied Sciences, K.R. Mangalam University, Gurugram, Haryana, India*

[2]*Shree S.K. Patel College of Pharmaceutical Education and Research, Ganpati University, Gujarat, India*

[3]*Nano Drug Delivery®, Raleigh-Durham, NC, USA*

ABSTRACT

Marine natural products (MNPs), which come from marine species and terrestrial plants, are a significant source of new drugs. Computational methodologies assist in the investigation and exploration of MNPs by repurposing current MNPs and targeting new metabolites based on genomic research. Natural substances with unique chemical properties are produced by marine organisms. These chemical characteristics are helpful in identifying new MNP uses. MNPs are gathered in chemical databases that enable their analysis and screening for biotechnologically relevant applications in order to fully exploit their potential applications. MNPs are currently poorly categorized in general chemical databases, nevertheless. The availability of free-access MNP databases has decreased, and their molecular diversity is still rather low when compared to paid-access databases. Recent developments in MNP extraction and separation methods are highlighted in this study. We will also discuss how attempts to mine the human genome are probably going to speed up the search for new antibacterial substances.

Handbook of Research in Marine Pharmaceutics: Exploring Oceanic Microbial Diversity for Human Health and Wellness. Md. Faiyazuddin & Khalid Rehman Hakeem (Eds.)

2.1 INTRODUCTION

In the early 1950s, the pharmaceutical industry began studying the nucleosides spongouridine and spongothymidine in the marine sponge *Cryptotethya crypta* (Bergmann & Feeney, 1950; 1951). The majority of natural product (NP) research has focused on marine species (Zhang, Li, Zhu, Gu, & Li, 2016). Marine organisms are substantial producers of secondary metabolites with a variety of structural properties and physiological activity because they are more likely than terrestrial life forms to display significant bioactivity and chemical novelty (Kong, Jiang, & Zhang, 2010). By using marine resources to generate natural medicines, many human illnesses have become more targeted and successful. The marine environment, which is home to various life forms with diverse physiologies and capacities for adaptation, is swiftly emerging as a top location for discovering new therapeutic leads. The search for new, possibly bioactive molecules in nature has a wide range of relevant natural sources. Even though oceans make up more than 70% of the earth's surface, research into marine ecosystems didn't start until the middle of the 1970s, when scuba diving and contemporary snorkeling were both created. Remotely operated vehicles were also not widely employed until the 1990s. Since 1990, more than 11,000 novel NPs have been reported to have originated from invertebrates, which make up around 60% of all marine organisms. Microorganisms such as marine cyanobacteria, marine fungal spores, and several other forms of microbes have attracted interest due to their biological diversity and ability to synthesize compounds with unmatched structures as a result of the continued exploitation of the marine environment. Today, microbes constitute a rich source of structurally related bioactive components and have created some of the most important active substances yet to be fully comprehended. Even more recently, it was found that a number of compounds that had previously been recovered from marine microbes like sponges and tunicates were metabolic byproducts of other adjacent bacteria.

A wide range of bioactivities, including chemotherapeutic, microtubules, antimetabolites, antibiotics, and anti-infectives, make marine natural products highly attractive as high-value constituents for pharmaceutical applications. As a result, more and more businesses are investing in this area. In a similar vein, the cosmetics business is increasingly looking to the ocean for fresh materials. In the past, objects to be applied to the human body for washing, beautification, encouraging beauty, or modifying the look without harming bodily structure or functions were referred to as cosmetics.

The US Food and Drug Administration (FDA) and the European Medicines Agency (EMA) have both approved at least eight related drugs in recent years, proving the seas have moved quite quickly in the field of marine natural drug development (Figure 2.1) (Martins et al., 2014).

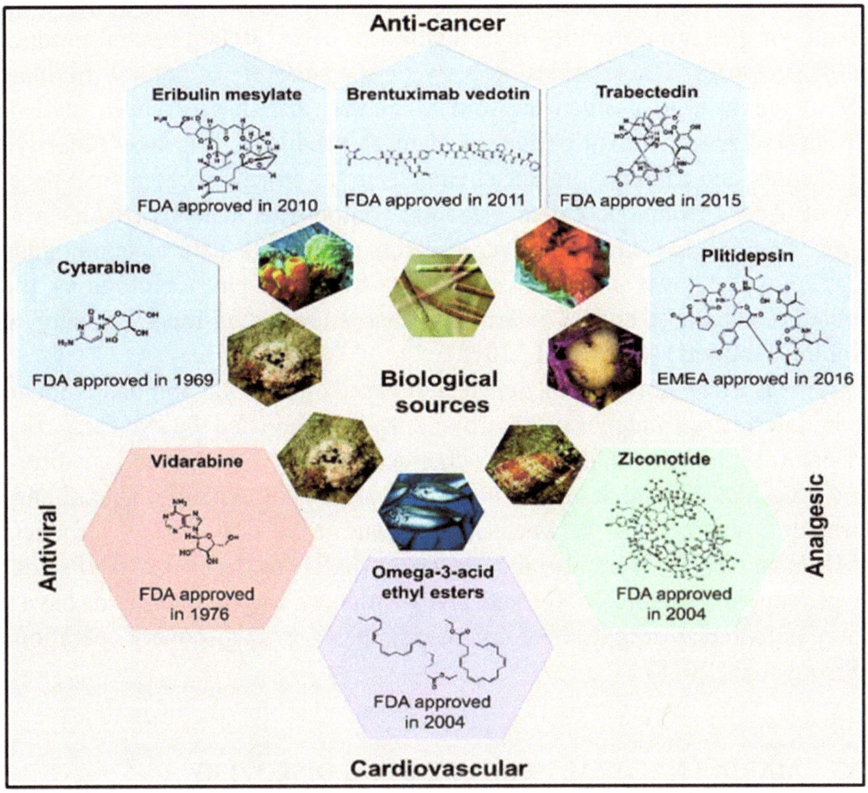

FIGURE 2.1 Eight medications have been approved by the FDA or EMEA, representing significant progress in the field of marine natural drug development.

Source: Reprinted from Pereira and Aires-de-Sousa, 2018. Copyright © 2018 by the authors. Licensee MDPI, Basel, Switzerland.http://creativecommons.org/licenses/by/4.0/

2.2 CHEMICAL SPACE FOR MARINE NATURAL PRODUCTS

One of the best sources of chemical inspiration for the development of new drugs is still natural products. Access to the structures of more than 440,000 NPs is possible because of public databases like NaprAlert, ChEMBL, PubChem, and Supernatural II. About 10% of them are readily available for purchase from enterprises and government research agencies, where they may be utilized for testing experiments. Numerous studies have documented comparisons between these chemical spaces and those created by medications and other types of small molecules in order to better comprehend the chemical space established by recognized and easily accessible NPs along with individual NP databases (Pereira, 2019). The current study reported the results of a cheminformatics investigation of over 130,000 natural product (NP) structures. The relationship between the structure of natural products (NPs) and the type of organism from which they arise has also been studied. To investigate the distribution of natural products, drug-like bioactive molecules, and common organic chemicals in the structural chemistry space, we utilized a technique called principal components analysis (Petersen & Amstutz, 2008). Comprehensive research employing natural compounds that are well-known and widely accessible from nine physical libraries, the Protein Data Bank, and 18 internet databases (including the Dictionary of Natural Products) (Chen et al., 2018).

This is a new era of medical research based on natural compounds found in both terrestrial and marine environments. We evaluated the two categories of natural products' major physicochemical properties, structural qualities, and drug-likeness and described how they varied evolutionarily. The primary structural distinctions between marine and terrestrial natural products (MNPs and TNPs) were also investigated. MNPs are larger than TNPs and, in general, less soluble. MNPs and TNPs with new drug-like scaffolds have a lot of potential as drug leads or drug candidates in drug discovery operations (Shang et al., 2018).

2.3 MARINE NATURAL PRODUCT DRUG DISCOVERY

The significance of natural products (NPs) in drug discovery, particularly their role in the creation of current medications, has received substantial documentation. In recent years, the study of drugs derived from marine natural resources has seen a rebirth. New chemical entities have entered the clinic for the treatment of pain, cancer, and other disease states as a result of the extensive collection, screening, and discovery of novel marine natural products. The peptide ziconotide (Prialt; Elan Pharmaceuticals), which was first discovered in a tropical cone snail, was the first marine-derived medication to be approved in the US in December 2004 for the treatment of pain. In October 2007, Trabectedin (Yondelis; PharmaMar) became the first marine anticancer drug to get EU approval. Natural substances have traditionally been used as a source of therapeutic compounds, particularly those derived from terrestrial plants and microorganisms (e.g., morphine from poppies and penicillins from fungi). Natural products have a huge influence on contemporary pharmaceutical discovery programs. In fact, pharmacologically active substances from plants and microorganisms serve as a crucial route for developing novel experimental medications (Molinski et al., 2009).

However, demand for marine natural products was on hold while the technology to capture the source species was improved. But by the late 1950s, the idea of pharmaceuticals from the ocean had aroused some curiosity. Oceans have a vast variety of organisms and makeup over 70% of the surface of the Earth. Marine invertebrates including sponges, jellyfish, anemones, and stony corals have been discovered to contain the highest concentrations of bioactive MNPs, with the Cnidaria phylum having the highest concentrations. The majority of marine natural products (MNPs) really appear to have their genuine source in the symbiotic microbes that live alongside them. In order to ensure the fullest development of their structural variety and bioactivity, the marine natural products are tested in the most diversified biological assays that are now practical, which are made simpler by emerging technology (Jiménez, 2018). Additionally, it has been shown that the marine environment is a very rich source of diverse natural products (NPs) with pertinent potency against parasites, viruses, and fungal drug-resistant disorders (Figure 2.2).

2.4 DATA CONTENT

2.4.1 DATA EXTRACTION AND CURATION

The majority of compound records were taken from the late D. John Faulkner's exceptional annual MNPs reports that were included in the National Production Report (Faulkner, 1984), a series that was kept up by the team of 11 Blunt JW (2003) and is currently 12. Carroll AR (2019).

The references were listed in Web of Science (https://apps.webofknowledge.com, 2018), and EndNote was used to extract the full text (version 9.0.0, Clarivate Analytics Inc., 2018). After human moderation, about 20,000 pieces from the 1960s through December 2018 were gathered and combined into one document repository. The bulk of these studies focused on the discovery of novel compounds, their biological activity, as well as any structural alterations or newly found stereochemistry of compounds that had already been studied.

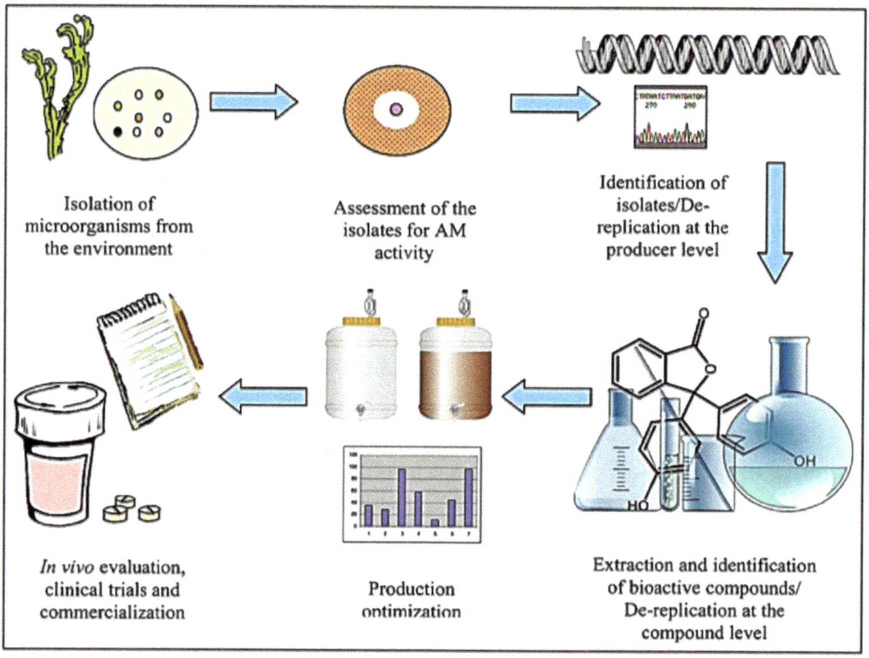

FIGURE 2.2 Broad process for finding naturally occurring, microbe-derived physiologically active chemicals like antimicrobials. The process begins with the separation of microorganisms from the environment, such as from the surfaces of marine eukaryotes, and is then followed by the screening of their antibacterial activity and the identification of the producing organism. After purification, the chemical structure of the bioactive molecule is revealed. For additional *in vivo* tests and product development, production optimization can be used to increase the yield of the desired chemical.

Source: Reprinted from Penesyan, Kjelleberg, & Egan, 2010. https://creativecommons.org/licenses/by/3.0/

These journals were manually mined for information on smaller molecules, including their chemical compositions, names of their constituent compounds, sources of origin, and other relevant details. After the structure of MNPs was utilized as a query to hunt for them in generic chemical databases, the scientific literature and patents of successful compounds were combined into a general document library (such as PubChem and ChEMBL). Using a taxonomy with over 4800 distinct categories based only on chemical structures and structural signals, we created a computer software (ClassyFire) that automatically classifies all known chemical compounds (Carroll et al., 2019). The nuclear magnetic resonance, infrared, ultraviolet-visible, and mass spectra of the structures were displayed if they were available from Wiley SpectraBase.

2.4.2 *CALCULATED PROPERTIES*

To assess each molecule's drug-likeness, certain physicochemical and pharmacokinetic properties were evaluated. Using RDKit, the following physicochemical properties were identified: molecular mass (also known as monoisotopic mass), polar surface area, quantitative estimate of drug-likeness QED weighted, aromatic rings, and heavy atoms. Pipeline Pilot ADMET models were utilized to predict pharmacokinetic traits such as blood-brain barrier permeability, intestinal absorption by humans, water solubility, CYP2D6 binding, hepatotoxicity, and plasma protein binding (Lyu et al., 2021).

2.4.3 *ORGANISM RESOURCES*

Terrestrial and marine species have been a consistent and expanding source of novel natural compounds, generating leads for various pharmacological, nutraceutical, and cosmeceutical applications. Accurate taxonomic and geographic information also supports meta-analytic statistical methods that apply biogeographic concepts to reveal unexplored areas and species groups for more effective NP discovery. Taxonomists historically identified species that produced novel NPs from samples using voucher specimens preserved in museums or institutional collections. Taxonomy has been continuously updated, especially with the advancement of molecular genetics. Screening platform technology has improved repeatability by enhancing the accuracy and precision of molecular descriptions, benefiting the identification of novel NPs. However, the biological entity that produces the target molecule is a fundamental aspect of natural product research. Therefore, details provided on the taxonomy and location of the source species' sample should be as precise as those regarding the chemical composition and biological activity of the reported NPs (Leal et al., 2016).

2.4.4 *BIOLOGICALLY ACTIVITY DATA*

Some biological activity information based on the original bioassays performed by the discoverers was obtained from the source document repository, although the majority of it was a succinct description of the pharmacological effects, such as "anti-fungal," "antiviral," "antioxidant," "anti-inflammatory," "antitumor," and "antimycotic" (Figure 2.3).

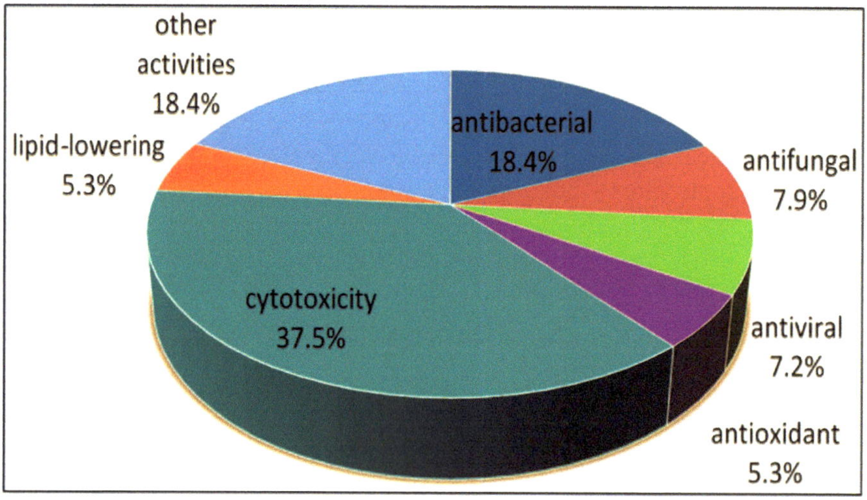

FIGURE 2.3 Novel compounds from marine sources that fall into the bioactive category.
Source: Adapted from Jin et al., 2016

The binding database is the name of the first informational database on measured protein-ligand affinities. It is designed to make access to focused data sets, such as the affinity data associated with a particular drug target, as well as large studies that benefit from the thoroughness of a significant and growing data collection, easier (Gilson, Liu, Baitaluk, Nicola, Hwang, & Chong, 2016).

The open-access database known as PubChem BioAssay is maintained by the NCBI, NLM, and NIH. It was founded in 2004 to serve as a central location for data generated by research in the fields of functional genomics, medicinal chemistry, and chemogenomics. Several NCBI-hosted biological and literature databases, such as PubMed, Protein, Gene, Taxonomy, etc., are linked to the PubChem BioAssay database. Research in several areas, including medicinal chemistry, drug discovery, pharmaceutical genomics, and informatics research has greatly benefited from the gathering of bioactivity and toxicity data in PubChem BioAssay. This was recently brought up in the analysis of how the public utilizes the resource PubChem (Wang et al., 2017) (Table 2.1).

2.5 EXPERT OPINION

Marine organisms, rather than terrestrial species and non-marine microorganisms, are now the most sought-after source of bioactive natural products

TABLE 2.1 Presents Statistics for Several Data Collections in the Comprehensive Marine Natural Products Databases (CMNPD)

Data Collection	Counting	Description of Data
Compounds	–	MNPs (Marine Natural Products)
Chemical entities	31 561[a]	Distinctive chemical structures
Compounds with 3D conformers	25 224	Apply the PubChem3D conformer models' filtering standards.
Biologically active compounds	15 774	compounds with information about their biological activity
Organisms	–	Source marine organisms
Kingdom	7	Taxonomic classification
Phyla	38	Taxonomic classification
Classes	93	Taxonomic classification
Orders	289	Taxonomic classification
Families	682	Taxonomic classification
Genera	1480	Taxonomic classification
Species	3354[b]	Taxonomic classification
Targets	2652	predefined targets in accordance with the ChEMBL target list
Isolated proteins	1122	Group of targets
Cell type	923	Group of targets
Microbes	459	Group of targets
Others	148	Group of targets
Bioactivities	72 343	Biological activity data
Brief information	15 980	Manual compilation of literature
Data from standardized experiments	56 363	ChEMBL Incorporation
Records	128 488	Scientific literature and patents
Scientific literature	119 543	Literature abstracts/citations
Patents	8945	Patent abstracts/citations

The CMNPD's statistics for numerous data sets (Lyu, Chen, Qiang, Liu, Wang, Zhang, & Liu, 2021).

[a]*Chemical entities include several versions of some chemicals.*

[b]*Due to the fact that several unidentified species have been grouped under the same category, this is the lowest estimate of the number of species.*

for medication development due to their unique chemical structures and potential biological activity. However, for the efficient use of marine natural resources, it is essential to develop sampling, separation, and improved spectrometric techniques as well as approaches for the structural characterization of active components. There are now 23 compounds in various phases of medicine development in the global pipeline for marine pharmaceutical trials, and despite numerous challenges, six distinct types of chemical structures from marine sources have already been authorized as pharmacological agents (Jiménez, 2018). China was one of the first nations to adopt marine medicine to heal illnesses. Over 110 marine plants and other marine herbal formulations have been described in ancient Chinese herbal texts including Shennong Bencaojin, Xinxiu Bencao, and Bencao Gangmu (including those for Chinese food therapy). There is a wealth of knowledge about marine plants and how to use them in medicine that has been gathered over thousands of years, and this knowledge has served as a crucial basis for contemporary research on marine medications. The number of newly found sea herbs has significantly increased since the middle of the 20th century as a result of special attention being given to traditional Chinese medicine. The pharmacology, chemistry, biology, and ecology of marine medicinal bioresources have also been the subject of extensive surveys that have yielded a wealth of information (Proksch, 2009).

Despite the fact that the area of MNP drug discovery has been completely developed in the 21st century, there are still a number of obstacles that prevent the production of medications from NPs, notably MNPs. One of the most crucial requirements is the reliable creation of systems employing fermentation procedures, chemical synthesis, or genetic techniques to produce a sustained supply of NPs. For instance, difficulties with data collection have hindered the anti-cancer clinical study employing the bissteroidal alkaloid (cephalostatin 1) (Pereira, 2019). The method employed to make marine natural products produced from macro- or microbes available for clinical studies has typically been the design and total or semi-synthetic chemical synthesis. However organic synthesis still poses a significant difficulty, especially for substances with a low level of complexity. Computer-aided synthesis planning (CASP) has utilized machine-learning techniques successfully. Around 50 years after the first study in this field, computer-aided synthetic routes of isolated NPs, such as tacamonidine, a pentacyclic alkaloid (not yet generated in the lab), have been published. This has allowed researchers to draw the conclusion that computers are presently capable of creating syntheses crucial to the routine practice of organic synthesis (Coley, Green, & Jensen, 2018; Szymkuć et al., 2016).

2.6 DATA ACCESS

2.6.1 WEB INTERFACE

A user-friendly interface made it possible to access the comprehensive marine natural products database (CMNPD) at the address (https://docs.cmnpd.org/about). This interface is based on the new ChEMBL online interface (https://www.ebi.ac.uk/chembl/), which serves as the blueprint for a free-to-use drug discovery database that intends to assemble knowledge and information on medicinal chemistry for use in pharmaceutical research and development (Mendez et al., 2019).

2.6.2 DATA BROWSING

Investigate four main categories of items (compounds, species, goal, and documentation) with unique Comprehensive Marine Natural Products Databases (CMNPD) IDs on the thorough list pages and specialist report card websites. The list pages as a whole include interactive filters that can be used to examine a portion of the original data in a chosen range for the filter property and display the distribution of the dataset in terms of a number of specific properties (e.g., molecular mass, target type, and organism class system). Additional details on each entry are provided on the report card pages, including name and categorization (for compounds, species, and targets), bibliometric data from books and patents (for documents), structure, computed properties, and biologically active compounds (for compounds), as well as internal and external links, such as PubChem to other report card websites (Lyu et al., 2021).

2.6.3 DATA QUERY

Data from the database of assessment outcomes is returned by the data query. There are quick search options on the toolbar located in the upper right corner of the homepage and on each individual page. The complete marine natural products database (CMNPD) free-text search allows users to enter any substance name, organism identification, target name, or search entity identifier. Based on the term mentioned below, the search field will present suitable options. In the online interface, users can query data.

1. **Free-Text Quick Search:** Users have the option of going directly to a chosen item's report card page or receiving a list of entities that match a given keyword (e.g., all organisms that match "Cnidaria") (e.g., organism Democnidari).
2. **Search In-Depth:** Query conditions and Boolean operators are merged. First, the grouped conditions' inner Boolean operation is carried out, and then, in ascending order, the outer Boolean operations.
3. **The Outcome of a Keyword Search For "Cnidaria":** The interactive filters on the left allow users to further refine their selection of entries.

2.6.4 DATA VISUALIZATION

The assessment system's visualization system, which is used to control visualization effects like the creation of theme maps and the generation of statistical graphs from the assessment data, is another essential element. A number of interactive data visualizations have been created that allow users to explore and understand the data stored in CMNPD using the JavaScript graphing libraries Plotly and ECharts (https://www.findbestopensource.com/product/apache-echarts) and Plotly (https://plotly.com/graphing-libraries/). In the sunburst charts, many taxonomic and structural classification hierarchies are displayed.

By allowing users to build network visualization by selecting a master node from these entries, vis-network (https://almende.github.io/vis/) provides a systematic analysis function that enables users to intuitively show the interaction between organisms, chemicals, and targets. Clicking on the node will allow you to see the complete entry's details.

2.7 DATA DEPOSITION

The extensive marine natural products database provides a deposit system that allows users to add new compounds, update information on existing compounds, and adjustments to existing data, all of which contribute to the quantity and quality of data.

Only data that has already been published is acceptable, and submissions must include references. Academics doing research on marine natural products (MNPs) are encouraged to submit new compounds to extensive marine natural products databases once the paper has been accepted by the journal (CMNPD).

When depositing a novel chemical entity, the MOL format is preferred over SMILES or InChI strings since it assures that the conformational structure expression matches that in the publication (Lyu et al., 2021).

2.8 APPLICATIONS OF MARINE NATURAL PRODUCTS

In the fields of biotechnology, bioengineering, and pharmaceutical sciences, marine natural products (MNPs) databases are becoming important.

2.8.1 PHARMACEUTICAL SCIENCES

In order to find new drugs, the pharmaceutical sciences have long used molecular databases. The creation of marine natural product databases in this field, such as Marin Lit, has also been spurred by the significant number of authorized medications based on MNPs. Additionally, this region is a pioneer in the industrial development of MNPs due to the great interest in novel marine chemicals in this region. Annual reports on novel MNPs are released by Blunt each year, which aid in maintaining the organization and up-to-dateness of new compound information.

With eight presently on the market that have been licensed for use and at least 12 more in clinical studies, MNPs have already demonstrated their effectiveness as medications. Biologically active MNPs may have features such as bioactive molecules, antitumor, antibacterial, antiviral, anti-inflammatory, and Alzheimer's. The strong demand for MNPs in the pharmaceutical sciences is a result of this achievement (Barbosa & Roque, 2019).

2.8.2 COSMETIC PRODUCT

Marine natural products have also been referred to as cosmetic product additives. It serves a wide range of uses in cosmetics, including the ability to protect the skin, whiten skin, prevent protozoan illnesses of the skin, treat atopic dermatitis, and reduce wrinkles (Barbosa & Roque, 2019).

2.8.3 AGRICULTURAL

Marine natural products from various sources showed insecticidal, herbicidal, and fungicidal action, which increased the likelihood of agricultural use. It has mostly been shown that a family of alkaloids is effective against fungi and insects (Barbosa & Roque, 2019).

2.8.4 FOOD INDUSTRY

A very broad range of applications for marine natural products is found in the food business. They can function as coloring agents, gelling agents, preservatives, stabilizers, and additives. MNPs have also been referred to as supplementary ingredients, dietary supplements, prebiotics, and nutraceuticals with the emergence of functional foods (Barbosa & Roque, 2019).

2.8.5 MATERIALS SCIENCE AND ENERGY

The enormous variety of biomaterials found in marine animals is correlated with the enormous variety of living forms found in the ocean. Thus, retaining compounds of interest from fishing byproducts as a green method of acquiring high-value chemicals is of interest. Diverse applications in ceramics, gels, and other biopolymers are available for marine-based materials science compounds. A novel method of producing biofuels, such as biodiesel and bioethanol, called MNPs has also proven significant (Da Silva & Rodrigues, 2014).

2.8.6 BIOTECHNOLOGY

It might be beneficial for several fields of biotechnology to utilize marine natural product databases more. It's possible that new uses for MNPs may be found as a result of the vast amounts of data present in these databases. As a result, it is crucial to have a single, open-access MNP database that links various sources of marine compounds with the appropriate molecular and physicochemical data. Compound dereplication might also benefit from improved MNP identification in databases, although marine life form compound and metabolite profiling are currently difficult (Barbosa & Roque, 2019).

2.9 CONCLUSION

New marine natural products from invertebrates (NMNPI) seek to promote social and ecological sustainability in order to boost economic growth and human well-being. There is a growing awareness that the chemical diversity of natural products is best suited to provide the fundamental scaffolding for future therapeutic drugs. In the drug development endeavor, there will be additional breakthroughs in the use of new natural products and chemical libraries based on natural products.

KEYWORDS

- **chemical databases**
- **drug discovery**
- **marine natural product**
- **potential applications**
- **terrestrial plants**
- **therapeutic drugs**

REFERENCES

Barbosa, A. J., & Roque, A. C. (2019). Free marine natural products databases for biotechnology and bioengineering. *Biotechnology Journal, 14*(11), 1800607.

Blunt, J. W., Copp, B. R., Munro, M. H., Northcote, P. T., & Prinsep, M. R. (2003). Marine natural products. *Natural Product Reports, 20*(1), 1–48.

Carroll, A. R., Copp, B. R., Davis, R. A., Keyzers, R. A., & Prinsep, M. R. (2019). Marine natural products. *Natural Product Reports, 36*(1), 122–173.

Chen, Y., Garcia de Lomana, M., Friedrich, N. O., & Kirchmair, J. (2018). Characterization of the chemical space of known and readily obtainable natural products. *Journal of Chemical Information and Modeling, 58*(8), 1518–1532.

Coley, C. W., Green, W. H., & Jensen, K. F. (2018). Machine learning in computer-aided synthesis planning. *Accounts of Chemical Research, 51*(5), 1281–1289.

Da Silva, V. C., & Rodrigues, C. M. (2014). Natural products: An extraordinary source of value-added compounds from diverse biomasses in Brazil. *Chemical and Biological Technologies in Agriculture, 1*(1), 1–6.

Djoumbou Feunang, Y., Eisner, R., Knox, C., Chepelev, L., Hastings, J., Owen, G., Fahy, E., Steinbeck, C., Subramanian, S., Bolton, E., Greiner, R., & Wishart, D. S. (2016). ClassyFire: Automated chemical classification with a comprehensive, computable taxonomy. *Journal of Cheminformatics, 8*(61), 61–81. https://doi.org/10.1186/s13321-016-0174-y.

Faulkner, D. J. (1984). Marine natural products: Metabolites of marine algae and herbivorous marine mollusks. *Natural Products Reports, 1*(1), 251–280.

Gilson, M. K., Liu, T., Baitaluk, M., Nicola, G., Hwang, L., & Chong, J. (2016). Binding DB in 2015: A public database for medicinal chemistry, computational chemistry, and systems

pharmacology. *Nucleic Acids Research, 44*(D1), D1045–D1053. https://doi.org/10.1093/nar/gkv1072.
Jiménez, C. (2018). Marine natural products in medicinal chemistry. *ACS Medicinal Chemistry Letters, 9*(10), 959–961.
Jin, L., Quan, C., Hou, X., & Fan, S. (2016). Potential pharmacological resources: Natural bioactive compounds from marine-derived fungi. *Marine Drugs, 14*(4), 76.
Kong, D. X., Jiang, Y. Y., & Zhang, H. Y. (2010). Marine natural products as sources of novel scaffolds: Achievement and concern. *Drug Discovery Today, 15*(21–22), 884–886.
Leal, M. C., Hilário, A., Munro, M. H., Blunt, J. W., & Calado, R. (2016). Natural product discovery needs improved taxonomic and geographic information. *Natural Product Reports, 33*(6), 747–750.
Lyu, C., Chen, T., Qiang, B., Liu, N., Wang, H., Zhang, L., & Liu, Z. (2021). CMNPD: A comprehensive marine natural products database towards facilitating drug discovery from the ocean. *Nucleic Acids Research, 49*(D1), D509–D515.
Martins, A., Vieira, H., Gaspar, H., & Santos, S. (2014). Marketed marine natural products in the pharmaceutical and cosmeceutical industries: Tips for success. *Marine Drugs, 12*(2), 1066–1101.
Mendez, D., Gaulton, A., Bento, A. P., Chambers, J., De Veij, M., Félix, E., Magariños, M. P., Mosquera, J. F., Mutowo, P., Nowotka, M., & Gordillo-Marañón, M. (2019). ChEMBL: Towards direct deposition of bioassay data. *Nucleic Acids Research, 47*(D1), D930–D940.
Molinski, T. F., Dalisay, D. S., Lievens, S. L., & Saludes, J. P. (2009). Drug development from marine natural products. *Nature Reviews Drug Discovery, 8*(1), 69–85.
Penesyan, A., Kjelleberg, S., & Egan, S. (2010). Development of novel drugs from marine surface-associated microorganisms. *Marine Drugs, 8*(3), 438–459.
Pereira, F.; Aires-de-Sousa, J. Computational Methodologies in the Exploration of Marine Natural Product Leads. Mar. Drugs 2018, 16, 236. https://doi.org/10.3390/md16070236
Pereira, F. (2019). Have marine natural product drug discovery efforts been productive and how can we improve their efficiency? *Expert Opinion on Drug Discovery, 14*(8), 717–722.
Petersen, F., & Amstutz, R. (Eds.). (2008). *Natural Compounds as Drugs,* Volume II (Vol. 66). Springer Science & Business Media.
Proksch, P. (2009). Chinese Marine Materia Medica. By Huashi Guan and Shuguang Wang. Shanghai Scientific and Technical Publishers, China Ocean Press, and Chemical Industry Press: Shanghai, Beijing, China, 2009; Hardback, 7064 pp; ¥ 2920; ISBN 978-7-5323-9958-1/R• 2707; ISBN 978-7-5323-9973-4/R• 2708; ISBN 978-7-1220-6012-9.
Shang, J., Hu, B., Wang, J., Zhu, F., Kang, Y., Li, D., Sun, H., Kong, D. X., & Hou, T. (2018). Cheminformatic insight into the differences between terrestrial and marine-originated natural products. *Journal of Chemical Information and Modeling, 58*(6), 1182–1193.
Szymkuć, S., Gajewska, E. P., Klucznik, T., Molga, K., Dittwald, P., Startek, M., Bajczyk, M., & Grzybowski, B. A. (2016). Computer-assisted synthetic planning: The end of the beginning. *Angewandte Chemie International Edition, 55*(20), 5904–5937.
Wang, Y., Bryant, S. H., Cheng, T., Wang, J., Gindulyte, A., Shoemaker, B. A., Thiessen, P. A., He, S., & Zhang, J. (2017). PubChem BioAssay: 2017 update. *Nucleic Acids Research, 45*(D1), D955–D963.
Zhang, G., Li, J., Zhu, T., Gu, Q., & Li, D. (2016). Advanced tools in marine natural drug discovery. *Current Opinion in Biotechnology, 42*, 13–23.

CHAPTER 3

Landscape on the Drug Development Strategies for Marine Products

SUMEL ASHIQUE,[1] AFZAL HUSSAIN,[2] PRASHANT KUMAR,[3]
NEERAJ MISHRA,[4] GAURAV GUPTA,[5–7] and KAMAL DUA[8,9]

[1]*School of Pharmaceutical Sciences, Lovely Professional University, Phagwara, Punjab, India*

[2]*Department of Pharmaceutics, College of Pharmacy, King Saud University, Riyadh, Saudi Arabia*

[3]*SRM Modinagar College of Pharmacy, SRMIST Delhi-NCR Campus, Modinagar, Ghaziabad, Uttar Pradesh, India*

[4]*Amity Institute of Pharmacy, Amity University Madhya Pradesh (AUMP), Gwalior, Madhya Pradesh, India*

[5]*School of Pharmacy, Suresh Gyan Vihar University, Jagatpura, Jaipur, Rajasthan, India*

[6]*Uttaranchal Institute of Pharmaceutical Sciences, Uttaranchal University, Dehradun, Uttarakhand, India*

[7]*Department of Pharmacology, Saveetha Dental College, Saveetha Institute of Medical and Technical Sciences, Saveetha University, Chennai, Tamil Nadu, India*

[8]*Faculty of Health, Australian Research Center in Complementary and Integrative Medicine, University of Technology Sydney, Ultimo, NSW, Australia*

[9]*Discipline of Pharmacy, Graduate School of Health, University of Technology, Sydney, Ultimo, NSW, Australia*

Handbook of Research in Marine Pharmaceutics: Exploring Oceanic Microbial Diversity for Human Health and Wellness. Md. Faiyazuddin & Khalid Rehman Hakeem (Eds.)

ABSTRACT

Ailments caused by illnesses alter patterns, and new ailments appear as a result of shifting environmental conditions. The resources already available for pharmaceuticals are already at capacity due to the rapid increase in the global population. Due to the rising needs of the global population, pharmaceutical producers are always searching for new sources to provide efficient and secure medications. Water covers 75% of the earth's surface, yet little is explored about the pharmacology of marine creatures, and much of it is still unknown. All across the world, they are heavily utilized in the creation of novel drugs. Research on these medications of marine origin is possible thanks to marine pharmacology. For these frequently complicated compounds to be successfully developed as drugs, a number of obstacles must be addressed, such as the supply issue and target identification. Nevertheless, strategies are available to get over these obstacles. The effectiveness of marine natural products as therapeutic lead depends on advancements in technologies such as sampling techniques, nanoscale NMR for structure analysis, total chemical synthesis, fermentation, and biotechnology. We provide examples of the high levels of innovation occurring in the sector of marine natural goods, which, in our opinion, will result in the release of a new wave of pharmaceuticals onto the market and into pharmacies in the future.

3.1 INTRODUCTION

In recent years, marine sources have received a lot of interest as prospective sources for natural products. The biopharmaceutical industry has a bright future due to the numerous chemicals that have been separated from marine sources and will be doing so in the future. Long recognized as a key source of new metabolites useful in the treatment of human diseases, and marine species. It has been stated that the maritime environment is a key source of chemical as well as biological diversity and that the oceans contain almost 300,000 different species, only a small portion of the total number of species still to be discovered. Each of these marine bioproducts has offered significant potential market value, and the ocean has a rich supply of ever more unique molecules with considerable potential as pharmaceuticals, nutraceuticals, cosmeceuticals, agrichemicals, and enzymes (James et al., 2012). The marine environment continues to be a gold mine for novel natural compounds that could be used as medication candidates. The ocean's wide range of temperatures, pressures, and nutrient concentrations allowed for significant speciation at entire phylogenetic levels, from bacteria to mammals. Many bioactive substances have been isolated from a variety of marine organisms, including sponges, soft corals, bryozoans, sea slugs, and tunicates (Fenical et al., 2006). Newer marine-derived microorganisms are promising sources of new bioactive substances that are crucial for the process of developing new drugs. In comparison to the amount of active bacteria extracted from seawater and sediments (5%), the reported percentage of pharmacologically active bacteria associated with marine animals (20%) and seaweeds (11%) is larger. The struggle for food and space among marine invertebrates' associated microbes may be the cause of their synthesis of bioactive compounds. The epibiotic defense of marine invertebrate hosts may depend on these bioactive substances, but they also have considerable industrial and medical uses. Current research to explore marine natural products is still in its embryonic stage due to a lack of ethnopharmacological history and the challenges associated with collecting marine species. More than 4200 unique compounds have been separated from shallow water to 900 m depths of the sea over the past 10 years thanks to the advancement in novel diving practices, remote-operated devices, and other tools that helped us overcome the challenges of collecting marine samples. The invention of high throughput screening (HTS) skills accelerated the assay speed and the drug discovery process using combinatorial chemistry. The discovery of new natural sources and creatures, such as those in the marine realm, which make up about 70% of the earth's outer surface and are the largest untapped source of riches, is linked to the return of natural products-initiated drug discovery (Montaser et al., 2011). Out of the 30 compounds in the global marine pharmaceutical clinical pipeline, including 22 drug candidates in phases III, II, or I of drug development, eight medications have received approval from the European Medicines Agency (EMEA), the Japanese Ministry of Health, the Australian Therapeutic Goods Administration (TGA), and the US FDA (Wiese et al., 2019; Newman et al., 2016; Gerwick et al., 2012). It's interesting to note that in the 21st century, marine-based natural products (MNPs) and derivatives make up six out of every eight approved pharmaceuticals (Pereira et al., 2018).

3.2 MARINE-BASED NATURAL PRODUCT DRUG DISCOVERY

The majority of bioactive MNPs have been found in marine-based invertebrates such as sponges, jellyfish, anemones, and rocky corals, with Porifera (like sponges) and *Cnidaria phyla* being the most abundant.

However, the true source of most MNPs appears to be the microorganisms that coexist with them in a close symbiotic relationship (Jimenez et al., 2018). Nearly all of the MNPs that have been licensed as medications or that are presently undergoing clinical trials are made synthetically by bacteria and cyanobacteria, such as salinosporamide A and brentuximab vedotin from cyanobacteria *Symploca* sp. (actinobacteria *Salinispora tropica*). Therefore, it is almost inevitable that substances derived from microorganisms will take over the MNP industry in the years to come. More than 75% of all antibiotics found to date have been extracted from terrestrial bacteria, making up the majority of commonly used antibiotics; though, antimicrobial compounds derived from marine sources have not yet reached the clinical trial stages (Newman et al., 2016, Choudhary et al., 2017). As mentioned above, compared to antibacterial macromolecules, there is a significant effort and amount of research going into the creation of innovative anti-cancer medicines. Nearly 50% of the bacterial isolates showed antibiotic potential against Gram-positive microorganisms (*Bacillus subtilis, Staphylococcus lentus*), Gram-negative strains (Escherichia coli), and mixed strains (*Bacillus subtilis, Staphylococcus lentus*) bacteria, according to Wiese et al. described in the investigations of antibiotic activity surveys conducted in their lab between 2008 and 2017 (Wiese et al., 2019). The authors used specimens of MNPs, such as polyketides along with cyclic peptides, developed from bacterial and marine-based fungi, to highlight the massive potential of marine significant microorganisms for the creation of new anti-infective medicines (Abdelmohsen et al., 2017; Andersen et al., 2018). Despite the fact that the key area of MNP-based drug discovery has been substantially adopted throughout the 21st century, there are still a number of barriers that limit the production of pharmaceuticals from NPs, notably MNPs. The reliable development of methods to manufacture a continuous supply of NPs through genetic, chemical, or fermentation approaches is one of the most important prerequisites.

3.3 ORIGINS AND ADVANCEMENTS OF MARINE CHEMICAL BIOLOGY

The chemistry of marine natural products has been fundamentally altered by the molecular revolution (Luesch et al., 2011). Over the past 10 years, new fields of science and inquiry that started to address long-standing issues in the field and offer novel suggestions for future research directions have emerged as a result of the eruption of various "omic" methodologies to genomics, proteomics, metabolomics, and transcriptomics now used by marine natural product specialists. Thanks to the integration of exhaustive natural product chemistry and modern molecular biology, scientists can now discuss fundamental queries about the biosynthetic capability of marine organisms, the synthetic functions of microbes in marine invertebrates under natural product chemistry, and the bioengineering perspective of marine drugs. In this part, we focused on recent advances in marine-based natural compound research that have been explored by modern omic tactics, with an emphasis on small molecule production.

3.4 DEVELOPMENT STRATEGIES

MNPs produced from macro-, or microorganisms have typically been made available for clinical studies through their scheme based on semi-synthetic as well as chemical synthesis. With some marine-derived medications having been successfully approved, marine natural products (MNPs) continue to be viable medicinal sources. However, it is never easy to find bioactive compounds in marine environments. Finding unique marine resources, resurrecting unfriendly creatures outside of the marine setting, characteristic novel compounds from recognized ones, identifying the role of MNPs, and maximizing their pharmaceutical utility are just a few of the challenges that must be solved. In this effort to motivate future MNP discovery pipeline initiatives, we examine several cutting-edge methods and procedures that can be used to address the aforementioned difficulties. In this prospective review, we will outline the cutting-edge methods and approaches that may be utilized in the area of natural drug development and exploring marine resources. It is not meant to be a complete and full treatment of the subject; rather, it aims to highlight the difficulties faced by marine drug development pipelines and serve as a catalyst for future work on the identification of MNPs and their advancement into clinical trials. We now have a better grasp of the genetic potential of these microbes with regard to their capacity to produce a wide range of chemical compounds. Many of the biosynthetic genes in bacteria are silent or are not translated in the laboratory. The bottleneck in drug development from microbial sources results from the fact that we have only discovered a small portion of the true biosynthetic variety of the chemicals of microbial origin (Pereira et al., 2019). Marine creatures unquestionably produce a wide variety of tiny compounds with intriguing chemical structures and powerful biological functions. Early research on biosynthesis was mostly based on isotope tracer experiments using a variety of marine microorganisms, algae, and invertebrates (Moore et al., 2005). The use of natural products in drug development has changed significantly over the past 30 years, with the big pharmaceutical corporations' involvement beginning to wane by the middle of the 1990s. Nevertheless, resourceful academics took advantage of the gap created by more extensive research and development initiatives, often working in conjunction with industry. In the previous five years, the field appears to have seen a resurgence. A large recognition that rivals approaches, including combinatorial chemistry, have failed to produce novel therapeutic leads in meaningful quantities has contributed to some of this and is partially what has motivated novel advances in analytical technologies, spectroscopy as well and high-throughput screening (Koehn et al., 2004). In the interim, focused efforts in medication development that were started in the 1970s and fundamental scientific research in the biochemistry and pharmacology of marine-based natural products have eventually resulted in the identification of marine-based drugs. Figure 3.1. describes various techniques used for the development of marine source-based formulations.

3.4.1 CO-CULTURE

In order to replicate the complex microbial habitat where they dwell in nature, two or more distinct bacteria are co-cultivated (also known as mixed fermentation). These bacterial communities yield bioactive secondary metabolites as a kind of shield because of the conflict over few resources and the competition for them. It is thought that the rivalry and hostility between these bacteria activate the biosynthetic genes that are dormant in opulent culture circumstances. Co-culturing is a potent experimental technique for activating quiet gene clusters or increasing the synthesis of substances that are already present constitutively. Growing microorganisms *in vitro* is a promising experimental strategy for increasing their chemical diversity. Competition and stress are induced in co-cultures, which increases the creation of new chemicals that are undetectable in pure cultures (Ka et al., 2014; Reen et al., 2015). In yeast extract-glucose media, the co-culture using two marine-based actinomycetes, ANAM-5 and AIAH10, increased cytotoxic and antifungal effects as well as the creation of one novel chemical (Haque et al., 2015). Here, the interaction between the two cultures may be what induces the synthesis of novel compounds produced from the marine brown-type alga *Rosenviagea* sp., *Pestalotia* sp. The co-cultures using marine-derived unrevealed Gram-negative bacterial strains of the genus *Thalassopia* (CNJ-328) resulted in chlorinated prenyl seco anthraquinone, pest alone, which had

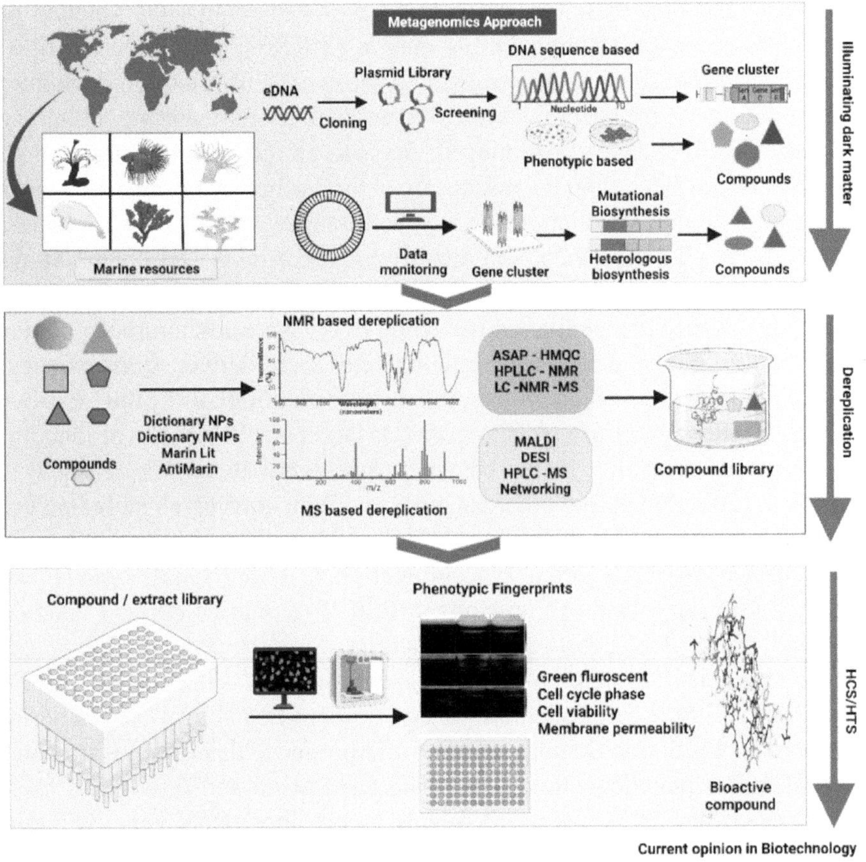

FIGURE 3.1 Several approaches for the development of marine source-based formulations.

antibacterial potential against methicillin-resistant *Staphylococcus aureus* (MRSA) as well as vancomycin-resistant Enterococcus faecium when 60 human cell lines from the National Cancer Institute were tested (Cueto et al., 2001). In a different study, a marine-derived fungus called *Libertella* sp. had an established (3-day-old) culture when a marine-proteobacterium called CNJ-328 encouraged the development of distinctive four primarane diterpenoids called libertellenone A-D, which were not developed by these cultures alone (CNL-523). These substances were cytotoxic to the human adenocarcinoma HCT-116 cell line, with libertellenone D being the most active of these (Oh et al., 2005). Emericella, a marine-derived fungus, and *Salinospora arenicola*, a marine actinomycete, were co-cultured to produce emericellamides A and B. In the co-culture, there was a 100-fold increase in their output (Oh et al., 2007). These substances had mild cytotoxic effects on HCT-116 cells and antibacterial action to counter the function of methicillin-resistant Staphylococcus aureus (MRSA). Due to interactions between different species and genera, Dusane et al. (2011) explored whether marine bacteria's antibacterial, biosurfactant and quorum-sensing inhibitory properties were enhanced or induced. *Bacillus* sp. (*B. thuringiensis* and *B. megaterium*) isolates from the marine alga Ulva californica were co-cultured to produce greater amounts of indole compounds and diketopiperazines, which improved antibiotic action against *B. megaterium* (Trischman et al., 2004). The Mediterranean sponge *Nocardiopsis* sp. RV163 and the Red Sea sponge *Dysidea avara, Actinokinospora* sp. EG49 were co-cultured in liquid-based culture media. These two actinomycetes were derived from sponges, and they produced several compounds that were identified using the LC-PDA and 1H-NMR fingerprinting techniques (Dashti et al., 2014). The production of istamycins was only increased when *Streptomyces tenjimariensis* was introduced 24 hours before other bacteria, resulting in research including the interaction between *S. tenjimariensis* and 53 unidentified marine microbes for stimulation of istamycin A and B accumulation by *S. tenjimariensis* (Slattery et al., 2001). Aspergillus fumigatus MR2012, a marine-derived fungus, and two hyper- and desert-adapted bacteria, *Streptomyces leeuwenhoekii* strains C34 and C58, were co-cultured, according to Wakefield et al. (2017), who reported the dual production of newly identified fungal and bacterial compounds. Luteoride D and pseurotin were produced using strains C34 and MR2012. The pentalenic acid was earliest extracted and discovered from this strain with C58, whereas all fungus metabolites under axenic culture were repressed. Chaxapeptin synthesis was also increased (Table 3.1).

3.4.3 METAGENOMICS STRATEGIES

Single-cell genome-based phylogenetic alignment (Rinke et al., 2013; Wilson et al., 2014) and innovative algorithm-based gene cluster network analysis (Cimermancic et al., 2014) are two remarkable techniques that were created to find bioresources (such as microorganisms) living in unexplored maritime environments. Metagenomics is a culture-independent technique to overcome the difficulties in retrieving the biosynthetic capability of the "uncultured bulk" of organisms (Brady et al., 2009). This method involves directly extracting the entire community DNA, also known as environmental DNA (referred to as eDNA) or metagenomic DNA, isolated from environmental samples like seawater, marine sand, or marine invertebrates. An appropriate cloning vector is then utilized to create a metagenomic library from the extracted sample DNA that comprises the genes necessary for the production of a specific NP, containing genes that encode the biosynthetic, regulatory, and resistance enzymes. For the purpose of utilizing the specific metagenomic eDNA for the creation of small molecules, two general methods are used: using both DNA unique sequence-based and function-based screening techniques. Marine uncultured bacteria research has made considerable use of DNA sequence-based metagenomics methods to identify clusters that produce tiny molecules. The cyanobactin class of NPs (ribosomal peptides) was found in 2011 by Eric Schmidt's team using PCR probes to search a sizable genomic DNA library of samples from the common ascidian Didemnum molle symbionts (Donia et al., 2011). Targeting functional genes that were preserved in a set of well-known or unidentified biosynthetic pathways, such as the resistance genes, NRPS, and PKS conserved motifs, requires the development of PCR-based sequence tag techniques.

TABLE 3.1 Marine-based Microbial Co-culture and Compounds Reported

Sl. No	Compounds Reported	Co-cultivated Microorganisms
1.	Aspergicin, Neoaspergillic acid, Ergosterol	*Aspergillus* sp., *Aspergillus* sp.
2.	Novel derivative of 8-Hydroxy-3-methyl-9-oxo9H-xanthene-1-carboxylic acid methyl ether	*Pestalotia* sp.
3.	Libertellenones A-D	*Libertella* sp. and *Thalassopia sp*
4.	Emericellamide A, Emericellamide B	*Salinospora Arenicola* and *Emericella* sp.
5.	Glionitrin A	*Aspergillus fumigatus, Sphingomonas* sp.
6.	Indole, Phe-Pro diketopiperazines	*S, sciuri, B. thuringensis* and *B. megaterium*
7	Istamycins	*Streptomyces tenjimariensis*
8	Derivative of Deoxyfunicone; 1,3,8-trihydroxy6-methylxanthen-9-one; 9R, 14S-epoxy-11-deoxyfunicone; 9S,14R-epoxy-11-deoxyfunicone	*Newer Penicillium* sp. *WC-29–5, Streptomyces fradiae*
9	Luteoride D	*Aspergillus fumigatus, Streptomyces leeuwenhoekii C34*
10	Chaxapeptin, pentalenic acid	*Streptomyces leeuwenhoekii C58 and Aspergillus fumigatus.*

Salinispora, a common marine Actinomycete genus, was genotyped using sequence tags in earlier applications to NP characterization (Prieto-Davó et al., 2013). By combining sequence tag interrogation and arraying eDNA as a library, Sean F. Brady and colleagues were able to successfully recover the whole pathways from a variety of biosynthetic systems. This platform for information technology is known as eSNaPD (Environmental Surveyor of NPD). Further heterologous expression of these biosynthetic clones from metagenomic libraries has produced novel bioactive compounds, such as the siderophores vibrioferrin and bisucaberin, which are obtained from the sea forms of the anticancer drug pederin, cyclic peptides generated from cyanobacteria, and fresh members of the peptides produced by ribosomal synthesis in the microviridin family from cyanobacteria in freshwater (Owen et al., 2013).

3.4.4 GENOME MINING APPROACHES

The rapid advancement of swift and economical next-generation DNA sequencing (NGS) technology has led to an unprecedented accumulation of genomic and metagenomics sequence data, opening up an innovative field in genomic study. Systematic investigations of genome sequences utilizing specific bioinformatic techniques undoubtedly contribute to the advancement of biological system knowledge and the sensible use of genomic code knowledge. The contrast between the small number of isolated compounds and the enormous biosynthetic capability and diversity validated by numerous organisms on the genomic level reveals the immense potential of genomic mining of novel NPs and the strategy it represents for accessing the enormous origin of novel, biologically active metabolites (Gross et al., 2009). By examining genomic data of DNA sequences that may encode enzymes involved in the production of particular compounds, the genome mining method effectively discovers known gene clusters/pathways and their homologs in the target genomes. The construction of ways for speedily purifying and structurally characterizing the compounds is further encouraged by the discovery of interesting essential components of NPs from unidentified clusters by the genome mining approach (Zerikly et al., 2009). Mutational biosynthesis entails inactivating genes within the cryptic biosynthetic gene cluster that are thought to be essential, stopping a biosynthetic pathway in situ, and then chromatographically comparing the metabolite outlines of the mutant within wild-type organisms. The whole biosynthetic gene cluster is cloned and expressed in a heterologous host in the heterologous biosynthesis strategy. Moreover, the compounds resulting from the ambiguous biosynthetic path can be recognized by describing the metabolites produced by the heterologous host carrying the vital biosynthetic gene cluster and relating these to the wild-type host.

3.4.5 HETEROLOGOUS BIOSYNTHESIS

A sizable collection of gene clusters with the potential to produce NPs has been made available through studies on metagenomics and genome mining. However, because linked genetic factors are either not expressed at all or only expressed at extremely low levels in laboratory settings, the majority of such biosynthetic paths are merely bioinformatically anticipated, making it difficult to get the relevant NPs. However, because the genetic and metabolic histories of most of those native organisms are unknown, in situ genetic alterations are typically not an option. The aforementioned situation specifically created the heterologous biosynthesis approach, which offers benefits in the directed manufacture of the encrypted product by shifting the inherent genetic pathway for a chosen NP to a substitute host. Three phases commonly make up the heterologous biosynthesis approach: locating and deciding the target biosynthetic gene groups; designing the route; assembling the gene group; and transferring it into a surrogate host; detecting the product and gene expressions (Zhang et al., 2011). In a nutshell, these targeted NP biosynthetic gene groups are selected and identified from gene sequence databases made available by the aforementioned genetic and computational methods. The target gene clusters are then carefully loaded and expressed in a suitable heterologous host using DNA engineering methods after being extracted from their original environment (Pfeifer et al., 2001). Quick gene modification methods are crucial for rearranging and remodeling target NP biosynthesis pathways in selected heterologous hosts. Also, numerous innovative *in vitro* and *in vivo* DNA assembly technologies, such as Gibson isothermal assembly (Gibson et al., 2009), the Golden Gate assembly technique (Engler et al., 2009), and others, have been created along with the conventional restriction enzyme cut or ligation method.

3.4.6 RAPID DE-REPLICATION OF NPS

MNPs are entering a novel "booming epoch," which is even equivalent to the "Golden Era of Antibiotics" in the mid-20th century, as new technologies are being created in marine springs-based structure isolation/characterization.

Natural product scientists have challenged the constantly rising issue that a lot of time is spent repeatedly discovering structures that have already been reported in the literature since the natural structures are quickly collected. To boost the practice of distinctive new compounds from well-known samples of blends, scientists have combined various fields of information in structural identifying and analyzing or developing completely novel methods, like computational spectrometry methods and hyphenated methods (recommended, utilizing the combination of two or more separation and detection approaches). Numerous public and private commercial databases, like PubChem, SpecInfo, Dictionary NPs, Dictionary MNPs, MarinLit, AntiMarin, etc., have been produced as a result (Gaudêncio et al., 2015).

3.4.7 MS-CENTERED HYPHENATED ANALYTICAL TECHNOLOGIES

Compound identification and characterization are essential early on in the pipelines for identifying natural therapeutic medicines. The workflow for analytically supported NP identification is known as an "essential control point" for compound dereplication. Through the process of dereplication, it may be possible to focus attention with additional time and resources on the candidates who have the most promise and avoid wasting early efforts on samples with low potential. In order to do speedy first-pass dereplication regarding screening of tasks, MS approaches were upgraded (Potterat et al., 2000). However, due to the limitations imposed by the complexity of some unique materials, like tissues and biofluids, it is difficult to apply existing MS techniques for analysis. Surface-based mass image analysis has evolved during the past 10 years, allowing for a significant reduction in sample preparation methods. Several matrix-based or matrix-free massive ionization methods, including desorption electrospray ionization (DESI), secondary ion mass spectrometry (SIMS), and matrix-assisted laser desorption/ionization (MALDI), are included in these methods (Nielsen et al., 2015). Although MS techniques are more accurately described as annotations than as definitive solutions, their usefulness could be significantly increased when combined with other analytical techniques (Lang et al., 2008).

3.5 COMPUTER-AIDED DRUG DESIGN (CADD)

To help the recent design of biological NP products, to assist in the *in vivo* & *in vitro* evaluation of isolated NPs and related extracts, and to virtually explore databases of known or planned NPs, computer calculation of biological actions of MNPs is necessary. Moreover, the areas of the chemical space that surround NPs are acknowledged as encouraging the development of novel therapeutic leads as a result of the millions of years of chemical structure evolution required for the best functioning of biochemical machinery (Ertl et al., 2008). The exploration of international networks between active substances and their targets, the simulating of interfaces concerning ligands and binding locations, and the establishment of SAR using newer NPs and MNPs are all examples of advances in computational methodologies that have been reported (Fang et al., 2017; Wu et al., 2017; Davis et al., 2015).

3.5.1 LIGAND-BASED (LB)

The search for novel lead compounds can be successful when collections of active molecule targets are established. In order to build QSARs, anticipate candidate properties, and give a novel idea for the creation of novel compounds, machine learning (ML) techniques have been developed. These algorithms use structural alignment for pharmacophore identification and virtual screening. Dinesh Kumar et al. predicted the targets designed for sporolides A and B including ligand-based pharmacophore selection towards well-known drugs and inhibitors. These reported NPs are polycyclic-based macrolides obtained from the necessary marine actinomycete and Salinispora tropica. It contains one aromatic ring, 6 H-bond acceptors, and one hydrophobic group, as discovered to constitute the eight pharmacophore features of sporolides A and B (Dineshkumar et al., 2014). The 3D models and a pharmacophore pattern with 400,000 known ligands were utilized to analyze the Community Binding Database. Further, *in-vitro* studies revealed that sporolide B considerably decreased the action of HIV-1 RT, indicating that it might be a potential treatment alternative for HIV and other associated retroviral viruses. The similar lab subsequently released a related computational investigation for the MNPs salinosporamide A, B, and C derived from the same origin and decided that the glucocorticoid receptor and methionine type aminopeptidase 2 could be novel drug targets, indicating potential anti-inflammatory and anti-cancer activities of salinosporamides (Dineshkumar et al., 2017). Shang et al. (2018) examined the dissimilarities between terrestrial and marine nanoparticles (NPs) using chemoinformatics approaches on a dataset encompassing 32,937 MNPs and 132,071 terrestrial NPs. The researchers discovered a pattern where MNPs had elongated rings and bigger chains, more halogens (specifically bromine), more nitrogen, and lower solubility. Because fewer MNP platforms are found in databases of well-known ligands, this is consistent with the fact that fewer MNPs have been used in drug discovery programs and points to the possibility of larger drug development potential for MNPs. Doliculide (26) is a marine natural product made by the sea hare *Dolabella auricularia.* Schneider et al. (2016) computationally discovered and biochemically established an anonymous, high-affinity macromolecular target of this MNP. The Doliculide, an NP-based product with potent actin-polymerizing and anticancer properties, and 134 intermediates and precursors of a total synthesis underwent automated target prediction using the SPiDER procedure. In order to create QSARs, which are built using experimental data that is currently available and molecular descriptors that encode structural information, linear regressions and machine learning methods are commonly used. Here, we discuss several current instances of QSAR models that have been utilized to calculate the biological actions and ADMET features of MNPs. The protein kinase B targets for additional structure-based investigations were chosen with the use of the QSAR approach, which was utilized to uncover pertinent structural properties. Later, a lead compound was chosen using ADMET predictions. In order to replicate transmembrane anion transport activity, Knight et al. (2016) used 43 synthesized variants of the marine alkaloid tambjamine. Philicity was measured using equations for two-, three-, and four-parameter linear models. The analysis of MNPs using QSAR approaches has also been prompted by the search for new antimalarial medications (Flores et al., 2018). The natural substance thiaplakortone-A, which was exposed in *Plakortis lita* (marine sponge) and is effective against chloroquine-resistant Plasmodium falciparum, was studied by Aswathy et al. (2017) using 42 analogs. Designing and screening novel virtual compounds required the development of numerous QSAR models, comprising 2D and 3D QSAR. The outcomes were integrated with computer-generated interfaces to *P. falciparum* calcium-dependent protein kinase 1.

3.5.2 STRUCTURE-BASED (SB)

Molecule docking is the main SB method for evaluating binding modes, estimating affinities to macromolecular targets, and helping to generate pharmacological leads. The method's application to MNPs is demonstrated in several recent articles (Kudryavtsev et al., 2014; Liu et al., 2015; Llorach et al., 2017; Botic et al., 2017). Liu et al. created, manufactured and tested the MNP tasiamide B as a BACE1 inhibitor, a potential target in

Alzheimer's disease. The acyclic peptide known as tasiamide B is composed of many amino acid residues and a statine-like unit. The examination of the structural-activity relationship (SAR) utilizing truncated derivatives led to the identification of a critical core structure and free carboxylic acid group for antagonistic action. A docking simulation (Liu et al., 2015) reinforced the results. Combining SB computational analysis and *in vitro* testing, the molecular target of selected 13 low M. Wt. MNPs commencing marine sponges and ascidians were identified. Some of them have the capacity to activate nicotinic acetylcholine receptors (nAChRs), as evidenced by their structural similarity to numerous cholinergic ligands (Kudryavtsev et al., 2014). Acetylcholinesterase (AChE) and Protein kinases are potential therapeutic targets in the fight against Alzheimer's disease (AD). The findings of these docking studies of meridians A-G were reported by Llorach-Pares et al. (2017) to support the future advancements of anti-AD drugs (indole alkaloids derived from tunicate Aplidium). The capacity of 12 pyrrole alkaloid derivatives to disrupt ATP binding cassette (ABC) carriers, which may be useful in overcoming cancer cells' multidrug resistance, was examined in a marine sponge called *Ianthella* sp. (Huang et al., 2014). Out of them—lamellarin O—was found to be an effective and specific inhibitor of the BCRP ABC transporter. A SAR analysis of the 12 MNPs and 6 created analogs, which was confirmed by in silico docking studies, revealed that a carboxylic ester, methoxy-acetophenone, and two phenolic residues are essential components of the inhibitory pharmacophore. Cen-Pacheco et al. (2018) employed molecular docking to recognize the various functions of two unique squalene derivatives produced from the red seaweed *Laurencia viridis* acting as inhibitors of Ser-Thr protein phosphatase type 2A (PP2A). Ebrahim and Sayed et al. (2016) reported on the examination of an MNP-based library that has 71 compounds with different scaffolds (e.g., macrolides, alkaloids, terpenes viz: sesquiterpenes, diterpenes, sesterterpenes, and triterpenes). Skariyachan et al. identified combinations from microbial symbionts linked with sea sponges as potential lead molecules opposite to the Ebola virus using a computational approach.

3.4.2 ARTIFICIAL INTELLIGENCE FOR MARINE NATURAL PRODUCT INNOVATION AND ADVANCEMENTS

The marine-based natural products are a significant source of drug discovery due to their unique structures and distinctive pharmacological activities. Due to the diversity and lavishness of natural products in the ocean, analyzing thousands of compounds using customary pharmacology and investigational models can be time-wasting and costly. Artificial intelligence (AI) plays a vital role in drug discovery. Artificial intelligence can contribute to multiple phases of drug development and is estimated to considerably reduce the time of drug development, improving drug development success rates. In a nutshell, target NP biosynthetic gene groups have been selected and further identified using gene sequence databases made available by the aforementioned genetic and computational methods. The target gene clusters are then carefully loaded and expressed in a suitable heterologous host using DNA engineering methods after being extracted from their original environment. Extensive computational techniques are employed to envisage the potential therapeutic assessment of marine natural products based on drug structures, drug targets, and drug interaction studies. Multi-target exploration of marine natural products can also be executed to determine whether drug targets play regulatory roles in disease pathways (García et al., 2022).

3.5 CONCLUSION

Over the past few decades, a lot of work has been done in this area. At least eight recently released medications, a large number of promising marine-derived compounds that are awaiting investigation, and dozens more therapeutic candidates that are currently undergoing clinical or preclinical testing are all the result of requests for medicinal substances from marine perseverance. Technological developments in gene sequencing and bioinformatics. To achieve the best results, a combination of these tools is typically engaged in successful situations. With the aim of illuminating the future path that will be taken by those who will advance MNPs into clinical use by outlining the key challenges and possible solutions in the marine natural drug discovery channels, we look back on these methodological and technical developments as well as effective case studies in the field of MNP here. The necessity to modify the MNP has decreased as a result of the adoption of these label-free methodologies in recognizing the biological targets of MNPs due to the normally small number of isolated compounds and typically complex architectures. It is possible to explore the biological role of MNPs towards genomic prospects using novel genomic techniques. For finding antibiotic targets, target-directed specific genome mining of producing organisms is exclusively promising. The more general profiling techniques support the target identification process through phenotypic screening and direct target identification.

ACKNOWLEDGMENT

Thanks to BIT, School of Pharmacy, Meerut for helping to draft this chapter.

CONFLICT OF INTEREST

All the authors declare no conflict of interest at all.

FUNDING

Not applicable

KEYWORDS

- **acetylcholinesterase**
- **development approaches**
- **future aspects**
- **marine formulations**
- **marine natural product**

REFERENCES

Abdelmohsen, U. R., Balasubramanian, S., & Oelschlaeger, T. A. (2017). Potential of marine natural products against drug-resistant fungal, viral, and parasitic infections. *Lancet Infectious Diseases*, *17*(2), e30–e41.

Andersen, R. J., Ntie-Kang, F., & Tietjen, I. (2018). Natural product-derived compounds in HIV suppression, remission, and eradication strategies. *Antiviral Research*, *158*, 63–77.

Aswathy, L., Jisha, R. S., Masand, V. H., Gajbhiye, J. M., & Shibi, I. G. (2017). Computational strategies to explore antimalarial thiazine alkaloid lead compounds based on an Australian marine sponge *Plakortis lita*. *Journal of Biomolecular Structure and Dynamics*, *35*(11), 2407–2429.

Botic, T., Defant, A., Zanini, P., Zuzek, M. C., Frangez, R., Janussen, D., & Sepcic, K. (2017). Discorhabdin alkaloids from Antarctic *Latrunculia* spp. sponges as a new class of cholinesterase inhibitors. *European Journal of Medicinal Chemistry*, *136*, 294–304.

Brady, S. F., Simmons, L., Kim, J. H., & Schmidt, E. W. (2009). Metagenomic approaches to natural products from free-living and symbiotic organisms. *Natural Product Reports, 26*(11), 1488–1503.
Cen-Pacheco, F., Perez Manriquez, C., Souto, L. M., Norte, M., Javier Fernandez, J., & Hernandez Daranas, A. (2018). Marine longilenes, oxasqualenoids with Ser-Thr protein phosphatase 2A inhibition activity. *Marine Drugs, 16*(4), 131.
Choudhary, A., Naughton, L. M., & Montanchez, I. (2017). Current status and prospects of marine natural products (MNPs) as antimicrobials. *Marine Drugs, 15*(9), 272.
Cimermancic, P., Medema, M. H., Claesen, J., Kurita, K., Brown, L. C., Mavrommatis, K., & Clardy, J. (2014). Insights into secondary metabolism from a global analysis of prokaryotic biosynthetic gene clusters. *Cell, 158*(2), 412–421.
Cueto, M., Jensen, P. R., Kauffman, C., Fenical, W., Lobkovsky, E., & Clardy, J. P. (2001). A new antibiotic produced by a marine fungus in response to bacterial challenge. *Journal of Natural Products, 64*(11), 1444–1446.
Dashti, Y., Grkovic, T., Abdelmohsen, U. R., Hentschel, U., & Quinn, R. J. (2014). Production of induced secondary metabolites by a co-culture of sponge-associated actinomycetes, *Actinokineospora* sp. EG49 and *Nocardiopsis* sp. RV163. *Marine Drugs, 12*(5), 3046–3059.
Davis, G. D. J., & Vasanthi, A. H. R. (2015). QSAR-based docking studies of marine algal anticancer compounds as inhibitors of protein kinase B (PKB). *European Journal of Pharmaceutical Sciences, 76*, 110–118.
Dineshkumar, K., Aparna, V., & Hopper, W. (2017). Ligand-based pharmacophore modeling and extended bioactivity prediction for salinosporamide A, B, and C from marine actinomycetes *Salinispora tropica. Combinatorial Chemistry & High Throughput Screening, 20*(1), 3–19.
Dineshkumar, K., Aparna, V., Madhuri, K. Z., & Hopper, W. (2014). Biological activity of sporolides A and B from *Salinispora tropica*: In silico target prediction using ligand-based pharmacophore mapping and in vitro activity validation on HIV-1 reverse transcriptase. *Chemical Biology & Drug Design, 83*(3), 350–361.
Donia, M. S., Ruffner, D. E., Cao, S., & Schmidt, E. W. (2011). Accessing the hidden majority of marine natural products through metagenomics. *ChemBioChem, 12*(8), 1230–1236.
Dusane, D. H., Matkar, P., Venugopalan, V. P., Kumar, A. R., & Zinjarde, S. S. (2011). Cross-species induction of antimicrobial compounds, biosurfactants, and quorum-sensing inhibitors in tropical marine epibiotic bacteria by pathogens and biofouling microorganisms. *Current Microbiology, 62*(3), 974–980.
Ebrahim, H. Y., & El Sayed, K. A. (2016). Discovery of novel antiangiogenic marine natural product scaffolds. *Marine Drugs, 14*(4), 57.
Engler, C., Gruetzner, R., Kandzia, R., & Marillonnet, S. (2009). Golden gate shuffling: A one-pot DNA shuffling method based on type II restriction enzymes. *PLoS One, 4*(5), e5553.
Ertl, P., Roggo, S., & Schuffenhauer, A. (2008). Natural product-likeness score and its application for prioritization of compound libraries. *Journal of Chemical Information and Modeling, 48*(1), 68–74.
Fang, J., Wu, Z., Cai, C., Wang, Q., Tang, Y., & Cheng, F. (2017). Quantitative and systems pharmacology. In-silico prediction of drug-target interactions of natural products enables new targeted cancer therapy. *Journal of Chemical Information and Modeling, 57*(11), 2657–2671.
Fenical, W., & Jensen, P. R. (2006). Developing a new resource for drug discovery: Marine actinomycete bacteria. *Nature Chemical Biology, 2*(12), 666-673.

Flores, M. C., Marquez, E. A., & Mora, J. R. (2018). Molecular modeling studies of bromopyrrole alkaloids as potential antimalarial compounds: A DFT approach. *Medicinal Chemistry Research*, *27*(4), 844–856.

García Márquez, F. P., Papaelias, M., & Marini, S. (2022). Artificial intelligence in marine science and engineering. *Journal of Marine Science and Engineering*, *10*(6), 711.

Gaudêncio, S. P., & Pereira, F. (2015). Dereplication: Racing to speed up the natural products discovery process. *Natural Product Reports*, *32*(6), 779–810.

Gerwick, W. H., & Moore, B. S. (2012). Lessons from the past and charting the future of marine natural products drug discovery and chemical biology. *Chemistry & Biology*, *19*(1), 85–98.

Gibson, D. G., Young, L., Chuang, R. Y., Venter, J. C., Hutchison, C. A., & Smith, H. O. (2009). Enzymatic assembly of DNA molecules up to several hundred kilobases. *Nature Methods*, *6*(5), 343–345.

Gross, H. (2009). Genomic mining–a concept for the discovery of new bioactive natural products. *Current Opinion in Drug Discovery & Development*, *12*(2), 207–219.

Haque, M. U., Haque, M. A., Sarker, A. K., & Islam, M. A. (2015). Comparative studies of secondary metabolites of co-culture and monoculture of two marine actinomycetes and their biological activities. *International Journal of Innovative Pharmaceutical Sciences and Research*, *3*(2), 35–45.

Huang, X. C., Xiao, X., Zhang, Y. K., Talele, T. T., Salim, A. A., Chen, Z. S., & Capon, R. J. (2014). Lamellarin O, a pyrrole alkaloid from an Australian marine sponge, *Ianthella* sp., reverses BCRP-mediated drug resistance in cancer cells. *Marine Drugs*, *12*(7), 3818–3837.

James, R. A., Vignesh, S., & Muthukumar, K. (2012). Marine drugs development and social implication. In *Coastal Environments: Focus on Asian Regions* (pp. 219–237). Springer, Dordrecht.

Jimenez, C. (2018). Marine natural products in medicinal chemistry. *ACS Medicinal Chemistry Letters*, *9*(10), 959–961.

Ka, A. K., & Keerthi, T. R. (2017). Co-culture as the novel approach for drug discovery from marine environments. *Novel Approaches in Drug Design and Development*, *2*, 1–4.

Knight, N. J., Hernando, E., Haynes, C. J. E., Busschaert, N., Clarke, H. J., Takimoto, K., & Gale, P. A. (2016). QSAR analysis of substituent effects on tambjamine anion transporters. *Chemical Science*, *7*(2), 1600–1608.

Koehn, F. E., & Carter, G. T. (2005). The evolving role of natural products in drug discovery. *Nature Reviews Drug Discovery*, *4*, 206–220.

Kudryavtsev, D., Makarieva, T., Utkina, N., Santalova, E., Kryukova, E., Methfessel, C., & Kasheverov, I. (2014). Marine natural products acting on the acetylcholine-binding protein and nicotinic receptors: From computer modeling to binding studies and electrophysiology. *Marine Drugs*, *12*(4), 1859–1875.

Lang, G., Mayhudin, N. A., Mitova, M. I., Sun, L., van der Sar, S., Blunt, J. W., & Munro, M. H. (2008). Evolving trends in the dereplication of natural product extracts: New methodology for rapid, small-scale investigation of natural product extracts. *Journal of Natural Products*, *71*(9), 1595–1599.

Liu, J., Chen, W., Xu, Y., Ren, S., Zhang, W., & Li, Y. (2015). Design, synthesis, and biological evaluation of tasiamide B derivatives as BACE1 inhibitors. *Bioorganic & Medicinal Chemistry*, *23*(8), 1963–1974.

Liu, J., Chen, W., Xu, Y., Ren, S., Zhang, W., & Li, Y. (2015). Design, synthesis, and biological evaluation of tasiamide B derivatives as BACE1 inhibitors. *Bioorganic & Medicinal Chemistry*, *23*(8), 1963–1974.

Llorach-Pares, L., Nonell-Canals, A., Sanchez-Martinez, M., & Avila, C. (2017). Computer-aided drug design applied to marine drug discovery: Meridianins as Alzheimer's disease therapeutic agents. *Marine Drugs, 15*(11), 366.
Llorach-Pares, L., Nonell-Canals, A., Sanchez-Martinez, M., & Avila, C. (2017). Computer-aided drug design applied to marine drug discovery: Meridianins as Alzheimer's disease therapeutic agents. *Marine Drugs, 15*(11), 366.
Luesch, H., Moore, R. E., Paul, V. J., Mooberry, S. L., & Corbett, T. H. (2001). Isolation of dolastatin 10 from the marine cyanobacterium *Symploca* species VP642 and total stereochemistry and biological evaluation of its analogue symplostatin 1. *Journal of Natural Products, 64*, 907–910.
Montaser, R., & Luesch, H. (2011). Marine natural products: A new wave of drugs? *Future Medicinal Chemistry, 3*(12), 1475–1489.
Moore, B. S. (2005). Biosynthesis of marine natural products: Microorganisms (Part A). *Natural Product Reports, 22*(5), 580–593.
Newman, D. J., & Cragg, G. M. (2016). Drugs and drug candidates from marine sources: An assessment of the current "State of play." *Planta Medica, 82*, 775–789.
Newman, D. J., & Cragg, G. M. (2016). Drugs and drug candidates from marine sources: An assessment of the current "State of play." *Planta Medica, 82*, 775–789.
Nielsen, K. F., & Larsen, T. O. (2015). The importance of mass spectrometric dereplication in fungal secondary metabolite analysis. *Frontiers in Microbiology, 6*, 71.
Oh, D. C., Jensen, P. R., Kauffman, C. A., & Fenical, W. (2005). Libertellenones A-D: Induction of cytotoxic diterpenoid biosynthesis by marine microbial competition. *Bioorganic & Medicinal Chemistry, 13*(17), 5267–5273.
Oh, D. C., Kauffman, C. A., Jensen, P. R., & Fenical, W. (2007). Induced production of emericellamides A and B from the marine-derived fungus *Emericella* sp. in competing co-culture. *Journal of Natural Products, 70*(4), 515–520.
Owen, J. G., Reddy, B. V., Ternei, M. A., Charlop-Powers, Z., Calle, P. Y., Kim, J. H., & Brady, S. F. (2013). Mapping gene clusters within arrayed metagenomic libraries to expand the structural diversity of biomedically relevant natural products. *Proceedings of the National Academy of Sciences, 110*(29), 11797–11802.
Pereira, F. (2019). Have marine natural product drug discovery efforts been productive and how can we improve their efficiency? *Expert Opinion on Drug Discovery, 14*(8), 717–722.
Pereira, F., & Aires-de-Sousa, J. (2018). Computational methodologies in the exploration of marine natural product leads. *Marine Drugs, 16*(7), 236.
Pfeifer, B. A., & Khosla, C. (2001). Biosynthesis of polyketides in heterologous hosts. *Microbiology and Molecular Biology Reviews, 65*(1), 106–118.
Potterat, O., Wagner, K., & Haag, H. (2000). Liquid chromatography–electrospray time-of-flight mass spectrometry for online accurate mass determination and identification of cyclodepsipeptides in a crude extract of the fungus *Metarrhizium anisopliae*. *Journal of Chromatography A, 872*(1–2), 85–90.
Prieto-Davó, A., Villarreal-Gómez, L. J., Forschner-Dancause, S., Bull, A. T., Stach, J. E., Smith, D. C., Rowley, D. C., & Jensen, P. R. (2013). Targeted search for actinomycetes from nearshore and deep-sea marine sediments. *FEMS Microbiology Ecology, 84*(3), 510–518.
Reen, F. J., Romano, S., Dobson, A. D., & O'Gara, F. (2015). The sound of silence: Activating silent biosynthetic gene clusters in marine microorganisms. *Marine Drugs, 13*(8), 4754–4783.
Rinke, C., Schwientek, P., Sczyrba, A., Ivanova, N. N., Anderson, I. J., Cheng, J. F., & Dodsworth, J. A. (2013). Insights into the phylogeny and coding potential of microbial dark matter. *Nature, 499*(7459), 431–437.

Schneider, G., Reker, D., Chen, T., Hauenstein, K., Schneider, P., & Altmann, K. H. (2016). Deorphaning the macromolecular targets of the natural anticancer compound doliculide. *Angewandte Chemie International Edition, 55*(44), 12408–12411.

Shang, J., Hu, B., Wang, J., Zhu, F., Kang, Y., Li, D., & Hou, T. (2018). A cheminformatic insight into the differences between terrestrial and marine-originated natural products. *Journal of Chemical Information and Modeling, 58*(5), 976–986.

Skariyachan, S., Acharya, A. B., Subramaniyan, S., Babu, S., Kulkarni, S., & Narayanappa, R. (2016). Secondary metabolites extracted from a marine sponge-associated *Comamonas testosteroni* and *Citrobacter freundii* as potential antimicrobials against MDR pathogens and hypothetical leads for VP40 matrix protein of Ebola virus: An in vitro and in silico investigation. *Journal of Biomolecular Structure and Dynamics, 34*(9), 1865–1883.

Slattery, M., Rajbhandari, I., & Wesson, K. (2001). Competition-mediated antibiotic induction in the marine bacterium *Streptomyces tenjimariensis. Microbial Ecology, 41*(2), 90–96.

Trischman, J. A., Oeffner, R. E., de Luna, M. G., & Kazaoka, M. (2004). Competitive induction and enhancement of indole and diketopiperazine in marine bacteria. *Marine Biotechnology, 6*(3), 215–220.

Wakefield, J., Hassan, H. M., Jaspars, M., Ebel, R., & Rateb, M. E. (2017). Dual induction of new microbial secondary metabolites by fungal bacterial co-cultivation. *Frontiers in Microbiology, 8*, 1284.

Wiese, J., & Imhoff, J. F. (2019). Marine bacteria and fungi as promising sources of new antibiotics. *Drug Development Research, 80*(1), 24–27.

Wilson, M. C., Mori, T., Rückert, C., Uria, A. R., Helf, M. J., Takada, K., & Rinke, C. (2014). An environmental bacterial taxon with a large and distinct metabolic repertoire. *Nature, 506*(7486), 58–62.

Wu, Z., Cheng, F., Li, J., Li, W., Liu, G., & Tang, Y. (2017). Sdtnbi: An integrated network and cheminformatics tool for systematic prediction of drug-target interactions and drug repositioning. *Briefings in Bioinformatics, 18*(2), 333–347.

Zerikly, M., & Challis, G. L. (2009). Strategies for the discovery of new natural products by genome mining. *ChemBioChem, 10*(4), 625–633.

Zhang, H., Boghigian, B. A., Armando, J., & Pfeifer, B. A. (2011). Methods and options for the heterologous production of complex natural products. *Natural Product Reports, 28*(1), 125–151.

CHAPTER 4

Marine Drug Delivery Systems

JASMIN HASSAN,[1] MD. FAIYAZUDDIN,[2] and MD. JASIM UDDIN[3]

[1]*Drug Delivery and Therapeutics Lab, London, United Kingdom*

[2]*School of Pharmacy, Al-Karim University, Katihar, Bihar, India*

[3]*Department of Pharmaceutical Technology, Faculty of Pharmacy, Universiti Malaya, 50603 Kuala Lumpur, Malaysia*

ABSTRACT

Since 1970, the ocean has caught the attention of researchers as a natural source of remedies. However, the ocean and its habitats are still a mystery to researchers to this date, and a treasury as well. For ages, we have turned to nature in search of medicines, and now approximately 60% of drugs in the market are from natural origins. In the time period from 1981 to 2014, a total of 1562 new chemical entities have been introduced to the market and only 21% of them are naturally derived. In this chapter, the authors have first provided a brief background history of the ocean as a natural source of drugs. Traditional applications and challenges in marine drug development have also been discussed alongside updated information regarding new approaches. The ocean serves as a solar-powered factory that provides diverse biochemical compounds such as alkaloids, lipids, peptides, polysaccharides, etc., that have the potential to treat diseases. In addition, different methods are used in the purification process of a potential biopharmaceutical compound, which requires biotechnology. Due to this, a multiple approach-based analysis on a molecular level is required to establish a marine-derived molecule as a therapeutic drug, and it has to go through phases of clinical trials for approval from the Food and Drug Administration (FDA) to be launched in the market as a pharmaceutical product. Applications of marine-derived molecules

Handbook of Research in Marine Pharmaceutics: Exploring Oceanic Microbial Diversity for Human Health and Wellness. Md. Faiyazuddin & Khalid Rehman Hakeem (Eds.)

that have therapeutic potentials like anticancer, antiviral, antibacterial, anti-inflammatory, vaccine adjuvants, etc., to treat different diseases have also been well-described by the authors. Overall, this chapter also provides lists of the marine-derived drugs that have received approvals from the FDA until 2020 and the drugs that are under clinical trials as well.

4.1 INTRODUCTION

The history behind each therapeutic has plenty of remarkable stories regarding the discovery of natural chemical compound-inspired drug development and treatment for diseases (Fenical, 1997; Shen, 2015). Since the very beginning of human civilization, we have always turned to nature as a source of pharmacologically active molecules (Choudhary et al., 2017; Li & Vederas, 2009). Even in recent years, we have been curious about natural sources of medicines, and the increasing number of publications (total 2274 in 2020 which was 287 in 2000) is proof of that (Catanesi et al., 2021). Terrain and ocean are two of those natural sources from which we get the maximum amount of remedies for treating diseases (Figure 4.1) (Fenical, 1997; Penesyan et al., 2010). Among them, the ocean and marine organisms have always appealed to us as an unprecedented entity of mystery (Galtsoff, 1934).

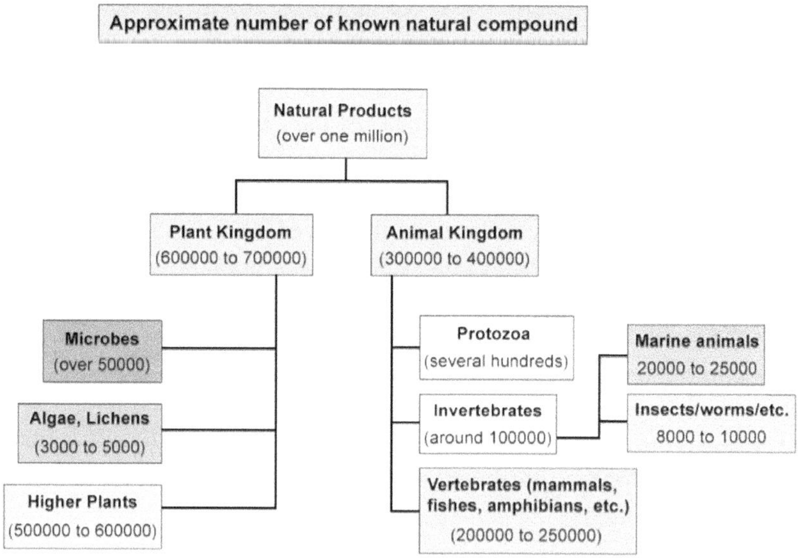

FIGURE 4.1 Summarized illustration of approximate numbers for known natural compounds to treat diseases.

Around 70% of Earth's surface is submerged in the ocean (Chen et al., 2020; Donia & Hamann, 2003; Penesyan et al., 2010), and represents an amazingly lavish source for diversified organisms owing to the distinct oceanic zones (Arthur James et al., 2014; Kijjoa & Sawangwong, 2004; Shen, 2015). In spite of the fact that the terrestrial environment represents a remarkable diversity of life, the ocean represents the significantly largest biodiversity on Earth (Donia & Hamann, 2003). With some overlap, 17 fundamental phyla of life happen on land while 32 happen in the sea, out of 34 (Kijjoa & Sawangwong, 2004). These diversified marine organisms render an abundance of marine natural active substances with varied novel structures and remarkable therapeutic values. Along with the structural variety, bioactive compounds isolated from marine organisms provide a broad range of pharmacological effects, and developed drugs from marine sources play a vital role as analgesic, anti-inflammatory, antiviral, antibacterial, antimalarial, antitumoral, antifungal, anticoagulant, and also in the treatment of cardiovascular diseases, neurological disorders, and cerebrovascular diseases (Choudhary et al., 2017; Donia & Hamann, 2003; Kijjoa & Sawangwong, 2004; Vignesh et al., 2011).

Due to being a potential source of nutrients and rich in sulfated polysaccharides (carbohydrates), various kinds of seaweeds have been eaten regularly by ancient maritime people, especially the Chinese and Japanese. Considering various pharmacological properties, seaweeds that contain sulfated polysaccharides have also been used traditionally as remedies to treat ailments (Ahmed et al., 2014; Ruggieri, 1976). In 1976, the Director of the New York Aquarium and Osborn Laboratories of Marine Sciences, George D. Ruggieri, published an article where he discussed a couple of unusual remedies generally used by ancient people (Romans, Chinese & Japanese) that were sourced from marine organisms (Ruggieri, 1976). Although the situation is different now in the evolution of time. Numerous researches have been conducted and are still ongoing to establish drugs obtained from marine natural products (MNPs) to treat different diseases (Donia & Hamann, 2003; Kijjoa & Sawangwong, 2004). However, it has not been an easy task to develop therapeutics from marine sources due to some inseparable challenges that come along with marine drug discovery. Firstly, insufficiency in the sustainable supply of drug candidates sometimes postpones or restrains additional research for drug development of many potential chemical compounds obtained from marine sources. However, to combat this issue, researchers have made an effort to feasibly synthesize analogs of those chemical derivatives with some modifications to gain the desired properties (Radjasa et al., 2011). Secondly, the identification of a potential bioactive material can be incorrect due to the inadequate yield of isolated marine compounds along with their structural complexity, which may cause difficulty in finding suitable analytical methods, subsequently obstructing the understanding of phenotypic variations (Maier, 2009; Papon et al., 2022). Thirdly, a major challenge for both analytical and biological evaluations is the presence of a tiny volume of minor metabolites in desired isolated compounds. To address this issue, *in silico* screening is applied to aid in understanding the minor drug candidates better (Iyer & Kadambi, 2011). A couple more challenges associated with marine drug development are a. development of universal methods for high-yield biosynthesis of potential compounds with small molecular weight, and b. genetic tool development to retrieve the *in vivo* potentiality of laboratory-cultured marine microorganisms. Additionally, feasible methods need to be developed to conduct biological assays for pure compounds, extracts, and fractions, which require improvement (Gerwick & Moore, 2012). Factors that limit marine drug development have been summarized in Figure 4.2.

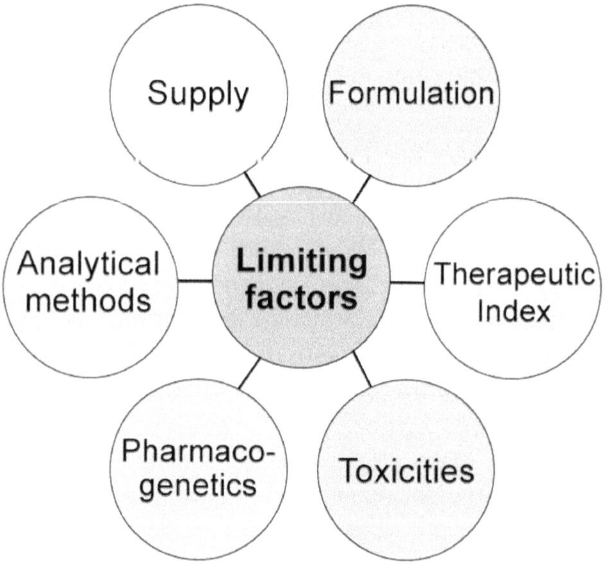

FIGURE 4.2 Marine drug development limiting factors.

Owing to the marine environment's chemical and physical conditions, nearly every class of marine organism shows various molecules with exceptional chemical structures (Kijjoa & Sawangwong, 2004). As we have mentioned before, apart from the chemical diversity, the ocean provides amazing pharmacological diversity as well. From the perspective of biodiversity, the ocean has the capacity to develop a natural pharmacy (Houghten, 1994; Kijjoa & Sawangwong, 2004; Montaser & Luesch, 2011; Ortholand & Ganesan, 2004). This leads us to the growing interest in pharmacologically active entities from marine sources (Cragg et al., 2006; Xiong et al., 2013). In 2022, Ahmed et al. published a paper where they presented data on the total number of search results containing the words marine bioactive compounds from the Scopus database, as shown in Figure 4.3 (Ahmed et al., 2022).

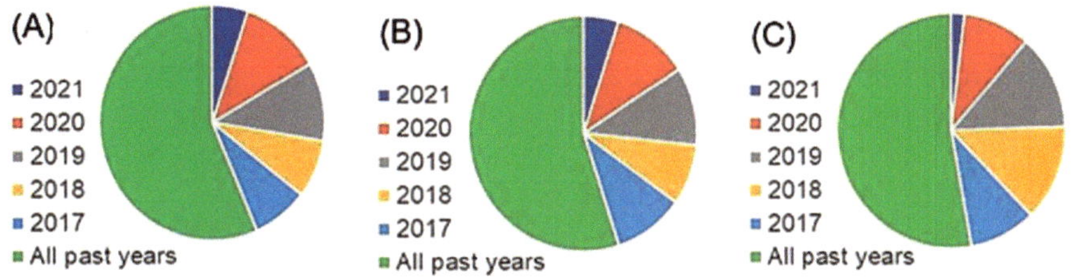

FIGURE 4.3 The number of articles search results obtained from the Scopus database from all years in a particular category of search terms – (A) bioactive compounds from marine sources; (B) bioactive peptides from marine sources; and (C) biomedical applications of marine peptides.

Source: Reproduced from: Ahmed et al. (2022). Copyright © 2022 by the authors; licensee MDPI, Basel, Switzerland. http://creativecommons.org/licenses/by/4.0/

In this chapter, the authors will be discussing the background history of medicines from marine resources, traditional applications of marine-derived drugs, pharmacological benefits of MNPs, and MNP-derived materials in pharmaceutical sciences extensively.

4.2 HISTORY OF MARINE DRUG ORIGIN

4.2.1 A NATURAL HABITAT OF PHARMACEUTICALS: MEDICINES FROM OCEAN

MNPs are an essential source of diversified structure and bioactive materials; quite a few of them have influenced the establishment of new therapeutic classes of drugs. All of these successes have had to deal with the challenges inherent to MNPs-derived drugs, e.g., insufficient resourcing of the candidate drug agent and structural complexity issues (Gerwick & Moore, 2012).

In the early days of the endeavor for MNP discovery, which was prior to the advent of SCUBA, the focus had been predominantly on the isolation and characterization of metabolites obtained from easily collected marine organisms, such as corals, red algae, and sponges (Iyer & Kadambi, 2011; Montaser & Luesch, 2011; Radjasa et al., 2011; Xiong et al., 2013). Moreover, it took around three years for researchers to completely attain the purpose of marine resources-based research in 1950 (Radjasa et al., 2011). This period from the 1940s to the 1950s is known as the "golden era" for MNPs-based drug discovery (Bérdy, 2005). It was after the revolution in the forties when almost all types of pharmacological activities of the MNPs had been discovered; actual progress toward success in drug discovery occurred steadily during the 1950s to 1970s (Bérdy, 2005; Radjasa et al., 2011). Since then, many MNPs with promising attributes as drug candidates have been discovered till now (Haefner, 2003; Kijjoa & Sawangwong, 2004). Some of them are enlisted in Table 4.1.

4.2.2 TRADITIONAL APPLICATIONS

From primordial times, the ocean has provided an environment that has led to numerous diversified interfaces, which have helped us obtain a favorable amount of pharmacologically active entities (Fox, 1960). To survive aberrant environmental states, such as photodynamic damage, extreme temperatures, oxidative stress, etc., marine organisms synthesize bioactive molecules that have also shown beneficial effects on human health according to recent research. Some of these have been listed in the above section in Table 4.1 (Catanesi et al., 2021; Pohnert, 2004).

From the very beginning, nature and remedies to treat diseases have had a strong connection, and therapeutic biomolecules have been used for thousands of years in traditional medicines. Marine organisms were widely used for their therapeutic values during the periods of ancient Greece and early Byzantium. For example, they were either consumed as fresh or dried flesh, or used in beverages, juices, broths, ointments, and powdered products (Catanesi et al., 2021; Voultsiadou Eleni, 2010). Classical Greek texts also mention some edible shellfish for their therapeutic attributes (Bhakuni & Rawat, 2005; Sennett et al., 2002; Voultsiadou et al., 2009).

Traditional Chinese medicine has also made a great deal of contribution to marine-derived drug research. Since the application of marine herbs and their derivatives belongs to a century-old tradition, the comprehensive knowledge regarding marine organisms and MNPs-derived drugs has been enriched

TABLE 4.1 Potential MNPs Derived Therapeutic Compounds

Sl. No.	Compound	Source	Application	References
1.	Hypothemycin	*Aigialas parvus*	Antiparasitic	Isaka et al. (2002)
2.	Kainic acid	*Digenea simplex*	Anthelmintic	Vignesh et al. (2011)
3.	Halovirs A-E	*Scytalidium* sp.	Antiviral	Rowley et al. (2003)
4.	Aplidine	*Aplidium albicans*	Anticancer	(Kijjoa & Sawangwong, (2004)
5.	SS-228Y	*Chainia* sp.	Antibacterial	Okazaki et al. (1975)
6.	Ara-C (cytarabine)	*Cryptotethya crypta*	Antileukemic	Vignesh et al. (2011)
7.	Keisslone	*Keissleriella* sp.	Antifungal	Thakur et al. (2008)
8.	Dolastatin 10	*Dolabella auricularia*	Anticancer	Kijjoa & Sawangwong (2004)
9.	Apratoxin A	*Lyngbya majuscule*	Antitumor	Luesch et al. (2001)
10.	Cephalosporins	*Cephalosporium acremonium*	Antibiotic	Vignesh et al. (2011)
11.	Hypoxisordarin	*Hypoxylon croceum*	Antifungal	Liu et al. (2003)
12.	Halichondrin B	*Halichondria okadai*	Anticancer	Kijjoa & Sawangwong (2004)
13.	Bryostatin	*Candidatus endobugula sertula*	Anticancer	Sudek et al. (2007)
14.	Ara-A	*Cryptotethya crypta*	Antiviral	Vignesh et al. (2011)
15.	Artemisinic acid	*Saccharomyces cerevisiae*	Antiparasitic	Martin et al. (2003)
16.	Mycaperoxide B	*Mycale*	Anticancer	Kijjoa & Sawangwong (2004)
17.	Ziconotide	*Conus magus*	Analgesic	Vignesh et al. (2011)
18.	Modiolides A-B	*Paraphaeospheria* sp.	Antibacterial	Tsuda et al. (2003)
19.	Kahalaide F	*Elysia rufescens*	Anticancer	Kijjoa & Sawangwong (2004)
20.	Ecteinascidin 743	*Ecteinascidia turbinate*	Anticancer	Haefner (2003); Vignesh et al. (2011)
21.	Shikimic acid	*Escherichia coli*	Antiviral	Chandran et al. (2003)

TABLE 4.1 *(Continued)*

Sl. No.	Compound	Source	Application	References
22.	Cyclodidemniserinol trisulfate	*Didemnum guttatum*	Antiviral (HIV)	Kijjoa & Sawangwong (2004)
23.	Dehydrodidemnine B	*Aplidium albicans*	Anticancer	Vignesh et al. (2011)
24.	Salinosporamide A	*Salinospora* sp.	Anticancer	Feling et al. (2003)
25.	Dithiocyanates	*Axinyssa*	Anti-infective	Kijjoa & Sawangwong 2004)
26.	GTS-10	*Amphiporus lactifloreus*	Alzheimer's disease	Vignesh et al. (2011)
27.	Tylactone	*Streptomyces venezuelae*	Antibiotic	Jung et al. (2006)
28.	Contignasterol	*Petrosia contignata*	Antiasthmatic	Kijjoa & Sawangwong 2004)
29.	Didemnim B	*Trididemnum solidum*	Antineoplastic/antiviral	Vignesh et al. (2011)
30.	Euniatin B	*Fusarium* sp.	Antibacterial	Christie et al. (1997)
31.	Spongistatin 1	*Hyrtios erecta*	Antifungal	Vignesh et al. (2011)
32.	Conotoxins	*Conus*	Analgesic	Kijjoa & Sawangwong, 2004)
33.	Manoalide	*Luffariella variabilis*	Anti-inflammatory	Vignesh et al. (2011)

due to monographs, local chronicles, primary medical prescriptions, folk formulas, and dietary suggestions. Conversely, unfairness occurred with the development of scientific methods and the dedication of scientists. All of this information and discoveries have collectively been compiled in the Chinese Marine Materia Medica, which is a kind of encyclopedia and represents the best collection of marine organisms such as algae, cyanobacteria, marine animals, and minerals. It is also contemplated as a pioneer in the new market drug development (Catanesi et al., 2021; Fu et al., 2016).

Seafoods, such as fish and marine invertebrates, have been highly recommended by ancient people since they are considered a rich source of omega-3 fatty acids to prevent chronic diseases and improve human health as well (Harris et al., 2009; Voultsiadou et al., 2009). However, some bizarre applications of marine organisms have also been done by ancient people. For example, an early Chinese pharmacopeia stated seaweeds as a treatment for edema, abscesses, etc., and a mixture of seven seaweeds used to be considered a vermifuge. Ancient Romans used to make medicine for toothache by mixing cinder from the burnt spine of a stingray with vinegar (Ruggieri, 1976). Aside from these, some beneficial knowledge has also been gained from the ancient history of marine organisms that helped today's revolution in the pharmaceutical industry, as mentioned before.

4.2.3 CHALLENGES IN MARINE DRUG DEVELOPMENT

Although the idea of marine organisms-derived drugs has ushered in a new era and created a unique field of research for scientists worldwide, the collection and supply of MNPs have always been challenging. This has made marine-derived drug discovery difficult to accomplish in earlier times (Bérdy, 2005; Donia & Hamann, 2003; Gerwick & Moore, 2012). This section will discuss these obstacles.

Due to the frequently changing environment, the same marine organism may produce different metabolites each time it is collected, even if it is from the same place in the ocean. Sometimes difficulties are faced because the microorganisms living in the marine animal produce the bioactive molecules, not the marine hosts (Jimeno, 2002; Martins et al., 2014; Molinski et al., 2009). An unsustainable supply of potential lead compounds may also cause issues sometimes, as they are present only in minimal quantities or are practically tough to isolate (Molinski et al., 2009; Summer et al., 2020). To undergo the phases of drug development, the required quantity of a lead compound might vary from a few grams to a kilogram. The unavailability of the desired compound in such ample quantities becomes a key issue sometimes (Martins et al., 2014). In addition, sample collection from spots deeper than 30 m is not possible by scuba diving, that's why trawling is done. Apart from being unselective, this technique sometimes causes sample damage and disturbance in benthic ecosystems. Moreover, it does not disclose information regarding the ambiance of the organisms (Martins et al., 2014; Montaser & Luesch, 2011). There are still quite a good number of unidentified species and strains and a lack of taxonomic knowledge for those marine species is also a major obstacle. Especially in the case of macroinvertebrates, because it is possible that many species might have been determined based on subtle morphological attributes, and an incorrect identification and characterization (Figure 4.4) may lead an entirely new drug development project to lose its authenticity (Benkendorff, 2013; Costello et al., 2006; Jiang et al., 2013). Attempted strategies in order to resolve this sustainable supply problem of marine organisms are mariculture and aquaculture. Yet the exceptional environment of the ocean makes the cultivation and maintenance of isolated samples challenging and sometimes impossible. For this reason, isolation and purification of the desired compound may need to be done from the specimens directly coming from the wild (Leal et al., 2012; Martins et al., 2014). Researchers have been trying to develop synthetic or semisynthetic analogs, more practicable derivatives, or a pharmacophore with less complexity that can generate the desired material later on. But these approaches are a challenge themselves (Radjasa et al., 2011; Suyama et al., 2011). Figure 4.4 shows the total number of MNPs misassigned in each five-year period until 2010.

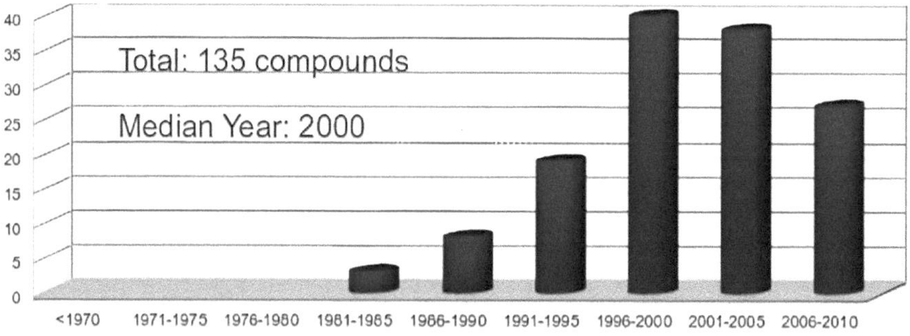

FIGURE 4.4 Total number of occurrences of misassignment of MNPs until 2010.

Source: Reproduced with permission from Suyama et al., 2011. Copyright 2011. Elsevier

Usually, high throughput screening (HTS) is used for the MNPs, and a high-quality library is required for this process, which contains all the information regarding crude extracts, purified fractions, and pure compounds. However, it may generate false negative results either due to the presence of a bioactive molecule in the crude extract in tiny amounts, or other compounds such as tannins might bind to the bioactive molecule, leading to blocking its activity, which is considered a drawback. Because of these reasons, in the recent past, the HTS approach used to be discouraged in new drug development projects (Butler, 2004; Gribbon & Andreas, 2005; Kingston, 2011). This issue is avoided by the screening of pre-fractionated libraries, which has turned out to be an effective technique (Carter, 2011).

Another challenge for MNPs-based drug development is the cost. Manufacturing and technology costs, often with poor yields, lead to an increase in per kilogram market cost and may cause the products to be economically unfeasible (Martins et al., 2014). Apart from these costs, marine organism collection costs are also present. One of the obstacles here is the fouling organisms of the ocean, which cause faster corrosion of equipment, affecting the safety and durability of marine facilities as well. In summary, this leads to consequential economic losses. To prevent this problem, the antifouling coating is a functional method and has quickly become a hotspot for research (Dafforn et al., 2011; Gu et al., 2020; Lejars et al., 2012; Satheesh et al., 2016; Schultz et al., 2010). These days, academia has also started to become aware of the market cost issue since it is evident that most discoveries are in the clinical pipeline and hardly reach the consumer market (Martins et al., 2014; Xiong et al., 2013). Some marine organisms are found in remote locations which require travel costs along with exclusive services for personnel's safety costs. To help readers understand the magnitude of these expenses, a small example can be given in terms of marine organism collection cost using a ship, which is around 14500 USD per day (Hale et al., 2002; Vignesh et al., 2011).

Approaches such as molecular docking and computational methods have helped in the cost minimization of marine-based new drug discovery to a great extent in the past few years (Chen et al., 2020; Pereira & Aires-de-Sousa, 2018).

4.3 HEALTH BENEFITS OF MARINE-DERIVED COMPOUNDS

It has already been mentioned that MNP-derived compounds have numerous health benefits. Until 2012, an approximate total of 121 products were commercially available in the market for their beneficial pharmacological properties, which have been collected from different marine organisms

(Figure 4.5) (Gerwick & Moore, 2012). In this section, we will discuss the health benefits separately for different MNPs.

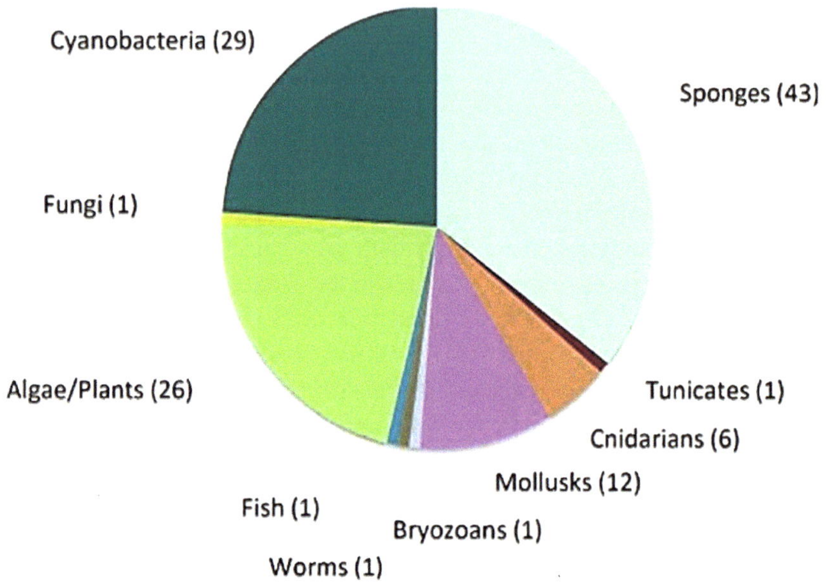

FIGURE 4.5 An approximate number of commercially available MNP-derived compounds with beneficial pharmacological properties from different marine organisms.

Source: Reproduced with permission from Gerwick & Moore, 2012. Copyright 2012. Elsevier.

4.3.1 POLYSACCHARIDES

Numerous varieties of marine polysaccharides (PS) convey their constant evolution to us (Laurienzo, 2010). During ancient times, the isolation of PSs using a hot aqueous process was the only method that had been applied, and insoluble compounds used to be separated through filtration. However, this process also had limitations; for example, the high viscosity of the crude mixture in the presence of solid residues made the filtration process troublesome and costly to conduct, which led the ancient method to become unappealing (Azevedo et al., 2015). As an alternative to the hot water method, the alkaline extraction method was adopted later for the extraction of PSs from the marine crude mixture, where impurities are dissolved in a basic solution which facilitates the separation process of PSs (McHugh, 2003). For example, the extraction process of carrageenan from marine seaweed has been illustrated in Figure 4.6 (Yashaswini et al., 2021).

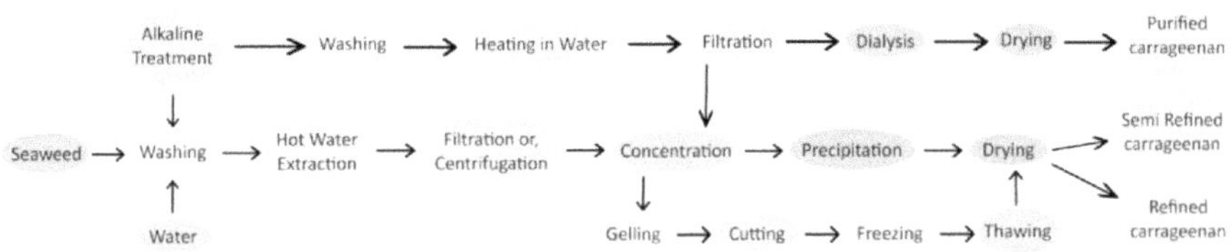

FIGURE 4.6 Simplified flowchart for the extraction process of carrageenan from seaweed via hot aqueous and alkaline treatment methods.

PS is mainly synthesized by photosynthetic marine organisms such as macroalgae, microalgae, and cyanobacteria (Ahmed et al., 2014; Raposo et al., 2015). Among them, both macroalgae and microalgae mostly produce sulfated polysaccharides (sPS). Seaweeds (red, green, brown) and cyanobacteria produce PS that provide a wide range of biomedical applications (Figure 4.7) (Lee et al., 2017; Raposo et al., 2015; Senni et al., 2011; Wijesekara et al., 2011; Wijesinghe & Jeon, 2012). In addition, the algal share in the world market is the largest among all marine organisms (Ahmed et al., 2014). The most studied PS obtained from marine organisms is macroalgae, also known as seaweed, and most of them are either green seaweeds (*Chlorophyta*), brown algae (*Chromophyta*), or red macroalgae (*Rhodophyta*) (Raposo et al., 2015).

Brown algae generally contain fucoidans. Oligosaccharides extracted from fucoidans via hydrolysis usually contain glucose, galactose, uronic acids, and other monosaccharides (MS) linked together with a main chain by various types of glycosidic bonds. This complex structure of fucoidans fails to form a linkage with some other types of compounds to facilitate the design of universal pharmaceuticals or drug-like substances for managing particular diseases (Raposo et al., 2015).

Red macroalgae contain a plethora of sulfated polysaccharides (sPS), among which galactans such as agaroids and carrageenans are present in large amounts. They consist of repeated units of 1,3-α-gal and 1,4-β-D-gal (Pomin, 2012) and 3,6-anhydrogal (3,6-Agal) (Usov, 2011) with various substituents of monosaccharides (xylose, mannose) and their derivatives (methoxy, sulfate, and pyruvate groups). The sulfation pattern categorizes carrageenan into different families, namely κ-carrageenan (C-4) and λ-carrageenan (C-2). Additionally, the L-isomer of agaroid differs from the D-isomer of carrageenan based on the rotation of gal in 1,3-linked compositions (Jiao et al., 2011).

In contrast, information about the structures and applications of green macroalgae is scarce. However, Wang et al. have published an insightful review article regarding the properties of sulfated polysaccharides obtained from a few species of chlorophytes. The structures of these sulfated polysaccharides are more diverse and complex, comprising glycosidic linkages among monomers and repeating di-units of aldobiuronic acid (glucuronic acid and iduronic acid) (L. Wang et al., 2014).

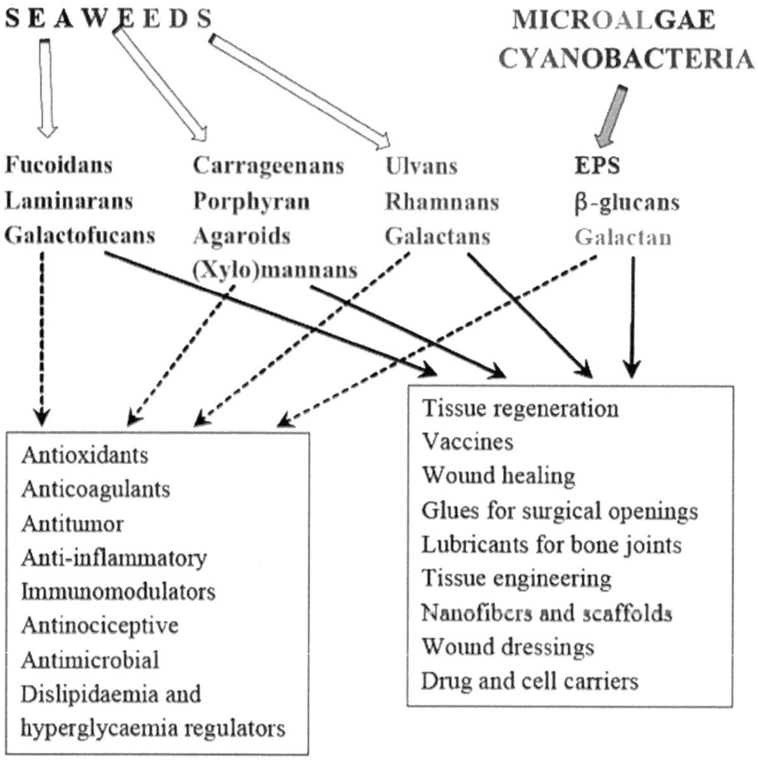

FIGURE 4.7 Pharmacological properties of different polysaccharides obtained from marine organisms.

Source: Reproduced from Raposo et al., 2015. Copyright © 2015 by the authors; licensee MDPI, Basel, Switzerland. http://creativecommons.org/licenses/by/4.0/

Many groups of researchers (Ale et al., 2011; Andersen et al., 1973; Cumashi et al., 2007; Duarte, Cardoso et al., 2001; Fedorov et al., 2013; Funami et al., 2007; Ghosh et al., 2009; Jiao et al., 2011; Li et al., 2006, 2008) have rigorously explored the structure of MSs, their linkage types, compared structural differences and similarities among them, and also studied their bioactive properties. Table 4.2 summarizes sources, structures, and applications for most of the PSs that have been obtained from marine organisms.

TABLE 4.2 Structure and Pharmacological Attributes of Marine-Derived PSS

Polysaccharides	Source(s)	Structure	Application(s)	References
S-fucan	*P. tetrastromatica, Okinawa mozuku, Leathesia difformis, Fucus spp. F. vesiculosus, Sargassum spp. Sargassum fusiforme, Ecklonia cava E. kurome*	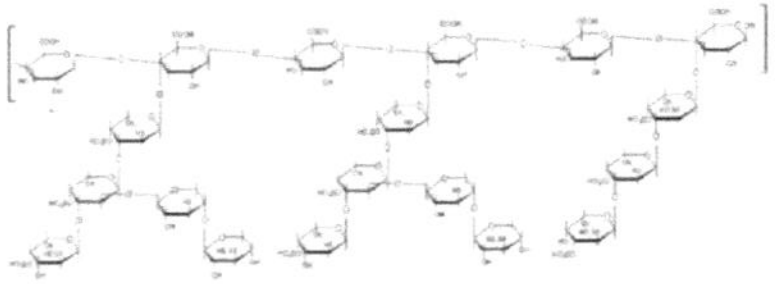	Antiviral, antitumor, anti-proliferative, immunomodulator, anti-inflammatory, antiadhesive, angiogenic, cardioprotective, gastroprotective.	Dantas-Santos et al. (2012); Hidari et al. (2008); Karmakar et al. (2009); Thomes et al. (2010); Wijesinghe & Jeon (2012)
S-galactofucan	*Lobophora variegate, D. menstrualis, D. mertensis, Spatoglossum schröederi, Adenocystis utricularis, Costaria costata, Saccharina japonica,*	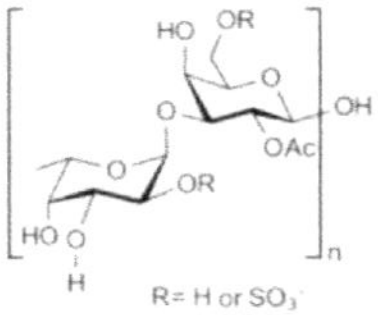	Antinociceptive, anti-thrombotic, anti-adhesive, peripheral anti-nociceptive, antioxidant, anti-inflammatory, anti-proliferative, anticoagulant.	Albuquerque et al. (2013); Costa et al. (2010); Medeiros et al. (2008); Paiva et al. (2011); Takano et al. (1995)
Porphyran	*P. yezoensis, Porphyra spp.,*	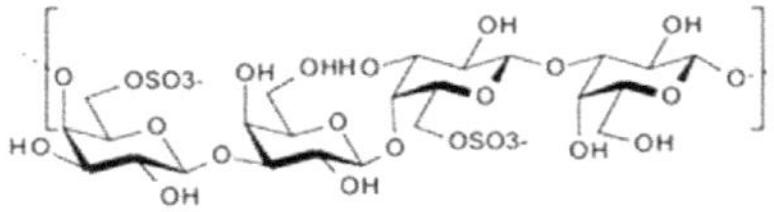	Regulates blood cholesterol, antitumor, hypotensive, immunomodulatory, and hypolipidemic.	Kwon & Nam (2006); Tsuge et al. (2007); Yoshizawa et al. (1993), 1995)
Fucoidan	*Fucus vesiculosus, C. novae-caledoniae, Eisenia bicyclis, S. horneri, Laminaria spp. S. patens, S. henslowianum, Lessonia vadose, Saccharina cichorioides, Undaria pinnatifida, Cystoseira indica.*	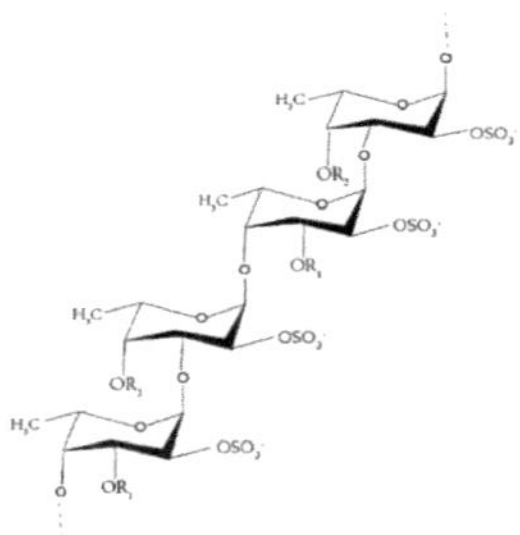	Anti-viral, anti-inflammatory, antitumor, anti-thrombotic, anti-proliferative, anti-angiogenic, anticoagulant, anti-metastatic, antioxidant, hyperlipidemic.	Chevolot et al. (1999); Cumashi et al. (2007); Holtkamp et al. (2009); Huang et al. (2010); Lee et al. (2008); Li et al. (2008); Xue et al. (2004); Yamamoto et al. (1984); Yoon et al. (2007)

TABLE 4.2 *(Continued)*

Polysaccharides	Source(s)	Structure	Application(s)	References
S-agarans	*Acanthophora spicifera, Bostrychia montagnei, Cryptopleura ramose, Gloiopeltis complanata, Aghardiella tenera, G.corticata.*	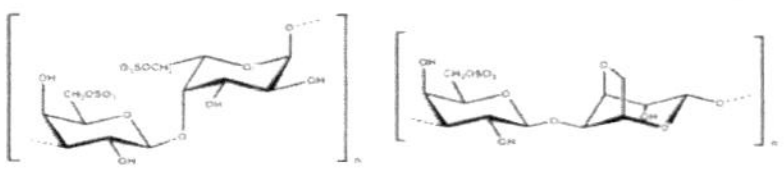	Antiviral	Carlucci et al. (1997); Clercq, (2000); Duarte et al. (2004); Duarte, Noseda et al. (2001); Mazumder et al. (2002); Takano et al. (1995); Witvrouw et al. (1994)
Agaroid-carrageenan	*Gymnogongrus tenuis, G. furcate.*	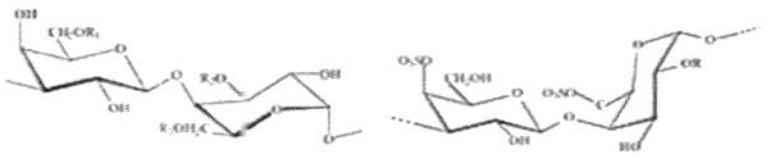	Anticoagulant	Perez Recalde et al. (2016); Takano et al. (1998)
S-λ-carrageenan	*Chondrus crispus, Phyllophora brodiei, S. pacifica.*	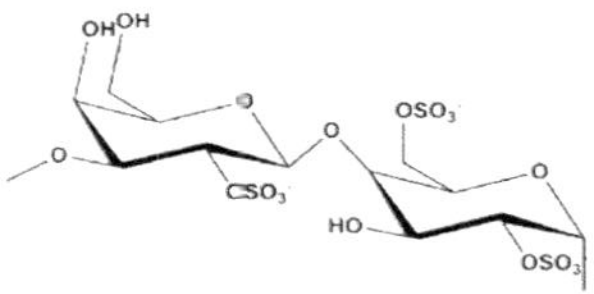	Antiviral, antithrombotic, anticoagulant.	Campo et al. (2009); Luescher-Mattli, (2005); Prajapati et al. (2014); Sen et al. (1994)
S-κ-carrageenan	*E. spinosa*	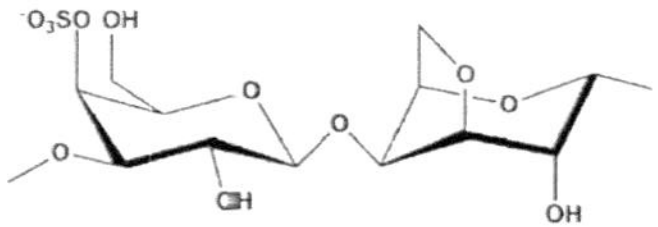	Antithrombotic, anticoagulant.	Campo et al. (2009); Kraan, (2012); Prajapati et al. (2014)
S-galactans	*Grateloupia indica, Schizymenia dubyi, S. binderi, Botryocladia occidentalis, G.corticata, Caulerpa pracemosa, U. lactuca,*	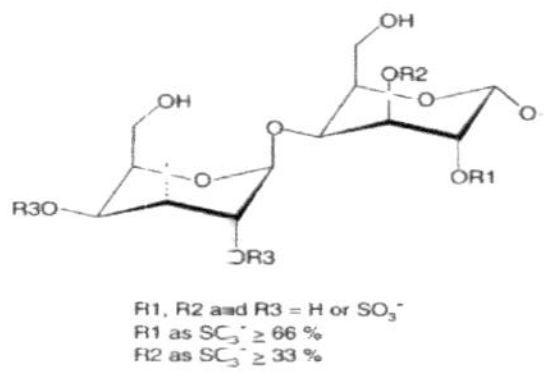	Antiviral, anticoagulant, antithrombotic, anti-venom.	Bourgougnon et al. (1994); Farias et al. (2000); Glauser et al. (2009); Ji et al. (2008); Sen et al. (1994); Toyama et al. (2010); Zúñiga et al. (2006)

TABLE 4.2 *(Continued)*

Polysaccharides	Source(s)	Structure	Application(s)	References
Xylomannan	*Sebdenia polydactyla, Nothogenia fastigiata.*	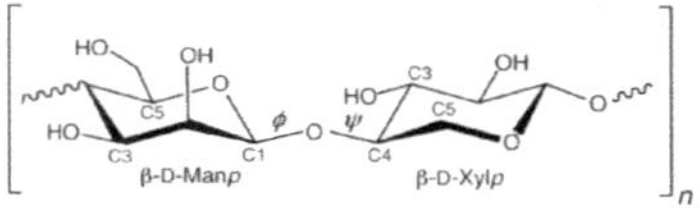	Antiviral, anticoagulant.	Damonte et al. (1994); Damonte et al. (1996); Ghosh et al. (2009); Kolender et al. (1997)
Rhamnan	*Monostroma latissimum, M. nitidum, Enteromorpha intestinalis*	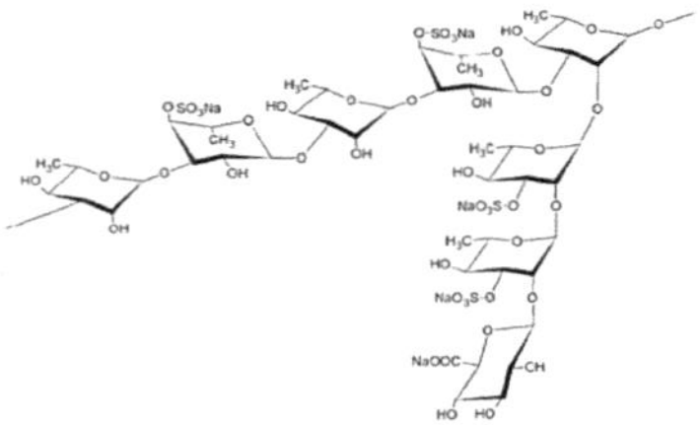	Immunomodulator, hepatoprotective, antithrombotic, antiviral, antitumor, anticoagulant.	Charles et al. (2007); Hayakawa et al. (2000); Jiao et al. (2009, 2010); Karnjanapratum & You, (2011); Qi et al. (2005, 2006); Terasawa et al. (2022)
S-ulvans and derivatives	*Ulva spp., U. pertusa, Eisenia bicyclis, Enteromorpha prolifera, Ulva reticulata, Ulva armoricana, Ulva rotunda.*	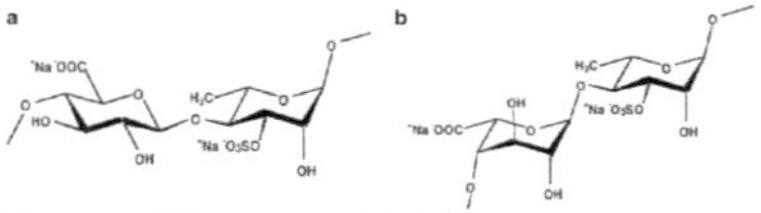 Note: a. A3s type and b. B3s type	Antioxidant, hepatoprotective, anti-proliferative, hypolipidemic, immunomodulator, hypocholesterolemia.	Kim et al. (2011); Lahaye et al. (1997); Lahaye & Robic (2007); Li et al. (2013); Pengzhan et al. (2003); Rao et al. (2004); Reisky et al. (2019); Takano et al. (1998); Teng et al. (2013); Usui et al. (1980); Xing et al. (2005)

4.3.2 FATTY ACIDS

Curiosity about the beneficial attributes of marine-derived fatty acids (FA) grew after observing that a few races, for example, Eskimos, have a lower incidence of thrombotic and atherosclerotic disorders. Aside from that, they also suffer less from inflammatory ailments like rheumatoid arthritis (Schmidt et al., 2001; Woods et al., 2000). Data obtained from human studies have shown that people who generally eat fish in their daily diet, such as Japanese and Eskimos, are less prone to develop prostate cancer than people who have a lower rate of fish intake (Lanier et al., 1989; Nobmann et al., 1992; Nutting et al., 1993).

Extended studies on cell lines and animal models have shown that docosahexaenoic acid (DHA) and eicosapentaenoic acid (EPA), these two omega-3 (ω3) fatty acids, are tumor suppressors (Connolly et al., 1997; Rose & Connolly, 1991). In 2013, Vaughan et al. suggested the potential for marine-derived polyunsaturated fatty acids (PUFAs) as an effective adjuvant to chemotherapeutics in their published literature (Vaughan et al., 2013).

Moreover, they are involved in partly inhibiting a good number of inflammation mediators such as prostaglandins, inflammatory cytokines, reactivity of T-helper cells, etc. (Calder, 2015; Vaughan et al., 2013). Furthermore, these two FAs have shown advantageous effects in treating different cardiac disorders to such an extent that they have been recommended in guidelines to patients for myocardial infarction management (Calder, 2012; Calder & Yaqoob, 2012; Innes & Calder, 2020; Nicholson et al., 2013; Saravanan et al., 2010; Schmidt et al., 2001).

ω3 FAs also exhibit a hypotriglyceridemic effect (Davidson, 2006). They decrease SREBP-1c (sterol regulatory element binding protein) levels via parallel suppression of hepatic lipogenesis, upregulation of FA oxidation in skeletal muscle and liver by activating peroxisome proliferator-activated receptor (PPAR), and also enhance metabolic flux of glucose into glycogen by downregulating the hepatocyte nuclear factor 4α (HNF-4α) (see more in Table 4.3) (Clarke, 2001; Jump & Clarke, 1999; Pawar et al., 2003; Pégorier et al., 2004; Price et al., 2000). This causes a redistribution of glucose on the way to oxidation from the triglyceride depot. As a result, the availability of the substrate required for VLDL (very low-density lipoprotein) synthesis decreases. Thus, ω3 FAs are one of the potent hypotriglyceridemic agents which act by concomitant upregulation of protein-encoding genes that trigger FA oxidation, and downregulation of protein-encoding genes that trigger lipid synthesis. These ω3 FAs may also work by increasing the depletion of postprandial chylomicron via lowering VLDL production and triggering lipoprotein lipase (LPL) activity. All these integrated effects aid in considering ω3 FAs as a beneficial tool in treating hypertriglyceridemia (Claudel et al., 2005; Hertz et al., 2001; Jump, 2002; Pan et al., 2004).

TABLE 4.3 Summarized Elaboration for the Effect of Ω3 FAs on Nuclear Receptors Involved in Regulation of Lipogenesis (Davidson, 2006)

Nuclear Receptor	Effects of ω3 FAs on Gene Regulation	Physicochemical Changes		
		Triglycerides	LDL	HDL
PPAR-α	↑	↓↓	↓	↑
HNF-4α	↓	↓↓	↔	↓
FXR	↑	↓↓	↑	↑
LXR	↓	↓↓	↓	↓
Net effects	↔	↓↓↓↓	↔	↔

Note: ↑ : increase; ↓ : decrease; ↔ : neutral effect; PPAR: peroxisome proliferator-activated receptor; HNF: hepatocyte nuclear factor; FXR: farnesol X receptor; LXR: liver X receptor; LDL: low-density lipoprotein; and HDL: high-density lipoprotein.

The quantity of the ω3 PUFA from marine sources varies depending on the species which has been mentioned in Table 4.4.

TABLE 4.4 An Estimated Amount of Ω3 PUFA in Marine-Derived Food

SL. No.	Source	g ω3 PUFA/100 g
1.	Salmon	1.0–2.0
2.	Tuna	0.5–1.6
3.	Shrimp	0.2–0.4
4.	Herring	1.2–3.1
5.	Cod	~0.2
6.	Trout	0.5–1.6
7.	Mackerel	1.8–5.3
8.	Halibut	0.5–1.0

**Dependent on factors like – season, region, and oceanic zone.*

Source: Information taken from: Hepburn et al. (1986); Innes & Calder (2020).

The ω3 PUFA belongs to an essential fatty acid family that has its first double bond at carbon atom number 3, which is counted as one of its structural characteristics (Figure 4.8) (Schmidt, 1997; Schmidt et al., 2001). The major marine-derived ω3 PUFAs are EPA (20:5n–3) and DHA (22:6n–3). They are also known as long-chain n–3 PUFA (Schmidt et al., 2001).

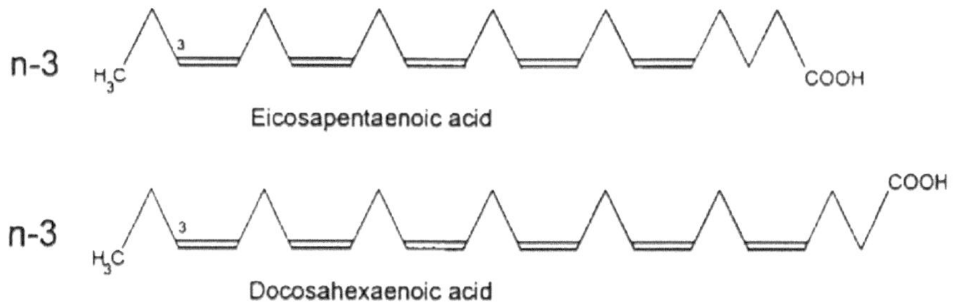

FIGURE 4.8 Structure of EPA and DHA.

Source: Reproduced with permission from Innes & Calder, 2020. Copyright © 2020 by the authors; licensee MDPI, Basel, Switzerland. http://creativecommons.org/licenses/by/4.0/

Microalgae are the predominant source of marine-derived ω3 PUFA, which is synthesized by an aerobic pathway inside the endoplasmic reticulum and chloroplast of an algal cell. Figure 4.9 explains the detailed mechanisms by which EPA and DHA are produced within algal cells in a simplified way. Despite the fact that microalgae are the leading primary makers of PUFA, higher marine ecological pyramids have metabolic pathways that can generate novel and eccentric PUFA. Besides, it has been seen that some organisms from lower tropical levels of the marine ecological system produce unique PUFA as well. Apart from microalgae, there are some other marine sources that provide PUFAs, such as mollusks, cephalopods, sponges (mostly Demospongiae), crustaceans, cnidarians, and tunicates (Monroig et al., 2013).

Filimonova and coworkers have recommended the use of marine FAs as the biochemical marker to evaluate contamination levels in marine ecosystems by studying the structural composition of FA chains (Filimonova et al., 2016).

4.3.3 COLLAGENS

Collagen, an intricate macro-protein that accounts for approximately 20–30% of all proteins, is the predominant element of membranes in the case of most living organisms, primarily serving structural functions such as cartilage and bones (Coppola et al., 2020; Geahchan et al., 2022; Müller & Müller, 2003; Szpak, 2011). To date, a total of 28 different kinds of collagens have been discovered, which have been chronologically denoted using Roman numerals I–XXVIII (Bou-Gharios et al., 2020; Halper, 2021; Karsdal, 2016; Salvatore et al., 2020). Typically, they consist of three helically-shaped helices containing about 1050 chains of amino acids (AAs) per helix. The fundamental structure of AA chains is the triplet [(Gly-X-Y) n], where glycine bonds to another two AAs with a repeated sequence (Figure 4.10). Generally, at position Y, proline or hydroxyproline is present, and apart from these two common AAs, there are approximately over 1000 AAs to be bonded. Ultimately, the AA chains are then oriented in primary, secondary, and tertiary structures in the shape of fibrils (Bächinger et al., 2010; Ferreira et al., 2012; Fidler et al., 2018; Salvatore et al., 2020; Silva et al., 2014; Sorushanova et al., 2019; Subhan et al., 2015).

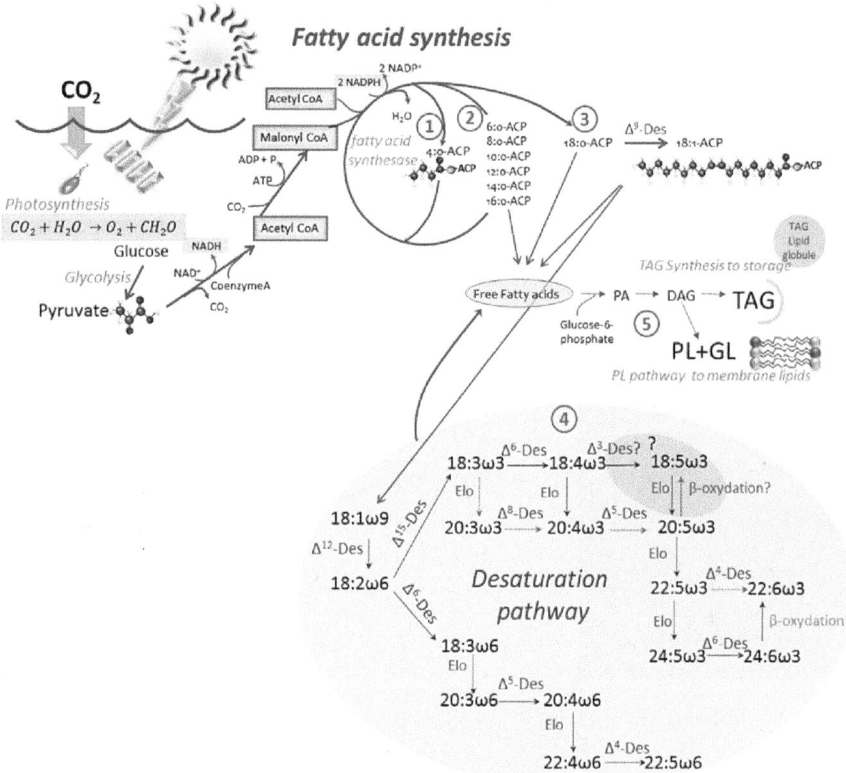

FIGURE 4.9 Simplified illustration of the aerobic pathway of PUFA synthesis in the phytoplankton cell.

Source: Modified from: Jónasdóttir (2019). Copyright © 2019 by the authors; licensee MDPI, Basel, Switzerland. http://creativecommons.org/licenses/by/4.0/

Collagens can also be classified on the basis of core structure along with infrastructural arrangements. Each of them has a certain alpha chain accompanying its core structure which makes a contribution to classify the type. They can be network-forming, fibril-forming, fibril-associated with interrupted triplex helix-like, and fibril-associated with interrupted triplex helix. Collagen types I, II, III, V, XI, XXIV, and XXVII fall under fibril-forming; collagen types IX, XII, XIV, and XX fall under fibril-associated with interrupted triplex helix; collagen types XVI, XIX, XXI, and XXII fall under fibril-associated with interrupted triplex helix-like; collagen type IV (basement membrane collagen); VI (beaded filament-forming collagen); VII (anchoring fibril-collagen); VIII and X (hexagonal network-collagen); XIII, XVII, XXIII, and XXV (transmembrane collagens); XV and XVIII (multiplexin collagen); XXVI and XXVIII (molecules with collagen-like domains) fall under network-forming. Each of them has its distinct function and offers high-order tissue composition (Coppola et al., 2020; Gordon & Hahn, 2010; Rest & Garrone, 1991; Senadheera et al., 2020; Silva et al., 2014; Stover & Verrelli, 2011; Szpak, 2011).

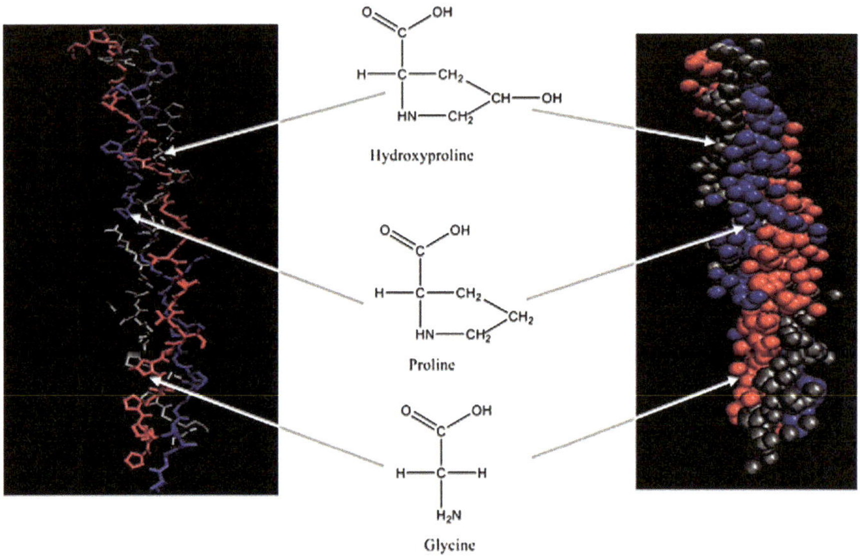

FIGURE 4.10 The helical structure of collagen.

Source: Reproduced from Senadheera et al., 2020. Copyright © 2020 by the authors; licensee MDPI, Basel, Switzerland. http://creativecommons.org/licenses/by/4.0/

Marine collagen (MC) is biocompatible, degradable, and has the capacity to absorb a considerable amount of water, making it a preferable compound for thickening, texturizing, and gel formation. Moreover, it has compelling properties regarding surface behavior that lead to its involvement in stabilization, protective colloid functions, film formation, emulsion, adhesion, cohesion, and foam formation. Aside from these, it is a good surfactant as well and has the penetrability for lipidic interfaces. Furthermore, it has some more excellent properties mentioned in Figure 4.11, and that is the reason the application of collagen is increasing in drug delivery devices and transdermal implants used in the pharmaceutical industry, as well as in cosmeceuticals, food, beverages, and nutraceuticals (Bourtoom, 2008; Hashim et al., 2015; In Jeong et al., 2007; Liu et al., 2022; Martucci & Ruseckaite, 2009; Song et al., 2006; Sorushanova et al., 2019; Wahyudi et al., 2016).

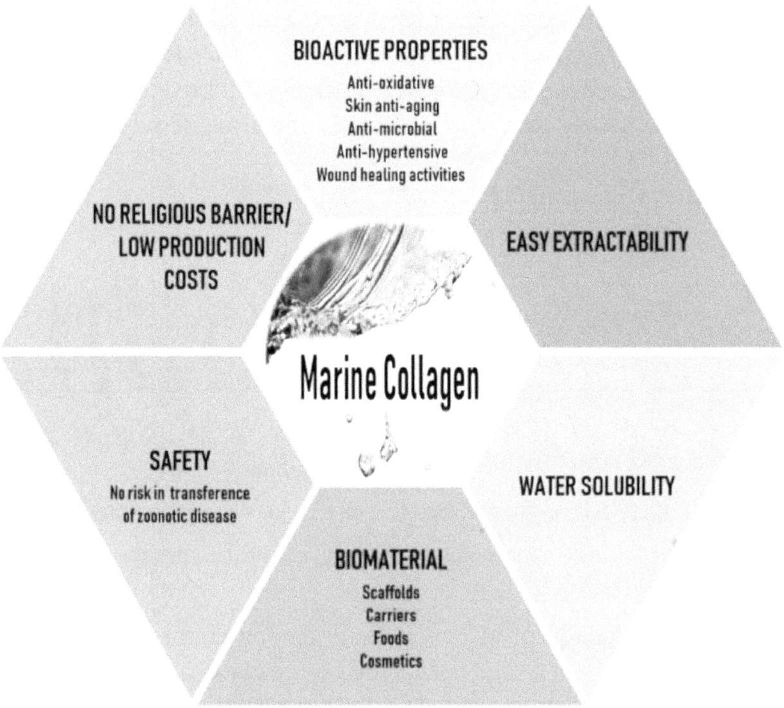

FIGURE 4.11 Properties and some of the applications of collagen.

Source: Reproduced from Lim et al., 2019. Copyright © 2019 by the authors; licensee MDPI, Basel, Switzerland. http://creativecommons.org/licenses/by/4.0/

MC is known for its feasibility in clinical trials because of weak immunogenicity, which reduces the chances of rejection if injected or ingested into a different animal's body. Despite its poor antigenicity, this attribute can be amplified by modification to suppress any immunogenic response (Coppola et al., 2020; Dong & Lv, 2016). Its unique properties also represent itself as a good source for biomaterials and bioplastics. In addition, marine-derived collagen does not carry transmissible diseases, which proves its safety for use in pharmaceuticals and increases interest among researchers, making it a popular area of research. Figure 4.12 shows the number of published articles regarding marine-derived collagens from the years 2000 to 2019, where data has been obtained from articles with keywords 'marine collagen' and 'fish collagen' indexed in Scopus. Particularly, fish biomass, jellyfish, starfish, sponges, and sharks contain a bit higher amount of collagen (Avila Rodríguez et al., 2018; Azizur Rahman, 2019; Felician et al., 2018; Langasco et al., 2017; Lee et al., 2001; Lim et al., 2019; Parisi et al., 2019). Apart from them, some other sources of collagen are prawns, octopus, cuttlefish, squid, sea urchin, mollusks, and sea anemones (Di Benedetto et al., 2014; Fleury et al., 2011; Kimura et al., 1981; Minamisako & Kimura, 1989; Nordwig et al., 1973; Tziveleka et al., 2017). Figure 4.13 summarizes the potential sources of marine-derived collagen.

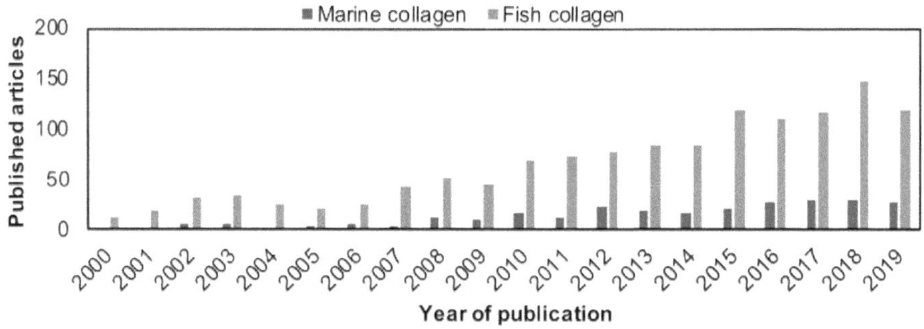

FIGURE 4.12 From 2000 to 2019 increasing research interest in marine derived collagen.
Source: Reproduced with permission from Salvatore et al., 2020. Copyright © 2020. Elsevier

Industrial production of MC relies to a greater extent on the extraction from sources, and lesser on the purification systems. Based on this, different methodologies have been developed for the purification of MC and its derivatives such as gelatin or peptides. Not only that, extraction protocols vary within a single marine source depending on the structural configuration of a specific body part, for example, skin, bone, cartilage, scales, bell, mesoglea, etc. (Chuaychan et al., 2017; Ikoma et al., 2003; Jankangram et al., 2016; Muralidharan et al., 2013; Yu et al., 2018). Keeping the distinctions aside, all of these procedures share five basic methodical steps (Figure 4.14) under the given operating temperature, 4–10°C (to avoid collagen denaturation) for all steps, which are: (i) isolation and cleaning; (ii) size minimization; (iii) elimination of non-collagenous components; (iv) collagen extraction by acidic or enzymatic treatment; and (v) salt precipitation and retrieval (Salvatore et al., 2020).

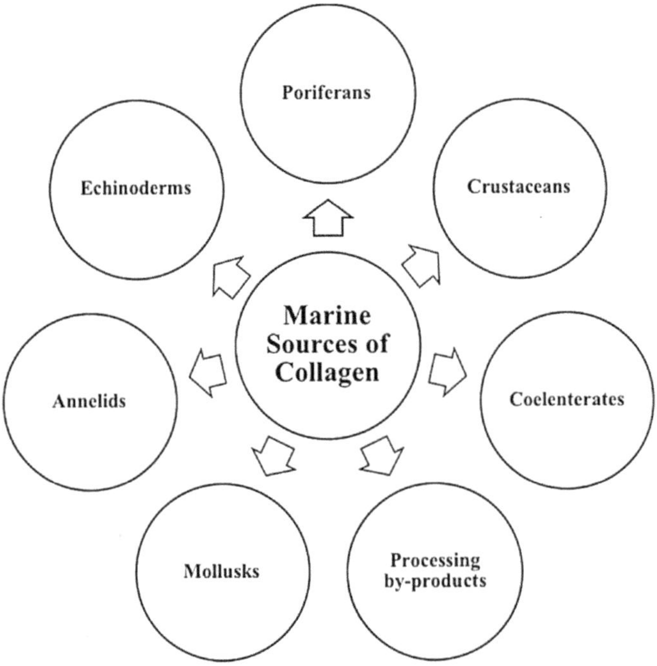

FIGURE 4.13 Potential marine sources of collagen.

Source: Reproduced from Senadheera et al., 2020. Copyright © 2020 by the authors; licensee MDPI, Basel, Switzerland. http://creativecommons.org/licenses/by/4.0/

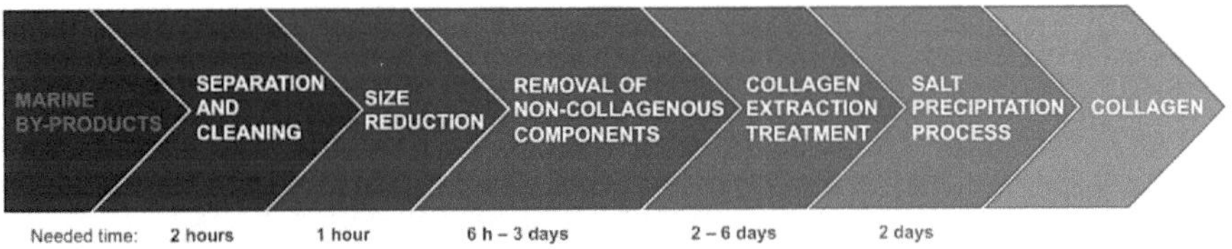

FIGURE 4.14 Schematic illustration for basic methodical steps to MC acquisition.

Source: Reproduced with permission from Salvatore et al., 2020. Copyright © 2020. Elsevier

4.3.4 PEPTIDES

Marine Peptides (MP) are one of the important MNPs which are distinct small parts of proteins that additionally act as a source of nitrogen and AAs along with varied physiological functions. They were first isolated and identified as cardiotonic, cardiotoxin, neurotoxin, antimicrobial, antitumor, and antiviral agents (Aneiros & Garateix, 2004; Anthony, 1974; Bernheimer et al., 1982;

Matsunaga et al., 1985; Norton et al., 1976; Zhou et al., 2013). These MPs are collected from fish, crabs, algae, mollusks, crustaceans, marine bacteria, and fungi. Depending on structural properties, AA sequences, and composition, they also have shown properties like antihypertensive, anti-proliferative, anticoagulant, anti-obesity, anti-aging, antioxidant, anti-diabetic, anti-cancer, strengthening immunity, cognitive function improvement, and chronic disease prevention, etc. (see more in Table 4.5) (Ahmed et al., 2021, 2022; Chai et al., 2016; Kim & Wijesekara, 2010; Masso-Silva & Diamond, 2014; Mayer et al., 2013; Ngo et al., 2012; Ryan et al., 2011; Sridhar et al., 2021; Xia et al., 2017).

Generally, peptides are first obtained from marine sources in order to further isolation and purification. After that, by using a bioassay-based fractionation procedure, they are fractionated. Then, a single bioactive peptide is isolated after purification. Finally, the obtained peptide extract is screened to find certain bioactivity (Figure 4.15). Typically, the use of appropriate proteolytic enzymes (e.g., trypsin, α-chymotrypsin, neutrase, papain, etc.) aids in the yield of biologically active peptides. The physicochemical parameters, i.e., temperature and pH of the protein solution, need to be adjusted for the optimization of the proteolytic enzyme's activity to get the maximum yield of the desired bioactive peptide. In the first step of peptide extraction, organic solvents such as methanol or ethyl acetate are used, and then the concentrated extract is dissolved in other solvents, for example, dichloromethane, carbon tetrachloride, or hexane (*see more in* Figure 4.16). After that, the partially purified extract is subjected to further purification using silica gel or size exclusion chromatography and washed out via polar solvents. Final purification is done via reversed-phase C18 HPLC (Aneiros & Garateix, 2004; Cheung et al., 2015; Du et al., 2014; Kim et al., 1997; Qian et al., 2007, 2008; Simpson et al., 1998; Sridhar et al., 2021; Yurchenko et al., 2014).

4.4 MARINE-DERIVED MATERIALS IN EMERGING DRUG DELIVERY SYSTEMS

4.4.1 TRANSDERMAL DRUG DELIVERY SYSTEM (TDDS)

In order to develop a new therapeutic drug delivery system, the transdermal route is usually considered an attractive approach apart from conventional drug delivery systems such as oral and parenteral drug delivery systems for administering pharmaceutical active ingredients. This is due to several advantages in some cases, including the avoidance of first-pass metabolism, reduction of unwanted effects, and improvement of patient compliance, making it a priority over conventional oral and intravenous techniques. However, one of the major limitations in this regard is the complex (brick

TABLE 4.5 List of Some Marine-Derived Peptides with Their Structure and Pharmacological Properties

Peptide	Source	Structure	No. of Natural AAs	Pharmacological Properties	References
Aminolipopeptides Trichoderins A, A1, and B	Sponge-derived fungus of *Trichoderma* sp.	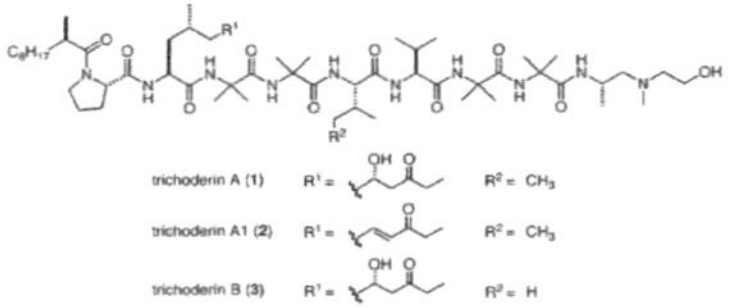	2	Antimycobacterial	Pruksakorn et al. (2010)
Callipeltins	*Sponge Callipelta* sp.	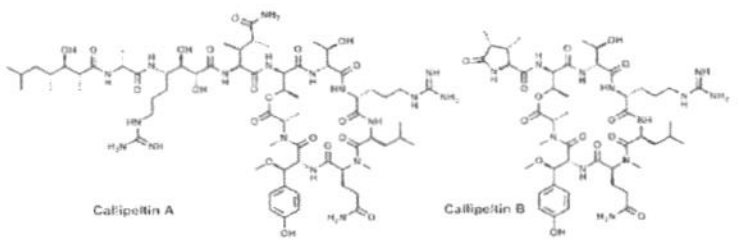	5	Anti-HIV	Kikuchi & Konno (2016); Pelay-Gimeno et al. (2013); Zampella et al. (1996)
Callyaerins A–G, I–M	Sponge *Callyspongia aerizusa*	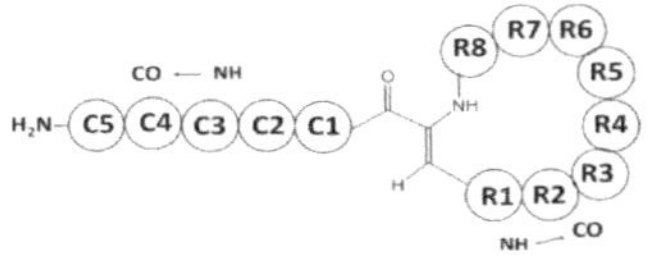	12	Antiviral, antimicrobial and antitubercular	Daletos et al. (2015); Sable et al. (2017)
Centrocins 1 and 2	Green sea urchin *Strongylocentrotus droebachiensis*		118	Antibacterial (broad spectrum)	Kang et al. (2015); Li et al. (2010)
Cm-p1 and Cm-p5	Snail *Cenchritis muricatus*	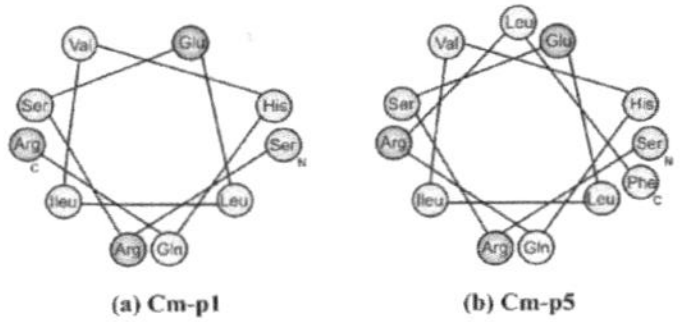	10	Antifungal	López-Abarrategui et al. (2012, 2015)

TABLE 4.5 *(Continued)*

Peptide	Source	Structure	No. of Natural AAs	Pharmacological Properties	References
Contulakin-G	Cone snail *Conus geographus*	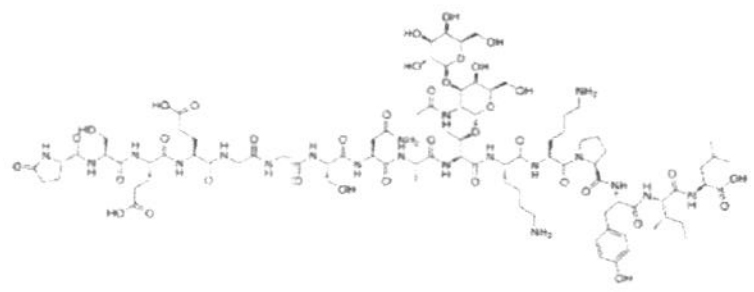	15	Analgesic	(Lee et al. (2015)
Crude peptide extract	Snail *Conus araneosus*	-	-	Sedative	Franklin & Rajesh (2015)
Cyclic depsipeptides Mirabamides	Sponge *Stelletta clavosa*	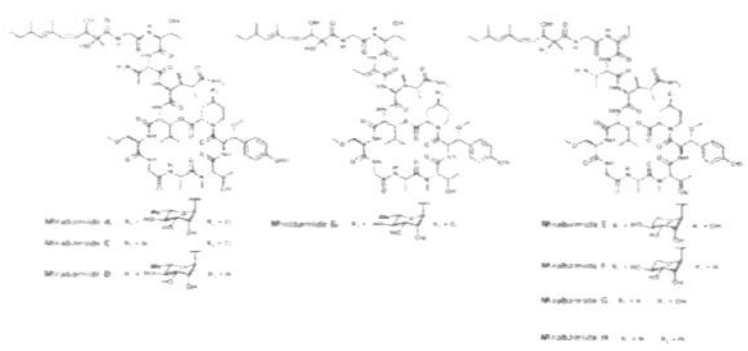	3	Antibacterial and anti-HIV	Pelay-Gimeno et al. (2013)
Cyclic lipopeptides, Maribasins A and B	Microorganism *Bacillus marinus* B-9987	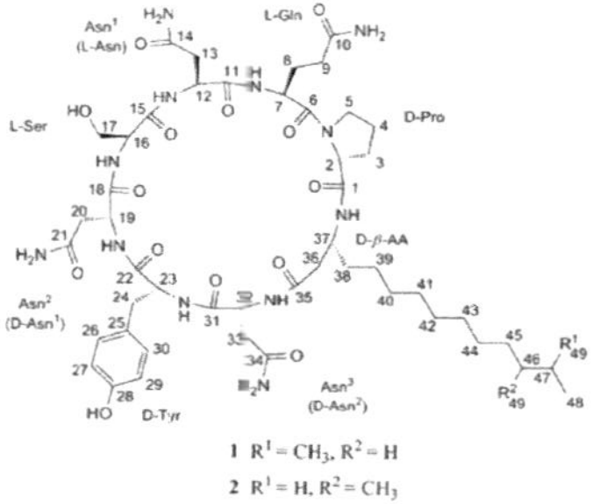	7	Antifungal (broad spectrum)	Zhang et al. (2010)

TABLE 4.5 *(Continued)*

Peptide	Source	Structure	No. of Natural AAs	Pharmacological Properties	References
Discodermin A	Sponge *Discodermia kiiensis*	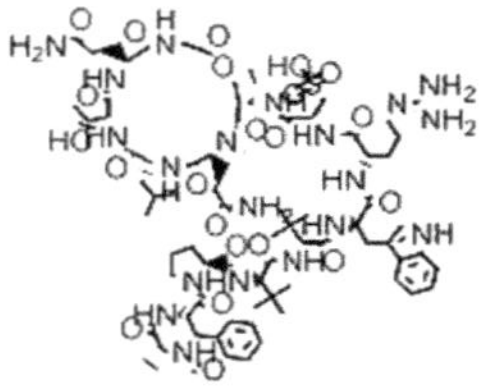	11	Antifungal and antibacterial (broad spectrum)	Otero-González et al. (2010)
EcDefensin	Estuary cod *Epinephelus coioides*	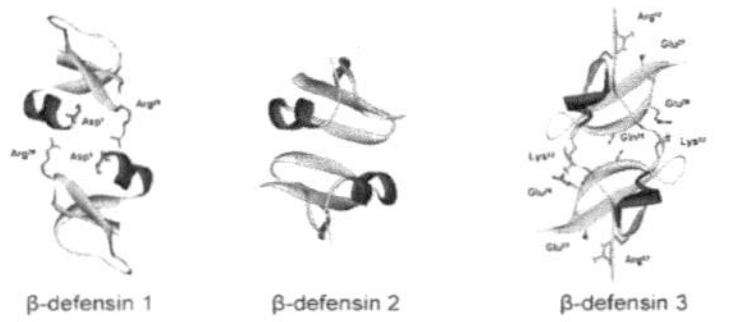	63	Antiviral (SGIV, Betanoda virus or NNV)	Guo et al. (2012)
Hemocyanin	Sea snail *Rapana thomasiana*	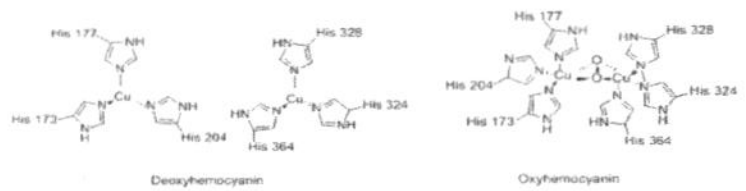	24	Antiviral (HSV)	Cheung et al. (2014)
HTP-1	Seahorse *Hippocampus trimaculatus*	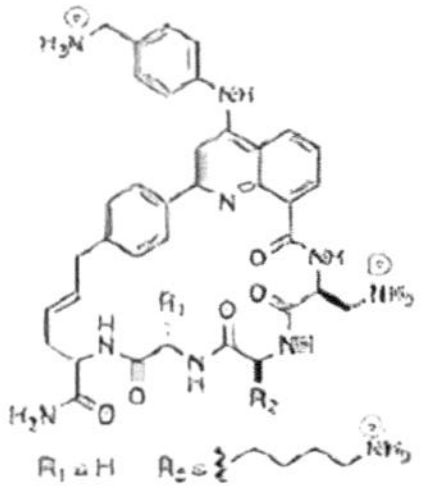	8	Neuroprotective	Pangestuti, Ryu et al. (2013)

TABLE 4.5 *(Continued)*

Peptide	Source	Structure	No. of Natural AAs	Pharmacological Properties	References
Kahalalide F Analogs	Mollusks *Sacoglossan* sp.	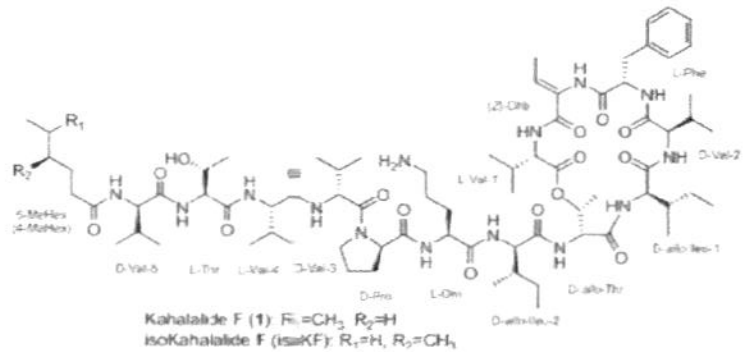	8	Antifungal and antitumor	Shilabin & Hamann (2011)
Koshikamide F–H	Sponges *Theonella swinhoei* and *Theonella cupola*	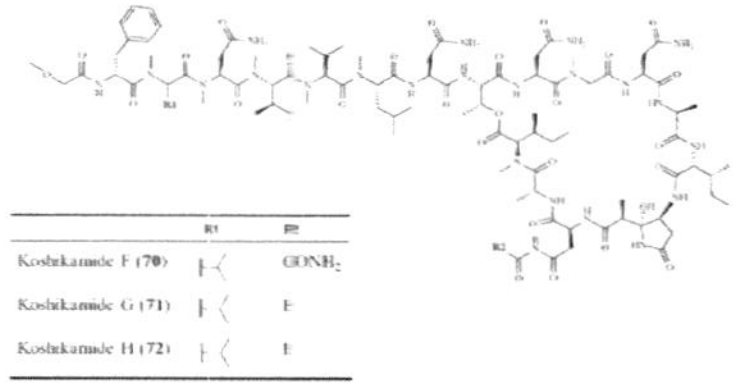	4	Anti-HIV	Kang et al. (2018); Plaza, Bifulco et al. (2010)
Neurotoxin AV3	Sea anemone *Anemonia viridis*	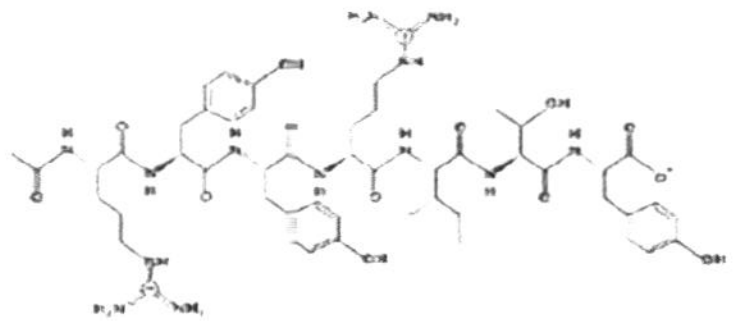	27	Modulation of voltage-gated sodium channel	Gur Barzilai et al. (2014)
Neurotensin	Cone snail *Conus geographus*	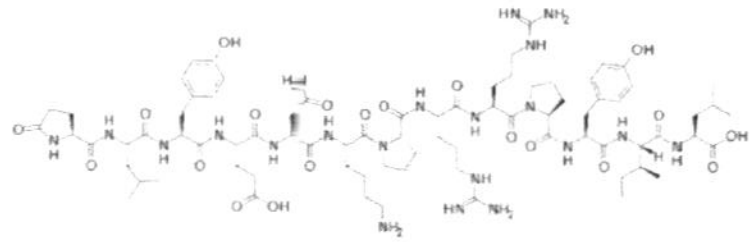	13	Analgesic	Lee et al. (2015)

TABLE 4.5 *(Continued)*

Peptide	Source	Structure	No. of Natural AAs	Pharmacological Properties	References
Peptide	Tunicate *Styela clava*	Ala-His-Ile-Ile-Ile	5	Antioxidant, cytotoxic, antihypertensive, hepatoprotective	Ko et al. (2016)
	Box jellyfish *Chiropsalmus quadrigatus*	Ala-Cys-Pro-Gly-Pro-Asn-Pro-Gly-Arg-Pro	10	Antihypertensive	So et al. (2016)
Terrelumamides A and B	Marine fungi *Aspergillus terreus*	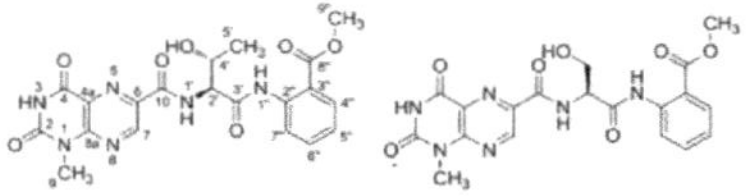 Terrelumamides A Terrelumamides B	1	Antidiabetic	You et al. (2015)
Thalassospiramides A, D and G	Bacteria *Thalassospira strain CNJ328*	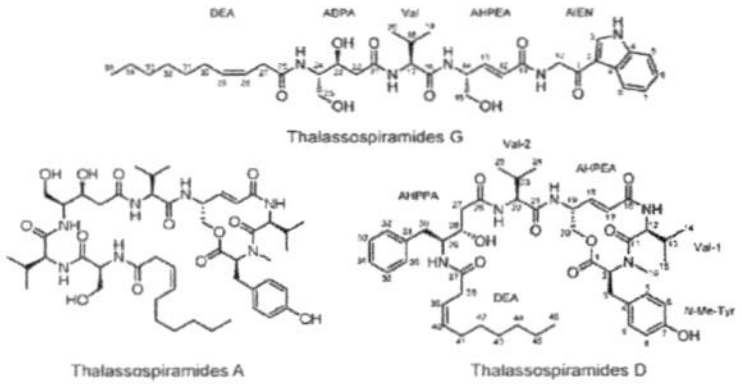	1, 4, 2	Anti-inflammatory	Um et al. (2013)

and mortar) structure of the skin, specifically the stratum corneum, which interferes with drug particle permeation. Improving drug permeability to overcome the skin barrier is challenging, leading to the necessity of developing innovative alternatives for TDDS (Costa Lima & Reis, 2018; Li et al., 2022; Moniz et al., 2021; Neupane et al., 2020; Sharma, 2019a). Recently, researchers have proposed two strategies: microneedles (MNs) and nanoparticles (NPs), which have shown the potential to increase drug permeability through the skin barrier (Moniz et al., 2021; Sharma, 2019a). In this section, we will discuss different marine-derived polymers used in MN fabrication and NP formulation.

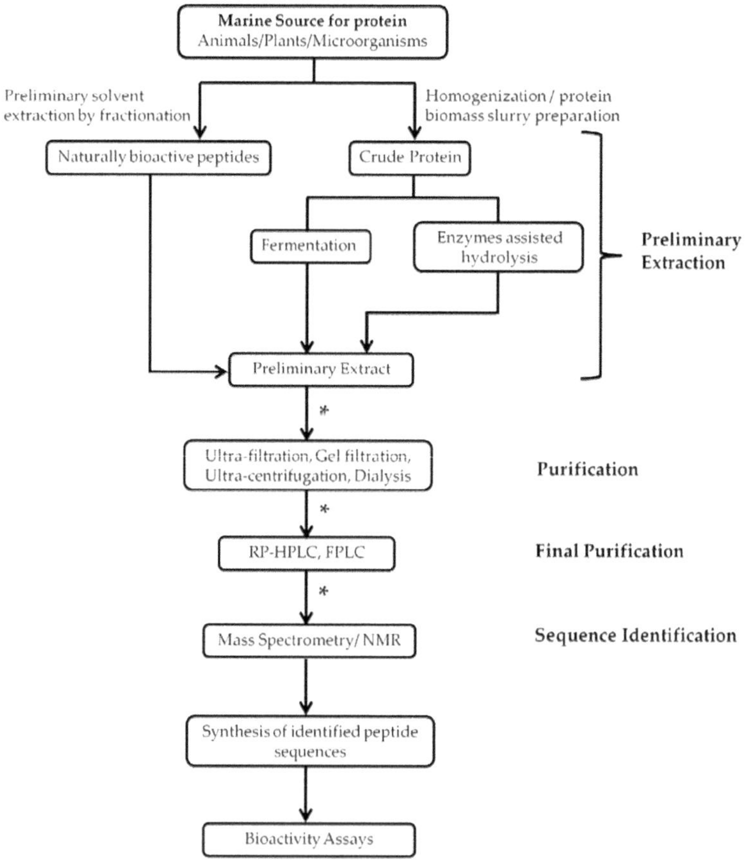

FIGURE 4.15 Steps involved in extraction, isolation, and identification of bioactive peptides from marine natural sources [*Steps in which bioassays have been done for screening].

Source: Reproduced from Sable et al., 2017. Copyright © 2017 by the authors; licensee MDPI, Basel, Switzerland. http://creativecommons.org/licenses/by/4.0/

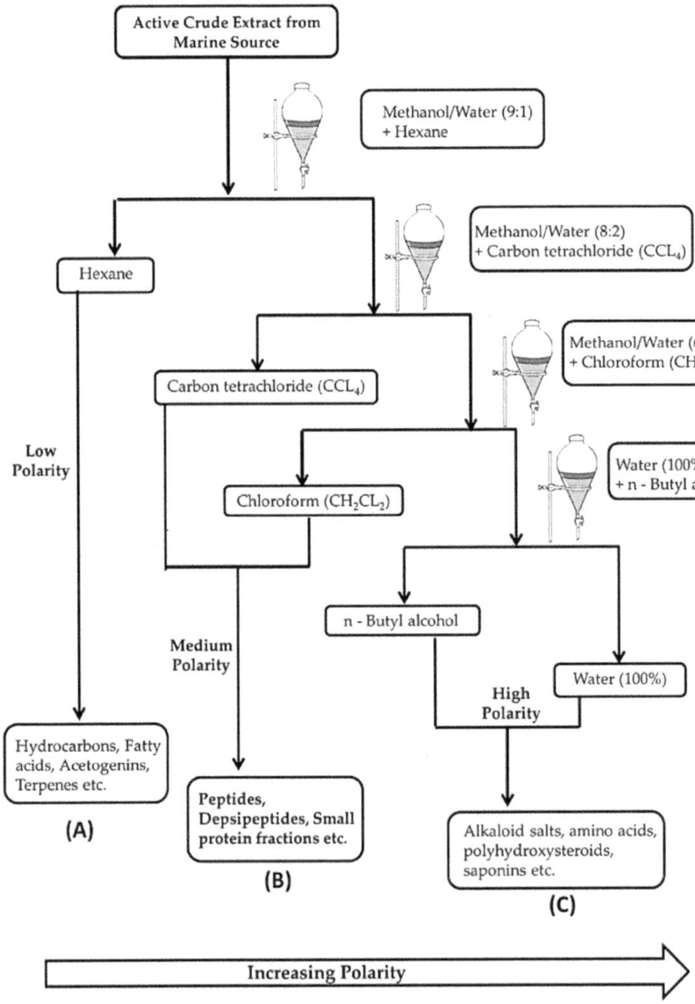

FIGURE 4.16 Schematic illustration of the solvent gradient-based isolation process of different chemical compounds of marine natural sources.

Source: Reproduced from Sable et al., 2017. Copyright © 2017 by the authors; licensee MDPI, Basel, Switzerland. http://creativecommons.org/licenses/by/4.0/

4.4.1.1 BIOPOLYMERS IN MNS FABRICATION

Due to the applicability in varied fields of treatment for different diseases, research regarding MNs has increased in the last two decades (Figure 4.17). From 1930 to 2021, a total of 3626 papers have been published according to the Scopus database, which includes reviews, conference papers, articles, letters, notes, and conference reviews (Biswas et al., 2017; Huang & Huang, 2019; Moniz et al., 2021).

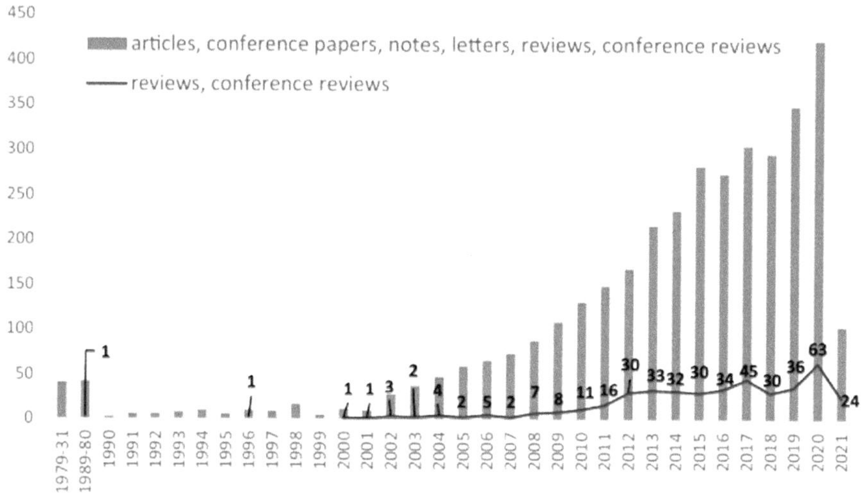

FIGURE 4.17 Rate of MN-related publications from 1930 to 2021.

Source: Reproduced with permission from Moniz et al., 2021. Copyright 2021. Elsevier.

MNs, a tiny therapeutic device, can be fabricated via various types of compounds (Hassan et al., 2022). Among them, marine-derived polymers are mostly used due to the beneficial properties discussed in section 3. Apart from those properties, MNP-derived polymers have shown adequate strength, availability, and cost efficiency. Moreover, polymeric MN fabrication and loading of a high amount of drugs are comparatively easy to be done (Juster et al., 2019; Kumar et al., 2022; Singh et al., 2019). Many techniques have been established to fabricate these MNs, such as thermal inkjet printing, stereolithography, micro-electromechanical systems, etc. (Hassan et al., 2022; Uddin et al., 2022; Zhang et al., 2009). The following marine-derived polymers are being used in MN fabrication right now.

4.4.1.1.1 Chitosan

Chitosan is one of the most popularly used polymers in biopharmaceutical sciences because of its availability. It is generally collected from the shells (exoskeleton) of arthropods and crustaceans. It is a crystalline, rigid, water-insoluble (at pH 7) polymer, which is a heteropolysaccharide. It consists of N-acetyl-D-glucosamine and D-glucosamine. Also, it is a partly

N-deacetylated analog of chitin (Figure 4.18). It is a natural cationic polymer due to the presence of free amino groups. Additionally, it can be dissolved in an acidic environment because of the addition of a proton (H^+) to free amino groups. The molecular weight (MW) of chitosan can be approximately 300 to 106 Da (Dabholkar et al., 2021; Fuhrmann & Zuberer, 2021; Husain et al., 2021; Moniz et al., 2021; Song et al., 2012). Various reactions can be facilitated by simply modifying amino groups present at the C-2 position and hydroxyl groups present at the C-3 and C-6 positions of chitosan. Moreover, among all the naturally derived PSs, chitosan is the only protonated PS that can interact with the negative charge of the outer layer of skin in facilitating the penetration of bioactive drug material. This modifiable quality of chitosan makes it a suitable polymer to be tailored according to the required mechanical strength and structural functions (Chatterjee et al., 2021; Kaczmarek et al., 2019; Li et al., 2022; Moniz et al., 2021).

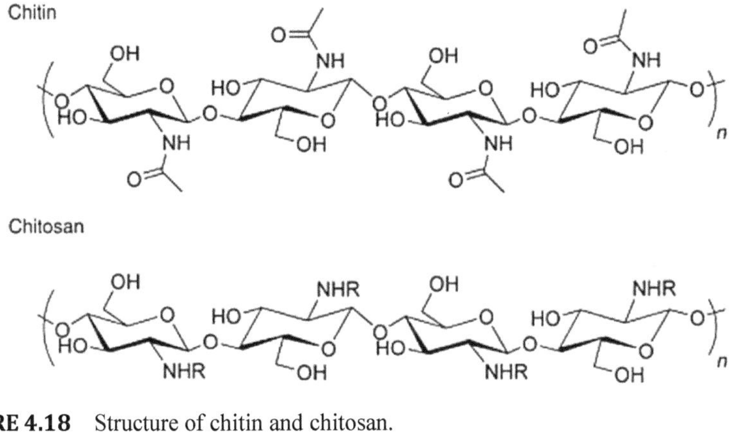

FIGURE 4.18 Structure of chitin and chitosan.

Chitosan is being widely used in many MN fabrication processes for the delivery of therapeutics, particularly in proteins, vaccines, anti-allergens, anti-cancer drugs, analgesics, anti-diabetic drug delivery, and wound healing purposes as well (Arshad, Hassan et al., 2019; Chi et al., 2020; Dathathri et al., 2020; Moniz et al., 2021; Yao et al., 2017). In 2013, Chen et al. conducted laboratory research where they successfully delivered ovalbumin protein via chitosan-made MNs that showed a comparatively higher antibody response in the rat body than conventional IM injections. They have also published a research work after that in 2018 based on the chitosan MN design and actuation mode. This group of researchers has published another of their experimental works where they have fabricated implantable MNs using chitosan in order to establish an extended-release vaccination process for the influenza virus and received the same outcome again in 2019 (Chen et al., 2013, 2018; Chen et al., 2019). Recently, Yang et al. conducted an *in-vitro* study where they delivered synergistic anti-cancer therapy for the treatment of melanoma using chitosan MNs, which exhibited a better antitumor effect in comparison to conventional injections (Yang et al., 2020).

Furthermore, aside from its non-toxic, biocompatible, hemostatic, and biodegradable properties, chitosan also has antibacterial and antifungal properties (Chen et al., 2019; Luo & Wang, 2014; Menchicchi et al., n.d.; Qin et al., 2020).

4.4.1.1.2 Alginate

Alginate is an anionic, biodegradable, marine-derived polysaccharide usually obtained from brown seaweed. It is an unbranched, linear copolymer comprising two types of blocks (i.e., G and M), α-L-guluronic acid (G) and β-D-mannuronic acid (M) binded together by the 1,4-glycosidic linkage in the varied ratio (G/M). Among all types of alginate, sodium alginate (MW = 32,000–400,000 g/mol) (Figure 4.19) is the most implemented one (Gombotz & Wee, 1998; Laurienzo, 2010; Lee et al., 2017; Li et al., 2022; Moniz et al., 2021).

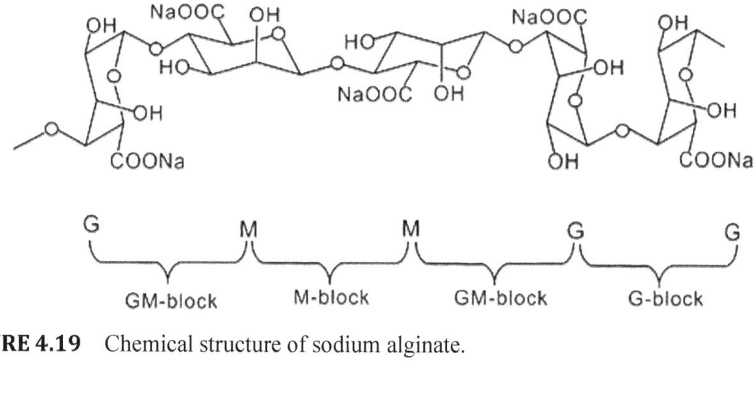

FIGURE 4.19 Chemical structure of sodium alginate.

Alginate with a higher G/M ratio exhibits wound healing properties by triggering cytokine generation via monocytes in the human body (Del

Gaudio et al., 2020; Varaprasad et al., 2020). Since it is comparatively cheaper, less toxic, biocompatible, and also shows mild gel-like properties in the presence of divalent cations e.g., Ca^{2+}, it is usually used to fabricate MNs (Gombotz & Wee, 1998; Lee et al., 2017). It has been used to engineer desired drug-loaded MNs such as insulin and vaccines (Chen et al., 2017; Moniz et al., 2021; Yu et al., 2017). In 2018, Zhang et al. published an experimental work where they fabricated alginate MNs to deliver insulin, which exhibited a better hypoglycemic effect in diabetic rats (Zhang et al., 2018). In 2019, a group of researchers fabricated Na-alginate MNs to deliver the *Bacillus* Calmette-Guérin (BCG) vaccine, which exhibited enhanced immunity in the treatment of tuberculosis (Arshad, Fatima et al., 2019).

4.4.1.1.3 Hyaluronic Acid

Hyaluronic Acid (HA) is an anionic, linear, non-sulfated glycosaminoglycan (GAG) which is one of the broadly used marine polymers, usually derived from crustaceans and bacteria. It is also known as hyaluronan (Papakonstantinou et al., 2012; Zhu et al., 2020). Structurally, it consists of N-acetyl-D-glucosamine and D-glucuronic acid which are linked by a glucuronidic β (1→3) bond (Figure 4.20) (Moniz et al., 2021; Weissman et al., 1954; Weissmann & Meyer, 1954).

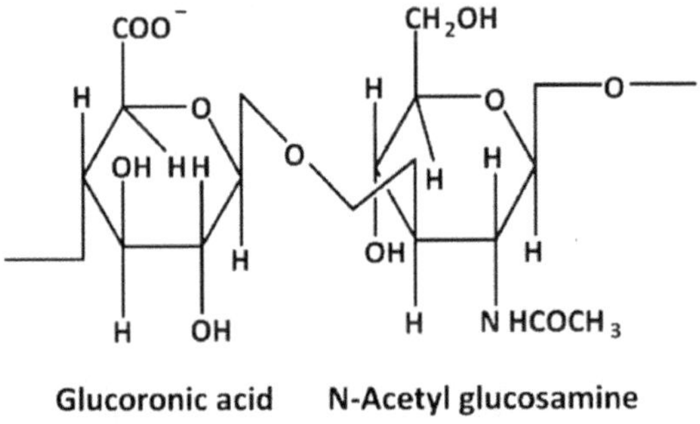

FIGURE 4.20 Structure of HA.

HA is also found in different body parts, usually in extracellular tissues. It is a vital element of animal skin as well, around 50% of the total amount found in the whole body (Collins & Birkinshaw, 2013; Juhlin, 1997; Mero

& Campisi, 2014; Reed et al., 1988; Tzellos et al., 2009). Due to its biocompatibility, high water affinity, non-toxicity, and non-immunogenicity, it has numerous biopharmaceutical applications in cancer therapy, cartilage regeneration, and ophthalmology (Highley et al., 2016; Hu et al., 2020; Misra et al., 2011; Tan et al., 2009; Zhang et al., 2020). It has also received authorization as a dermal filler from the Food and Drug Administration (FDA) (Becker et al., 2009).

In TDDS, HA is used in MN fabrication to deliver antidiabetic, antimicrobial, anti-inflammatory, anti-obesity, and anti-allergen medications; different anti-cancer therapies and vaccines, since it has an attribute of enhancing drug permeability by interacting with the stratum corneum (Dangol et al., 2017; Kim et al., 2018; Larrañeta et al., 2018; Leone et al., 2019; Matsuo, Hirobe et al., 2012; Matsuo, Yokota et al., 2012; Misra et al., 2011; Qiu et al., 2016; Ye, Yu et al., 2016; Ye, Wang et al., 2016; Yu et al., 2015). Usually, it is used after desired chemical modification along with another polymer as a conjugate for a better outcome, for example, in 2018, Chiu et al. engineered a composite MN using Na-hyaluronate and chitosan to deliver a vaccine. The tip of the MN was made of Na-hyaluronate and the base was made of chitosan. Here, the dissolvable Na-hyaluronate tip facilitated the swift release of the encapsulated antigen immediately after the insertion of MNs upon the skin, peaking the immune response, and the biodegradable chitosan slowly released the antigen until 4 weeks for prolonged immunity boost up (Figure 4.21) (Chiu et al., 2018). In 2016, Liu et al. published an experimental work where they delivered exendin-4 via MNs in GK/Slc rats which were fabricated using HA in the treatment of diabetes. The study showed a rapid release of the desired drug particle within 30s and the maximum amount of drug released within 5 mins after administration (Liu et al., 2016).

4.4.1.2 NANOPARTICLES (NP)

In reference to the properties mentioned in previous sections, marine-derived polymers are easily modifiable. They can be chemically modified via changing functional groups through the chemical reactions i.e., acetylation, carboxylation, and sulfation, which widens up their applications in biopharmaceutical fields. This has caught substantial attention from researchers worldwide to use these marine-derived polymers in nanomedicine (Figure 4.22) (Chen et al., 2020; Khanh et al., 2016; Sharma, 2019b).

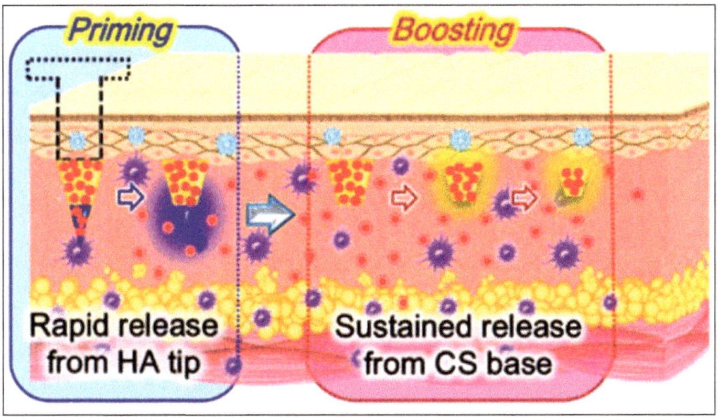

FIGURE 4.21 Release profile of vaccine from HA-CS-based composite MNs.

Source: Reproduced with permission from Chiu et al., 2018. Copyright 2018. American Chemical Society

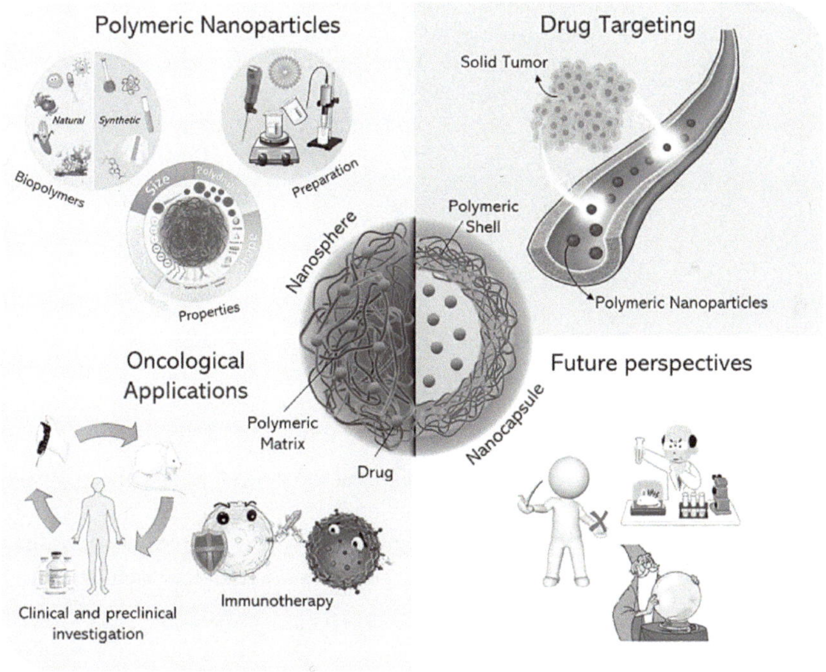

FIGURE 4.22 Overview of marine-derived nanoparticles. Sources, applications, scopes, and prospects.

Source: Reproduced from Gagliardi et al., 2021. Copyright © 2021 Gagliardi, Giuliano, Venkateswararao, Fresta, Bulotta, Awasthi and Cosco. https://creativecommons.org/licenses/by/4.0/

Due to the unique attributes i.e., nanometer size (diameter of particle= ~100 nm), targeting characteristics, higher surface-to-volume (sa/vol) ratio and a suitable amount of therapeutic release profile nanocarriers are being used widely in drug delivery of anti-cancer therapy (see more in Figure 4.23) (Gagliardi et al., 2021; Wicki et al., 2015; Yang et al., 2016).

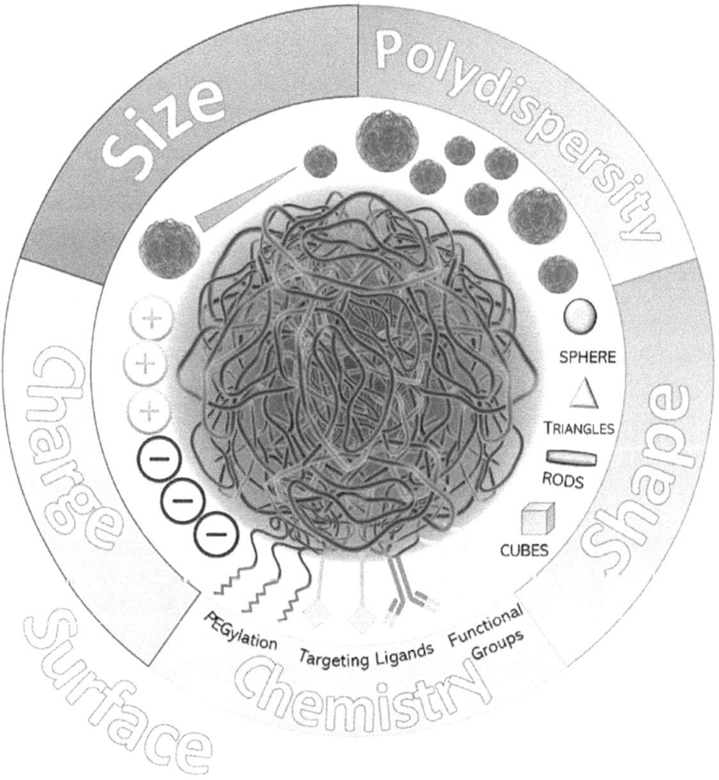

FIGURE 4.23 Schematic diagram of physicochemical attributes of marine-derived polymeric nanoparticles.

Source: Reproduced from Gagliardi et al., 2021. Copyright © 2021 Gagliardi, Giuliano, Venkateswararao, Fresta, Bulotta, Awasthi and Cosco. https://creativecommons.org/licenses/by/4.0/

Cyclodextrin is a chemically modified polymer that has been obtained via chemical reactions of a secondary metabolite (such as cyclosporine A, cephalosporium C, etc.) produced by the aquatic actinomycetes isolated from marine fungi (Bugni & Ireland, 2004; L. Guo et al., 2019).

A few years ago, a group of researchers engineered targeted cyclodextrin polymer-based nanoparticle CALAA-01, which contains siRNA (small interfering RNA). CALAA-01 is considered to be the first RNAi-based therapeutic made for experimental purposes to treat different types of cancers (Zuckerman et al., 2014).

Marine-derived polymer chitosan-based desired drug-encapsulated nanoparticles are widely used in sustained-release delivery systems, which can be easily degraded by various enzymes such as lipase, pectinase, cellulase, chitosanase, lysozymes, etc. (Taghipour-Sabzevar et al., 2019; Yang et al., 2014).

In 2009, Nicklas et al. engineered a marine sponge *Chondrosia reniformis*-derived collagen NP to deliver 17β-estradiol-hemihydrate via the transdermal route as a replacement for conventional hormone therapy. In their experimental work, they incorporated drug materials into collagen NPs and then placed them in a hydrogel. The authors reported the enhanced absorption and extended-release profile of estradiol-incorporated collagen NPs (Nicklas et al., 2009).

Another marine-derived polymer, fucoidan, is nowadays being used in anti-cancer nanotherapeutics. Several studies have mentioned the anti-tumor effects against various types of human tumors corresponding to the reaction with a biomolecule named P-selectin, which triggers metastasis usually by being expressed on the cancer cells (Lu et al., 2017; Shamay et al., 2016; Wu et al., 2016).

As mentioned previously in Section 4.4.1.1, marine-derived alginate contains 1,4-glycosidic linkage, which shows hypersensitivity to alkaline β-elimination and acid hydrolysis, making alginate easily dissolve inside the human body in a general physiological milieu where digestive enzymes are absent. Thus, framing alginate as an inexpensively modifiable marine-derived polymer leads to widened interactive capacity for use in nano-medicine development (Joye & McClements, 2014; Karlsson et al., 2018; Rottensteiner et al., 2014; Venkatesan et al., 2016).

4.5 MARINE-DERIVED BIOMOLECULES IN THE TREATMENT OF DIFFERENT DISEASES

We have discussed the structure, sources, original structural properties, modified structural properties, bio-pharmaceutical, and physicochemical properties of various biomolecules of marine origin until now. This section will discuss the mechanism of action for some of these marine-derived compounds in different disease treatments.

4.5.1 CARDIOPROTECTIVE

Cardiac disorder, also known as CVD (cardiovascular disease), is a cluster of disorders that adversely affect the heart and blood vessels connected to the heart by causing heart attack (coronary heart disease), heart failure, high blood pressure (hypertension), peripheral vascular disease, and stroke (cerebrovascular disease). It alone caused one-third of the total deaths worldwide in 1999 (Mayakrishnan et al., 2013; Rissanen et al., 2003). Some researchers reported a relationship between CVD and gut microflora based on the results of their experimental work. People with CVDs tend to be victims of changes in gut microbiota composition and diversity. Marine algal polysaccharides (MAPs) are known for their huge macromolecules and as suitable contenders for non-primate-derived SPs (Luthuli et al., 2019; Ma & Li, 2018; Magne et al., 2020). These MAPs get fermented (saccharolytic) by the gut microflora and produce SCFAs (short-chain fatty acids), mostly propionate, butyrate, and acetate at around a 20:20:60 molar ratio. Some of the SCFA-generating gut microbes are *Clostridium, Faecalibacterium, Roseburia, Eubacterium, Alloprevotella,* and *Bacteroides* (Martin-Gallausiaux et al., 2021; Scheppach, 1994; Zheng et al., 2020). It is usually predicted that CVD risk factor modulation can directly affect SCFA regulation; therefore, energy consumption, release of pancreatic peptide YY, enhanced lipid metabolism, blood pressure regulation, and at the same time slow down the adipose triglyceride lipase phosphorylation as well (Chambers et al., 2018). Most advantageous parts played by the SCFAs in the GI tract are facilitated via the mobilization of their GPR41 (G protein-coupled 41) receptor directly, which is also termed as FFAR2 (free FA receptor 2), along with GPR43/FFAR2 (for more details see Figure 4.24) (Kim et al., 2013). By enhancing GPR41 and GPR43 activity and SCFAs concentration, polysaccharides obtained from *Gracilaria lemaneiformis* have reduced obesity induced by a high-fat diet in mice, by restraining fatty acid accumulation in different organs, transforming dysbiosis of gut microorganisms, and increasing lipid metabolism (Sun et al., 2018). Furthermore, SCFAs can also trigger immune cells. They can upregulate the tight junction-related component genes and the family of mucin (MUC) gene expression in goblet cells to maintain the integrity of the gut environment, thus facilitating the reduction of CVD risk factors (Gan et al., 2022). In 2010, Thomesa et al. mentioned in their research article that fucoidan obtained from *Cladosiphon okamuranus* significantly reduced the enzyme markers associated with myocardial infarction induced by isoproterenol in rats (Mayakrishnan et al., 2013; Thomes et al., 2010).

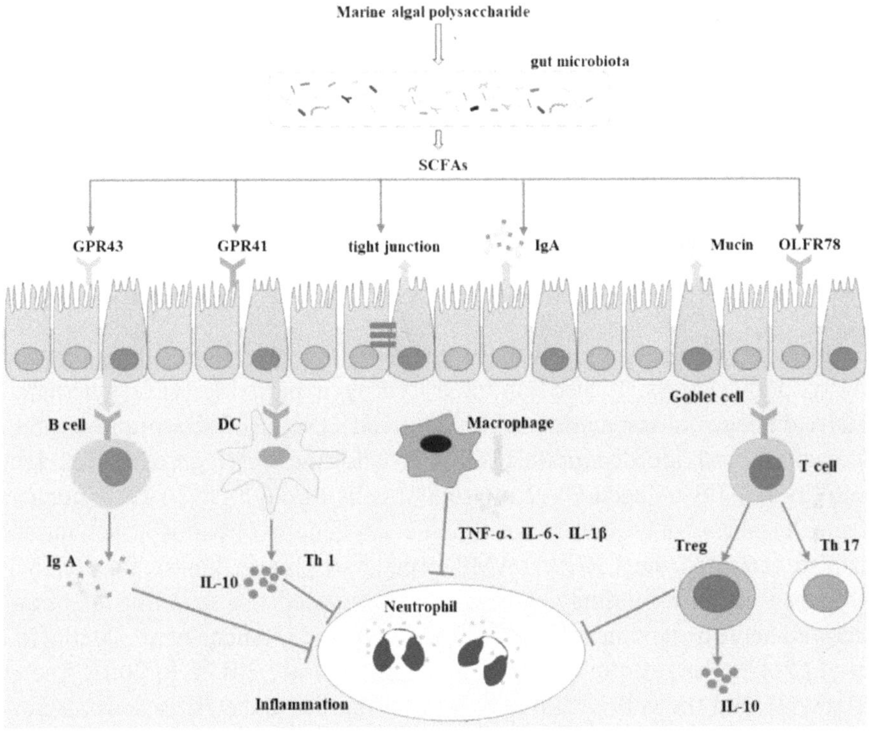

FIGURE 4.24 Production of SCFAs via fermentation of MAP by gut microbiota in the colon.

Source: Reproduced from Gan et al., 2022. Copyright © 2022 by the authors; licensee MDPI, Basel, Switzerland. http://creativecommons.org/licenses/by/4.0/

4.5.2 ANTI-INFLAMMATORY

Fucoxanthin is a tetraterpenoid which is a usual component of brown algae, also known as a carotenoid. This algal metabolite showed broader effectiveness in brain disorders e.g., Alzheimer's disease, traumatic brain injury, and ischemic stroke. It attenuates oxidative stress (OS) and inflammation in glial cells (Pangestuti, Vo et al., 2013; Xiang et al., 2017; Zhao et al., 2017). It has reduced inflammation in 42-amino acid β-amyloid (Aβ42) peptide-induced BV-2 microglial cells. This pharmacological activity happened by reduced secretion of proinflammatory mediators i.e., IL-1β (interleukin-1 beta), IL-6, TNF-α (tumor necrosis factor-alpha), and PGE_2 (prostaglandin E_2), along with the decreased expression of COX-2 (cyclooxygenase 2) and iNOS (inducible nitric oxide synthase), and reduced phosphorylation of the MAPK (mitogen-activated protein kinase) pathway (Hannan et al., 2020; Pangestuti, Vo et al.,

2013). In the case of Lipopolysaccharide (LPS)-induced BV-2 microglial cells, fucoxanthin showed a protective effect against neuroinflammation by reducing COX-2 and iNOS expression, as well as decreasing the secretion of inflammatory agents such as nitric oxide (NO), IL-6, TNF-α, and PGE_2. This inhibition affected the AP-1 (MAPKs-activating protein-1) pathways and the protein kinase B (Akt)/NF-κB (Hannan et al., 2020; Zhao et al., 2017).

4.5.3 ANTIOXIDANT

Sargassum siliquastrum-derived fucoxanthin prevented DNA damage caused by oxidative stress. It also inhibited hydrogen peroxide (H_2O_2)-mediated DNA damage in association with improved GSH production and expression of superoxide dismutase (SOD). In addition, it triggered antioxidant activity in LPS-induced BV-2 microglial cells through Nrf2/HO-1 (nuclear factor erythroid 2-related factor 2/heme oxygenase-1) pathway activation. It also activated the CREB (cAMP-dependent protein kinase PKA/cAMP response element-binding) pathway and enhanced the secretion of brain-derived neurotrophic factor (BDNF) neurotrophin (Hannan et al., 2020; Heo et al., 2008; Pangestuti, Vo et al., 2013; Zhao et al., 2017). In 2003, Lee et al. mentioned the antioxidant activity of a marine algae *Pelvetia siliquosa*-derived compound fucosterol. It increased antioxidant enzymes in cells, i.e., CAT (chloramphenicol acetyltransferase), SOD, and GPx (glutathione peroxidase) in rats. Moreover, it also prevented the sGPT (serum glutamate pyruvate transaminase; also known as alanine transferase) activity by 63.16% and sGOT (serum glutamic-oxaloacetic transaminase; also known as aspartate transaminase) activity by 25.57% (Lee et al., 2003). Furthermore, it inhibited oxidative stress-induced damage of human hepatoma HepG2 cells by increasing glutathione (GSH) levels (Choi et al., 2015).

4.5.4 ANTIVIRAL

Marine-derived antiviral compounds have numerous uses that also include: 1. chemotherapy for viral diseases of both humans and lower animals, and 2. biological control for contamination caused by human enteropathogenic viruses and dissemination in sewage-polluted waters. Macrolactin A is an antiviral compound obtained from marine bacteria, exhibiting preventive activity against human HIV replication, with the highest level of immunity (Gustafson et al., 1989; Yasuhara-Bell & Lu, 2010). Marine algal secondary metabolites have displayed widely varied *in vivo* and *in vitro* effects against various types of viruses, e.g., togaviruses (Semliki Forest virus, Sindbis virus), herpes viruses (HCMV, HSV-1, HSV-2), rhabdoviruses (VSV), paramyxoviruses (RSV), and both human and simian immunodeficiency viruses (HIV and SIV) (Gerber et al., 1958). Small molecules of polysaccharides obtained from red algae extracts showed antiviral effects against HSV and some other viruses. However, due to the wider range of unspecified responses against viral infection, the mentioned observations failed to generate interest among researchers (Ehresmann et al., 1977; Gerber et al., 1958; Richards et al., 1978; Witvrouw & De Clercq, 1997). Laminaran, a marine polysaccharide obtained from kelp, successfully prevents the adsorption of HIV on lymphocytes; therefore, it has been considered a good candidate for restraining the replication of HIV (W. Wang et al., 2012). Aside from that, numerous marine-derived compounds have shown antiviral activity against many viruses. Some of them are Human Immunodeficiency Virus (HIV), Enterovirus-71 (EV-71), Herpes Simplex Viruses (HSVs), Porcine Reproductive and Respiratory Syndrome Virus (PRRSV), Influenza Virus, Molluscum Contagiosum Virus (MCV), Respiratory Syncytial Virus (RSV), and Tobacco Mosaic Virus (TMV) (Moghadamtousi et al., 2015).

4.6 MARINE BIOTECHNOLOGY

There are many creatively effective strategies these days to find new MNP structures, including molecular networking, chemical prospecting using NMR, metabolomics, MALDI imaging, genome mining, and a genomisotopic system (Dona et al., 2016; Duncan et al., 2015; Gross et al., 2007; Helfrich et al., 2014; Schroeder et al., 2007; Spraker et al., 2020; Ziemert et al., 2016). A limitation of these structure-focused techniques is that they usually do not instantly connect with the biological activity of newly identified molecules, which hinders the identification of any physiological activity associated with a molecule that might be a candidate or subsequently misses the hypothesis due to an obscure screening process. By altering the procedure and utilizing either phenotypic analysis of an entire cell or *in-vitro* enzyme modifiable analysis done on crude marine-derived extract libraries as the primary screening process to find new MNPs, it provides the assurance that the candidate MNPs will have desired biochemical properties. This strategy has been used for ages and has proved its applicability (Chabner, 2016; Monks et al., 1997; Williams & Andersen, 2020).

From the past two decades, chemical genomics or chemical genetics has been considered an emerging new method for scientific inquiry in a sophisticated way, which aims to establish connectivity among a molecular target in a cell, a component's chemical structure, and a desired biochemical response (see more in Figure 4.25) (Burkard & Jallepalli, 2010; Cacace et al., 2017; Cong et al., 2012; Roemer et al., 2011).

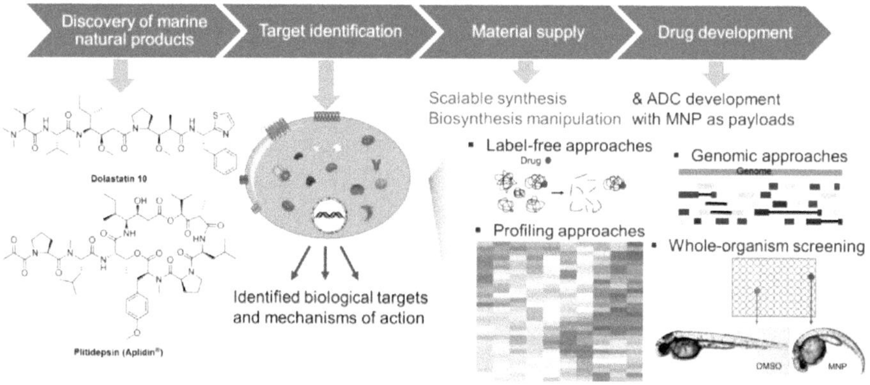

FIGURE 4.25 Drug development process from MNPs through target identification in cells.
Source: Reproduced with permission from Liang et al., 2019. Copyright © 2019. Elsevier.

4.7 PHARMACEUTICAL DRUGS OF MARINE ORIGIN

This section mentions the marine pharmaceuticals that have received approval from the FDA and those that are in the pipeline for approval.

4.7.1 PRECLINICAL PIPELINE

MNPs-derived components which are now under the pre-clinical phase of drug development due to promising pharmacological properties have been listed in Table 4.6.

4.7.2 CLINICAL PIPELINE

MNPs-derived components, which are now under the clinical phase of drug development due to promising pharmacological properties, have been listed in Table 4.7.

TABLE 4.6 List of Some MNP Derived Compounds Under Preclinical Pipeline with Their Potential Pharmacological Properties

Compound	Marine Source	Chemical Class	Chemical Structure	Drug Class	MOA	References
Agelasine D	Sponge	Terpene	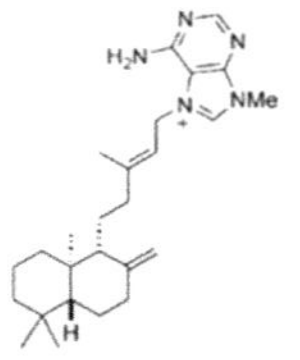 agelasine D	Antibacterial	Undefined	Hertiani et al. (2010)
Aqabamycin E	Bacterium	Alkaloid	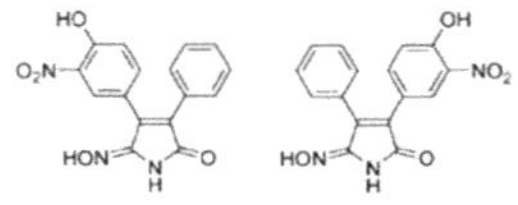 aqabamycin E	Antibacterial	Undefined	Al-Zereini et al. (2010)
Bacillistatin 1 and 2	Bacterium	Peptide	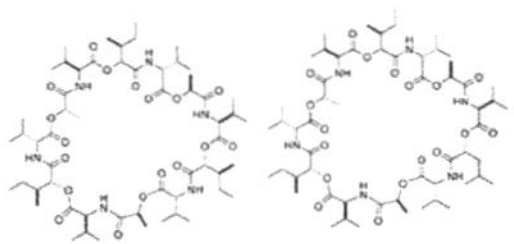 Bacillistatin 1 Bacillistatin 2	Antibacterial	Undefined	Pettit et al. (2009)
Bromophycolides	Alga	Terpene	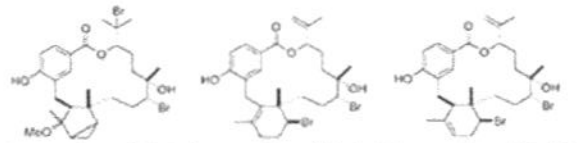 bromophycolide J bromophycolide M bromophycolide N 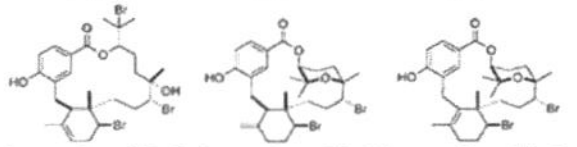 bromophycolide O bromophycolide P bromophycolide Q	Antibacterial	Undefined	Lane et al. (2009)

TABLE 4.6 *(Continued)*

Compound	Marine Source	Chemical Class	Chemical Structure	Drug Class	MOA	References
C. Cervi ornis diterpene	Alga	Terpene	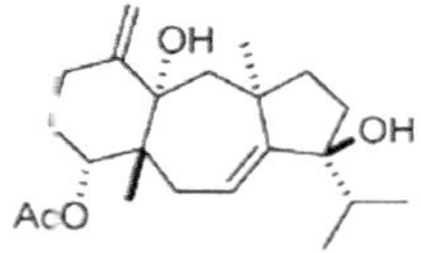	Antiprotozoal	Mitochondrial swelling & damage	Dos Santos et al. (2011)
C. hooperi diterpene	Sponge	Terpene	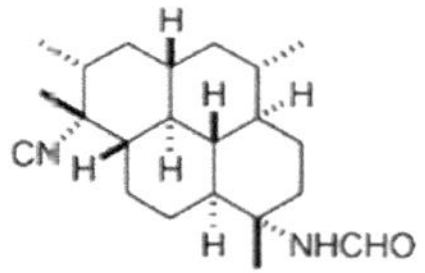	Antimalarial	Undefined	Wright et al. (2010); Wright & Lang-Unnasch, (2009)
Caboxamycin	Bacterium	Alkaloid	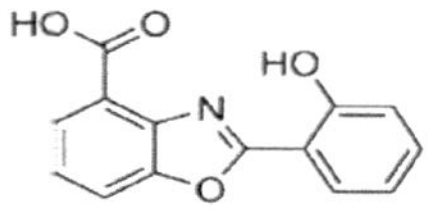	Antibacterial	Undefined	Hohmann et al. (2009)
Chrysophaentin A	Alga	Shikimate	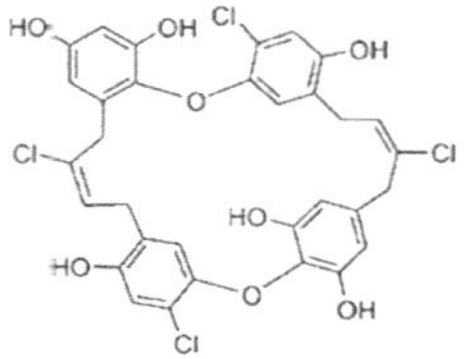	Antibacterial	Inhibit GTPase activity of FtsZ	Plaza, Keffer et al. (2010)
Convolutamine I	Bryozoan	Alkaloid	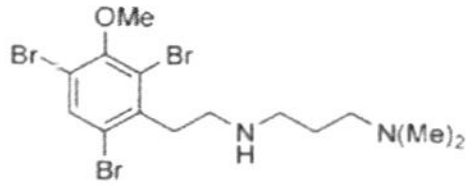	Antiprotozoal	Undefined	Davis et al. (2011)

TABLE 4.6 *(Continued)*

Compound	Marine Source	Chemical Class	Chemical Structure	Drug Class	MOA	References
Crossbyanol B	Cyanobacterium	Polyketide	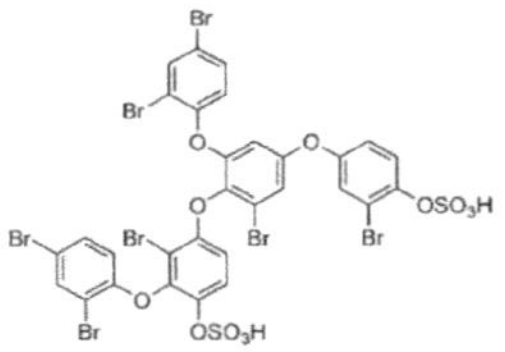	Antibacterial	Undefined	Choi et al. (2010)
Discorhabdin Z	Sponge	Alkaloid	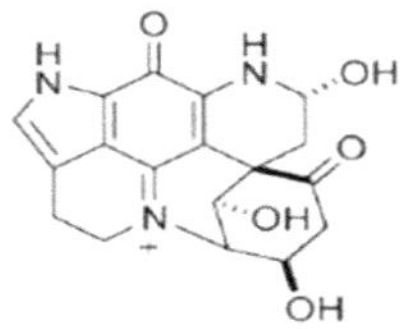	Antibacterial	Sortase A inhibition	Jeon et al. (2010)
Eusynstyelamide F	Bryozoan	Peptide	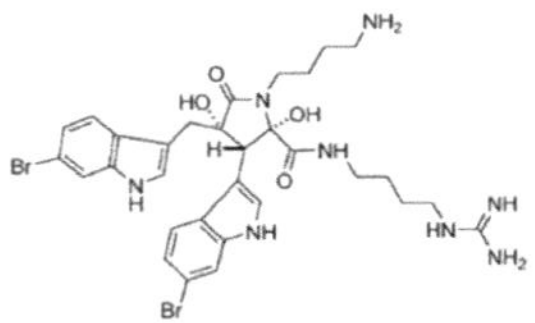	Antibacterial	Undefined	Tadesse et al. (2011)
Gallinamide A	Cyanobacterium	Peptide	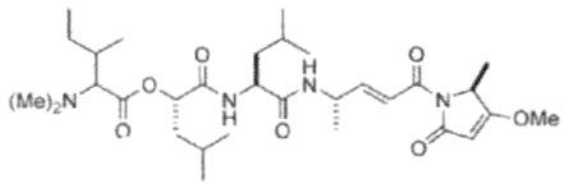	Antimalarial	Undefined	Linington et al. (2009)
Gracilioether B	Sponge	Polyketide	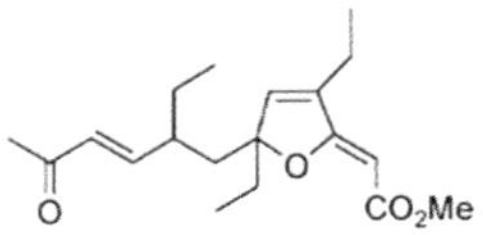	Antimalarial	Undefined	Ueoka et al. (2009)

TABLE 4.6 *(Continued)*

Compound	Marine Source	Chemical Class	Chemical Structure	Drug Class	MOA	References
H. salinus phenethylamine	Bacterium	Shikimate	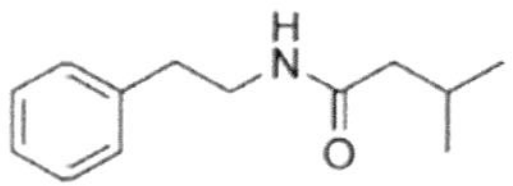	Antibacterial	Inhibit homoserine lactone receptor binding	Teasdale et al. (2009)
Homogentisic acid	Sponge	Shikimate	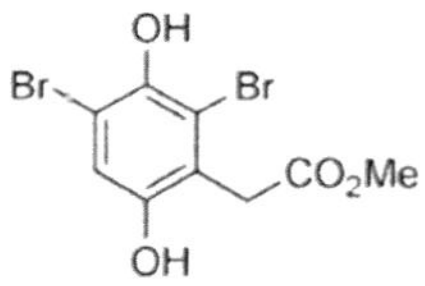	Antimalarial	Pfnek-1 enzyme inhibition	Lebouvier et al. (2009)
Hymenidin	Sponge	Alkaloid	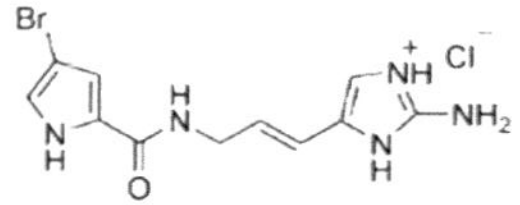	Antituberculosis	Undefined	Vicente et al. (2009)
Knightol	Coral	Terpene	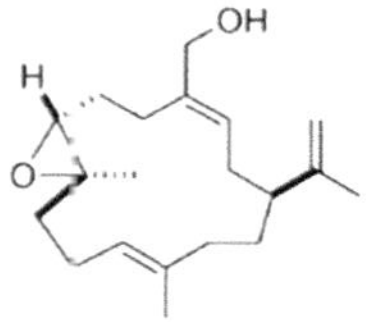	Antibacterial	Undefined	Tello et al. (2009)
Lyngbyoic acid	Cyanobacterium	Fatty acid		Antibacterial	Inhibit homoserine lactone receptor LasR	Kwan et al. (2011)
Manado peroxide C	Sponge	Polyketide	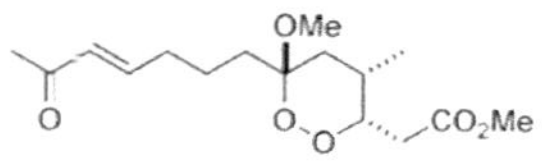	Antimalarial	Undefined	Fattorusso et al. (2010)

TABLE 4.6 *(Continued)*

Compound	Marine Source	Chemical Class	Chemical Structure	Drug Class	MOA	References
Manzamine A	Sponge	Alkaloid	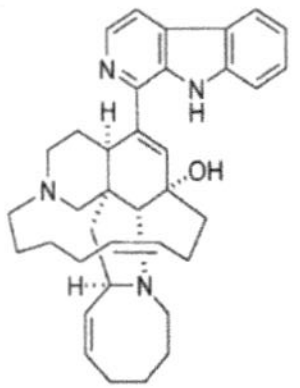	Antiviral	Early ICPO gene transcription inhibited	Palem et al. (2011)
MC21-B	Bacterium	Shikimate	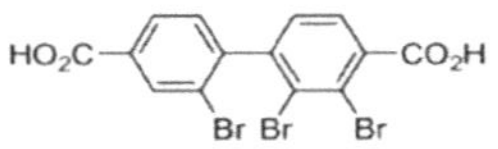	Antibacterial	Undefined	Isnansetyo & Kamei (2009)
Motualevic acid F	Sponge	Fatty acid	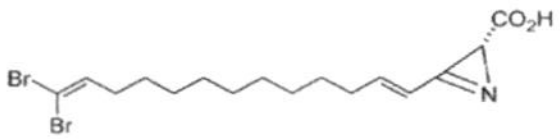	Antibacterial	Undefined	Keffer et al. (2009)
Nacardiopsis thiopeptide	Bacterium	Peptide	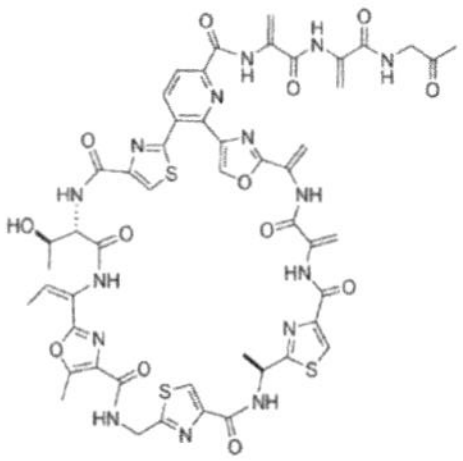	Antibacterial	Undefined	Engelhardt et al. (2010)
Neopetrosiamine A	Sponge	Alkaloid	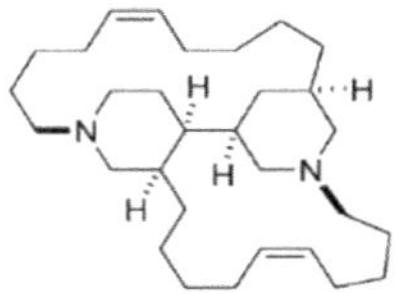	Antimalarial, Anti-tuberculosis	Undefined	Wei et al. (2010)

TABLE 4.6 *(Continued)*

Compound	Marine Source	Chemical Class	Chemical Structure	Drug Class	MOA	References
Plakortin	Sponge	Polyketide	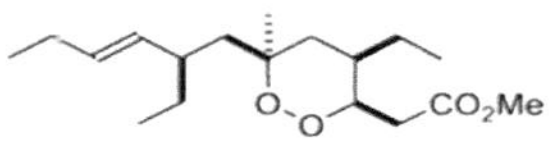	Antimalarial	Toxic carbon-radical	Taglialatela-Scafati et al. (2010)
Pseudopterosin U	Coral	Terpene	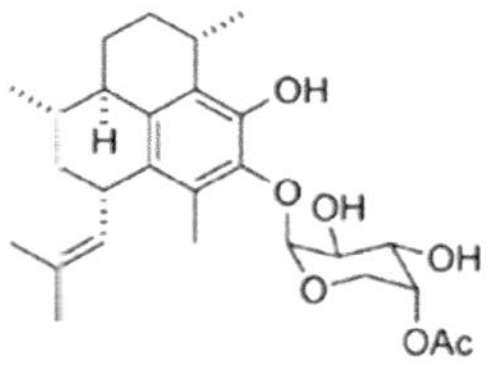	Antibacterial	Undefined	Correa et al. (2011)
Salinisporamycin	Bacterium	Polyketide	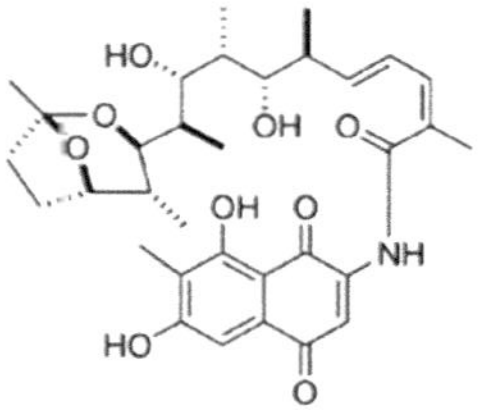	Antibacterial	Undefined	Matsuda et al. (2009)
Saadamycin	Bacterium	Polyketide	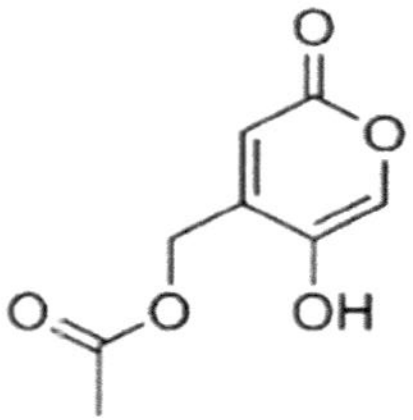	Antifungal	Undefined	El-Gendy & El-Bondkly 2010)

TABLE 4.6 *(Continued)*

Compound	Marine Source	Chemical Class	Chemical Structure	Drug Class	MOA	References
Theonellamide F	Sponge	Peptide	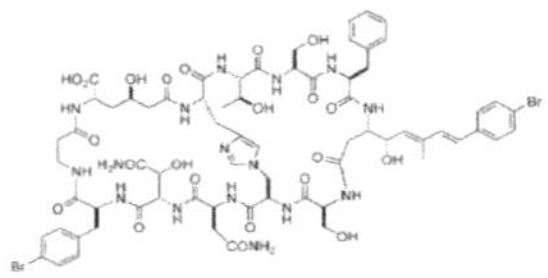	Antifungal	Activate 1,3-β-D-glucan synthesis	Nishimura et al. (2010)
Valinomycin	Bacterium	Peptide	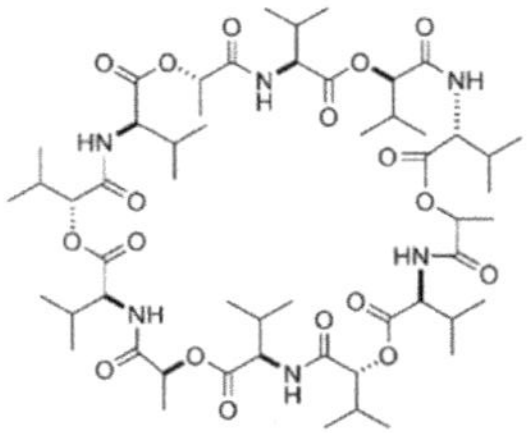	Antiprotozoal	Undefined	Pimentel-Elardo et al. (2010)
Xiamycin	Bacterium	Alkaloid	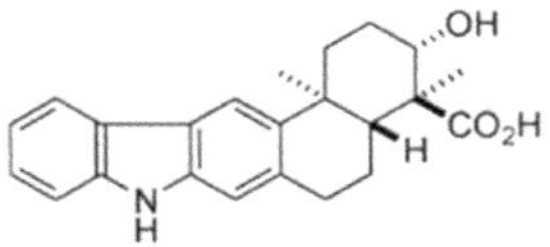	Antiviral	Selective inhibition of CCR5 tropic HIV	Ding et al. (2010)

TABLE 4.7 List of Some MNP-Derived Compounds Under Clinical Pipeline with Their Potential Pharmacological Properties (Malve, 2016)

Clinical Phase	MNP Derived Compound	Marine Source	Chemical Class	Drug Class
Phase I	Bryostatine 1	Bryozoa	Polyketide	Anticancer
	Elisidepsin	Mollusk	Despeptide	Anticancer
	Hemiasterlin (E7974)	Sponge	Tripeptide	Anticancer
	Pinatuzumab vedotin	Mollusk	Antibody-drug conjugates	Anticancer (non-Hodgkin lymphoma, chronic lymphocytic leukemia)
	Pseudoterosin	Soft coral	Diterpene glycoside	Wound healing
Phase II	DMXBA (GTS 21)	Worm	Alkaloid	Anti-Alzheimer's, anticognitive
Phase III	Plitidepsin	Tunicate	Depsipeptide	Anticancer
	Soblidotin	Bacterium	Peptide	Anticancer
	Tetrodotoxin	Pufferfish	Alkaloid	Chronic pain

4.7.3 APPROVED DRUGS

An array of MNP-derived molecules has received FDA approval to be launched in the market commercially on account of the pharmacological properties discussed throughout this chapter. An abundance of MNPs have been isolated from oceanic organisms. Although all of them have been purified, screened, and experimented upon, a tiny fraction of them have been legitimized to proceed to the clinical phase, and a few countable numbers among them successfully reached the market. MNPs-derived components that have received FDA approval due to their promising pharmacological properties have been listed in Table 4.8.

4.8 CONCLUSION

It's been about 3,500 million years since marine organisms appeared and 250 years of marine organism-based research. However, about 91% of them still lack accurate elaborative characterization. Marine species have the capability to survive unfavorable surroundings, which facilitates the production of varied secondary metabolites. From an ecological point of view, they develop secondary metabolic pathways to synthesize biomolecules in

TABLE 4.8 List of Some FDA Approved MNP Derived Compounds with Their Potential Application (Ahmed et al., 2022)

Name of the Compound	Source	Formula	Product Type	Elimination Half-Life	Bioavailability	Application(s)	Legal Status
Adcetris (Brentuximab vedotin)	*Dolabella auricularia*	$C_{6476}H_{9930}N_{1690}O_{2030}S_{40}$	D	Approximately 4 to 6 days	50–80%	Anti-cancer	Prescription-only
Bacitracin (Baciim)	*Bacillus subtilis*	$C_{66}H_{103}N_{17}O_{16}S$	N	Data is not available	Data is not available	Acute and chronic localized skin infections	Prescription-only for injection and OTC
Curacin A	*Lyngbya majuscula*	$C_{23}H_{35}NOS$	N	Data is not available	Data is not available	Anti-cancer	Not Available
Dactinomycin	*Streptomyces parvullus*	$C_{62}H_{86}N_{12}O_{16}$	D	36 h	Data is not available	Anti-cancer	Prescription-only
Dutasteride (Avodart)	–	$C_{27}H_{30}F_{6}N_{2}O_{2}$	S	4 to 5 Weeks	60%	Treatment of prostate cancer and hormone therapy	Prescription-only
Eribulin (Halaven)	Marine Sponge	$C_{40}H_{59}NO_{11}$	–	40 h	Data is not available	Anti-cancer	Prescription-only
Trabectedin (Yondelis)	Marine Tunicate	$C_{39}H_{43}N_{3}O_{11}S$	–	180 h	Data is not available	Anti-cancer (soft-tissue sarcoma and ovarian cancer)	Prescription-only
Ziconotide (intrathecal ziconotide)	Cone Snail	$C_{102}H_{172}N_{36}O_{32}S_{7}$	N	2.9–6.5 h	50%	Analgesics	Prescription-only

Note: Natural (N), synthetic (S), and Derivative (D).

order to adjust their way of living improve defense mechanisms to withstand severely hostile habitats, and create safety lines against predators, and other external factors, allowing themselves to survive. For this reason, marine-derived metabolites and their derivatives have taken up more space in pharmaceutical research over the past 10 years.

Since oceanic ambiance is not static, the possibility of resupplying the same marine organism mostly does not provide the same materials, known as a sustainable supply issue. Sometimes this issue also arises because the bioactive molecules are synthesized not by the macro-organism itself but by the microorganisms living inside it. The isolation of candidate compounds in these unusual cases is costly, which makes the new drug development project highly expensive to accomplish. To address the mentioned issues, approaches have been taken in recent years such as molecular docking, computational methods, etc., to mitigate both sustainable supply and cost issues.

In this chapter, a good number of marine-derived compounds with notable pharmacological values have been discussed. Some of them have already made their journey to the market, and some are yet to come since they are now under pre-clinical or clinical pipelines awaiting FDA approval. As mentioned, a couple of times throughout this chapter, the ocean represents a huge reservoir of biodiversity of materials with great economic values, rendering a never-ending area of research for scientists worldwide in the upcoming future.

KEYWORDS

- **Food and Drug Administration**
- **marine drug**
- **natural products**
- **natural sources**
- **pharmaceutical products**
- **therapeutics**

REFERENCES

Ahmed, A. B. A., Adel, M., Karimi, P., & Peidayesh, M. (2014). Pharmaceutical, cosmeceutical, and traditional applications of marine carbohydrates. In *Advances in Food and Nutrition Research* (Vol. 73, pp. 197–220). Academic Press. https://doi.org/10.1016/B978-0-12-800268-1.00010-X.

Ahmed, I., Asgher, M., Sher, F., Hussain, S. M., Nazish, N., Joshi, N., Sharma, A., Parra-Saldívar, R., Bilal, M., & Iqbal, H. M. N. (2022). Exploring marine as a rich source of bioactive peptides: Challenges and opportunities from marine pharmacology. *Marine Drugs*, *20*(3), 208. https://doi.org/10.3390/MD20030208.

Ahmed, S., Alam, W., Jeandet, P., Aschner, M., Alsharif, K. F., Saso, L., & Khan, H. (2022). Therapeutic potential of marine peptides in prostate cancer: Mechanistic insights. *Marine Drugs*, *20*(8), 466. https://doi.org/10.3390/MD20080466.

Ahmed, S., Mirzaei, H., Aschner, M., Khan, A., Al-Harrasi, A., & Khan, H. (2021). Marine peptides in breast cancer: Therapeutic and mechanistic understanding. *Biomedicine & Pharmacotherapy*, *142*, 112038. https://doi.org/10.1016/J.BIOPHA.2021.112038.

Albuquerque, I. R. L., Cordeiro, S. L., Gomes, D. L., Dreyfuss, J. L., Filgueira, L. G. A., Leit, E. L., Nader, H. B., & Rocha, H. A. O. (2013). Evaluation of anti-nociceptive and anti-inflammatory activities of a heterofucan from *Dictyota menstrualis*. *Marine Drugs*, *11*(8), 2722–2740. https://doi.org/10.3390/MD11082722.

Ale, M. T., Mikkelsen, J. D., & Meyer, A. S. (2011). Important determinants for fucoidan bioactivity: A critical review of structure-function relations and extraction methods for fucose-containing sulfated polysaccharides from brown seaweeds. *Marine Drugs*, *9*(10), 2106–2130. https://doi.org/10.3390/MD9102106.

Al-Zereini, W., Fotso Fondja Yao, C. B., Laatsch, H., & Anke, H. (2010). Aqabamycins A-G: Novel nitro maleimides from a marine Vibrio species. I. Taxonomy, fermentation, isolation, and biological activities. *The Journal of Antibiotics*, *63*(6), 297–301. https://doi.org/10.1038/ja.2010.34.

Andersen, N. S., Dolan, T. C. S., & Rees, D. A. (1973). Carrageenans. Part VII. Polysaccharides from *Eucheuma spinosum* and *Eucheuma cottonii*. The covalent structure of ι-carrageenan. *Journal of the Chemical Society, Perkin Transactions 1*, *1973*(0), 2173–2176. https://doi.org/10.1039/P19730002173.

Aneiros, A., & Garateix, A. (2004). Bioactive peptides from marine sources: Pharmacological properties and isolation procedures. *Journal of Chromatography B: Analytical Technologies in the Biomedical and Life Sciences*, *803*(1), 41–53. https://doi.org/10.1016/j.jchromb.2003.11.005.

Anthony, T. T. (1974). Sea snake venoms and neurotoxins. *Journal of Agricultural and Food Chemistry*, *22*(1), 36–43. https://doi.org/10.1021/JF60191A034.

Arshad, M. S., Fatima, S., Nazari, K., Ali, R., Farhan, M., Muhammad, S. A., Abbas, N., Hussain, A., Kucuk, I., Chang, M. W., Mehta, P., & Ahmad, Z. (2019). Engineering and characterization of BCG-loaded polymeric microneedles. *Journal of Drug Targeting*, *28*(5), 525–532. https://doi.org/10.1080/1061186X.2019.1693577.

Arshad, M. S., Hassan, S., Hussain, A., Abbas, N., Kucuk, I., Nazari, K., Ali, R., Ramzan, S., Alqahtani, A., Andriotis, E. G., Fatouros, D. G., Chang, M. W., & Ahmad, Z. (2019). Improved transdermal delivery of cetirizine hydrochloride using polymeric microneedles. *Daru: Journal of Faculty of Pharmacy, Tehran University of Medical Sciences*, *27*(2), 673–681. https://doi.org/10.1007/S40199-019-00301-3.

Arthur James, R., Vignesh, S., & Muthukumar, K. (2014). Marine drugs development and social implication. In *Coastal Environments: Focus on Asian Regions* (pp. 219–237). Springer Netherlands. https://doi.org/10.1007/978-90-481-3002-3_15.

Avila Rodríguez, M. I., Rodríguez Barroso, L. G., & Sánchez, M. L. (2018). Collagen: A review of its sources and potential cosmetic applications. *Journal of Cosmetic Dermatology*, *17*(1), 20–26. https://doi.org/10.1111/JOCD.12450.

Azevedo, G., Torres, M. D., Sousa-Pinto, I., & Hilliou, L. (2015). Effect of pre-extraction alkali treatment on the chemical structure and gelling properties of extracted hybrid carrageenan from *Chondrus crispus* and *Ahnfeltiopsis devoniensis*. *Food Hydrocolloids*, *50*, 150–158. https://doi.org/10.1016/J.FOODHYD.2015.03.029.

Azizur Rahman, M. (2019). Collagen of extracellular matrix from marine invertebrates and its medical applications. *Marine Drugs*, *17*(2), 118. https://doi.org/10.3390/MD17020118.

Bächinger, H. P., Mizuno, K., Vranka, J. A., & Boudko, S. P. (2010). Collagen formation and structure. *Ophthalmology*, *5*(1), 469–530. https://doi.org/10.2/JQUERY.MIN.JS.

Becker, L. C., Bergfeld, W. F., Belsito, D. V., Klaassen, C. D., Marks, J. G., Shank, R. C., Slaga, T. J., Snyder, P. W., & Andersen, F. A. (2009). Final report of the safety assessment of hyaluronic acid, potassium hyaluronate, and sodium hyaluronate. *International Journal of Toxicology*, *28*(4 Suppl), 5–67. https://doi.org/10.1177/1091581809337738.

Benkendorff, K. (2013). Natural product research in the Australian marine invertebrate *Dicathais orbita*. *Marine Drugs*, *11*(4), 1370–1398. https://doi.org/10.3390/MD11041370.

Bérdy, J. (2005). Bioactive microbial metabolites. *The Journal of Antibiotics*, *58*(1), 1–26. https://doi.org/10.1038/ja.2005.1.

Bernheimer, A. W., Avigad, L. S., & Lai, C. Y. (1982). Purification and properties of a toxin from the sea anemone *Condylactis gigantea*. *Archives of Biochemistry and Biophysics*, *214*(2), 840–845. https://doi.org/10.1016/0003-9861(82)90091-1.

Bhakuni, D. S., & Rawat, D. S. (2005). Bioactive marine natural products. In *Bioactive Marine Natural Products*, pp. 1–25. Springer Netherlands. https://doi.org/10.1007/1-4020-3484-9/COVER.

Biswas, S., Das, J., Barman, S., Shah, S. S., Gangopadhyay, M., Maiti, T. K., & Singh, N. D. P. (2017). Single component image guided 'on-demand' drug delivery system for early-stage prostate cancer. *Sensors and Actuators B: Chemical*, *244*, 327–333. https://doi.org/10.1016/J.SNB.2017.01.008.

Bou-Gharios, G., Abraham, D., & de Crombrugghe, B. (2020). Type I collagen structure, synthesis, and regulation. In *Principles of Bone Biology* (4th ed., pp. 295–337). Academic Press. https://doi.org/10.1016/B978-0-12-814841-9.00013-0.

Bourgougnon, N., Roussakis, C., Kornprobst, J. M., & Lahaye, M. (1994). Effects in vitro of sulfated polysaccharide from *Schizymenia dubyi* (*Rhodophyta*, *Gigartinales*) on a non-small-cell bronchopulmonary carcinoma line (NSCLC-N6). *Cancer Letters*, *85*(1), 87–92. https://doi.org/10.1016/0304-3835(94)90243-7.

Bourtoom, T. (2008). Review article edible films and coatings: Characteristics and properties. *International Food Research Journal*, *15*(3), 237–248. http://ifrj.upm.edu.my/15%20(3)%202008/01.%20Bourtoom,%20T.pdf (accessed on 7 June 2024).

Bugni, T. S., & Ireland, C. M. (2004). Marine-derived fungi: A chemically and biologically diverse group of microorganisms. *Natural Product Reports*, *21*(1), 143–163. https://doi.org/10.1039/B301926H.

Burkard, M. E., & Jallepalli, P. V. (2010). Validating cancer drug targets through chemical genetics. *Biochimica et Biophysica Acta*, *1806*(2), 251. https://doi.org/10.1016/J.BBCAN.2010.08.002.

Butler, M. S. (2004). The role of natural product chemistry in drug discovery. *Journal of Natural Products*, *67*(12), 2141–2153. https://doi.org/10.1021/NP040106Y.

Cacace, E., Kritikos, G., & Typas, A. (2017). Chemical genetics in drug discovery. *Current Opinion in Systems Biology*, *4*, 35–42. https://doi.org/10.1016/J.COISB.2017.05.020.

Calder, P. C. (2012). The role of marine omega-3 (n-3) fatty acids in inflammatory processes, atherosclerosis and plaque stability. *Molecular Nutrition & Food Research*, *56*(7), 1073–1080. https://doi.org/10.1002/MNFR.201100710.
Calder, P. C. (2015). Marine omega-3 fatty acids and inflammatory processes: Effects, mechanisms and clinical relevance. *Biochimica et Biophysica Acta (BBA) – Molecular and Cell Biology of Lipids*, *1851*(4), 469–484. https://doi.org/10.1016/J.BBALIP.2014.08.010.
Calder, P. C., & Yaqoob, P. (2012). Marine omega-3 fatty acids and coronary heart disease. *Current Opinion in Cardiology*, *27*(4), 412–419. https://doi.org/10.1097/HCO.0B013E328353FEBD.
Campo, V. L., Kawano, D. F., Silva, D. B. da, & Carvalho, I. (2009). Carrageenans: Biological properties, chemical modifications and structural analysis – A review. *Carbohydrate Polymers*, *77*(2), 167–180. https://doi.org/10.1016/J.CARBPOL.2009.01.020.
Carlucci, M. J., Scolaro, L. A., Errea, M. I., Matulewicz, M. C., & Damonte, E. B. (1997). Antiviral activity of natural sulphated galactans on herpes virus multiplication in cell culture. *Planta Medica*, *63*(5), 429–432. https://doi.org/10.1055/S-2006-957727/BIB.
Carter, G. T. (2011). Natural products and Pharma 2011: Strategic changes spur new opportunities. *Natural Product Reports*, *28*(11), 1783–1789. https://doi.org/10.1039/C1NP00033K.
Catanesi, M., Caioni, G., Castelli, V., Benedetti, E., D'angelo, M., & Cimini, A. (2021). Benefits under the sea: The role of marine compounds in neurodegenerative disorders. *Marine Drugs*, *19*(1), 24. https://doi.org/10.3390/MD19010024.
Chabner, B. A. (2016). NCI-60 cell line screening: A radical departure in its time. *Journal of the National Cancer Institute*, *108*(5), 1–7. https://doi.org/10.1093/JNCI/DJV388.
Chai, Q. Y., Yang, Z., Lin, H. W., & Han, B. N. (2016). Alkynyl-containing peptides of marine origin: A review. *Marine Drugs*, *14*(11), 216. https://doi.org/10.3390/MD14110216.
Chambers, E. S., Preston, T., Frost, G., & Morrison, D. J. (2018). Role of gut microbiota-generated short-chain fatty acids in metabolic and cardiovascular health. *Current Nutrition Reports*, *7*(4), 198–206. https://doi.org/10.1007/S13668-018-0248-8/FIGURES/1.
Chandran, S. S., Yi, J., Draths, K. M., Von Daeniken, R., Weber, W., & Frost, J. W. (2003). Phosphoenolpyruvate availability and the biosynthesis of shikimic acid. *Biotechnology Progress*, *19*(3), 808–814. https://doi.org/10.1021/BP025769P.
Charles, A. L., Chang, C. K., Wu, M. L., & Huang, T. C. (2007). Studies on the expression of liver detoxifying enzymes in rats fed seaweed (*Monostroma nitidum*). *Food and Chemical Toxicology*, *45*(12), 2390–2396. https://doi.org/10.1016/J.FCT.2007.06.014.
Chatterjee, S., Hui, P. C., Siu, W. S., Kan, C. W., Leung, P. C., Wanxue, C., & Chiou, J. C. (2021). Influence of pH-responsive compounds synthesized from chitosan and hyaluronic acid on dual-responsive (pH/temperature) hydrogel drug delivery systems of *Cortex Moutan*. *International Journal of Biological Macromolecules*, *168*, 163–174. https://doi.org/10.1016/J.IJBIOMAC.2020.12.035.
Chen, G., Seukep, A. J., & Guo, M. (2020). Recent advances in molecular docking for the research and discovery of potential marine drugs. *Marine Drugs*, *18*(11), 545. https://doi.org/10.3390/MD18110545.
Chen, M., Huang, S. F., Lai, K. Y., & Ling, M. H. (2013). Fully embeddable chitosan microneedles as a sustained release depot for intradermal vaccination. *Biomaterials*, *34*(12), 3077–3086. https://doi.org/10.1016/J.BIOMATERIALS.2012.12.041.
Chen, M., Lai, K. Y., Ling, M. H., & Lin, C. W. (2018). Enhancing immunogenicity of antigens through sustained intradermal delivery using chitosan microneedles with a patch-dissolvable design. *Acta Biomaterialia*, *65*, 66–75. https://doi.org/10.1016/J.ACTBIO.2017.11.004.

Chen, W., Tian, R., Xu, C., Yung, B. C., Wang, G., Liu, Y., Ni, Q., Zhang, F., Zhou, Z., Wang, J., Niu, G., Ma, Y., Fu, L., & Chen, X. (2017). Microneedle-array patches loaded with dual mineralized protein/peptide particles for type 2 diabetes therapy. *Nature Communications*, *8*(1), 1–11. https://doi.org/10.1038/s41467-017-01764-1.

Chen, X., Zhao, X., & Wang, G. (2020). Review on marine carbohydrate-based gold nanoparticles represented by alginate and chitosan for biomedical application. *Carbohydrate Polymers*, *244*, 116311. https://doi.org/10.1016/J.CARBPOL.2020.116311.

Chen, Y. H., Lai, K. Y., Chiu, Y. H., Wu, Y. W., Shiau, A. L., & Chen, M. C. (2019). Implantable microneedles with an immune-boosting function for effective intradermal influenza vaccination. *Acta Biomaterialia*, *97*, 230–238. https://doi.org/10.1016/J.ACTBIO.2019.07.048.

Cheung, R. C. F., Ng, T. B., & Wong, J. H. (2015). Marine peptides: Bioactivities and applications. *Marine Drugs*, *13*(7), 4006–4043. https://doi.org/10.3390/MD13074006.

Cheung, R. C. F., Wong, J. H., Pan, W. L., Chan, Y. S., Yin, C. M., Dan, X. L., Wang, H. X., Fang, E. F., Lam, S. K., Ngai, P. H. K., Xia, L. X., Liu, F., Ye, X. Y., Zhang, G. Q., Liu, Q. H., Sha, O., Lin, P., Ki, C., Bekhit, A. A., & Ng, T. B. (2014). Antifungal and antiviral products of marine organisms. *Applied Microbiology and Biotechnology*, *98*(8), 3475–3494. https://doi.org/10.1007/S00253-014-5575-0/TABLES/4.

Chevolot, L., Foucault, A., Chaubet, F., Kervarec, N., Sinquin, C., Fisher, A. M., & Boisson-Vidal, C. (1999). Further data on the structure of brown seaweed fucans: Relationships with anticoagulant activity. *Carbohydrate Research*, *319*(1–4), 154–165. https://doi.org/10.1016/S0008-6215(99)00127-5.

Chi, J., Zhang, X., Chen, C., Shao, C., Zhao, Y., & Wang, Y. (2020). Antibacterial and angiogenic chitosan microneedle array patch for promoting wound healing. *Bioactive Materials*, *5*(2), 253–259. https://doi.org/10.1016/J.BIOACTMAT.2020.02.004.

Chiu, Y. H., Chen, M. C., & Wan, S. W. (2018). Sodium hyaluronate/chitosan composite microneedles as a single-dose intradermal immunization system. *Biomacromolecules*, *19*(6), 2278–2285.https://doi.org/10.1021/ACS.BIOMAC.8B00441/SUPPL_FILE/BM8B00441_SI_001.PDF.

Choi, H., Engene, N., Smith, J. E., Preskitt, L. B., & Gerwick, W. H. (2010). Crossbyanols A-D, toxic brominated polyphenyl ethers from the Hawaiian bloom-forming cyanobacterium *Leptolyngbya crossbyana*. *Journal of Natural Products*, *73*(4), 517–522. https://doi.org/10.1021/NP900661G/SUPPL_FILE/NP900661G_SI_001.PDF.

Choi, J. S., Han, Y. R., Byeon, J. S., Choung, S. Y., Sohn, H. S., & Jung, H. A. (2015). Protective effect of fucosterol isolated from the edible brown algae, *Ecklonia stolonifera* and *Eisenia bicyclis*, on tert-butyl hydroperoxide- and tacrine-induced HepG2 cell injury. *Journal of Pharmacy and Pharmacology*, *67*(8), 1170–1178. https://doi.org/10.1111/JPHP.12404.

Choudhary, A., Naughton, L. M., Montánchez, I., Dobson, A. D. W., & Rai, D. K. (2017). Current status and future prospects of marine natural products (MNPs) as antimicrobials. *Marine Drugs*, *15*(9), 272. https://doi.org/10.3390/MD15090272.

Christie, S. N., McCaughey, C., McBride, M., & Coyle, P. V. (1997). Herpes simplex type 1 and genital herpes in Northern Ireland. *International Journal of STD & AIDS*, *8*(1), 68–69. https://pubmed.ncbi.nlm.nih.gov/9043990/ (accessed on 7 June 2024).

Chuaychan, S., Benjakul, S., & Kishimura, H. (2017). Characteristics and gelling property of gelatin from scale of spotted golden goatfish (*Parupeneus heptacanthus*). *Journal of Food Processing and Preservation*, *41*(5), e13139. https://doi.org/10.1111/JFPP.13139.

Clarke, S. D. (2001). Polyunsaturated fatty acid regulation of gene transcription: A molecular mechanism to improve the metabolic syndrome. *The Journal of Nutrition*, *131*(4), 1129–1132. https://doi.org/10.1093/JN/131.4.1129.

Claudel, T., Staels, B., & Kuipers, F. (2005). The Farnesoid X receptor: A molecular link between bile acid and lipid and glucose metabolism. *Arteriosclerosis, Thrombosis, and Vascular Biology, 25*(10), 2020–2031. https://doi.org/10.1161/01.ATV.0000178994.21828.A7.

Collins, M. N., & Birkinshaw, C. (2013). Hyaluronic acid-based scaffolds for tissue engineering—A review. *Carbohydrate Polymers, 92*(2), 1262–1279. https://doi.org/10.1016/J.CARBPOL.2012.10.028.

Cong, F., Cheung, A. K., & Huang, S. M. A. (2012). Chemical genetics-based target identification in drug discovery. *Annual Review of Pharmacology and Toxicology, 52*, 57–78. https://doi.org/10.1146/ANNUREV-PHARMTOX-010611-134639.

Connolly, J. M., Coleman, M., & Rose, D. P. (1997). Effects of dietary fatty acids on DU145 human prostate cancer cell growth in athymic nude mice. *Nutrition and Cancer, 29*(2), 114–119. https://doi.org/10.1080/01635589709514611.

Coppola, D., Oliviero, M., Vitale, G. A., Lauritano, C., D'Ambra, I., Iannace, S., & de Pascale, D. (2020). Marine collagen from alternative and sustainable sources: Extraction, processing and applications. *Marine Drugs, 18*(4), 214. https://doi.org/10.3390/MD18040214.

Correa, H., Aristizabal, F., Duque, C., & Kerr, R. (2011). Cytotoxic and antimicrobial activity of pseudopterosins and seco-pseudopterosins isolated from the octocoral *Pseudopterogorgia elisabethae* of San Andrés and Providencia Islands (Southwest Caribbean Sea). *Marine Drugs, 9*(3), 334–344. https://doi.org/10.3390/MD9030334.

Costa Lima, S. A., & Reis, S. (2018). Nanotechnological approaches in drug absorption through skin topical delivery. In *Nanoparticles in Life Sciences and Biomedicine* (pp. 171–194). https://doi.org/10.1201/9781351207355-7.

Costa, L. S., Fidelis, G. P., Cordeiro, S. L., Oliveira, R. M., Sabry, D. A., Câmara, R. B. G., Nobre, L. T. D. B., Costa, M. S. S. P., Almeida-Lima, J., Farias, E. H. C., Leite, E. L., & Rocha, H. A. O. (2010). Biological activities of sulfated polysaccharides from tropical seaweeds. *Biomedicine & Pharmacotherapy, 64*(1), 21–28. https://doi.org/10.1016/J.BIOPHA.2009.03.005.

Costello, M. J., Bouchet, P., Emblow, C. S., & Legakis, A. (2006). European marine biodiversity inventory and taxonomic resources: State of the art and gaps in knowledge. *Marine Ecology Progress Series, 316*, 257–268. https://doi.org/10.3354/MEPS316257.

Cragg, G. M., Newman, D. J., & Yang, S. S. (2006). Natural product extracts of plant and marine origin having antileukemia potential. The NCI experience. *Journal of Natural Products, 69*(3), 488–498. https://doi.org/10.1021/NP0581216.

Cumashi, A., Ushakova, N. A., Preobrazhenskaya, M. E., D'Incecco, A., Piccoli, A., Totani, L., Tinari, N., Morozevich, G. E., Berman, A. E., Bilan, M. I., Usov, A. I., Ustyuzhanina, N. E., Grachev, A. A., Sanderson, C. J., Kelly, M., Rabinovich, G. A., Iacobelli, S., & Nifantiev, N. E. (2007). A comparative study of the anti-inflammatory, anticoagulant, antiangiogenic, and antiadhesive activities of nine different fucoidans from brown seaweeds. *Glycobiology, 17*(5), 541–552. https://doi.org/10.1093/GLYCOB/CWM014.

Dabholkar, N., Gorantla, S., Waghule, T., Rapalli, V. K., Kothuru, A., Goel, S., & Singhvi, G. (2021). Biodegradable microneedles fabricated with carbohydrates and proteins: Revolutionary approach for transdermal drug delivery. *International Journal of Biological Macromolecules, 170*, 602–621. https://doi.org/10.1016/J.IJBIOMAC.2020.12.177.

Dafforn, K. A., Lewis, J. A., & Johnston, E. L. (2011). Antifouling strategies: History and regulation, ecological impacts and mitigation. *Marine Pollution Bulletin, 62*(3), 453–465. https://doi.org/10.1016/J.MARPOLBUL.2011.01.012.

Daletos, G., Kalscheuer, R., Koliwer-Brandl, H., Hartmann, R., De Voogd, N. J., Wray, V., Lin, W., & Proksch, P. (2015). Callyaerins from the marine sponge *Callyspongia aerizusa*: Cyclic peptides with antitubercular activity. *Journal of Natural Products, 78*(8), 1910–1925. https://doi.org/10.1021/ACS.JNATPROD.5B00266.

Damonte, E. B., Matulewicz, M. C., Cerezob, A. S., & Coto, C. E. (1996). Herpes simplex virus-inhibitory sulfated xylogalactans from the red seaweed *Nothogenia fastigiata. Chemotherapy, 42*(1), 57–64. https://doi.org/10.1159/000239422.

Damonte, E., Neyts, J., Pujol, C. A., Snoeck, R., Andrei, G., Ikeda, S., Witvrouw, M., Reymen, D., Haines, H., Matulewicz, M. C., Cerezo, A., Coto, C. E., & De Clerco, E. (1994). Antiviral activity of a sulphated polysaccharide from the red seaweed *Nothogenia fastigiata. Biochemical Pharmacology, 47*(12), 2187–2192. https://doi.org/10.1016/0006-2952 (94)90254-2.

Dangol, M., Kim, S., Li, C. G., Fakhraei Lahiji, S., Jang, M., Ma, Y., Huh, I., & Jung, H. (2017). Anti-obesity effect of a novel caffeine-loaded dissolving microneedle patch in high-fat diet-induced obese C57BL/6J mice. *Journal of Controlled Release, 265*, 41–47. https://doi.org/10.1016/J.JCONREL.2017.03.400.

Dantas-Santos, N., Almeida-Lima, J., Vidal, A. A. J., Gomes, D. L., Oliveira, R. M., Pedrosa, S. S., Pereira, P., Gama, F. M., & Rocha, H. A. O. (2012). Antiproliferative activity of fucan nanogel. *Marine Drugs, 10*(9), 2002–2022. https://doi.org/10.3390/MD10092002.

Dathathri, E., Lal, S., Mittal, M., Thakur, G., & De, S. (2020). Fabrication of low-cost composite polymer-based micro needle patch for transdermal drug delivery. *Applied Nanoscience, 10*(2), 371–377. https://doi.org/10.1007/S13204-019-01190-3.

Davidson, M. H. (2006). Mechanisms for the hypotriglyceridemic effect of marine omega-3 fatty acids. *The American Journal of Cardiology, 98*(4), 27–33. https://doi.org/10.1016/J.AMJCARD.2005.12.024.

Davis, R. A., Sykes, M., Avery, V. M., Camp, D., & Quinn, R. J. (2011). Convolutamines I and J, antitrypanosomal alkaloids from the bryozoan *Amathia tortusa. Bioorganic & Medicinal Chemistry, 19*(22), 6615–6619. https://doi.org/10.1016/J.BMC.2011.06.006.

De Clercq, E. (2000). Current lead natural products for the chemotherapy of human immunodeficiency virus (HIV) infection. *Medicinal Research Reviews, 20*(5), 323–349. https://pubmed.ncbi.nlm.nih.gov/10934347/.

Del Gaudio, P., Amante, C., Civale, R., Bizzarro, V., Petrella, A., Pepe, G., Campiglia, P., Russo, P., & Aquino, R. P. (2020). In situ gelling alginate-pectin blend particles loaded with Ac2-26: A new weapon to improve wound care armamentarium. *Carbohydrate Polymers, 227*, 115305. https://doi.org/10.1016/J.CARBPOL.2019.115305.

Di Benedetto, C., Barbaglio, A., Martinello, T., Alongi, V., Fassini, D., Cullorà, E., Patruno, M., Bonasoro, F., Barbosa, M. A., Carnevali, M. D. C., & Sugni, M. (2014). Production, characterization and biocompatibility of marine collagen matrices from an alternative and sustainable source: The sea urchin *Paracentrotus lividus. Marine Drugs, 12*(9), 4912–4933. https://doi.org/10.3390/MD12094912.

Ding, L., Münch, J., Goerls, H., Maier, A., Fiebig, H. H., Lin, W. H., & Hertweck, C. (2010). Xiamycin, a pentacyclic indolosesquiterpene with selective anti-HIV activity from a bacterial mangrove endophyte. *Bioorganic & Medicinal Chemistry Letters, 20*(22), 6685–6687. https://doi.org/10.1016/J.BMCL.2010.09.010.

Dona, A. C., Kyriakides, M., Scott, F., Shephard, E. A., Varshavi, D., Veselkov, K., & Everett, J. R. (2016). A guide to the identification of metabolites in NMR-based metabonomics/metabolomics experiments. *Computational and Structural Biotechnology Journal, 14*, 135–153. https://doi.org/10.1016/J.CSBJ.2016.02.005.

Dong, C., & Lv, Y. (2016). Application of Collagen Scaffold in Tissue Engineering: Recent Advances and New Perspectives. *Polymers, 8*(2), 42. https://doi.org/10.3390/POLYM 8020042.

Donia, M., & Hamann, M. T. (2003). Marine natural products and their potential applications as anti-infective agents. *The Lancet. Infectious Diseases, 3*(6), 338. https://doi.org/10.1016/ S1473-3099(03)00655-8.

Dos Santos, A. O., Britta, E. A., Bianco, E. M., Ueda-Nakamura, T., Dias Filho, B. P., Pereira, R. C., & Nakamura, C. V. (2011). 4-Acetoxydolastane Diterpene from the Brazilian Brown Alga Canistrocarpus cervicornis as Antileishmanial Agent. *Marine Drugs, 9*(11), 2369–2383. https://doi.org/10.3390/MD9112369.

Du, F. Y., Li, X. M., Zhang, P., Li, C. S., & Wang, B. G. (2014). Cyclodepsipeptides and Other O-Containing Heterocyclic Metabolites from Beauveria felina EN-135, a Marine-Derived Entomopathogenic Fungus. *Marine Drugs, 12*(5), 2816–2826. https://doi.org/10.3390/ MD12052816.

Duarte, M. E. R., Cardoso, M. A., Noseda, M. D., & Cerezo, A. S. (2001). Structural studies on fucoidans from the brown seaweed *Sargassum stenophyllum. Carbohydrate Research, 333*(4), 281–293. https://doi.org/10.1016/S0008-6215(01)00149-5.

Duarte, M. E. R., Cauduro, J. P., Noseda, D. G., Noseda, M. D., Gonçalves, A. G., Pujol, C. A., Damonte, E. B., & Cerezo, A. S. (2004). The structure of the agaran sulfate from *Acanthophora spicifera* (*Rhodomelaceae, Ceramiales*) and its antiviral activity. Relation between structure and antiviral activity in agarans. *Carbohydrate Research, 339*(2), 335–347. https://doi.org/10.1016/J.CARRES.2003.09.028.

Duarte, M. E. R., Noseda, D. G., Noseda, M. D., Tulio, S., Pujol, C. A., & Damonte, E. B. (2001). Inhibitory effect of sulfated galactans from the marine alga *Bostrychia montagnei* on herpes simplex virus replication in vitro. *Phytomedicine: International Journal of Phytotherapy and Phytopharmacology, 8*(1), 53–58. https://doi.org/10.1078/0944-7113-00007.

Duncan, K. R., Crüsemann, M., Lechner, A., Sarkar, A., Li, J., Ziemert, N., Wang, M., Bandeira, N., Moore, B. S., Dorrestein, P. C., & Jensen, P. R. (2015). Molecular networking and pattern-based genome mining improves discovery of biosynthetic gene clusters and their products from Salinispora species. *Chemistry & Biology, 22*(4), 460–471. https://doi. org/10.1016/J.CHEMBIOL.2015.03.010.

Ehresmann, D. W., Deig, E. F., Hatch, M. T., DiSalvo, L. H., & Vedros, N. A. (1977). Antiviral substances from California Marine Algae1. *Journal of Phycology, 13*(1), 37–40. https://doi. org/10.1111/J.1529-8817.1977.TB02883.X.

El-Gendy, M. M. A., & El-Bondkly, A. M. A. (2010). Production and genetic improvement of a novel antimycotic agent, Saadamycin, against Dermatophytes and other clinical fungi from Endophytic Streptomyces sp. Hedaya48. *Journal of Industrial Microbiology and Biotechnology, 37*(8), 831–841. https://doi.org/10.1007/S10295-010-0729-2.

Engelhardt, K., Degnes, K. F., Kemmler, M., Bredholt, H., Fjaervik, E., Klinkenberg, G., Sletta, H., Ellingsen, T. E., & Zotchev, S. B. (2010). Production of a new thiopeptide antibiotic, TP-1161, by a marine Nocardiovsis species. *Applied and Environmental Microbiology, 76*(15), 4969–4976. https://doi.org/10.1128/AEM.00741-10/SUPPL_FILE/ TP116_DISCOVERY_SUPPLEMENTALMATERIAL_REVISED_30_05_10_REV. DOC.

Farias, W. R. L., Valente, A. P., Pereira, M. S., & Mourão, P. A. S. (2000). Structure and anticoagulant activity of sulfated Galactans. *Journal of Biological Chemistry, 275*(38), 29299–29307. https://doi.org/10.1074/jbc.m002422200.

Fattorusso, C., Persico, M., Calcinai, B., Cerrano, C., Parapini, S., Taramelli, D., Novellino, E., Romano, A., Scala, F., Fattorusso, E., & Taglialatela-Scafati, O. (2010). Manadoperoxides A-D from the Indonesian sponge Plakortis cfr. simplex. Further insights on the structure-activity relationships of simple 1,2-dioxane antimalarials. *Journal of Natural Products, 73*(6), 1138–1145. https://doi.org/10.1021/NP100196B/SUPPL_FILE/NP100196B_SI_001.PDF.

Fedorov, S. N., Ermakova, S. P., Zvyagintseva, T. N., & Stonik, V. A. (2013). Anticancer and cancer preventive properties of Marine Polysaccharides: Some results and prospects. *Marine Drugs, 11*(12), 4876–4901. https://doi.org/10.3390/MD11124876.

Felician, F. F., Xia, C., Qi, W., & Xu, H. (2018). Collagen from Marine Biological Sources and Medical Applications. *Chemistry & Biodiversity, 15*(5), e1700557. https://doi.org/10.1002/CBDV.201700557.

Feling, R. H., Buchanan, G. O., Mincer, T. J., Kauffman, C. A., Jensen, P. R., & Fenical, W. (2003). Salinosporamide A: a highly cytotoxic proteasome inhibitor from a novel microbial source, a marine bacterium of the new genus salinospora. *Angewandte Chemie (International Ed. in English), 42*(3), 355–357. https://doi.org/10.1002/ANIE.200390115.

Fenical, W. (1997). New pharmaceuticals from marine organisms. *Trends in Biotechnology, 15*(9), 339–341. https://doi.org/10.1016/S0167-7799(97)01081-0.

Ferreira, A. M., Gentile, P., Chiono, V., & Ciardelli, G. (2012). Collagen for bone tissue regeneration. *Acta Biomaterialia, 8*(9), 3191–3200. https://doi.org/10.1016/J.ACTBIO.2012.06.014.

Fidler, A. L., Boudko, S. P., Rokas, A., & Hudson, B. G. (2018). The triple helix of collagens – An ancient protein structure that enabled animal multicellularity and tissue evolution. *Journal of Cell Science, 131*(7), 1–15. https://doi.org/10.1242/JCS.203950/57104.

Filimonova, V., Gonçalves, F., Marques, J. C., De Troch, M., & Gonçalves, A. M. M. (2016). Fatty acid profiling as bioindicator of chemical stress in marine organisms: A review. *Ecological Indicators, 67*, 657–672. https://doi.org/10.1016/J.ECOLIND.2016.03.044.

Fleury, C., Serpentini, A., Kypriotou, M., Renard, E., Galéra, P., & Lebel, J. M. (2011). Characterization of a Non-fibrillar-Related Collagen in the Mollusc *Haliotis tuberculata* and its biological activity on human dermal fibroblasts. *Marine Biotechnology, 13*(5), 1003–1016. https://doi.org/10.1007/S10126-011-9364-9/FIGURES/6.

Fox, D. L. (1960). Perspectives in marine biochemistry. *Annals of the New York Academy of Sciences, 90*(3), 617–621. https://doi.org/10.1111/J.1749-6632.1960.TB26409.X.

Franklin, J. B., & Rajesh, R. P. (2015). A sleep-inducing peptide from the venom of the Indian cone snail *Conus araneosus*. *Toxicon, 103*, 39–47. https://doi.org/10.1016/J.TOXICON.2015.06.017.

Fu, X. M., Zhang, M. Q., Shao, C. L., Li, G. Q., Bai, H., Dai, G. L., Chen, Q. W., Kong, W., Fu, X. J., & Wang, C. Y. (2016). Chinese Marine Materia Medica Resources: Status and potential. *Marine Drugs, 14*(3), 46. https://doi.org/10.3390/MD14030046.

Fuhrmann, J. J., & Zuberer, D. A. (2021). Carbon transformations and soil organic matter formation. *Principles and Applications of Soil Microbiology,* Third Edition, 327–361. https://doi.org/10.1016/B978-0-12-820202-9.00013-7.

Funami, T., Hiroe, M., Noda, S., Asai, I., Ikeda, S., & Nishinari, K. (2007). Influence of molecular structure imaged with atomic force microscopy on the rheological behavior of carrageenan aqueous systems in the presence or absence of cations. *Food Hydrocolloids, 21*(4), 617–629. https://doi.org/10.1016/J.FOODHYD.2006.07.013.

Gagliardi, A., Giuliano, E., Venkateswararao, E., Fresta, M., Bulotta, S., Awasthi, V., & Cosco, D. (2021). Biodegradable polymeric nanoparticles for drug delivery to solid tumors. *Frontiers in Pharmacology, 12*, 17. https://doi.org/10.3389/FPHAR.2021.601626/BIBTEX.

Galtsoff, P. S. (1934). The Mystery of the Ocean. *The Scientific Monthly, 39*(2), 172–175. https://www.jstor.org/stable/15786 (accessed on 7 June 2024).

Gan, R. Y., Hu, Y., Cheong, K. L., Yu, B., Chen, J., & Zhong, S. (2022). A Comprehensive review of the cardioprotective effect of marine algae polysaccharide on the gut microbiota. *Foods, 11*(22), 3550. https://doi.org/10.3390/FOODS11223550.

Geahchan, S., Baharlouei, P., & Rahman, M. A. (2022). Marine Collagen: A Promising Biomaterial for Wound Healing, Skin Anti-Aging, and Bone Regeneration. *Marine Drugs, 20*(1), 61. https://doi.org/10.3390/MD20010061.

Gerber, P., Dutcher, J. D., Adams, E. V., & Sherman, J. H. (1958). Protective effect of seaweed extracts for chicken embryos infected with influenza B or mumps virus. *Proceedings of the Society for Experimental Biology and Medicine, 99*(3), 590–593. https://doi.org/10.3181/00379727-99-24429.

Gerwick, W. H., & Moore, B. S. (2012). Lessons from the past and charting the future of marine natural products drug discovery and chemical biology. *Chemistry & Biology, 19*(1), 85. https://doi.org/10.1016/J.CHEMBIOL.2011.12.014.

Ghosh, T., Pujol, C. A., Damonte, E. B., Sinha, S., & Ray, B. (2009). Sulfated xylomannans from the red seaweed *Sebdenia polydactyla*: structural features, chemical modification and antiviral activity. *Antiviral Chemistry & Chemotherapy, 19*(6), 235–242. https://doi.org/10.1177/095632020901900603.

Glauser, B. F., Rezende, R. M., Melo, F. R., Pereira, M. S., Francischetti, I. M. B., Monteiro, R. Q., Rezaie, A. R., & Mourão, P. A. S. (2009). Anticoagulant activity of a sulfated galactan: Serpin-independent effect and specific interaction with factor Xa. *Thrombosis and Haemostasis, 102*(6), 1183–1193. https://doi.org/10.1160/TH09-04-0273.

Gombotz, W. R., & Wee, S. F. (1998). Protein release from alginate matrices. *Advanced Drug Delivery Reviews, 31*(3), 267–285. https://doi.org/10.1016/S0169-409X(97)00124-5.

Gordon, M. K., & Hahn, R. A. (2010). Collagens. *Cell and Tissue Research, 339*(1), 247–257. https://doi.org/10.1007/S00441-009-0844-4.

Gribbon, P., & Andreas, S. (2005). High-throughput drug discovery: What can we expect from HTS? *Drug Discovery Today, 10*(1), 17–22. https://doi.org/10.1016/S1359-6446(04)03275-1.

Gross, H., Stockwell, V. O., Henkels, M. D., Nowak-Thompson, B., Loper, J. E., & Gerwick, W. H. (2007). The genomisotopic approach: A systematic method to isolate products of orphan biosynthetic gene clusters. *Chemistry & Biology, 14*(1), 53–63. https://doi.org/10.1016/J.CHEMBIOL.2006.11.007.

Gu, Y., Yu, L., Mou, J., Wu, D., Xu, M., Zhou, P., & Ren, Y. (2020). Research strategies to develop environmentally friendly marine antifouling coatings. *Marine Drugs, 18*(7), 371. https://doi.org/10.3390/MD18070371.

Guo, L., Cao, X., Yang, S., Wang, X., Wen, Y., Zhang, F., Chen, H., & Wang, L. (2019). Characterization, solubility and antibacterial activity of inclusion complex of questin with hydroxypropyl-β-cyclodextrin. *3 Biotech, 9*(4), 123. https://doi.org/10.1007/S13205-019-1663-2.

Guo, M., Wei, J., Huang, X., Huang, Y., & Qin, Q. (2012). Antiviral effects of β-defensin derived from orange-spotted grouper (*Epinephelus coioides*). *Fish & Shellfish Immunology, 32*(5), 828–838. https://doi.org/10.1016/J.FSI.2012.02.005.

Gur Barzilai, M., Kahn, R., Regev, N., Gordon, D., Moran, Y., & Gurevitz, M. (2014). The specificity of Av3 sea anemone toxin for arthropods is determined at linker DI/SS2–S6 in the pore module of target sodium channels. *Biochemical Journal, 463*(2), 271–277. https://doi.org/10.1042/BJ20140576.

Gustafson, K., Roman, M., & Fenical, W. (1989). The Macrolactins, a novel class of antiviral and cytotoxic macrolides from a deep-sea marine bacterium. *Journal of the American Chemical Society, 111*(19), 7519–7524. https://doi.org/10.1021/JA00201A036.

Haefner, B. (2003). Drugs from the deep: Marine natural products as drug candidates. *Drug Discovery Today, 8*(12), 536–544. https://doi.org/10.1016/S1359-6446(03)02713-2.

Hale, K. J., Hummersone, M. G., Manaviazar, S., & Frigerio, M. (2002). The chemistry and biology of the bryostatin antitumour macrolides. *Natural Product Reports, 19*(4), 413–453. https://doi.org/10.1039/B009211H.

Halper, J. (2021). Basic structure, physiology, and biochemistry of connective tissues and extracellular matrix collagens. *Progress in Heritable Soft Connective Tissue Diseases (J. Halper (ed.); 2nd ed., Vol. 1348,* 5–43. Springer International Publishing. https://doi.org/10.1007/978–3-030–80614–9.

Hannan, M. A., Dash, R., Haque, M. N., Mohibbullah, M., Sohag, A. A. M., Rahman, M. A., Uddin, M. J., Alam, M., & Moon, I. S. (2020). Neuroprotective potentials of marine algae and their bioactive metabolites: Pharmacological insights and therapeutic advances. *Marine Drugs, 18*(7), 347. https://doi.org/10.3390/MD18070347.

Harris, M., Bruhn, C., Schor, D., KAPSAK, W. R., & BLAKISTONE, B. (2009). Communicating the net benefits of seafood consumption – IFT.org. *Food Technology Magazine, 63*(11), 1–37. https://www.ift.org/news-and-publications/food-technology-magazine/issues/2009/november/features/seafood-consumption (accessed on 7 June 2024).

Hashim, P., Ridzwan, M., Bakar, M. S., & Hashim, M. (2015). Collagen in food and beverage industries. *International Food Research Journal, 22*(1), 1–8. http://www.ifrj.upm.edu.my/22 (01) 2015/(1).pdf.

Hassan, J., Haigh, C., Ahmed, T., Uddin, M. J., & Das, D. B. (2022). Potential of microneedle systems for COVID-19 vaccination: Current trends and challenges. *Pharmaceutics, 14*(5), 1066. https://doi.org/10.3390/pharmaceutics14051066.

Hayakawa, Y., Hayashi, T., Lee, J. B., Srisomporn, P., Maeda, M., Ozawa, T., & Sakuragawa, N. (2000). Inhibition of thrombin by sulfated polysaccharides isolated from green algae. *Biochimica et Biophysica Acta (BBA) – Protein Structure and Molecular Enzymology, 1543*(1), 86–94. https://doi.org/10.1016/S0167-4838(00)00193-X.

Helfrich, E. J. N., Reiter, S., & Piel, J. (2014). Recent advances in genome-based polyketide discovery. *Current Opinion in Biotechnology, 29*(1), 107–115. https://doi.org/10.1016/J.COPBIO.2014.03.004.

Heo, S. J., Ko, S. C., Kang, S. M., Kang, H. S., Kim, J. P., Kim, S. H., Lee, K. W., Cho, M. G., & Jeon, Y. J. (2008). Cytoprotective effect of fucoxanthin isolated from brown algae Sargassum siliquastrum against H_2O_2-Induced cell damage. *European Food Research and Technology, 228*(1), 145–151. https://doi.org/10.1007/S00217-008-0918-7/FIGURES/6.

Hepburn, F. N., Exler, J., & Weihrauch, J. L. (1986). Provisional tables on the content of omega-3 fatty acids and other fat components of selected foods. *Journal of the American Dietetic Association, 86*(6), 788–793. https://doi.org/10.1016/S0002-8223(21)04023-2.

Hertiani, T., Edrada-Ebel, R. A., Ortlepp, S., van Soest, R. W. M., de Voogd, N. J., Wray, V., Hentschel, U., Kozytska, S., Müller, W. E. G., & Proksch, P. (2010). From anti-fouling to biofilm inhibition: New cytotoxic secondary metabolites from two Indonesian Agelas sponges. *Bioorganic & Medicinal Chemistry, 18*(3), 1297–1311. https://doi.org/10.1016/J.BMC.2009.12.028.

Hertz, R., Sheena, V., Kalderon, B., Berman, I., & Bar-Tana, J. (2001). Suppression of hepatocyte nuclear factor-4α by acyl-CoA thioesters of hypolipidemic peroxisome proliferators. *Biochemical Pharmacology, 61*(9), 1057–1062. https://doi.org/10.1016/S0006-2952(01)00578-0.

Hidari, K. I. P. J., Takahashi, N., Arihara, M., Nagaoka, M., Morita, K., & Suzuki, T. (2008). Structure and anti-dengue virus activity of sulfated polysaccharide from a marine alga. *Biochemical and Biophysical Research Communications, 376*(1), 91–95. https://doi.org/10.1016/J.BBRC.2008.08.100.

Highley, C. B., Prestwich, G. D., & Burdick, J. A. (2016). Recent advances in hyaluronic acid hydrogels for biomedical applications. *Current Opinion in Biotechnology, 40*, 35–40. https://doi.org/10.1016/J.COPBIO.2016.02.008.

Hohmann, C., Schneider, K., Bruntner, C., Irran, E., Nicholson, G., Bull, A. T., Jones, A. L., Brown, R., Stach, J. E. M., Goodfellow, M., Beil, W., Krämer, M., Imhoff, J. F., Süssmuth, R. D., & Fiedler, H. P. (2009). Caboxamycin, a new antibiotic of the benzoxazole family produced by the deep-sea strain Streptomyces sp. NTK 937. *The Journal of Antibiotics, 62*(2), 99–104. https://doi.org/10.1038/ja.2008.24.

Holtkamp, A. D., Kelly, S., Ulber, R., & Lang, S. (2009). Fucoidans and fucoidanases—focus on techniques for molecular structure elucidation and modification of marine polysaccharides. *Applied Microbiology and Biotechnology, 82*(1), 1–11. https://doi.org/10.1007/S00253-008-1790-X.

Houghten, R. A. (1994). Combinatorial libraries. Finding the needle in the haystack. *Current Biology: CB, 4*(6), 564–567. https://doi.org/10.1016/S0960-9822(00)00127-5.

Hu, Z., Hu, Z., Hu, Z., Wang, S., Dai, Z., Zhang, H., & Zheng, X. (2020). A novel theranostic nano-platform (PB@FePt–HA-g-PEG) for tumor chemodynamic–photothermal co-therapy and triple-modal imaging (MR/CT/PI) diagnosis. *Journal of Materials Chemistry B, 8*(24), 5351–5360. https://doi.org/10.1039/D0TB00708K.

Huang, L., Wen, K., Gao, X., & Liu, Y. (2010). Hypolipidemic effect of fucoidan from Laminaria japonica in hyperlipidemic rats. *Journal of Agricultural and Food Chemistry, 58*(4), 2229–2234. https://doi.org/10.1021/JF9042564.

Huang, S., & Huang, G. (2019). The dextrans as vehicles for gene and drug delivery. *International Journal of Biological Macromolecules, 127*, 1–10. https://doi.org/10.1016/J.IJBIOMAC.2019.01.123.

Husain, F. M., Khan, A., Khan, R. A., Siddique, J. A., Oves, M., Parwaz Khan, A. A., Ansari, M. O., & Cancar, H. D. (2021). Bio-based aerogels and their environment applications: an overview. In *Advances in Aerogel Composites for Environmental Remediation* (pp. 347–356). https://doi.org/10.1016/B978-0-12-820732-1.00018-7.

Ikoma, T., Kobayashi, H., Tanaka, J., Walsh, D., & Mann, S. (2003). Physical properties of type I collagen extracted from fish scales of Pagrus major and Oreochromis niloticas. *International Journal of Biological Macromolecules, 32*(3–5), 199–204. https://doi.org/10.1016/S0141-8130(03)00054-0.

In Jeong, S., Kim, S. Y., Cho, S. K., Chong, M. S., Kim, K. S., Kim, H., Lee, S. B., & Lee, Y. M. (2007). Tissue-engineered vascular grafts composed of marine collagen and PLGA fibers using pulsatile perfusion bioreactors. *Biomaterials, 28*(6), 1115–1122. https://doi.org/10.1016/J.BIOMATERIALS.2006.10.025.

Innes, J. K., & Calder, P. C. (2020). Marine Omega-3 (N-3) Fatty acids for cardiovascular health: An update for 2020. *International Journal of Molecular Sciences, 21*(4), 1362. https://doi.org/10.3390/IJMS21041362.

Isaka, M., Suyarnsestakorn, C., Tanticharoen, M., Kongsaeree, P., & Thebtaranonth, Y. (2002). Aigialomycins A-E, new resorcylic macrolides from the marine mangrove fungus

Aigialus parvus. *Journal of Organic Chemistry, 67*(5), 1561–1566. https://doi.org/10.1021/JO010930G.

Isnansetyo, A., & Kamei, Y. (2009). Anti-methicillin-resistant Staphylococcus aureus (MRSA) activity of MC21-B, an antibacterial compound produced by the marine bacterium Pseudoalteromonas phenolica O-BC30T. *International Journal of Antimicrobial Agents, 34*(2), 131–135. https://doi.org/10.1016/J.IJANTIMICAG.2009.02.009.

Iyer, U., & Kadambi, V. J. (2011). Antibody drug conjugates — Trojan horses in the war on cancer. *Journal of Pharmacological and Toxicological Methods, 64*(3), 207–212. https://doi.org/10.1016/J.VASCN.2011.07.005.

Jankangram, W., Chooluck, S., & Pomthong, B. (2016). Comparison of the properties of collagen extracted from dried jellyfish and dried squid. *African Journal of Biotechnology, 15*(16), 642–648. https://doi.org/10.4314/ajb.v15i16.

Jeon, J. E., Na, Z., Jung, M., Lee, H. S., Sim, C. J., Nahm, K., Oh, K. B., & Shin, J. (2010). Discorhabdins from the Korean Marine Sponge *Sceptrella sp. Journal of Natural Products, 73*(2), 258–262. https://doi.org/10.1021/NP9005629/SUPPL_FILE/NP9005629_SI_001.PDF.

Ji, H., Shao, H., Zhang, C., Hong, P., & Xiong, H. (2008). Separation of the polysaccharides in *Caulerpa racemosa* and their chemical composition and antitumor activity. *Journal of Applied Polymer Science, 110*(3), 1435–1440. https://doi.org/10.1002/APP.28676.

Jiang, X., Xiong, J., Song, Z., Morse, J. C., Jones, F. C., & Xie, Z. (2013). Is coarse taxonomy sufficient for detecting macroinvertebrate patterns in floodplain lakes? *Ecological Indicators, 27*, 48–55. https://doi.org/10.1016/J.ECOLIND.2012.11.015.

Jiao, G., Yu, G., Zhang, J., & Ewart, H. S. (2011). Chemical structures and bioactivities of sulfated polysaccharides from marine algae. *Marine Drugs, 9*(2), 196–223. https://doi.org/10.3390/MD9020196.

Jiao, L., Jiang, P., Zhang, L., & Wu, M. (2010). Antitumor and immunomodulating activity of polysaccharides from *Enteromorpha intestinalis. Biotechnology and Bioprocess Engineering, 15*(3), 421–428. https://doi.org/10.1007/S12257-008-0269-Z.

Jiao, L., Li, X., Li, T., Jiang, P., Zhang, L., Wu, M., & Zhang, L. (2009). Characterization and anti-tumor activity of alkali-extracted polysaccharide from *Enteromorpha intestinalis. International Immunopharmacology, 9*(3), 324–329. https://doi.org/10.1016/J.INTIMP.2008.12.010.

Jimeno, J. M. (2002). A clinical armamentarium of marine-derived anti-cancer compounds. *Anti-Cancer Drugs, 15*(9), 15–19 (supplementary file), https://pubmed.ncbi.nlm.nih.gov/12173490/ (accessed on 7 June 2024).

Jónasdóttir, S. H. (2019). Fatty acid profiles and production in marine phytoplankton. *Marine Drugs, 17*(3), 151. https://doi.org/10.3390/MD17030151.

Joye, I. J., & McClements, D. J. (2014). Biopolymer-based nanoparticles and microparticles: Fabrication, characterization, and application. *Current Opinion in Colloid & Interface Science, 19*(5), 417–427. https://doi.org/10.1016/J.COCIS.2014.07.002.

Juhlin, L. (1997). Hyaluronan in skin. *Journal of Internal Medicine, 242*(1), 61–66. https://doi.org/10.1046/J.1365-2796.1997.00175.X.

Jump, D. B. (2002). The biochemistry of n-3 polyunsaturated fatty acids. *The Journal of Biological Chemistry, 277*(11), 8755–8758. https://doi.org/10.1074/JBC.R100062200.

Jump, D. B., & Clarke, S. D. (1999). Regulation of gene expression by dietary fat. *Annual Review of Nutrition, 19*, 63–90. https://doi.org/10.1146/ANNUREV.NUTR.19.1.63.

Jung, W. S., Lee, S. K., Hong, J. S. J., Park, S. R., Jeong, S. J., Han, A. R., Sohng, J. K., Kim, B. G., Choi, C. Y., Sherman, D. H., & Yoon, Y. J. (2006). Heterologous expression of tylosin polyketide synthase and production of a hybrid bioactive macrolide in *Streptomyces venezuelae*. *Applied Microbiology and Biotechnology, 72*(4), 763–769. https://doi.org/10.1007/S00253-006-0318-5.

Juster, H., van der Aar, B., & de Brouwer, H. (2019). A review on microfabrication of thermoplastic polymer-based microneedle arrays. *Polymer Engineering & Science, 59*(5), 877–890. https://doi.org/10.1002/PEN.25078.

Kaczmarek, M. B., Struszczyk-Swita, K., Li, X., Szczęsna-Antczak, M., & Daroch, M. (2019). Enzymatic modifications of chitin, chitosan, and Chito oligosaccharides. *Frontiers in Bioengineering and Biotechnology, 7*(SEP), 243. https://doi.org/10.3389/FBIOE.2019.00243/BIBTEX.

Kang, H. K., Choi, M. C., Seo, C. H., & Park, Y. (2018). Therapeutic properties and biological benefits of marine-derived anticancer peptides. *International Journal of Molecular Sciences, 19*(3), 919. https://doi.org/10.3390/IJMS19030919.

Kang, H. K., Seo, C. H., & Park, Y. (2015). Marine peptides and their anti-infective activities. *Marine Drugs, 13*(1), 618–654. https://doi.org/10.3390/MD13010618.

Karlsson, J., Vaughan, H. J., & Green, J. J. (2018). Biodegradable polymeric nanoparticles for therapeutic cancer treatments. *Annual Review of Chemical and Biomolecular Engineering, 9*, 105–127. https://doi.org/10.1146/ANNUREV-CHEMBIOENG-060817-084055.

Karmakar, P., Ghosh, T., Sinha, S., Saha, S., Mandal, P., Ghosal, P. K., & Ray, B. (2009). Polysaccharides from the brown seaweed *Padina tetrastromatica*: Characterization of a sulfated fucan. *Carbohydrate Polymers, 78*(3), 416–421. https://doi.org/10.1016/J.CARBPOL.2009.04.039.

Karnjanapratum, S., & You, S. G. (2011). Molecular characteristics of sulfated polysaccharides from *Monostroma nitidum* and their in vitro anticancer and immunomodulatory activities. *International Journal of Biological Macromolecules, 48*(2), 311–318. https://doi.org/10.1016/J.IJBIOMAC.2010.12.002.

Karsdal, M. A. (2016). Biochemistry of collagens, laminins and elastin. In M. A. Karsdal (Ed.), *Biochemistry of Collagens, Laminins and Elastin*. 31–48. Elsevier. https://doi.org/10.1016/C2015-0-05547-2.

Keffer, J. L., Plaza, A., & Bewley, C. A. (2009). Motualevic acids A-F, antimicrobial acids from the sponge *Siliquariaspongia sp. Organic Letters, 11*(5), 1087–1090. https://doi.org/10.1021/OL802890B/SUPPL_FILE/OL802890B_SI_001.PDF.

Khanh, T. M. T., Van Vo, T., Duan, W., Ha-Lien Tran, P., & Truong-Dinh Tran, T. (2016). Perspectives of engineered marine derived polymers for biomedical nanoparticles. *Current Pharmaceutical Design, 22*(19), 2844–2856. https://www.ingentaconnect.com/content/ben/cpd/2016/00000022/00000019/art00008# (accessed on 7 June 2024).

Kijjoa, A., & Sawangwong, P. (2004). Drugs and cosmetics from the sea. *Marine Drugs, 2*(2), 73–82. https://doi.org/10.3390/MD202073.

Kikuchi, M., & Konno, H. (2016). Cytotoxic evaluation of natural and synthetic callipeltins: A revision of cytotoxicity of callipeltin B. *Bioscience, Biotechnology, and Biochemistry, 80*(6), 1066–1069. https://doi.org/10.1080/09168451.2016.1148581.

Kim, J. H., Shin, J. U., Kim, S. H., Noh, J. Y., Kim, H. R., Lee, J., Chu, H., Jeong, K. Y., Park, K. H., Kim, J. D., Kim, H. K., Jeong, D. H., Yong, T. S., Park, J. W., & Lee, K. H. (2018). Successful transdermal allergen delivery and allergen-specific immunotherapy using biodegradable microneedle patches. *Biomaterials, 150*, 38–48. https://doi.org/10.1016/J.BIOMATERIALS.2017.10.013.

Kim, J. K., Cho, M. L., Karnjanapratum, S., Shin, I. S., & You, S. G. (2011). In vitro and in vivo immunomodulatory activity of sulfated polysaccharides from *Enteromorpha prolifera*. *International Journal of Biological Macromolecules, 49*(5), 1051–1058. https://doi.org/10.1016/J.IJBIOMAC.2011.08.032.

Kim, M. H., Kang, S. G., Park, J. H., Yanagisawa, M., & Kim, C. H. (2013). Short-Chain Fatty Acids Activate GPR41 and GPR43 on Intestinal epithelial cells to promote inflammatory responses in mice. *Gastroenterology, 145*(2), 396–406.e10. https://doi.org/10.1053/J.GASTRO.2013.04.056.

Kim, S. K., & Wijesekara, I. (2010). Development and biological activities of marine-derived bioactive peptides: A review. *Journal of Functional Foods, 2*(1), 1–9. https://doi.org/10.1016/J.JFF.2010.01.003.

Kim, S. K., Jeon, Y. J., Byeun, H. G., Kim, Y. T., & Lee, C. K. (1997). Enzymatic Recovery of Cod Frame Proteins with Crude Proteinase from Tuna Pyloric Caeca. *Fisheries Science, 63*(3), 421–427.

Kimura, S., Takema, Y., & Kubota, M. (1981). Octopus skin collagen. Isolation and characterization of collagen comprising two distinct alpha chains. *Journal of Biological Chemistry, 256*(24), 13230–13234. https://doi.org/10.1016/S0021-9258(18)43033-5.

Kingston, D. G. I. (2011). Modern natural products drug discovery and its relevance to biodiversity conservation. *Journal of Natural Products, 74*(3), 496. https://doi.org/10.1021/NP100550T.

Ko, S. C., Kim, J. Il, Park, S. J., Jung, W. K., & Jeon, Y. J. (2016). Antihypertensive peptide purified from *Styela clava* flesh tissue stimulates glucose uptake through AMP-activated protein kinase (AMPK) activation in skeletal muscle cells. *European Food Research and Technology, 242*(2), 163–170. https://doi.org/10.1007/S00217-015-2526-7/FIGURES/5.

Kolender, A. A., Pujol, C. A., Damonte, E. B., Matulewicz, M. C., & Cerezo, A. S. (1997). The system of sulfated α (1 → 3) linked d-mannans from the red seaweed *Nothogenia fastigiata*: Structures, antiherpetic and anticoagulant properties. *Carbohydrate Research, 304*(1), 53–60. https://doi.org/10.1016/S0008-6215(97)00201-2.

Kraan, S. (2012). Algal polysaccharides, novel applications and outlook. *Carbohydrates – Comprehensive Studies on Glycobiology and Glycotechnology*, 352–489, https://doi.org/10.5772/51572.

Kumar, M., Jha, A., & Mishra, B. (2022). Marine biopolymers for transdermal drug delivery. *Marine Biomaterials*, 157–207. https://doi.org/10.1007/978–981–16–4787–1.

Kwan, J. C., Meickle, T., Ladwa, D., Teplitski, M., Paul, V., & Luesch, H. (2011). Lyngbyoic acid, a "tagged" fatty acid from a marine cyanobacterium, disrupts quorum sensing in *Pseudomonas aeruginosa*. *Molecular BioSystems, 7*(4), 1205–1216. https://doi.org/10.1039/C0MB00180E.

Kwon, M. J., & Nam, T. J. (2006). *Porphyran Induces Apoptosis Related Signal Pathway in AGS Gastric Cancer Cell Lines.* Life Sciences, *79*(20), 1956–1962. https://doi.org/10.1016/J.LFS.2006.06.031.

Lahaye, M., & Robic, A. (2007). Structure and function properties of Ulvan, a polysaccharide from green seaweeds. *Biomacromolecules, 8*(6), 1765–1774. https://doi.org/10.1021/BM061185Q/ASSET/IMAGES/MEDIUM/BM061185QN00001.GIF.

Lahaye, M., Brunel, M., & Bonnin, E. (1997). Fine chemical structure analysis of oligosaccharides produced by an ulvan-lyase degradation of the water-soluble cell-wall

polysaccharides from *Ulva sp.* (*Ulvales, Chlorophyta*). *Carbohydrate Research, 304*(3–4), 325–333. https://doi.org/10.1016/S0008-6215(97)00270-X.
Lane, A. L., Stout, E. P., Lin, A. S., Prudhomme, J., Le Roch, K., Fairchild, C. R., Franzblau, S. G., Hay, M. E., Aalbersberg, W., & Kubanek, J. (2009). Antimalarial bromophyeolides J-Q from the Fijian red alga *Callophycus serratus*. *Journal of Organic Chemistry, 74*(7), 2736–2742. https://doi.org/10.1021/JO900008W/SUPPL_FILE/JO900008W_SI_001.PDF.
Langasco, R., Cadeddu, B., Formato, M., Lepedda, A. J., Cossu, M., Giunchedi, P., Pronzato, R., Rassu, G., Manconi, R., & Gavini, E. (2017). Natural collagenic skeleton of marine sponges in pharmaceutics: Innovative biomaterial for topical drug delivery. *Materials Science and Engineering: C, 70*, 710–720. https://doi.org/10.1016/J.MSEC.2016.09.041.
Lanier, A. P., Bulkow, L. R., & Ireland, B. (1989). Cancer in Alaskan Indians, Eskimos, and Aleuts, 1969-83: implications for etiology and control. *Public Health Reports, 104*(6), 658.
Larrañeta, E., Henry, M., Irwin, N. J., Trotter, J., Perminova, A. A., & Donnelly, R. F. (2018). Synthesis and characterization of hyaluronic acid hydrogels crosslinked using a solvent-free process for potential biomedical applications. *Carbohydrate Polymers, 181*, 1194–1205. https://doi.org/10.1016/J.CARBPOL.2017.12.015.
Laurienzo, P. (2010). Marine Polysaccharides in Pharmaceutical Applications: An Overview. *Marine Drugs, 8*(9), 2435–2465. https://doi.org/10.3390/MD8092435.
Leal, M. C., Madeira, C., Brandão, C. A., Puga, J., & Calado, R. (2012). Bioprospecting of marine invertebrates for new natural products — A chemical and zoogeographical perspective. *Molecules, 17*(8), 9842. https://doi.org/10.3390/MOLECULES17089842.
Lebouvier, N., Jullian, V., Desvignes, I., Maurel, S., Parenty, A., Dorin-Semblat, D., Doerig, C., Sauvain, M., & Laurent, D. (2009). Antiplasmodial activities of homogentisic acid derivative protein kinase inhibitors isolated from a vanuatu marine sponge *Pseudoceratina sp. Marine Drugs, 7*(4), 640–653. https://doi.org/10.3390/MD7040640.
Lee, C. R., Grodzinsky, A. J., & Spector, M. (2001). The effects of cross-linking of collagen-glycosaminoglycan scaffolds on compressive stiffness, chondrocyte-mediated contraction, proliferation and biosynthesis. *Biomaterials, 22*(23), 3145–3154. https://doi.org/10.1016/S0142-9612(01)00067-9.
Lee, H. K., Zhang, L., Smith, M. D., Walewska, A., Vellore, N. A., Baron, R., McIntosh, J. M., White, H. S., Olivera, B. M., & Bulaj, G. (2015). A marine analgesic peptide, Contulakin-G, and neurotensin are distinct agonists for neurotensin receptors: Uncovering structural determinants of desensitization properties. *Frontiers in Pharmacology, 6*(FEB), 11. https://doi.org/10.3389/FPHAR.2015.00011.
Lee, S. H., Athukorala, Y., Lee, J. S., & Jeon, Y. J. (2008). Simple separation of anticoagulant sulfated galactan from marine red algae. *Journal of Applied Phycology, 20*(6), 1053–1059. https://doi.org/10.1007/S10811-007-9306-0.
Lee, S., Yeon, S. L., Sang, H. J., Sam, S. K., & Kuk, H. S. (2003). Anti-oxidant activities of fucosterol from the marine algae Pelvetia siliquosa. *Archives of Pharmacal Research, 26*(9), 719–722. https://doi.org/10.1007/BF02976680.
Lee, Y. E., Kim, H., Seo, C., Park, T., Lee, K. Bin, Yoo, S. Y., Hong, S. C., Kim, J. T., & Lee, J. (2017). Marine polysaccharides: therapeutic efficacy and biomedical applications. *Archives of Pharmacal Research, 40*(9), 1006–1020. https://doi.org/10.1007/S12272-017-0958-2.
Lejars, M., Margaillan, A., & Bressy, C. (2012). Fouling release coatings: A nontoxic alternative to biocidal antifouling coatings. *Chemical Reviews, 112*(8), 4347–4390. https://doi.org/10.1021/CR200350V.

Leone, M., Priester, M. I., Romeijn, S., Nejadnik, M. R., Mönkäre, J., O'Mahony, C., Jiskoot, W., Kersten, G., & Bouwstra, J. A. (2019). Hyaluronan-based dissolving microneedles with high antigen content for intradermal vaccination: Formulation, physicochemical characterization and immunogenicity assessment. *European Journal of Pharmaceutics and Biopharmaceutics, 134*, 49–59. https://doi.org/10.1016/J.EJPB.2018.11.013.

Li, B., Liu, S., Xing, R., Li, K., Li, R., Qin, Y., Wang, X., Wei, Z., & Li, P. (2013). Degradation of sulfated polysaccharides from *Enteromorpha prolifera* and their antioxidant activities. *Carbohydrate Polymers, 92*(2), 1991–1996. https://doi.org/10.1016/J.CARBPOL.2012.11.088.

Li, B., Lu, F., Wei, X., & Zhao, R. (2008). Fucoidan: Structure and Bioactivity. *Molecules, 13*(8), 1671–1695. https://doi.org/10.3390/MOLECULES13081671.

Li, B., Wei, X. J., Sun, J. L., & Xu, S. Y. (2006). Structural investigation of a fucoidan containing a fucose-free core from the brown seaweed, *Hizikia fusiforme. Carbohydrate Research, 341*(9), 1135–1146. https://doi.org/10.1016/J.CARRES.2006.03.035.

Li, C., Haug, T., Moe, M. K., Styrvold, O. B., & Stensvåg, K. (2010). Centrocins: Isolation and characterization of novel dimeric antimicrobial peptides from the green sea urchin, *Strongylocentrotus droebachiensis. Developmental & Comparative Immunology, 34*(9), 959–968. https://doi.org/10.1016/J.DCI.2010.04.004.

Li, J. W. H., & Vederas, J. C. (2009). Drug discovery and natural products: end of an era or an endless frontier? *Science, 325*(5937), 161–165. https://doi.org/10.1126/SCIENCE.1168243.

Li, J., Xiang, H., Zhang, Q., & Miao, X. (2022). Polysaccharide-based transdermal drug delivery. *Pharmaceuticals, 15*(5), 602. https://doi.org/10.3390/PH15050602.

Liang, X., Luo, D., & Luesch, H. (2019). Advances in exploring the therapeutic potential of marine natural products. *Pharmacological Research, 147*, 104373. https://doi.org/10.1016/J.PHRS.2019.104373.

Lim, Y. S., Ok, Y. J., Hwang, S. Y., Kwak, J. Y., & Yoon, S. (2019). Marine collagen as a promising biomaterial for biomedical applications. *Marine Drugs, 17*(8), 467. https://doi.org/10.3390/MD17080467.

Linington, R. G., Clark, B. R., Trimble, E. E., Almanza, A., Ureña, L. D., Kyle, D. E., & Gerwick, W. H. (2009). Antimalarial peptides from marine cyanobacteria: Isolation and structural elucidation of gallinamide A. *Journal of Natural Products, 72*(1), 14–17. https://doi.org/10.1021/NP8003529.

Liu, S., Lau, C. S., Liang, K., Wen, F., & Teoh, S. H. (2022). Marine collagen scaffolds in tissue engineering. *Current Opinion in Biotechnology, 74*, 92–103. https://doi.org/10.1016/J.COPBIO.2021.10.011.

Liu, S., Wu, D., Quan, Y. S., Kamiyama, F., Kusamori, K., Katsumi, H., Sakane, T., & Yamamoto, A. (2016). Improvement of transdermal delivery of exendin-4 using novel tip-loaded microneedle arrays fabricated from hyaluronic acid. *Molecular Pharmaceutics, 13*(1), 272–279. https://doi.org/10.1021/ACS.MOLPHARMACEUT.5B00765.

Liu, Z., Jensen, P. R., & Fenical, W. (2003). A cyclic carbonate and related polyketides from a marine-derived fungus of the genus Phoma. *Phytochemistry, 64*(2), 571–574. https://doi.org/10.1016/S0031-9422(03)00272-3.

López-Abarrategui, C., Alba, A., Silva, O. N., Reyes-Acosta, O., Vasconcelos, I. M., Oliveira, J. T. A., Migliolo, L., Costa, M. P., Costa, C. R., Silva, M. R. R., Garay, H. E., Dias, S. C., Franco, O. L., & Otero-González, A. J. (2012). Functional characterization of a synthetic hydrophilic antifungal peptide derived from the marine snail *Cenchritis muricatus. Biochimie, 94*(4), 968–974. https://doi.org/10.1016/J.BIOCHI.2011.12.016.

López-Abarrategui, C., McBeth, C., Mandal, S. M., Sun, Z. J., Heffron, G., Alba-Menéndez, A., Migliolo, L., Reyes-Acosta, O., García-Villarino, M., Nolasco, D. O., Falcão, R., Cherobim, M. D., Dias, S. C., Brandt, W., Wessjohann, L., Starnbach, M., Franco, O. L., & Otero-González, A. J. (2015). Cm-p5: An antifungal hydrophilic peptide derived from the coastal mollusk *Cenchritis muricatus* (Gastropoda: Littorinidae). *FASEB Journal, 29*(8), 3315–3325. https://doi.org/10.1096/FJ.14-269860.

Lu, K. Y., Li, R., Hsu, C. H., Lin, C. W., Chou, S. C., Tsai, M. L., & Mi, F. L. (2017). Development of a new type of multifunctional fucoidan-based nanoparticles for anticancer drug delivery. *Carbohydrate Polymers, 165*, 410–420. https://doi.org/10.1016/J.CARBPOL.2017.02.065.

Luesch, H., Yoshida, W. Y., Moore, R. E., Paul, V. J., & Corbett, T. H. (2001). Total structure determination of apratoxin A, a potent novel cytotoxin from the marine cyanobacterium *Lyngbya majuscula*. *Journal of the American Chemical Society, 123*(23), 5418–5423. https://doi.org/10.1021/JA010453J.

Luescher-Mattli, M. (2005). Algae, a possible source for new drugs in the treatment of HIV and other viral diseases. *Current Medicinal Chemistry – Anti-Infective Agents, 2*(3), 219–225. https://doi.org/10.2174/1568012033483051.

Luo, Y., & Wang, Q. (2014). Recent development of chitosan-based polyelectrolyte complexes with natural polysaccharides for drug delivery. *International Journal of Biological Macromolecules, 64*, 353–367. https://doi.org/10.1016/J.IJBIOMAC.2013.12.017.

Luthuli, S., Wu, S., Cheng, Y., Zheng, X., Wu, M., & Tong, H. (2019). Therapeutic Effects of Fucoidan: A Review on Recent Studies. *Marine Drugs, 17*(9), 487. https://doi.org/10.3390/MD17090487.

Ma, J., & Li, H. (2018). The role of gut microbiota in atherosclerosis and hypertension. *Frontiers in Pharmacology, 9*(SEP), 1082. https://doi.org/10.3389/FPHAR.2018.01082.

Magne, F., Gotteland, M., Gauthier, L., Zazueta, A., Pesoa, S., Navarrete, P., & Balamurugan, R. (2020). The Firmicutes/Bacteroidetes Ratio: A relevant marker of gut dysbiosis in obese patients? *Nutrients, 12*(5), 1474. https://doi.org/10.3390/NU12051474.

Maier, M. E. (2009). Structural revisions of natural products by total synthesis. *Natural Product Reports, 26*(9), 1105–1124. https://doi.org/10.1039/B809658A.

Malve, H. (2016). Exploring the ocean for new drug developments: Marine pharmacology. *Journal of Pharmacy & Bioallied Sciences, 8*(2), 83. https://doi.org/10.4103/0975-7406.171700.

Martin, V. J. J., Piteral, D. J., Withers, S. T., Newman, J. D., & Keasling, J. D. (2003). Engineering a mevalonate pathway in *Escherichia coli* for production of terpenoids. *Nature Biotechnology, 21*(7), 796–802. https://doi.org/10.1038/nbt833.

Martin-Gallausiaux, C., Marinelli, L., Blottière, H. M., Larraufie, P., & Lapaque, N. (2021). SCFA: Mechanisms and functional importance in the gut. *Proceedings of the Nutrition Society, 80*(1), 37–49. https://doi.org/10.1017/S0029665120006916.

Martins, A., Vieira, H., Gaspar, H., & Santos, S. (2014). Marketed marine natural products in the pharmaceutical and cosmeceutical industries: Tips for success. *Marine Drugs, 12*(2), 1066. https://doi.org/10.3390/MD12021066.

Martucci, J. F., & Ruseckaite, R. A. (2009). Biodegradation of three-layer laminate films based on gelatin under indoor soil conditions. *Polymer Degradation and Stability, 94*(8), 1307–1313. https://doi.org/10.1016/J.POLYMDEGRADSTAB.2009.03.018.

Masso-Silva, J. A., & Diamond, G. (2014). Antimicrobial peptides from fish. *Pharmaceuticals (Basel, Switzerland), 7*(3), 265–310. https://doi.org/10.3390/PH7030265.

Matsuda, S., Adachi, K., Matsuo, Y., Nukina, M., & Shizuri, Y. (2009). Salinisporamycin, a novel metabolite from *Salinispora arenicora. The Journal of Antibiotics, 62*(9), 519–526. https://doi.org/10.1038/ja.2009.75.

Matsunaga, S., Fusetani, N., & Konosu, S. (1985). Bioactive marine metabolites, IV. Isolation and the Amino acid composition of discodermin a, an antimicrobial peptide, from the marine sponge discodermia kiiensis. *Journal of Natural Products, 48*(2), 236–241. https://doi.org/10.1021/NP50038A006.

Matsuo, K., Hirobe, S., Yokota, Y., Ayabe, Y., Seto, M., Quan, Y. S., Kamiyama, F., Tougan, T., Horii, T., Mukai, Y., Okada, N., & Nakagawa, S. (2012). Transcutaneous immunization using a dissolving microneedle array protects against tetanus, diphtheria, malaria, and influenza. *Journal of Controlled Release, 160*(3), 495–501. https://doi.org/10.1016/J.JCONREL.2012.04.001.

Matsuo, K., Yokota, Y., Zhai, Y., Quan, Y. S., Kamiyama, F., Mukai, Y., Okada, N., & Nakagawa, S. (2012). A low-invasive and effective transcutaneous immunization system using a novel dissolving microneedle array for soluble and particulate antigens. *Journal of Controlled Release, 161*(1), 10–17. https://doi.org/10.1016/J.JCONREL.2012.01.033.

Mayakrishnan, V., Kannappan, P., Abdullah, N., & Ahmed, A. B. A. (2013). Cardioprotective activity of polysaccharides derived from marine algae: An overview. *Trends in Food Science & Technology, 30*(2), 98–104. https://doi.org/10.1016/J.TIFS.2013.01.007.

Mayer, A. M. S., Rodríguez, A. D., Taglialatela-Scafati, O., & Fusetani, N. (2013). Marine pharmacology in 2009–2011: Marine compounds with antibacterial, antidiabetic, antifungal, anti-inflammatory, antiprotozoal, antituberculosis, and antiviral activities; affecting the immune and nervous systems, and other miscellaneous mechanisms of action. *Marine Drugs, 11*(7), 2510–2573. https://doi.org/10.3390/MD11072510.

Mazumder, S., Ghosal, P. K., Pujol, C. A., Carlucci, M. J., Damonte, E. B., & Ray, B. (2002). Isolation, chemical investigation and antiviral activity of polysaccharides from *Gracilaria corticata* (Gracilariaceae, Rhodophyta). *International Journal of Biological Macromolecules, 31*(1–3), 87–95. https://doi.org/10.1016/J.IJBIOMAC.2002.11.005

McHugh, D. J. (2003). A guide to the seaweed industry. Food and Agriculture organization of the United Nations Rome, 2003 https://www.fao.org/3/y4765e/y4765e00.htm (accessed on 7 June 2024).

Medeiros, V. P., Queiroz, K. C. S., Cardoso, M. L., Monteiro, G. R. G., Oliveira, F. W., Chavante, S. F., Guimaraes, L. A., Rocha, H. A. O., & Leite, E. L. (2008). Sulfated galactofucan from Lobophora variegata: Anticoagulant and anti-inflammatory properties. *Biochemistry (Moscow), 73*(9), 1018–1024. https://doi.org/10.1134/S0006297908090095.

Menchicchi, B., Hensel, A., & Goycoolea, M. F. (n.d.). Polysaccharides as Bacterial Antiadhesive agents and "smart" constituents for improved drug delivery systems against *Helicobacter pylori* Infection. *Curr Pharm Des., 21*(33), 4888–906. doi: 10.2174/1381612821666150820104028.

Mero, A., & Campisi, M. (2014). Hyaluronic acid bioconjugates for the delivery of bioactive molecules. *Polymers, 6*(2), 346–369. https://doi.org/10.3390/POLYM6020346.

Minamisako, K., & Kimura, S. (1989). Characterization of muscle collagen from fleshy prawn *Penaeus chinensis. Comparative Biochemistry and Physiology Part B: Comparative Biochemistry, 94*(2), 349–353. https://doi.org/10.1016/0305-0491(89)90354-4.

Misra, S., Heldin, P., Hascall, V. C., Karamanos, N. K., Skandalis, S. S., Markwald, R. R., & Ghatak, S. (2011). Hyaluronan–CD44 interactions as potential targets for cancer therapy. *The FEBS Journal, 278*(9), 1429–1443. https://doi.org/10.1111/J.1742-4658.2011.08071.X.

Moghadamtousi, S. Z., Nikzad, S., Kadir, H. A., Abubakar, S., & Zandi, K. (2015). Potential antiviral agents from marine fungi: An overview. *Marine Drugs, 13*(7), 4520–4538. https://doi.org/10.3390/MD13074520.

Molinski, T. F., Dalisay, D. S., Lievens, S. L., & Saludes, J. P. (2009). Drug development from marine natural products. *Nature Reviews Drug Discovery, 8*(1), 69–85. https://doi.org/10.1038/NRD2487.

Moniz, T., Costa Lima, S. A., & Reis, S. (2021). Marine polymeric microneedles for transdermal drug delivery. *Carbohydrate Polymers, 266*, 118098. https://doi.org/10.1016/J.CARBPOL.2021.118098.

Monks, A., Scudiero, D. A., Johnson, G. S., Paull, K. D., & Sausville, E. A. (1997). The NCI anti-cancer drug screen: A smart screen to identify effectors of novel targets. *Anticancer Drug Design, 12*(7), 533–541. https://pubmed.ncbi.nlm.nih.gov/9365500/ (accessed on 7 June 2024).

Monroig, Ó., Tocher, D. R., & Navarro, J. C. (2013). Biosynthesis of polyunsaturated fatty acids in marine invertebrates: Recent advances in molecular mechanisms. *Marine Drugs, 11*(10), 3998–4018. https://doi.org/10.3390/MD11103998.

Montaser, R., & Luesch, H. (2011). Marine natural products: A new wave of drugs? *Future Medicinal Chemistry, 3*(12), 1475. https://doi.org/10.4155/FMC.11.118.

Müller, W. E. G., & Müller, M. (2003). The origin of metazoan complexity: Porifera as integrated animals. *Integrative and Comparative Biology, 43*(1), 3–10. https://doi.org/10.1093/ICB/43.1.3.

Muralidharan, N., Jeya Shakila, R., Sukumar, D., & Jeyasekaran, G. (2013). Skin, bone and muscle collagen extraction from the trash fish, leather jacket (*Odonus niger*) and their characterization. *Journal of Food Science and Technology, 50*(6), 1106–1113. https://doi.org/10.1007/S13197-011-0440-Y/FIGURES/4.

Neupane, R., Boddu, S. H. S., Renukuntla, J., Babu, R. J., & Tiwari, A. K. (2020). Alternatives to biological skin in permeation studies: Current trends and possibilities. *Pharmaceutics, 12*(2), 152. https://doi.org/10.3390/PHARMACEUTICS12020152.

Ngo, D. H., Vo, T. S., Ngo, D. N., Wijesekara, I., & Kim, S. K. (2012). Biological activities and potential health benefits of bioactive peptides derived from marine organisms. *International Journal of Biological Macromolecules, 51*(4), 378–383. https://doi.org/10.1016/J.IJBIOMAC.2012.06.001.

Nicholson, T., Khademi, H., & Moghadasian, M. H. (2013). The role of marine n-3 fatty acids in improving cardiovascular health: A review. *Food & Function, 4*(3), 357–365. https://doi.org/10.1039/C2FO30235G.

Nicklas, M., Schatton, W., Heinemann, S., Hanke, T., & Kreuter, J. (2009). Preparation and characterization of marine sponge collagen nanoparticles and employment for the transdermal delivery of 17β-estradiol-hemihydrate. *Journal of Microencapsulation, 35*(9), 1035–1042. https://doi.org/10.1080/03639040902755213.

Nishimura, S., Arita, Y., Honda, M., Iwamoto, K., Matsuyama, A., Shirai, A., Kawasaki, H., Kakeya, H., Kobayashi, T., Matsunaga, S., & Yoshida, M. (2010). Marine antifungal theonellamides target 3β-hydroxysterol to activate Rho1 signaling. *Nature Chemical Biology, 6*(7), 519–526. https://doi.org/10.1038/nchembio.387.

Nobmann, E. D., Byers, T., Lanier, A. P., Hankin, J. H., & Jackson, M. Y. (1992). The diet of Alaska Native adults: 1987-1988. *The American Journal of Clinical Nutrition, 55*(5), 1024–1032. https://doi.org/10.1093/AJCN/55.5.1024.

Nordwig, A., Nowack, H., & Hieber-Rogall, E. (1973). Sea anemone collagen: Further evidence for the existence of only one α-chain type. *Journal of Molecular Evolution, 2*(2), 175–180. https://doi.org/10.1007/BF01653997.

Norton, T. R., Shibata, S., Kashiwagi, M., & Bentley, J. (1976). Isolation and characterization of the cardiotonic polypeptide Anthopleurin-A from the sea anemone *Anthopleura xanthogrammica. Journal of Pharmaceutical Sciences*, *65*(9), 1368–1374. https://doi.org/10.1002/JPS.2600650927.

Nutting, P. A., Freeman, W. L., Risser, D. R., Helgerson, S. D., Paisano, R., Hisnanick, J., Beaver, S. K., Peters, I., Carney, J. P., & Speers, M. A. (1993). Cancer incidence among American Indians and Alaska Natives, 1980 through 1987. *American Journal of Public Health*, *83*(11), 1589–1598. https://doi.org/10.2105/AJPH.83.11.1589.

Okazaki, T., Kitahara, T., & Okami, Y. (1975). Studies on marine microorganisms. IV. A new antibiotic SS-228 Y produced by *Chainia* isolated from shallow sea mud. *The Journal of Antibiotics*, *28*(3), 176–184. https://doi.org/10.7164/ANTIBIOTICS.28.176.

Ortholand, J. Y., & Ganesan, A. (2004). Natural products and combinatorial chemistry: Back to the future. *Current Opinion in Chemical Biology*, *8*(3), 271–280. https://doi.org/10.1016/J.CBPA.2004.04.011.

Otero-González, A. J., Simas Magalhães, B., Garcia-Villarino, M., Ló pez-Abarrategui, C., Amaro Sousa, D., Campos Dias, S., & Luiz Franco, O. (2010). Antimicrobial peptides from marine invertebrates as a new frontier for microbial infection control. *The FASEB Journal*, *24*(5), 1320–1334. https://doi.org/10.1096/FJ.09-143388.

Paiva, A. A. D. O., Castro, A. J. G., Nascimento, M. S., Will, L. S. E. P., Santos, N. D., Araújo, R. M., Xavier, C. A. C., Rocha, F. A., & Leite, E. L. (2011). Antioxidant and anti-inflammatory effect of polysaccharides from *Lobophora variegata* on zymosan-induced arthritis in rats. *International Immunopharmacology*, *11*(9), 1241–1250. https://doi.org/10.1016/J.INTIMP.2011.04.003.

Palem, J. R., Bedadala, G. R., El Sayed, K. A., & Hsia, S. C. V. (2011). Manzamine A as a novel inhibitor of herpes simplex virus type-1 replication in cultured corneal cells. *Planta Medica*, *77*(1), 46–51. https://doi.org/10.1055/S-0030-1250093/BIB.

Pan, M., Cederbaum, A. I., Zhang, Y. L., Ginsberg, H. N., Williams, K. J., & Fisher, E. A. (2004). Lipid peroxidation and oxidant stress regulate hepatic apolipoprotein B degradation and VLDL production. *The Journal of Clinical Investigation*, *113*(9), 1277–1287. https://doi.org/10.1172/JCI19197.

Pangestuti, R., Ryu, B., Himaya, S., & Kim, S. K. (2013). Optimization of hydrolysis conditions, isolation, and identification of neuroprotective peptides derived from seahorse *Hippocampus trimaculatus*. *Amino Acids*, *45*(2), 369–381. https://doi.org/10.1007/S00726-013-1510-4/FIGURES/8.

Pangestuti, R., Vo, T. S., Ngo, D. H., & Kim, S. K. (2013). Fucoxanthin ameliorates inflammation and oxidative responses in microglia. *Journal of Agricultural and Food Chemistry*, *61*(16), 3876–3883. https://doi.org/10.1021/JF400015K/ASSET/IMAGES/MEDIUM/JF-2013-00015K_0009.GIF.

Papakonstantinou, E., Roth, M., & Karakiulakis, G. (2012). Hyaluronic acid: A key molecule in skin aging. *Dermato-Endocrinology*, *4*(3), 253. https://doi.org/10.4161/DERM.21923.

Papon, N., Copp, B. R., & Courdavault, V. (2022). Marine drugs: Biology, pipelines, current and future prospects for production. *Biotechnology Advances*, *54*, 107871. https://doi.org/10.1016/J.BIOTECHADV.2021.107871.

Parisi, J. R., Fernandes, K. R., Avanzi, I. R., Dorileo, B. P., Santana, A. F., Andrade, A. L., Gabbai-Armelin, P. R., Fortulan, C. A., Trichês, E. S., Granito, R. N., & Renno, A. C. M. (2019). Incorporation of collagen from marine sponges (spongin) into hydroxyapatite samples: Characterization and in vitro biological evaluation. *Marine Biotechnology*, *21*(1), 30–37. https://doi.org/10.1007/S10126-018-9855-Z/FIGURES/6.

Pawar, A., Botolin, D., Mangelsdorf, D. J., & Jump, D. B. (2003). The role of liver X receptor-α in the fatty acid regulation of hepatic gene expression. *Journal of Biological Chemistry, 278*(42), 40736–40743. https://doi.org/10.1074/jbc.M307973200.

Pégorier, J. P., Le May, C., & Girard, J. (2004). Control of gene expression by fatty acids. *The Journal of Nutrition, 134*(9), 2444S–2449S. https://doi.org/10.1093/JN/134.9.2444S.

Pelay-Gimeno, M., Tulla-Puche, J., & Albericio, F. (2013). "Head-to-Side-Chain" cyclodepsipeptides of marine origin. *Marine Drugs, 11*(5), 1693. https://doi.org/10.3390/MD11051693.

Penesyan, A., Kjelleberg, S., & Egan, S. (2010). Development of novel drugs from marine surface associated microorganisms. *Marine Drugs, 8*(3), 438–459. https://doi.org/10.3390/MD8030438.

Pengzhan, Y., Ning, L., Xiguang, L., Gefei, Z., Quanbin, Z., & Pengcheng, L. (2003). Antihyperlipidemic effects of different molecular weight sulfated polysaccharides from *Ulva pertusa* (Chlorophyta). *Pharmacological Research, 48*(6), 543–549. https://doi.org/10.1016/S1043-6618(03)00215-9.

Pereira, F., & Aires-de-Sousa, J. (2018). Computational methodologies in the exploration of marine natural product leads. *Marine Drugs, 16*(7), 236. https://doi.org/10.3390/MD16070236.

Perez Recalde, M., Canelón, D. J., Compagnone, R. S., Matulewicz, M. C., Cerezo, A. S., & Ciancia, M. (2016). Carrageenan and agaran structures from the red seaweed *Gymnogongrus tenuis*. *Carbohydrate Polymers, 136*, 1370–1378. https://doi.org/10.1016/J.CARBPOL.2015.10.007.

Pettit, G. R., Knight, J. C., Herald, D. L., Pettit, R. K., Hogan, F., Mukku, V. J. R. V., Hamblin, J. S., Dodson, M. J., & Chapuis, J. C. (2009). Antineoplastic agents. 570. Isolation and structure elucidation of bacillistatins 1 and 2 from a marine *Bacillus silvestris*. *Journal of Natural Products, 72*(3), 366–371. https://doi.org/10.1021/NP800603U/ASSET/IMAGES/MEDIUM/NP-2008-00603U_0005.GIF.

Pimentel-Elardo, S. M., Kozytska, S., Bugni, T. S., Ireland, C. M., Moll, H., & Hentschel, U. (2010). Anti-parasitic compounds from *Streptomyces sp.* strains isolated from Mediterranean sponges. *Marine Drugs, 8*(2), 373–380. https://doi.org/10.3390/MD8020373.

Plaza, A., Bifulco, G., Masullo, M., Lloyd, J. R., Keffer, J. L., Colin, P. L., Hooper, J. N. A., Bell, L. J., & Bewley, C. A. (2010). Mutremdamide A and koshikamides C-H, peptide inhibitors of HIV-1 entry from different *Theonella* species. *Journal of Organic Chemistry, 75*(13), 4344–4355. https://doi.org/10.1021/JO100076G/SUPPL_FILE/JO100076G_SI_001.PDF.

Plaza, A., Keffer, J. L., Bifulco, G., Lloyd, J. R., & Bewley, C. A. (2010). Chrysophaentins A-H, antibacterial bisdiarylbutene macrocycles that inhibit the bacterial cell division protein FtsZ. *Journal of the American Chemical Society, 132*(26), 9069–9077. https://doi.org/10.1021/JA102100H/SUPPL_FILE/JA102100H_SI_002.PDB.

Pohnert, G. (2004). Chemical defense strategies of marine organisms. In T. Kanatani (Ed.), *The Chemistry of Pheromones and Other Semiochemicals I* (1st ed., Vol. 239, pp. 179–219). Springer, Berlin, Heidelberg. https://doi.org/10.1007/B95453.

Pomin, V. H. (2012). Fucanomics and galactanomics: Marine distribution, medicinal impact, conceptions, and challenges. *Marine Drugs, 10*(4), 793–811. https://doi.org/10.3390/MD10040793.

Prajapati, V. D., Maheriya, P. M., Jani, G. K., & Solanki, H. K. (2014). RETRACTED: Carrageenan: A natural seaweed polysaccharide and its applications. *Carbohydrate Polymers, 105*(1), 97–112. https://doi.org/10.1016/J.CARBPOL.2014.01.067.

Price, P. T., Nelson, C. M., & Clarke, S. D. (2000). Omega-3 polyunsaturated fatty acid regulation of gene expression. *Current Opinion in Lipidology, 11*(1), 3–7. https://doi.org/10.1097/00041433-200002000-00002.

Pruksakorn, P., Arai, M., Kotoku, N., Vilchze, C., Baughn, A. D., Moodley, P., Jacobs, W. R., & Kobayashi, M. (2010). Trichoderins, novel aminolipopeptides from a marine sponge-derived *Trichoderma sp.*, are active against dormant mycobacteria. *Bioorganic & Medicinal Chemistry Letters*, *20*(12), 3658–3663. https://doi.org/10.1016/J.BMCL.2010.04.100.

Qi, H., Zhang, Q., Zhao, T., Chen, R., Zhang, H., Niu, X., & Li, Z. (2005). Antioxidant activity of different sulfate content derivatives of polysaccharide extracted from *Ulva pertusa* (Chlorophyta) in vitro. *International Journal of Biological Macromolecules*, *37*(4), 195–199. https://doi.org/10.1016/J.IJBIOMAC.2005.10.008.

Qi, H., Zhao, T., Zhang, Q., Li, Z., Zhao, Z., & Xing, R. (2006). Antioxidant activity of different molecular weight sulfated polysaccharides from *Ulva pertusa* Kjellm (Chlorophyta). *Journal of Applied Phycology*, *17*(6), 527–534. https://doi.org/10.1007/S10811-005-9003-9.

Qian, Z. J., Je, J. Y., & Kim, S. K. (2007). Antihypertensive effect of angiotensin I converting enzyme-inhibitory peptide from hydrolysates of bigeye tuna dark muscle, *Thunnus obesus*. *Journal of Agricultural and Food Chemistry*, *55*(21), 8398–8403. https://doi.org/10.1021/JF0710635.

Qian, Z. J., Jung, W. K., Byun, H. G., & Kim, S. K. (2008). Protective effect of an antioxidative peptide purified from gastrointestinal digests of oyster, *Crassostrea gigas*, against free radical induced DNA damage. *Bioresource Technology*, *99*(9), 3365–3371. https://doi.org/10.1016/J.BIORTECH.2007.08.018.

Qin, Y., Li, P., & Guo, Z. (2020). Cationic chitosan derivatives as potential antifungals: A review of structural optimization and applications. *Carbohydrate Polymers*, *236*, 116002. https://doi.org/10.1016/J.CARBPOL.2020.116002.

Qiu, Y., Li, C., Zhang, S., Yang, G., He, M., & Gao, Y. (2016). Systemic delivery of artemether by dissolving microneedles. *International Journal of Pharmaceutics*, *508*(1–2), 1–9. https://doi.org/10.1016/J.IJPHARM.2016.05.006.

Radjasa, O. K., Vaske, Y. M., Navarro, G., Vervoort, H. C., Tenney, K., Linington, R. G., & Crews, P. (2011). Highlights of marine invertebrate-derived biosynthetic products: Their biomedical potential and possible production by microbial associants. *Bioorganic & Medicinal Chemistry*, *19*(22), 6658. https://doi.org/10.1016/J.BMC.2011.07.017.

Rao, H. B. R., Sathivel, A., & Devaki, D. T. (2004). Antihepatotoxic nature of *Ulva reticulata* (Chlorophyceae) on acetaminophen-induced hepatotoxicity in experimental rats. *Journal of Medicinal Food*, *7*(4), 495–499. https://doi.org/10.1089/JMF.2004.7.495.

Raposo, D. J., Filomena, M., Morais, D., Bernardo, A. M., & De Morais, R. M. S. C. (2015). Marine polysaccharides from algae with potential biomedical applications. *Marine Drugs*, *13*(5), 2967–3028. https://doi.org/10.3390/MD13052967.

Reed, R. K., Lilja, K., & Laurent, T. C. (1988). Hyaluronan in the rat with special reference to the skin. *Acta Physiologica Scandinavica*, *134*(3), 405–411. https://doi.org/10.1111/J.1748-1716.1988.TB08508.X.

Reisky, L., Préchoux, A., Zühlke, M. K., Bäumgen, M., Robb, C. S., Gerlach, N., Roret, T., Stanetty, C., Larocque, R., Michel, G., Song, T., Markert, S., Unfried, F., Mihovilovic, M. D., Trautwein-Schult, A., Becher, D., Schweder, T., Bornscheuer, U. T., & Hehemann, J. H. (2019). A marine bacterial enzymatic cascade degrades the algal polysaccharide ulvan. *Nature Chemical Biology*, *15*(8), 803–812. https://doi.org/10.1038/s41589-019-0311-9.

Rest, M. Van Der, & Garrone, R. (1991). Collagen family of proteins. *The FASEB Journal*, *5*(13), 2814–2823. https://doi.org/10.1096/FASEBJ.5.13.1916105.

Richards, J. T., Kern, E. R., Glasgow, L. A., Overall, J. C., Deign, E. F., & Hatch, M. T. (1978). Antiviral activity of extracts from marine algae. *Antimicrobial Agents and Chemotherapy*, *14*(1), 24–30. https://doi.org/10.1128/AAC.14.1.24.

Rissanen, T. H., Voutilainen, S., Virtanen, J. K., Venho, B., Vanharanta, M., Mursu, J., & Salonen, J. T. (2003). Low intake of fruits, berries and vegetables is associated with excess mortality in men: The Kuopio Ischaemic Heart Disease Risk Factor (KIHD) Study. *The Journal of Nutrition, 133*(1), 199–204. https://doi.org/10.1093/JN/133.1.199.

Roemer, T., Davies, J., Giaever, G., & Nislow, C. (2011). Bugs, drugs and chemical genomics. *Nature Chemical Biology, 8*(1), 46–56. https://doi.org/10.1038/nchembio.744.

Rose, D. P., & Connolly, J. M. (1991). Effects of fatty acids and eicosanoid synthesis inhibitors on the growth of two human prostate cancer cell lines. *The Prostate, 18*(3), 243–254. https://doi.org/10.1002/PROS.2990180306.

Rottensteiner, U., Sarker, B., Heusinger, D., Dafinova, D., Rath, S. N., Beier, J. P., Kneser, U., Horch, R. E., Detsch, R., Boccaccini, A. R., & Arkudas, A. (2014). In vitro and in vivo biocompatibility of alginate dialdehyde/gelatin hydrogels with and without nanoscaled bioactive glass for bone tissue engineering applications. *Materials, 7*(3), 1957–1974. https://doi.org/10.3390/MA7031957.

Rowley, D. C., Kelly, S., Kauffman, C. A., Jensen, P. R., & Fenical, W. (2003). Haloviirs A-E, new antiviral agents from a marine-derived fungus of the genus *Scytalidium. Bioorganic and Medicinal Chemistry, 11*(19), 4263–4274. https://doi.org/10.1016/S0968-0896(03)00395-X.

Ruggieri, G. D. (1976). Drugs from the sea. *Science, 194(4264),* 491–497. https://doi.org/10.1126/SCIENCE.9691/ASSET/0FB69C6E-9194-4CEF-B593-8701CBB63BAC/ASSETS/SCIENCE.9691.FP.PNG.

Ryan, J. T., Ross, R. P., Bolton, D., Fitzgerald, G. F., & Stanton, C. (2011). Bioactive peptides from muscle sources: Meat and fish. *Nutrients, 3*(9), 765–791. https://doi.org/10.3390/NU3090765.

Sable, R., Parajuli, P., & Jois, S. (2017). Peptides, peptidomimetics, and polypeptides from marine sources: A wealth of natural sources for pharmaceutical applications. *Marine Drugs, 15*(4), 124. https://doi.org/10.3390/MD15040124.

Salvatore, L., Gallo, N., Natali, M. L., Campa, L., Lunetti, P., Madaghiele, M., Blasi, F. S., Corallo, A., Capobianco, L., & Sannino, A. (2020). Marine collagen and its derivatives: Versatile and sustainable bio-resources for healthcare. *Materials Science and Engineering: C, 113*, 110963. https://doi.org/10.1016/J.MSEC.2020.110963.

Saravanan, P., Davidson, N. C., Schmidt, E. B., & Calder, P. C. (2010). Cardiovascular effects of marine omega-3 fatty acids. *The Lancet, 376*(9740), 540–550. https://doi.org/10.1016/S0140-6736(10)60445-X.

Satheesh, S., Ba-Akdah, M. A., & Al-Sofyani, A. A. (2016). Natural antifouling compound production by microbes associated with marine macroorganisms – A review. *Electronic Journal of Biotechnology, 21*, 26–35. https://doi.org/10.1016/J.EJBT.2016.02.002.

Scheppach, W. (1994). Effects of short chain fatty acids on gut morphology and function. *Gut, 35*(1 Suppl), S35–S38. https://doi.org/10.1136/GUT.35.1_SUPPL.S35.

Schmidt, E. B. (1997). n–3 fatty acids and the risk of coronary heart disease. *Danish Medical Bulletin, 44*(1), 1–22. https://pubmed.ncbi.nlm.nih.gov/9062761/ (accessed on 7 June 2024).

Schmidt, E. B., Christensen, J. H., Aardestrup, I., Madsen, T., Riahi, S., Hansen, V. E., & Skou, H. A. (2001). Marine n-3 fatty acids: Basic features and background. *Lipids, 36*(1), S65–S68. https://doi.org/10.1007/S11745-001-0684-X.

Schroeder, F. C., Gibson, D. M., Churchill, A. C. L., Sojikul, P., Wursthorn, E. J., Krasnoff, S. B., & Clardy, J. (2007). Differential analysis of 2D NMR spectra: New natural products from a pilot-scale fungal extract library. *Angewandte Chemie International Edition, 46*(6), 901–904. https://doi.org/10.1002/ANIE.200603821.

Schultz, M. P., Bendick, J. A., Holm, E. R., & Hertel, W. M. (2010). Economic impact of biofouling on a naval surface ship. *Biofouling*, *27*(1), 87–98. https://doi.org/10.1080/08927014.2010.542809.

Sen, A. K., Das, A. K., Banerji, N., Siddhanta, A. K., Mody, K. H., Ramavat, B. K., Chauhan, V. D., Vedasiromoni, J. R., & Ganguly, D. K. (1994). A new sulfated polysaccharide with potent blood anti-coagulant activity from the red seaweed *Grateloupia indica*. *International Journal of Biological Macromolecules*, *16*(5), 279–280. https://doi.org/10.1016/0141-8130(94)90034-5.

Senadheera, T. R. L., Dave, D., & Shahidi, F. (2020). Sea cucumber derived type I collagen: A comprehensive review. *Marine Drugs*, *18*(9), 471. https://doi.org/10.3390/MD18090471.

Sennett, S. H., McCarthy, P. J., Wright, A. E., & Pomponi, S. A. (2002). Natural products from marine invertebrates: The Harbor Branch Oceanographic Institution experience. *Pharmaceutical News*, *9*(6), 483–488.

Senni, K., Pereira, J., Gueniche, F., Delbarre-Ladrat, C., Sinquin, C., Ratiskol, J., Godeau, G., Fischer, A. M., Helley, D., & Colliec-Jouault, S. (2011). Marine polysaccharides: A source of bioactive molecules for cell therapy and tissue engineering. *Marine Drugs*, *9*(9), 1664–1681. https://doi.org/10.3390/MD9091664.

Shamay, Y., Elkabets, M., Li, H., Shah, J., Brook, S., Wang, F., Adler, K., Baut, E., Scaltriti, M., Jena, P. V., Gardner, E. E., Poirier, J. T., Rudin, C. M., Baselga, J., Haimovitz-Friedman, A., & Heller, D. A. (2016). P-selectin is a nanotherapeutic delivery target in the tumor microenvironment. *Science Translational Medicine*, *8*(345), 345–374. https://doi.org/10.1126/SCITRANSLMED.AAF7374/SUPPL_FILE/8-345RA87_SM.PDF.

Sharma, M. (2019a). Transdermal and intravenous nano drug delivery systems: Present and future. In *Applications of Targeted Nano Drugs and Delivery Systems: Nanoscience and Nanotechnology in Drug Delivery* (pp. 499–550). Elsevier. https://doi.org/10.1016/B978-0-12-814029-1.00018-1.

Shen, B. (2015). A new golden age of natural products drug discovery. *Cell*, *163*(6), 1297–1300. https://doi.org/10.1016/J.CELL.2015.11.031.

Shilabin, A. G., & Hamann, M. T. (2011). In vitro and in vivo evaluation of select kahalalide F analogs with antitumor and antifungal activities. *Bioorganic & Medicinal Chemistry*, *19*(22), 6628–6632. https://doi.org/10.1016/J.BMC.2011.06.050.

Silva, T. H., Moreira-Silva, J., Marques, A. L. P., Domingues, A., Bayon, Y., & Reis, R. L. (2014). Marine origin collagens and its potential applications. *Marine Drugs*, *12*(12), 5881. https://doi.org/10.3390/MD12125881.

Simpson, B. K., Nayeri, G., Yaylayan, V., & Ashie, I. N. A. (1998). Enzymatic hydrolysis of shrimp meat. *Food Chemistry*, *61*(1–2), 131–138. https://doi.org/10.1016/S0308-8146(97)00131-3.

Singh, P., Carrier, A., Chen, Y., Lin, S., Wang, J., Cui, S., & Zhang, X. (2019). Polymeric microneedles for controlled transdermal drug delivery. *Journal of Controlled Release*, *315*, 97–113. https://doi.org/10.1016/J.JCONREL.2019.10.022.

So, P. B. T., Rubio, P., Lirio, S., Macabeo, A. P., Huang, H. Y., Corpuz, M. J. A. T., & Villaflores, O. B. (2016). In vitro angiotensin I converting enzyme inhibition by a peptide isolated from *Chiropsalmus quadrigatus* Haeckel (box jellyfish) venom hydrolysate. *Toxicon*, *119*, 77–83. https://doi.org/10.1016/J.TOXICON.2016.04.050.

Song, E. H., Shang, J., & Ratner, D. M. (2012). Polysaccharides. In *Polymer Science: A Comprehensive Reference, 10 Volume Set* (Vol. 9, pp. 137–155). https://doi.org/10.1016/B978-0-444-53349-4.00246-6.

Song, E., Yeon Kim, S., Chun, T., Byun, H. J., & Lee, Y. M. (2006). Collagen scaffolds derived from a marine source and their biocompatibility. *Biomaterials*, *27*(15), 2951–2961. https://doi.org/10.1016/J.BIOMATERIALS.2006.01.015.

Sorushanova, A., Delgado, L. M., Wu, Z., Shologu, N., Kshirsagar, A., Raghunath, R., Mullen, A. M., Bayon, Y., Pandit, A., Raghunath, M., & Zeugolis, D. I. (2019). The collagen suprafamily: From biosynthesis to advanced biomaterial development. *Advanced Materials*, *31*(1), 1801651. https://doi.org/10.1002/ADMA.201801651.

Spraker, J. E., Luu, G. T., & Sanchez, L. M. (2020). Imaging mass spectrometry for natural products discovery: A review of ionization methods. *Natural Product Reports*, *37*(2), 150–162. https://doi.org/10.1039/C9NP00038K.

Sridhar, K., Inbaraj, B. S., & Chen, B. H. (2021). Recent developments on production, purification and biological activity of marine peptides. *Food Research International*, *147*, 110468. https://doi.org/10.1016/J.FOODRES.2021.110468.

Stover, D. A., & Verrelli, B. C. (2011). Comparative vertebrate evolutionary analyses of type I collagen: Potential of COL1a1 gene structure and intron variation for common bone-related diseases. *Molecular Biology and Evolution*, *28*(1), 533–542. https://doi.org/10.1093/MOLBEV/MSQ221.

Subhan, F., Ikram, M., Shehzad, A., & Ghafoor, A. (2015). Marine collagen: An emerging player in biomedical applications. *Journal of Food Science and Technology*, *52*(8), 4703–4707. https://doi.org/10.1007/S13197-014-1652-8/TABLES/1.

Sudek, S., Lopanik, N. B., Waggoner, L. E., Hildebrand, M., Anderson, C., Liu, H., Patel, A., Sherman, D. H., & Haygood, M. G. (2007). Identification of the putative bryostatin polyketide synthase gene cluster from "Candidatus Endobugula sertula," the uncultivated microbial symbiont of the marine bryozoan *Bugula neritina*. *Journal of Natural Products*, *70*(1), 67–74. https://doi.org/10.1021/NP060361D.

Summer, K., Browne, J., Liu, L., & Benkendorff, K. (2020). Molluscan compounds provide drug leads for the treatment and prevention of respiratory disease. *Marine Drugs*, *18*(11), 570. https://doi.org/10.3390/MD18110570.

Sun, X., Duan, M., Liu, Y., Luo, T., Ma, N., Song, S., & Ai, C. (2018). The beneficial effects of *Gracilaria lemaneiformis* polysaccharides on obesity and the gut microbiota in high fat diet-fed mice. *Journal of Functional Foods*, *46*, 48–56. https://doi.org/10.1016/J.JFF.2018.04.041.

Suyama, T. L., Gerwick, W. H., & McPhail, K. L. (2011). Survey of marine natural product structure revisions: A synergy of spectroscopy and chemical synthesis. *Bioorganic & Medicinal Chemistry*, *19*(22), 6675. https://doi.org/10.1016/J.BMC.2011.06.011.

Szpak, P. (2011). Fish bone chemistry and ultrastructure: Implications for taphonomy and stable isotope analysis. *Journal of Archaeological Science*, *38*(12), 3358–3372. https://doi.org/10.1016/J.JAS.2011.07.022.

Tadesse, M., Tabudravu, J. N., Jaspars, M., Strøm, M. B., Hansen, E., Andersen, J. H., Kristiansen, P. E., & Haug, T. (2011). The antibacterial ent-eusynstyelamide B and eusynstyelamides D, E, and F from the arctic bryozoan *Tegella cf. spitzbergensis*. *Journal of Natural Products*, *74*(4), 837–841. https://doi.org/10.1021/NP100499C/SUPPL_FILE/NP100499C_SI_001.PDF.

Taghipour-Sabzevar, V., Sharifi, T., & Moghaddam, M. M. (2019). Polymeric nanoparticles as carrier for targeted and controlled delivery of anticancer agents. *Therapeutic Delivery*, *10*(8), 527–550. https://doi.org/10.4155/TDE-2019-0044.

Taglialatela-Scafati, O., Fattorusso, E., Romano, A., Scala, F., Barone, V., Cimino, P., Stendardo, E., Catalanotti, B., Persico, M., & Fattorusso, C. (2010). Insight into the mechanism of action of plakortins, simple 1,2-dioxane antimalarials. *Organic & Biomolecular Chemistry*, *8*(4), 846–856. https://doi.org/10.1039/B918600J.

Takano, R., Hayashi, K., Hara, S., & Hirase, S. (1995). Funoran from the red seaweed, Gloiopeltis complanata: polysaccharides with sulphated agarose structure and their precursor structure. *Carbohydrate Polymers*, *27*(4), 305–311. https://doi.org/10.1016/0144-8617(95)00070-4.

Takano, R., Iwane-Sakata, H., Hayashi, K., Hara, S., & Hirase, S. (1998). Concurrence of agaroid and carrageenan chains in funoran from the red seaweed *Gloiopeltis furcata* post. et ruprecht (cryptonemiales, rhodophyta). *Carbohydrate Polymers*, *35*(1–2), 81–87. https://doi.org/10.1016/S0144-8617(97)00230-0.

Tan, H., Chu, C. R., Payne, K. A., & Marra, K. G. (2009). Injectable in situ forming biodegradable chitosan–hyaluronic acid-based hydrogels for cartilage tissue engineering. *Biomaterials*, *30*(13), 2499–2506. https://doi.org/10.1016/J.BIOMATERIALS.2008.12.080.

Teasdale, M. E., Liu, J., Wallace, J., Akhlaghi, F., & Rowley, D. C. (2009). Secondary metabolites produced by the marine bacterium *Halobacillus salinus* that inhibit quorum sensing-controlled phenotypes in gram-negative bacteria. *Applied and Environmental Microbiology*, *75*(3), 567–572. https://doi.org/10.1128/AEM.00632-08.

Tello, E., Castellanos, L., Arevalo-Ferro, C., & Duque, C. (2009). Cembranoid diterpenes from the Caribbean Sea whip *Eunicea knighti*. *Journal of Natural Products*, *72*(9), 1595–1602. https://doi.org/10.1021/NP9002492.

Teng, Z., Qian, L., & Zhou, Y. (2013). Hypolipidemic activity of the polysaccharides from *Enteromorpha prolifera*. *International Journal of Biological Macromolecules*, *62*, 254–256. https://doi.org/10.1016/J.IJBIOMAC.2013.09.010.

Terasawa, M., Hiramoto, K., Uchida, R., & Suzuki, K. (2022). Anti-inflammatory activity of orally administered *Monostroma nitidum* rhamnan sulfate against lipopolysaccharide-induced damage to mouse organs and vascular endothelium. *Marine Drugs*, *20*(2), 121. https://doi.org/10.3390/MD20020121.

Thakur, N. L., Jain, R., Natalio, F., Hamer, B., Thakur, A. N., & Müller, W. E. G. (2008). Marine molecular biology: An emerging field of biological sciences. *Biotechnology Advances*, *26*(3), 233–245. https://doi.org/10.1016/J.BIOTECHADV.2008.01.001.

Thomes, P., Rajendran, M., Pasanban, B., & Rengasamy, R. (2010). Cardioprotective activity of *Cladosiphon okamuranus* fucoidan against isoproterenol induced myocardial infarction in rats. *Phytomedicine*, *18*(1), 52–57. https://doi.org/10.1016/J.PHYMED.2010.06.006.

Toyama, M. H., Toyama, D. O., Torres, V. M., Pontes, G. C., Farias, W. R. L., Melo, F. R., Oliveira, S. C. B., Fagundes, F. H. R., Diz Filho, E. B. S., & Cavada, B. S. (2010). Effects of low molecular weight sulfated galactan fragments from Botryocladia occidentalis on the pharmacological and enzymatic activity of sPLA2 from *Crotalus durissus* cascavella. *Protein Journal*, *29*(8), 567–571. https://doi.org/10.1007/S10930-010-9294-9.

Tsuda, M., Mugishima, T., Komatsu, K., Sone, T., Tanaka, M., Mikami, Y., & Kobayashi, J. (2003). Modiolides A and B, two new 10-membered macrolides from a marine-derived fungus. *Journal of Natural Products*, *66*(3), 412–415. https://doi.org/10.1021/NP0203943.

Tsuge, K., Okabe, M., Yoshimura, T., Sumi, T., Tachibana, H., & Yamada, K. (2007). Dietary effects of porphyran from *Porphyra yezoensis* on growth and lipid metabolism of Sprague-Dawley rats. *Food Science and Technology Research*, *10*(2), 147–151. https://doi.org/10.3136/FSTR.10.147.

Tzellos, T. G., Klagas, I., Vahtsevanos, K., Triaridis, S., Printza, A., Kyrgidis, A., Karakiulakis, G., Zouboulis, C. C., & Papakonstantinou, E. (2009). Extrinsic ageing in the human skin is associated with alterations in the expression of hyaluronic acid and its metabolizing enzymes. *Experimental Dermatology*, *18*(12), 1028–1035. https://doi.org/10.1111/J.1600-0625.2009.00889.X.

Tziveleka, L. A., Ioannou, E., Tsiourvas, D., Berillis, P., Foufa, E., & Roussis, V. (2017). Collagen from the marine sponges *Axinella cannabina* and *Suberites carnosus*: Isolation and morphological, biochemical, and biophysical characterization. *Marine Drugs*, *15*(6), 152. https://doi.org/10.3390/MD15060152.

Uddin, M. J., Hassan, J., & Douroumis, D. (2022). Thermal inkjet printing: Prospects and applications in the development of medicine. *Technologies*, *10*(5), 108. https://doi.org/10.3390/technologies10050108.

Ueoka, R., Nakao, Y., Kawatsu, S., Yaegashi, J., Matsumoto, Y., Matsunaga, S., Furihata, K., Van Soest, R. W. M., & Fusetani, N. (2009). Gracilioethers A-C, antimalarial metabolites from the marine sponge *Agelas gracilis*. *Journal of Organic Chemistry*, *74*(11), 4203–4207. https://doi.org/10.1021/JO900380F.

Um, S., Pyee, Y., Kim, E. H., Lee, S. K., Shin, J., & Oh, D. C. (2013). Thalassospiramide G, a new γ-amino-acid-bearing peptide from the marine bacterium Thalassospira sp. *Marine Drugs*, *11*(3), 611–622. https://doi.org/10.3390/MD11030611.

Usov, A. I. (2011). Polysaccharides of the red algae. *Advances in Carbohydrate Chemistry and Biochemistry*, *65*, 115–217. https://doi.org/10.1016/B978-0-12-385520-6.00004-2.

Usui, T., Asari, K., & Mizuno, T. (1980). Isolation of highly purified "fucoidan" from Eisenia bicyclis and its anticoagulant and antitumor activities. *Agricultural and Biological Chemistry*, *44*(8), 1965–1966. https://doi.org/10.1271/BBB1961.44.1965.

Varaprasad, K., Jayaramudu, T., Kanikireddy, V., Toro, C., & Sadiku, E. R. (2020). Alginate-based composite materials for wound dressing application: A mini review. *Carbohydrate Polymers*, *236*, 116025. https://doi.org/10.1016/J.CARBPOL.2020.116025.

Vaughan, V. C., Hassing, M. R., & Lewandowski, P. A. (2013). Marine polyunsaturated fatty acids and cancer therapy. *British Journal of Cancer*, *108*(3), 486–492. https://doi.org/10.1038/bjc.2012.586.

Venkatesan, J., Anil, S., Kim, S. K., & Shim, M. S. (2016). Seaweed polysaccharide-based nanoparticles: Preparation and applications for drug delivery. *Polymers*, *8*(2), 30. https://doi.org/10.3390/POLYM8020030.

Vicente, J., Vera, B., Rodríguez, A. D., Rodríguez-Escudero, I., & Raptis, R. G. (2009). Euryjanicin A: A new cycloheptapeptide from the Caribbean marine sponge *Prosuberites laughlini*. *Tetrahedron Letters*, *50*(32), 4571–4574. https://doi.org/10.1016/J.TETLET.2009.05.067.

Vignesh, S., Raja, A., & James, R. A. (2011). Marine drugs: Implication and future studies. *International Journal of Pharmacology*, *7*(1), 22–30. https://doi.org/10.3923/ijp.2011.22.30.

Voultsiadou Eleni, E. (2010). Therapeutic properties and uses of marine invertebrates in the ancient Greek world and early Byzantium. *Journal of Ethnopharmacology*, *130*(2), 237–247. https://doi.org/10.1016/J.JEP.2010.04.041.

Voultsiadou, E., Koutsoubas, D., & Achparaki, M. (2009). Bivalve mollusc exploitation in Mediterranean coastal communities: An historical approach. *Journal of Biological Research-Thessaloniki*, *12*, 1–10. https://users.auth.gr/users/3/0/052103/public_html/pdf/Bivalve exploitation 09.pdf (accessed on 7 June 2024).

Wahyudi, H., Reynolds, A. A., Li, Y., Owen, S. C., & Yu, S. M. (2016). Targeting collagen for diagnostic imaging and therapeutic delivery. *Journal of Controlled Release*, *240*, 323–331. https://doi.org/10.1016/J.JCONREL.2016.01.007.

Wang, L., Wang, X., Wu, H., & Liu, R. (2014). Overview on biological activities and molecular characteristics of sulfated polysaccharides from marine green algae in recent years. *Marine Drugs*, *12*(9), 4984–5020. https://doi.org/10.3390/MD12094984.

Wang, W., Wang, S. X., & Guan, H. S. (2012). The antiviral activities and mechanisms of marine polysaccharides: An overview. *Marine Drugs*, *10*(12), 2795–2816. https://doi.org/10.3390/MD10122795.

Wei, X., Nieves, K., & Rodríguez, A. D. (2010). Neopetrosiamine A, biologically active bis-piperidine alkaloid from the Caribbean Sea sponge *Neopetrosia proxima*. *Bioorganic & Medicinal Chemistry Letters*, *20*(19), 5905–5908. https://doi.org/10.1016/J.BMCL.2010.07.084.

Weissman, B., Meyer, K., Sampson, P., & Linker, A. (1954). Isolation of oligosaccharides enzymatically produced from hyaluronic acid. *Journal of Biological Chemistry*, *208*(1), 417–429. https://doi.org/10.1016/S0021-9258(18)65660-1.

Weissmann, B., & Meyer, K. (1954). The structure of hyalobiuronic acid and of hyaluronic acid from umbilical cord. *Journal of the American Chemical Society*, *76*(7), 1753–1757. https://doi.org/10.1021/JA01636A010.

Wicki, A., Witzigmann, D., Balasubramanian, V., & Huwyler, J. (2015). Nanomedicine in cancer therapy: Challenges, opportunities, and clinical applications. *Journal of Controlled Release*, *200*, 138–157. https://doi.org/10.1016/J.JCONREL.2014.12.030.

Wijesekara, I., Pangestuti, R., & Kim, S. K. (2011). Biological activities and potential health benefits of sulfated polysaccharides derived from marine algae. *Carbohydrate Polymers*, *84*(1), 14–21. https://doi.org/10.1016/J.CARBPOL.2010.10.062.

Wijesinghe, W. A. J. P., & Jeon, Y. J. (2012). Biological activities and potential industrial applications of fucose rich sulfated polysaccharides and fucoidans isolated from brown seaweeds: A review. *Carbohydrate Polymers*, *88*(1), 13–20. https://doi.org/10.1016/J.CARBPOL.2011.12.029.

Williams, D. E., & Andersen, R. J. (2020). Biologically active marine natural products and their molecular targets discovered using a chemical genetics approach. *Natural Product Reports*, *37*(5), 617–633. https://doi.org/10.1039/C9NP00054B.

Witvrouw, M., & De Clercq, E. (1997). Sulfated polysaccharides extracted from sea algae as potential antiviral drugs. *General Pharmacology: The Vascular System*, *29*(4), 497–511. https://doi.org/10.1016/S0306-3623(96)00563-0.

Witvrouw, M., Este, J. A., Mateu, M. Q., Reymen, D., Andrei, G., Snoeck, R., Ikeda, S., Pauwels, R., Bianchini, N. V., Desmyter, J., & De Clercq, E. (1994). Activity of a sulfated polysaccharide extracted from the red seaweed *Aghardhiella tenera* against human immunodeficiency virus and other enveloped viruses. *Antiviral Chemistry and Chemotherapy (United Kingdom)*, *5*(5), 297–303. https://doi.org/10.3/JQUERY-UI.JS.

Woods, R. K., Thien, F. C., & Abramson, M. J. (2000). Dietary marine fatty acids (fish oil) for asthma. *The Cochrane Database of Systematic Reviews*, *2019*(2), CD001283. https://doi.org/10.1002/14651858.CD001283.

Wright, A. D., & Lang-Unnasch, N. (2009). Diterpene formamides from the tropical marine sponge *Cymbastela hooperi* and their antimalarial activity in vitro. *Journal of Natural Products*, *72*(3), 492–495. https://doi.org/10.1021/NP800654W/ASSET/IMAGES/MEDIUM/NP-2008-00654W_0001.GIF.

Wright, A. D., McCluskey, A., Robertson, M. J., MacGregor, K. A., Gordon, C. P., & Guenther, J. (2010). Anti-malarial, anti-algal, anti-tubercular, anti-bacterial, anti-photosynthetic, and anti-fouling activity of diterpene and diterpene isonitriles from the tropical marine sponge *Cymbastela hooperi*. *Organic & Biomolecular Chemistry*, *9*(2), 400–407. https://doi.org/10.1039/C0OB00326C.

Wu, L., Sun, J., Su, X., Yu, Q., Yu, Q., & Zhang, P. (2016). A review about the development of fucoidan in antitumor activity: Progress and challenges. *Carbohydrate Polymers, 154,* 96–111. https://doi.org/10.1016/J.CARBPOL.2016.08.005.

Xia, E. Q., Zhu, S. S., He, M. J., Luo, F., Fu, C. Z., & Zou, T. Bin. (2017). Marine peptides as potential agents for the management of type 2 diabetes mellitus—A prospect. *Marine Drugs, 15*(4), 88. https://doi.org/10.3390/MD15040088.

Xiang, S., Liu, F., Lin, J., Chen, H., Huang, C., Chen, L., Zhou, Y., Ye, L., Zhang, K., Jin, J., Zhen, J., Wang, C., He, S., Wang, Q., Cui, W., & Zhang, J. (2017). Fucoxanthin inhibits β-amyloid assembly and attenuates β-amyloid oligomer-induced cognitive impairments. *Journal of Agricultural and Food Chemistry, 65*(20), 4092–4102. https://doi.org/10.1021/ACS.JAFC.7B00805/ASSET/IMAGES/MEDIUM/JF-2017-008054_0011.GIF.

Xing, R., Liu, S., Yu, H., Guo, Z., Li, Z., & Li, P. (2005). Preparation of high-molecular weight and high-sulfate content chitosans and their potential antioxidant activity in vitro. *Carbohydrate Polymers, 61*(2), 148–154. https://doi.org/10.1016/J.CARBPOL.2005.04.007.

Xiong, Z. Q., Wang, J. F., Hao, Y. Y., & Wang, Y. (2013). Recent advances in the discovery and development of marine microbial natural products. *Marine Drugs, 11*(3), 700–717. https://doi.org/10.3390/MD11030700.

Xue, C., Chen, L., Li, Z., Cai, Y., Lin, H., & Fang, Y. (2004). Antioxidative activities of low molecular fucoidans from kelp Laminaria japonica. *Developments in Food Science, 42*(C), 139–145. https://doi.org/10.1016/S0167-4501(04)80016-5.

Yamamoto, I., Takahashi, M., Tamura, E., Maruyama, H., & Mori, H. (1984). Antitumor activity of edible marine algae: Effect of crude fucoidan fractions prepared from edible brown seaweeds against L-1210 leukemia. *Hydrobiologia, 116*(1), 145–148. https://doi.org/10.1007/BF00027653.

Yang, B., Han, X., Ji, B., & Lu, R. (2016). Competition between tumor and mononuclear phagocyte system causing the low tumor distribution of nanoparticles and strategies to improve tumor accumulation. *Current Drug Delivery, 13*(8), 1261–1274. https://doi.org/10.2174/1567201813666160418105703.

Yang, P., Lu, C., Qin, W., Chen, M., Quan, G., Liu, H., Wang, L., Bai, X., Pan, X., & Wu, C. (2020). Construction of a core-shell microneedle system to achieve targeted co-delivery of checkpoint inhibitors for melanoma immunotherapy. *Acta Biomaterialia, 104,* 147–157. https://doi.org/10.1016/J.ACTBIO.2019.12.037.

Yang, Y., Wang, S., Wang, Y., Wang, X., Wang, Q., & Chen, M. (2014). Advances in self-assembled chitosan nanomaterials for drug delivery. *Biotechnology Advances, 32*(7), 1301–1316. https://doi.org/10.1016/J.BIOTECHADV.2014.07.007.

Yao, G., Quan, G., Lin, S., Peng, T., Wang, Q., Ran, H., Chen, H., Zhang, Q., Wang, L., Pan, X., & Wu, C. (2017). Novel dissolving microneedles for enhanced transdermal delivery of levonorgestrel: In vitro and in vivo characterization. *International Journal of Pharmaceutics, 534*(1–2), 378–386. https://doi.org/10.1016/J.IJPHARM.2017.10.035.

Yashaswini, D. G. V., Venkatesan, J., & Anil, S. (2021). Hydrocolloids from marine macroalgae: Isolation and applications. *Algae for Food: Cultivation, Processing and Nutritional Benefits,* 185–199. https://doi.org/10.1201/9781003165941-13.

Yasuhara-Bell, J., & Lu, Y. (2010). Marine compounds and their antiviral activities. *Antiviral Research, 86*(3), 231–240. https://doi.org/10.1016/J.ANTIVIRAL.2010.03.009.

Ye, Y., Wang, J., Hu, Q., Hochu, G. M., Xin, H., Wang, C., & Gu, Z. (2016). Synergistic transcutaneous immunotherapy enhances antitumor immune responses through delivery of checkpoint inhibitors. *ACS Nano*, *10*(9), 8956–8963. https://doi.org/10.1021/ACSNANO.6B04989/SUPPL_FILE/NN6B04989_SI_001.PDF.

Ye, Y., Yu, J., Wang, C., Nguyen, N. Y., Walker, G. M., Buse, J. B., & Gu, Z. (2016). Microneedles integrated with pancreatic cells and synthetic glucose-signal amplifiers for smart insulin delivery. *Advanced Materials*, *28*(16), 3115–3121. https://doi.org/10.1002/ADMA.201506025.

Yoon, S. J., Pyun, Y. R., Hwang, J. K., & Mourão, P. A. S. (2007). A sulfated fucan from the brown alga *Laminaria cichorioides* has mainly heparin cofactor II-dependent anticoagulant activity. *Carbohydrate Research*, *342*(15), 2326–2330. https://doi.org/10.1016/J.CARRES.2007.06.019.

Yoshizawa, Y., Enomoto, A., Todoh, H., Ametani, A., & Kaminogawa, S. (1993). Activation of murine macrophages by polysaccharide fractions from marine algae (*Porphyra yezoensis*). *Bioscience, Biotechnology, and Biochemistry*, *57*(11), 1862–1866. https://doi.org/10.1271/BBB.57.1862.

Yoshizawa, Y., Tsunehiro, J., Fukui, F., Ametani, A., Kaminogawa, S., Nomura, K., & Itoh, M. (1995). Macrophage stimulation activity of the polysaccharide fraction from a marine alga (*Porphyra yezoensis*): Structure-function relationships and improved solubility. *Bioscience, Biotechnology, and Biochemistry*, *59*(10), 1933–1937. https://doi.org/10.1271/BBB.59.1933.

You, M., Liao, L., Hong, S. H., Park, W., Kwon, D. I., Lee, J., Noh, M., Oh, D. C., Oh, K. B., & Shin, J. (2015). Lumazine peptides from the marine-derived fungus *Aspergillus terreus*. *Marine Drugs*, *13*(3), 1290–1303. https://doi.org/10.3390/MD13031290.

Yu, F., Zong, C., Jin, S., Zheng, J., Chen, N., Huang, J., Chen, Y., Huang, F., Yang, Z., Tang, Y., & Ding, G. (2018). Optimization of extraction conditions and characterization of pepsin-solubilised collagen from skin of giant croaker (*Nibea japonica*). *Marine Drugs*, *16*(1), 29. https://doi.org/10.3390/MD16010029.

Yu, J., Zhang, Y., Ye, Y., DiSanto, R., Sun, W., Ranson, D., Ligler, F. S., Buse, J. B., Gu, Z., & Hu, D. (2015). Microneedle-array patches loaded with hypoxia-sensitive vesicles provide fast glucose-responsive insulin delivery. *Proceedings of the National Academy of Sciences of the United States of America*, *112*(27), 8260–8265. https://doi.org/10.1073/PNAS.1505405112/SUPPL_FILE/PNAS.201505405SI.PDF.

Yu, W., Jiang, G., Zhang, Y., Liu, D., Xu, B., & Zhou, J. (2017). Polymer microneedles fabricated from alginate and hyaluronate for transdermal delivery of insulin. *Materials Science & Engineering. C, Materials for Biological Applications*, *80*, 187–196. https://doi.org/10.1016/J.MSEC.2017.05.143.

Yurchenko, A. N., Smetanina, O. F., Kalinovsky, A. I., Pushilin, M. A., Glazunov, V. P., Khudyakova, Y. V., Kirichuk, N. N., Ermakova, S. P., Dyshlovoy, S. A., Yurchenko, E. A., & Afiyatullov, S. S. (2014). Oxirapentyns F-K from the marine-sediment-derived fungus Isaria felina KMM 4639. *Journal of Natural Products*, *77*(6), 1321–1328. https://doi.org/10.1021/NP500014M/SUPPL_FILE/NP500014M_SI_002.CIF.

Zampella, A., D'Auria, M. V., Gomez Paloma, L., Casapullo, A., Minale, L., Debitus, C., & Henin, Y. (1996). Callipeltin A, an anti-HIV cyclic depsipeptide from the New Caledonian lithistida sponge *Callipelta* sp. *Journal of the American Chemical Society*, *118*(26), 6202–6209. https://doi.org/10.1021/JA954287P/SUPPL_FILE/JA6202.PDF.

Zhang, D. J., Liu, R. F., Li, Y. G., Tao, L. M., & Tian, L. (2010). Two new antifungal cyclic lipopeptides from Bacillus marinus B-9987. *Chemical and Pharmaceutical Bulletin*, *58*(12), 1630–1634. https://doi.org/10.1248/CPB.58.1630.

Zhang, P., Dalton, C., & Jullien, G. A. (2009). Design and fabrication of MEMS-based microneedle arrays for medical applications. *Microsystem Technologies*, *15*(7), 1073–1082. https://doi.org/10.1007/S00542-009-0883-5/FIGURES/8.

Zhang, Y., Jiang, G., Yu, W., Liu, D., & Xu, B. (2018). Microneedles fabricated from alginate and maltose for transdermal delivery of insulin on diabetic rats. *Materials Science and Engineering: C*, *85*, 18–26. https://doi.org/10.1016/J.MSEC.2017.12.006.

Zhang, Z., Suner, S. S., Blake, D. A., Ayyala, R. S., & Sahiner, N. (2020). Antimicrobial activity and biocompatibility of slow-release hyaluronic acid-antibiotic conjugated particles. *International Journal of Pharmaceutics*, *576*, 119024. https://doi.org/10.1016/J.IJPHARM.2020.119024.

Zhao, D., Kwon, S. H., Chun, Y. S., Gu, M. Y., & Yang, H. O. (2017). Anti-neuroinflammatory effects of fucoxanthin via inhibition of Akt/NF-κB and MAPKs/AP-1 pathways and activation of PKA/CREB pathway in lipopolysaccharide-activated BV-2 microglial cells. *Neurochemical Research*, *42*(2), 667–677. https://doi.org/10.1007/S11064-016-2123-6/FIGURES/6.

Zheng, L. X., Chen, X. Q., & Cheong, K. L. (2020). Current trends in marine algae polysaccharides: The digestive tract, microbial catabolism, and prebiotic potential. *International Journal of Biological Macromolecules*, *151*, 344–354. https://doi.org/10.1016/J.IJBIOMAC.2020.02.168.

Zhou, X., Liu, J., Yang, B., Lin, X., Yang, X. W., & Liu, Y. (2013). Marine natural products with anti-HIV activities in the last decade. *Current Medicinal Chemistry*, *20*(7), 953–973. https://doi.org/10.2174/0929867311320070009.

Zhu, J., Tang, X., Jia, Y., Ho, C. T., & Huang, Q. (2020). Applications and delivery mechanisms of hyaluronic acid used for topical/transdermal delivery – A review. *International Journal of Pharmaceutics*, *578*, 119127. https://doi.org/10.1016/J.IJPHARM.2020.119127.

Ziemert, N., Alanjary, M., & Weber, T. (2016). The evolution of genome mining in microbes – A review. *Natural Product Reports*, *33*(8), 988–1005. https://doi.org/10.1039/C6NP00025H.

Zuckerman, J. E., Gritli, I., Tolcher, A., Heidel, J. D., Lim, D., Morgan, R., Chmielowski, B., Ribas, A., Davis, M. E., & Yen, Y. (2014). Correlating animal and human phase Ia/Ib clinical data with CALAA-01, a targeted, polymer-based nanoparticle containing siRNA. *Proceedings of the National Academy of Sciences of the United States of America*, *111*(31), 11449–11454. https://doi.org/10.1073/PNAS.1411393111/SUPPL_FILE/PNAS.1411393111.SAPP.PDF.

Zúñiga, E. A., Matsuhiro, B., & Mejías, E. (2006). Preparation of a low-molecular weight fraction by free radical depolymerization of the sulfated galactan from *Schizymenia binderi* (*Gigartinales, Rhodophyta*) and its anticoagulant activity. *Carbohydrate Polymers*, *66*(2), 208–215. https://doi.org/10.1016/J.CARBPOL.2006.03.007.

CHAPTER 5

Marine Polysaccharides in Drug Delivery

MADHAVI PATEL,[1] BIJAL PRAJAPATI,[1] and BHUPENDRA PRAJAPATI[2,3]

[1]*Parul Institute of Pharmacy, Faculty of Pharmacy, Parul University, Vadodara, Gujarat, India*

[2]*Shree S. K. Patel College of Pharmaceutical Education and Research, Ganpat University, Gujarat, India*

[3]*Faculty of Pharmacy, Silpakorn University, Nakhon Pathom, Thailand*

ABSTRACT

Oceans are considered an immense spring of substances that find their application in biomedical and biotechnological fields. In developing a new system of formulation to be utilized in biomedical fields, the sea can be exploited as a renewable source to provide bio-compounds. Polysaccharides found inside the seabed are abundant, reducing extraction costs. The solubility issues in an aqueous medium and interaction with other bio-compounds are already at par. Algae in the sea are a rich source of polysaccharides like alginate and fucoidan. These biomaterials can be used as a starting source to develop various carrier devices for the delivery of drugs. Marine polysaccharides find their therapeutic application in inhibiting cancer, viral infections, and inflammation. Tissue engineering has also emerged as a current technology in which marine polysaccharides find their utilization. A particular element of marine polysaccharides is that they are cheap, biodegradable, and biocompatible. These characteristics make them an appropriate candidate for wound dressing excipients, wastewater treatment, and biosensor development. Chitosan and fucoidan have bioactive properties that can be combined with other polysaccharides to develop nanostructures. Marine plants and animals can produce marine bacteria, which shows that the marine polysaccharide

Handbook of Research in Marine Pharmaceutics: Exploring Oceanic Microbial Diversity for Human Health and Wellness. Md. Faiyazuddin & Khalid Rehman Hakeem (Eds.)

field is constantly evolving. The polysaccharides of interest are now produced *in vitro* due to advances in biological technologies with the help of various microorganisms tailoring the fabrication of biologically active compounds. The knowledge of these polysaccharides requires special attention towards pharmaceutical applications so that newer polysaccharides with desirable characteristics can be developed. Marine polysaccharides exhibit a lower cytotoxic behavior. Their recently studied benefits on health have led to their impediment as a key area of research in the biomedical and pharmaceutical fields. This chapter focuses on the applications of marine polysaccharides for developing novel drug delivery specific to tissues. Their availability in nature is abundant, which makes their profile a biodegradable component that can be used to investigate further applications to target therapeutic delivery with specificity.

5.1 INTRODUCTION TO MARINE POLYSACCHARIDES

Over 70% of the Earth's surface is shielded by oceans and is considered a huge ecosystem rich with a variety of biomass. The marine environment is known to contain diverse plants and animals loaded with different bioactive comprising various biological properties (Das, Lyla, & Khan, 2006; Cardoso, Costa, & Mano, 2016; McCarthy & Pomponi, 2004). Recently, biomaterials of marine origin are gaining interest and being tested for wide biomedicine applications due to their considerable noncytotoxic, biocompatible, and biodegradable properties. This has created a wide scope for the discovery of novel bioactive from the marine environment, which has opened boundless opportunities for the pharmaceutical industry in drug discovery and development (Molinski, Dalisay, Lievens, & Saludes, 2009; Silva, Duarte, Moreira-Silva, Mano, & Reis, 2012; Munro, Blunt, Dumdei, Hickford et al., 1999).

Polysaccharides are carbohydrates containing more than 10 sugar units, macromolecular compounds with high molecular weight and distributed to a large extent in plants, animals, and microorganisms. They form structural material and are major sources of energy and water. Furthermore, they are reported to exert several biological effects such as antiviral, anti-aging, anti-cancer, glucose-lowering, immunoregulation, antioxidants, and hepatoprotective (Schuerch, 1986; Cui, Ye, Wang, Li, Yao, & Qian, 2014; Zhang, Gao, Hu, Zhang, Sun, Rong, & Jia, 2017).

Recently, polysaccharides have gained much research attention due to their broad application in therapeutics and favorable characteristics (Lima-Tenório, Tenório-Neto, Guilherme, Garcia et al., 2015). Marine polysaccharides are known for their utilization in cosmetics and food on a large scale. However, in the past few years, more interest has been shown in incorporating marine polysaccharides in pharmaceutical sciences as a substance for the incorporation of bioactives (Laurienzo, 2010). Marine polysaccharides, due to their high compatibility, biodegradability, greater drug loading capacity, and non-toxicity, are a promising tool as a natural biopolymer for efficient drug delivery while also exhibiting unique therapeutic properties. Hence, they are being explored for the immobilization of biomolecules, tissue engineering, coating of stents, and various other biological therapies (Atanase, 2021; Zhong, Gao, Cheng, Liu, Wang, & Han, 2020; Atanase, Desbrieres, & Riess, 2017; Lee, Kim, Seo, Park et al., 2017).

5.2 SOURCES OF MARINE POLYSACCHARIDES

Polysaccharides in general are derived from different natural sources such as plants, animals, and microbes. Marine polysaccharides are obtained from similar natural marine sources where algae (seaweed) remain the major source of polysaccharides (Senni, Pereira, Gueniche, Delbarre-Ladrat et al., 2011). The polysaccharides are also derived via sulfation, acetylation, carboxymethylation, phosphorylation, and other chemical alterations to infer their structures with required changes. Such changes can bring favorable physicochemical characteristics with enhanced functions (Prashanth & Tharanathan, 2007). Seaweeds are abundantly occurring species that produce polysaccharides in the ocean and have provided many ingredients for drug delivery systems (DDs) (MacArtain, Gill, Brooks, Campbell, & Rowland, 2007). Apart from seaweeds, animal sources like cartilaginous fish tissue and crustaceans' skeletons, as well as microorganisms like prokaryotes, have been extracted and tested for various biological activities and their applications in the design of novel drug delivery systems (Senni, Pereira, Gueniche, Delbarre-Ladrat et al., 2011). Chitosan, carrageenan, fucoidan, laminarin, sodium alginate, sulfated polysaccharides, etc., are reported for their wide utilization in the development of different DDs.

5.3 CHARACTERISTICS OF MARINE POLYSACCHARIDES

Marine polysaccharides, like common polysaccharides, are carbohydrates with complex structures and are formed by the dehydration of many monosaccharides connected via glycosidic bonds. Homopolysaccharides contain a single type of monosaccharides, whereas heteropolysaccharides are composed of diverse monosaccharides (Habibi, Lucia, & Rojas, 2010; Muzzarelli, Greco, Busilacchi, Sollazzo, & Gigante, 2012). They find wide applications in the food and pharmaceutical industries due to their morphological and structural flexibility. They are water-soluble and can swell in water to give colloidal viscous solutions exhibiting pseudoplastic flow properties. Their wide applications are due to their innate functional characteristics like water holding, binding, thickening, gelling, stabilizing, etc., where, when desired, few tailor-made characteristics can be achieved for the polysaccharides for specific applications (Li & Nie, 2016). The molecular structure of the marine polysaccharides, due to the presence of amine, carboxyl, and hydroxyl-like multifunctional groups, facilitates their chemical or enzymatic modifications.

5.4 EXTRACTION AND PURIFICATION OF MARINE POLYSACCHARIDES

5.4.1 FROM SEAWEED

Polysaccharides are structural elements of the cell wall, and algae also possess polysaccharides in their cell wall. The conventional method of algal polysaccharide isolation consists of hot water extraction and removal of interfering materials such as lipids, coloring material, and low molecular weight compounds by methanol or chloroform. However, this method of extraction is time-consuming and requires high temperatures, and provides less yield (Zhang, Wang, Wang, Liu, Hou, & Zhang, 2010).

Advanced extraction techniques like microwave-assisted extraction are mostly preferred methods along with ultrasonic, supercritical fluid, ionic liquid, and enzyme-assisted extraction (Herrero & Ibanez, 2015; Martins, Vieira, Correia, Ferreira et al., 2016). Higher extraction yield with less extraction time, energy, and cost consumption are merits of microwave-assisted extraction when compared to conventional hot water extraction (Hahn, Lang, Ulber, & Muffler, 2012). Although providing a high yield, this technique may degrade or change the nature of polysaccharides under extraction. Similarly, laminarin from brown algae was extracted by ultrasonic extraction, providing a higher content than the conventional technique (Kadam, O'Donnell, Rai, Hossain, Burgess, Walsh, & Tiwari, 2015).

The obtained polysaccharides are in crude form and associated with impurities like protein, sulfate content, and other low molecular weight compounds. Further purification of this crude mixture is carried out by membrane separation, size exclusion, affinity and ion exchange chromatography, and precipitation by ethanol (Chakraborty, Miao, McDonald, & Chen, 2012; Ermakova, Men Shova, Vishchuk, Kim et al., 2013; Usoltseva, Anastyuk, Shevchenko, Zvyagintseva, & Ermakova, 2016).

5.4.2 FROM MICROORGANISM

According to interest, the particular strain of polysaccharide is obtained and placed for cultivation to obtain a pure culture in identical conditions, and this growing process of pure culture is referred to as fermentation. In comparison to fungi, bacteria are easy to cultivate, and after several sub-cultures, pure culture can be developed. Submerged and solid-state fermentation are two main types of fermentation processes followed for the development of exopolysaccharides in microorganisms (Correa, Brugnari, Bracht, Peralta, & Ferreira, 2016; Chao, Wu, & Lu, 2011).

Submerged fermentation employs liquid substrates to facilitate fermentation, while solid-based substrates are used for solid-state fermentation. Extraction of microbial polysaccharides includes confiscation of the biomass via centrifugation or filtration followed by evaporation to yield concentrated fluid and precipitation by alcohol. Further purification is performed by HPLC, size exclusion, gas-liquid, and anion exchange chromatography (Jindal & Khattar, 2018).

5.5 MARINE POLYSACCHARIDES IN DRUG DELIVERY SYSTEMS: MERITS AND DEMERITS

Marine polysaccharides, due to their diverse characteristics and applications, are highly advantageous for the development of DDs. However, they are obtained from natural sources where variation in their quality is a question mark, and also the scarcity of not having explored metabolism details is another drawback that needs to be dealt with (Yu, Shen, Song, & Xie, 2018; Haag & Kratz, 2006; García-González, Alnaief, & Smirnova, 2011). Figure 5.1 summarizes the merits and demerits of marine polysaccharides in DDs.

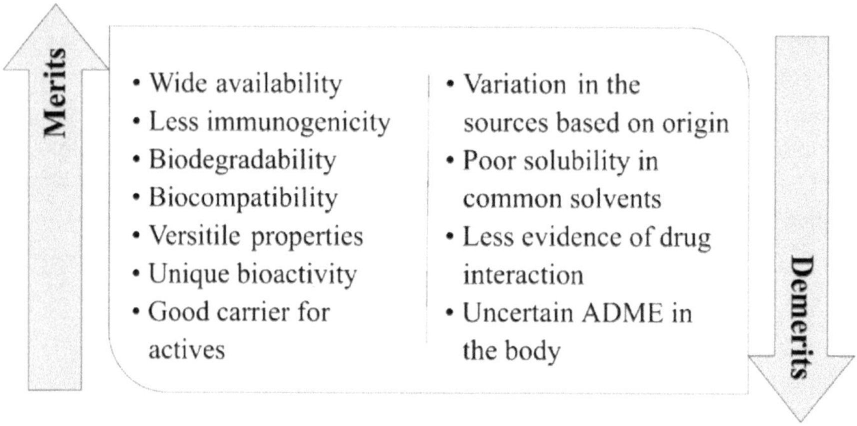

FIGURE 5.1 Merits and demerits of marine polysaccharides in DDs.

5.6 CLASSIFICATION OF MARINE POLYSACCHARIDES

Marine polysaccharides are broadly classified into plant, animal, and microbial polysaccharides based on the various sources they are obtained from. Figure 5.2 represents the different sources of marine polysaccharides. The red, brown, and green seaweeds are the major sources of marine plant-origin polysaccharides, comprising carrageenan and agar, fucoidan and algin, and ulvan, respectively. For marine animal polysaccharides, marine crustaceans are processed to extract chitin, cartilaginous bones of fish for chondroitin sulfate, starfish and sea cucumbers for sulfated polysaccharides, and scallops, abalones, and mollusks for glycosamines. Marine bacteria, fungi, or microalgae are processed to isolate microbial exopolysaccharides and structural polysaccharides (Sun, Ma, & Hu, 2021). Apart from sources, marine polysaccharides can also be categorized based on their electrostatic environment and their own sulfated or carboxylated structure. Carrageenans, fucoidan, alginate, hyaluronic acid, ulvan, dermatan sulfate, and chondroitin sulfate are polyanionic polysaccharides. Chitosan is prepared by deacetylation of chitin and is polycationic in nature, whereas agarose and chitin are neutral molecules. Alginate and hyaluronic acid are carboxylated, carrageenan and fucoidan are sulfated, whereas dermatan sulfate, chondroitin sulfate, and ulvan contain both types (Cardoso, Costa, & Mano, 2016).

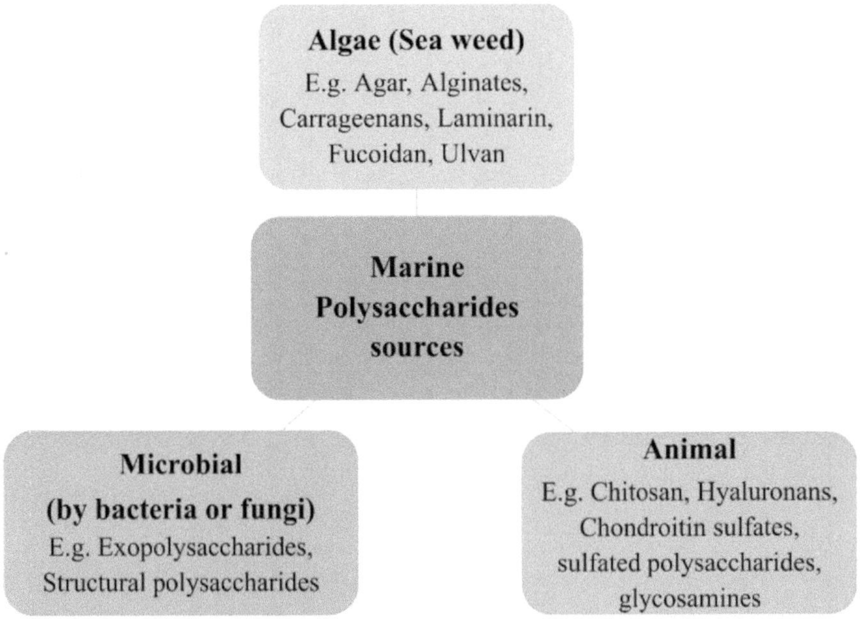

FIGURE 5.2 Various sources of marine polysaccharides.

5.7 MARINE POLYSACCHARIDES IN VARIOUS DRUG DELIVERY SYSTEMS

Marine-origin compounds are very important as they present vital features in the drug discovery field for good bioavailability and affinity to the target. Detailed information on several pharmacological actions exerted by different marine polysaccharides along with their incorporation in drug delivery systems is listed in Table 5.1. The chemical structure of a few marine-origin polysaccharides is displayed in Figure 5.3.

5.7.1 CHITOSAN

Chitosan is obtained from chitin and is a copolymer comprising glucosamine units and N-acetyl glucosamine units. Chitosan is a multipurpose basic material for the preparation of a varied range of products with applications extending

TABLE 5.1 Polysaccharides of Marine Origin

SL. No.	Name of Polysaccharide	Source	Application in DDs	Pharmacological Activity	References
1.	Alginate	Brown Algae (*Laminaria hyperborea, Ascophyllum nodosum, Macrocystis pyrifera, Sargassum* species)	Stabilizer, excipient, beads, particles, hydrogel matrices, microparticles	Glucose regulation, in dyslipidemia, and osteoarthritis, appetite regulation	Laurienzo (2010); Pereiira, & Cotas (2019); Gombotz, & Wee (1998); Sudhakar, Kuotsu, & Bandyopadhyay (2006); Lima, Sher, & Mano (2012); Yoo, & Park (2001); Tang, Chen, Chiang, Chen, Chang, & Chen (2015); Arlov, Oztürk, Steinwachs, Skjåk-Bræk, & Zenobi-Wong (2017).
2.	Fucoidan	Brown Algae (*Ascophyllum nodosum, Macrocystis pyrifera, Laminaria hyperborea*)	Nanoparticles, Microspheres, Controlled release of insulin	Anti-obesity, antitumor, antiviral, antithrombin, anticoagulant, anti-inflammatory, burn treatment, antidiabetic, tissue engineering	Li, Lu, Wei, & Zhao (2008); Pinheiro, Bourbon, Cerqueira, Maricato et al. (2015); Sezer & Akbuğa (2006); Anastyuk, Shevchenko, Ermakova, Vishchuk et al. (2012); Kim, Ravichandran, Khan, & Kim (2008); Alves, Sousa, & Reis (2013); Jeong, Kim, Jung, Park et al. (2013); Rabea, Badawy, Stevens, Smagghe, & Steurbaut (2003); Kim, Jeon, & Lee (2014)
3.	Carrageenan	Red Algae (*Chondrus crispus, Gigartina skottsbergii, Eucheuma denticulatum*)	Controlled and sustained release, tablet compressor	Immuno-modulator, Antiviral, Anticoagulant, antihyperlipidemic, Antioxidant, Antitumor, Antibacterial	Yermak, Kim, Titlynov, Isakov, & Solov'eva (1999); Ganta, Devalapally, Shahiwala, & Amiji (2008); Bonferoni, Rossi, Ferrari, Bertoni, Bolhuis, & Caramella (1998); Picker (1999); Panlasigui, Baello, Dimatangal, & Dumelod (2003); Silva, Dore, Marques, Nascimento et al. (2010); Rocha De Souza, Marques, Guerra Dore, Ferreira da Silva, Oliveira Rocha, & Leite (2007); Carlucci, Ciancia, Matulewicz, Cerezo, & Damonte (1999); Zhou, Sun, Xin, Zhang, Li, & Xu (2004); Schaeffer & Krylov (2000); Pereira & Costa-Lotufo (2012)

TABLE 5.1 *(Continued)*

SL. No.	Name of Polysaccharide	Source	Application in DDs	Pharmacological Activity	References
4.	Ulvan	Green Algae (*Ulva lactuca, Ulva reticulata Ulva faciata*)	Tissue engineering, Cosmetics	Antioxidant, Anti-viral, immune enhancer, anti-coagulant, anti-tumor, anti-hyperlipidemic	Alves, Sousa, & Reis (2013); Kidgell, Magnusson, De Nys, & Glasson (2019); Ahmed & Ahmed (2014); Qin, Du, Xiao, Li, & Gao (2002)
5.	Agar (Agarose)	Red algae (*Gelidium, Gracilaria, Pterocladia* species)	Hydrogels	Anti-bacterial, in diabetes, weight loss	Rocha De Souza, Marques, Guerra Dore, Ferreira da Silva, Oliveira Rocha, & Leite (2007); Svec, Tennikova, & Deyl (2003); Hoare & Kohane (2008)
6.	Chitosan (Deacetylated chitin)	Marine crustacean	Films and scaffolds, sustained release, nanoparticles	Antimicrobial (Alves, Sousa, & Reis (2013), anti-tumor, anti-inflammatory (Jeong, Kim, Jung, Park et al. (2013)	Hoare & Kohane (2008); Prabaharan & Mano (2004); Felt, Buri, & Gurny (1998); Chen & Abatangelo (2002); Jiang, Quan, Liao, & Wang (2006)
7.	Hyaluronans	Marine animals	Nanoparticles, liposomes, coating material, hydrogels	Wound cure (Rabea, Badawy, Stevens, Smagghe, & Steurbaut, (2003)), for arthritis (Kim, Jeon, & Lee, 2014)	Oyarzun-Ampuero, Brea, Loza, Torres, & Alonso (2009); Simões, Moreira, Fonseca, Düzgüneş, De Lima (2004); Grech, Mano, & Reis (2008); Santos, Alves, & Mano (2010); Teien, Abildgaard, & Höök (1976); Wang, Varghese, Sharma, Strehin et al. (2007)
8.	Chondroitin sulfates	Marine animals	Controlled release delivery	Normalize chemokines, cytokines, and Anti-coagulant (Yermak, Kim, Titlynov, Isakov, & Solov'eva (1999),	Santo, Gomes, Mano, & Reis (2012); Mourão & Pereira (1999); Mikami & Kitagawa (2013)

TABLE 5.1 *(Continued)*

SL. No.	Name of Polysaccharide	Source	Application in DDs	Pharmacological Activity	References
9.	Dermatan sulfate, Keratan sulfate, Heparan sulfate	Ray skin (glycosaminoglycans)	Targeted delivery, Peptide delivery	Anticoagulant, In osteoarthritis, anti-adhesive	Funderburgh (2002)
10.	Exopolysac-charides (Eps)	Microbial exopolysaccharides from bacteria, cyanobacteria, fungi, etc. (*Halomonas* species, *Alteromonas* species, *Idiomarina* species)	Emulsifying and moisture-retaining agent, biocompatible nanoparticles	Hypotensive, Antihyperlipidemic, Antidiabetic	Squillaci, Finamore, Diana, Restaino et al. (2016); Mehta, Sidhu, Pinnaka, & Roy Choudhury (2014); Vecina, Oliveira, Araujo, Baggio, Torello, Saad, & De Souza Queiroz (2014); Parnell, & Reimer (2012)
11.	Laminarin	Brown algae	Nanoparticles, gel, microparticles, hydrogel	Antioxidants, hypoglycemic, antihyperlipidemic	Zargarzadeh, Amaral, Custódio, & Mano (2020); Yu, Wang, Guo, Zhao, & Chen (2019); Castanheira, Correia, Rodrigues, & Mano (2020); Jia, Yang, Wang, Liu, & Xie (2014); Lee, Kim, Jang, Cho, & Choi (2011)
12.	Porphyran	Red algae	Nanoparticles, Capping, and reducing agent	Antibacterial, antioxidant, hepatoprotective, antihyperlipidemic, anticancer, antiviral, anticoagulant, hypotensive	Khoshnevisan, Daneshpour, Barkhi, Gholami, Samadian, & Maleki (2018); Venkatpurwar, Shiras, & Pokharkar (2011); Xu, Huo, Xu, Zhu, Wu, Wei, & Zhou (2021); Correc, Hehemann, Czjzek, & Helbert (2011); Bhatia, Sharma, Sharma, Kavale et al. (2008)

from the medical, food industry, pharmaceutical, and agriculture, along with environmental pollution protection. The reactive hydroxyl and amino groups existing in chitosan are responsible for many useful properties including biocompatibility, polyelectrolyte, gel-forming, antioxidant, antimicrobial, and metal chelating with easy processability (Di, Liang, Shen, Pei, Wu, & He, 2022). Chitin, the second majorly available natural polymer in the industry, is considered after cellulose. Chitin is a biological polymer that originates in many animals like the exoskeletons of many crustaceans, mollusks, algae, insects, and cell walls of fungi (Sur, Rathore, Dave, Reddy, Chouhan, & Sadhu, 2019; Abedini, Ebrahimi, Roozbehani, Domb, & Hosseinkhani, 2018).

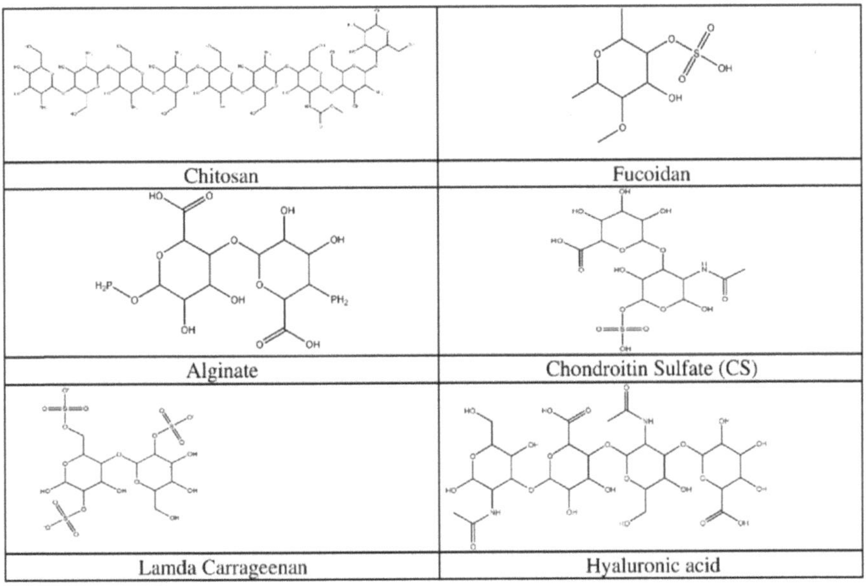

FIGURE 5.3 Polysaccharides of marine origin.

The proportion of chitin in the said animals generally ranges from 15 to 30% in crab cuticles, 20 to 30% in crustacean exoskeletons, 30 to 40% in shrimp cuticles, 5 to 25% in insect cuticles, and 2 to 44% in fungi cell walls. The exoskeletons of crustaceans, specifically prawns, crab, and shrimp shells, are largely available as waste resulting from food processing units. Chitosan is mainly derived from them for industrial applications. Conversely, it is more obtainable as a byproduct from the breeding of cocoons by the silk industry, a byproduct resulting from protein extraction from insects in the food industry or animal feed industry, and by fungi fermentation. Chitosan is classically converted from chitin via different chemical or enzymatic treatments for use in industrial or research applications.

Chitin is composed of 1–4 linked 2-acetamido 2-deoxy β-D-glucopyranose. Chitosan possesses amino groups that appear upon solubilization in diluted acids ($pH < 6$), differing from chitin, and forms complexes with various metal ions. Hence, it is found applications for wastewater treatment and purification. A deacetylated form of chitin, chitosan possesses a heterogeneous structure.

Chitosan has protonated groups such as amino groups, making it a polycation that consequently forms ionic complexes with a wide range of synthetic or natural anionic substances like proteins, lipids, DNA, and a few negatively charged non-natural polymers, for instance, polyacrylic acid. Chitosan is the only naturally occurring positively charged material in polysaccharides.

The molecular weight of chitosan usually lies between 300 to 1000 kDa based on the source of chitin.

Chitosan is biodegradable, biocompatible, hydrophilic, nontoxic, has good bioavailability, and selective permeability for water, along with exceptional chemical resistance. Chitosan is employed in the formation of gels, films, microparticles, nanoparticles, and beads for the incorporation of actives. Due to its unique biodegradability property, chitosan gets decomposed *in vivo* into amino sugars that can be certainly absorbed by the human body. Hence, chitosan and its derivatives have been mostly studied and employed as many drug carriers (Shariatinia, 2019).

5.7.2 FUCOIDAN

Fucoidan is a polysaccharide that was first extracted and isolated from marine brown seaweed in 1913 with dilute acetic acid as an extraction solvent and followed by a further purification process (Kylin, 1913). Fucoidan is constructed of 1→3-linked α-L-fucopyranosyls or of interchanging 1→3-linked and 1→4-linked residues that may be sulfate substituted. Fucoidan is a general term covering a different cluster of fucose-containing sulfated polysaccharides based on different extraction methods. Their composition originates from various brown seaweeds reported. Thus, fucose-containing sulfated polysaccharides is a highly preferred terminology over only fucoidan. This sulfated polysaccharide has varied chemical structures due to the different extraction methods utilized which affect the compositions as well.

The compositions along with the physical and chemical features of fucoidans are greatly affected by variations in source species, season, age, and very importantly their geographical location (Lim & Aida, 2017).

Fucoidan is a sulfated polysaccharide built of fucose linked α-(1→2)-, α-(1→3)-, and/or α-(1→4)-glycosidic bonds. It is also reported to contain other monosaccharides, like glucose, galactose, rhamnose, mannose, xylose, and uronic acids. The wide molecular weight it ranges from 7 kDa to as high as 2379 kDa based on the species, collection season, and original geographical conditions. The proportion of its sulfate content ranges between 5 to 38% (Lim, Aida, Schiehser, Rosenau, & Böhmdorfer, 2019; Bilan, Grachev, Ustuzhanina, Shashkov, Nifantiev, & Usov, 2004). It shows biological activities as an antiviral, antibacterial, antitumor, antioxidant, anti-inflammatory, and anticoagulant (Ale, Maruyama, Tamauchi, Mikkelsen, & Meyer, 2011).

5.7.3 ALGINATE

Alginate is a naturally occurring polysaccharide primarily made of mannuronic acid and guluronic acid. It is a structural constituent present in brown seaweed of the Phaeophyceae family and is also available in the form of capsular polysaccharides in a few soil bacteria (*Azotobacter* and *Pseudomonas*).

The amount of alginate in such algae can provide up to 40% of its dry weight. The chemical organization and sequence of guluronic acid units and mannuronic acid units present in alginate may differ extensively among various species and in different fragments of the algae. The composition and sequence of the monomer unit of the polysaccharide are also influenced by the season, especially when the algae are harvested, their growth location, and the age of the tissue mass used for the generation of alginate (Da Silva, Vidart, Da Silva, Gimenes, & Vieira, 2017).

Alginic acid and its salts contain a family of unbranched binary copolymers made up of the radicals of α-L-guluronic acids and β-D-mannuronic acid joined with (1→4)-bonds chemically. The polymeric chain of alginate comprises homopolymeric polyguluronic and polymannuronic chains (blocks) intermingled with intermittent residues of these acids.

Currently, various commercial alginates are prepared by extraction from algal sources only. Whereas, microbial fermentation and post-polymerization of alginates are of paramount importance in high-tech field applications within biotechnology and pharmacy. The process of isolation of alginates obtained from brown algae biomass consists of simple stages of preliminary extraction by processing with hydrochloric acid, washing, filtration, and alkali neutralization. From this solution, sodium alginate is precipitated using alcohol and for higher purity, generally, it is re-precipitated in the same way.

Alginic acid forms alginates of several types by binding with polyvalent metals. All the carboxyl groups bound with the cations in complete alginates are insoluble in water. In the case of incomplete alginates, they are either soluble or insoluble in water. Conversely, complete alginates composed of univalent metals are water-soluble and give viscous sticky solutions when dissolved. It comprises potassium, sodium, magnesium, or ammonium ions.

Alginate, the polymer showing a polyelectrolyte nature, is dignified as nontoxic, non-immunogenic, biodegradable, and biocompatible. Alginate can hold water and possess gelling, stabilizing, and viscosifying properties. Its physical properties considerably regulate the stability and drug release rate of the formed gels, and also the function of cells that are encapsulated within alginate gels (Lee & Mooney, 2012). If the guluronic acid content is less than 20–25%, then alginate gel is not formed. While alginates with a low ratio of mannuronide to guluronide produce a stiff gel, and if the ratio is high, the gel is more elastic. When the polysaccharide molecular weight increases up to 400–500 kDa, the strength of the gel increases linearly (Haug, 1964; Khotimchenko, Kovalev, Savchenko, & Ziganshina, 2001).

5.7.4 CHONDROITIN SULFATE (CS)

Repeating disaccharide units of N-acetylgalactosamine and glucuronic acid are linked by β-(1→3) glycosidic bonds and are sulfated in different carbon positions to form chondroitin sulfate. It is a glycosaminoglycan comprising 20–100 repeating units. CS has substantial heterogeneity in structure and length that differs according to the position of the sulfate group (Vázquez, Rodríguez-Amado, Montemayor, Fraguas, Del Pilar González, & Murado, 2013; Malavaki, Mizumoto, Karamanos, & Sugahara, 2008).

CS is extracted from numerous terrestrial and marine animal sources, such as cartilage and fish bones. The concentration and composition of CS differ by source organism and the tissue in which it is present. CS from these marine sources contains varied chain lengths and many sulfated disaccharides. The presence of specific sulfation patterns makes the synthesis of CS challenging and complex. CS can be extracted using biosynthesis and chemical techniques to synthesize a specific structure, sulfation pattern, and molecular weight. Also, biological fermentation utilizing different fungi and bacteria is employed for the production of CS (He, Fu, Li, Jones, Linhardt, & Koffas, 2015; Jin, Zhang, Yuan, Kang, Du, & Chen, 2016; Schiraldi, Cimini, & De Rosa, 2010). From an industrial viewpoint, the isolation method of CS includes various stages of chemical hydrolysis of cartilage, proteoglycan core breakdown, removal of proteins, and retrieval and purification of CS (Rodén, Baker, Cifonelli, & Mathews, 1972; Chascall, Calabro, Midura, & Yanagishita, 1994; Sumi, Ohba, Ikegami, Shibata et al., 2002). In recent times, other alternative methods for the isolation of CS have been reported to substitute the classical methods, which include digestion of proteins and cartilage mediated by enzymes, selective precipitations using alcohol, resuspension, neutralization by salt solutions, and separation based on the molecular weight using ultrafiltration diafiltration techniques (Lignot, Lahogue, & Bourseau, 2003; Murado, Fraguas, Montemayor, Vázquez, & González, 2010).

5.7.5 CARRAGEENAN

Carrageenan, a sulfated polysaccharide extracted and purified from red macroalgae *Eucheuma, Chondrus, Gigartina, Hypnea, Sarconema, Agardhiella, Furcellaria, Solieria,* and *Iridaea* species. It is water-soluble with a linear chain of partially sulfated galactans (Oliyaei, Moosavi-Nasab, & Mazloomi, 2022; Pacheco-Quito, Ruiz-Caro, & Veiga, 2020). Carrageenan has a wide molecular weight ranging from 20 to 40 kDa on the lower side up to 200 to 800 kDa on the higher side depending on the algae origin and the preparation method adopted for carrageenan (McKim, Willoughby, Sr Blakemore, & Weiner, 2019). Different grades of carrageenan are available, among which the kappa, iota, and lambda have broad applications in food products and pharmaceutical industries. A few other types of carrageenan also exist like mu (μ), nu (ν), Teta (θ), and beta (β), which are classified based on their solubility in potassium chloride (Geonzon, Descallar, Du, Bacabac, & Matsukawa, 2020). Carrageenan and its various derivatives are highly biocompatible, biodegradable, and anionic, comprising reactive functional groups, and are utilized in different areas for a wide range of applications (Khan, Raza, Mehboob, Kadir et al., 2020). Commercial production of kappa carrageenan, lambda carrageenan, and iota carrageenan is carried out by extraction from *algae Kappaphycus alvarezii, Gigartina,* or *Chondrus* species, and *Eucheuma denticulatum*, respectively (Khotimchenko, Kovalev, Savchenko,

& Ziganshina, 2001). Structurally, carrageenan possesses repeating units of D-galactose and 3,6-anhydro-galactose units linked together (McKim, Willoughby, Sr Blakemore, & Weiner, 2019). The strength of anionic properties in various carrageenan types is about 22 to 35% of the sulfation degree. Their anionic nature significantly impacts solubility and gel strength (Necas & Bartosikova, 2013).

Carrageenan mainly acts as a gel-forming, thickening, and stabilizing agent in foods and green edible films (Huang, Luo, Liu, Dong et al., 2020; Roy & Rhim, 2020). In biomedical applications, it serves as an agent for encapsulating actives (Chen, Han, Zhao, Tang, & Wang, 2019; Yew, Shameli, Mohamad, Lee, & Teow, 2020). Therapeutically, carrageenan is explored for its antioxidant and anti-hyperglycemic potential (Rafiquzzaman, Ahmed, Lee, Noh, Jo, & Kong, 2016). A study reveals the influence of carrageenan food supplements on patients with cardiovascular disease, resulting in the reversal of elevated lipid profile and demonstrating moderate modulation of immune markers. It has also exerted antitumor, immunomodulatory, antiviral, and anticoagulant effects (Groult, Cousin, Chot-Plassot, Maura et al., 2019; Cicinskas, Kalitnik, Karetin, Mohan Ram, Achary, & Kravchenko, 2020; Song, Peng, Wang, Liu et al., 2020). Carrageenan can serve as a prebiotic and is found to effectively regulate the composition of gut microorganisms (Qiu, Aweya, Liu, Liu, Tang, Zhang, & Cheong, 2022).

5.7.6 HYALURONIC ACID (HA)

HA is a polysaccharide made up of repeating disaccharide units comprising N-acetyl-D-glucosamine and D-glucuronic acid (Abdallah, Fernández, Matias, & Do Rosário Bronze, 2020). Other naturally occurring glycosaminoglycans are smaller in size than HA, which has 100 to 20,000 repeating units with a molecular weight between 105 and 108 Da (Laurent & Fraser, 1992; Sadhasivam & Muthuvel, 2014). HA is found in abundance within the human body in the umbilical cord, embedded in an intracellular matrix of connective tissues, in fluids like the synovial fluid (1400 to 3600 μg/mL), and the vitreous humor (Fakhari & Berkland, 2013). Commercially, extraction of HA is carried out from the vitreous humor of Swordfish and Sharks, Mollusc bivalve, and the liver of Stingray and Tuna fish (Murado, Montemayor, Cabo, Vázquez & González, 2012; Kanchana, Arumugam, Giji, & Balasubramanian, 2013). Chemical hydrolysis of the tissue is a must to confirm the disruption of the proteoglycan core, which will then be subjected to the elimination of proteins for recovery of the glycosaminoglycan.

The purification method generally employed is ultrafiltration-diafiltration. Here, the separation process is size-based to eliminate impurities and then concentrate the polysaccharide in solution. Also, dialysis and ion exchange techniques are the purification techniques employed for HA (Choi, Choi, Kim, Lee, Noh, & Kim, 2014; Opdensteinen, Clodt, Müschen, Filiz, & Buyel, 2019).

5.7.7 ULVAN

Ulvan was first introduced in 1993 as the sulfated rhamnoglycuronans isolated and purified from *Ulva* species (Kidgell, Magnusson, De Nys, & Glasson, 2019). Basically, ulvan is a sulfated polysaccharide principally composed of uronic acids and sulfated from members of the order Ulvales. Hence, a repeating disaccharide-like structure of ulvan chiefly comprised of a uronic acid is generally linked to a sulfated neutral sugar (Ray & Lahaye, 1995). Ulvan is a cell wall polysaccharide of *Ulva* species algae. It contributes 9 to 36% of the Ulva biomass and mostly consists of sulfated rhamnose, uronic acids, glucuronic acid and iduronic acid, and xylose (Quemener, Lahaye, & Bobin-Dubigeon, 1997; Lahaye & Robic, 2007). The source of algal species, the ecophysiology they reside in as well as handling processes to synthesize ulvan and biomass affect the composition of ulvan. α- and β-(1,4)-linked monosaccharides such as xylose, rhamnose, glucuronic, and iduronic acid by specific repeating disaccharide units create the ulvan backbone (Paradossi, Cavalieri, Pizzoferrato, & Liquori, 1999). While the mole ratio of basic sugars in ulvan from a specific source and batch is defined, the molecular structure can be altered via depolymerization and the addition or removal of functional groups. Hence, molecular weights and degree of sulfation (2.3 to 40%) of ulvan extracts differ widely which exerts a large influence on physicochemical properties and biological activities. The molecular structure of ulvan gets altered by polymerization and by performing removal or addition of sulfate esters (Wang, Hu, Nie, Yu, & Xie, 2016; Qi, Zhao, Zhang, Li, Zhao, & Xing, 2005; Pengzhan, Quanbin, Hong, Xizhen, & Zhien, 2004).

Ulvan is highly practiced in biomedical applications like tissue engineering, wound dressing, regenerative medicine, and biofilm prevention. Also, Ulvan is beneficial as an antiviral, anticoagulant, antihyperlipidemic, anticancer, antioxidant, as well as immunostimulatory (Alves, Sousa, & Reis, 2013; Kidgell, Magnusson, De Nys, & Glasson, 2019; Ahmed & Ahmed, 2014; Qin, Du, Xiao, Li, & Gao, 2002).

5.8 APPLICATIONS OF MARINE POLYSACCHARIDES

5.8.1 AS MULTIFUNCTIONAL EXCIPIENTS

Marine polysaccharides have many valuable applications in the development of pharmaceuticals, with an exceptional impact on the development of multifunctional excipients. This results in the improvement of bioavailability as well as applications in drug delivery. The anti-cancer property of marine polysaccharides has applications as an excipient in anti-cancer drug formulation (Saravanan, Bhaskar, Rao, & Dhanaraju, 2003).

5.8.2 BINDER

As a good binder, marine polysaccharides used in the manufacturing of tablets employ the wet granulation method, where binders in the form of a solution or as solid dry powder are followed by the addition of a granulating agent. Marine polysaccharides with a higher concentration of hydroxyl groups usually show a greater water-binding capacity, making the wet granulation process easier. Bioadhesive marine polysaccharides are found to have very good cohesive strength as well as high binding strength, allowing them to remain attached to the actives under proper tensional conditions (Paolucci, Fasulo, & Volpe, 2015).

5.8.3 DILUENT

Marine polysaccharides find their applications as fillers or diluents to increase the bulk of solid dosage forms, particularly when they contain a low concentration of active pharmaceutical ingredients. For example, Chitosan is used as a diluent, binder, and disintegrating agent (De Jesus Raposo, De Morais, & De Morais, 2014).

5.8.4 DISINTEGRANTS

The breakdown of tablet formulations into smaller particles is essential for faster dissolution of the dosage form. Disintegrants are added to the tablet, and tablets with disintegrants dissolve more rapidly than those without them. Increasing porosity, wettability, capillary action, and wicking are the mechanisms by which marine polysaccharides behave through swelling when they come in contact with aqueous fluids. This aids in breaking the tablet into granular mass and then into smaller particles (Thomas & Kim, 2013).

5.8.5 DISSOLUTION ENHANCERS

Dissolution enhancers are necessary to improve the dissolution of poorly soluble drugs. Marine polysaccharides may function to enhance dissolution, reducing cohesive forces that hold a solid dosage form together and helping to break it into smaller granules, thereby increasing the active surface area for dissolution (Laurienzo, 2010).

5.8.6 DRUG DELIVERY CARRIERS

Extensive utilization of marine polysaccharides is found to produce carriers for drug delivery. Properties like biocompatibility, non-toxicity, biodegradability, and stimuli responsiveness make them suitable sources for the construction of composite devices with the loading and release of drugs in a controlled manner (Dhanaraju, Elizabeth, & Gunasekaran, 2011). For example, alginate is found as and in the multifunctional excipient, hydrogel matrices, stabilizers, microparticles, beads, etc. Multifunctional excipient applications of fucoidans in microspheres, fucospheres, nanoparticles, insulin-controlled release. Chitosan from marine chitin has applications in nanoparticles, beads, membranes, scaffolds, and films, regenerative medicine, and tissue engineering, for release in an accelerated manner and sustained release. Chondroitin sulfate for controlled release and anticancer drug delivery. Carrageenans from red algae in tablet compression, controlled release, sustained release, fast release, and pH-sensitive delivery systems. Hyaluronan has applications in the preparation of hydrogels, nanoparticles, microparticles, coating material, and liposomes.

5.8.7 AS THERAPEUTIC AGENTS

With precise therapeutic effects, marine polysaccharides can be integrated in the form of excipients in the pharmaceutical formulation, used for the same diseased condition. Carrageenans are found with anticoagulant, anticancer activity, anti-inflammatory activity, antioxidant, immuno-modulator, antibacterial, anti-hyperlipidemic, and antiviral properties along with their use in dengue, hepatitis A, and HIV (Carlucci, Ciancia, Matulewicz, Cerezo, & Damonte, 1999; Silva, Dore, Marques, Nascimento et al., 2010; Rocha De Souza, Marques, Guerra Dore, Ferreira Da Silva, Oliveira Rocha, & Leite, 2007; Hejazi & Amiji, 2003). Fucoidans have anti-tumor, anti-inflammatory, anti-coagulant, anti-thrombin, anti-adhesive, and anti-viral activity and are useful in burn treatment (Li, Lu, Wei, & Zhao, 2008; Pinheiro, Bourbon, Cerqueira, Maricato et al., 2015; Sezer, & Akbuğa, 2006; Anastyuk, Shevchenko, Ermakova, Vishchuk et al., 2012; Kim, Ravichandran, Khan, & Kim, 2008; Alves, Sousa, & Reis, 2013; Jeong, Kim, Jung, Park et al., 2013; Rabea, Badawy, Stevens, Smagghe, & Steurbaut, 2003; Kim, Jeon, & Lee, 2014). Alginate has anti-cystic fibrosis activity and anti-hypertensive efficacy. Chitosan is associated with antimicrobial, anti-inflammatory, anti-tumor activity, and antifungal properties. Chondroitin sulfate is useful as an anti-coagulant, a supplement for arthritis, and keratan sulfate for osteo-arthritis. Agarose has therapeutic effects for glucose intolerance in type 2 diabetes mellitus, anti-bacterial activity, and weight loss. Hyaluronans are utilized for wound treatment and as a supplement for arthritis.

5.8.8 IN THE DEVELOPMENT OF NANO FORMULATIONS

Marine polysaccharides, with good biodegradability, biocompatibility, low cost, nontoxicity, and abundance, the nanomaterials based on them have garnered attention as one of the most important research areas recently. Nowadays, medications based on marine nanoparticles or biopolymer nanoparticles, micelles, metals and metal oxides, liposomes, and dendrimers are actively being considered for combating various diseases, including cancer, tissue engineering, gene and drug delivery, and wound healing. Marine polysaccharides can be easily processed to convert into nanoparticles, microparticles, nanofibers, beads, gels, scaffolds, and sponges, showcasing numerous biomedical applications in the field of nanobiotechnology. Chitosan has potential applications for nasal and gastrointestinal delivery systems due to its strong electrostatic interaction with mucus.

The cationic property of chitosan is utilized in nanoformulations. Chitosan combined with gelatin serves as a polymer for the preparation of pH-sensitive microspheres, which have been reported to release the drug substance in acidic media only (Hejazi & Amiji, 2003; Illum, Jabbal-Gill, Hinchcliffe, Fisher, & Davis, 2001). Protein drugs can be encapsulated in the pH-sensitive hydrogel of chitosan to achieve controlled drug release (Chen, Wu, Mi, Lin, Yu, & Sung, 2004).

Nanofibers can be easily prepared with chitosan, which can be cast into films from its acidic aqueous solutions. Electrospinning to prepare nanofibers from acetic acid solutions of chitosan has been reported. Microparticles of alginates for better bioavailability and reduced side effects compared with a drug in conventional formulations were reported for tuberculosis (Geng, Kwon, & Jang, 2005; Sharma, Khuller, & Garg, 2003). Alginate nanocomposites as a sustained release system for Vitamin B1 and B6 were described (Kevadiya, Joshi, Patel, Ingole, Mody, & Bajaj, 2010). For many years, wound-healing dressings of alginates have been effectively utilized for the cleansing of secreting wounds and cuts (Gilchrist & Martin, 1983). Surface-coated nanostructures of polysaccharides have been achieved by the addition of chitosan to heparin and hyaluronic acid (Boddohi, Almodóvar, Zhang, Johnson, & Kipper, 2010). Tailored nanocomposites of HA due to their antibacterial properties, with chitosan, have been reported for chronic ulcer dressing and wound healing dressing (Abdel-Mohsen, Hrdina, Burgert, Abdel-Rahman et al., 2013).

Various types of DDs formed by the utilization of marine polysaccharides are represented in Figure 5.4.

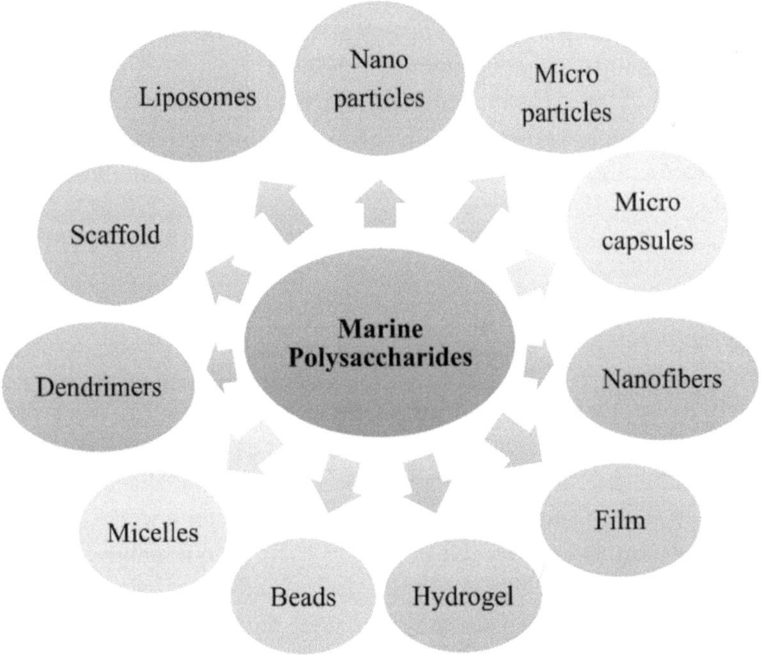

FIGURE 5.4 Marine polysaccharides in different drug delivery systems.

5.8.9 IN BIOMEDICAL AND TISSUE ENGINEERING

Biomedical applications of chitosan are found in artificial skins and surgical sutures for wound healing approved by the FDA (Croisier & Jérôme, 2013; Rinaudo, 2006; Aranaz, Harris, & Heras, 2010; Thanou, Verhoef, & Junginger, 2001). Other applications include artificial blood vessels, drug delivery tissue engineering, and controlled drug release (Prabaharan & Mano, 2004; Di Martino, Sittinger, & Risbud, 2005). Also, for ocular contact lenses and eye humor fluid, topical for bandages, sponges, and burn dressings (Şenel & McClure, 2004). Chitosan demonstrated its utilization in blood cholesterol maintenance, as an anti-inflammatory, and for tumor inhibition. Other uses include dental plaque inhibition, anti-viral, bone healing treatment, hemostatic agent, and wound healing (Ueno, Mori, & Fujinaga, 2001; Ma, Kotaki, Inai, & Ramakrishna, 2005). It has periodontal as well as orthopedic applications too (Khor & Lim, 2003). In biomedical and biotechnological applications, alginate gels are used for the control of viruses and other living cells (Ito & Chang, 1992; Kalyanasundaram, Feinstein, Nicholson, Leong, & Garver, 1999). Bioartificial endocrine glands are created by utilizing alginate gel as implantation material like in the thyroid follicles and islets of Langerhans (Bürgi-Saville, Reut, Gerber, Peter et al., 1998). Oral vaccines with antigen-coated microspheres of alginate are reported for the development of immunity against rotavirus pathogens (Bowersock, HogenEsch, Suckow, Guimond et al., 1999). Carrageenan in combination with synthetic polymers has biomedical applications. A microwave irradiation process for blends of κ-carrageenan-graft-PVP has been reported (Prasad, Mehta, Meena, & Siddhanta, 2006). HA has been widely utilized in polysaccharides in tissue culture scaffolds in the biomedical field (Collins & Birkinshaw, 2013).

5.9 CONCLUSION

Natural polysaccharides are found in wide applications in drug delivery systems where marine source polysaccharides have contributed majorly. These polysaccharides, due to their biocompatible and biodegradable characteristics, are found as highly suitable substances for the design and development of controlled or targeted release DDs. Due to their unique nature, tailor-made development of polysaccharides with desired characteristics has made it possible to incorporate a wide range of drug substances for efficient drug delivery. Such biomaterials and their derivatives serve as versatile excipients for many pharmaceuticals. With the development of high-end extraction and purification processes, the field of marine polysaccharides has gradually evolved. Not only as multifunctional excipients for DDs, but many marine polysaccharides have also demonstrated potential therapeutic substances such as anticancer, antidiabetic, antiviral, antiobesity, anti-inflammatory, antioxidant, anticoagulant, etc. In the fields of biomedical tissue engineering, wound accessories, scaffold development, and protein and peptide delivery, marine polysaccharides have exhibited eye-catching applications. Nanoparticulate drug delivery for targeted and controlled release formulations has strengthened its base after the initiation of the use of marine polysaccharides in their fabrication. More and more experiments incorporating marine polysaccharides and their derivatives will explore a way for designing new and specific drug delivery systems for therapy.

5.10 FUTURE PERSPECTIVES

Nature has gifted many sources to humans, and marine polysaccharides are one such precious resource for drug delivery. The inherent properties of marine polysaccharides have facilitated the design and development of various drug delivery systems, including nanoparticles, through the incorporation of different types of drugs, proteins, or peptides. The production and utilization of marine polysaccharides in various drug delivery systems are not easy. Several challenges, such as large-scale production feasibility, controlled release of drugs within their therapeutic window, effective targeted drug delivery, and route of administration, need to be addressed when incorporating marine polysaccharides into drug delivery systems. Advanced extraction, purification, and chemical modification technologies pave the way for marine polysaccharides to be converted into highly compatible forms for embedding bioactive substances. The synthesis of new derivatives based on existing ones may open the door to exploring the therapeutic and excipient values of marine polysaccharides. However, the metabolism of such substances is yet to be fully explored to understand their functionalities. More clinical evidence needs to be generated to strengthen the application of these potential substances in the biomedical and tissue engineering fields.

ACKNOWLEDGMENT

The authors are thankful to Shree S. K. Patel College of Pharmaceutical Education and Research, Ganpat University, Gujarat, India, and Parul Institute of

Pharmacy, Parul University, Vadodara, Gujarat for extending support to carry out this work. Dr. Prajapati, extends his sincere appreciation to the Faculty of Pharmacy, Silpakorn University, Thailand, for their generous support that enabled the completion of this work.

CONFLICT OF INTEREST

The authors declare no competing interests.

KEYWORDS

- **biodegradable**
- **biomaterial**
- **drug delivery systems**
- **fucoidan**
- **marine polysaccharides**
- **targeted delivery**

REFERENCES

Abdallah, M. M., Fernández, N., Matias, A. A., & do Rosário Bronze, M. (2020). Hyaluronic acid and Chondroitin sulfate from marine and terrestrial sources: Extraction and purification methods. *Carbohydrate Polymers, 243*, 116441.

Abdel-Mohsen, A. M., Hrdina, R., Burgert, L., Abdel-Rahman, R. M., Hašová, M., Šmejkalová, D., Kolář, M., Pekar, M., & Aly, A. S. (2013). Antibacterial activity and cell viability of hyaluronan fiber with silver nanoparticles. *Carbohydrate Polymers, 92*(2), 1177–1187.

Abedini, F., Ebrahimi, M., Roozbehani, A. H., Domb, A. J., & Hosseinkhani, H. (2018). Overview on natural hydrophilic polysaccharide polymers in drug delivery. *Polymers for Advanced Technologies, 29*(10), 2564–2573.

Ahmed, O. M., & Ahmed, R. R. (2014). Anti-proliferative and apoptotic efficacies of ulvan polysaccharides against different types of carcinoma cells in vitro and in vivo. *Journal of Cancer Science and Therapy, 6*, 202–208.

Ale, M. T., Maruyama, H., Tamauchi, H., Mikkelsen, J. D., & Meyer, A. S. (2011). Fucoidan from Sargassum sp. and *Fucus vesiculosus* reduces cell viability of lung carcinoma and melanoma cells in vitro and activates natural killer cells in mice in vivo. *International Journal of Biological Macromolecules, 49*(3), 331–336.

Alves, A., Sousa, R. A., & Reis, R. L. (2013). In vitro cytotoxicity assessment of ulvan, a polysaccharide extracted from green algae. *Phytotherapy Research, 27*(8), 1143–1148.

Anastyuk, S. D., Shevchenko, N. M., Ermakova, S. P., Vishchuk, O. S., Nazarenko, E. L., Dmitrenok, P. S., & Zvyagintseva, T. N. (2012). Anticancer activity in vitro of a fucoidan from the brown alga *Fucus evanescens* and its low-molecular fragments, structurally characterized by tandem mass-spectrometry. *Carbohydrate Polymers, 87*(1), 186–194.

Aranaz, I., Harris, R., & Heras, A. (2010). Chitosan amphiphilic derivatives. Chemistry and applications. *Current Organic Chemistry, 14*(3), 308–330.

Arlov, O., Oztürk, E., Steinwachs, M., Skjåk-Bræk, G., & Zenobi-Wong, M. (2017). Biomimetic sulphated alginate hydrogels suppress IL-1B-induced inflammatory responses in human chondrocytes. *European Cells and Materials, 33*, 76.

Atanase, L. I. (2021). Micellar drug delivery systems based on natural biopolymers. *Polymers, 13*(3), 477.

Atanase, L. I., Desbrieres, J., & Riess, G. (2017). Micellization of synthetic and polysaccharides-based graft copolymers in aqueous media. *Progress in Polymer Science, 73*, 32–60.

Bhatia, S., Sharma, A., Sharma, K., Kavale, M., Chaugule, B., Dhalwal, K., & Mahadik, K. (2008). Novel algal polysaccharides from marine source: Porphyran. *Pharmacognosy Reviews, 2*(4), 271.

Bilan, M. I., Grachev, A. A., Ustuzhanina, N. E., Shashkov, A. S., Nifantiev, N. E., & Usov, A. I. (2004). A highly regular fraction of a fucoidan from the brown seaweed *Fucus distichus* L. *Carbohydrate Research, 339*(3), 511–517.

Boddohi, S., Almodóvar, J., Zhang, H., Johnson, P. A., & Kipper, M. J. (2010). Layer-by-layer assembly of polysaccharide-based nanostructured surfaces containing polyelectrolyte complex nanoparticles. *Colloids and Surfaces B: Biointerfaces, 77*(1), 60–68.

Bonferoni, M. C., Rossi, S., Ferrari, F., Bertoni, M., Bolhuis, G. K., & Caramella, C. (1998). On the employment of λ carrageenan in a matrix system. III. Optimization of a λ carrageenan–HPMC hydrophilic matrix. *Journal of Controlled Release, 51*(2–3), 231–239.

Bowersock, T. L., HogenEsch, H., Suckow, M., Guimond, P., Martin, S., Borie, D., Torregrosa, S., Park, H., & Park, K. (1999). Oral vaccination of animals with antigens encapsulated in alginate microspheres. *Vaccine, 17*(13–14), 1804–1811.

Bürgi-Saville, M. E., Reut, B., Gerber, H., Peter, H. J., Paulsson, M., Kaempf, J., Simon, F., Marti, U., & Bürgi, U. (1998). Alginate gel culture allows the retention of extracellular matrix and follicular structure of rat thyroid tissue but does not lead to the formation of follicles by FRTL-5 cells. *Thyroid, 8*(12), 1147–1155.

Cardoso, M. J., Costa, R. R., & Mano, J. F. (2016). Marine origin polysaccharides in drug delivery systems. *Marine Drugs, 14*(2), 34.

Carlucci, M. J., Ciancia, M., Matulewicz, M. C., Cerezo, A. S., & Damonte, E. B. (1999). Antiherpetic activity and mode of action of natural carrageenans of diverse structural types. *Antiviral Research, 43*(2), 93–102.

Castanheira, E. J., Correia, T. R., Rodrigues, J. M., & Mano, J. F. (2020). Novel biodegradable laminarin microparticles for biomedical applications. *Bulletin of the Chemical Society of Japan, 93*(6), 713-719.

Chakraborty, M., Miao, C., McDonald, A., & Chen, S. (2012). Concomitant extraction of bio-oil and value-added polysaccharides from *Chlorella sorokiniana* using a unique sequential hydrothermal extraction technology. *Fuel, 95*, 63–70.

Chao, C. H., Wu, H. J., & Lu, M. K. (2011). Promotion of fungal growth and underlying physiochemical changes of polysaccharides in *Rigidoporus ulmarius*, an edible Basidiomycete mushroom. *Carbohydrate Polymers, 85*(3), 609–614.

Chascall, V., Calabro, A., Midura, R. J., & Yanagishita, M. (1994). [24] Isolation and characterization of proteoglycans. In *Methods in Enzymology* (Vol. 230, pp. 390–417). Academic Press.

Chen, S. C., Wu, Y. C., Mi, F. L., Lin, Y. H., Yu, L. C., & Sung, H. W. (2004). A novel pH-sensitive hydrogel composed of N, O-carboxymethyl chitosan and alginate cross-linked by genipin for protein drug delivery. *Journal of Controlled Release, 96*(2), 285–300.

Chen, W. J., & Abatangelo, G. (2002). Functions of hyaluronan in wound repair. *Hyaluronan, 2,* 147–156.

Chen, X., Han, W., Zhao, X., Tang, W., & Wang, F. (2019). Epirubicin-loaded marine carrageenan oligosaccharide capped gold nanoparticle system for pH-triggered anticancer drug release. *Scientific Reports, 9*(1), 1–10.

Choi, S., Choi, W., Kim, S., Lee, S. Y., Noh, I., & Kim, C. W. (2014). Purification and biocompatibility of fermented hyaluronic acid for its applications to biomaterials. *Biomaterials Research, 18*(1), 1–10.

Cicinskas, E., Kalitnik, A. A., Karetin, Y. A., Mohan Ram, M. S., Achary, A., & Kravchenko, A. O. (2020). Immunomodulating properties of carrageenan from *Tichocarpus crinitus. Inflammation, 43*(4), 1387–1396.

Collins, M. N., & Birkinshaw, C. (2013). Hyaluronic acid-based scaffolds for tissue engineering—A review. *Carbohydrate Polymers, 92*(2), 1262–1279.

Correa, R. C., Brugnari, T., Bracht, A., Peralta, R. M., & Ferreira, I. C. (2016). Biotechnological, nutritional and therapeutic uses of Pleurotus spp. (*Oyster mushroom*) related with its chemical composition: A review on the past decade findings. *Trends in Food Science & Technology, 50*, 103–117.

Correc, G., Hehemann, J. H., Czjzek, M., & Helbert, W. (2011). Structural analysis of the degradation products of porphyran digested by *Zobellia galactanivorans* β-porphyranase A. *Carbohydrate Polymers, 83*(1), 277–283.

Croisier, F., & Jérôme, C. (2013). Chitosan-based biomaterials for tissue engineering. *European Polymer Journal, 49*(4), 780–792.

Cui, Y., Ye, Q., Wang, H., Li, Y., Yao, W., & Qian, H. (2014). Hepatoprotective potential of Aloe vera polysaccharides against chronic alcohol-induced hepatotoxicity in mice. *Journal of the Science of Food and Agriculture, 94*(9), 1764–1771.

da Silva, T. L., Vidart, J. M., da Silva, M. G., Gimenes, M. L., & Vieira, M. G. (2017). Alginate and sericin: Environmental and pharmaceutical applications. In *Biological Activities and Application of Marine Polysaccharides* (pp. 57–86). Elsevier.

Das, S., Lyla, P. S., & Khan, S. A. (2006). Marine microbial diversity and ecology: Importance and future perspectives. *Current Science, 90*(10), 1325–1335.

de Jesus Raposo, M. F., De Morais, A. M., & De Morais, R. M. (2014). Bioactivity and applications of polysaccharides from marine microalgae. *Polysaccharides: Bioactivity and Biotechnology, 1,* 1683–1727.

Dhanaraju, M. D., Elizabeth, S., & Gunasekaran, T. (2011). Triamcinolone-loaded glutaraldehyde cross-linked chitosan microspheres: Prolonged release approach for the treatment of rheumatoid arthritis. *Drug Delivery, 18*(3), 198–207.

Di Martino, A., Sittinger, M., & Risbud, M. V. (2005). Chitosan: A versatile biopolymer for orthopaedic tissue engineering. *Biomaterials, 26*(30), 5983–5990.

Di, X., Liang, X., Shen, C., Pei, Y., Wu, B., & He, Z. (2022). Carbohydrates used in polymeric systems for drug delivery: From structures to applications. *Pharmaceutics, 14*(4), 739.

Ermakova, S., Men'shova, R., Vishchuk, O., Kim, S. M., Um, B. H., Isakov, V., & Zvyagintseva, T. (2013). Water-soluble polysaccharides from the brown alga Eisenia bicyclis: Structural characteristics and antitumor activity. *Algal Research, 2*(1), 51–58.

Fakhari, A., & Berkland, C. (2013). Applications and emerging trends of hyaluronic acid in tissue engineering, as a dermal filler and in osteoarthritis treatment. *Acta Biomaterialia, 9*(7), 7081–7092.

Felt, O., Buri, P., & Gurny, R. (1998). Chitosan: a unique polysaccharide for drug delivery. *Drug Development and Industrial Pharmacy, 24*(11), 979–993.

Funderburgh, J. L. (2002). Keratan sulfate biosynthesis. *IUBMB Life, 54*(4), 187–194.

Ganta, S., Devalapally, H., Shahiwala, A., & Amiji, M. (2008). A review of stimuli-responsive nanocarriers for drug and gene delivery. *Journal of Controlled Release, 126*(3), 187–204.

García-González, C. A., Alnaief, M., & Smirnova, I. (2011). Polysaccharide-based aerogels—Promising biodegradable carriers for drug delivery systems. *Carbohydrate Polymers, 86*(4), 1425–1438.

Geng, X., Kwon, O. H., & Jang, J. (2005). Electrospinning of chitosan dissolved in concentrated acetic acid solution. *Biomaterials, 26*(27), 5427–5432.

Geonzon, L. C., Descallar, F. B., Du, L., Bacabac, R. G., & Matsukawa, S. (2020). Gelation mechanism and network structure in gels of carrageenans and their mixtures viewed at different length scales–a review. *Food Hydrocolloids, 108*, 106039.

Gilchrist, T., & Martin, A. M. (1983). Wound treatment with Sorbsan—an alginate fiber dressing. *Biomaterials, 4*(4), 317–320.

Gombotz, W. R., & Wee, S. (1998). Protein release from alginate matrices. *Advanced Drug Delivery Reviews, 31*(3), 267–285.

Grech, J. M., Mano, J. F., & Reis, R. L. (2008). Chitosan beads as templates for layer-by-layer assembly and their application in the sustained release of bioactive agents. *Journal of Bioactive and Compatible Polymers, 23*(4), 367–380.

Haag, R., & Kratz, F. (2006). Polymer therapeutics: Concepts and applications. *Angewandte Chemie International Edition, 45*(8), 1198–1215.

Habibi, Y., Lucia, L. A., & Rojas, O. J. (2010). Cellulose nanocrystals: Chemistry, self-assembly, and applications. *Chemical Reviews, 110*(6), 3479–3500.

Hahn, T., Lang, S., Ulber, R., & Muffler, K. (2012). Novel procedures for the extraction of fucoidan from brown algae. *Process Biochemistry, 47*(12), 1691–1698.

Haug, A. (1964). Composition and properties of alginates. *Norwegian Institute of Seaweed Research*, 1–123.

He, W., Fu, L., Li, G., Jones, J. A., Linhardt, R. J., & Koffas, M. (2015). Production of chondroitin in metabolically engineered *E. coli*. *Metabolic Engineering, 27*, 92–100.

Hejazi, R., & Amiji, M. (2003). Chitosan-based gastrointestinal delivery systems. *Journal of Controlled Release, 89*(2), 151–165.

Herrero, M., & Ibanez, E. (2015). Green processes and sustainability: An overview on the extraction of high added-value products from seaweeds and microalgae. *The Journal of Supercritical Fluids, 96*, 211–216.

Hoare, T. R., & Kohane, D. S. (2008). Hydrogels in drug delivery: Progress and challenges. *Polymer, 49*(8), 1993–2007.

Huang, X., Luo, X., Liu, L., Dong, K., Yang, R., Lin, C., Song, H., Li, S., & Huang, Q. (2020). Formation mechanism of egg white protein/κ-Carrageenan composite film and its application to oil packaging. *Food Hydrocolloids, 105*, 105780.

Illum, L., Jabbal-Gill, I., Hinchcliffe, M., Fisher, A. N., & Davis, S. S. (2001). Chitosan as a novel nasal delivery system for vaccines. *Advanced Drug Delivery Reviews, 51*(1–3), 81–98.

Ito, Y., & Chang, T. M. (1992). In vitro study of multicellular hepatocyte spheroids formed in microcapsules. *Artificial Organs, 16*(4), 422–427.

Jeong, Y. T., Kim, Y. D., Jung, Y. M., Park, D. C., Lee, D. S., Ku, S. K., & Lee, J. Y. (2013). Low molecular weight fucoidan improves endoplasmic reticulum stress-reduced insulin sensitivity through AMP-activated protein kinase activation in L6 myotubes and restores lipid homeostasis in a mouse model of type 2 diabetes. *Molecular Pharmacology, 84*(1), 147–157.

Jia, X., Yang, J., Wang, Z., Liu, R., & Xie, R. (2014). Polysaccharides from Laminaria japonica show hypoglycemic and hypolipidemic activities in mice with experimentally induced diabetes. *Experimental Biology and Medicine, 239*(12), 1663–1670.

Jiang, G. B., Quan, D., Liao, K., & Wang, H. (2006). Novel polymer micelles prepared from chitosan grafted hydrophobic palmitoyl groups for drug delivery. *Molecular Pharmaceutics, 3*(2), 152–160.

Jin, P., Zhang, L., Yuan, P., Kang, Z., Du, G., & Chen, J. (2016). Efficient biosynthesis of polysaccharides chondroitin and heparosan by metabolically engineered Bacillus subtilis. *Carbohydrate Polymers, 140*, 424–432.

Jindal, N., & Khattar, J. S. (2018). Microbial polysaccharides in food industry. In *Biopolymers for Food Design* (pp. 95–123). Academic Press.

Kadam, S. U., O'Donnell, C. P., Rai, D. K., Hossain, M. B., Burgess, C. M., Walsh, D., & Tiwari, B. K. (2015). Laminarin from Irish brown seaweeds *Ascophyllum nodosum* and *Laminaria hyperborea*: Ultrasound assisted extraction, characterization and bioactivity. *Marine Drugs, 13*(7), 4270–4280.

Kalyanasundaram, S., Feinstein, S., Nicholson, J. P., Leong, K. W., & Garver, R. I. (1999). Coacervate microspheres as carriers of recombinant adenoviruses. *Cancer Gene Therapy, 6*(2), 107–112.

Kanchana, S., Arumugam, M., Giji, S., & Balasubramanian, T. (2013). Isolation, characterization and antioxidant activity of hyaluronic acid from marine bivalve mollusc *Amussium pleuronectus* (Linnaeus, 1758). *Bioactive Carbohydrates and Dietary Fibre, 2*(1), 1–7.

Kevadiya, B. D., Joshi, G. V., Patel, H. A., Ingole, P. G., Mody, H. M., & Bajaj, H. C. (2010). Montmorillonite-alginate nanocomposites as a drug delivery system: Intercalation and in vitro release of vitamin B1 and vitamin B6. *Journal of Biomaterials Applications, 25*(2), 161–177.

Khan, M. U., Raza, M. A., Mehboob, H., Kadir, M. R., Abd Razak, S. I., Shah, S. A., Iqbal, M. Z., & Amin, R. (2020). Development and in vitro evaluation of κ-carrageenan based polymeric hybrid nanocomposite scaffolds for bone tissue engineering. *RSC Advances, 10*(66), 40529–40542.

Khor, E., & Lim, L. Y. (2003). Implantable applications of chitin and chitosan. *Biomaterials, 24*(13), 2339–2349.

Khoshnevisan, K., Daneshpour, M., Barkhi, M., Gholami, M., Samadian, H., & Maleki, H. (2018). The promising potentials of capped gold nanoparticles for drug delivery systems. *Journal of Drug Targeting, 26*(7), 525–532.

Khotimchenko, Y. S., Kovalev, V. V., Savchenko, O. V., & Ziganshina, O. A. (2001). Physical–chemical properties, physiological activity, and usage of alginates, the polysaccharides of brown algae. *Russian Journal of Marine Biology, 27*(1), S53–S64.
Kidgell, J. T., Magnusson, M., de Nys, R., & Glasson, C. R. (2019). Ulvan: A systematic review of extraction, composition and function. *Algal Research, 39*, 101422.
Kim, M. J., Jeon, J., & Lee, J. S. (2014). Fucoidan prevents high-fat diet-induced obesity in animals by suppression of fat accumulation. *Phytotherapy Research, 28*(1), 137–143.
Kim, S. K., Ravichandran, Y. D., Khan, S. B., & Kim, Y. T. (2008). Prospective of the cosmeceuticals derived from marine organisms. *Biotechnology and Bioprocess Engineering, 13*(5), 511–523.
Kylin, H. (1913). Biochemistry of sea algae. *H. Z. Physiol. Chem., 83*, 171–197.
Lahaye, M., & Robic, A. (2007). Structure and functional properties of ulvan, a polysaccharide from green seaweeds. *Biomacromolecules, 8*(6), 1765–1774.
Laurent, T. C., & Fraser, J. R. (1992). Hyaluronan. *FASEB Journal: Official Publication of the Federation of American Societies for Experimental Biology, 6*(7), 2397–2404.
Laurienzo, P. (2010). Marine polysaccharides in pharmaceutical applications: An overview. *Marine Drugs, 8*(9), 2435–2465.
Lee, K. Y., & Mooney, D. J. (2012). Alginate: Properties and biomedical applications. *Progress in Polymer Science, 37*(1), 106–126.
Lee, S. J., Kim, C. W., Jang, H. J., Cho, S. Y., & Choi, J. W. (2011). Anti-hyperlipidemia and anti-arteriosclerosis effects of Laminaria japonica in Sprague-Dawley rats. *Fisheries and Aquatic Sciences, 14*(4), 235–241.
Lee, Y. E., Kim, H., Seo, C., Park, T., Lee, K. B., Yoo, S. Y., Hong, S. C., Kim, J. T., & Lee, J. (2017). Marine polysaccharides: Therapeutic efficacy and biomedical applications. *Archives of Pharmacal Research, 40*(9), 1006–1020.
Li, B., Lu, F., Wei, X., & Zhao, R. (2008). Fucoidan: Structure and bioactivity. *Molecules, 13*(8), 1671–1695.
Li, J. M., & Nie, S. P. (2016). The functional and nutritional aspects of hydrocolloids in foods. *Food Hydrocolloids, 53*, 46–61.
Lignot, B., Lahogue, V., & Bourseau, P. (2003). Enzymatic extraction of chondroitin sulfate from skate cartilage and concentration-desalting by ultrafiltration. *Journal of Biotechnology, 103*(3), 281–284.
Lim, S. J., & Aida, W. M. (2017). Extraction of sulfated polysaccharides (fucoidan) from brown seaweed. In *Seaweed polysaccharides* (pp. 27–46). Elsevier.
Lim, S. J., Aida, W. M., Schiehser, S., Rosenau, T., & Böhmdorfer, S. (2019). Structural elucidation of fucoidan from *Cladosiphon okamuranus* (*Okinawa mozuku*). *Food Chemistry, 272*, 222–226.
Lima, A. C., Sher, P., & Mano, J. F. (2012). Production methodologies of polymeric and hydrogel particles for drug delivery applications. *Expert Opinion on Drug Delivery, 9*(2), 231–248.
Lima-Tenório, M. K., Tenório-Neto, E. T., Guilherme, M. R., Garcia, F. P., Nakamura, C. V., Pineda, E. A., & Rubira, A. F. (2015). Water transport properties through starch-based hydrogel nanocomposites responding to both pH and a remote magnetic field. *Chemical Engineering Journal, 259*, 620–629.
Ma, Z., Kotaki, M., Inai, R., & Ramakrishna, S. (2005). Potential of nanofiber matrix as tissue-engineering scaffolds. *Tissue Engineering, 11*(1–2), 101–109.

MacArtain, P., Gill, C. I., Brooks, M., Campbell, R., & Rowland, I. R. (2007). Nutritional value of edible seaweeds. *Nutrition Reviews, 65*(12), 535–543.

Malavaki, C., Mizumoto, S., Karamanos, N., & Sugahara, K. (2008). Recent advances in the structural study of functional chondroitin sulfate and dermatan sulfate in health and disease. *Connective Tissue Research, 49*(3–4), 133–139.

Martins, M., Vieira, F. A., Correia, I., Ferreira, R. A., Abreu, H., Coutinho, J. A., & Ventura, S. P. (2016). Recovery of phycobiliproteins from the red macroalga *Gracilaria* sp. using ionic liquid aqueous solutions. *Green Chemistry, 18*(15), 4287–4296.

McCarthy, P. J., & Pomponi, S. A. (2004). A search for new pharmaceutical drugs from marine organisms. *Marine Biology, 22*, 1–2.

McKim, J. M., Willoughby, J. A., Sr, Blakemore, W. R., & Weiner, M. L. (2019). Clarifying the confusion between poligeenan, degraded carrageenan, and carrageenan: a review of the chemistry, nomenclature, and in vivo toxicology by the oral route. *Critical Reviews in Food Science and Nutrition, 59*(19), 3054–3073.

Mehta, A., Sidhu, C., Pinnaka, A. K., & Roy Choudhury, A. (2014). Extracellular polysaccharide production by a novel osmotolerant marine strain of *Alteromonas macleodii* and its application towards biomineralization of silver. *PLoS One, 9*(6), e98798.

Mikami, T., & Kitagawa, H. (2013). Biosynthesis and function of chondroitin sulfate. *Biochimica et Biophysica Acta (BBA)-General Subjects, 1830*(10), 4719–4733.

Molinski, T. F., Dalisay, D. S., Lievens, S. L., & Saludes, J. P. (2009). Drug development from marine natural products. *Nature Reviews Drug Discovery, 8*(1), 69–85.

Mourão, P. A., & Pereira, M. S. (1999). Searching for alternatives to heparin: sulfated fucans from marine invertebrates. *Trends in Cardiovascular Medicine, 9*(8), 225–232.

Munro, M. H., Blunt, J. W., Dumdei, E. J., Hickford, S. J., Lill, R. E., Li, S., Battershill, C. N., & Duckworth, A. R. (1999). The discovery and development of marine compounds with pharmaceutical potential. *Progress in Industrial Microbiology, 35*, 15–25.

Murado, M. A., Fraguas, J., Montemayor, M. I., Vázquez, J. A., & González, P. (2010). Preparation of highly purified chondroitin sulphate from skate (*Raja clavata*) cartilage by-products: Process optimization including a new procedure of alkaline hydroalcoholic hydrolysis. *Biochemical Engineering Journal, 49*(1), 126–132.

Murado, M. A., Montemayor, M. I., Cabo, M. L., Vázquez, J. A., & González, M. P. (2012). Optimization of extraction and purification process of hyaluronic acid from fish eyeball. *Food and Bioproducts Processing, 90*(3), 491–498.

Muzzarelli, R. A., Greco, F., Busilacchi, A., Sollazzo, V., & Gigante, A. (2012). Chitosan, hyaluronan and chondroitin sulfate in tissue engineering for cartilage regeneration: A review. *Carbohydrate Polymers, 89*(3), 723–739.

Necas, J., & Bartosikova, L. (2013). Carrageenan: a review. *Veterinarni Medicina, 58*(6), 187–205.

Oliyaei, N., Moosavi-Nasab, M., & Mazloomi, S. M. (2022). Therapeutic activity of fucoidan and carrageenan as marine algal polysaccharides against viruses. *3 Biotech, 12*(7), 1–5.

Opdensteinen, P., Clodt, J. I., Müschen, C. R., Filiz, V., & Buyel, J. F. (2019). A combined ultrafiltration/diafiltration step facilitates the purification of cyanovirin-N from transgenic tobacco extracts. *Frontiers in Bioengineering and Biotechnology, 6*, 206.

Oyarzun-Ampuero, F. A., Brea, J., Loza, M. I., Torres, D., & Alonso, M. J. (2009). Chitosan–hyaluronic acid nanoparticles loaded with heparin for the treatment of asthma. *International Journal of Pharmaceutics, 381*(2), 122–129.

Pacheco-Quito, E. M., Ruiz-Caro, R., & Veiga, M. D. (2020). Carrageenan: drug delivery systems and other biomedical applications. *Marine Drugs, 18*(11), 583.

Panlasigui, L. N., Baello, O. Q., Dimatangal, J. M., & Dumelod, B. D. (2003). Blood cholesterol and lipid-lowering effects of carrageenan on human volunteers. *Asia Pacific Journal of Clinical Nutrition, 12*(2), 158–165.

Paolucci, M., Fasulo, G., & Volpe, M. G. (2015). Employment of marine polysaccharides to manufacture functional biocomposites for aquaculture feeding applications. *Marine Drugs, 13*(5), 2680–2693.

Paradossi, G., Cavalieri, F., Pizzoferrato, L., & Liquori, A. M. (1999). A physico-chemical study on the polysaccharide ulvan from hot water extraction of the macroalga Ulva. *International Journal of Biological Macromolecules, 25*(4), 309–315.

Parnell, J. A., & Reimer, R. A. (2012). Prebiotic fiber modulation of the gut microbiota improves risk factors for obesity and the metabolic syndrome. *Gut Microbes, 3*(1), 29–34.

Pengzhan, Y., Quanbin, Z., Hong, Z., Xizhen, N., & Zhien, L. (2004). Preparation of polysaccharides in different molecular weights from *Ulva pertusa* Kjellm (Chlorophyta). *Chinese Journal of Oceanology and Limnology, 22*(4), 381–385.

Pereira, L., & Cotas, J. (2019). Alginates—A general overview, alginates recent uses of this natural polymer; introductory chapter: Alginates—A overview. In *IntechOpen*, pp. 1–16.

Pereira, R. C., & Costa-Lotufo, L. V. (2012). Bioprospecting for bioactives from seaweeds: potential, obstacles and alternatives. *Revista Brasileira de Farmacognosia, 22*, 894–905.

Picker, K. M. (1999). Matrix tablets of carrageenans. I. A compaction study. *Drug Development and Industrial Pharmacy, 25*(3), 329–337.

Pinheiro, A. C., Bourbon, A. I., Cerqueira, M. A., Maricato, É., Nunes, C., Coimbra, M. A., & Vicente, A. A. (2015). Chitosan/fucoidan multilayer nanocapsules as a vehicle for controlled release of bioactive compounds. *Carbohydrate Polymers, 115*, 1–9.

Prabaharan, M., & Mano, J. F. (2004). Chitosan-based particles as controlled drug delivery systems. *Drug Delivery, 12*(1), 41–57.

Prasad, K., Mehta, G., Meena, R., & Siddhanta, A. K. (2006). Hydrogel-forming agar-graft-PVP and κ-carrageenan-graft-PVP blends: Rapid synthesis and characterization. *Journal of Applied Polymer Science, 102*(4), 3654–3663.

Prashanth, K. H., & Tharanathan, R. N. (2007). Chitin/chitosan: Modifications and their unlimited application potential—An overview. *Trends in Food Science & Technology, 18*(3), 117–131.

Qi, H., Zhao, T., Zhang, Q., Li, Z., Zhao, Z., & Xing, R. (2005). Antioxidant activity of different molecular weight sulfated polysaccharides from *Ulva pertusa* Kjellm (Chlorophyta). *Journal of Applied Phycology, 17*(6), 527–534.

Qin, C., Du, Y., Xiao, L., Li, Z., & Gao, X. (2002). Enzymic preparation of water-soluble chitosan and their antitumor activity. *International Journal of Biological Macromolecules, 31*(1–3), 111–117.

Qiu, S. M., Aweya, J. J., Liu, X., Liu, Y., Tang, S., Zhang, W., & Cheong, K. L. (2022). Bioactive polysaccharides from red seaweed as potent food supplements: A systematic review of their extraction, purification, and biological activities. *Carbohydrate Polymers, 275*, 118696.

Quemener, B., Lahaye, M., & Bobin-Dubigeon, C. (1997). Sugar determination in ulvans by a chemical-enzymatic method coupled to high performance anion exchange chromatography. *Journal of Applied Phycology, 9*(2), 179–188.

Rabea, E. I., Badawy, M. E., Stevens, C. V., Smagghe, G., & Steurbaut, W. (2003). Chitosan as antimicrobial agent: Applications and mode of action. *Biomacromolecules, 4*(6), 1457–1465.

Rafiquzzaman, S. M., Ahmed, R., Lee, J. M., Noh, G., Jo, G., & Kong, I. S. (2016). Improved methods for isolation of carrageenan from *Hypnea musciformis* and its antioxidant activity. *Journal of Applied Phycology, 28*(2), 1265–1274.

Ray, B., & Lahaye, M. (1995). Cell-wall polysaccharides from the marine green alga Ulva "rigida" (*Ulvales, Chlorophyta*). Extraction and chemical composition. *Carbohydrate Research, 274*(8), 251–261.

Rinaudo, M. (2006). Chitin and chitosan: Properties and applications. *Progress in Polymer Science, 31*(7), 603–632.

Rocha de Souza, M. C., Marques, C. T., Guerra Dore, C. M., Ferreira da Silva, F. R., Oliveira Rocha, H. A., & Leite, E. L. (2007). Antioxidant activities of sulfated polysaccharides from brown and red seaweeds. *Journal of Applied Phycology, 19*(2), 153–160.

Rocha de Souza, M. C., Marques, C. T., Guerra Dore, C. M., Ferreira da Silva, F. R., Oliveira Rocha, H. A., & Leite, E. L. (2007). Antioxidant activities of sulfated polysaccharides from brown and red seaweeds. *Journal of Applied Phycology, 19*(2), 153–160.

Rodén, L., Baker, J. R., Cifonelli, J. A., & Mathews, M. B. (1972). [7] Isolation and characterization of connective tissue polysaccharides. In *Methods in Enzymology* (Vol. 28, pp. 73–140). Academic Press.

Roy, S., & Rhim, J. W. (2020). Preparation of gelatin/carrageenan-based color-indicator film integrated with shikonin and propolis for smart food packaging applications. *ACS Applied Bio Materials, 4*(1), 770–779.

Sadhasivam, G., & Muthuvel, A. (2014). Isolation and characterization of hyaluronic acid from marine organisms. *Advances in Food and Nutrition Research, 72*, 61–77.

Santo, V. E., Gomes, M. E., Mano, J. F., & Reis, R. L. (2012). Chitosan–chondroitin sulphate nanoparticles for controlled delivery of platelet lysates in bone regenerative medicine. *Journal of Tissue Engineering and Regenerative Medicine, 6*(S3), s47–s59.

Santos, J. R., Alves, N. M., & Mano, J. F. (2010). New thermo-responsive hydrogels based on poly (N-isopropylacrylamide)/hyaluronic acid semi-interpenetrated polymer networks: Swelling properties and drug release studies. *Journal of Bioactive and Compatible Polymers, 25*(2), 169–184.

Saravanan, M., Bhaskar, K., Rao, G. S., & Dhanaraju, M. D. (2003). Ibuprofen-loaded ethylcellulose/polystyrene microspheres: an approach to get prolonged drug release with reduced burst effect and low ethylcellulose content. *Journal of Microencapsulation, 20*(3), 289–302.

Savage, P. E. (2012). Algae under pressure and in hot water. *Science, 338*(6110), 1039–1040.

Schaeffer, D. J., & Krylov, V. S. (2000). Anti-HIV activity of extracts and compounds from algae and cyanobacteria. *Ecotoxicology and Environmental Safety, 45*(3), 208–227.

Schiraldi, C., Cimini, D., & De Rosa, M. (2010). Production of chondroitin sulfate and chondroitin. *Applied Microbiology and Biotechnology, 87*(4), 1209–1220.

Schuerch, C. (1986). *Encyclopaedia of Polymer Science and Engineering* (2nd ed., Vol. 13, pp. 87–162). John Wiley & Sons, New York.

Şenel, S., & McClure, S. J. (2004). Potential applications of chitosan in veterinary medicine. *Advanced Drug Delivery Reviews, 56*(10), 1467–1480.

Senni, K., Pereira, J., Gueniche, F., Delbarre-Ladrat, C., Sinquin, C., Ratiskol, J., Godeau, G., Fischer, A. M., Helley, D., & Colliec-Jouault, S. (2011). Marine polysaccharides: A

source of bioactive molecules for cell therapy and tissue engineering. *Marine Drugs, 9*(9), 1664–1681.
Sezer, A. D., & Akbuğa, J. (2006). Fucosphere—New microsphere carriers for peptide and protein delivery: Preparation and in vitro characterization. *Journal of Microencapsulation, 23*(5), 513–522.
Shariatinia, Z. (2019). Pharmaceutical applications of chitosan. *Advances in Colloid and Interface Science, 263*, 131–194.
Sharma, S., Khuller, G. K., & Garg, S. K. (2003). Alginate-based oral drug delivery system for tuberculosis: Pharmacokinetics and therapeutic effects. *Journal of Antimicrobial Chemotherapy, 51*(4), 931-938.
Silva, F. R., Dore, C. M., Marques, C. T., Nascimento, M. S., Benevides, N. M., Rocha, H. A., & Leite, E. L. (2010). Anticoagulant activity, paw edema and pleurisy induced carrageenan: Action of major types of commercial carrageenans. *Carbohydrate Polymers, 79*(1), 26–33.
Silva, F. R., Dore, C. M., Marques, C. T., Nascimento, M. S., Benevides, N. M., Rocha, H. A., & Chavante, S. F. (2010). Anticoagulant activity, paw edema and pleurisy induced carrageenan: Action of major types of commercial carrageenans. *Carbohydrate Polymers, 79*(1), 26–33.
Silva, T. H., Duarte, A. R., Moreira-Silva, J., Mano, J. F., & Reis, R. L. (2012). Biomaterials from marine-origin biopolymers. In *Biomimetic Approaches for Biomaterials Development* (pp. 1–23), Wiley-VCH.
Simões, S., Moreira, J. N., Fonseca, C., Düzgüneş, N., & De Lima, M. C. (2004). On the formulation of pH-sensitive liposomes with long circulation times. *Advanced Drug Delivery Reviews, 56*(7), 947–965.
Sokolova, E. V., Bogdanovich, L. N., Ivanova, T. B., Byankina, A. O., Kryzhanovskiy, S. P., & Yermak, I. M. (2019). λ-Carrageenan oligosaccharides of distinct anti-heparanase and anticoagulant activities inhibit MDA-MB-231 breast cancer cell migration. *Marine Drugs, 17*(3), 140.
Song, S., Peng, H., Wang, Q., Liu, Z., Dong, X., Wen, C., Ai, C., Zhang, Y., Wang, Z., & Zhu, B. (2020). Inhibitory activities of marine sulfated polysaccharides against SARS-CoV-2. *Food & Function, 11*(9), 7415–7420.
Squillaci, G., Finamore, R., Diana, P., Restaino, O. F., Schiraldi, C., Arbucci, S., & Morana, A. (2016). Production and properties of an exopolysaccharide synthesized by the extreme halophilic archaeon *Haloterrigena turkmenica. Applied Microbiology and Biotechnology, 100*(2), 613–623.
Sudhakar, Y., Kuotsu, K., & Bandyopadhyay, A. K. (2006). Buccal bioadhesive drug delivery—a promising option for orally less efficient drugs. *Journal of Controlled Release, 114*(1), 15–40.
Sumi, T., Ohba, H., Ikegami, T., Shibata, M., Sakaki, T., Sallay, I., & Park, S. S. (2002). Method for the preparation of chondroitin sulfate compounds. *United States Patent US 6,342,367.*
Sun, Y., Ma, X., & Hu, H. (2021). Marine polysaccharides as a versatile biomass for the construction of nano drug delivery systems. *Marine Drugs, 19*(6), 345.
Sur, S., Rathore, A., Dave, V., Reddy, K. R., Chouhan, R. S., & Sadhu, V. (2019). Recent developments in functionalized polymer nanoparticles for efficient drug delivery system. *Nano-Structures & Nano-Objects, 20*, 100397.

Svec, F., Tennikova, T. B., & Deyl, Z. (Eds.). (2003). *Monolithic Materials: Preparation, Properties and Applications*. Elsevier. 1st Edition, Volume 67 – April 29, 2003. eBook ISBN: 9780080536972.

Tang, H. C., Chen, W. C., Chiang, C. W., Chen, L. Y., Chang, Y. C., & Chen, C. H. (2015). Differentiation effects of platelet-rich plasma concentrations on synovial fluid mesenchymal stem cells from pigs cultivated in alginate complex hydrogel. *International Journal of Molecular Sciences, 16*(8), 18507–18521.

Teien, A. N., Abildgaard, U., & Höök, M. (1976). The anticoagulant effect of heparan sulfate and dermatan sulfate. *Thrombosis Research, 8*(6), 859–867.

Thanou, M., Verhoef, J. C., & Junginger, H. E. (2001). Oral drug absorption enhancement by chitosan and its derivatives. *Advanced Drug Delivery Reviews, 52*(2), 117–126.

Thomas, N. V., & Kim, S. K. (2013). Beneficial effects of marine algal compounds in cosmeceuticals. *Marine Drugs, 11*(1), 146–164.

Ueno, H., Mori, T., & Fujinaga, T. (2001). Topical formulations and wound healing applications of chitosan. *Advanced Drug Delivery Reviews, 52*(2), 105–115.

Usoltseva, R. V., Anastyuk, S. D., Shevchenko, N. M., Zvyagintseva, T. N., & Ermakova, S. P. (2016). The comparison of structure and anticancer activity in vitro of polysaccharides from brown algae *Alaria marginata* and *A. angusta*. *Carbohydrate Polymers, 153*, 258–265.

Vázquez, J. A., Rodríguez-Amado, I., Montemayor, M. I., Fraguas, J., del Pilar González, M., & Murado, M. A. (2013). Chondroitin sulfate, hyaluronic acid and chitin/chitosan production using marine waste sources: Characteristics, applications and eco-friendly processes: A review. *Marine Drugs, 11*(3), 747–774.

Vecina, J. F., Oliveira, A. G., Araujo, T. G., Baggio, S. R., Torello, C. O., Saad, M. J., & de Souza Queiroz, M. L. (2014). Chlorella modulates insulin signaling pathway and prevents high-fat diet-induced insulin resistance in mice. *Life Sciences, 95*(1), 45–52.

Venkatpurwar, V., Shiras, A., & Pokharkar, V. (2011). Porphyran capped gold nanoparticles as a novel carrier for delivery of anticancer drug: in vitro cytotoxicity study. *International Journal of Pharmaceutics, 409*(1–2), 314–320.

Wang, D. A., Varghese, S., Sharma, B., Strehin, I., Fermanian, S., Gorham, J., & Elisseeff, J. H. (2007). Multifunctional chondroitin sulphate for cartilage tissue–biomaterial integration. *Nature Materials, 6*(5), 385–392.

Wang, J., Hu, S., Nie, S., Yu, Q., & Xie, M. (2016). Reviews on mechanisms of in vitro antioxidant activity of polysaccharides. *Oxidative Medicine and Cellular Longevity*, pp. 1–13.

Xu, Y. Y., Huo, Y. F., Xu, L., Zhu, Y. Z., Wu, Y. T., Wei, X. Y., & Zhou, T. (2021). Resveratrol-loaded ovalbumin/*Porphyra haitanensis* polysaccharide composite nanoparticles: Fabrication, characterization and antitumor activity. *Journal of Drug Delivery Science and Technology, 66*, 102811.

Yermak, I. M., Kim, Y. H., Titlynov, E. A., Isakov, V. V., & Solov'eva, T. F. (1999). Chemical structure and gel properties of carrageenans from algae belonging to the Gigartinaceae and Tichocarpaceae, collected from the Russian Pacific coast. In *Sixteenth International Seaweed Symposium* (pp. 555–562). Springer, Dordrecht.

Yew, Y. P., Shameli, K., Mohamad, S. E., Lee, K. X., & Teow, S. Y. (2020). Green synthesized montmorillonite/carrageenan/Fe_3O_4 nanocomposites for pH-responsive release of protocatechuic acid and its anticancer activity. *International Journal of Molecular Sciences, 21*(14), 4851.

Yoo, H. S., & Park, T. G. (2001). Biodegradable polymeric micelles composed of doxorubicin conjugated PLGA–PEG block copolymer. *Journal of Controlled Release, 70*(1–2), 63–70.

Yu, Y., Shen, M., Song, Q., & Xie, J. (2018). Biological activities and pharmaceutical applications of polysaccharide from natural resources: A review. *Carbohydrate Polymers, 183*, 91–101.

Yu, Y., Wang, B., Guo, C., Zhao, F., & Chen, D. (2019). Protoporphyrin IX-loaded laminarin nanoparticles for anticancer treatment, their cellular behavior, ROS detection, and animal studies. *Nanoscale Research Letters, 14*(1), 1–2.

Zargarzadeh, M., Amaral, A. J., Custódio, C. A., & Mano, J. F. (2020). Biomedical applications of laminarin. *Carbohydrate Polymers, 232*, 115774.

Zhang, C., Gao, Z., Hu, C., Zhang, J., Sun, X., Rong, C., & Jia, L. (2017). Antioxidant, antibacterial and anti-aging activities of intracellular zinc polysaccharides from – SH-05. *International Journal of Biological Macromolecules, 95*, 778–787.

Zhang, Z., Wang, F., Wang, X., Liu, X., Hou, Y., & Zhang, Q. (2010). Extraction of the polysaccharides from five algae and their potential antioxidant activity in vitro. *Carbohydrate Polymers, 82*(1), 118–121.

Zhong, H., Gao, X., Cheng, C., Liu, C., Wang, Q., & Han, X. (2020). The structural characteristics of seaweed polysaccharides and their application in gel drug delivery systems. *Marine Drugs, 18*(12), 658.

Zhou, G., Sun, Y., Xin, H., Zhang, Y., Li, Z., & Xu, Z. (2004). In vivo antitumor and immunomodulation activities of different molecular weight lambda-carrageenans from *Chondrus ocellatus*. *Pharmacological Research, 50*(1), 47–53.

CHAPTER 6

Marine Biopolymers in Smart Oral Delivery of Pharmaceuticals

SUNNY R. SHAH,[1] PRASHANT L. PINGALE,[2] SUDARSHAN SINGH,[3] TITILOPE JOHN JAYEOYE,[4] ASHVIN V. DUDHREJIYA,[1] and BHUPENDRA PRAJAPATI[5]

[1]*B. K. Mody Government Pharmacy College, Gujarat Technological University, Rajkot, Gujarat, India*

[2]*GES's Sir Dr. M. S. Gosavi College of Pharmaceutical Education and Research, Nashik, Maharashtra, India*

[3]*Faculty of Pharmacy, Chiang Mai University, Chiang Mai, Thailand*

[4]*Department of Chemistry, Faculty of Science, Chulalongkorn University, Bangkok, Thailand*

[5]*Shree S. K. Patel College of Pharmaceutical Education and Research, Ganpat University, Kherva, Gujarat, India*

ABSTRACT

Water covers nearly three-fourths of the globe's surface with the availability of a wide range of marine organisms that supply or are used in the development of valuable biomaterials including lipids, proteins, and polysaccharides. The seaweed polysaccharides are considered the most abundant and widely available in the ocean, from the tidal level to free-floating or anchored. Marine microorganisms such as bacteria, microalgae, and seaweed represent a large source of valuable biodegradable polymers. Biopolymers originating from a marine source are a diverse and versatile class of inactive pharmaceuticals that have great potential in the

Handbook of Research in Marine Pharmaceutics: Exploring Oceanic Microbial Diversity for Human Health and Wellness. Md. Faiyazuddin & Khalid Rehman Hakeem (Eds.)

development of the economy. Marine polymers have gained significant interest among researchers and pharmaceutical formulators due to their high potential in replacing existing synthetic materials, both non-biodegradable and biodegradable. Moreover, the fortification of natural bioactives within marine-originated biopolymers has equally gained interest due to synergistic effects. Marine polymers are now significantly used in the development of several medical products including medical materials, cosmetics, clothing fabrics, water treatment chemicals, industrial plastics, absorbents, biosensors, food packaging, and delivering smart pharmaceuticals. Smart oral pharmaceuticals are products that are prepared using cutting-edge technology to deliver the required active content that can produce improved therapeutic efficacy with reduced side effects. This chapter highlights the potential sources of novel marine-based polymers with recent expansion in the processing of different novel natural polymers from a marine source to develop multifunctional excipients and valorization of waste biomass to produce biopolymers that can effectively deliver pharmaceuticals.

6.1 INTRODUCTION

The mainstay of therapeutic treatment for millennia has been traditional oral drug delivery systems such as tablets, capsules, solutions, and suspensions. If the chemist is in charge of establishing and determining the drug's pharmacological action, the formulation scientist is responsible for developing an appropriate dosage form that strikes a balance between bioavailability and patient compliance. However, the umbrella of limitations associated with conventional oral drug delivery, such as inadequate dissolution, poor absorption, drug exposure to non-target organs, dose-dependent toxicity, etc., limits the drug's bioavailability, which in turn restricts formulation scientists' ability to take advantage of the potential patient compliance for oral delivery (Wen, Jung, & Li, 2015). While many alternative drug delivery methods have been investigated, including intravenous, intramuscular, topical, transdermal, etc., to increase the bioavailability of drugs, oral drug delivery is still the most accepted method due to patient compliance. The drug release modification for oral delivery can be achieved with the help of a suitable formulation strategy to improve the bioavailability of biopharmaceutical classification III and IV drugs.

The use of biopolymers in the development of dosage forms has always proven beneficial, especially for the treatment of chronic diseases. It is simple to alter the molecular weight, content, purification intensity, attached functional groups, and molecular conformation of biopolymers (Barcelos, Vespermann, Pelissari, & Molina, 2019). Bioceramics (Bongio, Van Den Beucken, Leeuwenburgh, & Jansen, 2010), polysaccharides, enzymes, peptides, lipids, pigments, toxins, and products made from algae are only a few examples of materials that are frequently employed in biological and biomedical applications. The market for biomaterials offers enormous opportunities for entrepreneurs, but developing new and improved goods at competitive rates will require intensive study (Priyan Shanura Fernando, Kim, Kim, & Jeon, 2018). Marine biotechnology industries are growing as a result of the isolation, characterization, and application of marine biomaterials. The versatility and biocompatibility profile of the polymer are the primary factors considered while selecting polymers for the development of drug delivery systems. These requirements are met by naturally occurring polysaccharides, which are also inexpensive and accessible (Manna & Jana, 2022). However, marine sources are somewhat underutilized for drug discovery despite having rich sources of many chemical compounds and a high level of biodiversity.

6.2 PHYSIOLOGY OF THE GASTROINTESTINAL TRACT

The largest barrier preventing pathogenic invaders from entering the body is the gut mucosa barrier. It keeps poisons and toxins from entering the bloodstream by filtering them out. However, the delivery of medicinal drugs orally or to the brain is also impeded by these barriers. Many of these elements can be considered in order to improve drug delivery and overcome these numerous obstacles (Homayun, Lin, & Choi, 2019). Designing strategies to improve the oral bioavailability of potential medications requires knowledge of the structure and biological characteristics of the intestinal mucosa barrier. Before entering the systemic circulation, the medication must partition across the cell membranes and cross the cytoplasm of the cell. The medicine must also get past biological obstacles like enzymes that can break it down throughout this process. Before getting absorbed, for instance, the drug must partition across the cell membranes and cross the cytoplasm of the cell.

6.2.1 BIOLOGICAL BARRIERS

For a drug to be therapeutically active, it has to bypass all the barriers in the gastrointestinal tract and reach the site of action. However, the luminal side of the gastrointestinal tract is covered in an aqueous mucus layer, and the goblet cells secrete the glycoprotein-rich mucus. Before a drug molecule can overcome the intestinal mucosa barrier's epithelial cell layer, it must first pass through this mucus layer. Depending on its physicochemical characteristics, a drug molecule can pass the mucosal surfaces in many different ways. Drugs that enter cells through the transcellular pathway frequently balance solubility and hydrophobicity efficiently. The paracellular pathway is a passageway between cells that connects to nearby cell membranes at intercellular junctions. The entire lateral side of the cells is covered with an 80 nm long, convoluted canal of intercellular connections. Small chemicals and ions can pass through the intestinal mucosa barrier using this channel. Three sections make up the intercellular junction: desmosomes, adherens junctions, and tight junctions (Vanslembrouck, Chen, Larabell, & Van Hengel, 2022). Drugs with hydrophobicity and the ability to partition into cell membranes are more likely to enter the intestinal mucosa through the transcellular pathway. Molecules cannot pass freely through the paracellular pathway due to tight interactions. The molecules can partition into the cellular membranes from the basolateral to the apical side, allowing them to passively diffuse across transcellular channels. Efflux pumps can prevent medication molecules from passively diffusing across cells. Drugs or drug conjugates have also been transported through biological barriers via nutrient transporters. Transporters can help medicinal molecules traverse cells via receptor-mediated transcellular pathways (Yazdani, Jaldin-Fincati, Pereira, & Klip, 2019).

6.2.2 PHYSICOCHEMICAL BARRIERS TO DRUG DELIVERY

Drugs must be released from their dosage forms in order to be absorbed in the GI tract; the released drug dose must be in a solution form or should be able to dissolve in the GI fluid. Additionally, the medicine must be dissolved and pass through the intestinal wall. The water solubility and intestinal epithelial membrane permeability of pharmaceuticals are the major determinants of GI absorption. Drugs' poor water solubility and membrane permeability, among other physicochemical characteristics, provide a significant barrier to effective oral medication administration to humans. Drugs' low chemical and biological stability, as well as physiological obstacles including pH, efflux transporters, and metabolic enzymes, can further impede absorption. Some medications might also cause nausea and localized irritability. A complete understanding of the physicochemical properties, GI permeability, biological barriers, pharmacokinetics, and pharmacodynamics of medicines is required for the development of oral drug delivery systems.

6.2.3 METABOLIC BARRIERS TO DRUG DELIVERY

Drug metabolism can affect oral bioavailability in addition to solubility and permeability. The BDDCS provides insights into how diet affects drug absorption, as well as details on how drug absorption, elimination, and transport interact. According to the BDDCS, the primary route of elimination affects drug permeability (Benet, Hosey, Ursu, & Oprea, 2016). Drug interactions with transporters are not expected to be significant for Class 1 BDDCS medicines due to their high solubility and extensive metabolism. Therefore, the bioavailability of such medications should not be significantly impacted by high-fat meals. Due to their insolubility, Class 2 BDDCS medicines, which are poorly soluble and highly metabolized, may experience significant transporter effects, primarily efflux transporter effects. As a result of the intestinal suppression of efflux pumps such as P-gp transporters, high-fat diets may enhance their bioavailability (Bocci, Oprea, & Benet, 2022). A decrease or elimination of the impact of high-fat meals and a large minimization of other drug transporter interactions may result from dosage form modifications that greatly improve the solubility of BDDCS class 2 medications. Due to their low permeability, Class 3 BDDCS medicines are thought to be particularly susceptible to the effects of uptake transporters. Because fatty diets block intestinal uptake transporters, they can decrease the bioavailability of many medications. The main effect, however, will rely on the level of transporter inhibition as well as the substrate's relative affinity for the transporters if a medication is a substrate for transporters (influx or efflux). Predicting the impact of a high-fat meal on drug absorption is challenging for class 4 BDDCS agents since interactions between class 2 and 3 substances can occur (Cheng & Wong, 2020). In contrast, when fatty diet effects do occur, they are typically characterized by an increase in the drug's bioavailability, which is brought about by a combination of the drug's improved GI solubilization and the suppression of efflux transporters.

6.2.4 BIOCHEMICAL BARRIERS TO DRUG DELIVERY

Individual variances in enzyme activity (inhibition or induction), genetic polymorphisms, and even disease states can have a significant impact on how drugs are metabolized (Carvalho Henriques, Yang, Lapetina, Carr et al., 2020). Enzymes from mammalian and microbial origins are found in the intestine. The mammalian enzymes are present in the lumen and the enterocytes and the microflora's enzymes are found in the ileum and colon (Thursby & Juge, 2017). Digestive enzymes released by the pancreas, such as lipases, amylase, and peptidases, such as chymotrypsin and trypsin, as well as those originating from the colon's intestinal flora, which are primarily present in the lower GI tract, generally cause intestinal metabolism to occur. In addition, the first-pass metabolism, which consists of intracellular and brush-border metabolism, is carried out on the surface of the enterocyte by enzymes found in the brush-border membrane. The small intestine is where brush-border metabolism primarily takes place. The brush border metabolism involves the peptidases isomaltase, alkaline phosphatase, sucrose, and others (Patricia & Dhamoon, 2019). There may be a first-pass metabolic limit to oral absorption. The enterocytes are the site of intracellular metabolism, which mostly involves phase-I metabolizing enzymes such as cytochrome P450 enzymes like CYP3A4, various phase-II conjugating enzymes involved in processes like sulfation and glucuronidation, as well as additional enzymes like esterases (Phang-Lyn & Llerena, 2022). Although pre-absorptive metabolism occurs in the intestinal epithelium, it can also serve as a significant site for the delivery of ester-type prodrugs like aspirin. The primary metabolic barrier, in addition to the intestinal epithelium, is the hepatic first-pass metabolism. Membrane transporters can be divided into two groups: uptake transporters and efflux transporters. They help move drugs and endogenous substances into or out of cells. Therefore, membrane transporters play a crucial role in determining the bioavailability, distribution, and absorption of oral drugs. While efflux transporters are members of the ABC superfamily, the primary uptake transporters that allow xenobiotic medicines to enter cells are members of the solute carrier superfamily. Efflux transporters, such as the bile salt export pump (BSEP), Pgp, MRP1–6, and BCRP, are highly expressed in the liver and gut. The majority of these membrane transporters pump substrates against a gradient of concentration by using ATP. P-gp is mostly found in the brush border surface of enterocytes in the small intestine and a major portion of the colon, where it serves as a defense barrier against foreign substances. In mature enterocytes, CYP3A4 co-localizes with P-gp and shows overlapping substrate specificity. As a result, the majority of substrate medicines may be digested by pumping them through P-gp from enterocytes into the lumen before they can be reabsorbed into cells, extending their time in the presence of CYP3A4. Many medications' bioavailability is constrained by this method. Additionally, it can result in drug interactions, particularly when medicines are designed to inhibit P-gp or CYP3A4 (Järvinen, Deng, Kiander, Sinokki, Kidron, & Sjöstedt, 2022).

6.3 STRATEGIES TO IMPROVE ORAL DRUG DELIVERY

A variety of techniques have been investigated to increase drug penetration across biological barriers. The chemical alteration of drug entities like prodrugs and peptidomimetics is one way. A formulation that improves medication penetration via biological barriers is a different approach. By designing the most effective formulation, drug molecules' intestinal absorption can be increased. The inclusion of ion pairing and complexation molecules (Ristroph & Prud'homme, 2019), non-surfactant membrane penetration enhancers, surfactant adjuvants, or mixtures of these additives have all been proposed as ways to improve drug absorption for peptides (Maher, Brayden, Casettari, & Illum, 2019). To increase drug permeability, it has been examined to include tight junction perturbers such as cytoskeletal agents, oxidants, hormones, calcium chelators, and bacterial toxins in the formulation. Because of their prolonged retention at the targeted mucosal barrier, mucoadhesives are used in another innovative delivery system to improve drug delivery; lectins have been found as promising peptide carriers in an oral mucoadhesive system (Zhu, Chen, Paul, Lu, Wu, & Qi, 2021). For peptides, it has also been proposed that coadministration with enzyme-involved metabolism inhibitors will improve oral absorption. Protein-protein interactions at intercellular junctions can be changed to boost the paracellular penetration of molecules through biological barriers. Peptides derived from tight junction proteins like occludin and claudin can alter the intercellular junctions of biological barriers (Buckley & Turner, 2018).

The low rate of absorption of hydrophobic pharmaceuticals is mostly due to the lack of effective oral dosage forms for drugs with poor water solubility. Food impact, gastrointestinal discomfort, slow onset of action, lack of dose proportionality, and substantial intra- and inter-subject variability are other characteristics associated with limited bioavailability. A significant factor limiting the oral bioavailability of hydrophobic medicines is drug solubility. To design formulations of poorly soluble medications, it is important to carefully screen factors including surfactant selection, particle size reduction, and salt selection. Historically, a mix of surfactants has been used to increase drug absorption when taken orally. The hydrophilic and lipophilic groups of surfactants' hydrophilic head and hydrophobic tail aid in the localization of drug molecules at the interface, reducing interfacial tension (Ghosh, Ray,

& Pramanik, 2020). Surfactants boost the solubility and permeability of medications by briefly loosening tight intracellular junctions, which is one of the processes by which they increase bioavailability (Liu, Sun, Cheng et al., 2021). Surfactants can become a safety problem when used in larger quantities, therefore this must be carefully considered. Other methods, such as micro/nanonization, can also significantly increase a drug's bioavailability (Pattnaik, Swain, & Ramakrishna, 2022). Drugs' particle size is significantly reduced using these approaches, which enhances their surface area and, in turn, the dissolution rate.

Orally delivered drug molecules must successfully navigate through a number of delivery-related obstacles in order to be absorbed. The gastro-intestinal epithelium layer is a significant barrier to medication absorption. Depending on the physicochemical characteristics of the drug molecules, the drug molecules can either flow through or between the cells. According to recent research, the intestine's metabolism significantly hinders the absorption of drugs. The issue is complicated by the coordinated activity of these drug-metabolizing enzymes and efflux mechanisms. Although crossing the intestinal epithelial layer is difficult, pharmaceutical researchers and medicinal chemists propose to overcome these obstacles by developing novel strategies that optimize pharmacological activity and improve drug delivery.

6.4 MARINE-BASED POLYMERS USED IN DELIVERING ORAL PHARMACEUTICALS

Marine species have a variety of unique compounds with a broad spectrum of biological properties and bioactivity. The high biodiversity of marine compounds and the ease of their extraction and purification methods have led to a recent increase in interest in several scientific fields that research their wide range of applications. Marine biomaterials' non-cytotoxic qualities, biodegradability, and biocompatibility make them suitable for a variety of biomedical applications. The pharmaceutical business is built on these biological characteristics, which have led to the discovery of a wide variety of novel bioactive substances having pharmacological effects (Montuori, De Pascale, & Lauritano, 2022). Due to their anti-tumoral effects, certain of these substances, including polypeptides derived from tunicates and sponges, have been investigated for the treatment of cancer. Many of these substances, like Aplidin and Ecteinascidin, have already been tested in clinical studies (Jha, Kumar, & Mishra, 2022).

The food and cosmetics sectors mostly use marine polysaccharides, but the medicinal sciences also use them extensively where there is growing interest in using them as materials to incorporate bioactive compounds. The majority of marine polysaccharides come from marine algae, although they can also be found in animal tissues including cartilaginous fish tissue and crab bones. Additionally, some polysaccharides from marine microorganisms, such as some prokaryotes, can be isolated. The large complex category known as marine polysaccharides is made up of several macromolecules with varied biological properties. Numerous chemical and biological characteristics of polysaccharides include biocompatibility, biodegradability, adhesiveness, and the capacity to create hydrogels. Sulfated polysaccharides are one of the numerous marine polysaccharide groups that stand out (Taher, Omer, Hamed, Ali, Tamer, & Eldin, 2019). They exhibit bioactivities such as antioxidant, anticoagulant, anticancer, antiviral, anti-allergic, anti-adhesive, anti-angiogenic, and anti-inflammatory properties in comparison to other marine polysaccharides. An extensive examination of one of these DDS materials resulted in the creation of novel chemical modification approaches designed to enhance biological activity and change their affinity for certain drugs. Using chemical methods or interactions with other synthetic or natural polymers, it has been feasible to improve the efficacy of medication release as well as the capacity to include pharmaceuticals. The potential for novel, less invasive, and more individualized therapy approaches has sparked enthusiasm for the investigation of DDSs for marine polysaccharides with therapeutic properties. Several studies demonstrated that marine-originated biomaterials permit the loading of lower pharmacological dosages, which may significantly reduce the adverse effects brought on by the medicines (Hasnain, Ahmed, Alkahtani, Milivojevic et al., 2021).

In order to expand the use of DDSs as diagnostic tools, these materials may be utilized as signaling markers to direct the delivery of a carrier to a specific site. These methods may also have a significant positive impact on gene therapy, which is frequently constrained by the health risks related to viral vectors. Genetic material and other therapeutic ingredients being encapsulated, in contrast, have been demonstrated to have many benefits due to biomaterials' stability, protection, increased solubility, prolonged release, biocompatibility, and in certain cases, biodegradability (Nayak & Hasnain, 2020). The primary focus of this review is the use of marine polysaccharides as the foundation for oral DDSs. Alginate, chitosan, carrageenan, and hyaluronan are the primary marine polysaccharides utilized in or being researched for the pharmaceutical industry (also known as hyaluronic acid). We will discuss the many modifications and adaptations that may be made to these biopolymers to achieve goals like medication delivery and protection, stimuli responsiveness, and targeting ability (Biswas, Jony, Nandy et al., 2021).

6.4.1 CHITIN AND CHITOSAN

One of the most prevalent naturally occurring polymers in our biosphere, chitin, is the source of the linear polysaccharide known as chitosan. Deacetylating chitin yields chitosan, a substance containing both N-acetyl-D-glucosamine and D-glucosamine residues (acetylated and deacetylated units). Enzymes like chitinase and lysozyme can break down chitosan as well as chitin. Chitin is often transformed into soluble derivatives such as chitosan, a material that dissolves in acidic conditions, and carboxymethyl chitin because it is not water-soluble. Since they are sensitive to pH fluctuations, the amine groups in chitosan can either be positively charged in an acidic environment or neutral in an alkaline environment range. Among the marine polysaccharides, chitosan is one that is most widely utilized and studied for biological purposes. It is used to make films, membranes, and scaffolds for tissue engineering and regenerative medicine, as well as nanoparticles, beads, and capsules for controlled drug delivery systems (Ansari, Murteza, Ahsan, Hasnain, & Nayak, 2022).

Chitosan demonstrates significant antibacterial activity against pathogenic microorganisms, which is a desirable quality for making wound-infection-preventing films. It also exhibits anti-inflammatory and anti-tumor properties (Singh, Nwabor, Syukri, & Voravuthikunchai, 2021). Chitosan is a great choice for excipients in DDSs and for building devices that must come into contact with biological environments because of all its biological qualities (Sun, Liu, Omer, Yang, & Ouyang, 2019). Electrostatic interactions between a bioactive chemical and a polysaccharide determine the stability, protection, and acceleration or delaying of drug release in chitosan-based DDSs (Singh, Nwabor, Sukri, Wunnoo et al., 2022). The shape, size, density, degree of cross-linking, level of chitosan acetylation, and physicochemical characteristics of the bioactive agent may all have an impact on the release profile and rate of biomolecules from chitosan-based carriers (Singha & Basu, 2022). Different ways of release are possible, including passive diffusion, DDS erosion, and release from the surfaces of the DDSs. The level of deacetylation in chitosan may be used as a degradation control strategy. For chitosan-based carriers, an alternative release mechanism is utilized called enzyme degradation.

By applying various chemical changes to chitosan, it is also feasible to boost the binding capability of medications that are insufficiently water-soluble. Chitosan can be chemically modified to increase release effectiveness and provide further benefits, including pharmaceutical stability and protection. Grafting carboxymethyl cyclodextrin onto hydroxypropyl chitosan can be done using a water-soluble carbodiimide, which is created when chitosan and propylene epoxide combine under alkaline conditions (EDC). Drugs that are hydrophobic can be encapsulated because HPCH includes hydrophobic groups. These DDSs may also be pH-responsive due to the free amine groups, which may be protonated at lower pH values. The results of testing using ketoprofen as a model hydrophobic drug showed that this chitosan derivative has significant potential as a biodegradable delivery strategy for hydrophobic pharmaceuticals in a pH-sensitive controlled release. Additionally, it has been shown that chitosan retains its pH sensitivity while becoming more soluble in water when thiol groups are added (Lemos, De Souza, & Fajardo, 2021).

Chitosan has the ability to react with other substances, which may help with sustained release, cargo stabilization, and cargo protection. Chitosan has been shown in numerous trials to improve and prolong the absorption of hydrophilic medicines administered orally and intravenously. Through the use of PEG, chitosan may be chemically altered to improve its biocompatibility, specifically by reducing its toxicity and enhancing protein adsorption, cell adhesion, growth, and proliferation. According to researchers, chitosan-PEG nano-capsules for the oral administration of peptides have increased intestinal absorption and decreased cytotoxicity (Layek & Das, 2021).

Using electrostatic complexation, chitosan-based capsules can also be made, and in certain situations, they can respond to stimuli outside of pH. One example of this is the production of LbL microcapsules utilizing negatively charged elastin-like recombinants (ELRs), and recombinant polypeptides with built-in reactivity to temperature. For the delivery of active compounds, novel thermosensitive ELR/chitosan microcapsules were created. Studies using cells also revealed that such structures had low cytotoxicity. Despite the lack of research on these microcapsules' pH response, these findings strongly suggest that chitosan can link with stimuli-responsive biomaterials that can come from a wide range of sources, some unusual like genetically altered polypeptides. One of the most common mechanisms in smart DDSs is thermal responses, although it is questionable if their sensitivity would be adequate to treat conditions like a normal fever, when fluctuations in body temperature are as minor as 1–2°C, for instance. Furthermore, not everyone's body temperature is quite the same. Due to the need for trigger-based administration and the requirement that triggers work within a specific range, combining two or more physiological characteristics could provide disease treatment (Bharathi, Ganesh, Harini, Vatsala et al., 2022).

6.4.2 ALGINATE

Alginate is produced from *Laminaria digitata* and *Ascophyllum nodosum,* respectively. Alginate is completely safe, biocompatible, and extremely bioavailable. Alginate is one of the biopolymers with the broadest range of biomedical applications because of these key benefits. Alginate typically functions as a stabilizing ingredient in pharmaceutical formulations and is utilized as an excipient in several DDSs (Sudarshan Singh, Ozioma F Nwabor, Wanatsanan Chulrik, Chutima Jansakun, & Phuangthip Bhoopong, 2022). Alginate is soluble in neutral and alkaline environments because it has charged carboxyl groups at pH values greater than 3–4 (He, Shang, Liu, & Yuan, 2020). In contrast to the stomach's acidic environment, which interferes with the alginate carrier's stability, the intestine's alkaline pH promotes this biopolymer's solubility and drug release. Such pH sensitivity encourages greater protection for medications that preferentially absorb in the intestinal tract. Alginate is a suitable biomaterial for the building of DDSs as a result of its solubility and sensitivity to pH. Alginate is used more frequently than other biopolymers due to its biocompatibility, low toxicity, high bioavailability, and less expensive extraction and purification processes, and capacity to be processed into hydrogel matrices, beads, and particles (Iravani & Varma, 2022). Alginate is additionally utilized as an excipient in pharmaceutical tablets to enhance the drug's stability and level of protection. The form of alginate most frequently used in the pharmaceutical sector to create tablets is sodium alginate, particularly when the medicine is not soluble in water. The medication release can be prolonged by using sodium alginate.

Alginate is utilized to create microparticles that can contain various bioactive substances, including proteins (Omer, Ahmed, El-Subruiti, Khalifa, & Eltaweil, 2021). Large amounts of medication can be held in alginate microparticles, which also help to preserve the payload against proteolytic attack. Several processes, including variations in temperature and pH, the use of biodegradable polymers, enzymatic degradation, and other methods that react to chemical and physical stimuli, might cause a bioactive ingredient to be released from the carrier (Dey, Roy, Hasnain, & Nayak, 2020). Since these factors can change greatly, they are challenging to regulate and program.

New microparticle release mechanisms, however, have been created that rely on completely regulated external stimuli, including ultrasound triggering.

An alginate type vibrates when treated with ultrasonic waves. Microparticles created by researchers were found to have a per-fluorocarbon breakthrough capacity. After 15 minutes of exposure, the results revealed that these microparticles had been disrupted, proving that these systems are potential DDSs that can be externally controlled by sound. The usual morphologies of alginate processing using a range of methods, such as calcium cross-linked encapsulation, multiple-phase emulsion, and emulsion, are microcapsules. Oral DDSs can be made hybrid and more flexible by combining alginate with other biomaterials through electrostatic interactions, chemical modification, or cross-linking. Albumin is one of the model proteins that are most frequently employed in controlled release research, and chitosan/alginate-PEG complex capsules are trustworthy examples of how to encapsulate this kind of protein (Essa, Elebyary, Abdelquader, El Maghraby, & Elkordy, 2021). To increase the adaptability of these systems, new spherical structures made of other kinds of synthetic materials can be constructed. Alginate and poly(N-isopropyl acrylamide) (PNIPAAm) can be used to create devices that can deliver biomolecules with dual stimuli-responsive dependence. This is possible because PNIPAAm has thermosensitive features (Anand & Rajinikanth, 2021).

Chitosan-alginate-PNIPAAm beads were found to have reduced release rates with decreasing temperatures in studies utilizing indomethacin as a model medication. The fact that the same thing occurs when the pH is dropped shows that the pH and temperature may both be changed in order to control the permeability of the particles. Alginate may also be complexed with natural polymers like chitosan by a process akin to polymer conjugation to improve the protection of the cargo during oral delivery, such as when administering insulin (Hariyadi & Islam, 2020). Alginate can be used to build capsules for cell encapsulation that are frequently connected with cytotherapy treatments or just to make cellular micro-cultures in more complicated systems where it is impossible to use a normal bioreactor. Moreover, a recent study indicated that chitosan-PVA-Alginate fortified metallic nanoparticles cosmeceutical peel-off mask multifarious application in reduction of skin aging and efficacy against pathogenic microorganisms responsible for the development of acne vulgaris (Singh, Chunglok, Nwabor, Ushir, Singh, & Panpipat, 2022).

In this context, a novel method for creating alginate-based capsules for the integration of various cell types has been presented. Layers of alternating liquid alginate and chitosan were utilized to cover the cells and confine them. To create anchoring sites that assist cell viability, poly (lactic acid) microparticles were combined with the cells in the encapsulation process. The results showed how effective these capsules are as culture systems and how highly viable the encapsulated cells are. Since it would enable the encapsulation of several distinct cell types along with other biomolecules like growth factors and different cytokines, it holds great potential not just for cell culture but also for other medical uses (Putri, Picchioni, Harjanto, & Chalid, 2021).

6.4.3 CARRAGEENAN

Red algae have a sulfated polysaccharide called carrageenan, which has the molecular structure (AB)n sequence, which is a linear series of alternate galactose residues. These polymers might or might not include sulfates. Alternating -(13) (unit A) and -(14) (unit B) glycosidic linkages hold them together. Carrageenans are classified as polyanions because of the negative charge that the sulfated groups provide them (Qureshi, Nayak, Maji, Kim, Banerjee, & Pal, 2019). According to the number of sulfate groups they contain, one, two, or three, carrageenans are divided into three groups: kappa, iota, and lambda. The simple extraction method involves submerging the raw material in an alkaline solution until a gel forms. The gel is then submerged in warm water and heated throughout the extraction process to 74°C. Additional purification stages, including dialysis, can be carried out based on the type of carrageenan and the required level of purity (Pacheco-Quito, Ruiz-Caro, & Veiga, 2020).

The process is finished by filtering, precipitation, and drying. While types are regularly derived from the genera *Kappaphycus* and *Eucheuma*, respectively, they are typically collected from algae in the *Gigartinaceae* family. The quantity of sulfated groups present affects the ability to gel. Carrageenans can gel when exposed to cations, but carrageenan cannot due to its high sulfation level. Many businesses have employed gelation capabilities, including the food industry (using carrageenan as an emulsifier and stabilizer), the cosmetics industry, and the pharmaceutical industry (Joy, Vigneshkumar, John, & George, 2021). Carrageenan is not frequently used as an excipient in the pharmaceutical business, unlike other biomaterials of marine origin, therefore information about their uses, properties, and functions is scarce. As an illustration, a study was done where two forms of carrageenan (and their potential to form tablets) were examined in terms of compression behavior. According to the findings, both carrageenans are suitable excipients for regulated release. As a result of their anticoagulant, anticancer, immunomodulatory, anti-hyperlipidemic, and antioxidant capabilities, carrageenans are also used in a variety of biological applications. They are also protective against certain viruses, fungi, and bacteria (Guan, Li, & Mao, 2017).

Because of the latter, carrageenans have been proposed as potential treatments for respiratory illnesses like the infamous avian flu, and they are now being investigated for their ability to destroy a variety of viruses like those that cause dengue fever, hepatitis A, HIV, and herpes. Carrageenan and its derivatives have varying degrees of toxicity, according to studies, although they pose no threat to patients' health. These qualities make carrageenan a viable biomaterial for biological purposes (Kalsoom Khan, Saba, Nawazish, Akhtar et al., 2017). Carrageenan's ability to gelify and dissolve in water, as well as other physicochemical properties, determine how it is used as an excipient in oral administration systems. Carrageenan's degree of sulfation, which affects its mechanical properties and rate of dissolution, has a big influence on how much weight it can support. These elements may impact the cargo's discharge, either slowing it down or speeding it up. It is feasible to exert further by joining or conjugating with different polymers and regulating the medication release profile, irrespective of other conditions like pH or the type of carrageenan. There are also other polymers added, such as the temperature-sensitive semi-synthetic polymer HPMC, that can address issues with pH erosion and offer the medication greater protection, encouraging a sustained release that is pH independent (Jafari, Farahani, Sedighi, Rabiee, & Savoji, 2021). While pH-responsive polymers might be conjugated, the opposite reaction (i.e., pH-triggered breakdown) could be preferred. The pH may be changed to influence both the loading and release processes of carrageenan/alginate interpenetrated networks. When drugs and genes are delivered using stimuli-responsive materials, the carrier may also operate as an active trigger and therapy optimizer, opening up new possibilities for drug and gene delivery. When building nano-carriers, temperature-sensitive materials can encourage a regulated release at temperatures higher than 37°C. A method like this can be beneficial in circumstances as typical as a fever. However, in conditions of hyperthermia, where the drug would be available in a confined location, different nanocarriers may be used (Lim, Ooi, Tey, & Chan, 2017).

In the pharmaceutical sector, carrageenan is often utilized as a raw material to create DDSs, capsules for cells used in cell therapies, and items requiring the regeneration of cartilage. Given that carrageenan hydrogels are effective settings for encasing a range of human cell types and achieving chondrogenic differentiation, their use can be used as a vehicle for the regulated delivery of biomolecules, particularly for cargo stability. Due to the differentiation already described as well as the ease of injection and possibility for usage as growth factor reservoirs, this approach showed promise for cartilage regeneration methods (Vijayakumar, Saravanakumar, Malaikozhundan, Divya et al., 2020).

Hydrogels made of carrageenan and other marine-derived materials have also demonstrated their suitability as effective tools for cell encapsulation. Utilizing novel processes, spherical beads as well as fibrillar hydrogels composed of carrageenan and alginate have been created. Fibrillar hydrogels were created via wet spinning, and they show better capacity for application as cell carriers in drug delivery systems. Carrageenan possesses biological qualities; hence, it is anticipated that carrageenan-based devices would make good DDSs for the administration of both bioactive substances and cells for cytotherapies (Mobarak, Jumaah, Ghani, Abdullah, & Ahmad, 2015).

6.4.4 HYALURONIC ACID

The glycosaminoglycan family includes hyaluronan. GAGs are repeating disaccharide units of N-acetylated hexosamine and uronic acid that are linear, negatively charged heteropolysaccharides. It is a linear polysaccharide made up of 13 and 14 alternately arranged glycosidic linkages that join the D-glucuronic chain disaccharide and N-acetyl-D-glucosamine units. A significant part of the extracellular matrix and dispersion in cartilage, vitreous humor, and synovial fluid is hyaluronan. Hyaluronan plays a significant part in a number of biological processes and is superior to other materials because of its high viscoelasticity for a number of biomedical applications. It is specifically engaged in cell proliferation, differentiation, and migration as well as tissue regeneration. Because it is present in the synovial fluid of joints, hyaluronan is a biological marker that can be employed to identify disorders related to rheumatoid arthritis (Carrion, Nasrollahzadeh, Sajjadi, Jaleh, Soufi, & Iravani, 2021).

Hyaluronan is now produced in vast amounts using a variety of processes and materials, including bacterial fermentation. Hyaluronan can also be extracted straight from sources like cartilage and the vitreous humor of many fish species that live in the ocean. Hydrolases like hyaluronidase, which dissolve the glycosidic link between two residues, have an impact on its biodegradability. Because it is present in a variety of bodily biological fluids, hyaluronan can be utilized as a biomarker to monitor its mobility in those fluids (Prasathkumar & Sadhasivam, 2021). For spherical structures, hyaluronan has also been employed as a covering material. Hyaluronan and chitosan multilayers can be alternately adsorbed onto cross-linked chitosan spheres as templates. The results demonstrated that the release from the microspheres was more sustained than the release from the uncoated cores when gentamycin sulfate was utilized as a model medication for *in-vitro* release. These findings demonstrate that LbL coating can provide cargo stabilization and, for appropriate lows, increased sustained release. Hyaluronan coatings on liposomes are also created using spherical templates (Cazorla-Luna, Martín-Illana, Notario-Pérez, Ruiz-Caro, & Veiga, 2021). The intracellular administration of peptides is one practical use of such pH sensitivity. Hyaluronan-coated liposomes for the administration of anticancer medications were introduced by scientists. In this instance, the covering shielded the liposome and its contents from attacks by bloodstream proteins. When a liposome enters the extracellular matrix of a tumor, where hyaluronidase breaks down the outer layer of hyaluronan, it is exposed to cytoplasmic pH changes that allow the intracellular drug release (Yazdi, Taghizadeh, Taghizadeh, Stadler et al., 2020).

During *in vivo* studies, high anticancer activity was also discovered. The capacity of hyaluronan to interact with many proteins is an intriguing characteristic. Hyaluronan has membrane receptors; hence, this ability can be useful as a diagnostic tool. Here, cell growth is accompanied by a robust expression of the receptor CD44. Using hyaluronan devices, it is possible to identify an elevated amount of CD44 expression as a valuable marker for the early detection of cancer (Mallakpour, Azadi, & Hussain, 2021). Bioactive compounds created by a number of construction methods can be stored using hyaluronan hydrogels. Due to the capacity of certain of these devices to produce fluorescence, Hyaluronan-based nanoparticles that interact with metals, including gold, were broadly used as indicators for disorders such as cancer as well as rheumatoid arthritis (Grosso & de-Paz, 2021).

6.4.5 MARINE POLYMERS FORM MICROALGAE AND MARINE CREATURE WASTE

The three main evolutionary groupings of seaweeds, Rhodophyta (red seaweeds), *Phaeophyceae* (brown seaweeds), and Chlorophyceae, are macroscopic, multicellular algae (green). This biological variety exhibits significant chemical diversity with unique polymers within each category. Although a thorough analysis of these polymers is outside the purview of this chapter, a brief summary of the major types of polymers that these organisms make is provided. These species' distinctive structural and non-structural polysaccharides, which can account for over 70% of dry weight, are of interest as polymers. Agars and carrageenans from red seaweeds are among those that are heavily exploited for industrial use, with alginates from brown seaweeds being of special economic importance as hydrogels (Claverie, McReynolds, Petitpas, Thomas, & Fernandes, 2020). Except for storage polysaccharides, which are found in or connected to the cell wall and resemble the starch found in plants, they are all polysaccharides. Although their physical characteristics (such as microfibril diameter) and the enzymes involved in their formation differ, cellulose is the most prevalent material across all taxa (Kuznetsova, Andryukov, Besednova, Zaporozhets, & Kalinin, 2020).

Matrix polysaccharides are virtually always sulfated in seaweeds, a trait that is not seen in terrestrial plants. Sulfation has a crucial impact on these biopolymers' biological activities; a thorough examination of their bioactivity may lead to the development of custom molecules with excellent therapeutic potential. From a biomimetic standpoint, it is very interesting to combine the biological capabilities of these biopolymers with their chemical characteristics (structure, derivatization with small molecules). The ability to distribute medicines, proteins, and/or nucleic acids, as well as the ability to provide receptor- or enzyme-mediated cell-matrix interactions, are all crucial properties that these materials must preserve (Silva, Alves, Ferreira, Oliveira et al., 2012). This is because they support the structure of microalgae mechanically. The incorporation of certain cellulose nanofibrils in polysaccharide matrices results in tissue that is stronger, more extensible, and more fatigue-resistant. This is an example of adaptive viscoelastic behavior. Recently, it was discovered that calcareous coralline algae synthesize chitin and proteins that resemble collagen. These two biopolymers likely contribute to the biomineralization of the algal skeletons. By mimicking naturally occurring biopolymer combinations, materials with tunable qualities, such as hydrophobicity, stiffness, mechanical resistance, etc., may be created (Rasmussen & Morrissey, 2007).

Marine collagens, such as those found in fish bone, skin, scales, and cartilage, have a smaller particle size and lower molecular weight than pig or bovine collagen, making them more accessible. Additionally, they may absorb substances into the body up to 1.5 times more effectively, and their blood circulates more quickly (Himaya & Kim, 2015). Collagen obtained from marine sources is analogous to collagen from regular bovine and pig sources with regard to amino acid content as well as biocompatibility. Fish by-products including skin, scales, and fishbone, which are consumed often worldwide and produce a significant amount of waste, can be used to make fish collagen (Heo, Oh, & Jung,).

Collagen from fish skin is excellent for cutaneous applications since it has no tendency to irritate the skin and has a good water-holding capacity (around 6% of its weight when exposed to 63% humidity for 24 hours) (Santiesteban-Romero, Martínez-Ruiz, Sosa-Hernández, Parra-Saldívar, & Iqbal, 2022). Collagen obtained from codfish skin denatured at a temperature of about 16°C, which may be related to the species' habitat, collagen from the skin of two teleost species and two Chondrichthyes species has been studied. Although tilapia, catfish, pomfret, and mackerel skin may be extracted for 74 hours at a low extraction temperature (just under 13.26°C), it only produces 2.27% of the dry mass (Coppola, Oliviero, Vitale, Lauritano, D'Ambra, Iannace, & De Pascale, 2020).

Fish scales make up a sizable portion of the trash produced by the fish processing industry. Recent research indicates that fish scale collagen has properties similar to type I collagen, which has two $\alpha 1$ chains and one $\alpha 2$ chain. The collagen from the scales of tilapia may have a high denaturation temperature (57.9–79.0°C) because of its large amino acid content as well as superior inter/intrachain connections (Van der Waals interactions, ionic bonds, dipole-dipole bonds, and hydrogen bonds) (Hickey, 2012). However, fresh carp scales barely reached the denaturation temperature of collagen at 32°C. Due to lower extraction yields (0.13%) compared to fish skin (4.3%), collagen from the scales of mackerel, pomfret, tilapia, and catfish required a higher extraction temperature (range 16.6–19.03°C) and longer extraction time. Additionally, it has been shown that fish scale collagen has good water absorption (13.3%) and retention properties (15%), making it appropriate for use in therapeutic as well as medicinal therapies (Mourelle, Gómez, & Legido, 2017). Studying the collagen in the bones of the tilapia, it was discovered that its denaturation temperature is 32.5°C (Bedi, Srivastava, Srivastava, Mahapatra, Dkhar, Chandra, & Srivastava, 2022). The collagen in the bones of the Lutjanus species, according to researchers, has anticancer properties. The collagen yields 4.5% and has 8.815 mg of protein per milliliter. One of the main methods for removing collagen from fishbone is high-intensity pulsed electric fields (PEF). In research, semi-bionic extraction and PEF treatments were used to extract collagen, chondroitin, and calcium from discarded fish bones (Hosseini, Ramezanzade, & McClements, 2021). The study examined the physical-chemical composition and antioxidant properties of silvertip shark collagen (*Carcharhinus albimarginatus*). This cartilage was used to extract type 2 gelatin, type 2 acid-soluble collagen, and type 2 pepsin-soluble collagen. Type 2 gelatin had a denaturation temperature of 32.5°C, greater than that of type 2 gelatin and PSC, two additional collagens. Compared to type 2 ASC and type 2 gelatin, PSC showed greater reducing power and antioxidant activity. Due to their greater antioxidant activity, collagens from silvertip sharks could be useful in biomedical applications (Wang, Zhou, Calvo-Lerma, Liu, Collado, & Barba, 2022) (Figure 6.1).

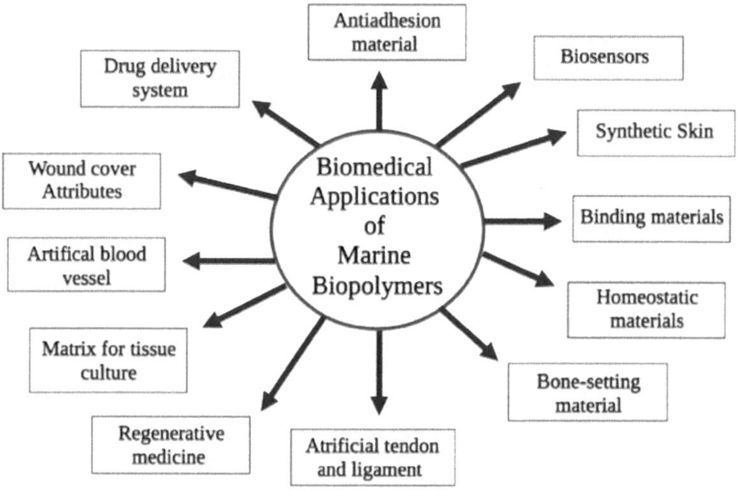

FIGURE 6.1 Biomedical applications of marine-based biopolymers.

6.5 APPLICATION OF MARINE-BASED POLYMERS IN THE FABRICATION OF NOVEL ORAL DOSAGE FORMS

The use of marine-origin polysaccharides has increased due to their abundance in availability, affordability, biocompatibility, and biodegradability. These features encourage scientists to utilize marine-based polymers in the fabrication of novel drug delivery systems, tissue engineering, cancer therapy, wound dressing, biosensors, and the treatment of wastewater. Considering the physicochemical properties, their usage in oral drug delivery has been elaborated.

6.5.1 NANO-EMULSIONS

To address significant issues with traditional drug delivery systems, a cutting-edge method of drug delivery has been developed. Nanoemulsions are emulsions produced in nanometer sizes to enhance the delivery of pharmacologically active substances (Manna, Jana, Mandal, & Bhowmik, 2022). These are isotropic systems that are thermodynamically stable when two immiscible liquids are combined into one phase using an emulsifying agent like a surfactant and a co-surfactant. Nanoemulsions typically have droplet sizes between 20 and 200 nm. The size and shape of the particles scattered in the continuous phase are the primary distinctions between an emulsion and a nanoemulsion (Hosseini, Ramezanzade, & McClements, 2021). Chitosan is a flexible marine biopolymer with several beneficial properties, such as non-toxicity, ease of modification, antibacterial activity, biodegradability, and biocompatibility. These qualities make it suitable for use in meat-based food systems, which are particularly susceptible to spoilage due to their richness in protein, micronutrients, and moisture. As a substitute for synthetic compounds, which are believed to have toxicological effects, bioactive components from plant extracts combined with chitosan are excellent determinants for meat preservation. This is due to the natural origin of these components (Rahnama, Anvar, Ahari, & Kazempoor, 2021).

Natural marine polymer materials are used extensively as medical polymer materials and porous support materials because of their great biocompatibility and lack of toxicity due to their ability to break down into metabolites in the body (Ozogul, Elabed, Ceylan, Ocak, & Ozogul, 2021). Alginate, a marine biopolymer, is negatively charged, while chitosan is positively charged. Ionic bonding causes the two components to combine and create a rather stable structure. It is well known that beta-glucan has excellent skin penetration, skin moisturizing and soothing benefits, and burn as well as wound healing properties in addition to contributing to the durability of materials. According to several studies, beta-glucan nanoemulsion-infused alginate-chitosan hydrogel patches were prepared to test the antibacterial effectiveness (Jana, Gandhi, & Roy, 2020).

6.5.2 DENDRIMERS

Following the current epidemic caused by the SARS-CoV-2 virus, scientists from all around the world have focused their study on all aspects of this novel threat to civilization. However, some preventative measures and supportive treatments can reduce viral transmission without a particular treatment intervention, as shown by factors like the fundamentals in the epidemiology of viral infections, reproduction, rate of case deaths, as well as incubation time are important factors (Anusha, Arasu, Al-Dhabi, & Raj, 2022). In this overview, the SARS-CoV-2 coronavirus life cycle in a human host cell and infection prevention methods are briefly discussed. The antiviral characteristics of synthetic and natural polymers, including cyclodextrin, carrageenan, polyethyleneimine, nanoparticles, hydrophobically modified chitosan, glycosylated chitosan, amine-based dendrimers, are highlighted (Ramanjooloo, Andersen, & Bhaw-Luximon, 2021).

Because of their spherical shape and hollow interior core, dendrimers are three-dimensional macromolecules that are extensively branched and have nano-sized dimensions. Other molecules can be coated with the empty internal cavity's macromolecular cavity, which prevents further oxidation and clustering of those molecules (Ahmad, Ahmad, Khan, Akhtar, Singh, & Ahmad, 2018). Dendrimers are useful as diagnostic tools, antibacterial, anticancer, and biological uses because of their special features. Given that their moiety has a substantial amount of surface area, different compositions, and groups, dendrimers have a distinct branching structure that gives them a spherical form. Such groups can be seen as models for drug grafting and immobilization since they represent the functional active areas (Noreen, Ma, Saeed, Pervaiz et al., 2022).

On the dendrimer's external surface, branching units bind covalently to the central core and terminal functional groups which have an atom or multifunctional molecule at its center. Monodispersing macromolecules with distinct sizes and molecular masses are called dendrimers. They possess well-controlled rheological qualities as a result of their predictable shape and size. It is possible to modify the properties of the terminal groups of dendrimers, such as their solubility, miscibility, hydrophilicity or lipophilicity, and acidity (Nwuche, Igbokwe, Ajagbe, & Onwosi, 2021). Small molecules that can fit inside their cavities and interact through charge interactions, hydrogen bonds, and lipophilicity, such as medicines, metals, and imaging moieties, can be encapsulated by dendrimers, which is one of their most significant features. Medication may become more stable, have less potential for toxicity, and release more quickly when they are enclosed in dendrimers (Guimarães, 2021).

When compared to proteins, dendrimers are less expensive, have superior biocompatibility, and offer precisely regulated huge macromolecular surfaces. The dendrimers serve as a covering and can enclose other biologically active substances as well as conjugate molecules (Shahcheraghi, Golchin, Sadri, Tabari, Borhanifar, & Makani, 2022). Moreover, the insertion of a number of chemical groups, including basic, hydrogen-bonding-capable, acidic, hydrophobic, as well as charges, enables the customization and control of the chemical characteristics and structural design of the dendrimer's branches, core, and surface groups. One of the most crucial physicochemical characteristics that improves the *in vitro* and *in vivo* behaviors of dendrimers is variation in their composition, design, and attributes (Maheshwari, Kuche, Advankar, Soni et al., 2018). An amine-terminated PAMAM dendrimer known as

G4-PAMAM (polyamidoamine) dendrimer is employed as a covering for the nano Ni substance, which has strong antimicrobial properties against a variety of bacteria. In ambient settings, the nano Ni material often experiences rapid oxidation and aggregates. Ni NPs can be coated with dendrimers, which will protect them from further oxidation and aggregation. When tested against the bacteria E. coli, the chemical reaction displayed by these dendrimer-encapsulated Ni nanoparticles disrupts the cell wall, preventing bacterial growth (Kerry, Rout, Das, Fraceto, Paramithiotis, & Patra, 2019).

6.5.3 SOLID-LIPID NANOPARTICLES

Many researchers have proposed solid lipid nanoparticles (SLNs) as biopolymer-based nanocarriers of medicinal drugs and diagnostics. The benefits of these colloidal carriers as a drug delivery method include the potential for regulated drug release and drug targeting, protection of the molecule against chemical deterioration, lack of carrier biotoxicity and biodegradation, and no production issues on a large scale. When given orally, one benefit of nanoparticles is that they can be absorbed by a variety of methods. The transport of the medicine in the capsule can be improved by SLNs (Tang, Partridge, Woolley, Pilmer, & Lim, 2022). This is due to a number of mechanisms, including muco-adhesion, nanoparticle internalization, and penetration-boosting impact. Due to its exceptional biocompatibility, biodegradability, as well as ability to alter the physicochemical characteristics of nanoparticles, chitosan has been widely employed as a coating material for a variety of nanoparticles. Chitosan-coated lipid nanoparticles for solid dosage forms are being developed (Sánchez-Dengra, Bermejo, González-Álvarez, & González-Álvarez, 2022). With the use of a high-pressure hot homogenization process, we manufactured SLNs either with or without chitosan coatings. By examining these nanoparticles' entrapment effectiveness, zeta potential, particle size, TEM, as well as *in-vitro* drug release, their characteristics were determined (Carducci, Casadei, Mariani, & Barbosa, 2019).

According to a different study, chitosan is significantly utilized to enhance SLNs' interaction with the intestinal mucosa to improve drug absorption. Chitosan, produced by the deacetylation of chitin, is a biocompatible and biodegradable substance with mucoadhesive qualities that may help drugs be absorbed due to prolonged interaction with the intestinal epithelium (Tabassum & Ahmad, 2021). Additionally, it has been noted that chitosan improves SLN stability and reduces compound loss, is a powerful permeability enhancer due to its capacity to reversibly change tight junctions, and can give SLNs stealth qualities when combined with Brij-78. Chitosan and lipid-based colloids have recently been combined to create novel entities that combine the benefits of both vectors and do away with some of the limitations of lipid carriers, such as the abrupt release of drugs once supplied to humans and the difficulty of incorporating particular drugs (Sridhar, Inbaraj, & Chen, 2021).

6.5.4 PEPTIDES AND PROTEIN

Over the past few decades, edible seaweeds have been viewed as prospective sources of both novel biologically active chemicals and necessary nutrients for humans. Food-derived peptides that are biologically active can have a variety of effects, such as opiate-like, mineral-binding, immunomodulatory, antibacterial, antioxidant, antithrombotic, and blood pressure-lowering ones (Zaky, Simal-Gandara, Eun, Shim, & Abd El-Aty, 2021). Different sources of animal and plant protein have been found to contain bioactive peptides, with milk peptides being by far the most well-known. By acting as ACE-inhibitory peptides, certain physiologically active peptides have the potential to prevent hypertension and its harmful effects (Zheng, Wang, Zhuang, Li et al., 2020). The importance of ACE-inhibitory peptides derived from food lies in the fact that they satisfy the demands for naturalness and safety, although they are less effective than synthetic medications like captopril (Patil, Usman, Zhang, Mehmood, Zhou, Teng, & Li, 2022).

There are now two prominent pharmaceuticals made from marine peptides on the market: ziconotide and brentuximab vedotin, which are both natural marine peptides and peptide derivatives (Sukmarini, 2021). Ziconotide [Prialt®], a peptide first found in a tropical cone snail, was the first substance produced from marine life to be authorized for use in the United States in December 2004 to alleviate pain (Zhao & Antunes, 2022). The FDA later authorized the use of another sea hare-derived marine peptide medication [Adcetris®] for the treatment of cancer in 2011 (Giddings & Newman, 2019). Moreover, the first bioactive peptides, such as neurotoxic, cardiotonic peptide, antiviral, and antitumor peptides, and cardiotoxin and antimicrobial peptides, were found and isolated in marine animals. Apart from this, plitidepsin and glembatumumab are used in the treatment of cancer, Dermochlorella® DG containing purified oligopeptide from algae used in cosmetics, Katsuobushi oligopeptide used as an antihypertensive agent, homophymine A – an anti-HIV cyclodepsipeptide, dolastatin 10 and dolastatin 15 used as potent anti-cancer agents, hemiasterlin used as an antimitotic agent, monomethyl auristatin E used as an antineoplastic agent, didemnin B having potent anti-tumor activity, Pardaxin used to treat Methicillin-Resistant Staphylococcus aureus Infection, etc., are some of the well-known peptides derived from marine sources (Calado, Leal, Gaspar, Santos et al., 2018).

Overall, marine resources offer a plentiful supply for the extraction of peptides. These peptides have numerous uses in both medicines and nutraceuticals. More peptides will be found as technology develops and tools to look for marine sources become available (Elnahtawy, Elshafei, & Elzoghby, 2022). Additionally, it is now possible to extract pure peptides from natural resources due to advancements in the sensitivity of analytical techniques. A synthesis plan can be created to synthesize these peptides on a wide scale for pharmaceutical uses once the peptides acquired from marine sources have been characterized. Some of these active peptides can also be altered for intravenous or oral delivery. Therefore, marine sources can be a useful supply of nutrition, medications, and nutraceuticals in addition to being a rich source of food (Šimat, Elabed, Kulawik, Ceylan et al., 2020).

6.5.6 MARINE PEPTIDES SECLUSION AND PROCESSING

Due to the interaction between protein and polysaccharides in the seaweed matrix, isolating protein from seaweeds can be challenging. As a result, methods for protein precipitation such as isoelectric point, ammonium sulfate saturation, and trichloroacetic acid are not entirely appropriate for seaweeds (Cermeño, Kleekayai, Amigo-Benavent, Harnedy-Rothwell, & FitzGerald, 2020). To address this issue, multiple methods are suggested, including proteolytic treatment of the whole seaweed followed by filtration and dialysis or alginate lyase treatment of the seaweed matrix to produce an enriched protein precipitate that is recovered by centrifugation (Zayed & Ulber, 2020).

A typical method for finding marine bioactive (linear or cyclic) peptides involves first extracting the peptides from the sources. A single bioactive peptide is produced after the extract has been purified, fractionated under the guidance of a bioassay, and screened for particular bioactivity (Akbarian, Khani, Eghbalpour, & Uversky, 2022). In the initial phase of cyclic peptide extraction, organic solvents like methanol or ethyl acetate are typically utilized. The extracted material is then concentrated and divided among several solvents, such as hexane, carbon tetrachloride, or dichloromethane. Size exclusion chromatography, reverse-phase HPLC, and ion exchange chromatography are frequently used in the general process for linear peptides (Qamar, Torres, Parekh, & Falconer, 2021). An exclusion chromatography column is used to size-fractionate the aqueous extract first. A cation/anion exchange chromatography column is used to apply the concentrated target fraction, which is then loaded and eluted according to the peptide's basicity. Final purification will employ reverse-phase C18 HPLC with an additional step of re-chromatography on the same column for removing contaminants (Li, Jian, & Fu, 2019).

6.5.7 MUCOADHESIVE DRUG DELIVERY SYSTEM

The attachment of two materials, at least one of which is a mucosal surface, is known as mucoadhesion. Mucosal medication delivery has drawn a lot of interest during the past few decades. The nature of the mucosal tissue and the physicochemical characteristics of the polymeric formulation are just two examples of the many variables that affect a dosage form's capacity to adhere to mucous membranes. The mucus layer, which covers the mucous membrane, is a viscoelastic gel that is 95% water in composition (Yermak, Davydova, & Volod'ko, 2022). The mucin glycoproteins (mucins), which are in charge of the mucus layer's viscoelastic characteristics and interactions with its surroundings, are its most crucial functional constituents. They are in charge of giving the mucous membranes their gel-like texture and anti-adhesive qualities. By increasing the residence duration at the distribution site, mucoadhesive polymers may be a useful strategy for increasing the bioavailability of the active ingredient (Wagner, Wheeler, & Ribbeck, 2018). Due to their suitability for numerous drug delivery applications that rely on their mucoadhesive qualities, polysaccharides are among the most extensively investigated natural macromolecules (Singh & Bothara, 2014). Although the underlying mechanisms are not fully understood, several natural polymers have mucoadhesive characteristics. Some of the natural polymers extensively used in the pharmaceutical industry and as discussed in the preceding section such as chitosan, sodium alginate (alginic acid), carrageenan, hyaluronic acid, etc., have been derived from marine sources.

6.6 MARINE BIOPOLYMERS IN SMART DELIVERY OF PHARMACEUTICALS

Oceans are a vast source of natural pharmaceuticals. The exploitation of the sea as a renewable source of biodegradable compounds with wide biotechnological, biomedical, and pharmaceutical applications. Marine polysaccharides are among the most abundant biomaterials in the seas, besides their solubility behavior in aqueous solvent extraction media, and their interaction with other bio-compounds. Several marine polysaccharides are explored in the fabrication of conventional drug delivery systems. Perhaps these polysaccharides are now significantly utilized in the development of smart drug delivery.

6.6.1 ELECTROSPUN NANOFIBERS

An innovative technological platform for the production of nonwoven nanofibrous scaffolds with increased surface area, high encapsulation efficacy, intrinsic interconnectivity, and structural resemblance to the natural extracellular matrix has been made possible by the development of electrospinning (Iliou, Kikionis, Ioannou, & Roussis, 2022). For biomedical applications like tissue engineering, drug delivery, and wound healing, marine biopolymer-based electrospun nanofibrous scaffolds with multifunctional features and customizable mechanical properties are presently receiving a lot of interest. Because they have unique qualities like high encapsulation efficacy, inherent interconnectedness, and structural similarity to the extracellular matrix (ECM), they can be employed as scaffolds to encourage cell development and as carriers for bioactive substances. The most effective and popular method for creating ultrafine polymeric fibers is electrospinning (Behere & Ingavle, 2022).

Along with well-known polymers from marine sources as shown in Figure 6.2. Halomonas maura, a halophilic bacterium, produces the exopolysaccharide known as mauran (Banerjee, Sarkar, Cuadros-Orellana, & Bandopadhyay, 2019). It is a sulfated polysaccharide that primarily consists of mannose, galactose, glucose, and glucuronic acid. It also contains a significant amount of uronic acid. Mauran aqueous solutions exhibit viscoelastic, thixotropic, and pseudoplastic behavior and are quite viscous over a wide pH range (Iliou, Kikionis, Ioannou, & Roussis, 2022). Mauran and PVA were successfully combined to create electrospun nanofibers with a homogeneous structure and small average diameter. *In vitro* testing of the nanofibers' biocompatibility was performed on murine fibroblast cells and mesenchymal stem cells. Investigations on cell adhesion revealed the scaffolds' ability to foster cell adhesion, migration, proliferation, and differentiation, suggesting they have great potential for tissue engineering and other biomedical applications (Raveendran, Dhandayuthapani, Nagaoka, Yoshida, Maekawa, & Kumar, 2013).

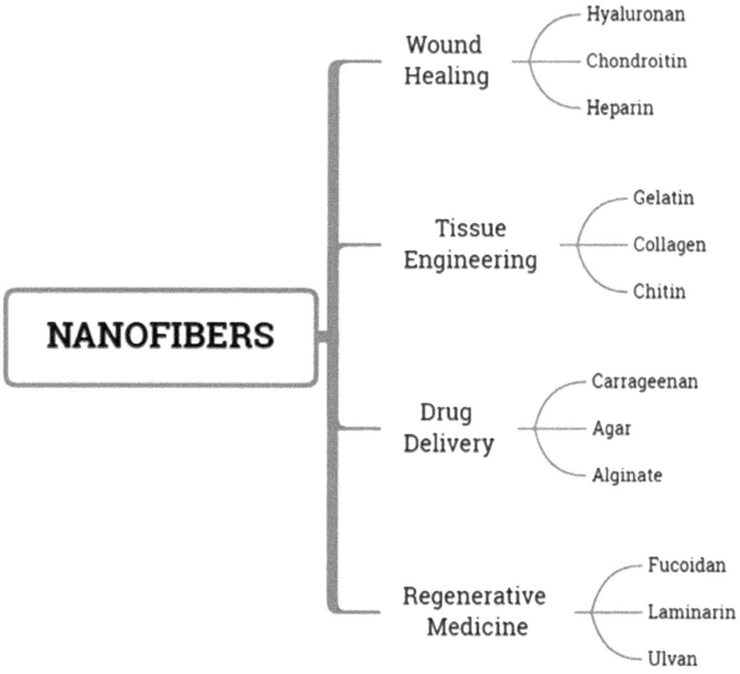

FIGURE 6.2 Various pharmaceutical applications of nanofibers.

6.6.2 *THREE-DIMENSIONAL PRINTED ORAL DRUG DELIVERY USING MARINE POLYMERS*

The first ever FDA authorized 3D printed medicine product, Spritam®, is an orodispersible dosage form, comprising either 250, 500, 750, or 1000 mg of the antiepileptic medication levetiracetam (Souto, Campos, Filho, Teixeira et al., 2019). The use of 3D printing technology to formulate drug delivery systems is growing, and the nanomedicine industry seems to find it particularly appealing. In order to safeguard the anticancer medication camptothecin, systems that were 3D printed with chitosan-based polymeric micelles were then loaded with the medicine to delay its release in the upper GI tract and prevent it; the chitosan micelles are released at a colonic pH. Medication intestinal permeability significantly increased in 3D printed devices (Almeida, Linares, Mora-Castaño, Casas, Caraballo, & Sarmento, 2021). Due to its biocompatibility, biodegradability, and cell adhesion, alginate-based hydrogel inks are frequently utilized in printing.

Kondiah et al. investigated the production of TPGS-PLGA NPs that were placed within an alginate-gelatin hydrogel matrix that was 3D-(bio) printed in order to address the poor oral bioavailability of HDMTX. A controlled release of HDMTX after oral administration was achieved by using 3D (bio) printing to have better control over the system design parameters through physicochemical and physico-mechanical adjustments (Kondiah, Rants'o, Makhathini, Mdanda, & Choonara, 2022).

6.7 CONCLUSION

Growing world population and urbanization contribute to the increased production of marine waste, which deposits in landfills, developing several health-associated issues. Meanwhile, the growing shortage of pharmaceutical resources and materials as a result of over-consumption and unsustainable production processes has shifted attention to the development of sustainable resources from marine waste. Therefore, the valorization of marine waste using cutting-edge technology has created a way to develop a sustainable and multifarious excipient for pharmaceutical and food industry applications. Moreover, these marine-based polymers are significantly exploited in the development of mucoadhesive, additive manufacturing, proteins, peptides, etc. Although several polymers have been explored, further studies can be conducted to develop new excipients from marine creature bones and their cost implications for valorization on a large scale, including life-cycle assessment analysis and techno-economic modeling.

KEYWORDS

- **bile salt export pump**
- **carrageenan**
- **chitosan**
- **elastin-like recombinants**
- **oral pharmaceuticals**
- **poly(N-isopropyl acrylamide)**
- **sea waste valorization**

REFERENCES

Ahmad, U., Ahmad, Z., Khan, A. A., Akhtar, J., Singh, S. P., & Ahmad, F. J. (2018). Strategies in development and delivery of nanotechnology based cosmetic products. *Drug research, 68*(9), 545–552.

Akbarian, M., Khani, A., Eghbalpour, S., & Uversky, V. N. (2022). Bioactive Peptides: Synthesis, Sources, Applications, and Proposed Mechanisms of Action. *International Journal of Molecular Sciences, 23*(3), 1445.

Almeida, A., Linares, V., Mora-Castaño, G., Casas, M., Caraballo, I., & Sarmento, B. (2021). 3D printed systems for colon-specific delivery of camptothecin-loaded chitosan micelles. *European Journal of Pharmaceutics and Biopharmaceutics, 167*, 48–56.

Anand, S., & Rajinikanth, P. (2021). Alginate-based nanomaterials in drug delivery applications. In *Biopolymer-Based Nanomaterials in Drug Delivery and Biomedical Applications* (pp. 339–364). Elsevier.

Ansari, M. T., Murteza, S., Ahsan, M. N., Hasnain, M. S., & Nayak, A. K. (2022). Chitosan as a responsive biopolymer in drug delivery. In *Chitosan in Drug Delivery* (pp. 389–410). Elsevier.

Anusha, J., Arasu, M. V., Al-Dhabi, N. A., & Raj, C. J. (2022). Nanomaterials in electrochemical biosensors and their applications. In *Emerging Nanomaterials for Advanced Technologies* (pp. 487–516). Springer.

Banerjee, A., Sarkar, S., Cuadros-Orellana, S., & Bandopadhyay, R. (2019). Exopolysaccharides and biofilms in mitigating salinity stress: The biotechnological potential of halophilic and soil-inhabiting PGPR microorganisms. In *Microorganisms in saline environments: Strategies and functions* (pp. 133–153). Springer.

Barcelos, M. C. S., Vespermann, K. A. C., Pelissari, F. M., & Molina, G. (2019). Current status of biotechnological production and applications of microbial exopolysaccharides. *Critical Reviews in Food Science and Nutrition, 60*(9), 1475–1495. https://doi.org/10.1080/10408398.2019.1575791.

Bedi, N., Srivastava, D. K., Srivastava, A., Mahapatra, S., Dkhar, D. S., Chandra, P., & Srivastava, A. (2022). Marine biological macromolecules as matrix material for biosensor fabrication. *Biotechnology and Bioengineering, 119*(9), 2046–2063.

Behere, I., & Ingavle, G. (2022). In vitro and in vivo advancement of multifunctional electrospun nanofiber scaffolds in wound healing applications: Innovative nanofiber designs, stem cell approaches, and future perspectives. *Journal of Biomedical Materials Research Part A, 110*(3), 443–461.

Benet, L. Z., Hosey, C. M., Ursu, O., & Oprea, T. I. (2016). BDDCS, the Rule of 5 and drug ability. *Advanced Drug Delivery Reviews, 101*, 89–98. https://doi.org/10.1016/j.addr. 2016.05.007.

Bharathi, R., Ganesh, S. S., Harini, G., Vatsala, K., Anushikaa, R., Aravind, S., Abinaya, S., & Selvamurugan, N. (2022). Chitosan-based scaffolds as drug delivery systems in bone tissue engineering. *International Journal of Biological Macromolecules, 222*(Part A), 132–153.

Biswas, M. C., Jony, B., Nandy, P. K., Chowdhury, R. A., Halder, S., Kumar, D., Ramakrishna, S., Hassan, M., Ahsan, M. A., & Hoque, M. E. (2021). Recent advancement of biopolymers and their potential biomedical applications. *Journal of Polymers and the Environment, 30*(51), 51–24.

Bocci, G., Oprea, T. I., & Benet, L. Z. (2022). State of the art and uses for the biopharmaceutics drug disposition classification system (BDDCS): New additions, revisions, and citation references. *AAPS J, 24*(2), 37. https://doi.org/10.1208/s12248-022-00687-0.

Bongio, M., van den Beucken, J. J. J. P., Leeuwenburgh, S. C. G., & Jansen, J. A. (2010). Development of bone substitute materials: From 'biocompatible' to 'instructive.' *Journal of Materials Chemistry, 20*, 8747. https://doi.org/10.1039/c0jm00795a.

Buckley, A., & Turner, J. R. (2018). Cell biology of tight junction barrier regulation and mucosal disease. *Cold Spring Harbor Perspectives in Biology, 10*(6), a029314. https://doi.org/10.1101/cshperspect.a029314.

Calado, R., Leal, M. C., Gaspar, H., Santos, S., Marques, A., Nunes, M. L., & Vieira, H. (2018). How to succeed in marketing marine natural products for nutraceutical, pharmaceutical and cosmeceutical markets. In *Grand challenges in marine biotechnology* (pp. 317–403). Springer.

Carducci, F., Casadei, B. R., Mariani, P., & Barbosa, L. R. (2019). X-ray characterization of pharmaceutical and cosmetic lipidic nanoparticles for cutaneous application. *Current Pharmaceutical Design, 25*(22), 2364–2374.

Carrion, C. C., Nasrollahzadeh, M., Sajjadi, M., Jaleh, B., Soufi, G. J., & Iravani, S. (2021). Lignin, lipid, protein, hyaluronic acid, starch, cellulose, gum, pectin, alginate and chitosan-based nanomaterials for cancer nano therapy: Challenges and opportunities. *International Journal of Biological Macromolecules, 178*, 193–228.

Carvalho Henriques, B., Yang, E. H., Lapetina, D., Carr, M. S., Yavorskyy, V., & Hague, J., et al. (2020). How can drug metabolism and transporter genetics inform psychotropic prescribing? *Frontiers in Genetics, 11*, 491895. https://doi.org/10.3389/fgene.2020.491895.

Cazorla-Luna, R., Martín-Illana, A., Notario-Pérez, F., Ruiz-Caro, R., & Veiga, M.-D. (2021). Naturally occurring polyelectrolytes and their use for the development of complex-based mucoadhesive drug delivery systems: an overview. *Polymers, 13*, 2241.

Cermeño, M., Kleekayai, T., Amigo-Benavent, M., Harnedy-Rothwell, P., & FitzGerald, R. J. (2020). Current knowledge on the extraction, purification, identification, and validation of bioactive peptides from seaweed. *Electrophoresis, 41*(19), 1694–1717.

Cheng, L., & Wong, H. (2020). Food effects on oral drug absorption: Application of physiologically-based pharmacokinetic modeling as a predictive tool. *Pharmaceutics, 12*(7), 672. https://doi.org/10.3390/pharmaceutics12070672.

Claverie, M., McReynolds, C., Petitpas, A., Thomas, M., & Fernandes, S. C. (2020). Marine-derived polymeric materials and biomimetics: An overview. *Polymers, 12*, 1002.

Coppola, D., Oliviero, M., Vitale, G. A., Lauritano, C., D'Ambra, I., Iannace, S., & de Pascale, D. (2020). Marine collagen from alternative and sustainable sources: Extraction, processing and applications. *Marine Drugs, 18*(4), 214.

Dey, S., Roy, S., Hasnain, M. S., & Nayak, A. K. (2020). Grafted alginates in drug delivery. In *Alginates in Drug Delivery* (pp. 71–100). Elsevier.

Elnahtawy, A. I., Elshafei, N. S., & Elzoghby, A. O. (2022). Marine Polymer-Based Nanocarriers for Drug Delivery Applications. In *Marine Biomaterials* (pp. 15–59). Springer.

Essa, E. A., Elebyary, T. T., Abdelquader, M. M., El Maghraby, G. M., & Elkordy, A. A. (2021). Smart liquids for oral controlled drug release: An overview of alginate and non-alginate-based systems. *Journal of Drug Delivery Science and Technology, 61*, 102211.

Ghosh, S., Ray, A., & Pramanik, N. (2020). Self-assembly of surfactants: An overview on general aspects of amphiphiles. *Biophysical Chemistry, 265*, 106429. https://doi.org/10.1016/j.bpc.2020.106429.

Giddings, L.-A., & Newman, D. J. (2019). Bioactive compounds from extremophilic marine fungi. In *Fungi in Extreme Environments: Ecological Role and Biotechnological Significance* (pp. 349–382). Springer.

Grosso, R., & de-Paz, M.-V. (2021). Thiolated-Polymer-Based Nanoparticles as an Avant-Garde Approach for Anticancer Therapies—Reviewing Thiomers from Chitosan and Hyaluronic Acid. *Pharmaceutics, 13*, 854.

Guan, J., Li, L., & Mao, S. (2017). Applications of carrageenan in advanced drug delivery. In *Seaweed Polysaccharides* (pp. 283–303). Elsevier.

Guimarães, T. M. T. (2021). Alginate-derived microneedle devices for transdermal delivery of rivastigmine-application in skin mimetic model, *Pharmaceutics*, *14*(4), 752.

Hariyadi, D. M., & Islam, N. (2020). Current status of alginate in drug delivery. *Advances in Pharmacological and Pharmaceutical Sciences, 2020*, pp. 1–16.

Hasnain, M. S., Ahmed, S. A., Alkahtani, S., Milivojevic, M., Kandar, C. C., Dhara, A. K., & Nayak, A. K. (2020). Biopolymers for drug delivery. In *Advanced Biopolymeric Systems for Drug Delivery* (pp. 1–29). Springer.

He, L., Shang, Z., Liu, H., & Yuan, Z. X. (2020). Alginate-based platforms for cancer-targeted drug delivery. *BioMed Research International, 2020*, pp. 1–17.

Heo, S. Y., Oh, G. W., & Jung, W. K. (2020). 3D printed marine biomaterials composites for bone tissue engineering. In S. K. Kim (Ed.), *Encyclopedia of Marine Biotechnology* (Vol. 2, pp. 1299–1314). Springer.

Hickey, R. M. (2012). Extraction and characterization of bioactive carbohydrates with health benefits from marine resources: Macro-and microalgae, cyanobacteria, and invertebrates. In S. Kim (Ed.), *Marine Bioactive Compounds* (pp. 159–172). Springer.

Himaya, S., & Kim, S. K. (2015). Marine nutraceuticals. In S. K. Kim (Ed.), *Springer Handbook of Marine Biotechnology* (pp. 995–1014). Springer.

Homayun, B., Lin, X., & Choi, H.-J. (2019). Challenges and recent progress in oral drug delivery systems for biopharmaceuticals. *Pharmaceutics, 11*(3), 129. https://doi.org/10.3390/pharmaceutics11030129.

Hosseini, S. F., Ramezanzade, L., & McClements, D. J. (2021). Recent advances in nanoencapsulation of hydrophobic marine bioactives: Bioavailability, safety, and sensory attributes of nano-fortified functional foods. *Trends in Food Science & Technology, 109*, 322–339.

Iliou, K., Kikionis, S., Ioannou, E., & Roussis, V. (2022). Marine biopolymers as bioactive functional ingredients of electrospun nanofibrous scaffolds for biomedical applications. *Marine Drugs, 20*(6), 314.

Iravani, S., & Varma, R. S. (2022). Alginate-Based Micro- and Nanosystems for Targeted Cancer Therapy. *Marine Drugs, 20*, 598.

Jafari, A., Farahani, M., Sedighi, M., Rabiee, N., & Savoji, H. (2021). Carrageenans for tissue engineering and regenerative medicine applications: A review. *Carbohydrate Polymers, 281*(1), 119045.

Jana, S., Gandhi, A., & Roy, C. (2020). Marine biomaterials-based systems: Delivery of therapeutics. In S. K. Kim (Ed.), *Encyclopedia of Marine Biotechnology* (Vol. 2, pp. 1141–1174). Springer.

Järvinen, E., Deng, F., Kiander, W., Sinokki, A., Kidron, H., & Sjöstedt, N. (2022). The role of uptake and efflux transporters in the disposition of glucuronide and sulfate conjugates. *Frontiers in Pharmacology, 12*, 802539. https://doi.org/10.3389/fphar.2022.802539.

Jha, A., Kumar, M., Mishra, B. (2022). Marine biopolymer-based anticancer drug delivery systems. *Marine Biomaterials*, 351–401.

Joy, R., Vigneshkumar, P., John, F., & George, J. (2021). Hydrogels based on carrageenan. In *Plant and Algal Hydrogels for Drug Delivery and Regenerative Medicine* (pp. 293–325). Elsevier.

Kalsoom Khan, A., Saba, A. U., Nawazish, S., Akhtar, F., Rashid, R., Mir, S., & Pervaiz, F. (2017). Carrageenan based bionanocomposites as drug delivery tool with special emphasis on the influence of ferromagnetic nanoparticles. *Oxidative medicine and cellular longevity, 2017*, pp.1–13.

Kerry, R. G., Rout, J. R., Das, G., Fraceto, L. F., Paramithiotis, S., & Patra, J. K. (2019). Applications of nanotechnology in food and agriculture. In *Food Molecular Microbiology* (pp. 223–249). CRC Press.

Kondiah, P. P., Rants'o, T. A., Makhathini, S. S., Mdanda, S., & Choonara, Y. E. (2022). An Oral 3D Printed PLGA-Tocopherol PEG Succinate Nanocomposite Hydrogel for High-Dose Methotrexate Delivery in Maintenance Chemotherapy. *Biomedicines, 10*(10), 1470.

Kuznetsova, T. A., Andryukov, B. G., Besednova, N. N., Zaporozhets, T. S., & Kalinin, A. V. (2020). Marine algae polysaccharides as basis for wound dressings, drug delivery, and tissue engineering: A review. *Journal of Marine Science and Engineering, 8*, 481.

Layek, B., & Das, S. (2021). Chitosan-based nanomaterials in drug delivery applications. In *Biopolymer-Based Nanomaterials in Drug Delivery and Biomedical Applications* (pp. 185–219). Elsevier.

Lemos, T. S., de Souza, J. F., & Fajardo, A. R. (2021). Magnetic microspheres based on pectin coated by chitosan towards smart drug release. *Carbohydrate Polymers, 265*, 118013.

Li, W., Jian, W., & Fu, Y. (2019). Basic sample preparation techniques in LC-MS bioanalysis: Protein precipitation, liquid–liquid extraction, and solid-phase extraction. In *Sample Preparation in LC-MS Bioanalysis* (pp. 1–30), Elsevier.

Lim, H. P., Ooi, C. W., Tey, B. T., & Chan, E. S. (2017). Controlled delivery of oral insulin as part using pH-responsive alginate/κ-carrageenan composite hydrogel beads. *Reactive and Functional Polymers, 120*, 20–29.

Liu, J., Sun, Y., Cheng, M., Liu, Q., Liu, W., Gao, C., Feng, J., Jin, Y., & Tu, L. (2021). Improving oral bioavailability of luteolin nanocrystals by surface modification of sodium dodecyl sulfate. *AAPS PharmSciTech, 22*, pp. 133. https://doi.org/10.1208/s12249-021-02012-y.

Maher, S., Brayden, D. J., Casettari, L., & Illum, L. (2019). Application of permeation enhancers in oral delivery of macromolecules: An update. *Pharmaceutics, 11*(1), 41. https://doi.org/10.3390/pharmaceutics11010041.

Maheshwari, R., Kuche, K. N., Advankar, A., Soni, N., Raval, N., Sharma, P. A., & Tekade, R. K. (2018). Natural ingredients/botanical extracts for the nutraceutical industry. In *Flavors for Nutraceutical and Functional Foods* (pp. 75–122). CRC Press.

Mallakpour, S., Azadi, E., & Hussain, C. M. (2021). Recent advancements in synthesis and drug delivery utilization of polysaccharides-based nanocomposites: The important role of nanoparticles and layered double hydroxides. *International Journal of Biological Macromolecules, 193*, 183–204.

Manna, S., & Jana, S. (2022). Marine Polysaccharides in tailor-made drug delivery. *Current Pharmaceutical Design, 28*(9), 1046–1066. https://doi.org/10.2174/1381612828666220328122539.

Manna, S., Jana, S., Mandal, D., & Bhowmik, M. (2022). Marine biopolymer-based oral in situ gel: Drug delivery application. In S. K. Kim (Ed.), *Marine Biomaterials* (pp. 139–155). Springer.

Mobarak, N., Jumaah, F., Ghani, M., Abdullah, M., & Ahmad, A. (2015). Carboxymethyl carrageenan-based biopolymer electrolytes. *Electrochimica Acta, 175*, 224–231.

Montuori, E., de Pascale, D., & Lauritano, C. (2022). Recent discoveries on marine organism immunomodulatory activities. *Marine Drugs, 20*(4), 422.

Mourelle, M. L., Gómez, C. P., & Legido, J. L. (2017). The potential use of marine microalgae and cyanobacteria in cosmetics and thalassotherapy. *Cosmetics, 4*(4), 46.
Nayak, A. K., & Hasnain, M. S. (Eds.). (2020). *Advanced Biopolymeric Systems for Drug Delivery*. Springer, Nature Switzerland AG; 1st ed. 2020 edition (12 July 2020). Hardcover: 386 pp., ISBN: 978–3030469221 https://www.springerprofessional.de/en/advanced-biopolymeric-systems-for-drug-delivery/18172594.
Noreen, S., Ma, J. X., Saeed, M., Pervaiz, F., Hanif, M. F., Ahmed, B., & Madni, A. (2022). Natural polysaccharide-based biodegradable polymeric platforms for transdermal drug delivery system: A critical analysis. *Drug Delivery and Translational Research, 12,* 1–18.
Nwuche, C. O., Igbokwe, V. C., Ajagbe, D. D., & Onwosi, C. O. (2021). Nanoparticles, biosurfactants and microbes in bioremediation. In *Rhizomicrobiome Dynamics in Bioremediation* (pp. 162–179). CRC Press.
Omer, A. M., Ahmed, M. S., El-Subruiti, G. M., Khalifa, R. E., & Eltaweil, A. S. (2021). pH-sensitive alginate/carboxymethyl chitosan/aminated chitosan microcapsules for efficient encapsulation and delivery of diclofenac sodium. *Pharmaceutics, 13*, 338.
Ozogul, F., Elabed, N., Ceylan, Z., Ocak, E., & Ozogul, Y. (2021). Nano-technological approaches for plant and marine-based polysaccharides for nano-encapsulations and their applications in food industry. In *Advances in Food and Nutrition Research* (Vol. 97, pp. 187–236). Elsevier.
Pacheco-Quito, E.-M., Ruiz-Caro, R., & Veiga, M.-D. (2020). Carrageenan: drug delivery systems and other biomedical applications. *Marine Drugs, 18*, 583.
Patil, P. J., Usman, M., Zhang, C., Mehmood, A., Zhou, M., Teng, C., & Li, X. (2022). An updated review on food-derived bioactive peptides: Focus on the regulatory requirements, safety, and bioavailability. *Comprehensive Reviews in Food Science and Food Safety, 21*(8), 1732–1776.
Patricia, J. J., & Dhamoon, A. S., (2022) Physiology, Digestion. In: StatPearls [Internet]. Treasure Island (FL): StatPearls Publishing; 2024 January. Available from: https://www.ncbi.nlm.nih.gov/books/NBK544242/.
Pattnaik, S., Swain, K., & Ramakrishna, S. (2022). Optimal delivery of poorly soluble drugs using electrospun nanofiber technology: Challenges, state of the art, and future directions. *WIREs Nanomedicine and Nanobiotechnology, 15*(2), e1859. https://doi.org/10.1002/wnan.1859.
Phang-Lyn, S., & Llerena, V. A., (2023). Biochemistry, Biotransformation. In: StatPearls [Internet]. Treasure Island (FL): StatPearls Publishing; 2024 Jan-. Available from: https://www.ncbi.nlm.nih.gov/books/NBK544353/.
Prasathkumar, M., & Sadhasivam, S. (2021). Chitosan/Hyaluronic acid/Alginate and an assorted polymers loaded with honey, plant, and marine compounds for progressive wound healing—Know-how. *International Journal of Biological Macromolecules, 186*, 656–685.
Priyan Shanura Fernando, I., Kim, K.-N., Kim, D., & Jeon, Y.-J. (2018). Algal polysaccharides: Potential bioactive substances for cosmeceutical applications. *Critical Reviews in Biotechnology, 39*(1), 99–113. https://doi.org/10.1080/07388551.2018.1503995.
Putri, A. P., Picchioni, F., Harjanto, S., & Chalid, M. (2021). Alginate modification and lectin-conjugation approach to synthesize the mucoadhesive matrix. *Applied Sciences, 11*, 11818.
Qamar, S., Torres, Y. J., Parekh, H. S., & Falconer, J. R. (2021). Extraction of medicinal cannabinoids through supercritical carbon dioxide technologies: A review. *Journal of Chromatography B, 1167*, 122581.

Qureshi, D., Nayak, S. K., Maji, S., Kim, D., Banerjee, I., & Pal, K. (2019). Carrageenan: a wonder polymer from marine algae for potential drug delivery applications. *Current Pharmaceutical Design, 25*, 1172–1186.

Rahnama, M., Anvar, S. A., Ahari, H., & Kazempoor, R. (2021). Antibacterial effects of extracted corn zein with garlic extract-based nanoemulsion on the shelf life of Vannamei prawn (*Litopenaeus vannamei*) at refrigerated temperature. *Journal of Food Science, 86*(11), 4969–4990.

Ramanjooloo, A., Andersen, R. J., & Bhaw-Luximon, A. (2021). Marine sponge-derived/inspired drugs and their applications in drug delivery systems. *Future Medicinal Chemistry, 13*(5), 487–504.

Rasmussen, R. S., & Morrissey, M. T. (2007). Marine biotechnology for production of food ingredients. *Advances in Food and Nutrition Research, 52*, 237–292.

Raveendran, S., Dhandayuthapani, B., Nagaoka, Y., Yoshida, Y., Maekawa, T., & Kumar, D. S. (2013). Biocompatible nanofibers based on extremophilic bacterial polysaccharide, Mauran from *Halomonas maura*. *Carbohydrate Polymers, 92*(2), 1225–1233.

Ristroph, K. D., & Prud'homme, R. K. (2019). Hydrophobic ion pairing: Encapsulating small molecules, peptides, and proteins into nanocarriers. *Nanoscale Advances, 1*(10), 4207–4237. https://doi.org/10.1039/c9na00308h.

Sánchez-Dengra, B., Bermejo, M., González-Álvarez, I., & González-Álvarez, M. (2022). Marine-based biopolymers for central nervous system drug delivery. In *Marine Biomaterials* (pp. 317–349). Springer.

Santiesteban-Romero, B., Martínez-Ruiz, M., Sosa-Hernández, J. E., Parra-Saldívar, R., & Iqbal, H. M. (2022). Microalgae photo-protectants and related bio-carriers loaded with bioactive entities for skin applications—An insight of microalgae biotechnology. *Marine Drugs, 20*(9), 487.

Shahcheraghi, N., Golchin, H., Sadri, Z., Tabari, Y., Borhanifar, F., & Makani, S. (2022). Nano-biotechnology, an applicable approach for sustainable future. *3 Biotech, 12*(1), 1–24.

Silva, T. H., Alves, A., Ferreira, B., Oliveira, J. M., Reys, L., Ferreira, R., Sousa, R., Silva, S. S., Mano, J., & Reis, R. (2012). Materials of marine origin: A review on polymers and ceramics of biomedical interest. *International Materials Reviews, 57*, 276–306.

Šimat, V., Elabed, N., Kulawik, P., Ceylan, Z., Jamroz, E., Yazgan, H., Özogul, F. (2020). Recent advances in marine-based nutraceuticals and their health benefits. *Marine Drugs, 18*(12), 627.

Singh, S., & Bothara, S. B. (2014). *Manilkara zapota* (Linn.) Seeds: A potential source of natural gum. *International Scholarly Research Notices, 2014*, 1–10.

Singh, S., Chunglok, W., Nwabor, O. F., Ushir, Y. V., & Singh, S. (2022). Hydrophilic biopolymer matrix antibacterial peel-off facial mask functionalized with biogenic nanostructured material for cosmeceutical applications. *Journal of Polymers and the Environment, 30*, 938–953.

Singh, S., Nwabor, O. F., Sukri, D. M., Wunnoo, S., Dumjun, K., Lethongkam, S., Kusolphat, P., Hemtanon, N., Klinprathum, K., & Sunghan, J. (2022). Poly (vinyl alcohol) copolymerized with xanthan gum/hypromellose/sodium carboxymethyl cellulose dermal dressings functionalized with biogenic nanostructured materials for antibacterial and wound healing application. *International Journal of Biological Macromolecules, 216*, 235–250.

Singh, S., Nwabor, O. F., Syukri, D. M., & Voravuthikunchai, S. P. (2021). Chitosan-poly (vinyl alcohol) intelligent films fortified with anthocyanins isolated from *Clitoria ternatea* and *Carissa carandas* for monitoring beverage freshness. *International Journal of Biological Macromolecules, 182*, 1015–1025.

Singha, I., & Basu, A. (2022). Chitosan based injectable hydrogels for smart drug delivery applications. *Sensors International, 3*, 100168.

Souto, E. B., Campos, J., Filho, S., Teixeira, M., Martins-Gomes, C., Zielinska, A., & Silva, A. (2019). 3D printing in the design of pharmaceutical dosage forms. *Pharmaceutical Development and Technology, 24*(9), 1044–1053.

Sridhar, K., Inbaraj, B. S., & Chen, B. H. (2021). Recent developments on production, purification and biological activity of marine peptides. *Food Research International, 147*, 110468.

Sudarshan Singh, W. C., Nwabor, O. F., Chulrik, W., Jansakun, C., & Bhoopong, P. (2022). Porous biodegradable sodium alginate composite fortified with *Hibiscus sabdariffa* L. calyx extract for the multifarious biological applications and extension of climacteric fruit shelf-life. *Journal of Polymers and the Environment, 31, 1–17,* https://link.springer.com/article/10.1007/s10924–022–02596-x.

Sukmarini, L. (2021). Drug development from peptide-derived marine natural products. In *Proceedings of the IOP Conference Series: Materials Science and Engineering* (p. 012063), IOP Publishing Ltd, https://iopscience.iop.org/article/10.1088/1757–899X/1011/1/012063.

Sun, X., Liu, C., Omer, A., Yang, L.-Y., & Ouyang, X.-K. (2019). Dual-layered pH-sensitive alginate/chitosan/kappa-carrageenan microbeads for colon-targeted release of 5-fluorouracil. *International Journal of Biological Macromolecules, 132*, 487–494.

Tabassum, H., & Ahmad, I. Z. (2021). Nanoformulations loaded with microalgal bioactive compounds for disease therapy. In *Bioprospecting Algae for Nanosized Materials* (pp. 229–260). Springer.

Taher, M., Omer, A. M., Hamed, A., Ali, A., Tamer, T., & Eldin, M. (2019). Development of smart alginate/chitosan grafted microcapsules for colon site-specific drug delivery. *Egyptian Journal of Chemistry, 62*, 1037–1045.

Tang, E. K., Partridge, G. J., Woolley, L. D., Pilmer, L., & Lim, L. Y. (2022). Effects of formulation on the palatability and efficacy of in-feed praziquantel medications for marine finfish aquaculture. *Marine Drugs, 20*(6), 323.

Thursby, E., & Juge, N. (2017). Introduction to the human gut microbiota. *Biochemical Journal, 474*(11), 1823–1836. https://doi.org/10.1042/BCJ20160510.

Vanslembrouck, B., Chen, J.-H., Larabell, C., & van Hengel, J. (2022). Microscopic visualization of cell-cell adhesion complexes at micro and nanoscale. *Frontiers in Cell and Developmental Biology, 10*, 819534. https://doi.org/10.3389/fcell.2022.819534.

Vijayakumar, S., Saravanakumar, K., Malaikozhundan, B., Divya, M., Vaseeharan, B., Durán-Lara, E. F., & Wang, M. H. (2020). Biopolymer K-carrageenan wrapped ZnO nanoparticles as drug delivery vehicles for anti MRSA therapy. *International journal of biological macromolecules, 144*, 9–18.

Wagner, C., Wheeler, K., & Ribbeck, K. (2018). Mucins and their role in shaping the functions of mucus barriers. *Annual Review of Cell and Developmental Biology, 34*, 189–215.

Wanat, K. (2020). Biological barriers, and the influence of protein binding on the passage of drugs across them. *Molecular Biology Reports, 47*(4), 3221–3231. https://doi.org/10.1007/s11033-020-05361-2.

Wang, M., Zhou, J., Calvo-Lerma, J., Liu, Y., Collado, M. C., & Barba, F. J. (2022). Effects of marine bioactive compounds on gut ecology based on in vitro digestion and colonic fermentation models. *Nutrients, 14*(12), 3307.

Wen, H., Jung, H., & Li, X. (2015). Drug delivery approaches in addressing clinical pharmacology-related issues: Opportunities and challenges. *AAPS J, 17*(6), 1327–1340. https://doi.org/10.1208/s12248-015-9814-9.

Yazdani, S., Jaldin-Fincati, J. R., Pereira, R. V. S., & Klip, A. (2019). Endothelial cell barriers: Transport of molecules between blood and tissues. *Traffic, 20*(5), 390–403. https://doi.org/10.1111/tra.12645.

Yazdi, M. K., Taghizadeh, A., Taghizadeh, M., Stadler, F. J., Farokhi, M., Mottaghitalab, F., Zarrintaj, P., Ramsey, J. D., Seidi, F., & Saeb, M. R. (2020). Agarose-based biomaterials for advanced drug delivery. *Journal of Controlled Release, 326*, 523–543.

Yermak, I. M., Davydova, V. N., & Volod'ko, A. V. (2022). Mucoadhesive Marine Polysaccharides. *Marine Drugs, 20*(9), 522.

Zaky, A. A., Simal-Gandara, J., Eun, J. B., Shim, J. H., & Abd El-Aty, A. M. (2021). Bioactivities, applications, safety, and health benefits of bioactive peptides from food and by-products: A review. *Frontiers in Nutrition, 8*, 815640. doi: 10.3389/fnut.2021.815640.

Zayed, A., & Ulber, R. (2020). Fucoidans: Downstream processes and recent applications. *Marine Drugs, 18*(3), 170.

Zhao, Y., & Antunes, A. (2022). Biomedical potential of the neglected molluscivorous and vermivorous *Conus* species. *Marine Drugs, 20*(2), 105.

Zheng, Y., Wang, X., Zhuang, Y., Li, Y., Shi, P., Tian, H., Li, X., & Chen, X. (2020). Isolation of novel ACE-inhibitory peptide from naked oat globulin hydrolysates in silico approach: Molecular docking, in vivo antihypertension and effects on renin and intracellular endothelin-1. *Journal of Food Science, 85*(5), 1328–1337.

Zhu, Q., Chen, Z., Paul, P. K., Lu, Y., Wu, W., & Qi, J. (2021). Oral delivery of proteins and peptides: Challenges, status quo and future perspectives. *Acta Pharmaceutica Sinica B, 11*(10), 2416–2448. https://doi.org/10.1016/j.apsb.202.

CHAPTER 7

Marine-Polysaccharides as Nano-Drug Delivery Vehicles

POOJA JAIN, RAHMUDDIN KHAN, BUSHRA AHMED, AYESHA SIDDIQUI, AHSAN ALI, MOHD. AAMIR MIRZA, and ZEENAT IQBAL

Department of Pharmaceutics, School of Pharmaceutical Education and Research (SPER), Jamia Hamdard, New Delhi, India

ABSTRACT

Current trends in the application of nanomaterials are emerging in the nano-biotechnological sector for the development of medicines. The use of marine-origin polysaccharides has increased in recent research because they are abundant, cheap, biocompatible, and biodegradable. These features motivate researchers to exploit marine-origin polysaccharides for nanotechnology as drug delivery systems for tissue engineering, chemotherapy, wound dressing, and as biosensors. Considering the physicochemical and bioactive characteristics of fucoidan and chitosan, a variety of nanostructures have been formulated with these polysaccharides. This review provides an outline of these marine polysaccharides, including their sources, chemical structure, biological properties, and nanomedicine applications; their combination as nanoparticles with descriptions of the most used production methods; and their physicochemical and biological properties applied to the design of nanoparticles to deliver several classes of compounds.

Handbook of Research in Marine Pharmaceutics: Exploring Oceanic Microbial Diversity for Human Health and Wellness. Md. Faiyazuddin & Khalid Rehman Hakeem (Eds.)

7.1 INTRODUCTION

The most prevalent macromolecular polymers in nature are polysaccharides, which come from renewable sources including algae, plants, and microorganisms like fungi and bacteria.

Due to their breadth and diversity, the marine environment is one of the most significant sources of naturally occurring bioactive substances. The majority of the earth's surface, or around 71%, is made up of oceans. The development of novel systems with several biotechnological and medical applications is aided by sea exploration as a renewable source of natural chemicals (Zhang, Li, Zhu, Gu, & Li, 2016). There are numerous natural sources of polysaccharides, including pectin and gums from plants, dextran from microbes, and chitosan from animals (Liu, Jiao, Wang, Zhou, & Zhang, 2008; Ngwuluka, 2018). These polysaccharides can be seen as an essential functional element required for several physiological and biological processes, including immune system regulation, anti-cancer, anti-hyperglycemic, and antioxidant properties. Microbial polysaccharides have long been used in biotechnology and the health sciences (Gopinath, Saravanan, Al-Maleki, Ramesh, & Vadivelu, 2018; Sutherland, 1998). The high biodiversity and ease of extraction and purification of marine chemicals have led to a significant increase in the creation of marine-based drugs during the past few decades. Additionally, they are frequently biodegradable and biocompatible (Molinski, Dalisay, Lievens, & Saludes, 2009). A significant percentage of marine biodiversity is still being studied since it is challenging to replicate the marine milieu in a lab setting (Ganesan, 2008). The current tendency is to develop novel methods to enhance the discovery and development of marine pharmaceuticals, mainly to enhance extraction yields, purification methods, and cultivation methods (Molinski, Dalisay, Lievens, & Saludes, 2009) or apply nanotechnological methods to resolve physicochemical problems with marine chemicals. Due to the low bioavailability of this phenomenon, new strategies must be devised to fully utilize these chemicals in their possible usage in the future. Holothurin, a toxin that Nigrelli isolated from the marine organism Actinopyga agassizi, was the first chemical of marine origin to be physiologically active (Arpicco, Battaglia, Brusa, Cavalli, Chirio, Dosio et al., 2016). In mice, holothurin showed some anticancer properties. Since then, researchers have been looking for helpful natural and therapeutic compounds in marine life. Numerous polysaccharides that have been extracted from marine algae in recent years have garnered a lot of interest for use in pharmaceuticals, cosmetics, and functional foods. A particular class of biomacromolecules known as polysaccharides is used as a structural element in the cell walls of marine algae. Numerous pharmacological effects, such as anticoagulant, antioxidant, anticancer, and immunomodulatory, are frequently connected to the polysaccharides present in marine algae (Brown, Allsopp, Magee, Gill, Nitecki, Strain et al., 2014; Moore & Tischer, 1964).

Although there is increasing interest in employing marine polysaccharides as the building blocks for bioactive molecules in pharmaceutical sciences, their primary applications are in the food and cosmetics sectors (Rioux, Beaulieu, & Turgeon, 2017). Most marine polysaccharides come from marine algae, while they can also be found in animal tissues like cartilaginous fish tissue and crab bones. Polysaccharides exhibit a variety of chemical and biological properties, including biocompatibility, biodegradability, adhesiveness, and the capacity to produce hydrogels. Some of these polysaccharides can also be isolated from marine bacteria.

Sulfated polysaccharides stand out among the numerous marine polysaccharides (Hamed, Özogul, Özogul, & Regenstein, 2015). They display bioactivities such as antioxidant (Laurienzo, 2010), anticoagulant (Senni, Pereira, Gueniche, Delbarre-Ladrat, Sinquin, Ratiskol et al., 2011), anticancer (Ngo & Kim, 2013), antiviral (Costa, Fidelis, Cordeiro, Oliveira, Sabry D De, Câmara et al., 2010), antiallergic (Wijesekara, Pangestuti, & Kim, 2011), anti-adhesive, anti-angiogenic, and anti-inflammatory properties as compared to other marine polysaccharides. A thorough analysis of one of these drug delivery system (DDS) materials led to the development of unique chemical modification techniques intended to harness this biological activity and alter the material's affinity for medications. Chemical procedures or interactions with other natural or synthetic polymers can improve the efficacy of drug release and the ability to incorporate pharmaceuticals (Barahona, Chandía, Encinas, Matsuhiro, & Zúñiga, 2011). Marine polysaccharides and their derivatives make fantastic building blocks for DDSs. For instance, hyaluronic acid can be used to target the overexpressed CD44 receptor on tumor cells (HA). Strong antibacterial activity can be seen in chitosan (CS) (Liu, Chen, Shang, & Yin, 2019). Marine polysaccharides also have a range of functional groups, including hydroxyl, amino, and carboxylic acid groups, that are active. These active areas can be used to chemically alter marine polysaccharides, expanding their potential applications (Wang, Zhao, Han, Bai, Li, Liu et al., 2015). DDSs based on marine polysaccharides can benefit from several nanoscale systems' advantages as well as polysaccharide features. This chapter is an attempt to apprise the readers about the various biomaterials used in the development of drug delivery systems, marine polysaccharides as drug delivery systems, fucoidan-chitosan nanoparticles, and the method of preparations thereof.

7.2 BIOMATERIALS

Introducing biocompatible and biodegradable drug carriers is one strategy to avoid or minimize the adverse effects that might result in medication delivery to cells with enhanced efficiency and performance throughout the health rehabilitation process. Due to the adverse effects associated with the application and delivery of traditional medications, the hunt for controlled drug release has required the need for materials that can carry pharmaceuticals to the target location without difficulty or issue during and after delivery. As a result of low drug concentrations at the delivery site brought on by drug metabolism, degradation, and transport away from the target, it is also possible that the drug release rate is so rapid that therapeutic activities are no longer effective (Schneider, Langer, Loveday, & Hair, 2017). Extended therapeutic windows require materials that can allow drug release without altering or deteriorating over time. Apart from allowing surface modification with ligands, the design parameters of biomaterials used for drug carriers should also protect pharmaceuticals against rapid breakdown and/or degeneracy within the target location. This is accomplished by encapsulating enough drugs in the biomaterial to achieve efficient healing, sustaining drug stability for effective therapeutics through body transport and at the target site while preserving biological activity, ensuring a predictable release rate in the therapeutic period ranging from days to years. Biomaterials and their degradation products must be biocompatible and nontoxic within the body (Bertrand, Wu, Xu, Kamaly, & Farokhzad, 2014). The FDA has approved PLGA, poly (lactic acid) (PLA), and polyglycolic acid (PGA) materials as micro-particle depot systems because they are adaptable in adjusting material biodegradation time and are biocompatible with nontoxic natural degradation products. Doxorubicin, a chemotherapeutic, is well encapsulated in the liposomal form with a PEG coat. In comparison to the usage of free medication, this improves the drug's circulatory half-life and tumor absorption while also lowering its toxicological activity in patients (Miller, Zhang, Kos, Xiong, Zhou, Perelman et al., 2017). Marqibo, a liposomal encapsulating vincristine for the treatment of uncommon leukemia, is another authorized nanoparticle drug carrier (Silverman & Deitcher, 2013). Abraxane is an albumin-bound paclitaxel nanoparticle used to treat breast cancer (Ferrari, 2005). The Gliadel wafer, which consists of dime-sized wafers made of the chemotherapeutic agent carmustine and a polymer matrix made of poly (carboxyphenoxy-propane/sebacic acid), is surgically inserted into the brain post-tumor resection and used as an adjunct to surgery in patients with recurrent glioblastoma multiforme (Brem, Piantadosi, Burger, Walker, Selker, Vick et al., 1995; Brem, Mahaley, Vick, Black, Schold, Burger et al., 1991).

7.2.1 SMART BIOMATERIALS

For the development of next-generation precision pharmaceuticals, innovative "smart" or responsive biomaterials with the capacity to exploit and respond to these processes are in high demand. Environmentally responsive materials are attractive, particularly *in vivo,* since they may be used to modulate medication release, cell adhesiveness, mechanical characteristics, or permeability. Stimulants such as pH, temperature, and light can cause these environmental changes (Jeong, Kim, & Bae, 2012; Shimoboji, Larenas, Fowler, Hoffman, & Stayton, 2003). The pH-sensitive polymers that are sensitive to gastrointestinal tract pH fluctuation have been created to successfully transfer medications from the stomach to the small intestine. pH-sensitive hydrogels made from poly (methacrylic acid) grafted with poly (ethylene glycol) (PMAA-g-PEG) swell in reaction to pH, for instance, as food is broken down into nutritious compounds in the stomach (pH 2) and then absorbed in the small intestine (pH 7) (Kim, La Flamme, & Peppas, 2003). Orally administered insulin-loaded PMAA-g-PEG gels to diabetic mice resulted in a significant drop in glucose levels because protein activity is preserved in acidic and digestive enzyme conditions (Lowman, Morishita, Kajita, Nagai, & Peppas, 1999). Biomaterials for on-demand and pulsatile drug administration have been created that are remotely activated by stimuli such as visible light, near-infrared (NIR) light, ultrasound, electric currents, and magnetic fields. Thus, smart biomaterials that respond to a variety of environmental stimuli (e.g., temperature, pressure, pH, enzymes, glucose), biological signals, or pathological abnormalities for actuating drug release have been developed (Griset, Walpole, Liu, Gaffey, Colson, & Grinstaff, 2009; Kopeček, 2003; Peppas, Hilt, Khademhosseini, & Langer, 2006; Yoshida, Uchida, Kaneko, Sakai, Kikuchi, Sakurai et al., 1995).

7.2.2 BIOMATERIAL FROM MARINE ORIGIN AS A DRUG DELIVERY SYSTEM

Biomaterials play an important role in human health research. The marine environment is vast and shows promising bioactivity and chemical innovation.

It also exhibits biodegradability and biocompatibility, making it an important source for extracting secondary metabolites (Barbosa, Coutinho, Costa Lima, & Reis, 2019). Marine biomaterials consist of simple polymers to complex polysaccharides. This is a newly emerging area of research with significant applications (Haefner, 2003). Recently, researchers have paid considerable attention to marine-derived biomaterials for various biomedical applications. It has been observed that marine pigments are obtained from sources such as different fishes, mollusks, sponges, mangroves, seawater, etc. Marine biomaterials are generally biodegradable and biocompatible, showing biological properties that aid in the discovery of a wide range of novel bioactive compounds, such as metallic nanoparticles, which present challenges in the fields of cosmetics and bionanotechnology. Metallic nanoparticles have already been synthesized using physical or chemical methods, so in order to minimize the toxic effects of chemicals, there is a need to shift to marine sources to produce the safest form of nanoparticles. Therefore, it is evident that marine biomaterials have gained a strong market position in recent days, attracting various marine researchers and proving to be beneficial from an industrial perspective. In order to adapt to the eco-friendly and green environment, the process involves interlinking microbial biotechnology with nanotechnology (Manivasagan, Nam, & Oh, 2016). The goal is to create new techniques and methods for the progress of marine drug discovery and development, to improve the yield extraction, purification process, and cultivation methods, or to implement nanotechnological approaches to overcome physicochemical issues of marine compounds. The isolation, characterization, and application of marine biomaterials play an important role in developing the marine biotechnology industry. Recent advances in technologies help to isolate bioactive materials from the marine source in an effective manner.

A drug delivery system (DDS) can be defined as a formulation or a device that enables the introduction of a therapeutic substance into the human body. We are very much aware that the drug delivery system can be changed in order to achieve the therapeutic level, sustained release in a body with minimum side effects by using various biomaterials from marine origin with different densities, different charges, and side chain functionality is considered an advancement in drug delivery system. Marine-derived polymers are biocompatible and have surface characteristics and mechanical strength that are vital features for a successful scaffold to be used as a drug delivery system. Marine biomaterials can be classified into three main groups: polysaccharides, proteins, and lipids. Polysaccharides are polymeric carbohydrate structures formed by repeated monosaccharide units that are attached to each other by glycosidic bonds. Biomaterials are known to offer various advantages for the delivery of genetic material or therapeutic agents. Several types of DDS have been developed so far, namely, lipid-, phospholipid, and polymer-based, which can be considered to optimize drugs' physicochemical features as well as *in vivo* efficacy (Manivasagan, Nam, & Oh, 2016).

7.2.3 BIOMATERIALS IN TISSUE ENGINEERING AND OTHER BIOMEDICAL APPLICATIONS

In recent years, it has been seen that biomaterials derived from marine sources play a crucial role in providing a template and extracellular environment to support regenerative cells and promote tissue regeneration. This is evident in various biomedical applications, including cancer therapy, tissue engineering, drug delivery, catalysis, sensors, electronic materials, and wastewater treatment. Fucoidan and chitosan-composed biomaterials have been used for bone regeneration. Similarly, in wound healing, a solution of the compound is used in the form of hydrogels. Particularly, chitosan is observed to be used in the proliferation and activation of inflammatory responses, while fucoidan shows significant gel contraction, suggesting that both are helpful in wound healing.

7.3 BIOMATERIALS AS DRUG DELIVERY SYSTEMS

7.3.1 BIOCERAMIC

Bioceramics, a special set of fully, partially, or non-crystalline ceramics (e.g., calcium phosphates, bioactive glasses, and glass-ceramics) that are designed for the repair and reconstruction of diseased parts of the body, have high potential as scaffold materials. Another name that can be given to bioceramics is a ceramic material that has been used for skeletal repair or medical implants. Traditionally, bioceramics have been used to fill and restore bone and dental defects (repair of hard tissues). However, the basic idea behind their successful use in recent pharmaceutical development is because of ceramics' inherent attributes, such as their high levels of chemical resistance, compressive strength, and wear resistance. There have been many developments in the class of bioceramics, including bioglass, A-W ceramic-glass, calcium phosphates, plasma spray HA coating, fillers composites in the form of calcium phosphates, etc. More recently, this category of biomaterials has also revealed promising applications in the field of soft-tissue engineering. The plan involves the proper knowledge of the quantity to be used in bioceramics, which still requires a deep understanding of the biological system as there are still doubts regarding the exact bonding of bone minerals and collagen. Therefore, in order to better understand the design of bioceramics, it is very necessary to understand the bone response to a particular ion (Baino, Novajra, Vitale-Brovarone, 2015).

7.3.2 NATURAL PRODUCTS

Nowadays, it is very common to use naturally derived biomaterials in multiple regenerative medicine applications. Natural biomaterials can be categorized into the following subtypes: (i) protein-based biomaterials; (ii) polysaccharide-based biomaterials; (iii) glycosaminoglycan-derived biomaterials, and tissue/organ-derived biomaterials. Some of the earliest uses of natural biomaterials date back to 3000 BC when the Egyptians used coconut shells to repair injured skulls and pieces of wood as false teeth. When we consider protein-based natural biomaterials like collagen, it plays a major role in maintaining the structural integrity of the extracellular matrix and is also used in films, beads, meshes, sponges, etc. Another important protein-based natural biomaterial is gelatin, which is highly biodegradable and derived from fish, porcine, and bovine sources. Due to its versatile nature, gelatin is widely used in cardiovascular and hepatic tissue engineering (Elamparithi, Punnoose, Paul, & Kuruvilla, 2017). Polysaccharide-based biomaterials include cellulose, one of the most abundant natural polymers. Cellulose derivatives such as carboxymethylcellulose, cellulose nitrate, cellulose acetate, and cellulose xanthate are gaining importance in pharmaceutical formulations and tissue engineering (Brovold, Almeida, Pla-Palacín, Sainz-Arnal, Sánchez-Romero, Rivas et al., 2018).

7.3.3 FATTY ACIDS

Fishes and other microorganisms were used to produce long-chain omega-3 polyunsaturated fatty acids (PUFAs), such as eicosapentaenoic acid and docosahexaenoic acid that are helpful in providing a cure in case of cardiovascular diseases. A major component is omega-3 fatty acid obtained from oily fish. Omega fatty acids can be studied under two categories, i.e., omega-3 fatty acid and omega-6 fatty acid, depending on the difference in the placement of the double bond. Both fatty acids are used in prostaglandin production, which is very helpful in tumor formation and inflammation. It has been observed that the angiogenic cascade is well-seen as a chemopreventive agent. Omega-3 fatty acids, in particular, eicosapentaenoic acid (EPA) and docosahexaenoic acid (DHA) found principally in oily fish, have potent anti-inflammatory, anti-apoptotic, anti-proliferative, and antiangiogenic effects observed (Spencer, Mann, Metcalfe, Webb, Pollard, Spencer et al., 2009). Clinical trials demonstrated that patients with acute myocardial infarction who took omega-3 fatty acids exhibited less adverse left ventricular remodeling and non-infarct myocardial fibrosis (Wan, Qin, Lei, Li, Meng, Fang et al., 2021).

7.3.4 MARINE PIGMENTS

Marine pigments are emerging as important in the fields of medicine, food, and cosmetics due to the side effects observed with the use of synthetic compounds. Marine pigments not only provide color but also consist of diverse chemical components such as terpenoids and melanin. Carotenoids are biologically active secondary metabolites ranging from orange to yellow pigments obtained from marine organisms. They also exhibit bioactivities like anticancer and antioxidant properties and serve as food colorants. Another important use of marine pigments is in biomedical applications such as drug delivery, which is a major challenge to deliver drugs to a specific site without affecting healthy tissue. Various derivatives of porphyrins, including polymer porphyrin conjugates, are used in drug delivery due to their biodegradability and better stability, allowing them to reach the targeted site with minimal side effects on nearby tissues (Kim & Venkatesan, 2013). In addition to drug delivery, other biomedical applications include photothermal therapy, photodynamic therapy, and photoacoustic imaging. Marine pigments are considered an emerging field in human healthcare systems, although there are still gaps in understanding their *in vitro* and *in vivo* applications. The future of marine pigment usage holds many more applications that need to be explored in the fields of food, cosmetics, and medicine.

7.3.5 MARINE COLLAGEN

A marine biomaterial is an important source acquired from the environment in the expanding area of biomedicine. Collagens appear to be abundant in the fisheries and mariculture of certain fish species. Marine fish collagen-based nanomaterials (i.e., collagen gels, scaffolds, sponges, films, membranes, and composites) have a wide range of applications, including drug delivery, wound healing, wound dressing, tissue engineering (e.g., bone, cartilage, dental, vascular, and skin tissues), and therapeutics for skin aging, diabetes, and obesity. This structural protein has been found to be very useful as a bone graft in engineering, and in the field of dentistry, and it is also believed that a patch with 3D printing loaded with an anticancer drug is used in its treatment. It is also used in the cosmeceutical industries due to its non-irritating properties, as the material shows promising absorption properties (Khrunyk, Lach, Petrenko, & Ehrlich, 2020).

7.4 APPLICATIONS OF MARINE BIOMATERIALS

Marine-derived biomaterials are abundant and biocompatible with humans. However, the field of marine biomaterials is still in its early stages, and several biomaterials must be identified and described for a variety of biological, biomedical, as well as industrial applications as shown in Table 7.1 (Barbosa, Coutinho, Costa Lima, & Reis, 2019; Manivasagan, Bharathiraja, Santha Moorthy, Mondal, Seo, Dae Lee et al., 2018). Various polysaccharides suited for biomedical uses are prevalent in marine creatures. Natural biomaterials such as fucoidan and chitosan have significant physicochemical and biological characteristics with proven therapeutic benefits.

TABLE 7.1 Applications of Marine Biomaterial

Sl. No.	Marine Biomaterial	Activity Shown	Advantages
1.	Peptides/chitosan	Antioxidant and anti-inflammatory activity	Helpful in cardiac problems as well as in cancer
2.	Toxins, pigments, sulfated polysaccharides	Anticancer activity	Designing of new anticancer drug
3.	Sulfated polysaccharides, fucoidan, carrageenan	Anticoagulant activity	A formulation consisting of anticoagulant agents
4.	Phlorotannin, polysaccharides, and lectins	Anti-HIV activity	*In vitro*, anti-HIV activity was observed
5.	Collagen, chitosan, hyaluronic acid, corals, algae (diatoms)	Tissue engineering activity	In replacement of infectious organ

7.4.1 MARINE POLYSACCHARIDES AS DRUG DELIVERY SYSTEM

Many different natural resources can be found in the water, which is a vast and endless supply. Marine polysaccharides are used as drug delivery systems in nanotechnology and tissue engineering, cancer treatments, wound dressing, biosensors, and even water purification. The most researched marine polysaccharides are chitin, chitosan, alginates, hyaluronic acid, fucoidan, carrageenan, agarose, and Ulva (Barbosa, Coutinho, Costa Lima, & Reis, 2019).

The drug delivery system (DDS) may fully regulate the distribution of pharmaceuticals in the organism in terms of time, space, and dosage, leading to an improvement in drug utilization effectiveness as well as cost and harmful side effect savings. The nano-drug delivery system (NDDS) can operate at the subcellular level *in vivo* due to its small size. The marine polysaccharide-based DDS is a carrier for various medicinal preparations that combine the benefits of polysaccharide materials and nanotechnology (Sun, Ma, & Hu, 2021).

Marine polysaccharides are known for their benefits in biocompatibility, biodegradability, minimal immunogenicity, and high natural availability (Venkatesan, Anil, Kim, & Shim, 2016). It is simple to modify a molecular structure by chemical, biological, or enzymatic means since the molecular backbone contains multifunctional groups (such as hydroxyl, carboxyl, and amine) (Sun, Ma, & Hu, 2021). Marine polysaccharides are used due to their physicochemical properties and pharmacological effects, e.g., anti-coagulant, anti-inflammatory, anti-tumor, etc.

Although numerous marine polysaccharide-based products, including dermal fillers and wound dressings, have been used clinically (e.g., hydrogels, nanofiber mats), the majority of these marine polysaccharide-based products are employed *in vitro* or particular parts of the body (Hariyadi & Islam, 2020). The carrier often needs to have a distinct structure and metabolism in the body. Materials made from marine polysaccharides have a disadvantage in this aspect because the season and location of production will inevitably affect the molecular weight and structure of the marine polysaccharide source. Industrial refining can produce items of great quality and stability, but it typically comes at a significant cost premium. The interaction between the drug molecules, polysaccharides, and the human body, including absorption, distribution, metabolism, and excretion, must also be thoroughly studied due to the ambiguity surrounding the selection of model pharmaceuticals in most research (Cardoso, Costa, & Mano, 2016).

7.4.1.1 MARINE BIOINSPIRED NANOPARTICLES FOR TREATING VIRAL INFECTION

Virus particles are much smaller than living cells. A pathogenic virus has been shown to be the cause of several viral infections. Numerous disorders brought on by viruses increase the death rate (Akram, Tahir, Shah, Mahmood, Altaf, Ahmad et al., 2018). Polyphenols, alkaloids, lipids, carbohydrates, steroids, terpenoids, exopolysaccharides, polyketides, zoanthoxanthins, and peptides are among the antiviral substances made by marine organisms. One of the metabolites produced by marine organisms is phlorotannins, which interfere with viral attachment, penetration, and replication (Riccio, Ruocco, Mutalipassi, Costantini, Zupo, Coppola et al., 2020). The silver nanoparticles derived from the seaweed *U. lactuca* show cytotoxic efficacy against vector-borne pathogens, i.e., *Aedes aegypti* and *Culex pipiens*, due to their small particle size. These silver nanoparticles, synthesized from *U. lactuca*, exhibit more effective action than conventional insecticides. Moreover, these silver nanoparticles adhere to the insect cuticle, penetrate the cell, and interfere with further cell processes (El-Sheekh, Shabaan, Hassan, & Morsi, 2022).

The interaction of viral proteins with silver nanoparticles triggers poliovirus inhibition. The silver nanoparticles derived from *Dictyosphaerium* sp., a freshwater microalga, have major antiviral activity against the Newcastle disease virus. The viral glycoprotein envelope is linked to the silver nanoparticles, preventing further virus penetration (Khalid, Khalid, Ahmed, Hanif, Ismail, & Janjua, 2017).

7.4.1.2 MARINE BIOINSPIRED NANOPARTICLES FOR TREATING NON-INFECTIOUS DISEASES

Inflammation is the natural response of the body which is caused by tissue injury, infection, and genetic modifications (Conte, Marturano, Peluso, Calarco, & Cerruti, 2017). Inflammation is caused when the immune system becomes active and releases a variety of inflammatory mediators such as bradykinins and prostaglandins (Assali, Shawahna, Dayyeh, Shareef, & Alhimony, 2018). Therefore, controlling inflammation and lowering prostaglandin levels, helps to avoid chronic diseases (Kaulmann & Bohn, 2014). One of the therapeutic methods for inflammatory conditions involves the use of steroids and nonsteroidal anti-inflammatory drugs to reduce inflammation and inflammatory mediators. Synthetic anti-inflammatory drugs have many side effects. It is therefore necessary to use marine-based organisms with great biological activity to produce therapeutically effective nanoparticles with no side effects. *Galaxaura elongata, Turbinaria ornata*, and other macroalgal species that create silver nanoparticles, as well as *Enteromorpha flexuosa*, have demonstrated significant anti-inflammatory activity through membrane stability.

Silver nanoparticles help to reduce prostaglandins by producing inhibitory proteins that denature cyclooxygenase and 5-lipoxygenase. ZnO nanoparticles wrapped in Kappa-carrageenan have established 82% anti-inflammatory efficiency (Vijayakumar, Saravanakumar, Malaikozhundan, Divya, Vaseeharan, Durán-Lara et al., 2020). Due to the high surface area to volume ratio, these nanoparticles are more effective than bulk materials in inhibiting cytokines and inflammatory coenzymes.

The gold nanoparticles derived from jellyfish extract show important deadly effects on HeLa cancer cells. *Rhodotorula mucilaginosa* derived from Cu_2O nanoparticles shows anticancer activity (Hassabo, Ibrahim, Ali, & Emam, 2022).

7.4.2 PREPARATION OF MARINE POLYSACCHARIDE-BASED NANOPARTICLES (NPS)

Self-assembly and covalent crosslinking are two methods that can be used to create marine polysaccharide-based nanoparticles (NPs). Non-covalent bonds play a role in the formation of self-assembled NPs. Ionic interactions, hydrophobic interactions, and polyelectrolyte complexation are the key driving forces (Chollet, Saboural, Chauvierre, Villemin, Letourneur, Chaubet et al., 2016). Covalent crosslinking can be accomplished in a variety of ways, including chemical crosslinking and radiation crosslinking (Tziveleka, Sapalidis, Kikionis, Aggelidou, Demiri, Kritis et al., 2020). The NPs produced can be nanogels, micelles, or vesicles, depending on the synthesis technique (Agnihotri, Mallikarjuna, & Aminabhavi, 2004).

Various types of nanoparticles are prepared using marine polysaccharides, e.g., seaweed polysaccharide-based nanoparticles, and Fucoidan-based nanoparticles (Venkatesan, Anil, Kim, & Shim, 2016). Fucoidan has been studied for the biosynthesis of metal nanoparticles, such as the green synthesis of gold nanoparticles, and it has been considered for cancer treatment.

Chitosan-Based Nanoparticles. Chitosan is an excellent material used in biological environments and for designing drug delivery systems. Chitosan can be used as a DDS to create stable solid dosage forms for oral delivery, hydrogels, nanoparticles, and colloidal systems for ocular application, mucoadhesive characteristics for nasal spray formulations and buccal drug delivery, and polymer robustness for vaginal tablets. It can also be mixed with antigens to create vaccines and, when refined, used in the parenteral route due to its harmless nature (Barbosa, Coutinho, Costa Lima, & Reis, 2019; Luo & Wang, 2014).

7.4.2.1 ALGINATE NANOPARTICLES

Alginate is a natural polysaccharide that is derived from brown seaweeds and has been widely investigated and used for biomedical applications (Chollet, Saboural, Chauvierre, Villemin, Letourneur, Chaubet et al., 2016). Researchers have found that alginate-based nanoparticles have numerous applications in insulin delivery, antifungal, and antitubercular drugs (Tziveleka, Sapalidis, Kikionis, Aggelidou, Demiri, Kritis et al., 2020). The insulin-loaded nanoparticles using alginate show a loading capacity of 14.3%. Similarly, loading insulin in alginate-dextran nanospheres through nanoemulsion dispersion resulted in an increase in encapsulation efficiency of up to 82.5%.

7.4.2.2 CARRAGEENAN-BASED NANOPARTICLES

The main source of carrageenan is red algae such as *Kappaphycus* sp. and *Eucheuma* sp. Based on the extraction procedure, carrageenan is classified into three types: kappa, iota, and lambda (Agnihotri, Mallikarjuna, & Aminabhavi, 2004). Kappa carrageenan is hard and firm, whereas iota carrageenan is elastic and soft by nature. There are various applications of carrageenan, e.g., in nanotechnological, biological, and pharmaceutical fields. The carrageenan nanoparticles are created in hydrophilic conditions, with a very gentle procedure that prohibits the use of organic solvents and other harmful chemical processes. This procedure produces appropriate nanoparticles that can deliver drugs in a sustained and controlled manner and are ideal candidates for biological applications (Luo & Wang, 2014).

7.4.3 CHARACTERISTICS OF MARINE POLYSACCHARIDES

Marine polysaccharides have been used in various areas of biomaterials, food, cosmetics, and pharmacology due to their wide range of pharmaceutical properties, including antioxidant, anti-inflammatory, antiallergic, antitumor, antiobesity, antidiabetic, anticoagulant, antiviral, immunomodulatory, cardio-protective, and antihepatopathy activities (De La Fuente, Seijo & Alonso, 2008).

Polysaccharides are a class of carbohydrates with complex and large molecular structures. Dehydration of many monosaccharide molecules results in the formation of polysaccharides. Glycosidic bonds connect the structural constituents of polysaccharides. Common glycosidic bonds include α-1,4-, β-1,4-, and α-1,6-glycosidic bonds. The structural unit can be joined together to form either a straight chain or a branched chain. In the branched chain, the connecting point is frequently made up of 1,6-glycosidic bonds; the straight chain is typically connected by 1,4-glycosidic bonds (such as starch) or 1,4-glycosidic bonds (such as cellulose) (Wang, Varghese, Sharma, Strehin, Fermanian, Gorham et al., 2007).

Marine-derived polysaccharides have a variety of therapeutic uses since they are harmless compounds, nontoxic, and physiologically inactive. Additionally, most of them can create hydrogels due to their exceptional hydrophilicity, which enables them to attach to proteins and other substances. Alginate, porphyran, fucoidan, and chitin, as well as their derivatives, have all been discovered and extracted from a variety of sources, particularly seaweed and crustaceans (Table 7.2). These marine polysaccharides have been identified as possible anti-allergic drugs in recent years due to their ability to boost innate immunity (De La Fuente, Seijo, & Alonso, 2008).

7.4.3.1 FUCOIDAN CHITOSAN NANOPARTICLES

Recent research has boosted the usage of marine-origin polysaccharides since they are ample, economical, biocompatible, and biodegradable, and have low immunogenicity. Such nanostructures are easily producible, conjugable, and complex with proteins or other bioactive molecules. Figure 7.1 represents a complex of Fucoidan and Chitosan-forming nanoparticles. They can be altered to create gels or interconnected polymeric networks. Their physicochemical and biological qualities make them helpful in nanotechnologies, such as drug delivery systems, tissue regeneration, cancer treatment, or wound dressing; and even water purification. In addition to having significant biological properties like biocompatibility, and anti-inflammatory activity, as well as adhesive and antimicrobial activity, marine polysaccharides can display a variety of chemical structures.

TABLE 7.2 Marine Polysaccharides from Various Sources

Sl. No	Polysaccharides from Marine Algae	Source	Uses	Pharmacological Activity	Reference
1.	Alginates	Brown Algae	One of the natural polymers used frequently in protein- and drug-delivery systems is alginate. Alginate is used in a variety of pharmaceutical formulations as a stabilizing agent as well as in applications for wound healing and tissue engineering. This is one of the most popular uses for alginate.	–	Hariyadi & Islam (2020)
2.	Carrageenan's	Red algae	In the pharmaceutical sector, carrageenan is typically employed as a raw ingredient for the creation of DDSs, cell capsules for cell treatments, and applications involving cartilage regeneration. It may be a good option to use hydrogels made of carrageenan as a vehicle for the controlled delivery of biomolecules, especially if the cargo needs to be stabilized.	Anticoagulant, Anti-tumor, Immuno-modulator, anti-hyperlipidemic, Antioxidant, Antibacterial, viral, bird flu, dengue, Hepatitis A, HIV	Cardoso, Costa, & Mano (2016)
3.	Fucoidans	Brown algae	Many types of brown algae contain the sulfated polysaccharide known as fucoidan. Fucoidan is a raw ingredient that can be utilized to create DDSs, just like other marine polysaccharides. Fucospheres, also known as fucoidan DDSs, are often created by electrostatic interactions between chitosan and fucoidan DDSs and have been proposed as potential burn therapies.	Anti-tumor, anti-thrombin, anti-coagulant, anti-inflammatory, anti-adhesive, and anti-viral, burn treatment	Sezer & Akbuğa (2006)

TABLE 7.2 *(Continued)*

Sl. No	Polysaccharides from Marine Algae	Source	Uses	Pharmacological Activity	Reference
4.	Ulvans	Green algae	Ulva is used to make the Sulfated polysaccharide known as ulvan. Nanocarriers. it is also used in the construction of membranes, due to electrostatic interactions with other cationic polymers.	Anti-viral, anti-oxidant, anti-coagulant, anti-tumor, anti-hyperlipidemic and immune system enhancer	Alves, Pinho, Neves, Sousa, & Reis (2012)
5.	Chitosan's	Shrimp and Crab	It is particularly useful in the creation of nanoparticles, beads, and capsules for controlled drug delivery systems as well as membranes, films, and scaffolds for tissue engineering and regenerative medicine.	Antimicrobial, anti-tumor, and inflammatory	Agnihotri, Mallikarjuna, & Aminabhavi (2004)
6.	Hyaluronans	Fish eye and mussels.	Hyaluronan can combine with polycations like chitosan to produce nanoparticles and microspheres, much like other polyanions can. In order to create nanoparticles for the delivery of genes to the cornea and conjunctiva.	Wound treatment, supplement for arthritis	De La Fuente, Seijo, & Alonso (2008)
7.	Chondroitin Sulphates	Fish cartilage	This polysaccharide has been utilized in the pharmaceutical industry as an active ingredient in medications with anticoagulant qualities, as a preventative measure against arthritis, and as hydrogels for the regeneration of cartilage tissue.	Anti-coagulant, supplement for arthritis	Wang, Varghese, Sharma, Strehin, Fermanian, Gorham et al. (2007)

FIGURE 7.1 Fucoidan-chitosan nanoparticles are complex of two polysaccharides i.e., fucoidan and chitosan molecules.

Fucoidan is a type of fucose-rich sulfated marine polysaccharide that has multiple fascinating biological actions (Figure 7.2). The family of fucoidans also includes ascophyllan (xylofucoglycuronan and xylofucomanuronan) and sargassans (glycuronofucogalactan) (Wang, Xing, Cao, Ji, Liang, & Song, 2019). It is primarily derived from brown seaweed (*Phaeophycean*) in the body wall of some marine invertebrates such as sea cucumber (*Holothuroidae*) and the egg jelly coat of sea urchins (*Echinoidea*). They were first discovered in 1913 by Kylin in brown algae (*Ascophyllum nodosum, Fucus vesiculosus*, *Laminaria digitata*, and *Laminaria saccharina*). Since then, fucoidans have been identified in 70 more species of brown algae. Some common sources of fucoidan are *Cladosiphon okamuranus, Cladosiphon okamuranus Tokida, Costaria costata, Ecklonia cava, Eisenia bicyclis, Fucus* sp., *Laminaria japonica, Macrocystis pyrifera, Sargassum* sp., *Saccharina japonica, Turbinaria conoides,* and *Undaria pinnatifida*, etc. (Barbosa, Coutinho, Costa Lima, & Reis, 2019). Fucoidan's anticoagulant action may be related directly to sulfate concentration and orientation, molecular size, and sugar composition.

Chitosan is a biopolymer derived from chitin. Chitosan has excellent biodegradability and biocompatibility. Chitosan (where the pKa is approximately 6.5) is soluble at low pH levels and insoluble at high pH levels because of its amino group. Chitosan also has the novel characteristic of mucoadhesion and enhances paracellular drug transport via the transient opening of the tight junction between epithelial cells. Therefore, chitosan has been widely studied in mucosal drug delivery, including in oral, buccal, and nasal delivery systems. Chitosan can also form complexes with polyanions like fucoidan. Thus, biodegradable polymers are excellent for medicinal and pharmaceutical formulation because of their low cost, accessibility, and simplicity of complexation. Figures 7.3 and 7.4 represent the chemical structure of fucoidan and chitosan, respectively (Berteau & Mulloy, 2003).

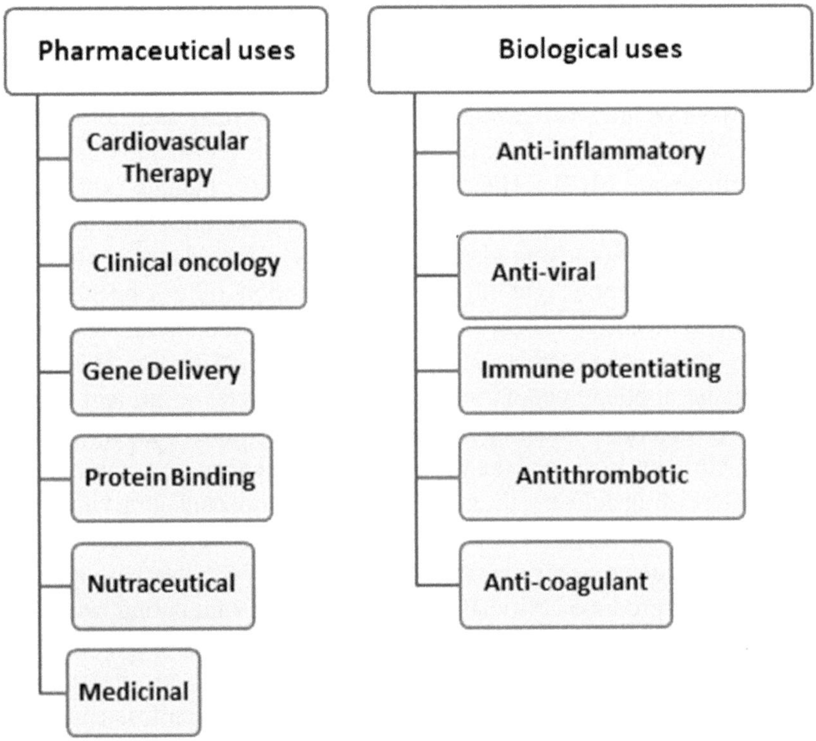

FIGURE 7.2 Pharmaceutical and biological applications of fucoidan-chitosan nanoparticles.

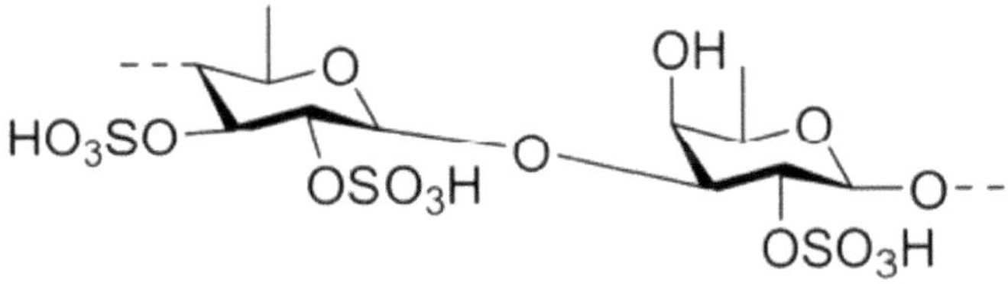

FIGURE 7.3 Chemical structure of fucoidan.

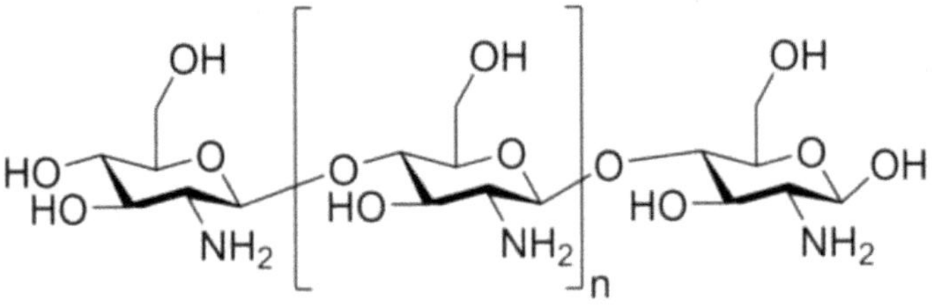

FIGURE 7.4 Chemical structure of chitosan.

Fucoidan is a promising therapeutic candidate for pharmaceutical use. It is a biological compound that shows antimutagenic, antitumor, anti-inflammatory, and anticoagulant effects. The biological activities of fucoidan can be classified into two categories, namely classical and non-classical activities (Wang, Xing, Cao, Ji, Liang, & Song, 2019; Vishchuk, Ermakova, & Zvyagintseva, 2013). The classical activities include antioxidant, antitumor, anticoagulant, antithrombotic, antiviral, immune potentiating, and anti-inflammatory. Some non-classical activities of fucoidans are alleviation of metabolic syndrome, gut protection, promotion of angiogenesis, and boosting bone health. Since fucoidans act as toxicity modulators and growth promoters, there are possible applications in aquaculture too. The field of nanomedicine applies Fucoidan-Chitosan nanoparticles in protein binding, small drug delivery, imaging, anticoagulation, gene delivery, clinical oncology, both diagnosis and therapeutics, tissue regeneration, angiogenesis, non-parenteral drug delivery, drug delivery to the brain, ocular tract infections, and cardiovascular therapy. In the biomedical field, fucoidan and chitosan are used for tissue engineering and wound healing. Fucoidan has significant capabilities to produce artificial bone, including stimulating osteogenic genes and boosting the synthesis of the fibrillar collagen matrix, in contrast to chitosan, which may be utilized to change cell adhesion, proliferation, and differentiation. Also, the conjugation of fucoidan with chitosan improved the antibacterial activity (Huang & Li, 2014). Due to its multiple fascinating biological actions, fucoidan has been the subject of significant research over the last 10 years. Recently, interest in fucoidans has increased due to the quest for novel medications. Numerous fucoidan structures have been deciphered in recent years, and several facets of their biological function have been clarified. Anticoagulant activity of the fucoidans could be directly related to the sulfate content and position, molecular weight, and sugar composition (Kuznetsova, Besednova, Mamaev, Momot, Shevchenko, & Zvyagintseva, 2003). Fucoidan-chitosan nanoparticles have been intensively investigated for oral administration of active medicinal substances, utilizing chitosan's virtue of greater contact with the mucus layer and, as a result, extended resistance periods at the absorption site. Low molecular weight positively charged chitosan and negatively charged oligofucoidan could self-assemble to form NPs. The size of the NPs plays a major role in the cellular uptake. The smaller NPs have a larger surface area, which results in increased interactions with the biological cell membrane. Many studies have reported that oligofucoidan nanoparticles in the range of 100–274 nm show enhanced antitumor activity in the tumor region. The pharmaceutical and biological applications of such nanoparticles are summarized in Figure 7.2.

The most commonly used production methods include ultrasonication of fucoidan and chitosan solutions, dropping method, ionic crosslinking of fucoidan and chitosan strands, drug-polymer conjugates, etc. The most popular preparation technique is polyelectrolyte complexation, which initially uses the polyelectrolyte self-assembly method. Using this method, different weight ratios can be combined by mixing chitosan and fucoidan solutions before sonicating the mixture at room temperature. This technique makes it possible to produce nanoparticles between 100 and 300 nm with a 90% encapsulation efficiency. Using the dropping method, it was determined how the pH of the chitosan solution and the mass ratio of fucoidan to chitosan affected the production of fucoidan-chitosan nanoparticles. When chitosan's pH was raised to 3.69, the nanoparticles tended to enlarge. In order to create the nanoparticles, this method calls for the dropwise addition of various fucoidan solutions into the chitosan solution while stirring continuously for 30 min. Ionic bonds that connect one polymer chain to another allow for the ionic crosslinking method to be used to prepare fucoidan-chitosan nanoparticles. Fucoidan solutions are added to chitosan solutions in this procedure by flush mixing with a pipette tip vibrating ultrasonically in an ice bath. The dropping technique can also be used for complex fucoidan-chitosan nanoparticles. Using drug-polymer conjugates, and fucoidan-chitosan nanoparticles for drug delivery are also possible. In this case, the drug is conjugated with the polymers that make up the nanoparticle by direct covalent linkage or noncovalent interactions, and typically, this conjugation takes place by a stable or cleavable linker, especially if the drug's primary objective is to reach the conjugate's intracellular target.

It is noteworthy that different synthesis methods produce nanoparticles made of the same polysaccharides with different physicochemical properties, giving these nanosystems a wide range of applications in pharmaceutics as well as in cosmetics and food products (Barbosa, Coutinho, Costa Lima, & Reis, 2019).

The physicochemical characterization of these nanoparticles includes the pH-responsive profile determination, morphology analysis using transmission electron microscopy, and characterization in terms of size, zeta potential, polydispersity, determination of the entrapment efficiency and drug loading, Fourier-transform infrared spectroscopy evaluation, determination of the antioxidant activity and *in-vitro* release assays, etc. Then, the stability of drug-loaded fucoidan/chitosan nanoparticles in biorelevant media and the antioxidant activity of the same should be validated as per the drug delivery system (Barbosa, Coutinho, Costa Lima, & Reis, 2019).

7.5 CONCLUSION

An important source of polysaccharides suitable for biomedical applications is marine organisms. Natural biomaterials with known therapeutic effects include fucoidan and chitosan, which display significant physicochemical and biological characteristics. However, there are some restrictions, like the purification as well as the hydrophilic nature still need to be improved. The use of fucoidan and chitosan as nanoparticles for pharmaceutical applications, specifically for utilizing many biological drugs that have poor aqueous solubility and permeability and less bioavailability, has been the subject of numerous studies. These marine-origin polysaccharides can be chemically altered to increase their bioactivity or bioavailability. Although each of these polysaccharides has unique properties, their combination has been used to enhance their biomedical applications.

ACKNOWLEDGMENT

Author Pooja Jain is thankful to CSIR for providing financial assistance in the form of SRF [09/0591 (11905)/2021 (EMR-I)].

KEYWORDS

- **chitosan**
- **drug delivery system**
- **fucoidan**
- **marine polysaccharides**
- **poly (lactic acid)**
- **polyglycolic acid**

REFERENCES

Agnihotri, S. A., Mallikarjuna, N. N., & Aminabhavi, T. M. (2004). Recent advances on chitosan-based micro- and nanoparticles in drug delivery. *Journal of Controlled Release, 100*, 5–28.

Agnihotri, S. A., Mallikarjuna, N. N., & Aminabhavi, T. M. (2004). Recent advances on chitosan-based micro- and nanoparticles in drug delivery. *Journal of Controlled Release, 100*, 5–28.

Akram, M., Tahir, I. M., Shah, S. M. A., Mahmood, Z., Altaf, A., Ahmad, K., et al. (2018). Antiviral potential of medicinal plants against HIV, HSV, influenza, hepatitis, and coxsackievirus: A systematic review. *Phytotherapy Research, 32*, 811–822.

Alves, A., Pinho, E. D., Neves, N. M., Sousa, R. A., & Reis, R. L. (2012). Processing ulvan into 2D structures: Cross-linked ulvan membranes as new biomaterials for drug delivery applications. *International Journal of Pharmaceutics, 426*, 76–81.

Arpicco, S., Battaglia, L., Brusa, P., Cavalli, R., Chirio, D., Dosio, F., et al. (2016). Recent studies on the delivery of hydrophilic drugs in nanoparticulate systems. *Journal of Drug Delivery Science and Technology, 32*, 298–312.

Assali, M., Shawahna, R., Dayyeh, S., Shareef, M., & Alhimony, I. A. (2018). Dexamethasone-diclofenac loaded polylactide nanoparticles: Preparation, release and anti-inflammatory activity. *European Journal of Pharmaceutical Sciences, 122*, 179–184.

Baino, F., Novajra, G., & Vitale-Brovarone, C. (2015). Bioceramics and scaffolds: A winning combination for tissue engineering. *Frontiers in Bioengineering and Biotechnology, 3*, 1–17, 202–219. https://doi.org/10.3389/fbioe.2015.00202.

Barahona, T., Chandía, N. P., Encinas, M. V., Matsuhiro, B., & Zúñiga, E. A. (2011). Antioxidant capacity of sulfated polysaccharides from seaweeds: A kinetic approach. *Food Hydrocolloids, 25*, 529–535.

Barbosa, A. I., Coutinho, A. J., Costa Lima, S. A., & Reis, S. (2019). Marine polysaccharides in pharmaceutical applications: Fucoidan and chitosan as key players in the drug delivery match field. *Marine Drugs, 17*(12), 654.

Berteau, O., & Mulloy, B. (2003). Sulfated fucans, fresh perspectives: Structures, functions, and biological properties of sulfated fucans and an overview of enzymes active toward this class of polysaccharide. *Glycobiology, 13*, 29R–40R.

Bertrand, N., Wu, J., Xu, X., Kamaly, N., & Farokhzad, O. C. (2014). Cancer nanotechnology: The impact of passive and active targeting in the era of modern cancer biology. *Advanced Drug Delivery Reviews, 66*, 2–25.

Best, S. M., Porter, A. E., Thian, E. S., & Huang, J. (2008). Bioceramics: Past, present and for the future. *Journal of the European Ceramic Society, 28*, 1319–1327.

Brem, H., Mahaley, M. S., Vick, N. A., Black, K. L., Schold, S. C., Burger, P. C., et al. (1991). Interstitial chemotherapy with drug polymer implants for the treatment of recurrent gliomas. *Journal of Neurosurgery, 74*, 441–446.

Brem, H., Piantadosi, S., Burger, P., Walker, M., Selker, R., Vick, N., et al. (1995). Placebo-controlled trial of safety and efficacy of intraoperative controlled delivery by biodegradable polymers of chemotherapy for recurrent gliomas. *The Lancet, 345*, 1008–1012.

Brovold, M., Almeida, J. I., Pla-Palacín, I., Sainz-Arnal, P., Sánchez-Romero, N., Rivas, J. J., et al. (2018). Naturally-derived biomaterials for tissue engineering applications. In H. J. Chun, K. Park, C. H. Kim, & G. Khang (Eds.), *Novel Biomaterials for Regenerative Medicine* (pp. 421–449). Springer. https://doi.org/10.1007/978-981-13-0947-2_23.

Brown, E. M., Allsopp, P. J., Magee, P. J., Gill, C. I., Nitecki, S., Strain, C. R., et al. (2014). Seaweed and human health. *Nutrition Reviews, 72*, 205–216.

Cardoso, M. J., Costa, R. R., & Mano, J. F. (2016). Marine origin polysaccharides in drug delivery systems. *Marine Drugs, 14*, 1–27.

Cardoso, M. J., Costa, R. R., & Mano, J. F. (2016). Marine origin polysaccharides in drug delivery systems. *Marine Drugs, 14*, 1–27.

Chollet, L., Saboural, P., Chauvierre, C., Villemin, J. N., Letourneur, D., Chaubet, F., et al. (2016). Fucoidans in nanomedicine. *Marine Drugs, 14*, 1–24.

Costa, L. S., Fidelis, G. P., Cordeiro, S. L., Oliveira, R. M., Sabry, D. A., Câmara, R. B. G., et al. (2010). Biological activities of sulfated polysaccharides from tropical seaweeds. *Biomedicine & Pharmacotherapy, 64*, 21–28.

De La Fuente, M., Seijo, B., & Alonso, M. J. (2008). Novel hyaluronic acid-chitosan nanoparticles for ocular gene therapy. *Investigative Ophthalmology and Visual Science, 49*, 2016–2024.

De La Fuente, M., Seijo, B., & Alonso, M. J. (2008). Novel hyaluronic acid-chitosan nanoparticles for ocular gene therapy. *Investigative Ophthalmology and Visual Science, 49*, 2016–2024.

Elamparithi, A., Punnoose, A. M., Paul, S. F. D., & Kuruvilla, S. (2017). Gelatin electrospun nanofibrous matrices for cardiac tissue engineering applications. *International Journal of Polymeric Materials and Polymeric Biomaterials, 66*, 20–27.

El-Sheekh, M. M., Shabaan, M. T., Hassan, L., & Morsi, H. H. (2022). Antiviral activity of algae biosynthesized silver and gold nanoparticles against Herps Simplex (HSV-1) virus in vitro using cell-line culture technique. *International Journal of Environmental Health Research, 32*, 1–12.

Ferrari, M. (2005). Cancer nanotechnology: Opportunities and challenges. *Nature Reviews Cancer, 5*, 161–171.

Ganesan, A. (2008). The impact of natural products upon modern drug discovery. *Current Opinion in Chemical Biology, 12*, 306–317.

Gopinath, V., Saravanan, S., Al-Maleki, A. R., Ramesh, M., & Vadivelu, J. (2018). A review of natural polysaccharides for drug delivery applications: Special focus on cellulose, starch and glycogen. *Biomedicine & Pharmacotherapy, 107*, 96–108.

Griset, A. P., Walpole, J., Liu, R., Gaffey, A., Colson, Y. L., & Grinstaff, M. W. (2009). Expansile nanoparticles: Synthesis, characterization, and in vivo efficacy of an acid-responsive polymeric drug delivery system. *Journal of the American Chemical Society, 131*, 2469–2471.

Haefner, B. (2003). Drugs from the deep: Marine natural products as drug candidates. *Drug Discovery Today, 8*, 536–544.

Hamed, I., Özogul, F., Özogul, Y., & Regenstein, J. M. (2015). Marine bioactive compounds and their health benefits: A review. *Comprehensive Reviews in Food Science and Food Safety, 14*, 446–465.

Hariyadi, D. M., & Islam, N. (2020). Current status of alginate in drug delivery. *Advances in Pharmacological and Pharmaceutical Sciences, 2020*, 1–16. https://doi.org/10.1155/2020/8880857.

Hassabo, A. A., Ibrahim, E., Ali, B., & Emam, H. E. (2022). Anticancer effects of biosynthesized Cu2O nanoparticles using marine yeast. *Biocatalysis and Agricultural Biotechnology, 39*. Elsevier Ltd.

Huang, Y.-C., & Li, R.-Y. (2014). Preparation and Characterization of Antioxidant Nanoparticles Composed of Chitosan and Fucoidan for Antibiotics Delivery. *Marine Drugs, 12*, 4379–4398.

Jeong, B., Kim, S. W., & Bae, Y. H. (2012). Thermosensitive sol–gel reversible hydrogels. *Advanced Drug Delivery Reviews, 64*, 154–162.

Kaulmann, A., & Bohn, T. (2014). Carotenoids, inflammation, and oxidative stress–implications of cellular signaling pathways and relation to chronic disease prevention. *Nutrition Research, 34*, 907–929.

Khalid, M., Khalid, N., Ahmed, I., Hanif, R., Ismail, M., & Janjua, H. A. (2017). Comparative studies of three novel freshwater microalgae strains for synthesis of silver nanoparticles: Insights of characterization, antibacterial, cytotoxicity and antiviral activities. *Nanomedicine: Nanotechnology, Biology and Medicine, 29*, 1851–1863.

Khrunyk, Y., Lach, S., Petrenko, I., & Ehrlich, H. (2020). Progress in modern marine biomaterials research. *Marine Drugs, 18*, 589.

Kim, B., La Flamme, K., & Peppas, N. A. (2003). Dynamic swelling behavior of pH-sensitive anionic hydrogels used for protein delivery. *Journal of Applied Polymer Science, 89*, 1606–1613.

Kim, S. K. (2013). Introduction to Marine Biomaterials. In S.-K. Kim & J. Venkatesan (Eds.), *Marine Biomaterials* (pp. 26–39). CRC Press.

Kim, S. K., & Venkatesan, J. (2013). Introduction to marine biomaterials. Marine Biomaterials: Characterization, Isolation and Application, 3-16.

Kopeček, J. (2003). Smart and genetically engineered biomaterials and drug delivery systems. *European Journal of Pharmaceutical Sciences, 20*, 1–16.

Kuznetsova, T. A., Besednova, N. N., Mamaev, A. N., Momot, A. P., Shevchenko, N. M., & Zvyagintseva, T. N. (2003). Anticoagulant activity of fucoidan from brown Algae *Fucus evanescens* of the Okhotsk Sea. *Bulletin of Experimental Biology and Medicine, 136*, 471–473.

Laurienzo, P. (2010). Marine polysaccharides in pharmaceutical applications: An overview. *Marine Drugs, 8*, 2435–2465.

Liu, Y.-L., Chen, D., Shang, P., & Yin, D.-C. (2019). A review of magnet systems for targeted drug delivery. *Journal of Controlled Release, 302*, 90–104.

Liu, Z., Jiao, Y., Wang, Y., Zhou, C., & Zhang, Z. (2008). Polysaccharides-based nanoparticles as drug delivery systems. *Advanced Drug Delivery Reviews, 60*, 1650–1662.

Lowman, A. M., Morishita, M., Kajita, M., Nagai, T., & Peppas, N. A. (1999). Oral delivery of insulin using pH-responsive complexation gels. *Journal of Pharmaceutical Sciences, 88*, 933–937.

Luo, Y., & Wang, Q. (2014). Recent development of chitosan-based polyelectrolyte complexes with natural polysaccharides for drug delivery. *International Journal of Biological Macromolecules, 64*, 353–367.

Manivasagan, P., Bharathiraja, S., Santha Moorthy, M., Mondal, S., Seo, H., & Lee, K. D., et al. (2018). Marine natural pigments as potential sources for therapeutic applications. *Critical Reviews in Biotechnology, 38*, 745–761.

Manivasagan, P., Nam, S. Y., & Oh, J. (2016). Marine microorganisms as potential biofactories for synthesis of metallic nanoparticles. *Critical Reviews in Microbiology, 42*, 1007–1019.

Miller, J. B., Zhang, S., Kos, P., Xiong, H., Zhou, K., Perelman, S. S., et al. (2017). Non-viral CRISPR/Cas gene editing in vitro and in vivo enabled by synthetic nanoparticle co-delivery of Cas9 mRNA and sgRNA. *Angewandte Chemie, 129*(5), 1079–1083.

Molinski, T. F., Dalisay, D. S., Lievens, S. L., & Saludes, J. P. (2009). Drug development from marine natural products. *Nature Reviews Drug Discovery, 8*, 69–85.

Moore, B., & Tischer, R. (1964). Extracellular polysaccharides of algae: Effects on life-support systems. *Science, 145*, 586–587.

Ngo, D.-H., & Kim, S.-K. (2013). Sulfated polysaccharides as bioactive agents from marine algae. *International Journal of Biological Macromolecules, 62*, 70–75.

Ngwuluka, N. C. (2018). Responsive polysaccharides and polysaccharides-based nanoparticles for drug delivery. In A. S. H. Makhlouf & N. Y. Abu-Thabit (Eds.), *Stimuli Responsive Polymeric Nanocarriers for Drug Delivery Applications, Volume 1* (pp. 531–554). Woodhead Publishing. [Online]. Available from: https://www.sciencedirect.com/science/article/pii/B9780081019979000230 (accessed on 7 June 2024).

Peppas, N. A., Hilt, J. Z., Khademhosseini, A., & Langer, R. (2006). Hydrogels in biology and medicine: From molecular principles to bionanotechnology. *Advanced Materials, 18*, 1345–1360.

Riccio, G., Ruocco, N., Mutalipassi, M., Costantini, M., Zupo, V., Coppola, D., et al. (2020). Ten-year research update review: Antiviral activities from marine organisms. *Biomolecules, 10*, 1–36.

Rioux, L. E., Beaulieu, L., & Turgeon, S. L. (2017). Seaweeds: A traditional ingredient for new gastronomic sensation. *Food Hydrocolloids, 68*, 255–265.

Rysz, J., Gluba-Brzózka, A., Franczyk, B., Jabłonowski, Z., & Ciałkowska-Rysz, A. (2017). Novel biomarkers in the diagnosis of chronic kidney disease and the prediction of its outcome. *International Journal of Molecular Sciences, 18*(8), 1702.

Schneider, C., Langer, R., Loveday, D., & Hair, D. (2017). Applications of ethylene vinyl acetate copolymers (EVA) in drug delivery systems. *Journal of Controlled Release, 262*, 284–295.

Senni, K., Pereira, J., Gueniche, F., Delbarre-Ladrat, C., Sinquin, C., Ratiskol, J., et al. (2011). Marine polysaccharides: A source of bioactive molecules for cell therapy and tissue engineering. *Marine Drugs, 9*, 1664–1681.

Sezer, A. D., & Akbuğa, J. (2006). Fucosphere – New microsphere carriers for peptide and protein delivery: Preparation and in vitro characterization. *Journal of Microencapsulation, 23*, 513–522.

Shimoboji, T., Larenas, E., Fowler, T., Hoffman, A. S., & Stayton, P. S. (2003). Temperature-induced switching of enzyme activity with smart polymer–enzyme conjugates. *Bioconjugate Chemistry, 14*, 517–525.

Silverman, J. A., & Deitcher, S. R. (2013). Marqibo® (vincristine sulfate liposome injection) improves the pharmacokinetics and pharmacodynamics of vincristine. *Cancer Chemotherapy and Pharmacology, 71*, 555–564.

Spencer, L., Mann, C., Metcalfe, M., Webb, M., Pollard, C., Spencer, D., et al. (2009). The effect of omega-3 FAs on tumour angiogenesis and their therapeutic potential. *European Journal of Cancer, 45*, 2077–2086.

Sun, Y., Ma, X., & Hu, H. (2021). Marine polysaccharides as a versatile biomass for the construction of nano drug delivery systems. *Marine Drugs, 19*, 1–15.

Sutherland, I. W. (1998). Novel and established applications of microbial polysaccharides. *Trends in Biotechnology, 16*, 41–46.

Tziveleka, L. A., Sapalidis, A., Kikionis, S., Aggelidou, E., Demiri, E., Kritis, A., et al. (2020). Hybrid sponge-like scaffolds based on ulvan and gelatin: Design, characterization and evaluation of their potential use in bone tissue engineering. *Materials, 13*, 1–20.

Venkatesan, J., Anil, S., Kim, S. K., & Shim, M. S. (2016). Seaweed polysaccharide-based nanoparticles: Preparation and applications for drug delivery. *Polymers, 8*, 1–25.

Vijayakumar, S., Saravanakumar, K., Malaikozhundan, B., Divya, M., Vaseeharan, B., Durán-Lara, E. F., et al. (2020). Biopolymer K-carrageenan wrapped ZnO nanoparticles as drug delivery vehicles for anti MRSA therapy. *International Journal of Biological Macromolecules, 144*, 9–18.

Vishchuk, O. S., Ermakova, S. P., & Zvyagintseva, T. N. (2013). The fucoidans from brown algae of Far-Eastern seas: Anti-tumor activity and structure–function relationship. *Food Chemistry, 141*, 1211–1217.

Wan, M., Qin, W., Lei, C., Li, Q., Meng, M., Fang, M., et al. (2021). Biomaterials from the sea: Future building blocks for biomedical applications. *Bioactive Materials, 6*, 4255–4285.

Wang, D. A., Varghese, S., Sharma, B., Strehin, I., Fermanian, S., Gorham, J., et al. (2007). Multifunctional chondroitin sulphate for cartilage tissue-biomaterial integration. *Nature Materials, 6*, 385–392.

Wang, D. A., Varghese, S., Sharma, B., Strehin, I., Fermanian, S., Gorham, J., et al. (2007). Multifunctional chondroitin sulphate for cartilage tissue-biomaterial integration. *Nature Materials, 6*, 385–392.

Wang, Y., Xing, M., Cao, Q., Ji, A., Liang, H., & Song, S. (2019). Biological activities of fucoidan and the factors mediating its therapeutic effects: A review of recent studies. *Marine Drugs, 17*, 183.

Wang, Y., Zhao, Q., Han, N., Bai, L., Li, J., Liu, J., et al. (2015). Mesoporous silica nanoparticles in drug delivery and biomedical applications. *Nanomedicine: Nanotechnology, Biology and Medicine, 11*, 313–327.

Wijesekara, I., Pangestuti, R., & Kim, S. K. (2011). Biological activities and potential health benefits of sulfated polysaccharides derived from marine algae. *Carbohydrate Polymers, 84*, 14–21.

Yoshida, R., Uchida, K., Kaneko, Y., Sakai, K., Kikuchi, A., Sakurai, Y., et al. (1995). Comb-type grafted hydrogels with rapid deswelling response to temperature changes. *Nature, 374*, 240–242.

Zhang, G., Li, J., Zhu, T., Gu, Q., & Li, D. (2016). Advanced tools in marine natural drug discovery. *Current Opinion in Biotechnology, 42*, 13–23.

CHAPTER 8

Marine-Derived Biomaterials: An Ignored Building Block in Regenerative Medicine and Tissue Engineering Applications

SOURAV MOHANTO,[1] ADRIJA BHUNIA,[1] ROKEYA SULTANA,[2] J. C. DARSHAN,[3] and MOHAMMED GULZAR AHMED[1]

[1]*Department of Pharmaceutics, Yenepoya Pharmacy College and Research Center, Yenepoya (Deemed to be University), Mangalore, Karnataka, India*

[2]*Department of Pharmacognosy, Yenepoya Pharmacy College and Research Center, Yenepoya (Deemed to be University), Mangalore, Karnataka, India*

[3]*Department of Pharmacy Practice, Yenepoya Pharmacy College and Research Center, Yenepoya (Deemed to be University), Mangalore, Karnataka, India*

ABSTRACT

Just like one-third of the healthy human body, the Earth's surface is shrouded by approximately 71% of water, and the ocean bears approximately 95% of the water on top of that. Several potential biomaterials have been derived from marine sources in the last decade, which are advantageous for biomedical applications. Marine sources-derived biomaterials have experienced various stages of discovery and development, including coral bone graft substitute formation to polysaccharide-based biomaterials to materialize various targeted delivery systems or carrier systems for tissue engineering applications. Despite several biological properties of marine-based biomaterials, still safety and effectiveness of these biomaterials remain questioned by several biomedical scientists. The significant limitations of marine-based

Handbook of Research in Marine Pharmaceutics: Exploring Oceanic Microbial Diversity for Human Health and Wellness. Md. Faiyazuddin & Khalid Rehman Hakeem (Eds.)

biomaterials are challenging to address due to restraining quality control requirements, i.e., immunogenicity, physicochemical structure, mechanical integrity, toxicity assessment, etc. Hence, this chapter primarily concentrated on the optimized extraction and isolation techniques of marine biomaterials for higher yields, structure elucidation, and physicochemical, and biological properties, and also summarized recent explicatory investigations as effective biomaterial biomedical applications. Therefore, future constraints and perspectives of marine-based biomaterials in drug delivery science have also been exclusively summarized in this chapter. The chapter also covers the critical aspects of available marine biomaterials for various biomedical and tissue engineering applications with the scope of future studies in other applications.

8.1 INTRODUCTION

Just like one-third of the healthy human body, the Earth's surface is shrouded by approximately 71% of water, and the ocean bears about 95% of the water on top of that. The marine ecosystem is the most expansive ecosystem on the planet, estimated to have approximately 2.2 million species in marine biology, although 91% still await identification and standard characterization. Marine species are classified into six kingdoms, i.e., bacteria, protozoans, chromista, fungi, plants, and animals (Macha et al., 2019b). Currently, formulation scientists have yielded more concentration to marine biological materials, mainly some biological macromolecules and their derivatives, such as chitin, alginate, and collagen in biomedical applications. They do not endure religious constraints for mammals, and zoonotic transmission risk, and further enclose enhanced biocompatibility, bioavailability, biodegradation rate, biological activity, and mechanical strength in tissue engineering applications (Zhang et al., 2022). Marine origins exemplify a natural choice to furnish adequate biomaterials and compounds to be exploited in developing biomaterials for tissue engineering and medical devices, which can achieve the injunction for tissue regeneration or regenerative medicine. The extracellular matrix (ECM) usually demonstrates equivalent features, i.e., cell viability, proliferation, angiogenesis, etc., with marine-derived biomaterials for sustaining and rebuilding living tissues (Carvalho et al., 2020). Furthermore, marine-derived biomaterials have also displayed low or no cytotoxicity, low immunogenicity, and higher biodegradation rates in several *in vitro* models. The most commonly derived marine biomaterials are proteins and polysaccharides in nature. Polysaccharides include chitin, chitosan, carrageenan, fucoidan, alginate, glycosaminoglycan, etc., whereas proteins include mainly collagen abundantly found in the sea resource (Romano et al., 2022). Figure 8.1 illustrates various marine-based biomaterials and their potential therapeutic effect in biomedical and tissue engineering applications.

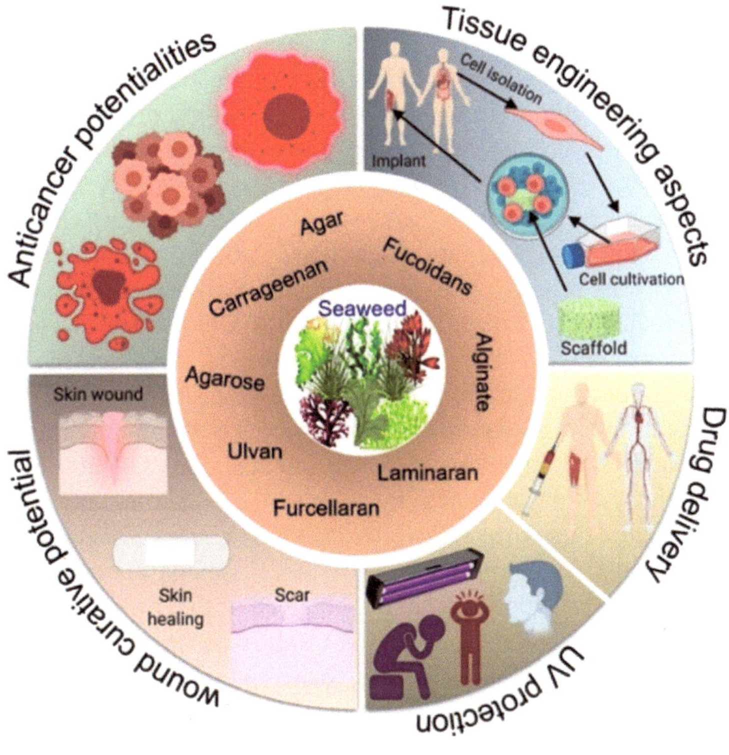

FIGURE 8.1 Schematic illustration described the various marine-based biomaterials applications in the biomedical and tissue regeneration field.

Marine-derived biomaterials have passed several developmental stages since first extracted in the form of coral bone graft substitutes, polysaccharide-based polymers, rapid advances in tissue regeneration, nanotechnology, and three-dimensional (3D) printing, and later as antimicrobial agents, regenerative matrix, and drug delivery systems/carriers. In the diversity, approximately 25,000 biologically active compounds are obtainable from microorganisms with the highest possibility of yielding biomaterials in regenerative medicine.

The evolution and utilization of marine biological materials are a substantial priority in material research and development. Several biomaterials (e.g., exopolysaccharides, fatty acids, carotenoids, mycosporine, etc.) with diverse physicochemical characteristics describe sustainability, low-cost production, and natural alternatives for tissue engineering applications. Moreover, several polysaccharides have been employed for antiviral, antimicrobial, antioxidant, anti-inflammatory, antithrombotic, and antitumor properties (Wan et al., 2021). These biomaterials can be further incorporated for the delivery of therapeutic molecules via scaffolds, wound dressings, and drug carrier materials for tissue engineering products, treatment, and repair of articular cartilage tissues (Carvalho et al., 2021a).

During the initial stages of marine biomaterial development, chitosan was found to be a natural polysaccharide significantly utilized in tissue engineering and drug delivery science. The structural modifications of chitosan-based biomaterials showed enhanced effectiveness as antioxidant and antitumor agents. The pseudo-natural polycationic substance, i.e., chitosan, further formation of the complex with other biomaterials showed improved drug release (controlled), drug encapsulation efficiency, cell/enzyme immobilization, and drug/gene carriers. Marine-derived chitosan biomaterials also possess higher biodegradability, antibacterial activity, hydrophilic properties, and the ability to assemble complexes when investigated *in vitro* (Younes & Rinaudo, 2015). In the later stages of marine biomaterial development, Carrageenans, the family of linear sulfated polysaccharides obtained from red seaweeds, showed attractive anti-inflammatory, antimicrobial, anti-hyperlipidemic, antioxidant, antitumor, and anticoagulant effects in tissue engineering application. The various physicochemical assessments further ratified the changes in the molecular weight, site sulfate group, and complete oxidation or partial oxidation of β-D-galactopyranose units exemplified improved biological characteristics, i.e., anticoagulant in nerve tissue engineering (Qureshi et al., 2019). The linear polysaccharides, i.e., Glycosaminoglycans (GAGs), are found in most animal tissues' extracellular matrix (ECM) and cell surface, free or bound with proteins. The advancement allowed marine biomaterials to enter the world of regenerative medicine, such as tissue and cartilage regeneration, due to their ability to restrain proteoglycans and promote growth factors, which are further crucial for cell proliferation and attachment (Valcarcel et al., 2017). The sulfated polysaccharide fucoidan (FD) extracted from brown algae has also been demonstrated to retain bio-multifunctional attributes, i.e., facilitating collagen formation, angiogenesis, specific binding of growth factors, antioxidant, anti-inflammatory, and immunomodulatory activities (Hao et al., 2021). These advancements in marine biomaterials have assembled it of significant prominence as a merging contender in the field of regenerative medicine & tissue engineering.

This chapter discusses the abundant marine biomaterials and their utilization in regenerative medicine for tissue engineering applications, specifically in bone, cartilage, skin, liver, etc. The chapter also exclusively summarizes the various extraction or isolation process limitations concerning marine biomaterials. Later in the chapter, a comprehensive summary is assembled of the different polysaccharides, proteins, and secondary metabolites derived from marine sources and their successful utilization in tissue engineering applications. The chapter also focuses on the various animal models and clinical standpoints to assess marine biomaterials and their limitations. The chapter concludes with closing remarks on the limits of marine biomaterials in drug delivery science and the conceivable tools to overcome the limitations in tissue engineering applications.

8.2 CHALLENGES OF MARINE BIOMATERIALS DURING EXTRACTION AND ISOLATION PROCESS

Chitin and its derivatives are mostly found in lower-level ocean sources, such as fungi, algae, and arthropods, exhibiting macromolecular polysaccharides in nature. The conventional method involves the production of chitin from the exoskeleton of arthropods (e.g., crabs and shrimp), which may further form complexes with proteins (Pakizeh et al., 2021; Younes & Rinaudo, 2015). Several empirical studies on tissue engineering understand the utilization of chitin due to enhanced biocompatibility, anti-inflammatory, hemostatic, and analgesic properties, promotion of growth factors, and standardized utilization techniques (Kaur & Dhillon, 2015). The industrial-scale production of marine-derived chitin involves the use of acid and alkali, resulting in serious environmental pollution. The removal of proteins and fat pigments is also taken into consideration during the extraction process (Kaur & Dhillon, 2015; Pakizeh et al., 2021). These limitations lead to an increase in production costs, acetylation of chitin, and a decrease in the physicochemical properties of chitin, which can be rectified through the use of proteolytic enzymes (e.g., pepsin, trypsin, papain, alkaline protease) and mild treatments with ethylenediaminetetraacetic acid (EDTA) and citric acid (Kaur & Dhillon, 2015; Pakizeh et al., 2021). Furthermore, low efficiency, deproteinization, and cost-effectiveness are identified as limitations during the enzymatic treatment extraction of chitin, which can be addressed by combining weak alkali treatment with an enzymatic approach (Vázquez et al., 2017; Zhang et al., 2022).

Collagen, a biological macromolecule of ocean sources (mainly in fishes), contains almost 25–30% protein. The isolation technique includes washing, extraction, and milling of the bioresource samples, followed by chemical treatment/precursor to remove/dissolve non-collagenous proteins, pigments, or fats. Based on different regions, the composition and physical characteristics of the extracellular matrix (ECM) are different for marine animals. The extraction methodology for marine collagen must include the preparation process, such as enzymatic and acid solutions. Marine collagen has led to advancements in wound healing, dental tissue engineering, and many more developments. Despite several advantages, it carries certain limitations for clinical trials of the mentioned applications, and further development is significant for nutraceuticals, which need further consideration (Lim et al., 2019). According to recent investigational statements, the extraction and isolation of different collagens from various sources can be processed via acid-base extraction, salt-mediated extraction, hot water extraction, and fermentation, followed by the recently advanced microwave-assisted extraction process. The morphological characteristics of collagen fiber and cross-linking capacity with other biomaterials may also vary due to different extraction, isolation, and purification techniques. It is also crucial to remove the maximum impurities from the extracted collagen, which may lead to partially modified collagen. Commonly used purification methods include centrifugation, ion exchange chromatography, dialysis, electrophoresis, and salting-out (Zhang et al., 2022).

Sodium alginate is classified as linear anionic polysaccharides, usually derived from marine-based Sargassum of brown algae and bacterial strains. The extraction process includes various steps such as pre-treatment via 0.1% formaldehyde to evade alginate pigments, further addition of alkaline solution to the gathered insoluble alginic acid, and sodium alginate is obtained in aqueous solution, and finally, filtering. Considering its application in the field of tissue engineering it still counts several limitations, such as biocompatibility in the form of hydrogels; continuous and sequential release of the drugs, which are yet to be solved, and improvement of the mechanical properties (H. Zhang et al., 2021). Alginates present on the cell wall of algae were conventionally extracted by adding an alkaline solution (NaOH, Na_2CO_3) after pre-treatment with dilute acid. After the filtration of the extract, $CaCl_2$ is incorporated into the filtrate to form an aggregation of alginate. The alginate obtained is further reacted with dilute hydrochloric acid, dispersed by sodium carbonate, filtered, and suspended the particles via ethanol. Finally, the powdered sodium alginate is obtained after the purification and transformation. At present, several advanced techniques have been employed, i.e., ultrasonic-assisted extraction (UAE), microwave-assisted extraction (MAE), pressurized liquid extraction (PLE), and enzyme extraction (EAE), to obtain higher yields. The activation of the raw materials via enzyme promotes a slow rate of degradation of alginate by acid. The specificity of the enzyme promotes the destruction of the cell wall making alginate dissolve more and also increasing the yield, reducing the solvent usage, and promoting efficiency (Zhang et al., 2022).

The sulfated polysaccharide-based high molecular weight carrageenan (CG) can also be derived from several marine sources, including marine algae (Family: Rhodophyceae, Examples: eucheuma, hypnea, chondrus, etc.) and has been utilized as biomaterials for tissue regeneration. The extraction process of CG includes pre-treatment, extraction, and precipitation initiated via chemical reagent, decantation, and drying. At first, the impurities in red algae are removed via pre-treatment, followed by using the hot alkali extraction technique, leading to CG being released from cells. After the treatment, the CG is clarified and obtained in powder form. Commonly, for the extraction of CG, the "alcohol precipitation method" is used. The physicochemical properties of CG make it insoluble in organic solvents such as methanol and ethanol. To precipitate CG from the solution, the concentrated form is kept in alcohol, promoting the collection of dried and desired size of CG by solvent evaporation. Several complex methods, such as using enzymes or fungi, and biological methods, decrease carrageenan's toxicity. In recent years, advanced extraction methods, i.e., MAE, UAE, DES, etc., have encountered superior benefits such as gelation and biological activity conditions of CGs, and improvement of the physicochemical characteristics (L. Li et al., 2014; Pacheco-Quito et al., 2020).

Fucoidan is a complex polysaccharide biofunctional material that can be sourced from marine-derived brown algae and invertebrates (e.g., sea urchins, trepang, etc.). It comprises l-fucose and sulfate groups into l-fucose-4-sulfate, which is a primary monosaccharide compound (Wang et al., 2019). The structural diversity is affected by the extraction of fucoidan because of the development of a specific amount of bias, resulting in a change in structure, charge, molecular weight, and chemical composition (January et al., 2019). Fucoidan can be extracted by several methods, such as dilute alkali, microwave-assisted treatment, dilute acid, enzyme-assisted, and hot water. The methods' requirement of higher temperature, solvent, or extraction time has led to the proposal of some developed polysaccharide extraction techniques that can provide higher yield and sustain the natural fucoidan structure. Some techniques that improve the extraction yield include a combination of DES with subcritical water extraction for better yield, UAE method for higher MW, and homogenization method for the extraction of fucoid for outstanding reactive oxygen species (ROS) scavenging property (Nguyen et al., 2020; Yuan & Macquarrie, 2015).

Ulvan, a seaweed polysaccharide, promotes extraction methods such as physical assistance, solvent, or other methods. The physicochemical properties of ulvan give an idea about the extraction method to be used via the interaction of the cell wall. The acid extraction induces nearly the entire desulfation of isolated ulvan; the purified enzymatic methods support an adequate SO_3 substituent level. The destruction and rapid dissolving of Ulvan can be caused by UAE, whereas it cannot destroy the intracellular structure of the same. The extraction process can reduce or enhance the degradation of ulvan, though the specific mechanism of the structure-activity relationship between them and the biological activity needs to be further studied (Cindana Mo'o et al., 2020; Tziveleka et al., 2019). The conventional extraction process of laminarin is a time-consuming and low-yield method. So, a new extraction technology is needed to enhance the extraction rate and yield. For example, for the extraction of laminarin, UAE stands better than solid-liquid extraction promoting high MW; also, the best extraction efficiencies for laminarin (87.6%) can be achieved by DES from marine kelp. In addition, various physicochemical and biological approaches have been utilized to alter its morphology to improve the bioactivity of laminarin. Catarina et al. developed cross-linked photolaminarins via acrylate groups, enhancing their utilization in biomedicine. Despite several strategies, little research has been successfully reported due to batch-to-batch variation of extraction and non-available standard characterization techniques. The structural activity relationship (SAR) and biofunctionality should be investigated further (Custódio et al., 2016; Zargarzadeh et al., 2020).

8.3 BIOMATERIALS FROM MARINE SOURCES FOR BIOMEDICAL AND TISSUE REGENERATIVE APPLICATION

Biomaterials can be objects developed for biological systems to evaluate, treat, boost, or substitute any body organ, tissue, or function. Biomaterials have traditionally been considered inert and do not interact with the host's biological systems. Materials from biological origins (e.g., wood) have a diversified account as biomaterials and have been usually used to restore tissues from disease or trauma (e.g., prosthetics). In the early 20th century, biomaterials were substituted by synthetic ceramics, polymers, and metal alloys because of enhanced activity and several other properties. However, biomaterials still have several limitations. The development of current biomaterials is paying attention to simulating multiple functions of the extracellular matrices of tissues, as it can restrain the responses in a well-defined manner, owing to inherent biocompatibility (Wan et al., 2021). The marine ecosystem, with around 2.2 million species, is one of the leading ecosystems on Earth. The ocean consists of water and is a profuse source of various biomaterials for humankind. Marine organisms are an immense foundation of new biologically active compounds. These compounds are exclusive because the aqueous environment requires a sky-scraping claim for specific and potent bioactive molecules. Ocean, a profuse resource of materials, is of great importance to humans as marine biomaterials. They are a pioneering emerging area of research with multiple applications. The brink of 91% of these species still looks out for detection and exploration (Wan et al., 2021). Researchers have paid far-reaching interest to marine-derived biomaterials for a range of applications. These resources have marvelous prospects for biomaterials of higher value. Last decade biomaterials commenced from aquatic organisms had a great rise in number. The field is slowly expanding. Marine biomaterials have experienced innumerous spans of innovation and advancement, varying from polysaccharide-based biomaterials to coralline bone grafts. Marine-derived biomaterials lack human pathogens on the whole (Mora et al., 2011). At present, research on marine biomaterials has undergone rejuvenation. Technological uprising has dramatically improved the efficacy of utilizing these renewable natural resources (Russo et al., 2015).

Marine organisms retain remarkable bioactive properties. The organisms have been used as biomimetic templates in biomedical applications, either straight or modified. In the year 1967, holothurin came into existence as the very first marine biomaterial. It was the first one to be experimented with *in an in-vivo* cancer model as an anti-tumor agent (Khrunyk et al., 2020). The discovery stimulated extensive research on marine biomaterials. It resembles natural bone and retains excellent biocompatibility, substantial osteoconductivity, and osteogenic potential (Nigrelli et al., 1967). Marine biomaterials also generate ways towards future *in vivo* studies of the biomaterial. It encloses a scope for habitats with colossal microbial biodiversity. Around 25,000 biologically active compounds were acknowledged (Demers et al., 2002). Bioactive materials are mainly developed by bacteria, fungi, and fungi-like protists. Marine microorganisms result in carotenoids, fatty acids, chitosan, mycosporine-like amino acids, exopolysaccharides, and mycosporine representing low-cost, sustainable, rapid development of alternatives for natural molecules for anti-aging, skin whitening, UV protection for the body, face, and hair care (Chiroff et al., 1975). Fungi and marine bacteria extract polyketides, alkaloids, peptides, and polysaccharides, which have been broadly studied (Corinaldesi et al., 2017).

8.3.1 POLYSACCHARIDES AND GLYCANS FROM MARINE ORGANISMS

Carbohydrates, called saccharides, comprise carbon, hydrogen, and oxygen molecules. Carbohydrates are the most abundant and essential biomolecule component for innumerable natural products and hence have drawn attention to numerous health benefits. Amid carbohydrates, polysaccharides exhibit several aquatic organisms' most productive bioactive substances. Marine micro and macro-organisms are an outstanding source of carbohydrates with various applications in biofunctional properties. These polysaccharides can be obtained from marine organisms cheaply compared to plants. Hence, in the modern era marine organisms have been recognized as advanced contributors to novel therapeutic agents. The polysaccharides are henceforth considered biocompatible with no toxicity. Carrageenan, chitosan, fucoidan, and alginate, indicated physicochemical and biological properties that immersed enormous concentrations from biomedical and pharmaceutical industries (Huang et al., 2018). Aquatic polysaccharides (mostly chitosan, alginate, carrageenan, chitin, and fucoidan) also work on cell proliferation and cycle, modulate different metabolic pathways, and copious pharmaceutical activities, i.e., antiviral, anticoagulant, immuno-stimulatory, antioxidative, anticancer, and antibacterial effects.

Polysaccharides have numerous advantages for human health and have also been designed into probable cosmeceuticals and nutraceuticals (Grice & Mariottini, 2018). In recent decades, glycans (or carbohydrate- or sugar-based compounds) have contributed outstandingly to developing novel therapeutics. The glycans are grouped depending on the isolated marine organism, namely, bacteria (Wan et al., 2021), chromists (Mora et al., 2011), plants (Russo et al., 2015), and animals (Khrunyk et al., 2020). Several investigational studies reported that glycans from marine sources possess several therapeutic activities against cytomegalovirus, Epstein-Barr virus, dengue virus, encephalomyocarditis virus, African swine fever virus, hepatitis C virus, human immunodeficiency virus, human cytomegalovirus, herpes simplex virus, human papillomavirus, human rhinovirus, influenza virus, murine sarcoma virus, Japanese encephalitis virus, murine leukemia virus, Newcastle disease virus, parainfluenza virus, Semliki Forest virus, respiratory syncytial virus, tobacco mosaic virus, viral hemorrhagic septicemia virus, vaccinia virus, varicella-zoster virus, and vesicular stomatitis virus, etc. (Cui et al., 2018).

8.3.1.1 CHITIN AND CHITOSAN

Chitosan and chitin are polysaccharide-based compounds, derivatives of glucose. They complement each other based on physicochemical properties. The contrast between them is due to the presence and absence of free amine groups (Gao et al., 2019). They are essential and novel biological compounds due to their emphasis on compatibility (low immunogenicity and non-toxicity) and biodegradation. Their positive impact on human and animal health has made them attractive. Chitosan and chitin are natural polysaccharides obtained from microorganisms, insects, plants, and animals (Tsurkan et al., 2021), summarizing the various sources in Table 8.1.

TABLE 8.1 Sources of Chitin and Chitosan from Marine Organisms (Nigrelli et al., 1967)

Sea Animals	Insects	Micro-organisms
Crustaceans	Scorpions	Green algae
Coelenterates	Brachiopods	Yeast (β-type)
Annelida	Cockroaches	Fungi (cell walls)
Mollusca	Spiders	Mycelia penicillium
Lobster	Beetles	Chytridiaceae
Shrimp	Ants	Ascomydes
Prawn	–	Blastocladiaceae
Krill	–	Spores
Crab	–	–

Chitosan, a deacetylated derivative of chitin, originates from the shells of insects, crabs, bone plates of squids, shrimps, and microorganisms. Chitosan present in the cell wall of fungi is in free form for one and covalent bond form to β-glucan in another. The polymers can be easily made by mycelia present in industrial waste or by mycelia obtained from the fungi medium attained from industrial by-products (Laurienzo, 2010). Chitosan biomaterial is also chemically represented as a copolymer of α-(1→4) glucosamine $(C_6H_{11}O_4N)_n$, having diverse numbers of N-acetyl groups. It comprises erratically distributed beta 1–4 bonds between N-acetyl-D-glucosamine and D-glucosamine (Senni et al., 2011).

Chitin is extrapolated from the Greek term 'chiton.' Chitin, a long carbon chain polymer of N-acetylglucosamine, is an amide result of glucose (Kady, 2019). It includes a variety of polysaccharides that appear as a component of a fungal cell wall, in the radulae of mollusks, in the scales of fish, in the exoskeleton of arthropods, and in cephalopod beaks (Kim & Thomas, 2007). Chitin stands as the second most affluent polysaccharide after cellulose. Chitin and cellulose could be analyzed in the presence of crystalline nanofibrils of whiskers. The component has several applications in biotechnological, medicine, and industries (Loo et al., 2022) as summarized in Table 8.2.

8.3.1.2 CARRAGEENAN

Carrageenans or carrageenans are polysaccharides obtained from red edible seaweeds and are natural linear sulfated polysaccharides. The content of carrageenan in seaweeds varies depending on the seaweed species and marine biology, ranging from 20–40% of dry weight (Peters, 1972) (Figure 8.2).

A large amount of distinguished and crucial red seaweed Chondrus crispus (Irish moss) is commonly used for fabricating hydrophilic colloids to produce carrageenan. Chondrus crispus (Irish moss) grows attached to the rocks and is a dark red parsley-like plant. Carrageenans are commercially significant hydrophilic colloids (water-soluble gums) (Torres et al., 2021; Yang et al., 2013). These biomaterials possess half-ester sulfate moieties and are anionic. Commercial carrageenans are available as the steady salt of sodium, potassium, and calcium or as a mixture. The cations association and confirmed sugar units of the polymer chain define the physical properties of carrageenans. Protein reactivity expressed by gelling and non-gelling carrageenans exhibits polymer regularity (Synowiecki & Al-Khateeb, 2003). Carrageenan is a multifunctional component that behaves differently in water and milk systems. In water, it shows distinctive hydrocolloid properties of thickening and gelling, while in milk systems, it also has the property of reacting with proteins to furnish additional stabilizing abilities. Carrageenan can uniquely form various gel textures at room temperature: rigid or elastic, clear or turbid, tough or tender, heat-stable or thermally reversible, and low or high melting/gelling temperatures (Carrageenan – Uses, Side Effects, and More, n.d.). Carrageenan can be categorized by structure, chemical, or production or process properties such as:

TABLE 8.2 Applications of Marine-Derived Chitin and Chitosan Biomaterials and Their Characteristics (Younes & Rinaudo, 2015)

SI. No	Name	Application	Characteristics/Advantages
1.	Chitin	Biomedical and pharmaceutical materials	Sponges and bandages for the treatment of wounds and suture threats
2.		Agriculture	Defensive-inducing mechanisms in plants
3.		Water engineering	For decontaminated plutonium-containing wastewater and water containing methyl mercury acetate
4.	Chitosan	Biomedical and pharmaceutical materials	Antimicrobial properties, treating major burns, preparation of artificial skins, surgical sutures, contact lenses, blood dialysis membranes, artificial blood vessels, anticoagulant, antitumor, osteoarthritis. Ophthalmology, antioxidant.
5.		Cosmetics	Skin and hair care products
6.		Tissue engineering	Cell growth proliferation in tracheal cartilage
7.		Agriculture	Seed and food coating fertilizer and fungicides
8.		Enzymatic, Chromatographic media, and analytical	Immobilization of enzymes Gel permeation Chromatography
9.		Food and feed additives	Clarification and deacidification of food and beverages, Dietary fiber
10.		Water engineering	Wastewater treatment, recovery of metal ions and pesticides
11.	Chitin and Chitosan	Medicine and pharmacy	Wound healing activity, antiulcer activity, burns treatment.
12.		Agriculture	Controlling parasitic nematodes in soil
13.			

- **According to Structure**
  KAPPA – stiff and friable gel, High gel strength, thermo-reversible, and shows syneresis.
  IOTA – Thixotropic, elastic gel, thermo-reversible, no syneresis,
  LAMBDA – High viscosity, cold soluble, non-gelling.
- **According to the Production Process**
  SEMI-REFINED – Low degree of purity with. an opaque gel containing high cellulose and fiber levels.
  REFINED –Transparent and clear gel with a high degree of purity (Buck et al., 2006).

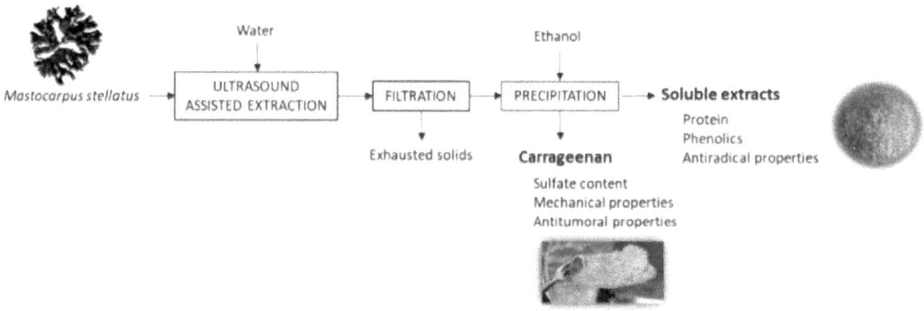

FIGURE 8.2 Flow diagram of the ultrasound-assisted extraction of carrageenan and bioactive from *Mastocarpus stellatus* species of red algae closely related to Irish moss, belonging to the *Phyllophoraceae* family.

Source: Reprinted with permission from Torres et al., (2021). Copyright © 2021 by the authors. Licensee MDPI, Basel, Switzerland. http://creativecommons.org/licenses/by/4.0/

The hydrophilic and anionic properties are applied for the application of carrageenans. They are used for thickening, suspending, or gelling; they are added to bodying, dispersion, syneresis control, binding, and emulsion stabilization and are predominantly used for application in foods and dairy. The properties of stabilizing, thickening, and gelling are considerably utilized in the food industry (Yegappan et al., 2018).

8.3.1.3 GLYCOSAMINOGLYCANS

Glycosaminoglycans (GAGs) are essential complexes of heteropolysaccharides spread in mammals' tissues. It comprises a basic structure of repetitive disaccharide units with a hexosamine (glycosamine or galactosamine) linked to a non-nitrogenated sugar, uronic acid (D-glucuronic acid or L-iduronic acid), or a galactose. It initiates several biological processes by the regulation of different protein factors (Norman Stanley, 1987). The GAGs found in marine organisms differ from terrestrial organisms, especially in terms of molecular weight and sulfation. They are long-chain polysaccharides divided into chondroitin sulfate, dermatan sulfate, heparan sulfate, heparin sulfate, and one non-sulfate group hyaluronic acid. They are acquired from animal tissue by costly lower-yield extraction process; regardless, they are positively valued polysaccharides and used in the biomedical market (Pacheco-Quito et al., 2020). GAGs composition of disaccharide, sulfation pattern, and chain length significantly vary depending upon the species and extraction factors. It includes antioxidant, immunomodulatory, antiproliferative, anti-inflammatory, antiviral, anticoagulant properties, and neuroprotective, serving as therapeutic agents modulating biological processes (Campo et al., 2009).

8.3.1.4 ALGINATE

Alginate acquired from brown seaweed is a naturally occurring anionic polymer. It has been analyzed and applied for several biomedical applications due to its biocompatibility, low toxicity, temperate gelation by an accumulation of divalent cations such as Ca^{2+}, and relatively low cost (Yegappan et al., 2018). Alginate is a biomaterial with applications in biomedical science because of its encouraging properties, including ease of gelation and biocompatibility. Alginate hydrogels are specifically used in drug delivery, tissue engineering, and wound healing applications. The gels sustain structural similarity to the extracellular matrices of tissues and hence can be used for critical roles (Zhang et al., 2010).

8.3.1.5 FUCOIDAN

Fucoidan developed from diverse species of brown algae is a long-chain sulfated polysaccharide. Commercially available fucoidan is normally extracted from the seaweed species *Cladosiphon okamuranus, Laminaria japonica, Fucus vesiculosus,* and *Undaria pinnatifida*. It has a rich record of medicinal and therapeutic use. Numerous health benefits of fucoidan include anti-viral, anti-cancer, wound healing, anti-inflammation, immune modulation, digestive health, and anti-aging applications (Zhang, 2010).

8.3.2 PROTEINS AND PEPTIDES FROM MARINE ORGANISMS

Marine proteins and peptides have pronounced prospects in developing different fields of pharmaceutical products, nutraceuticals, and cosmeceuticals. Peptides and proteins obtained from marine sources are considered harmless and economical. The use of protein and peptide-based drugs has recently been increasing for the treatment of various diseases by aiding multiple functions, such as antioxidants, anticancer drugs, antimicrobials, and anticoagulants. In addition, marine sources (macroalgae, fish, shellfish, and bivalves) may comprise definite proteins and peptides. Peptides are known for numerous biotic activities, including antiviral, antimicrobial, and antitumor activities, and as toxins (Cheung et al., 2015).

8.3.2.1 COLLAGEN

Collagen is a category of protein. It is the most abundant structural protein in animals. Consequently, aquatic organisms and their byproducts can be ecological, eco-friendly sources of collagen. A structural protein makes up the framework of cells and tissues. Collagen peptides are used as antioxidants to reduce the damage of free radicals and inflammation (both associated with chronic diseases and aging) in skin cells. In the cosmetic industry, aquatic collagen is one of the trendiest types, as it causes less inflammation than animal sources and is less likely to contain contaminants (Synowiecki & Al-Khateeb, 2003). Marine collagen is similar in amino acid content and biocompatibility to that of animals of porcine or bovine origin. Additionally, type I collagen is abundant in aquatic organisms. Collagen is an important component of pharmaceutical or food products and acts on connective tissues of the skin, ligaments, tendons, and teeth (Ricard-Blum, 2011). Collagen peptides are used as nutritional supplements to restore bone density in malnutrition or other degenerative diseases. The beneficial effects of collagen peptides on skin elasticity, such as reducing wrinkles or increasing collagen deposition in the dermis, have also been demonstrated in various clinical studies. In the cosmetic industry, marine collagen is effectively used for treating burns, wounds, and ulcers, as well as for antimicrobial protection, heat loss from injured tissue, and moisture prevention (Alexandra Benisek, 2021).

8.3.2.2 GELATIN

Gelatin is a colorless, flavorless, and translucent food component generally found in the collagen of animal body parts. It is rubbery when moist and brittle when dry. After hydrolysis, it is referred to as hydrolyzed collagen, collagen hydrolysate, and collagen peptides. It is usually used as a shell of drug or vitamin capsules, beverages, medications, and gelling agents in food, photographic films, papers, and cosmetics (Djagny et al., 2001).

8.3.2.3 MARINE ENZYMES

Marine microorganisms must withstand habitats characterized by exciting salinity, temperature, or pressure conditions. Their enzymes present associated structures, *viz.* thermostability or half stability, which are appealing for practical applications. The specific features accessible by marine enzymes can be beneficial for enlightening or developing new procedures and yields. The usage of marine enzymes can result in value-added or novel methods or products due to their nature. Thus, these enzymes can display the required, eventually novel, catalytic activity under operational conditions of temperature or salinity. This may provide an economical edge in either process or product development and readiness for the market. The microbially produced enzymes are important since microorganisms are easier and less expensive to cultivate than mammalian or plant cells and provide a higher yield and productivity in the targeted enzyme (Trincone, 2013; Zhang & Kim, 2010).

8.3.3 SECONDARY METABOLITES PRODUCED BY MARINE SOURCES AS BIOMATERIALS

Marine macro and microorganisms generate a dizzying array of secondary metabolites, which include alkaloids, steroids, terpenes, peptides, polyketides, and porphyrins. Marine secondary metabolites usually are molecules with varying molecular weights between 100 to 1000 Daltons and, unlike primary marine metabolites, are often established to be exclusive to an organism or a specific taxonomic group of the marine source. Other than marine macroorganisms like algae, sponges, or corals, specifically marine fungi and bacteria have been revealed to produce novel secondary metabolites with distinctive and varied chemical structures that may hold the key to developing novel drugs or drug leads (Srinivasan et al., 2021). Marine secondary metabolites are available in prokaryotic and eukaryotic microorganisms, unicellular bacteria (e.g., *Bacillus* sp., *Pseudomonas* sp.), eukaryotic fungi (e.g., *Penicillium* sp., *Aspergillus* sp.), filamentous actinomycetes (e.g., *Streptomyces* sp.), and terrestrial plants being the center of versatile producers. A potential clarification of why organisms produce a high assortment of bioactive secondary marine metabolites is that these molecules furnish producers with a specific advantage against competing organisms and act as an adaptation to ecological conditions. In addition, secondary metabolites can harbor numerous advantageous properties for human health, such as antioxidant, antiviral, antibacterial, antidiabetic, antitumor, anticoagulant, anti-inflammatory, and antihypertensive activities (Fahmy & Abdel-Tawab, 2021).

8.3.3.1 BROMOPHENOLS

Bromophenols (BPs), a group of brominated compounds, are found in marine organisms and are believed to be the reason for the typical sea-like taste and flavor. Numerous bromophenols are present in marine polychaetes and algae, such as *Padina arborescens*, the brown macroalgae *Sargassum siliquastrum,* and *Lobophora variegata*, with high concentrations up to 7000 ng/g (Kotterman et al., 2003). Marine algae contain various bromophenols that exhibit a range of biological activities, including antioxidant, antimicrobial, anticancer, antidiabetic, and anti-thrombotic effects. Bromophenols are widely used in pesticides, herbicides, wood protectants, and flame retardants, leading to frequent contact with these chemical products in daily life (Hassenklöver et al., 2006).

8.3.3.2 MYCOSPORINES AND MYCOSPORINE-LIKE AMINO ACIDS (MMAS)

Mycosporines and mycosporine-like amino acids are small secondary metabolites produced by organisms living in environments with high levels of sunlight, typically marine environments. For example, lichens, fungi, algae, and cyanobacteria produce them primarily when exposed to solar ultraviolet radiation. There are over 30 natural types, such as mycosporine-glycine, mycosporine-2-glycine, shinorine, palythine, porphyra, etc., known for their UV-absorbing and antioxidative properties. These amino acids are mainly found in marine species like phytoplankton, cyanobacteria, fungi, algae, and coral, serving as a natural defense against UV radiation (Oren & Gunde-Cimerman, 2007).

Mycosporines and mycosporine-like amino acids (MAAs) are colorless, water-soluble, UV-absorbing compounds characterized by small molecular mass and high molar extinction coefficients. They are small molecular-weight water-soluble molecules absorbing UV emission in the wavelength range of 310–365 nm. A wide range of microorganisms accumulates them, including prokaryotic (cyanobacteria), eukaryotic (microalgae, yeasts, and fungi), and a variety of marine macroalgae, corals, and other marine life forms (Singh et al., 2008). The role that MAAs play as sunscreen complexes to guard against harm by injurious levels of UV radiation is well established. However, accumulating evidence suggests that MAAs may have additional functions:

- They may help as antioxidant molecules scavenging toxic oxygen radicals.
- They are generally earning as compatible solutes following salt stress.
- Their formation will be induced by certain organisms' desiccation or thermal stress.
- They usually have a role as an accessory light-harvesting pigment in photosynthesis or as an intracellular nitrogen reservoir.
- They go well for helping in fungal reproduction
- MAAs represent high potential in biotechnological and cosmetics applications. Indeed, their UV-absorbing properties allow the creation of products derived from natural photoprotectors, basically harmless to the environment and efficient against UV damage (Rastogi & Incharoensakdi, 2014).

8.4 MARINE-DERIVED BIOMATERIALS IN TISSUE ENGINEERING APPLICATION

8.4.1 BONE

Surgery or prosthetic processes are the primary treatment worldwide for bone-related defects and fractures, such as osteoporosis, osteoarthritis, accidental cases, trauma, etc. The increasing age and poor lifestyles in developing countries have led to severe bone disorders. This disorder has made it the second most transplanted biological component after blood.

Medical techniques, such as xenograft, autograft, and allograft, show delayed recovery rates with other morbidities. Current clinical remedies reported for bone defects (e.g., autologous and allogeneic transplantations) using autografts and allografts demonstrate restrictions and painful complications. The structure and morphology of bone make it quite challenging for drug delivery and absorption for formulation scientists worldwide. The limitations of synthetic materials create a pressing need to investigate new materials to deliver drugs or act as carriers (Borciani et al., 2022; Lalzawmliana et al., 2019).

Several biopolymers, such as chitosan, alginate, collagen, etc., derived from marine bioresources, have proven their effectiveness for biomedical and tissue engineering applications. 3D-based scaffolds used as drug delivery carriers, developed from marine resources, have abundant mechanical and biological properties (i.e., bioactivity, biodegradability, renewability) for bone tissue engineering. In a recent study, Machałowski et al. developed a chitin-based biocomposite system deposited with calcium carbonate by applying loaded cultivated mollusk from *Cornu Aspersum* and chitinous matrix from *Aplysina fistularis* via chemical acid-base treatment for tissue engineering applications. The study report confirmed that the mechanical resistance and improved surface roughness were almost four times better for biofunctionalized scaffold when tested with human fetal osteoblast cell lines for the first time. The cell line study revealed the potential utilization and delivery of active molecules via a marine-derived matrix system (Machałowski et al., 2022). Another recent investigation by Alves AL and Team concluded the anti-inflammatory potentials of collagen and gelatin from blue sharks and codfish in the bone marrow, respectively, depicted via graphical illustration in Figure 8.3(i). The *in vivo* study in C57BL/6 mice after peritoneal injection of biomaterials resulted in high expression of Arginase 1, low expression of pro-inflammatory cytokines, nitric oxide synthase, etc., via shark collagen due to less immunogenicity. The increase in pro-anti-inflammatory expression, i.e., IL-10 and Arginase, and neutrophil recruitment in the abdominal cavity of mice was also reported, along with the biodegradability on-site after 24 hours of injection. Hence, the activation of M2 macrophages for the anti-inflammatory microenvironment by marine-derived collagen materials was proved, except for the functionalized collagen, which can be used for regenerative medicine based on clinical investigation (Alves et al., 2022). Furthermore, Correia et al. also developed blended chitosan collagen derived from fish skins attached via catechol groups. The characteristics of functional biomaterials improved, i.e., adhesiveness, biodegradability, and absorption, for application in tissue engineering. The composite membranes were compared regarding surface morphology, mechanical, and adhesive properties towards chitosan and blended with unmodified collagen. The membranes were assessed on cell viability, metabolic activity, and morphology on L929 fibroblastic cells. The study hence proved the improved adhesive, mechanical properties, and enhancement of cell attachment and proliferation of the functionalized collagen with catechol groups membrane, which can be utilized for the restoration of the morphological and functional properties of injured tissues (Correia et al., 2022).

The structural properties of bone improved by collagen have always drawn attention toward application in bone tissue engineering. Recently, several investigations proved that cellulose nanofibrils (TNC) composite with living cells from Tunicate exhibited higher mechanical strength in bone and cartilage tissue repair, similar to collagen. To understand its properties, Apelgren et al. investigated the mechanical and biological properties of an *in vivo* model of Norwegian fjords sourced Tunicates (TNC) derived from aquaculture. The study reported that the hydrogel comprised higher tensile shear, viscosity, and tunable length (2.4–3 μm) of TNC formed through a microfluidizer, further recommended for printability. The study further concluded that post-print cross-linking with alginate stabilized the hydrogel after 30 days. Moreover, the *in vivo* study model concluded no negative remarks on necrosis, further infection, invasion of neutrophil granulocytes, or extensive fibrosis via the hydrogel system. Moreover, the histochemical analysis confirmed the construction of lymphocytes, macrophages, etc., in the TNC constructs and PTFE controls. The suggested results open up a new path for the biomedical application of TNC structure for tissue engineering (Apelgren et al., 2022). Several recent investigations based on marine-derived biomaterials i.e., polydopamine/nano-hydroxyapatite (Yang et al., 2022), Poly (methyl methacrylate)/chitosan from Oyster shell (Rekha Sahoo et al., 2022), calcium phosphate cement from Mytilus edulis (Z. Xu et al., 2022), Whitlockite (WH) powder from Plankton (Baek et al., 2022), Polycaprolactone/Collagen/Carbonated Hydroxyapatite-based scaffold from fish bones (Kim et al., 2022), chitosan/collagen/β-glycerophosphate from fucoidan (Ohmes et al., 2022), etc., also showed potential interesting outcomes in the treatment of bone tissue engineering.

8.4.2 CARTILAGE

Synthetic biomaterials commonly associated with several limitations produce considerable challenges in cartilage tissue engineering. To address the challenges, Han et al. recently modified calcium silicate (CS) nanoparticles with natural fish scale (FS) with improved mechanical strength in tendon repair surgery. The study reported that the scaffolds of FS-CS NPs indicated excellent mechanical strength (125.05 MPa) and tangential stress (14.16 MJ/m^3) resistance compared to the natural tendon, leading to cuff tendon repairment. Furthermore, the bone marrow mesenchymal stem cells (BMSCs) activated by the BMP-2/Smad/Runx2 pathway enhanced the biomechanical function and bone interface regeneration due to the excellent bioactivity (Han et al., 2023), depicted in Figure 8.3(ii). Currently, hydrogel as a drug delivery system has attained the limelight in cartilage tissue regeneration. Hydrogel delivery system for tissue engineering via compressive and absorption methodology covers unique techniques while compacting biopolymeric solution and revoking excess water. Recently, Carvalho et al. developed an eco-friendly marine hydrogel using compressive and absorption methodology using collagen (derived from jellyfish and blue shark skin), fucoidan (derived from brown algae), and squid pens at different ratios. The physicochemical assessment exhibited excellent water uptake, a higher degradation rate with porous structure along with 18% mass loss of the porous structure over 30 days. The *in vitro* cell line study on the chondrocyte cells (ATDC5) confirmed increased cell encapsulation efficiency, cell proliferation, non-cytotoxic behavior, cell viability, etc., assuring the enriched microenvironment with a higher ratio of various marine-derived polymers suitable for biomedical articular cartilage therapies (Carvalho et al., 2022). Tunicates are marine-derived sulfated polysaccharide-based biomaterial enhanced with extracellular matrix, making it an ideal choice of biomaterial for cartilage tissue regeneration. To utilize its properties, Govindharaj et al. developed 3D printed hydrogels via tunicates derived from *P. nigra* for obtaining decellularized extracellular matrix (dECM) loaded with human mesenchymal stem cells (hMSCs) for cartilage tissue engineering, summarized the development via graphical representation in Figure 8.3(iii). The *in vitro* study confirmed the improved biocompatibility, metabolic activity, cell attachment, proliferation, and chondrogenic differentiation. The 3D printed scaffolds with biocompatibility and non-toxicity helped to bridge the gaps effectively with increased mechanical strength for cartilage regeneration (Govindharaj et al., 2022). The decellularized extracellular matrix (dECM) with stem cells derived from marine bioresources indicates a prominent effect on injured tendon tissue repair compared to the decellularized extracellular matrix (ECM). Recently, Liu et al. constructed a natural pore structure consisting of decellularized tilapia fish skin (DTFS) rich in collagen along with 1-ethyl-3-(3-dimethylaminopropyl)-carbodiimide (EDC) and N-hydroxysuccinimide (NHS) for improvement in the mechanical strength as depicted in Figure 8.3(iv) (A). The assessment of the cell line study of DTFS confirmed no cytotoxicity, promotion of cell viability, proliferation, and tenogenic differentiation of tendon-derived stem cells (TDSCs) as depicted in Figure 8.3(iv) (B). The effectiveness of collagen-based scaffolds in a rat Achilles tendon defect model made it be further considered as novel therapeutics for cartilage tissue repair (Liu et al., 2022). Several recent investigations on marine biomaterials such as collagen type II from porcine skin (Wu et al., 2021), *Prionace glauca* skin collagen (Diogo et al., 2021), alginate/dopamine/silk fibroin (Zhang et al., 2021), Gracilariasalicornia sourced →3)-2-*O*-methyl-*O*-6-sulfonato-β-d-galactopyranosyl-(1→4)-2-*O*-methyl-3,6-anhydro-α-d-galactopyranose-(1→/GS-AgNP (Chakraborty et al., 2021), *Acipenser schrenckii*-derived chondroitin sulfate (Wang et al., 2020), etc., have shown enough potential for cartilage tissue repair with no or minimal toxicity, higher cell proliferation and viability, collagen deposition *in vitro* and *in vivo* models.

8.4.3 CARDIAC TISSUE

Several cardiovascular diseases or disorders, i.e., atherosclerosis, stroke, heart attack, arrhythmia, heart valve problems, etc., specifically commence due to high lipid deposition in the blood vessel or clot formation in the blood vessel and hence to treat these conditions natural biomaterials came into the picture enclosing higher biocompatibility, low cytotoxicity, mechanical strength, increased biodegradation rate, etc. Several marine-derived lipidic substances, i.e., Eicosapentaenoic acid (EPA) and Docosahexaenoic acid (DHA), predominantly exhibited pharmacological actions in the human body in the 1900s. Several statistical reports also proffered that the high intake of marine resources, especially fish lowers the incidence of cardiovascular occurrence due to the deposition of omega-3 fatty acids (PUFA) in the body (Macha et al., 2019a). The marine-derived biomaterials possessing antioxidant properties that hinder reactive oxygen species (ROS) in pathological tissues are an immaculate preference for a disease condition, except the constraints include low biocompatibility, the bioavailability of antioxidant material, etc. To improve these limitations, Wang et al. developed a multi-omics approach to design thrombospondin type I (TSRL) protein from several (i.e., 1262 numbers of proteins) adhesive components and further utilized as a recombinant protein with a yield of 500 mg L^{-1}. The biocompatible TSRL macromolecule might self-assemble into a moisture-resistant coating via a Ca^{2+}-mediated reaction or redox-responsive hydrogels with tunable physical possessions. The study showed excellent protection against oxidative stress

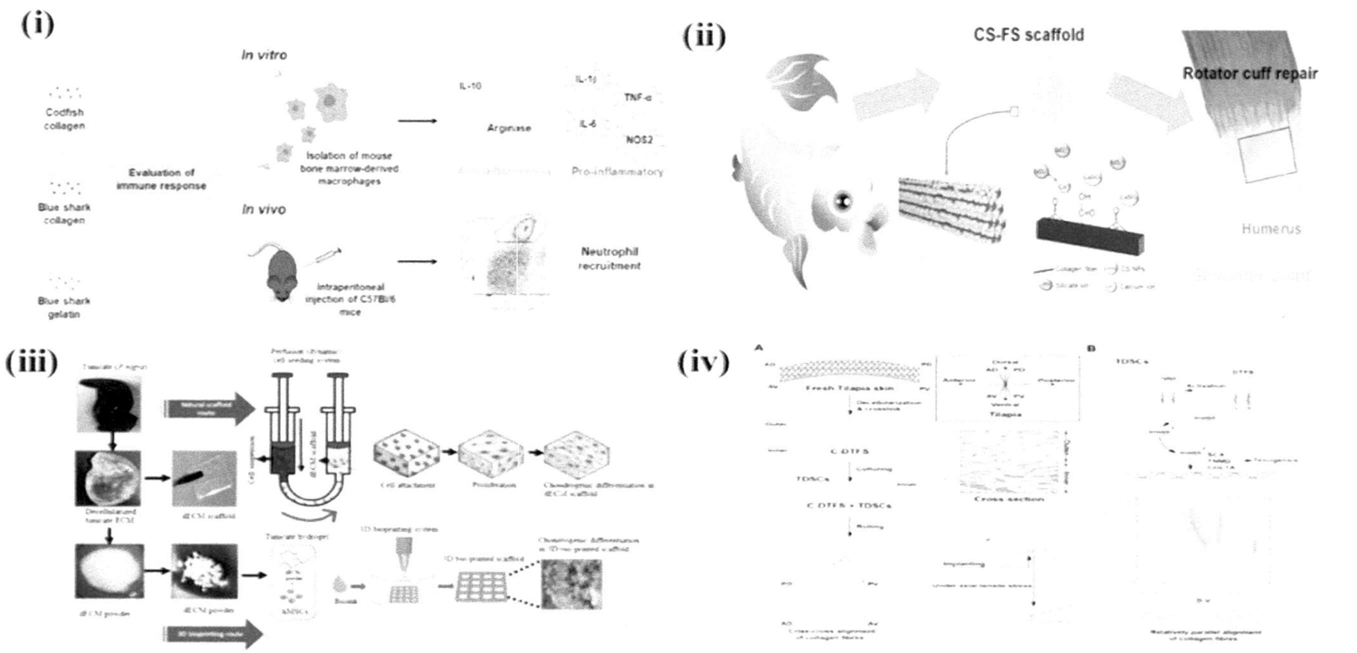

FIGURE 8.3 (i) Graphical representation of anti-inflammatory potentials of collagen and gelatin from blue sharks and codfish in the bone marrow regeneration via increasing in pro-anti-inflammatory expression and neutrophil recruitment (ii) demonstrated modified calcium silicate (CS) nanoparticles loaded natural fish scale (FS) scaffold (CS-FS) with higher mechanical strength in tendon repair surgery via activation of BMP-2/Smad/Runx2 pathway, showed improved bioactivity (iii) Graphical representation of 3D printed hydrogels via tunicates derived from P. nigra for obtaining decellularized extracellular matrix (dECM) loaded with human mesenchymal stem cells (hMSCs) for cartilage tissue engineering (iv) (A) Schematic illustration of decellularized tilapia fish skin (DTFS) rich in collagen along with 1-ethyl-3-(3-dimethyl aminopropyl)-carbodiimide (EDC) and N-hydroxysuccinimide (NHS) for improvement in the mechanical strength (B) Molecular mechanism underlying DTFS mediated tendon derivative stem cells proliferation and viability.

Source: i. Reprinted with permission from Alves et al., 2022. Copyright ©2022 Elsevier. ii. Reprinted with permission from Han et al., 2023. Copyright ©2023 Elsevier. iii. Reprinted with permission from Govindharaj et al., 2022. Copyright ©2022 Elsevier. iv. Reprinted with permission from Liu et al., 2022. Copyright ©2022 Elsevier.

via stronger 1,1-diphenyl-2-picrylhydrazyl (DPPH) radical scavenging rates of TSRL-based hydrogels than ascorbic acid and glutathione (GSH), summarized in Figure 8.4(i). The TSRL-based hydrogel was further operated on mice's skin and documented suppression of the degradation rate of elastic fibers and collagen-induced by ultraviolet radiation B (UVB), directing to augmented epidermal hyperplasia, further establishing the enhanced antioxidant activity. Therefore, the study findings ascertained that the first ever marine invertebrate possesses adhesive characteristics, further indicating the significant application in tissue engineering and cosmeceuticals (Wang et al., 2022).

Numerous functional nanomaterials with antioxidant attributes from marine sources also exhibit the potential for reducing vascular damage caused by ROS and avoiding off-target effects. To investigate further, Liu et al. assembled chitosan-based fucoidan nanoparticles (CFNs) derived from marine organisms, further bound with P-selectin, and showed anti-inflammatory characteristics by activating platelets, inhibiting leukocyte recruitment, and rolling on platelets. The authors summarized the *in vitro* study outcomes of CFNs, which demonstrated ROS scavenging abilities and binding affinity towards plaque (shown in Figure 8.4(ii)). However, the targeted delivery of CFNs as therapeutic molecules against atherosclerotic plaque needs to be more clearly identified. This study may open up unexplored opportunities for using marine-derived biomaterials for cardiac tissue repair, but at the same time, molecular-level investigations of marine derivatives are equally essential to establish their significance further (Liu et al., 2022). In another experiment, sodium alginate/chitosan-loaded clam heparinoid DG1 and curcumin (Cur) nanoparticles were found to enhance oral antithrombotic activity. The physicochemical characterization of the nanoparticles revealed a matrix structure via the polymers to encapsulate a high amount of DG1 and Cur, resulting in a smaller hydrodynamic diameter of the nanoparticles, high loading capacity, and a negative zeta potential (–20 to –30 mV), ensuring the stability of the NPs. The *in vitro* acute pulmonary embolism murine model further demonstrated that the polymeric matrix loaded DG1/Cur-NPs showed enhanced oral antithrombotic activity of DG1, but Cur alone can't achieve the same effect (Chen et al., 2022). Similarly, in a recent study, Zhou et al. also developed alginate/melanin NP-based hydrogel from cuttlefish ink and ocean algae, respectively, proposed for hydrogel development. The MNPs/Alg hydrogel was further investigated for cardiac tissue repair through regulation of the MI microenvironment and ROS induction of cardiomyocytes. Additionally, the developed hydrogel was found to be influential in the MI rat model by promoting polarization to regenerative M2 macrophages, as observed significantly (Zhou et al., 2021). The literature also reveals limited investigations on marine-based biomaterials in cardiac tissue regeneration or engineering due to challenges associated with limited animal models and toxicity assessment models.

8.4.4 LIVER

Tissue engineering utilizing biomaterial scaffolds has captivated many scientists and doctors because it strives to reconstruct living tissues or replace lost or damaged tissues. Furthermore, the engineered tissues can facilitate or eliminate the need for organ replacement. Thus, an extracellular matrix (ECM) should be necessarily prompted for the cells/tissues to adhere, proliferate, and differentiate as they are in their aboriginal tissue. Generally, artificial ECMs have been engaged in tissue engineering due to preferred cell types in contact with an appropriate environment, delivering mechanical aid until the newly developed tissues are structurally stable and current and store bio-signaling molecules. Naturally derived polymers have excellent biological properties, such as cell adhesion, mechanical properties, biodegradability, and biocompatibility, which correspond to synthetic polymers. However, they have some limitations, such as reproducibility on a large scale, complexity of purification, and tribulation in stemming degradation time. The design of natural polymer-based artificial ECMs for liver tissue engineering is critical because the liver plays essential roles in our body, such as metabolism, storage, synthesis, and release of carbohydrates, lipids, vitamins, proteins, inactivation of endogenous and exogenous substances and activation of precursor molecules. Also, hepatocytes are anchorage-dependent parenchymal cells within the liver that lose their liver-specific conditioning and viability in the lack of an ideal artificial ECM (Jiang et al., 2015).

In recent years, there has been a significant amount of empirical research on the development of novel biomaterials from marine sources. However, the biocompatibility and toxicity of these biomaterials remain a challenge for most formulation developers in tissue engineering applications. Therefore, to understand the toxicity assessment, Sumayya and Muraleedhara Kurup (2018) researched the *in vivo* biocompatibility and toxicity of biocomposite scaffolds loaded with hydroxyapatite/alginate/chitosan/fucoidan (HACF) in a rat model. The hematology and serum biochemistry investigation confirmed the activation of pro-inflammatory enzyme markers, tracked by lipid peroxidation, was common in scaffold-based implanted rats. The implanted scaffold showed a modulatory effect on stress signaling and inflammatory cascades, confirming the compatibility of the biomaterials in liver tissue engineering. In another study, Hiller et al. optimized the biocomposite loaded with alginate/gelatin/extracellular matrix from marine resources to bioprint human HepaRG liver cells using a pneumatic extrusion printer. The tissue model that resulted underwent testing to confirm its suitability for research into human adenovirus 5 (hAdV5) infection and transduction by an adeno-associated virus (AAV) vector. The study resulted in improved printability, cell viability, and sustained metabolic processes of the printed HepaRG cells using bioink consisting of alginate (0.5 mg/mL) and gelatin (1 mg/mL). The AAV vectors of serotype 6 successfully suppressed cyclophilin B through RNA induction after efficiently transducing the tissue models, summarized via graphical representation in Figure 8.4(iii). Furthermore, the 3D-printed device also allowed for adequate adenoviral replication, enabling generous research in virus biology and the creation of new antiviral drugs (Hiller et al., 2018).

To comprehend the cytocompatibility of marine-derived biomaterials in intestinal replacement for massive bowel resections, Shabafrooz et al. scrutinized hyaluronic acid (HA)/gelatin/collagen-based scaffolds on Caco-2 cell line *in vitro*. The characterization study defined the porous structure of the biomaterial-based scaffolds, further improving the swelling and degradation rate by adding HA. The incorporation of 15% HA on scaffolds with the composites (Gelatin: collagen, 9:1) resulted in the enhancement of structural characteristics and cytocompatibility on the Caco-2 cell line, whereas the addition of 25% HA could further improve the physiochemical features of the scaffolds, including cell viability, proliferation, etc., which may be necessary for hepatocyte tissue recovery (Shabafrooz et al., 2014). The utilization of marine biomaterials has increased in several tissue engineering sectors due to remarkable biological and unexplored biochemical properties. However, the limited clinical applications of the same biomaterials lead to non-industrialization and problems associated with large-scale extraction or isolation techniques. Therefore, in recent years, there have been very few investigations related to marine biomaterials in liver tissue injury or regenerative systems, but *in vitro*, studies could provide a valuable opportunity to develop market-oriented products with marine biomaterials for liver tissue engineering (Zhang et al., 2022).

8.4.5 PERIODONTAL

Periodontal regeneration incorporates interdisciplinary strategies for re-establishing the cementum, alveolar bone, and periodontal ligament surrounding teeth. Periodontal regeneration needs the right neovascularization

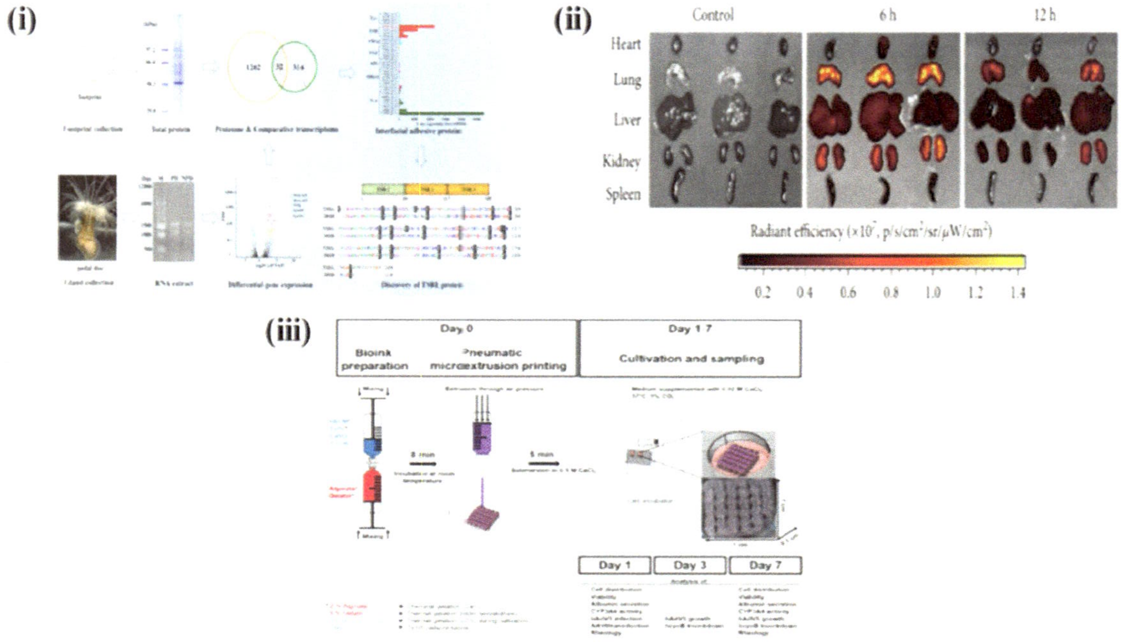

FIGURE 8.4 (i) Discovery of the TSRL protein from complicated sessile marine organism adhesives. SDS-PAGE exhibited multiple protein bands in footprints, suggesting the presence of unwanted ectc dermal cellular proteins. In parallel, comparative transcriptome analysis identified 314 transcripts that were differentially upregulated in the pedal disc (PD: pedal disc; NPD: remaining body tissues). Among the upregulated genes, 32 were identified in the footprints. The bar graph of candidate expression levels in pedal discs shows that c69606_g2_i4 (named TSRL) exhibited elevated expression. The TSRL protein contained a high portion of cysteine residues (13%), which was hypothesized to endow the protein with antioxidant properties. Sequence alignment further showed that the TSRL protein had high sequence identity (46%) with an ECM component (3R6B, human TSR), indicating satisfactory biocompatibility for tissue regeneration. (ii) Chitosan-based fucoidan nanoparticles (CFNs)/P-selectin accumulation and showed anti-inflammatory characteristics in mice via ex vivo fluorescence imaging+? (iii) Schematic representation of 3D printing associated cell-laden hybrid alginate/gelatin/human extracellular matrix (hECM) via a pneumatic extrusion printer on various days. *Detailed explanation of the function that the components of the bioink have.

Source: i. Reprinted with permission from Wang et al., 2022 Copyright ©2022 Elsevier. ii. Reprinted with permission from Liu, et al., 2022. Copyright ©2022 Hindawi. iii. Reprinted with permission from Hiller et al., 2018. Copyright ©2018 MDPI.

niches, signal molecules, proliferation, and differentiation of the regeneration-related cells. It has been noted that several biomaterials, technologies, and methods can create settings conducive to regeneration. Several innovative approaches for developing bone/tissue regenerative scaffolds, injectable collagen-based hydrogel, collagen sponges, etc., from marine biomaterials have received significant attention in periodontal regeneration. The regeneration process in periodontal tissue engineering is mainly initiated with angiogenesis, proliferation, and differentiation of cell-attached, sufficient molecular signaling pathways toward the damaged tissue can be countered via marine-based collagen biomaterials sourced from animal to plant. Additionally, adequate blood supply is mandatory as the signal molecules in the regeneration process are required. Several investigations further promoted the recruitment of IL-8, insulin-like growth factor 1 (IGF-1), VEGF, platelet-derived growth factor AA (PDGF-AA), monocyte chemoattractant protein 1 (MCP-1), etc., which further promote the induction of endothelial lineage during neovascularization process (Binlateh et al., 2022), depicted via Figure 8.5(i). Recently, Kibar et al. investigated partially purified marine-derived exopolysaccharide (EPS) produced by *Weissella cibaria* EIR/P2 for periodontal regeneration. The advanced analytical instruments confirmed the presence of glucose, dextran, and other polysaccharide derivatives with a molecular mass of 8×10^6 Da interpreted by gel permeation/size exclusion chromatography. The marine biomaterial was further investigated for *in vitro* antibacterial activity against Streptococcus mutans and reported 70% biofilm inhibition via dextran (at 50 mg/mL). The cell line-based MTT assay further confirmed the viability and proliferation of human periodontal ligament fibroblast cells (hPDLFCs). Therefore, *W. cibaria*-elicited dextran could be an adequate multifunctional biomaterial in other tissue engineering applications if investigated *in vivo* and clinical aspects (Kibar et al., 2020). In another study, Echazu MIA et al. formulated *Larrea divaricata* Cav. extract loaded in pH-responsive biopolymer-silica (Chi-CMC-SiO_2) composites for treating gingivoperiodontal diseases. The *in vitro* investigations of the composite system reported maximum encapsulation efficiency of the extract, extract release within 4 days, and a perfect microenvironment for biomineralization, followed by swelling, thermal behavior, and mucoadhesiveness. To understand the cytotoxicity behavior of the extract and composite system, the authors investigated a cytotoxicity study on 3T3 fibroblast cell lines and showed eminent proliferation via extract-loaded composites. Therefore, this study may establish the importance of bioactive molecules from marine sources, which could be further proven in periodontal tissue regeneration if investigated in animal models (Alvarez Echazú et al., 2018). The recent investigations on marine or plant-derived biomaterials application in periodontal tissue regeneration have been reviewed less due to stability issues of periodontal tissue for long-term use, alveolar bone loss which can't be restored, Sharpey's fiber structure, fast dissolution rate, and poor structural stability, etc., and hence novel methods have to be incorporated for a prerequisite to achieve periodontal tissue regeneration, both structurally and functionally (Binlateh et al., 2022; Sheikh et al., 2017; Varghese et al., 2020).

8.4.6 SKIN

Peptides/proteins administered via the oral route are advantageous as they extend excellent efficacy and have fewer side effects. Recently, self-assembling supramolecular active polypeptides (APs) derived from marine shellfish meat for *in vitro* and *in vivo* (via oral route of administration) were prepared by controllable enzymatic (animal protease) hydrolysis technology. The *in vivo* study further demonstrated that the oral administration of APs could inhibit inflammation, and promote fibroblast proliferation and revascularization, leading to epithelialization, collagen deposition, and hair follicle regeneration to achieve scarless healing, as pictured in Figure 8.5(ii). The results confirmed that the *in vivo* study in mice promoted dermal wound healing that could be utilized for the treatment of skin wounds (Yang et al., 2022). Wound healing is generally hindered due to microbial infections and slow regression of inflammation. Lu et al. derived the skin of tilapia fish loaded into a gelatin-fucose gum-tannic acid (Gel & Fuc-TA) based hydrogel system for wound dressing and tissue regeneration application. The hydrogel exhibited an excellent network structure, antibacterial and antioxidant activities, cell compatibility, and swelling and release properties. Furthermore, it can also facilitate the platelet endothelial cell adhesion molecule-1 (CD31), enhance collagen deposition, alpha-smooth muscle actin (α-SMA), and vascular endothelial growth factor (VEGF), and directly promote/accelerate wound repair. The hydrogel further improved the expression of microbiome-related proinflammatory factors, such as lipopolysaccharide (LPS), Toll-like receptor 2 (TLR2), and Toll-like receptor 4 (TLR4), which further lowered IL-6, IL-1β, TNF-α. The hydrogel may also help increase the Arg-1, IL-10, and TGF-β, leading to the reduction of inflammatory response. Hence, they concluded that the green cross-linked hydrogel might help promote wound healing (Lu et al., 2022).

Similarly, Rozan HE et al. prepared a hydrogel by loading diatom biosilica (DBs) and doxycycline (DOXY) coated with hydroxybutyl chitosan (HBC) for wound healing. The hydrogel composite showed significant inhibitory activity against *E. coli* (inhibition rate: 98%) and *S. aureus* (inhibition rate: 100%). The *in vitro* study on L929 cells confirmed minimum cytotoxicity, indicating excellent biocompatibility of the composite hydrogel, whereas the *in vivo* report established the wound re-epithelialization and acceleration of healing, as shown in Figure 8.5(iii). After the 12th day of treatment with the hydrogel, the wound closure was found to be $99.4 \pm 0.4\%$ with proper collagen deposition, and the presence of neovascularization indicates the potential of HBC/DBs/DOXY hydrogel in wound healing (Rozan et al., 2022). Jana et al. summarized the two significant factors for the promotion of wound healing, i.e., angiogenesis and ECM re-construction under diabetic conditions, and formulated electrospun microfibers by combining Rohu fish skin-derived collagen (Fcol) with ion-doped bioactive glass (BAG). The microfibrous mats exhibited a higher damping factor and non-uniform tan-delta value due to viscoelastic properties as measured by a nano-DMA test. The maximum tensile strength and toughness are observed with bioactive glass microfibers, while microfibers demonstrate the lowest values with cobalt dopant. *In vitro* study on human dermal fibroblasts (HDFs) cultured over the microfibers for 48 h showed excellent cell-cell and cell-material interactions. For *in vivo*, a rabbit model with a full-thickness diabetic wound was used with the mats, revealing the wound's early healing. The microfibers-treated injuries led to a significant increase in the density of blood vessels by CD-31 than control, Duoderm rather than Fcol/BAG treated wounds. The fish collagen/ion-doped bioactive glass microfibers treated wounds indicated mature collagen deposition and excellent ECM remodeling signifying the positivity in diabetic wound healing (Jana et al., 2022).

Among several other wounds, burn wounds are one of the most common skin injuries with a high mortality rate due to the affected surface area and severity of the wound. The healing process occurs naturally, but prolonged healing periods might lead to chronic wounds. For rapid recovery and prevention of impairments, several dressings have been designed. The high biocompatibility, biodegradability, and lack of toxicity of marine biopolymers have led to the development of novel biomedical systems. Figure 8.5(iv) demonstrates a second-degree burn model of SKH-1 hairless mice treated with the formulated electrospun nanofibrous matrices made up of ulvan and marine gelatin at various compositions for the characterization of wound healing efficacy. The synergistic effect of hydrolyzed collagen/ulvan with silver nanoparticles was also evaluated. The healing efficiency for burn wounds depends on histopathological analysis and parameters such as transepidermal water

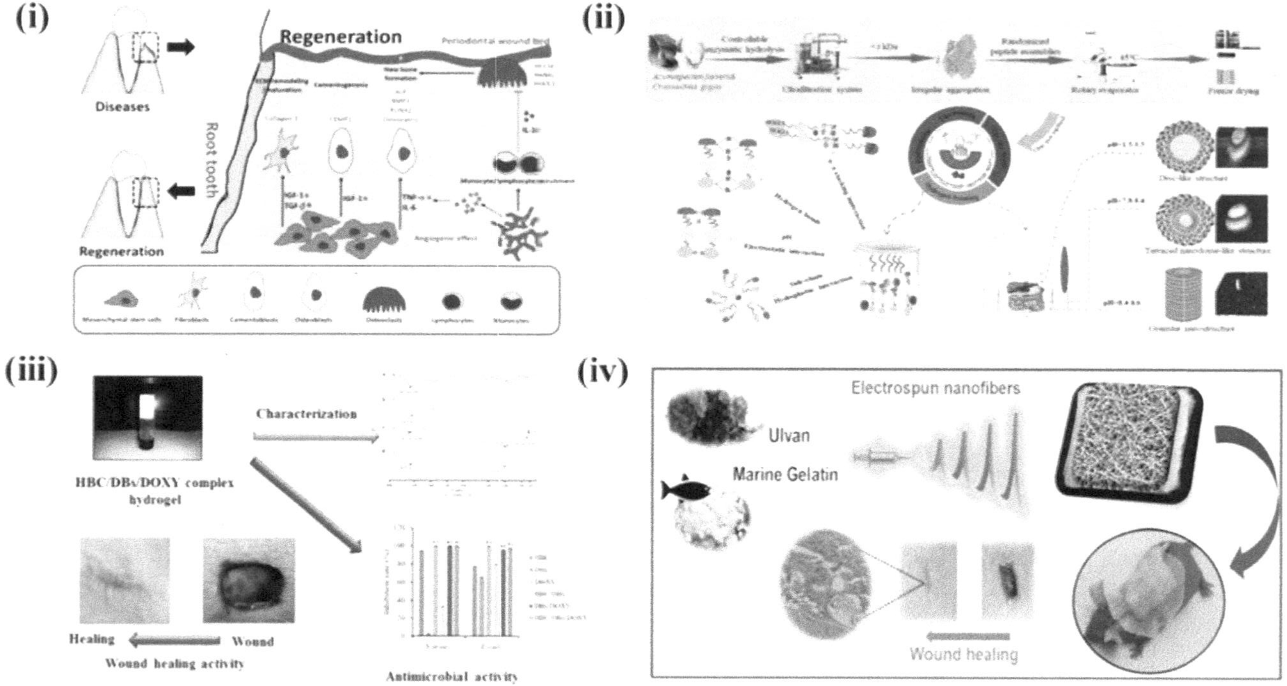

FIGURE 8.5 (i) Schematic illustration of Mesenchymal stem cells (MSC) initiated mechanisms of periodontal regeneration process. MSC induced angiogenic effect, secretion of inflammatory cytokines (TNF-a and IL-6) and growth factors (IGF-1 and TGF-b), and recruitment of immune cells (monocytes and lymphocytes). The immune cells release IL-10 to suppress osteoclast activation in homeostatic maintenance of new bone formation. (ii) Schematic illustration of self-assembling supramolecular (APs) peptide from marine shellfish for skin tissue regeneration. (iii) Graphical representation of Diatom biosilica (DBs)/hydroxybutyl chitosan (HBC)/doxycycline (DOXY) loaded hydrogel for wound healing. (iv) Graphical representation of Ulvan/marine gelatin loaded electrospun nanofibrous matrices application in second-degree burn wound model on SKH-1 hairless mice.

Source: i. Reprinted with permission from Binlateh et al., 2022. Copyright ©2022 MDPI. ii. Reprinted with permission from Yang, et al., 2022. Copyright ©2022 Elsevier. iii. Reprinted with permission from Rozan et al., 2022. Copyright ©2022 Elsevier. iv. Reprinted with permission from Terezaki et al., 2022. Copyright ©2022 Elsevier.

loss, wound area, skin thickness, texture, and hemoglobin. Hence, the study reported that ulvan and marine gelatin-composed patches develop a rapid rate of wound contraction in the early stages by suppressing inflammation, acquiring uniform wound closure, and exhibiting effective therapies for burn wounds (Terezaki et al., 2022). Cyanobacteria-produced exopolysaccharides (EPS) are anionic complex biomolecules with possible biomedical applications. In the current study, *Nostoc* sp. strains (e.g., PCC7936 and PCC7413) produced EPS and were evaluated as unique biomaterials for wound dressings. The culture medium introduced with acetate ions stimulated EPS production, attaining PCC7413 and PCC7936. The culture mediums depicted nine monosaccharide residues with MW > 1000 kDa, yielding changes in the molar percentages. The anionic nature and sulfate groups in both EPS were confirmed by results exhibited in FTIR and DLS, which are essential features of biomedical applications. The MTT assay and scratch assay further confirmed the increased biocompatibility and fibroblast migration and proliferation of hydrogels in PCC7936. Hence, *Nostoc* EPS hydrogels indicated favorable properties for application in skin injuries (Alvarez et al., 2021).

8.4.7 OTHER APPLICATIONS OF TISSUE ENGINEERING

Several past investigations have shown the efficacy of marine biomaterials as 3D printable subjects which can enable the precise fabrication of complex structures consisting of multiple cells for delivery. Collagen, which is the most common marine-derived biomaterial, was seen to be incorporated with gelatin to improve the mechanical properties, but the degradation rate seems to be unfit if this process has been done (Adamiak & Sionkowska, 2020). Traditional approaches have also further shown cytotoxicity due to the presence of crosslinking agents, whereas physical methods have also reported denaturation as the most common limitation. Therefore, Sanz et al. tried to develop photo-cross-linkable collagen methacrylate (ColMA) via coupling of methacrylate groups to the side chains of collagen for better biocompatibility and mechanical strength in tissue engineering applications. Therefore, this preliminary study could further promote the coaxial printing of neural and skeletal muscle cell cultures for neuromuscular junction (NMJ) regeneration with low cost, scalability, and customization for a wide range of drug screening to study neuromuscular junction physiology (Sanz et al., 2020).

The tissue regeneration in the corneal endothelium has shown great challenges. Recent advancement was also performed on patients via Descemet stripping automated endothelial keratoplasty (DSAEK) but due to transfusion limitations, it has been rejected by the donor. Therefore, Li et al. developed a fish-derived collagen membrane loaded with CECs for corneal endothelial transplantation. The extraction of collagen from fish was initiated with decellularization, and decalcification, followed by purification and washing. The analytical instruments further gave precise results on the microsporic structure and thermal resistance behavior of the cellulose. The cell line study confirmed the viability and proliferation, and no cytotoxicity was reported when evaluating various parameters on LDH and WST-1. The study also reported an *In vivo* model experiment on rabbit anterior chambers and showed excellent water content, biocompatibility, and transparency after investigation. Therefore, based on the above results, it may be concluded that FSCM/CECs have high potential efficacy for endothelial keratoplasty transplantation (Li et al., 2020). In another study, salmon-derived collagen-based gel loaded with human umbilical vein endothelial cells (HUVECs) was developed to understand the effect of culturing cells/tissues for vascular tissue engineering. The cell-containing gel was further evaluated for DNA content, MTT assay, and physicochemical characterization demonstrated the cell number increased with culture time under static conditions. The study has also shown that IL-8 and IL-6 productions of the cells significantly increased under stretch conditions than static conditions. Therefore, this approach can be further investigated for an *in vivo* model to understand the proliferation of the cells in biological conditions (Kanayama et al., 2008). Currently, there are very limited investigations made on marine-derived biomaterials on corneal tissue regeneration due to limitations associated with the incorporation of therapeutic cells into the adhesive biomaterial. Further limitations also include the suturing of corneal implants. Therefore, to address these issues, Koivusalo et al. developed dopamine-loaded hydrazone crosslinked hyaluronic acid (HA-DOPA) based hydrogel surface modified via proteins/peptides to increase retention time. The hASCs encapsulated in HA-DOPA hydrogel exhibited excellent tissue adhesion upon implantation in a porcine corneal organ culture model. This approach further bypasses the suturing process in the cornea, while the delivery of stem cells can be possible via this excellent next-generation biomaterial if investigated at the clinical level as well (Koivusalo et al., 2019).

8.6 CLINICAL PERSPECTIVE OF MARINE BASED BIOMATERIALS IN TISSUE ENGINEERING APPLICATION

Although *in vitro* models have made significant strides toward simulating the complexity and cellular interaction found in tissues, *in vivo,* testing is still necessary to carefully examine newly developed marine-based materials' biological performance when implanted in a natural system. It is essential to characterize this reaction at the cellular and molecular levels properly, and this has been part of substantial research over the past few decades (Costa-Pinto et al., 2016). Animal models are incredibly beneficial in the biomedical field, particularly for demonstrating specific natural responses, such as cell biology/delivery, genetics, immune responses to infection and cancer, drug and vaccine development, tissue engineering, and regenerative medicine using novel biomaterials. Due to the phylogenetic similarity between humans and other animals, the studies conducted *in vivo* provide a chance to examine and forecast any implanted or injectable material's reaction to a clinical scenario that mimics human physiology in a controlled environment (Carvalho et al., 2021a). Various animals, including rats, mice, horses, rabbits, primates, pigs, cats, and dogs, among others, have been employed in this context. Mainly when testing biomaterials using the TERM approach, it is essential to take into account several critical factors, including the biomaterial's shape and size, concerning the chosen animal model, including age, species, sex, genetic background, etc., ethical aspects, quality, & certification of the animal facility where the surgical procedures will be carried out (Costa-Pinto et al., 2016; Wan et al., 2021).

Holothurin was first used as an anti-tumor drug in *in-vivo* cancer models in 1967 after being isolated from marine animals. This led to the beginning of the usage of marine biomaterials. As a result of this procedure, four significant findings have sparked extensive studies on marine biomaterials. Studies on polysaccharide-based polymers, coral bone graft alternatives, and the utilization of collagen in tissue regeneration are included in the first and second phases, respectively. The final stage of the development of marine-derived biomaterials is marine-derived composites. At this point, cutting-edge strategies like nanotechnology and three-dimensional (3D) printing have piqued scientists' tremendous interest (Mazlum et al., 2022; Sharifian & Homaei, 2022). Before the biomaterials are approved for use in clinical settings, they must first undergo an initial screening process that verifies both their safety and the safety of any degraded products produced when they come in contact with host tissue (Azab et al., 2007). This assessment has historically been carried out using subcutaneous, intraperitoneal, and intramuscular rodent (mice and rats) implant models (Costa-Pinto et al., 2016). The *in vivo* technique entails developing lesions in the cartilage tissue, such as on the knees of animals, and applying the developed biomaterial from the marine source to the injured tissue for a predetermined time period (commonly 8–12 weeks). The end goal entails killing the animals and assessing the marine-based structures' capacity for regeneration as well as monitoring immune/inflammatory responses, vascularization, and the rate of biomaterial degradation through currently employed methods, such as X-ray microtomography and histological investigation, among others (Carvalho et al., 2021a; You et al., 2018).

Ionic/chemical gelation and freeze-drying have been utilized most often in the *in vivo* studies, which is similar to what was shown for *in vitro* studies, followed by 3D bioprinting, which employs alginate, chitosan, and collagen; has significantly increased over the past 10 years (Bianchi et al., 2017; Carvalho et al., 2021a). The most conclusive findings to date highlight the ability of these systems to repair damaged articular cartilage and demonstrate their capacity to sustain the fundamental needs of cells, including the ability to support proliferation and differentiation. However, as they allow for less intrusive and painful treatments, it is required to confirm the biomaterials with a few more *in vivo* investigations using animal models which are complex, and to develop the systems that can be used as injectable materials (Carvalho et al., 2021a; Yang et al., 2019). Using subcutaneous, intramuscular, and intraperitoneal murine models, the impact of anti-inflammatory and inflammatory mediators on host tissue at the implant site can be evaluated. These mediators include nitric oxide, tumor necrosis factor-alpha (TNF-a), interferon-gamma (IFN-gamma), vascular endothelial growth factor (VEGF), and fibroblast growth factor (FGF). Inflammatory tissue response, foreign body reaction (FBR), macrophage phagocytic potential, potential giant cell development, creation of fibrotic capsule, and width, as well as vascularization in the neighboring tissue at the site of implantation are typically examined factors to be accessed via animal models (Carvalho et al., 2021a; Yang et al., 2019).

A study reported that Sinularin (80 mg/kg, intraplantar) exhibited analgesic benefits in rats subcutaneously administered. It also reduced carrageenan-induced spinal neuro-inflammation, microglial and astrocyte activation, and iNOS upregulation in the lumbar spinal cord's dorsal horn. In addition to that, the administration of sinularin (80 mg/kg) demonstrated its analgesic action by upregulating TGF-1 and inhibiting the leukocyte infiltration caused by carrageenan. At 10 M, the C18 terpene nanolobatolide from the soft coral Sinularian anolobata Verseveldt reduced the iNOS buildup in microglial cells treated with INF to 45.5%. Another study examined the effects of marine sponge scaffolds impregnated with Insulin-like Growth Factor-1 and Bone Morphogenetic Protein-2 on the *in vivo* performance over a period of 90 days. The researchers have shown that this specific group of sea sponges could be an excellent natural resource of cristoballite with an acicular and extremely high aspect ratio that is non-cytotoxic and biocompatible. Their findings indicated that compared to bare coral scaffolds, growth factor-laden sponge skeletons could speed up early bone development and implant penetration during *in vivo* characterizations (SEM, histology, radiography, and fluorochrome labeling). This study also suggested that a novel source of natural bio-scaffold for repairing bone deformities is a marine sponge's skeleton. The sea sponge skeleton's microstructure, size of a pore, permeability, interconnected pore network, and composition contribute to the process of osteoblastic cellular attachment. Its role as a novel scaffold source for the regeneration of bone and as a local carrier of growth factors for bone repair is indicated by the existence of osteoinductive and conductive frameworks as well as by the structural diversity and the abundance of natural marine sponge skeletons (Bhattacharya et al., 2013; Salgado et al., 2011). Three groups of male Wistar rats (12 weeks, 300–350 g) were randomly assigned. Animals were divided into two subgroups with various euthanasia times (15 and 45 days). The animals were kept in a controlled setting with the availability of food and water, a constant temperature of 22°C, and 12-hour light/dark cycles. Rats were given anesthesia for the surgical operations using a mixture of fentanyl (0.05 mg/kg), xylazine (8 mg/kg), acepromazine (1 mg/kg), and ketamine (80 mg/kg). The skulls of the immobilized animals were shaved, cleaned, and povidone-iodine disinfected. A skin incision was made to expose the bone, and the periosteum was removed using aseptic procedures. Under extensive saline irrigation, a trephine drill defect measuring 8 mm was made in the parietal area. In the generated defect, the implants were inserted using a randomization strategy. The skin was then sutured with nylon after the wound was closed with biodegradable Vicryl 5-0. In conclusion, this investigation showed that BS/SPG scaffolds were biocompatible as well as capable of supporting bone growth in a significant bone defect of the rat. Additionally, BS/SPG altered VEGF's immune expression (Parisi et al., 2020). Similarly, a 3D bioprinting tissue scaffold system was studied in which Alginate was obtained from the Human mesenchymal stem cells (hMSCs) and was encapsulated, which showed a significant improvement of the chondro-inductive potential and also promoted the chondrogenesis by MSCs (Delplace et al., 2021). The studies have also shown that alginate obtained from the Chondrocyte cells (ATDC5) was encapsulated, it did not affect the biological behavior of cells and contributed to cell proliferation (Izadifar et al., 2016). However, when the same encapsulated Alginate from the Human chondrocyte cells showed high viability and proliferative capacity with lower apoptosis levels after bioprinting. Alginate obtained from the

Chondrocytic cells and human osteoblast cells (MG63) was encapsulated to show the printed cells' viability maintained for up to 7 days with almost nil decrease (Shim et al., 2012).

The spinning technique was carried out for the marine biomaterials, and the researchers conducted *in vivo* studies. The Polyethylene Glycol-Polyvinyl alcohol-alginate was obtained from the type B synoviocyte cells (SFB), which were similar to fibroblasts, and also from the Chondrocyte cells and underwent seeding to show cytocompatibility and the promotion of chondrogenic differentiation with improved cell adhesion (Chen et al., 2010). Alginate-hyaluronic acid was procured from the Human immortalized chondrocyte cell line (C28/I2) and was encapsulated to show spherical morphology, and cytocompatible (Mohabatpour et al., 2016). Alginate extracted from the chondrocyte cells (HTB-94) was assembled to undergo seeding, which exhibited good uniformity, structural integrity, and cellular compatibility. The alginate-based scaffolds are cheap and easy to synthesize, process, and scale up (Bhattarai et al., 2006). Supercritical fluids were used as a tissue-scaffolds system; shark collagen was extracted from chondrocyte cells (ATDC5), and seeding was accomplished to obtain an allowance of cell adhesion, growth, and proliferation as a result (Fernandes-Silva et al., 2013).

The ionic/chemical gelation was preferred as the tissue-scaffolds system for assessing marine-origin biomaterials, which were studied *in vivo*. Polyvinyl alcohol (PVA)/chitosan and alginate were implanted into rabbits, and the ability of the blended biomaterial to repair damaged tissue is significantly superior to that of the single polymer system (Dashtdar et al., 2015). Carrageenan/fibrin/hyaluronic acid was injected into mice's bodies to show that it can repair damaged tissue and has the advantage of a minimally invasive technique (Pereira et al., 2009). Hydroxyapatite-alginate was administered into mice by implantation technique to show that *in vitro* and *in vivo* studies stimulate chondrocytes to secrete mineralized matrix (You et al., 2018). Rabbits were used as the animal model for nano hydroxyapatite-chitosan-gelatin was administered with an injectable type of surgery that provided good osteoinduction for the regeneration of subchondral bone lesions and preserved the native cartilage (Wang et al., 2017). The rabbits underwent an implantation type of surgery with alginate, which showed the induction of regeneration of damaged knee tissues (Mata et al., 2017). Alginate, which was administered to the white rabbit, also underwent implantation surgery to show that biopolymer-containing growth factors can generate significantly more new cartilage (Stevens et al., 2004). Dogs were taken as a proper model for the study of Agarose by implantation, which showed that there were no obvious biocompatibility issues or cytological evidence of an inflammatory or immune response (Ng et al., 2010). Agarose injected into rabbits showed histological and morphological evaluation revealing a considerable improvement in restoration capacity (Rahfoth et al., 1998). Again, mice were used as an animal model, which showed cartilaginous tissue that is stable and regenerates *in vivo* after subcutaneous implantation surgery with agarose (Vinardell et al., 2012).

For the subsequent study, the freeze-drying technique developed a tissue scaffold system for the *in vivo* studies. Albino rats were preferred as the primary model for the hydroxyapatite-alginate-chitosan-fucoidan biomaterials composite analysis and were operated by implantation method to ensure that there was no observation of inflammation around the implants and biocompatibility (Song et al., 2006). The rat model underwent implantation with poly (vinyl alcohol)--carrageenan, which promoted cell growth and vascular integration within scaffolds and ensured an adequate nutrient supply for the survival of neo-tissues (Song et al., 2006). Collagen with EDC/NHS crosslinking was implanted into the bodies of the mice, and the scaffold showed a safe, cytocompatible material that supports tissue regeneration (Korpayev et al., 2020). In another study, mice were once again chosen for the study of chitosan-collagen after implantation surgery in which the capability of the extracted cells to differentiate into adipogenic, osteogenic, and chondrogenic lineages was observed (Yang et al., 2019). Similarly, Chitosan-silk fibroin was implanted into a rat model after surgery to observe that the histological studies revealed hyaline-like cartilage regeneration (Bermueller et al., 2013). Collagen was implanted into the rat, which offered good biocompatibility, safety, and no cytotoxic side effects; there was just a small amount of visible local inflammation. Furthermore, Collagen introduced into the rabbit's body showed repair of cartilage (Apelgren et al., 2018).

The 3D printing was carried out for various biomaterials, administered into animal models, and observed for the outcomes. Nano-fibrillated cellulose-alginate was implanted into a mouse model to monitor that the structure permitted the biomaterials and surrounding tissues to be integrated *in vivo* without risk (Apelgren et al., 2018). One more study with the same animal model and surgical methods was observed with the production of new cartilage tissues, increased chondrocyte proliferation, and good proliferative capacity (Apelgren et al., 2017). Again, a similar approach has been investigated and showed that future human reconstructive surgery would benefit from the structure's exceptional mechanical realism (Möller et al., 2017). Chitosan-oxidized chondroitin sulfate was administered into a mouse model by implantation type of surgery which exhibited favorable biocompatibility, encapsulated cells were highly viable, and a low inflammatory gene expression by macrophages was observed (Chen et al., 2022). The rabbit model was introduced with Alginate-gelatin-hydroxyapatite, showing that even after six months the quality of the mended cartilage compared to that of the unrepaired tissue defects around healthy cartilage with the help of osteochondral scaffold (Yang et al., 2020). The higher ECM and GAG content in the engineered cartilage was observed without an adverse tissue response after implantation of PCL-alginate into the mice's body (Kundu et al., 2015).

The effects of marine hydrolysates on fat metabolism *in vivo* have also been studied in several recent investigations. In mice fed a high-calorie diet, MC peptides taken orally were found to drastically reduce body fat and weight. Peroxisome proliferator-activated receptor (PPAR) expression was particularly high. In addition to acting as a lipid sensor that regulates fatty acid burn, this nuclear hormone receptor is crucial in modifying insulin sensitivity. Therefore, its overexpression caused the metabolism of fatty acids to rise, which then caused fat to burn. Additionally, serum levels of triglycerides, total cholesterol, and low-density lipoproteins (LDL) fell while rising for high-density lipoproteins (HDL) (Salvatore et al., 2020). Additionally, MC peptides can inhibit the angiotensin I-converting enzyme (ACE), which suggests lowering blood pressure and hyperlipidemia. Marine hydrolysates were found to increase adiponectin and decrease leptin in type 2 diabetes-induced rats. The study further confirms that MC peptides can improve insulin sensitivity by controlling the release of adipocytokines, i.e., a family of inflammatory mediators associated with diabetes. The most significant amount of MC peptides that can be delivered is unrestricted, but it should be noted that an excessive peptide concentration could harm internal organs. Studies on mice also showed that high-dose collagen peptide injection (10%) considerably decreased body weight while significantly increasing the volume of internal organs such as the liver, thymus, spleen, and kidney, indicating the hyperplasia of the critical immunological and metabolic organs. Overall, clinical investigations on MC food supplements have been carried out based on favorable preclinical results and the absence of any deleterious or toxic consequences upon correct peptide dosage. Given the excellent absorption of MC derivatives through the intestinal wall, oral administration of fish collagen formulations is favored. A randomized clinical trial (NCT03872297) using overweight adults is currently evaluating the effects of Naticol®, a widely available fish type I collagen peptide, on body weight and fat (Salvatore et al., 2020).

According to several *in vivo* studies, marine biomaterials, especially collagen obtained from different aquatic species and processed into various shapes (such as films, sponges, beads, and ointments), can be considered reliable bioresources for medical purposes. Marine biomaterial has enhanced cell renewal, wound healing, and inflammation reduction, whether used in its native, denatured, or peptide form. It also helps with calcium and other mineral absorption, inhibits bacterial growth, functions as an antioxidant, and aids in controlling body weight and type 2 diabetes (Zhu et al., 2010). The proven usefulness of native MC in pharmaceutical products seems to be considerably difficult, even though it is realistic to expect an increase in MC demand in industries like nutraceuticals, and cosmetics where peptides and gelatin are mainly utilized. This is primarily because improving the reproducibility and safety of the finished products calls for strict control over batch-to-batch consistency. Additionally, more research is required to understand marine collagen's immunological reaction, its bioactivity, as well as the underlying mechanisms of action, which are still poorly understood (Salvatore et al., 2020; Zhu et al., 2017). The applications of various marine-derived biomaterials and their possible mechanisms in tissue regenerative systems have been summarized in Table 8.3.

8.7 CHALLENGES OF MARINE BIOMATERIALS IN THE DELIVERY SYSTEM

The tissue engineering application primarily focuses on the regeneration of the injured tissue or supports the injured tissue via supplying viable nutrients, growth factors, etc., in various major organs. The development of a delivery system with synthetic biomaterials showed multiple limitations, and hence, natural-based marine biomaterials came into conception due to their distinctive biological and mechanical characteristics similar to regular hard or soft tissues to support the injured tissues (Carvalho et al., 2021b; Macha et al., 2019a; Romano et al., 2022; Wan et al., 2021; Zhang et al., 2022). The limitations associated with marine biomaterials began from its large-scale extraction or isolation method development, followed by removing the impurities process and stability, non-availability of *In vivo* models, cell line investigations, etc., limited the utilization of marine biomaterials in tissue engineering applications. The marine biomaterials consist of several macromolecules that retain immunogenicity and toxicity in significant organs if the experimentations were not done at an initial stage. Moreover, batch-to-batch variations of the yield contents and regulatory guidelines make it

TABLE 8.3 Applications of Various Marine Derived Biomaterials, Their Source, *In Vivo* Test Models, and Probable Mechanisms in Tissue Engineering

Biomaterial	**Source**	**Uses**	**Test model**	**Mechanism**	**Reference**
Phlorotannin	Brown alga *Ecklonia cava*	Treatment of tracheal stenosis caused by prolonged endotracheal intubation	Rabbit	• Inhibition of proliferation of human tracheal fibroblasts, treated with transforming growth factor β1	Lee et al. (2020)
Collagen	Jellyfish	Bone tissue regeneration	Wistar rats	• After interacting with collagen, the four main integrins (α1β1, α2β1, α10β1, and α11β1) can control osteoclast adherence to the bone, as well as create an acidic environment that promotes bone deterioration. • osteoclast-associated receptors (OSCAR) and leukocyte-associated Ig-like receptor-1, collagen has recently been demonstrated to be a ligand (LAIR-1)	Flaig et al. (2020)
Collagen	Salmon	Bone tissue regeneration	Mouse	• In mouse pre-osteoblastic cells (MC3T3-E1) osteoblasts, collagen in salmon skin can also significantly up-regulate the gene expression of various collagen-modifying enzymes, which benefits osteoblasts.	Xu et al. (2021)
Alginate	Brown seaweeds	Endodontics	*In vitro* (Cell line)	• Biological gutta-percha for regenerative endodontic therapy. • Human apical papilla stem cells were able to distribute, endure, multiply, and develop into osteoblast-like cells with calcified osseous extracellular matrix, offering a good root canal healing environment.	Lin et al. (2022)

TABLE 8.3 *(Continued)*

Biomaterial	Source	Uses	Test model	Mechanism	Reference
Chitosan	Shells of arthropods such as crab, shrimp and lobster	Wound healing	*In vitro* (Cell line)	• The highest levels of transforming growth factor-1 (TGF-1) and smooth muscle actin (SMA) were found in the overall sponge sample, but matrix metalloproteinase-1 expression was most stimulated by the CMCS hydrogel (MMP-1). • Additionally, the CMCS sponge dressing showed the best results for wound closure and accelerating wound repair *In vivo*, with the quickest epithelization and best healing results due to the epidermis's well-organized structure and the dermis's well-placed collagen.	Wang et al. (2020)
Coralline	Coralline algae	Hard tissue remodeling	*In vitro* (Cell line)	• The biphasic osteochondral region had a good integration of the bio-fabricated matrix. After 4 months, new tissue had replaced the biomaterial, which had first undergone resorption and then been replaced by the development of bone and cartilage. • Without the addition of new factors, ossification was increased and induced.	Vago et al. (2002)
Collagen	Echinoderms	Guided Tissue Regeneration (GTR)	*In vitro* (Cell line)	• Mesenchymal stem cells in PDL can develop into fibroblasts, osteoblasts, and cementoblasts, injured periodontal tissue does not successfully recover under conventional treatment due to endogenous regeneration processes. • Healing through protracted junctional epithelial development frequently leads to endogenous regeneration.	Ferrario et al. (2017)

TABLE 8.3 *(Continued)*

Biomaterial	Source	Uses	Test model	Mechanism	Reference
Corals	Coral	Repair of cranial bone defects	Canine models	• Expanded, osteogenically induced autologous ASCs from canine subcutaneous fat were seeded on coral scaffolds. • Parietal bone was damaged in both directions, creating 20 mm x 20 mm full-thickness defects. • ASC-coral constructs (experimental group) or just coral alone were used to repair the flaws.	Cui et al. (2007)

more unaffordable for research even though the earth consists of more than 70% water. As the marine biomaterials may consist of several biological macromolecules hence the fabrication process in the delivery system is also tricky or unknown due to the non-availed standardization protocol (Carvalho et al., 2021b; Macha et al., 2019a; Romano et al., 2022; Wan et al., 2021; Zhang et al., 2022). The authors believed that the marine-derived biomaterials consist of several macromolecules, and hence, the identifications of the macromolecules have yet to be appropriately investigated. So, to conceive and characterize those biomaterials for physicochemical properties via FTIR, DSC, TGA, etc., is also challenging to comprehend. Moreover, the release of active drug molecules from the composite system based on the marine biomaterials may also be impacted by the existence of additional elements in the biomaterial, hindering the application process in tissue engineering, especially in bone, periodontal, corneal, etc. Developing a drug delivery system doesn't require any particular method or technique, but understanding the materials to be selected for incorporation followed by characterization may require further detailed studies on these biomaterials. Some challenges can be overcome in the near future by completing the optimized isolation and standardization techniques, which can be attainable via biomass valorization beneath the notion of circular economy, followed by the formation of optimized procedure/yields tools that can be scaled up to overcome the current blockage of reproductivity issues while ensuring the accessibility at competitive prices for the clinical health market (Carvalho et al., 2021b).

8.8 CONCLUSION AND FUTURE PROSPECTIVE

In previous years, several marine-derived biomaterials have been attempted to extract/isolate and successfully characterized at a laboratory hierarchy. Several varieties of chemical classes, such as steroids from echinoderms, terpenes from sponges, alkaloids from tunicates, etc., have been revealed to have potential effects as anti-inflammatory agents, anti-proliferative agents for tumor cells, improving collagen deposition in various tissue engineering applications, food products, cosmeceuticals, etc., which further develop tremendous appeal among scientists. Despite having several biological and physicochemical characteristics of marine biomaterials, so far a very minimum number of developments have been reached to the consumer due to harvesting of aquatic organisms in an extreme environment, lower yield values of the natural products, obtaining a sustainable supply of the marine bioresources, standardization issues of the marine biomolecules, purification, ecological issues, regulatory guidelines, and insufficient investment by pharmaceutical companies causing inconvenience to afford in drug delivery research. The aquatic vertebrates also possess several bioactive molecules, but the challenges remain the same, especially the lower yield values. Furthermore, marine-based animals cannot be accumulated in large portions as repeated re-collection can lead to the depletion of the species, which should be considered for early screening assays. Therefore, the establishment of ex-situ, inland culturing methods for species of interest can be an initial stage of solution, leading to sustainable research development on the establishment of the transgenic animal cell line expressing the same level of molecules like marine source, standardization of biosynthetic pathways, etc., can overcome this issue at a general level. Although some edible species are widely cultivated (e.g., bivalves, holothurians, solitary ascidians), other species (e.g., sponges, cnidarians, and some echinoids) may be challenging to maintain under laboratory prerequisites for an extended duration, to obtain the release of gametes and complete the life cycle in captivity (Atanasov et al., 2021; Jaspars et al., 2016; Romano et al., 2022).

To date, the laboratory-scale investigation has been witnessed primarily in Europe and North America compared to the rest of the world. However, the most noteworthy coastal region, i.e., the Indo-Pacific and Oceania region, is still untouched. Further exploration in this region may provide a significant boost to the healthcare industry (Leal et al., 2012). In early 2010, metabolomic investigations progressively identified several marine invertebrates with biological applications. Metabolomics can inform researchers about the metabolite composition of a natural extract and provide insights into how metabolites change during different physiological states. Additionally, it can help in discovering new chemical scaffolds for creating novel synthetic medications. Metabolomics has also enabled the distinction of MNPs created by the associated microbiota, sequestered from symbionts, or synthesized by the animal (Atanasov et al., 2021; Romano et al., 2022). Safety and immunogenicity are significant concerns in the development, characterization, and application of marine-derived biomaterials for human use. Since most marine biomaterials may contain numerous active molecules, proper physicochemical characterization is essential for further investigation and incorporation of these biomaterials into tissue engineering applications (Zhang et al., 2022). Intradisciplinary research with various subjects which can further conclude the formation and structural properties of biomaterials and their mechanism of conversion is highly required. Nevertheless, following this multidisciplinary and multi-scale approach to investigations on marine biomaterials opens up the mere question of "why" behind every reason, which leads to deeper research investigations and can avail financial assistance from the government or non-government organizations to make it market-ready research (Wan et al., 2021).

KEYWORDS

- **biomaterials**
- **biomedical**
- **drug delivery**
- **marine**
- **polysaccharides**
- **tissue engineering**

REFERENCES

Adamiak, K., & Sionkowska, A. (2020). Current methods of collagen cross-linking: Review. *International Journal of Biological Macromolecules, 161*, 550–560. https://doi.org/10.1016/j.ijbiomac.2020.06.075.

Alexandra Benisek. (2021). What Is Marine Collagen? Radiance by Webmd. https://www.webmd.com/beauty/marine-collagen (accessed on 7 June 2024).

Alvarez Echazú, M. I., Olivetti, C. E., Peralta, I., Alonso, M. R., Anesini, C., Perez, C. J., Alvarez, G. S., & Desimone, M. F. (2018). Development of pH-responsive biopolymer-silica composites loaded with *Larrea divaricata* Cav. extract with antioxidant activity. *Colloids and Surfaces B: Biointerfaces, 169*, 82–91. https://doi.org/10.1016/j.colsurfb.2018.05.015.

Alvarez, X., Alves, A., Ribeiro, M. P., Lazzari, M., Coutinho, P., & Otero, A. (2021). Biochemical characterization of *Nostoc* sp. exopolysaccharides and evaluation of potential use in wound healing. *Carbohydrate Polymers, 254*, 117303. https://doi.org/10.1016/j.carbpol.2020.117303.

Alves, A. L., Costa-Gouveia, J., Vieira de Castro, J., Sotelo, C. G., Vázquez, J. A., Pérez-Martín, R. I., Torrado, E., Neves, N., Reis, R. L., Castro, A. G., & Silva, T. H. (2022). Study of the immunologic response of marine-derived collagen and gelatin extracts for tissue engineering applications. *Acta Biomaterialia, 141*, 123–131. https://doi.org/10.1016/j.actbio.2022.01.009.

Apelgren, P., Amoroso, M., Lindahl, A., Brantsing, C., Rotter, N., Gatenholm, P., & Kölby, L. (2017). Chondrocytes and stem cells in 3D-bioprinted structures create human cartilage in vivo. *PloS One, 12*(12), e0189428. https://doi.org/10.1371/journal.pone.0189428.

Apelgren, P., Amoroso, M., Säljö, K., Lindahl, A., Brantsing, C., Stridh Orrhult, L., Gatenholm, P., & Kölby, L. (2018). Skin Grafting on 3D Bioprinted Cartilage Constructs In Vivo.

Plastic and Reconstructive Surgery. Global Open, 6(9), e1930. https://doi.org/10.1097/GOX.0000000000001930.

Apelgren, P., Sämfors, S., Säljö, K., Mölne, J., Gatenholm, P., Troedsson, C., Thompson, E. M., & Kölby, L. (2022). Biomaterial and biocompatibility evaluation of tunicate nanocellulose for tissue engineering. *Biomaterials Advances, 137*, 212828. https://doi.org/10.1016/j.bioadv.2022.212828.

Atanasov, A. G., Zotchev, S. B., Dirsch, V. M., & Supuran, C. T. (2021). Natural products in drug discovery: advances and opportunities. *Nature Reviews. Drug Discovery, 20*(3), 200–216. https://doi.org/10.1038/s41573-020-00114-z.

Azab, A. K., Doviner, V., Orkin, B., Kleinstern, J., Srebnik, M., Nissan, A., & Rubinstein, A. (2007). Biocompatibility evaluation of crosslinked chitosan hydrogels after subcutaneous and intraperitoneal implantation in the rat. *Journal of Biomedical Materials Research. Part A, 83*(2), 414–422. https://doi.org/10.1002/jbm.a.31256.

Baek, J. W., Park, H., Kim, K. S., Chun, S. K., & Kim, B. S. (2022). Marine Plankton-Derived Whitlockite Powder-Based 3D-Printed Porous Scaffold for Bone Tissue Engineering. *Materials (Basel, Switzerland), 15*(10), pp. 3413. https://doi.org/10.3390/ma15103413.

Bermueller, C., Schwarz, S., Elsaesser, A. F., Sewing, J., Baur, N., von Bomhard, A., Scheithauer, M., Notbohm, H., & Rotter, N. (2013). Marine collagen scaffolds for nasal cartilage repair: prevention of nasal septal perforations in a new orthotopic rat model using tissue engineering techniques. *Tissue Engineering. Part A, 19*(19–20), 2201–2214. https://doi.org/10.1089/ten.TEA.2012.0650.

Bhattacharya, R., Kundu, B., N., Nandi, S. K., & Basu, D. (2013). Systematic approach to treat chronic osteomyelitis through localized drug delivery system: bench to bedside. *Materials Science & Engineering. C, Materials for Biological Applications, 33*(7), 3986–3993. https://doi.org/10.1016/j.msec.2013.05.036.

Bhattarai, N., Li, Z., Edmondson, D., & Zhang, M. (2006). Alginate-Based Nanofibrous Scaffolds: Structural, Mechanical, and Biological Properties. *Advanced Materials, 18*(11), 1463–1467. https://doi.org/10.1002/adma.200502537.

Bianchi, M., Pisciotta, A., Bertoni, L., Berni, M., Gambardella, A., Visani, A., Russo, A., de Pol, A., & Carnevale, G. (2017). Osteogenic sifferentiation of hDPSCs on biogenic bone apatite thin films. *Stem Cells International, 2017*, 3579283. https://doi.org/10.1155/2017/3579283.

Binlateh, T., Thammanichanon, P., Rittipakorn, P., Thinsathid, N., & Jitprasertwong, P. (2022). Collagen-based biomaterials in periodontal regeneration: current applications and future perspectives of plant-based collagen. *Biomimetics, 7*(2), 34. https://doi.org/10.3390/biomimetics7020034.

Borciani, G., Fischetti, T., Ciapetti, G., Montesissa, M., Baldini, N., & Graziani, G. (2022). Marine biological waste as a source of hydroxyapatite for bone tissue engineering applications. *Ceramics International, 49*(2), 1572–1584. https://doi.org/10.1016/j.ceramint. 2022.10.341.

Buck, C. B., Thompson, C. D., Roberts, J. N., Müller, M., Lowy, D. R., & Schiller, J. T. (2006). Carrageenan is a potent inhibitor of papillomavirus infection. *PLoS Pathogens, 2*(7), e69. https://doi.org/10.1371/journal.ppat.0020069.

Campo, V., Kawano, D., Silva, D., & Ivone, C. (2009). Carrageenans: Biological properties, chemical modifications and structural analysis–A review. *Carbohydrate Polymers, 77*, 167–180. https://doi.org/10.1016/j.carbpol.2009.01.020.

Carrageenan–Uses, Side Effects, and More. (2021). WebMD. https://doi.org/https://www.webmd.com/vitamins/ai/ingredientmono-710/carrageenan (accessed on June 22, 2024).

Carvalho, D. N., López-Cebral, R., Sousa, R. O., Alves, A. L., Reys, L. L., Silva, S. S., Oliveira, J. M., Reis, R. L., & Silva, T. H. (2020). Marine collagen-chitosan-fucoidan cryogels as cell-laden biocomposites envisaging tissue engineering. *Biomedical Materials (Bristol, England), 15*(5), 55030. https://doi.org/10.1088/1748-605X/ab9f04.

Carvalho, D. N., Reis, R. L., & Silva, T. H. (2021a). Marine origin materials on biomaterials and advanced therapies to cartilage tissue engineering and regenerative medicine. *Biomaterials Science, 9*(20), 6718–6736. https://doi.org/10.1039/D1BM00809A.

Carvalho, D. N., Williams, D. S., Sotelo, C. G., Pérez-Martín, R. I., Mearns-Spragg, A., Reis, R. L., & Silva, T. H. (2022). Marine origin biomaterials using a compressive and absorption methodology as cell-laden hydrogel envisaging cartilage tissue engineering. *Biomaterials Advances, 137*, 212843. https://doi.org/10.1016/j.bioadv.2022.212843.

Chakraborty, K., Antony, T., & Dhara, S. (2021). Marine macroalgal polygalactan-built nanoparticle construct for osteogenesis. *Biomacromolecules, 22*(5), 2197–2210. https://doi.org/10.1021/acs.biomac.1c00270.

Chen, G. L., Cai, H. Y., Chen, J. P., Li, R., Zhong, S. Y., Jia, X. J., Liu, X. F., & Song, B. B. (2022). Chitosan/Alginate Nanoparticles for the Enhanced Oral Antithrombotic Activity of Clam Heparinoid from the Clam *Coelomactra antiquata*. *Marine Drugs, 20*(2), 136. https://doi.org/10.3390/md20020136.

Chen, W. H., Chen, Q. W., Chen, Q., Cui, C., Duan, S., Kang, Y., Liu, Y., Liu, Y., Muhammad, W., Shao, S., Tang, C., Wang, J., Wang, L., Xiong, M. H., Yin, L., Zhang, K., Zhang, Z., Zhen, X., Feng, J., Chen, X. (2022). Biomedical polymers: synthesis, properties, and applications. *Science China. Chemistry, 65*(6), 1010–1075. https://doi.org/10.1007/s11426-022-1243-5.

Chen, Y., Bilgen, B., Pareta, R. A., Myles, A. J., Fenniri, H., Ciombor, D. M., Aaron, R. K., & Webster, T. J. (2010). Self-assembled rosette nanotube/hydrogel composites for cartilage tissue engineering. *Tissue Engineering. Part C, Methods, 16*(6), 1233–1243. https://doi.org/10.1089/ten.TEC.2009.0400.

Cheung, R. C. F., Ng, T. B., & Wong, J. H. (2015). Marine Peptides: Bioactivities and Applications. *Marine Drugs, 13*(7), 4006–4043. https://doi.org/10.3390/md13074006.

Chiroff, R. T., White, E. W., Weber, K. N., & Roy, D. M. (1975). Tissue ingrowth of Replamineform implants. *Journal of Biomedical Materials Research, 9*(4), 29–45. https://doi.org/10.1002/jbm.820090407.

Cindana Mo'o, F. R., Wilar, G., Devkota, H. P., & Wathoni, N. (2020). Ulvan, a Polysaccharide from Macroalga Ulva sp.: A Review of Chemistry, Biological Activities and Potential for Food and Biomedical Applications. *Applied Sciences, 10*(16), 5488. https://doi.org/10.3390/app10165488.

Corinaldesi, C., Barone, G., Marcellini, F., Dell'Anno, A., & Danovaro, R. (2017). Marine microbial-derived molecules and their potential use in cosmeceutical and cosmetic products. *Marine Drugs, 15*(4), 118. https://doi.org/10.3390/md15040118.

Correia, C., Sousa, R. O., Vale, A. C., Peixoto, D., Silva, T. H., Reis, R. L., Pashkuleva, I., & Alves, N. M. (2022). Adhesive and biodegradable membranes made of sustainable catechol-functionalized marine collagen and chitosan. *Colloids and Surfaces B: Biointerfaces, 213*, 112409. https://doi.org/10.1016/j.colsurfb.2022.112409.

Costa-Pinto, A., Santos, T. C., Neves, N. M., & Reis, R. L. (2016). Testing Natural Biomaterials in Animal Models. In *Biomaterials from Nature for Advanced Devices and Therapies* (pp. 562–579). John Wiley & Sons, Ltd. https://doi.org/10.1002/9781119126218.ch30.

Cui, J., Li, Y., Wang, S., Chi, Y., Hwang, H., & Wang, P. (2018). Directional preparation of anticoagulant-active sulfated polysaccharides from *Enteromorpha prolifera* using artificial neural networks. *Scientific Reports, 8*(1), 3062. https://doi.org/10.1038/s41598-018-21556-x.

Cui, L., Liu, B., Liu, G., Zhang, W., Cen, L., Sun, J., Yin, S., Liu, W., & Cao, Y. (2007). Repair of cranial bone defects with adipose derived stem cells and coral scaffold in a canine model. *Biomaterials, 28*(36), 5477–5486. https://doi.org/10.1016/j.biomaterials.2007.08.042.

Custódio, C. A., Reis, R. L., & Mano, J. F. (2016). Photo-Cross-Linked Laminarin-Based Hydrogels for Biomedical Applications. *Biomacromolecules, 17*(5), 1602–1609. https://doi.org/10.1021/acs.biomac.5b01736.

Dashtdar, H., Murali, M. R., Abbas, A. A., Suhaeb, A. M., Selvaratnam, L., Tay, L. X., & Kamarul, T. (2015). PVA-chitosan composite hydrogel versus alginate beads as a potential mesenchymal stem cell carrier for the treatment of focal cartilage defects. *Knee Surgery, Sports Traumatology, Arthroscopy: Official Journal of the ESSKA, 23*(5), 1368–1377. https://doi.org/10.1007/s00167-013-2723-5.

Delplace, V., Boutet, M. A., Le Visage, C., Maugars, Y., Guicheux, J., & Vinatier, C. (2021). Osteoarthritis: From upcoming treatments to treatments yet to come. *Joint Bone Spine, 88*(5), 105206. https://doi.org/10.1016/j.jbspin.2021.105206.

Demers, C., Hamdy, C. R., Corsi, K., Chellat, F., Tabrizian, M., & Yahia, L. (2002). Natural coral exoskeleton as a bone graft substitute: A review. *Bio-Medical Materials and Engineering, 12*(1), 15–35.

Diogo, G. S., Carneiro, F., Freitas-Ribeiro, S., Sotelo, C. G., Pérez-Martín, R. I., Pirraco, R. P., Reis, R. L., & Silva, T. H. (2021). Prionace glauca skin collagen bioengineered constructs as a promising approach to trigger cartilage regeneration. *Materials Science and Engineering: C, 120*, 111587. https://doi.org/10.1016/j.msec.2020.111587.

Djagny, V. B., Wang, Z., & Xu, S. (2001). Gelatin: a valuable protein for food and pharmaceutical industries: review. *Critical Reviews in Food Science and Nutrition, 41*(6), 481–492. https://doi.org/10.1080/20014091091904.

Fahmy, N. M., & Abdel-Tawab, A. M. (2021). Isolation and characterization of marine sponge-associated Streptomyces sp. NMF6 strain producing secondary metabolite(s) possessing antimicrobial, antioxidant, anticancer, and antiviral activities. *Journal, Genetic Engineering & Biotechnology, 19*(1), 102. https://doi.org/10.1186/s43141-021-00203-5.

Fernandes-Silva, S., Moreira-Silva, J., Silva, T. H., Perez-Martin, R. I., Sotelo, C. G., Mano, J. F., Duarte, A. R. C., & Reis, R. L. (2013). Porous hydrogels from shark skin collagen crosslinked under dense carbon dioxide atmosphere. *Macromolecular Bioscience, 13*(11), 1621–1631. https://doi.org/10.1002/mabi.201300228.

Ferrario, C., Leggio, L., Leone, R., Di Benedetto, C., Guidetti, L., Coccè, V., Ascagni, M., Bonasoro, F., La Porta, C. A. M., Candia Carnevali, M. D., & Sugni, M. (2017). Marine-derived collagen biomaterials from echinoderm connective tissues. *Marine Environmental Research, 128*, 46–57. https://doi.org/10.1016/j.marenvres.2016.03.007.

Flaig, I., Radenković, M., Najman, S., Pröhl, A., Jung, O., & Barbeck, M. (2020). In Vivo analysis of the biocompatibility and immune response of jellyfish collagen scaffolds and its suitability for bone regeneration. *International Journal of Molecular Sciences, 21*(12), 4518. https://doi.org/10.3390/ijms21124518.

Gao, Y., Zhang, L., & Jiao, W. (2019). Marine glycan-derived therapeutics in China. *Progress in Molecular Biology and Translational Science, 163*, 113–134. https://doi.org/10.1016/bs.pmbts.2019.02.006.

Govindharaj, M., Hashimi, N. Al, Soman, S. S., Kanwar, S., & Vijayavenkataraman, S. (2022). 3D bioprinting of human mesenchymal stem cells in a novel tunic decellularized ECM bioink for cartilage tissue engineering. *Materialia, 23*, 101457. https://doi.org/10.1016/j.mtla.2022.101457.

Grice, I. D., & Mariottini, G. L. (2018). Glycans with Antiviral Activity from Marine Organisms. *Results and Problems in Cell Differentiation, 65*, 439–475. https://doi.org/10.1007/978-3-319-92486-1_20.

Han, F., Li, T., Li, M., Zhang, B., Wang, Y., Zhu, Y., & Wu, C. (2023). Nano-calcium silicate mineralized fish scale scaffolds for enhancing tendon-bone healing. *Bioactive Materials, 20*, 29–40. https://doi.org/10.1016/j.bioactmat.2022.04.030.

Hao, Y., Zheng, W., Sun, Z., Zhang, D., Sui, K., Shen, P., Li, P., & Zhou, Q. (2021). Marine polysaccharide-based composite hydrogels containing fucoidan: Preparation, physicochemical characterization, and biocompatible evaluation. *International Journal of Biological Macromolecules, 183*, 1978–1986. https://doi.org/10.1016/j.ijbiomac.2021.05.190.

Hassenklöver, T., Predehl, S., Pilli, J., Ledwolorz, J., Assmann, M., & Bickmeyer, U. (2006). Bromophenols, both present in marine organisms and in industrial flame retardants, disturb cellular Ca2+ signaling in neuroendocrine cells (PC12). *Aquatic Toxicology (Amsterdam, Netherlands), 76*(1), 37–45. https://doi.org/10.1016/j.aquatox.2005.09.004.

Hiller, T., Berg, J., Elomaa, L., Röhrs, V., Ullah, I., Schaar, K., Dietrich, A.-C., Al-Zeer, M., Kurtz, A., Hocke, A., Hippenstiel, S., Fechner, H., Weinhart, M., & Kurreck, J. (2018). Generation of a 3D liver model comprising human extracellular matrix in an alginate/gelatin-based bioink by extrusion bioprinting for infection and transduction studies. *International Journal of Molecular Sciences, 19*(10), 3129. https://doi.org/10.3390/ijms19103129.

Huang, H., Song, Y., Li, X., Wang, X., Ling, C., Qin, X., Zhou, Z., Li, Q., Wei, X., & Ju, J. (2018). Abyssomicin monomers and dimers from the marine-derived *Streptomyces koyangensis* SCSIO 5802. *Journal of Natural Products, 81*(8), 1892–1898. https://doi.org/10.1021/acs.jnatprod.8b00448.

Izadifar, Z., Chang, T., Kulyk, W., Chen, X., & Eames, B. F. (2016). Analyzing biological performance of 3d-printed, cell-impregnated hybrid constructs for cartilage tissue engineering. *Tissue Engineering. Part C, Methods, 22*(3), 173–188. https://doi.org/10.1089/ten.TEC.2015.0307.

Jana, S., Datta, P., Das, H., Jaiswal, S., Ghosh, P. R., Lahiri, D., Kundu, B., & Nandi, S. K. (2022). Copper and cobalt doped bioactive glass-fish dermal collagen electrospun mat triggers key events of diabetic wound healing in full-thickness skin defect model. *Journal of the Mechanical Behavior of Biomedical Materials, 134*, 105414. https://doi.org/10.1016/j.jmbbm.2022.105414.

January, G. G., Naidoo, R. K., Kirby-McCullough, B., & Bauer, R. (2019). Assessing methodologies for fucoidan extraction from South African brown algae. *Algal Research, 40*, 101517. https://doi.org/10.1016/j.algal.2019.101517.

Jaspars, M., De Pascale, D., Andersen, J. H., Reyes, F., Crawford, A. D., & Ianora, A. (2016). The marine biodiscovery pipeline and ocean medicines of tomorrow. *Journal of the Marine Biological Association of the United Kingdom, 96*(1), 151–158. https://doi.org/10.1017/S0025315415002106.

Jiang, T., Singh, B., Choi, Y. J., Akaike, T., & Cho, C. S. (2015). Liver tissue engineering using functional marine biomaterials. In S. K. Kim (Ed.), *Functional Marine Biomaterials* (pp. 91–106). Woodhead Publishing. https://doi.org/10.1016/B978-1-78242-086-6.00006-6.

Kady, E. El. (2019). Chitin, Chitosan and Glucan, Properties and Applications. *World Journal of Agriculture and Soil Science, 3*(1), 1–19. https://doi.org/10.33552/WJASS.2019.03.000553.

Kanayama, T., Nagai, N., Mori, K., & Munekata, M. (2008). Application of elastic salmon collagen gel to uniaxial stretching culture of human umbilical vein endothelial cells. *Journal of Bioscience and Bioengineering, 105*(5), 554–557. https://doi.org/10.1263/jbb.105.554.

Kaur, S., & Dhillon, G. S. (2015). Recent trends in biological extraction of chitin from marine shell wastes: a review. *Critical Reviews in Biotechnology, 35*(1), 44–61. https://doi.org/10.3109/07388551.2013.798256.

Khrunyk, Y., Lach, S., Petrenko, I., & Ehrlich, H. (2020). Progress in Modern Marine Biomaterials Research. *Marine Drugs, 18*(12), 589. https://doi.org/10.3390/md18120589.

Kibar, H., Arslan, Y. E., Ceylan, A., Karaca, B., Haliscelik, O., & Kiran, F. (2020). *Weissella cibaria* EIR/P2-derived exopolysaccharide: A novel alternative to conventional biomaterials targeting periodontal regeneration. *International Journal of Biological Macromolecules, 165*, 2900–2908. https://doi.org/10.1016/j.ijbiomac.2020.10.106.

Kim, K. W., & Thomas, R. L. (2007). Antioxidative activity of chitosan's with varying molecular weights. *Food Chemistry, 101*(1), 308–313. https://doi.org/10.1016/j.foodchem.2006.01.038.

Kim, S. C., Heo, S. Y., Oh, G. W., Yi, M., & Jung, W. K. (2022). A 3D-Printed Polycaprolactone/marine collagen scaffold reinforced with carbonated hydroxyapatite from fish bones for bone regeneration. *Marine Drugs, 20*(6), 344. https://doi.org/10.3390/md20060344.

Koivusalo, L., Kauppila, M., Samanta, S., Parihar, V. S., Ilmarinen, T., Miettinen, S., Oommen, O. P., & Skottman, H. (2019). Tissue adhesive hyaluronic acid hydrogels for suture less stem cell delivery and regeneration of corneal epithelium and stroma. *Biomaterials, 225*, 119516. https://doi.org/10.1016/j.biomaterials.2019.119516.

Korpayev, S., Kaygusuz, G., Şen, M., Orhan, K., Oto, Ç., & Karakeçili, A. (2020). Chitosan/collagen based biomimetic osteochondral tissue constructs: A growth factor-free approach. *International Journal of Biological Macromolecules, 156*, 681–690. https://doi.org/10.1016/j.ijbiomac.2020.04.109.

Kundu, J., Shim, J. H., Jang, J., Kim, S. W., & Cho, D. W. (2015). An additive manufacturing-based PCL-alginate-chondrocyte bioprinted scaffold for cartilage tissue engineering. *Journal of Tissue Engineering and Regenerative Medicine, 9*(11), 1286–1297. https://doi.org/10.1002/term.1682.

Lalzawmliana, V., Anand, A., Mukherjee, P., Chaudhuri, S., Kundu, B., Nandi, S. K., & Thakur, N. L. (2019). Marine organisms as a source of natural matrix for bone tissue engineering. *Ceramics International, 45*(2, Part A), 1469–1481. https://doi.org/10.1016/j.ceramint.2018.10.108.

Laurienzo, P. (2010). Marine polysaccharides in pharmaceutical applications: an overview. *Marine Drugs, 8*(9), 2435–2465. https://doi.org/10.3390/md8092435.

Leal, M. C., Puga, J., Serôdio, J., Gomes, N. C. M., & Calado, R. (2012). Trends in the discovery of new marine natural products from invertebrates over the last two decades–where and what are we bioprospecting? *PloS One, 7*(1), e30580. https://doi.org/10.1371/journal.pone.0030580.

Lee, H., Jung, M., Ko, S. C., Heo, S. Y., Kang, H. W., Kim, S., Hwang, C., Lee, K., Oak, C., Jung, M., Oh, J., Park, W., Choi, I. W., & Jung, W. K. (2020). Fabrication and biological activity of polycaprolactone/phlorotannin endotracheal tube to prevent tracheal stenosis: An in vitro and in vivo study. *Journal of Biomedical Materials Research Part B: Applied Biomaterials, 108*, 1046–1056. https://doi.org/10.1002/jbm.b.34456.

Li, K. Y., Pan, H. A., Chen, K. H., Kuo, T. L., Chou, C. H., Liang, Y. J., & Lin, F. H. (2020). Fish-Scale Collagen Membrane Seeded with Corneal Endothelial Cells as Alternative Graft for Endothelial Keratoplasty Transplantation. *ACS Biomaterials Science & Engineering, 6*(5), 2570–2577. https://doi.org/10.1021/acsbiomaterials.9b00562.

Li, L., Ni, R., Shao, Y., & Mao, S. (2014). Carrageenan and its applications in drug delivery. *Carbohydrate Polymers, 103*, 1–11. https://doi.org/10.1016/j.carbpol.2013.12.008.

Lim, Y. S., Ok, Y. J., Hwang, S. Y., Kwak, J. Y., & Yoon, S. (2019). Marine collagen as a promising biomaterial for biomedical applications. *Marine Drugs, 17*(8), 467, https://doi.org/10.3390/md17080467.

Lin, G. S. S., Cher, C. Y., Goh, Y. H., Chan, D. Z. K., Karobari, M. I., Lai, J. C. H., & Noorani, T. Y. (2022). An insight into the role of marine biopolymer alginate in endodontics: A review. *Marine Drugs, 20*(8), 539. https://doi.org/10.3390/md20080539.

Liu, M., Zhang, Y., Ma, X., Zhang, B., Huang, Y., Zhao, J., Wang, S., Li, Y., Zhu, Y., Xiong, J., He, T., Wang, Y., Han, W., Yang, K., Bi, X., Liu, Y., & Zhang, H. (2022). Synthesis and characterization of fucoidan-chitosan nanoparticles targeting p-selectin for effective atherosclerosis therapy. *Oxidative Medicine and Cellular Longevity, 2022*, 1–27. https://doi.org/10.1155/2022/8006642.

Liu, Z., Yu, M. Z., Peng, H., Liu, R. T., Lim, T., Zhang, C. Q., Zhu, Z. Z., & Wei, X. J. (2022). Decellularized tilapia fish skin: A novel candidate for tendon tissue engineering. *Materials Today Bio, 17*, 100488. https://doi.org/10.1016/j.mtbio.2022.100488.

Loo, H. L., Goh, B. H., Lee, L. H., & Chuah, L. H. (2022). Application of chitosan-based nanoparticles in skin wound healing. *Asian Journal of Pharmaceutical Sciences, 17*(3), 299–332. https://doi.org/10.1016/j.ajps.2022.04.001.

Lu, Y., Zhu, X., Hu, C., Li, P., Zhao, M., Lu, J., & Xia, G. (2022). A fucoidan-gelatin wound dressing accelerates wound healing by enhancing antibacterial and anti-inflammatory activities. *International Journal of Biological Macromolecules, 223*, 36–48. https://doi.org/10.1016/j.ijbiomac.2022.10.255.

Macha, I. J., Ben-Nissan, B., & Müller, W. H. (2019). Marine-based biomaterials for tissue engineering applications. In A. H. Choi & B. Ben-Nissan (Eds.), *Marine-derived biomaterials for tissue engineering applications* (pp. 99–111). Springer Singapore. https://doi.org/10.1007/978-981-13-8855-2_5.

Machałowski, T., Idaszek, J., Chlanda, A., Heljak, M., Piasecki, A., Święszkowski, W., & Jesionowski, T. (2022). Naturally prefabricated 3D chitinous skeletal scaffold of marine demosponge origin, biomineralized ex vivo as a functional biomaterial. *Carbohydrate Polymers, 275*, 118750. https://doi.org/10.1016/j.carbpol.2021.118750.

Mata, M., Milian, L., Oliver, M., Zurriaga, J., Sancho-Tello, M., de Llano, J. J. M., & Carda, C. (2017). In vivo articular cartilage regeneration using human dental pulp stem cells cultured in an alginate scaffold: A preliminary study. *Stem Cells International, 2017*, 8309256. https://doi.org/10.1155/2017/8309256.

Mazlum, Y., Yazıcı, M., Naz, M., & Sayin, S. (2022). Marine biomaterials and their applications. In *Theory and Research in Agriculture, Forestry and Aquaculture Sciences* (pp. 103–124).

Mohabatpour, F., Karkhaneh, A., & Sharifi, A. M. (2016). A hydrogel/fiber composite scaffold for chondrocyte encapsulation in cartilage tissue regeneration. *RSC Advances, 6*(86), 83135–83145. https://doi.org/10.1039/C6RA15592H.

Möller, T., Amoroso, M., Hägg, D., Brantsing, C., Rotter, N., Apelgren, P., Lindahl, A., Kölby, L., & Gatenholm, P. (2017). In vivo chondrogenesis in 3D bioprinted human cell-laden hydrogel constructs. *Plastic and Reconstructive Surgery. Global Open, 5*(2), e1227. https://doi.org/10.1097/GOX.0000000000001227.

Mora, C., Tittensor, D. P., Adl, S., Simpson, A. G. B., & Worm, B. (2011). How many species are there on Earth and in the ocean? *PLOS Biology, 9*(8), 1–8. https://doi.org/10.1371/journal.pbio.1001127.

Ng, K. W., Lima, E. G., Bian, L., O'Conor, C. J., Jayabalan, P. S., Stoker, A. M., Kuroki, K., Cook, C. R., Ateshian, G. A., Cook, J. L., & Hung, C. T. (2010). Passaged adult chondrocytes can form engineered cartilage with functional mechanical properties: A canine model. *Tissue Engineering. Part A, 16*(3), 1041–1051. https://doi.org/10.1089/ten.TEA.2009.0581.

Nguyen, T. T., Mikkelsen, M. D., Tran, V. H. N., Trang, V. T. D., Rhein-Knudsen, N., Holck, J., Rasin, A. B., Cao, H. T. T., Van, T. T. T., & Meyer, A. S. (2020). Enzyme-assisted fucoidan extraction from brown macroalgae *Fucus distichus* subsp. evanescens and *Saccharina latissima*. *Marine Drugs, 18*(6), 296. https://doi.org/10.3390/md18060296.

Nigrelli, R. F., Stempien, M. F. J., Ruggieri, G. D., Liguori, V. R., & Cecil, J. T. (1967). Substances of potential biomedical importance from marine organisms. *Federation Proceedings, 26*(4), 1197–1205.

Norman Stanley. (1987). Chapter 3 – Production, properties and uses of carrageenan. *Food and Agriculture Organization of the United Nations*. https://www.fao.org/4/x5822e/x5822e05.htm (accessed on 7 June 2024).

Ohmes, J., Saure, L. M., Schütt, F., Trenkel, M., Seekamp, A., Scherließ, R., Adelung, R., & Fuchs, S. (2022). Injectable thermosensitive chitosan-collagen hydrogel as a delivery system for marine polysaccharide fucoidan. *Marine Drugs, 20*(6), 402. https://doi.org/10.3390/md20060402.

Oren, A., & Gunde-Cimerman, N. (2007). Mycosporines and mycosporine-like amino acids: UV protectants or multipurpose secondary metabolites? *FEMS Microbiology Letters, 269*(1), 1–10. https://doi.org/10.1111/j.1574-6968.2007.00650.x.

Pacheco-Quito, E. M., Ruiz-Caro, R., & Veiga, M. D. (2020). Carrageenan: Drug delivery systems and other biomedical applications. *Marine Drugs, 18*(11), 583. https://doi.org/10.3390/md18110583.

Pakizeh, M., Moradi, A., & Ghassemi, T. (2021). Chemical extraction and modification of chitin and chitosan from shrimp shells. *European Polymer Journal, 159*, 110709. https://doi.org/10.1016/j.eurpolymj.2021.110709.

Parisi, J. R., Fernandes, K. R., Aparecida do Vale, G. C., de França Santana, A., de Almeida Cruz, M., Fortulan, C. A., Zanotto, E. D., Peitl, O., Granito, R. N., & Rennó, A. C. M. (2020). Marine spongin incorporation into Biosilicate® for tissue engineering applications: An in vivo study. *Journal of Biomaterials Applications, 35*(2), 205–214. https://doi.org/10.1177/0885328220922161.

Pereira, R. C., Scaranari, M., Castagnola, P., Grandizio, M., Azevedo, H. S., Reis, R. L., Cancedda, R., & Gentili, C. (2009). Novel injectable gel (system) as a vehicle for human articular chondrocytes in cartilage tissue regeneration. *Journal of Tissue Engineering and Regenerative Medicine, 3*(2), 97–106. https://doi.org/10.1002/term.145.

Peters, W. (1972). Occurrence of chitin in mollusca. *Comparative Biochemistry and Physiology Part B: Comparative Biochemistry, 41*(3), 541–550. https://doi.org/10.1016/0305-0491(72)90117-4.

Qureshi, D., Nayak, S. K., Maji, S., Kim, D., Banerjee, I., & Pal, K. (2019). Carrageenan: A wonder polymer from marine algae for potential drug delivery applications. *Current Pharmaceutical Design, 25*(11), 1172–1186. https://doi.org/10.2174/1381612825666190425190754.

Rahfoth, B., Weisser, J., Sternkopf, F., Aigner, T., von der Mark, K., & Bräuer, R. (1998). Transplantation of allograft chondrocytes embedded in agarose gel into cartilage defects of rabbits. *Osteoarthritis and Cartilage, 6*(1), 50–65. https://doi.org/10.1053/joca.1997.0092.

Rastogi, R. P., & Incharoensakdi, A. (2014). Analysis of UV-absorbing photoprotectant mycosporine-like amino acid (MAA) in the cyanobacterium Arthrospira sp. CU2556.

Photochemical & Photobiological Sciences: Official Journal of the European Photochemistry Association and the European Society for Photobiology, 13(7), 1016–1024. https://doi.org/10.1039/c4pp00013g.
Rekha Sahoo, D., Biswal, T., & Acharya, S. K. (2022). Synthesis, characterization and biomedical applications of Poly (methyl methacrylate)/Chitosan (derived from oyster shell powder). *Materials Today: Proceedings*. https://doi.org/10.1016/j.matpr.2022.10.280.
Ricard-Blum, S. (2011). The collagen family. *Cold Spring Harbor Perspectives in Biology, 3*(1), a004978. https://doi.org/10.1101/cshperspect.a004978.
Romano, G., Almeida, M., Varela Coelho, A., Cutignano, A., Gonçalves, L. G., Hansen, E., Khnykin, D., Mass, T., Ramšak, A., Rocha, M. S., Silva, T. H., Sugni, M., Ballarin, L., & Genevière, A. M. (2022). Biomaterials and bioactive natural products from marine invertebrates: From basic research to innovative applications. *Marine Drugs, 20*(4), 219. https://doi.org/10.3390/md20040219.
Rozan, H. E., Wu, G., Zhou, Z., Li, Q., Sharaf, M., & Chen, X. (2022). The complex hydrogel based on diatom biosilica and hydroxybutyl chitosan for wound healing. *Colloids and Surfaces B: Biointerfaces, 216*, 112523. https://doi.org/10.1016/j.colsurfb.2022.112523.
Russo, P., Kisialiou, A., Lamonaca, P., Moroni, R., Prinzi, G., & Fini, M. (2015). New drugs from marine organisms in Alzheimer's disease. *Marine Drugs, 14*(1), 5. https://doi.org/10.3390/md14010005.
Salgado, P. C., Sathler, P. C., Castro, H. C., Alves, G. G., Oliveira, A. M. de, Oliveira, R. C. de, Maia, M. D. C., Rodrigues, C. R., Coelh, P. G., Fuly, A., Cabral, L. M., & Granjeiro, J. M. (2011). Bone remodeling, biomaterials and technological applications: Revisiting basic concepts. *Journal of Biomaterials and Nanobiotechnology, 02*(03), 318–328. https://doi.org/10.4236/jbnb.2011.23039.
Salvatore, L., Gallo, N., Natali, M. L., Campa, L., Lunetti, P., Madaghiele, M., Blasi, F. S., Corallo, A., Capobianco, L., & Sannino, A. (2020). Marine collagen and its derivatives: Versatile and sustainable bio-resources for healthcare. *Materials Science and Engineering: C, 113*, 110963. https://doi.org/10.1016/j.msec.2020.110963.
Sanz, B., Albillos Sanchez, A., Tangey, B., Gilmore, K., Yue, Z., Liu, X., & Wallace, G. (2020). Light cross-linkable marine collagen for coaxial printing of a 3D model of neuromuscular junction formation. *Biomedicines, 9*(1), 16. https://doi.org/10.3390/biomedicines9010016.
Senni, K., Pereira, J., Gueniche, F., Delbarre-Ladrat, C., Sinquin, C., Ratiskol, J., Godeau, G., Fischer, A.-M., Helley, D., & Colliec-Jouault, S. (2011). Marine polysaccharides: A source of bioactive molecules for cell therapy and tissue engineering. *Marine Drugs, 9*(9), 1664–1681. https://doi.org/10.3390/md9091664.
Shabafrooz, V., Mozafari, M., Köhler, G. A., Assefa, S., Vashaee, D., & Tayebi, L. (2014). The effect of hyaluronic acid on bio functionality of gelatin-collagen intestine tissue engineering scaffolds. *Journal of Biomedical Materials Research Part A, 102*(9), 3130–3139. https://doi.org/10.1002/jbm.a.34984.
Sharifian, S., & Homaei, A. (2022). Marine-derived polysaccharides: Prospects for future pharmaceuticals and drug delivery systems. In S. Jana & S. Jana (Eds.), *Marine Biomaterials: Drug Delivery and Therapeutic Applications* (pp. 403–453). Springer Nature Singapore. https://doi.org/10.1007/978-981-16-4787-1_12.
Sheikh, Z., Hamdan, N., Ikeda, Y., Grynpas, M., Ganss, B., & Glogauer, M. (2017). Natural graft tissues and synthetic biomaterials for periodontal and alveolar bone reconstructive applications: A review. *Biomaterials Research, 21*(1), 9. https://doi.org/10.1186/s40824-017-0095-5.

Shim, J.-H., Lee, J.-S., Kim, J. Y., & Cho, D.-W. (2012). Bioprinting of a mechanically enhanced three-dimensional dual cell-laden construct for osteochondral tissue engineering using a multi-head tissue/organ building system. *Journal of Micromechanics and Microengineering, 22*(8), 85014. https://doi.org/10.1088/0960-1317/22/8/085014.

Singh, S. P., Kumari, S., Rastogi, R. P., Singh, K. L., & Sinha, R. P. (2008). Mycosporine-like amino acids (MAAs): Chemical structure, biosynthesis and significance as UV-absorbing/screening compounds. *Indian Journal of Experimental Biology, 46*(1), 7–17.

Song, E., Yeon Kim, S., Chun, T., Byun, H. J., & Lee, Y. M. (2006). Collagen scaffolds derived from a marine source and their biocompatibility. *Biomaterials, 27*(15), 2951–2961. https://doi.org/10.1016/j.biomaterials.2006.01.015.

Srinivasan, R., Kannappan, A., Shi, C., & Lin, X. (2021). Marine bacterial secondary metabolites: A treasure house for structurally unique and effective antimicrobial compounds. *Marine Drugs, 19*(10), 530. https://doi.org/10.3390/md19100530.

Stevens, M. M., Marini, R. P., Martin, I., Langer, R., & Prasad Shastri, V. (2004). FGF-2 enhances TGF-beta1-induced periosteal chondrogenesis. *Journal of Orthopaedic Research: Official Publication of the Orthopaedic Research Society, 22*(5), 1114–1119. https://doi.org/10.1016/j.orthres.2003.12.021.

Sumayya, A. S., & Muraleedhara Kurup, G. (2018). Biocompatibility of subcutaneously implanted marine macromolecules cross-linked bio-composite scaffold for cartilage tissue engineering applications. *Journal of Biomaterials Science, Polymer Edition, 29*(3), 257–276. https://doi.org/10.1080/09205063.2017.1413759.

Synowiecki, J., & Al-Khateeb, N. A. (2003). Production, properties, and some new applications of chitin and its derivatives. *Critical Reviews in Food Science and Nutrition, 43*(2), 145–171. https://doi.org/10.1080/10408690390826473.

Terezaki, A., Kikionis, S., Ioannou, E., Sfiniadakis, I., Tziveleka, L.-A., Vitsos, A., Roussis, V., & Rallis, M. (2022). Ulvan/gelatin-based nanofibrous patches as a promising treatment for burn wounds. *Journal of Drug Delivery Science and Technology, 74*, 103535. https://doi.org/10.1016/j.jddst.2022.103535.

Torres, M. D., Flórez-Fernández, N., & Dominguez, H. (2021). Ultrasound-assisted water extraction of Mastocarpus stellatus carrageenan with adequate mechanical and antiproliferative properties. *Marine Drugs, 19*(5), 280. https://doi.org/10.3390/md19050280.

Trincone, A. (2013). Angling for uniqueness in enzymatic preparation of glycosides. *Biomolecules, 3*(2), 334–350. https://doi.org/10.3390/biom3020334.

Tsurkan, M. V, Voronkina, A., Khrunyk, Y., Wysokowski, M., Petrenko, I., & Ehrlich, H. (2021). Progress in chitin analytics. *Carbohydrate Polymers, 252*, 117204. https://doi.org/10.1016/j.carbpol.2020.117204.

Tziveleka, L.-A., Ioannou, E., & Roussis, V. (2019). Ulvan, a bioactive marine sulphated polysaccharide as a key constituent of hybrid biomaterials: A review. *Carbohydrate Polymers, 218*, 355–370. https://doi.org/10.1016/j.carbpol.2019.04.074.

Vago, R., Plotquin, D., Bunin, A., Sinelnikov, I., Atar, D., & Itzhak, D. (2002). Hard tissue remodeling using biofabricated coralline biomaterials. *Journal of Biochemical and Biophysical Methods, 50*(2–3), 253–259. https://doi.org/10.1016/s0165-022x(01)00235-4.

Valcarcel, J., Novoa-Carballal, R., Pérez-Martín, R. I., Reis, R. L., & Vázquez, J. A. (2017). Glycosaminoglycans from marine sources as therapeutic agents. *Biotechnology Advances, 35*(6), 711–725. https://doi.org/10.1016/j.biotechadv.2017.07.008.

Varghese, D. M. G., George V., D. T., Thomas, D. N. G., Pyas, D. A. M., & Maymol Francis, D. A. (2020). Marine-based biomaterials in dental regeneration. *International Journal of*

Innovative Research in Medical Science, 5(10 SE-Review Article), 443–448. https://doi.org/10.23958/ijirms/vol05-i10/965.
Vázquez, J. A., Noriega, D., Ramos, P., Valcarcel, J., Novoa-Carballal, R., Pastrana, L., Reis, R. L., & Pérez-Martín, R. I. (2017). Optimization of high purity chitin and chitosan production from Illex argentinus pens by a combination of enzymatic and chemical processes. *Carbohydrate Polymers, 174*, 262–272. https://doi.org/10.1016/j.carbpol.2017.06.070.
Vinardell, T., Sheehy, E. J., Buckley, C. T., & Kelly, D. J. (2012). A comparison of the functionality and in vivo phenotypic stability of cartilaginous tissues engineered from different stem cell sources. *Tissue Engineering. Part A, 18*(11–12), 1161–1170. https://doi.org/10.1089/ten.TEA.2011.0544.
Wan, M., Qin, W., Lei, C., Li, Q., Meng, M., Fang, M., Song, W., Chen, J., Tay, F., & Niu, L. (2021). Biomaterials from the sea: Future building blocks for biomedical applications. *Bioactive Materials, 6*(12), 4255–4285. https://doi.org/10.1016/j.bioactmat.2021.04.028.
Wang, B., Liu, W., Xing, D., Li, R., Lv, C., Li, Y., Yan, X., Ke, Y., Xu, Y., Du, Y., & Lin, J. (2017). Injectable nanohydroxyapatite-chitosan-gelatin micro-scaffolds induce regeneration of knee subchondral bone lesions. *Scientific Reports, 7*(1), 16709. https://doi.org/10.1038/s41598-017-17025-6.
Wang, D., Zhang, N., Meng, G., He, J., & Wu, F. (2020). The effect of form of carboxymethyl-chitosan dressings on biological properties in wound healing. *Colloids and Surfaces B: Biointerfaces, 194*, 111191. https://doi.org/10.1016/j.colsurfb.2020.111191.
Wang, L., Zhang, X., Xu, P., Yan, J., Zhang, Y., Su, H., Sun, C., Lu, Q., & Liu, W. (2022). Exploration of sea anemone-inspired high-performance biomaterials with enhanced antioxidant activity. *Bioactive Materials, 10*, 504–514. https://doi.org/10.1016/j.bioactmat.2021.08.021.
Wang, T., Zhang, S., Ren, S., Zhang, X., Yang, F., Chen, Y., & Wang, B. (2020). Structural characterization and proliferation activity of chondroitin sulfate from the sturgeon, Acipenser schrenckii. *International Journal of Biological Macromolecules, 164*, 3005–3011. https://doi.org/10.1016/j.ijbiomac.2020.08.110.
Wang, Y., Xing, M., Cao, Q., Ji, A., Liang, H., & Song, S. (2019). Biological activities of fucoidan and the factors mediating its therapeutic effects: A review of recent studies. *Marine Drugs, 17*(3), 183. https://doi.org/10.3390/md17030183.
Wu, Z., Korntner, S. H., Mullen, A. M., Skoufos, I., Tzora, A., & Zeugolis, D. I. (2021). In the quest of the optimal tissue source (porcine male and female articular, tracheal and auricular cartilage) for the development of collagen sponges for articular cartilage. *Biomedical Engineering Advances, 1*, 100002. https://doi.org/10.1016/j.bea.2021.100002.
Xu, N., Peng, X. L., Li, H. R., Liu, J. X., Cheng, J. S. Y., Qi, X. Y., Ye, S. J., Gong, H. L., Zhao, X. H., Yu, J., Xu, G., & Wei, D. X. (2021). Marine-derived collagen as biomaterials for human health. In *Frontiers in Nutrition, 8*, 702108.
Xu, Z., Zhu, Z., Chen, H., Han, L., Shi, P., Dong, X., Wu, D., Du, M., & Li, T. (2022). Application of a Mytilus edulis-derived promoting calcium absorption peptide in calcium phosphate cements for bone. *Biomaterials, 282*, 121390. https://doi.org/10.1016/j.biomaterials.2022.121390.
Yang, F., Zhao, D., Zhang, K., Wang, Z., Wang, Y., Wu, C., Cui, S., Guo, T., Chen, L., & Chen, J. (2022). Oral delivery of marine shellfish supramolecule peptides for skin wound healing. *Colloids and Surfaces B: Biointerfaces, 216*, 112592. https://doi.org/10.1016/j.colsurfb.2022.112592.
Yang, S., Qian, Z., Liu, D., Wen, N., Xu, J., & Guo, X. (2019). Integration of C-type natriuretic peptide gene-modified bone marrow mesenchymal stem cells with chitosan/silk

fibroin scaffolds as a promising strategy for articular cartilage regeneration. *Cell and Tissue Banking, 20*(2), 209–220. https://doi.org/10.1007/s10561-019-09760-z.

Yang, Y., Geng, Y., Liu, M., Liu, K., Lv, X., Yu, H., Pan, P., Chen, L., & Chen, J. (2022). Facile mussel-inspired polymerization to facilitate biomimetic in situ homogeneous mineralization for bone regeneration. *Composites Part B: Engineering, 247*, 110325. https://doi.org/10.1016/j.compositesb.2022.110325.

Yang, Y., Karakhanova, S., Werner, J., & Bazhin, A. V. (2013). Reactive oxygen species in cancer biology and anticancer therapy. *Current Medicinal Chemistry, 20*(30), 3677–3692. https://doi.org/10.2174/0929867311320999165.

Yang, Y., Yang, G., Song, Y., Xu, Y., Zhao, S., & Zhang, W. (2020). 3D Bioprinted Integrated Osteochondral Scaffold-Mediated Repair of Articular Cartilage Defects in the Rabbit Knee. *Journal of Medical and Biological Engineering, 40*(1), 71–81. https://doi.org/10.1007/s40846-019-00481-y.

Yegappan, R., Selvaprithiviraj, V., Amirthalingam, S., & Jayakumar, R. (2018). Carrageenan based hydrogels for drug delivery, tissue engineering and wound healing. *Carbohydrate Polymers, 198*, 385–400. https://doi.org/10.1016/j.carbpol.2018.06.086.

You, F., Chen, X., Cooper, D. M. L., Chang, T., & Eames, B. F. (2018). Homogeneous hydroxyapatite/alginate composite hydrogel promotes calcified cartilage matrix deposition with potential for three-dimensional bioprinting. *Biofabrication, 11*(1), 15015. https://doi.org/10.1088/1758-5090/aaf44a.

Younes, I., & Rinaudo, M. (2015). Chitin and chitosan preparation from marine sources. Structure, properties and applications. *Marine Drugs, 13*(3), 1133–1174. https://doi.org/10.3390/md13031133.

Yuan, Y., & Macquarrie, D. (2015). Microwave assisted extraction of sulfated polysaccharides (fucoidan) from Ascophyllum nodosum and its antioxidant activity. *Carbohydrate Polymers, 129*, 101–107. https://doi.org/10.1016/j.carbpol.2015.04.057.

Zargarzadeh, M., Amaral, A. J. R., Custódio, C. A., & Mano, J. F. (2020). Biomedical applications of laminarin. *Carbohydrate Polymers, 232*, 115774. https://doi.org/10.1016/j.carbpol.2019.115774.

Zhang, C., & Kim, S.-K. (2010). Research and application of marine microbial enzymes: status and prospects. *Marine Drugs, 8*(6), 1920–1934. https://doi.org/10.3390/md8061920.

Zhang, F.-X., Liu, P., Ding, W., Meng, Q.-B., Su, D.-H., Zhang, Q.-C., Lian, R.-X., Yu, B.-Q., Zhao, M.-D., Dong, J., Li, Y.-L., & Jiang, L.-B. (2021). Injectable Mussel-Inspired highly adhesive hydrogel with exosomes for endogenous cell recruitment and cartilage defect regeneration. *Biomaterials, 278*, 121169. https://doi.org/10.1016/j.biomaterials.2021.121169.

Zhang, F., Zhang, Z., & Linhardt, R. J. (2010). Chapter 3 – Glycosaminoglycans. In R. D. Cummings & J. M. Pierce (Eds.), *Handbook of Glycomics* (pp. 59–80). Academic Press. https://doi.org/10.1016/B978-0-12-373600-0.00003-2.

Zhang, H., Cheng, J., & Ao, Q. (2021). Preparation of Alginate-Based Biomaterials and Their Applications in Biomedicine. *Marine Drugs, 19*(5), 264. https://doi.org/10.3390/md19050264.

Zhang, H., Wu, X., Quan, L., & Ao, Q. (2022). Characteristics of Marine Biomaterials and Their Applications in Biomedicine. *Marine Drugs, 20*(6), 372, https://doi.org/10.3390/md20060372.

Zhang, L. (2010). Glycosaminoglycan (GAG) Biosynthesis and GAG-Binding Proteins. In L. Zhang (Ed.), *Glycosaminoglycans in Development, Health and Disease* (Vol. 93, pp. 1–17). Academic Press. https://doi.org/10.1016/S1877-1173(10)93001-9.

Zhou, J., Liu, W., Zhao, X., Xian, Y., Wu, W., Zhang, X., Zhao, N., Xu, F. J., & Wang, C. (2021). Natural melanin/alginate hydrogels achieve cardiac repair through ROS scavenging and macrophage polarization. *Advanced Science (Weinheim, Baden-Wurttemberg, Germany), 8*(20), e2100505. https://doi.org/10.1002/advs.202100505.

Zhu, C.-F., Li, G.-Z., Peng, H.-B., Zhang, F., Chen, Y., & Li, Y. (2010). Treatment with marine collagen peptides modulates glucose and lipid metabolism in Chinese patients with type 2 diabetes mellitus. *Applied Physiology, Nutrition, and Metabolism = Physiologie Appliquee, Nutrition et Metabolisme, 35*(6), 797–804. https://doi.org/10.1139/H10-075.

Zhu, C., Zhang, W., Mu, B., Zhang, F., Lai, N., Zhou, J., Xu, A., Liu, J., & Li, Y. (2017). Effects of marine collagen peptides on glucose metabolism and insulin resistance in type 2 diabetic rats. *Journal of Food Science and Technology, 54*(8), 2260–2269. https://doi.org/10.1007/s13197-017-2663-z.

CHAPTER 9

Marine Biopolymer-Based Gastro-Intestinal Mucoadhesive Drug Delivery Systems

OLIVIA SEN,[1] SREEJAN MANNA,[1] and SOUGATA JANA[2]

[1]*Department of Pharmaceutical Technology, Brainware University, Barasat, Kolkata, West Bengal, India*

[2]*Department of Health and Family Welfare, Directorate of Health Services, Kolkata, West Bengal, India*

ABSTRACT

In recent times, marine biopolymers are getting more importance as novel biomaterials for developing site specific drug delivery systems. These biopolymers have excellent biocompatibility, biodegradable nature, low cost, and non-toxic properties. Mucoadhesive delivery strategy has proven to be one of the efficient delivery systems for target specific drug delivery. The mucoadhesive behavior of several marine biopolymers, including alginate, carrageenan and chitosan are investigated by pharmaceutical researchers to fabricate gastro-intestinal mucoadhesive systems. These biopolymers have the ability to produce gels, and thereby control the release of drugs for a long duration. Therefore, these biopolymers facilitate localization of drugs in the particular gastrointestinal tract (GIT). Different approaches can be used to prepare efficient mucoadhesive drug delivery systems using these marine biopolymers which can also help maintain the different obstacles like gastric pH and gastric motility which has a huge impact on gastric retention time and drug delivery. Here, the authors have described several classes of marine biopolymers and their applications in developing mucoadhesive drug delivery.

Handbook of Research in Marine Pharmaceutics: Exploring Oceanic Microbial Diversity for Human Health and Wellness. Md. Faiyazuddin & Khalid Rehman Hakeem (Eds.)

9.1 INTRODUCTION

Oral route of administration is a very common and promising delivery route because of its convenient delivery (Kwon et al., 2016; Ma et al., 2018; Maharjan et al., 2016). However, this oral delivery also has several disadvantages. Because of its physiological variability and other factors, inappropriate absorption of drugs is the major drawback of oral delivery (Hu et al., 2018; Leal et al., 2017). Gastric residence time is another important parameter. Whereas oral delivery exhibits quick gastric emptying time and less bioavailability. But nowadays gastric retention has acquired lots of attention (Choi et al., 2014; Banerjee et al., 2016; Araújo et al., 2017). Mucoadhesive drug delivery systems are preferred because they can prolong the gastric emptying time at the targeted site of action, offer prolonged drug release, and reduce the degradation of drugs at various sites of the body (Butnarasu et al., 2019; Cook et al., 2017; Mackie et al., 2017). They can stick to mucus membranes and facilitate the improvement of the therapeutic index of drugs. They also increase bioavailability, and cost-effectiveness which may lead to great patient compliance (Menchicchi et al., 2018; Rossi et al., 2018; Albarkah et al., 2015). Adhesion is a process which involves adhering two surfaces to each other. Adhesion The process of fixing two surfaces with one another is known as "adhesion." The term "mucoadhesion" refers to adhesion that can take place at the mucosal membrane (Russo et al., 2016; Abodinar et al., 2016).

The mucosal layer is the most frequent biological substrate, hence the term mucoadhesion. A gel-like substance known as mucin often covers the body's internal organs, including the nose, mouth, rectum, vagina, and gastrointestinal tract (GIT) (Sosnik et al., 2014; Bassi et al., 2017). Since mucin is found throughout the physiological system, all mucoadhesives should be able to interact with the mucin layer to provide the proper adhesion (Woertz et al., 2013; Khutoryanskiy et al., 2011). Mucus has binding properties since it can adhere to the mucous membrane. The mucus layer, which covers the mucous membrane, is a viscoelastic gel that is 95% water in composition (Shaikh et al., 2011; Carvalho et al., 2010). The viscoelastic characteristics and interactions with its surroundings are for mucin glycoproteins (mucins), which are the most important constituents of the mucus layer. This mucin gives a gel-like texture and anti-adhesive characteristics to the mucous membrane (Punitha et al., 2010). Typically, mucoadhesive polymers are hydrophilic nets with a large number of polar functional groups. In order to generate weakly crosslinked networks, such polymers interact with mucus not only via diffusion but also through secondary chemical linkages. Some mucoadhesive polymers can alter the tight connections between cells, increasing the permeability of the drug through the epithelium as well as the retention time of the dosage form at the injection site (Singh et al., 2012; Bassi et al., 2018). The polymer must have particular structural features, such as strong hydrogen bonding groups, strong anionic or cationic charges, high molecular weight (MW), chain flexibility, and surface energy characteristics that enable spreading to mucus, in order to demonstrate good mucoadhesive properties (Chaudhari et al., 2014; Esim et al., 2020). For polymers with charged groups or nonionic functional groups or charged groups capable of forming hydrogen bonds with mucus, excellent mucoadhesive activity is often seen. Another crucial factor is the creation of hydrogen bonds between the mucosal layer and the functional groups of the polymers (Palem et al., 2016). In general, the adhesion is stronger the stronger the hydrogen bonding. Numerous polymer-based characteristics, including the degree of crosslinking, the chain length, and the presence of different functional groups, have an impact on the success and strength of mucoadhesion bonding. The hydroxyl, carboxyl, and amino functional groups are in charge of these interactions. The ability of carboxylic groups to form hydrogen bonds with the oligosaccharide chains of mucins has frequently been linked to the mucoadhesiveness of weakly anionic carboxyl-containing polymers, such as poly(acrylic through the epithelium as well as the retention time of the dosage form at the injection site (Singh et al., 2012; Bassi et al., 2018). The ability of carboxylic groups to form hydrogen bonds with the oligosaccharide chains of mucins has frequently been linked to poly-(methacrylic acid), carboxymethylcellulose, sodium alginate, and poly[(maleic acid)-co-(vinyl methyl ether)] (Pham et al., 2021). In order for mucoadhesion to take place, it was demonstrated that polymers needed to have functional groups capable of forming hydrogen bonds above the critical concentration and that the polymer chains needed to be flexible enough to form as many hydrogen bonds as feasible. Mucoadhesive polymers can help to increase the bioavailability of the active pharmaceutical ingredient by enhancing the gastric retention time at the targeted site. Mucoadhesive polymers have a significant role in a number of areas, including the gastrointestinal tract, vagina, rectal lumen, buccal cavity, and nasal cavity (Manna et al., 2022b; Paul et al., 2022). Carbohydrate hydrophilic biopolymers have gained popularity as drug delivery systems in recent years among mucoadhesive polymers. Since their mucoadhesive qualities make them attractive for a variety of drug delivery applications, polysaccharides are among the most extensively researched natural macromolecules. These polymers are usually hydrophilic, stable, nontoxic, and biodegradable in nature. They exhibit reactive functional groups which are responsible for mucoadhesive characteristics. Sodium carboxymethyl cellulose, chitosan, and pectin show very good mucoadhesive properties (Hamdan et al., 2020; Cook et al., 2017). Marine biopolymers also have good mucoadhesive properties. Marine carbohydrate biopolymers have drawn interest because of their biodegradability and biocompatibility, abundant functional groups, powerful water absorption, and advantageous physicochemical features (Brown et al., 2014; Cardoso et al., 2016; Claverie et al., 2020). Various marine biopolymers like chitosan, carrageenan, alginate, and their utilization in formulating mucoadhesive drug delivery can be discussed here in this review.

9.2 CLASSIFICATION OF MARINE BIOPOLYMERS

Classification of various marine biopolymers is shown in Figure 9.1

9.2.1 BIOPOLYMERS DERIVED FROM MARINE ALGAE

The principal source of polysaccharides is algae among a large number of marine organisms (Manivasagan & Oh, 2016). Among them, red macroalgae are the main useful sources of polysaccharides. But polysaccharides can also be extracted from green or brown macroalgae (Ruocco et al., 2016; Venkatesan et al., 2015). Seaweeds are multicellular marine algae. They are also another source of polysaccharides. Currently, these marine algae are very useful for producing cosmetics and food supplements.

9.2.1.1 ALGINATES

Alginates can be obtained from brown seaweeds, such as *Ascophyllum nodosum, Laminaria digitata, Laminaria hyperborea, Laminaria japonica,* and *Macrocystis pyrifera* (Lee & Mooney, 2012). Alginates are also extracted from bacterial biosynthesis from Pseudomonas and Azotobacter species (Aprilliza et al., 2017; Siddhesh et al., 2012). Alginate is biocompatible, less toxic, and has high bioavailability. It consists of consecutively organized two blocks including β-D-mannuronate (M) and (1–4)-linkages of α-L-guluronate (G) monomers (Hecht & Srebnik, 2016). According to the geographic source variation, the ratio of the M and G blocks can be altered (Najdahmadi et al., 2018). After cross-linking with bivalent or trivalent metal ions, the ionic character of alginate makes it easier to produce hydrogel matrices (Karmakar et al., 2022; Manna et al., 2022c). Due to the presence of a carboxyl group, which aids in protecting drug molecules at basic and neutral pH, alginate is reported to have a pH-responsive nature (Orivea et al., 2006; Becker & Brandon, 2001; O'Brien, 2011). Alginates are easy to extract and purify, so they can be used in various pharmaceutical and food industries. Alginates have very good biodegradability, biocompatibility, and less cytotoxicity, making them suitable for formulating hydrogel matrices, in situ gels, microparticulate formulations, nano-formulations, and transdermal preparations (Manna et al., 2022d; Jana et al., 2016; Jana et al., 2013b).

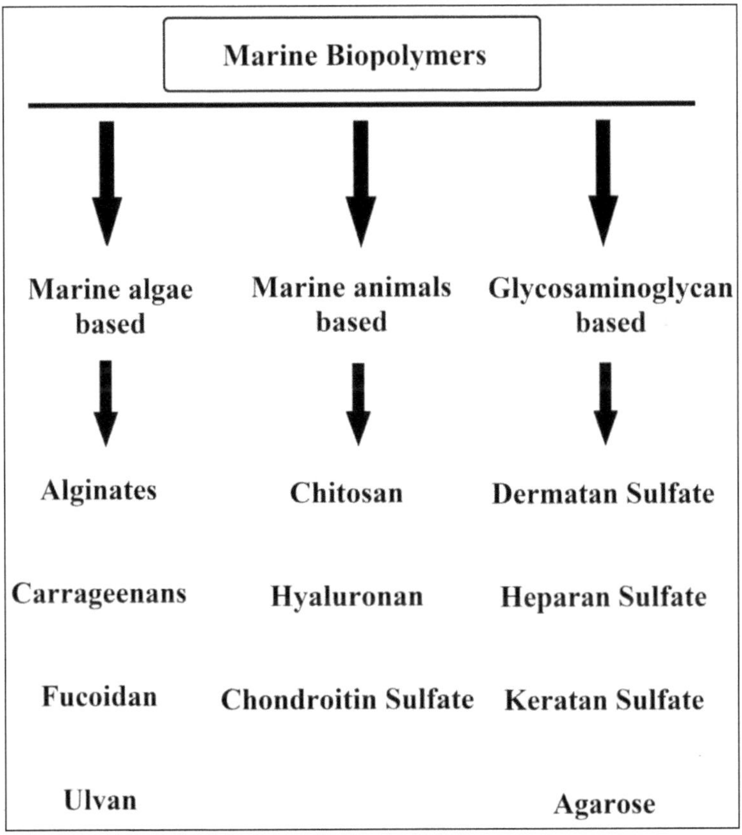

FIGURE 9.1 Classification of marine biopolymers.

9.2.1.2 CARRAGEENANS

Carrageenans (CG) are hydrophilic linear sulfated high molecular weight polysaccharides that are isolated from different red algae (Zia et al., 2017). It is an anionic polysaccharide since it includes 15% to 40% ester-sulfate. According to their sulfate concentration, method of extraction, and solubility, CG can be divided into six fundamental forms: kappa-CG, iota-CG, lambda-CG, mu-CG, nu-CG, beta-CG, and theta-CG (Necas et al., 2013). Therefore, *Kappaphycus alvarezii* and *Eucheuma denticulatum* are used to extract carrageenans (Rhein-Knudsen et al., 2015). Iota-CG is primarily made from Eucheuma denticulatum, also named Eucheuma spinosum, while kappa-carrageenan is primarily derived from *Kappaphycus alvarezii,* commonly called *Eucheuma cottonii* in the marketplace. CG is made up of d-galactose and 3,6-anhydro-galactose that are connected by alternating 1,3- and 1,4-glycosidic linkages (Jiao et al., 2011). Kappa and iota types of carrageenans are responsible for antiviral, antibacterial, anti-hyperlipidemic, anti-tumor, and immunomodulatory properties. Antioxidant properties are obtained only from Kappa-CG (Gómez-Ordóñez et al., 2014; Inic-Kanada et al., 2018). Lambda-CG is responsible for anticoagulant, antithrombotic, antitumor, and immunomodulatory activity. Carrageenan can be widely used as a biomaterial in biomedical applications in different drug delivery strategies, e.g., hydrogels, films, tablets, beads, pellets, suppositories, microparticles, nanoparticles, and tissue engineering (Manna et al., 2021; Liu et al., 2015).

9.2.1.3 FUCOIDAN

A long-chain sulfated polysaccharide called fucoidan is present in some types of brown algae. The seaweed species *Laminaria japonica, Fucus vesiculosus, Cladosiphon okamuranus,* and *Undaria pinnatifida* are frequently used to extract the fucoidan used in commercial products (Fitton et al., 2015). Animal species, including the sea cucumber, have also been found to contain different types of fucoidan. It consists of 44.1% fucose, 26.3% sulfate, 31.1% ash, and a small amount of amino glucose (Luthuli et al., 2019). Numerous studies have recently shown that fucoidans exhibit various biological actions, including antioxidant, antiviral, anticoagulant, anticancer, anticoagulant, and anti-inflammatory characteristics. Additionally, it is used for uropathy, gastroprotective properties, and renal treatment (Li et al., 2008). Fucospheres combined with chitosan are utilized to create a beneficial burn treatment by electrostatic interaction (van Weelden et al., 2019).

9.2.1.4 ULVANS

Ulvans are isolated from marine microalgae Ulva and Enteromorpha genus. It is a sulfated marine polysaccharide. Ulvans can be extracted by depolymerization of organic solvent on feedstock (Melcher et al., 2017). It consists of different types of sugar moieties like xylose, glucose, iduronic acid, glucuronic acid, etc. Molecular weight, electron density, and charge distribution in Ulvan can vary with the percentage of sugar moieties present in its structure (Marco et al., 2021; Fei et al., 2021).

9.2.2 BIOPOLYMERS DERIVED FROM MARINE ANIMALS

There are various marine sources of polysaccharides other than algae. Marine animals are the primary source of glycosaminoglycans (GAG) and chitin derivatives which are the source of marine polysaccharides (Claverie et al., 2020).

9.2.2.1 CHITOSAN

Chitosan is obtained from chitin, which can be isolated from the exoskeleton of numerous marine animals like shrimp, lobsters, and crabs (Younes & Rinaudo, 2015). The deacetylation process is used to transform chitin into chitosan. Acetylated units (D-glucosamine residues) and deacetylated units (N-acetyl-D-glucosamine) may coexist depending on the grade of deacetylation (Cheung et al., 2015). Chitosan behaves cationically in aqueous solution and is soluble in acidic pH. This characteristic of chitosan helps in the development of electrostatic complexes with polymers that have opposing charges (Manna et al., 2022a; Jana et al., 2013a). Chitosan has reportedly recently been found to exhibit a variety of biological characteristics, such as anticancer, antioxidant, and antibacterial behavior (Abd El-Hack et al., 2021). Chitosan is a biocompatible and biodegradable marine polymer, which contributes to its widespread use in the pharmaceutical and biomedical industries. In addition to these, chitosan has a wide variety of uses in the food processing industry. Additionally, it has shown potential as a tissue engineering and wound healing scaffold (Jana & Sen, 2017; Kim et al., 2008).

9.2.2.2 HYALURONAN

Hyaluronan is a non-sulfated polyanionic polysaccharide also referred to as hyaluronic acid. It is a natural mucoadhesive linear polysaccharide consisting of alternating N-acetylglucosamine and glucuronic acid units (Prydz et al., 2015; Gupta et al., 2019). With its very good physicochemical characteristics, such as non-toxicity, biocompatibility, biodegradability, and non-immunogenicity, hyaluronic acid has a wide variety of applications. It is a great tool for biomedical uses like osteoarthritis tissue engineering, ocular surgery, plastic surgery, and drug delivery (Papakonstantinou et al., 2012; Ghiselli et al., 2017).

9.2.2.3 CHONDROITIN SULFATES

Chondroitin sulfates are long unbranched sulfated glycosaminoglycans containing marine polysaccharides. They are primarily isolated from bovine and porcine cartilage. Additionally, they can also be obtained from other sea animals, such as salmon, sea cucumbers, sharks, whales, and mollusks. They are made up of N-acetylgalactosamine attached to glucuronic acid disaccharide (Henrotin et al., 2010; Mishra & Ganguli, 2021). Sulfated GAGs can also be used as anticoagulants, to stop bleeding, and prevent arthritis (Djerbal et al., 2017; Henrotin et al., 2010).

9.2.3 BIOPOLYMERS OBTAINED FROM GLYCOSAMINOGLYCANS

Glycosaminoglycans have low bioavailability and can be difficult to extract. As a result, glycosaminoglycans have limited applications in the pharmaceutical field. However, some polysaccharides, such as keratan sulfate, dermatan sulfate, heparan sulfate, and agarose, have found applications in the pharmaceutical industry (Merry, 2021).

- **Dermatan Sulfate:** A glycosaminoglycan termed dermatan sulfate, formerly known as a mucopolysaccharide, is mostly present in the skin but is also found in blood vessels, heart valves, tendons, and lungs (Westergren-Thorsson et al., 1991). Additionally, it is known as chondroitin sulfate B and is written as $C_{14}H_{21}NO_{15}S$. This carbohydrate is composed of N-acetyl galactosamine and iduronic acid, containing linear polymers of disaccharide units. Dermatan sulfate may play a role in fibrosis, cardiovascular disease, cancer, and wound healing (Rostand & Esko, 1997).

- **Heparan Sulfate:** Heparan sulfate is a linear polysaccharide glycosaminoglycan containing a repetitive unit of D-glucuronic acid and iduronic acid in combination with either sulfated or acetylated D-glucosamine (Li et al., 2016). Depending on the sulfated molecules, the biological properties of heparan sulfate may change (Wang et al., 2021; Brito et al., 2014).
- **Keratan Sulfates:** This marine polysaccharide is formed from -(1-4) linked N-acetyl glucosamine units and galactose residues. Keratan sulfate is divided into three types based on its protein binding ability, namely class I, class II, and class III, which are found in corneal cartilages, small cartilages, and nerve tissue precipitates, respectively (Caterson et al., 2018; Tai et al., 1997).
- **Agarose:** Agarose can be isolated from the cell wall of red seaweed. It is a polysaccharide of repetitive units of agarobiose. Agarose has been used in the production of hydrogels and nanoparticle delivery systems for a variety of pharmacological purposes (Khodadadi et al., 2020).

9.3 MECHANISM OF MUCOADHESION

Mucoadhesion comprises wetting, adsorption, and interpenetration of polymer chains, which is a complex process. The following stages are involved in mucoadhesion, i.e., the contact stage and consolidation stage as shown in Figure 9.2. In the contact stage, the mucoadhesive and mucosal membranes come into contact with each other through wetting and swelling phenomena. The consolidation stage involves the penetration of mucoadhesive materials into the interface of the mucoadhesion (Andrews et al., 2009).

9.3.1 MUCOADHESION THEORIES

Mucoadhesion is a complicated process. Various theories have been suggested to describe the mechanisms of mucoadhesion. These theories are wetting theory, diffusion theory, fracture theory, electronic theory, and adsorption theory (Shaikh et al., 2011).

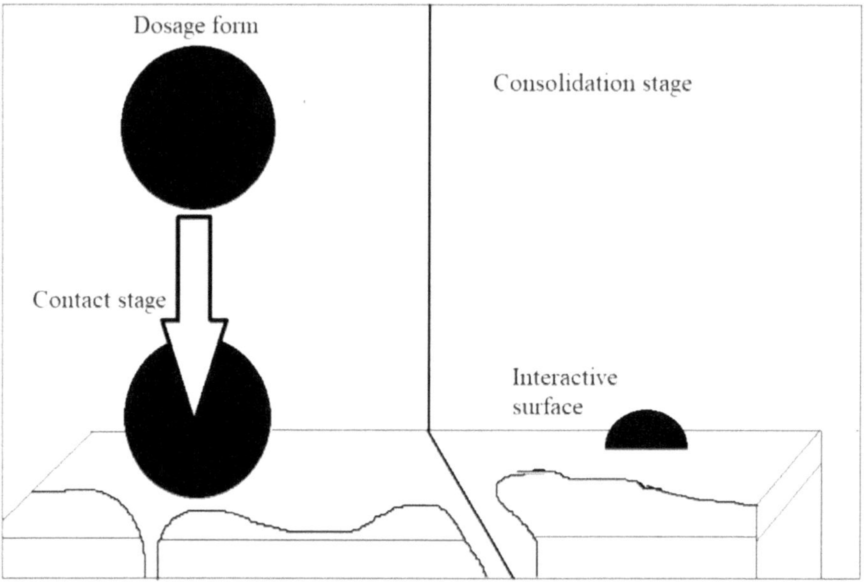

FIGURE 9.2 Mechanism of mucoadhesion.

9.3.1.1 WETTING THEORY

The wetting theory applies to liquid systems that exhibit surface affinity and spread across them. The contact angle is one of the measuring techniques that can be used to determine this affinity. According to the general norm, affinity increases with decreasing contact angle. For adequate spreadability, the contact angle should be equal to or nearly equal to zero. As shown in the equation below, the spreadability coefficient, SAB, can be determined from the variation between the surface energies γB and γA and the interfacial energy γAB (Hägerström et al., 2003).

$$SAB = \gamma B - \gamma A - \gamma AB$$

The greater the individual surface energy of mucous and device in relation to the interfacial energy, the greater the adhesion work, WA, i.e., the greater the energy required to segregate the two phases.

$$WA = \gamma A + \beta B - \gamma AB$$

9.3.1.2 DIFFUSION THEORY

The formation of a semi-permanent adhesive link between polymer chains and mucus can be explained by diffusion theory as shown in Figure 9.3. This theory proposes that the degree of polymer chain penetration enhances the adhesive force. The flexibility, diffusion coefficient, and nature of the adhesive chain, movement, and contact time all affect how much penetration occurs. Between 0.2 and 0.5 m of interpenetration, an effective bioadhesive bond must be created. The interpenetration depth between polymer chains and mucus can be calculated using the equation below: $l = (tDb)^{1/2}$ is the adhesion formula, where t is the contact time and Db is the mucoadhesive material's mucus diffusion coefficient (Smart et al., 2005).

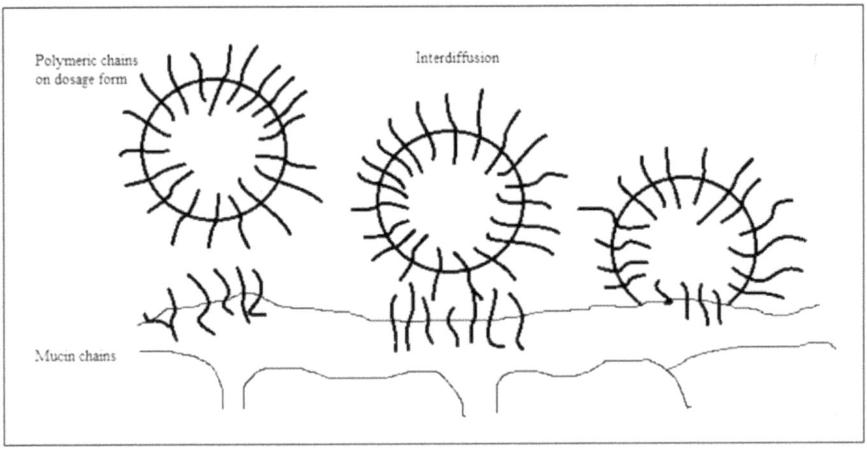

FIGURE 9.3 Mechanism of mucoadhesion according to diffusion theory.

9.3.1.3 FRACTURE THEORY

This theory explains the force needed for the separation of two surfaces after adhesion. Through the following equation, the fracture strength is equivalent to the adhesive strength. This theory is useful for the study of bioadhesion by tensile apparatus (Dodou et al., 2005).

$$\sigma = (E \times \varepsilon/L)^{1/2}$$

where, σ is the fracture strength; ε is the fracture energy; E is the young modulus of elasticity; L is the critical crack length.

9.3.1.4 ELECTRONIC THEORY

According to this theory, an electrical double layer forms at the mucoadhesive interface as a result of electrons being transferred between the mucoadhesive polymer and the mucin glycoprotein network. For example, the interaction between positively charged polymers like chitosan and the negatively charged mucosal surface becomes adhesive upon hydration and provides an intimate contact between the dosage form and absorbing tissue (Dodou et al., 2005).

9.3.1.5 ADSORPTION THEORY

In accordance with this theory, the surface forces between the chemical structures at the two surfaces lead to sticking after the initial contact between the two surfaces. Polar molecules or groups can reorient at the interface. Adhesion can lead to chemisorption if it is very strong. According to the notion, one or more secondary forces—including van der Waals forces, hydrogen bonds, and hydrophobic bonds—are ultimately responsible for adherence to tissue (Dodou et al., 2005).

9.4 APPLICATIONS OF MARINE BIOPOLYMERS IN MUCOADHESIVE DRUG DELIVERY

The applications of several marine biopolymer-based mucoadhesive drug delivery are briefly discussed here as shown in Table 9.1. The various drug delivery approaches for gastrointestinal mucoadhesive systems employing marine biopolymers are shown in Figure 9.4.

9.4.1 ALGINATE-BASED MUCOADHESIVE DRUG DELIVERY

Diós et al. designed sodium alginate-based mucoadhesive metronidazole buoyant tablets for *H. pylori* eradication. According to statistical analysis, tablets made of sodium alginate, low-substituted hydroxypropyl cellulose, and sodium bicarbonate demonstrated good *in vitro* floating and dissolution characteristics. The prepared floating tablets showed superiority in drug release study compared to conventional metronidazole. The mucoadhesive and rheological characteristics were deemed good (Diós et al., 2015).

TABLE 9.1 Applications of Several Marine Biopolymer-Based Mucoadhesive Drug Delivery

Sl. No.	Delivery Strategy	Polymeric composition	Special Features	Drug	References
1.	Alginate-based mucoadhesive drug delivery system	Alginate + hydroxypropyl cellulose	Mucoadhesive floating tablets	Metronidazole	Diós et al. (2015)
		Sodium alginate + pectin	Mucoadhesive beads	Ciprofloxacin hydrochloride	Jain et al. (2016)
		Sodium alginate + HPMC K4M	Mucoadhesive Floating beads	Metoprolol Tartrate	Biswas & Sahoo (2016)
		Sodium alginate + Linum usitatissimum mucilage	Mucoadhesive microspheres	Metformin HCl	Ghumman et al. (2020)
		Sodium alginate + ispaghula husk mucilage	Mucoadhesive beads	Glibenclamide	Nayak et al. (2013)
		Sodium alginate + Carboxymethylcellulose	Mucoadhesive microbeads	*Lactobacillus kefiri*	Demitri et al. (2017)
		Sodium alginate + sterculia gum	Mucoadhesive Floating beads	Risperidone	Bera et al., 2015)
		Sodium alginate + *Bletilla striata*	Mucoadhesive microspheres	–	Wu et al. (2020)
2.	Chitosan based mucoadhesive drug delivery	Chitosan + Sodium alginate	Mucoadhesive microcapsules	Probiotics	(Ta et al. (2021)
		Chitosan	Mucoadhesive floating beads	Emodin	Chen et al. (2019)
		Chitosan	Mucoadhesive films	Ascorbic acid	Kristó et al. (2022)
		Chitosan + Polycaprolactone	Mucoadhesive matrix tablets	Milnacipran HCl	Hussain et al. (2020)
		Chitosan + Carbopol 934 + HPMC K4M	Gastro retentive mucoadhesive tablets	Tranexamic acid	Agarwal et al. (2015)
		Thiolated chitosan	Mucoadhesive nanoparticles	α-mangostin	Samprasit et al. (2021)
		Chitosan + gelatin	Mucoadhesive buccul tablets	Propranolol hydrochloride	Abruzzo et al. (2015)
		Chitosan	Mucoadhesive microspheres	Benzydamine hydrochloride	Jiang et al. (2017)

TABLE 9.1 *(Continued)*

Sl. No.	Delivery Strategy	Polymeric composition	Special Features	Drug	References
3.	Carrageenan-based mucoadhesive drug delivery	Carrageenan + chitosan	Mucoadhesive beads	Echinochrome A	Yermak et al. (2021)
		Carboxymethylated kappa-carrageenan	Mucoadhesive microparticulate system	Insulin	Leong et al. (2011)
		Carboxymethylated iota-carrageenan + chitosan	Mucoadhesive nanoparticles	Insulin	Sahoo et al. (2017)
		Carboxymethylated iota-carrageenan + chitosan	Mucoadhesive nanoparticles	Insulin	Sahoo et al. (2019)
		κ-carrageenan + hydroxypropyl cellulose	In situ mucoadhesive gel	–	Vigani et al. (2019)
		κ-carrageenan + hydroxypropyl cellulose	In situ mucoadhesive gel	Hydroalcoholic extract of *Hibiscus sabdariffa*	Vigani et al. (2019)
		κ-carrageenan + polyethylene glycol 600 + poloxamer	Mucoadhesive buccal wafer	–	Kianfar et al. (2013)
4.	Agarose-based mucoadhesive drug delivery	Agarose + Carbopol 934P	Mucoadhesive gel	Benzocaine + Tibezonium iodide	Syed et al. (2022)
		Agarose + guar gum	Mucoadhesive buccal film	Zolmitriptan succinate	Jillani et al. (2022)

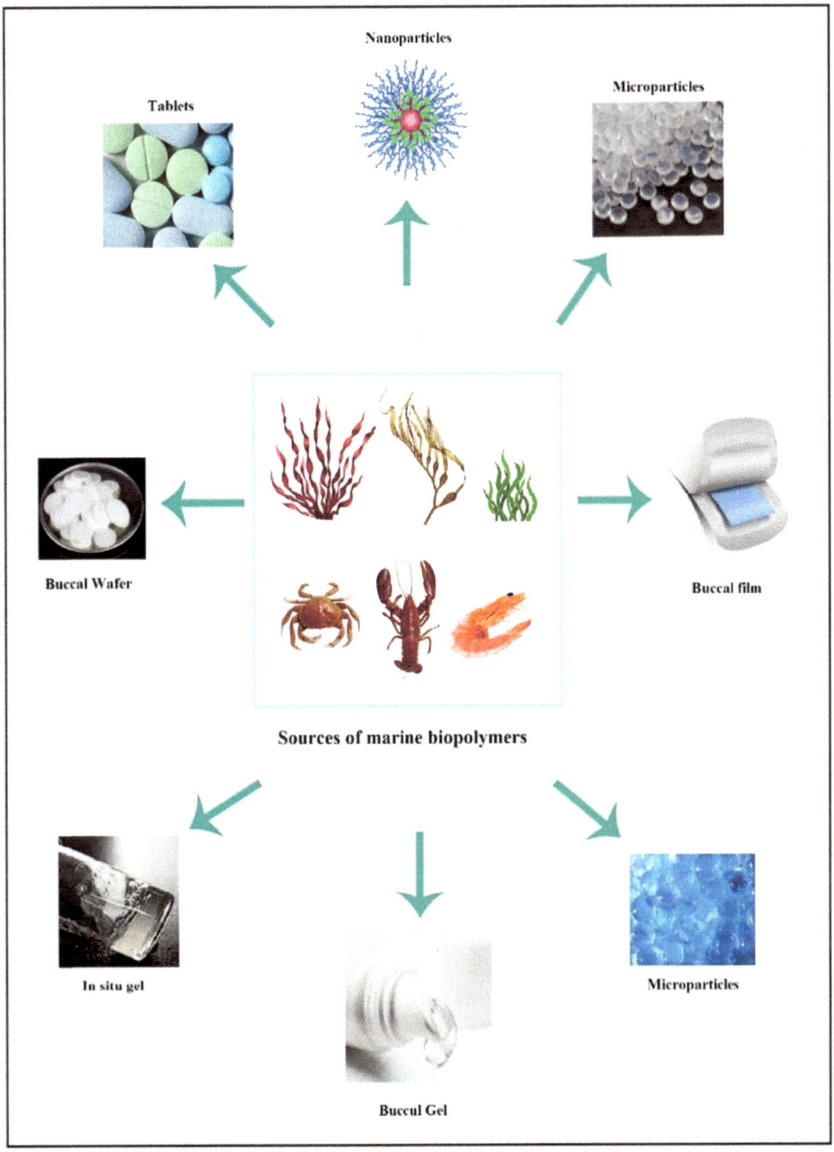

FIGURE 9.4 Different drug delivery strategies for gastrointestinal mucoadhesive systems employing marine biopolymers.

Jain et al. (2016) prepared a drug-resin complex with resinous matter and ciprofloxacin hydrochloride using activated sodium polystyrene sulfonate. The prepared drug-resin complex was added to an alginate-pectin polymer complex, which was further crosslinked with a calcium chloride solution to prepare mucoadhesive beads. These ionically cross-linked beads were fabricated using a 21-gauge needle. Subsequently, the beads were filtered, washed with deionized water, and dried at 37°C.

Biswas and Sahoo prepared alginate-based mucoadhesive metoprolol tartrate floating beads using the ionotropic gelation technique. Tapioca starch-containing beads showed a significant improvement in entrapment efficiency and sustained release behavior of the drug. Beads coated with HPMC revealed that the drug release extended up to 11 hours. Pharmacokinetic assessment using a rabbit model showed that the floating beads had considerably higher oral bioavailability (Biswas & Sahoo, 2016).

Ghumman et al. developed metformin HCl incorporated *Linum usitatissimum* mucilage and alginate-containing mucoadhesive microspheres through the ionotropic gelation method. Different ratios of polymers were used to analyze the *in vitro* release and encapsulation efficiency of the drug for 12 hours. All formulations had sustainable drug entrapment efficiency. SEM and FTIR analyses were also conducted. The pH of the buffer medium affected the swelling characteristics of the formulations. These microspheres were reported to have good mucoadhesivity (Ghumman et al., 2020).

Nayak et al. prepared glibenclamide-loaded ispaghula husk mucilage and alginate mucoadhesive beads through the ionic gelation method. The SEM study reported rough surfaces with characteristic large wrinkles and cracks. Drug entrapment efficiency was more than 90% for the optimized formulation with excellent mucoadhesive properties on biological membranes. The drug release data indicated a controlled release of glibenclamide following a zero-order release mechanism. *In vivo* studies showed a significant antidiabetic effect after oral administration in diabetes-induced rats (Nayak et al., 2013).

Demitri et al. formulated an encapsulation technique to isolate aerosols of alginate-containing solution and cross-linked them with $CaCl_2$-containing *Lactobacillus kefiri* for the preparation of alginate microbeads. Carboxymethylcellulose was also added to alginate, which possesses excellent mucoadhesivity with the mucosal surface. The SEM micrographs of the crosslinked and non-crosslinked microbeads can be differentiated by their porous and smooth surfaces, respectively. The study indicated that the incorporation of $CaCl_2$ in the bead formulation can organize the molecules of alginate to create a surface with a hard shell around the microbeads (Demitri et al., 2017).

Bera et al. developed risperidone-loaded oil-entrapped alginate floating beads coated with alginate and sterculia gum gel membrane. Beads were formulated by ionic gelation method. Entrapment efficiency and release of the drug were studied for up to 8 hours. Drug entrapment efficiency and cumulative drug release of optimized formulations were 83.73% ± 0.81% and 70.84% ± 0.27% after 8 hours, respectively. These beads exhibited slow release of the drug, excellent buoyancy, and good *ex vivo* mucoadhesion. The drug release followed the Korsmeyer-Peppas mechanism for risperidone-containing coated and uncoated beads. Surface morphology study, drug-excipient compatibility, and drug crystallinity study of optimized beads were investigated by SEM, FTIR, and P-XRD, respectively (Bera et al., 2015).

Wu et al. developed *Bletilla striata* combined alginate-based microspheres for enhancement of the mucoadhesion effect and retention in the GIT for a prolonged period of time. The prepared microspheres were spherical shaped and discrete. The inclusion of *Bletilla striata* made the microspheres more flexible and less stiff. After being blended with *Bletilla striata*, the swelling ability, mucin adsorption capacity, and rate of retention on the gastric mucosa of the sodium alginate matrix were improved. The mucoadhesion test reported that *Bletilla striata* and alginate microspheres remained in rats' stomachs for a longer period of time than the alginate microspheres. The *Bletilla striata* and alginate-based microspheres with improved mucoadhesion are strongly recommended as a possible drug carriers in the development of the gastroretentive drug delivery system (Wu et al., 2020).

9.4.2 CHITOSAN BASED MUCOADHESIVE DRUG DELIVERY

Ta et al. developed chitosan-coated alginate-based mucoadhesive microcapsules of probiotics using the electrospray method for gastrointestinal delivery. Microcapsules with a spherical shape and a size range of 30–600 μm were created using electro-spraying; however, the addition of chitosan expanded their size to 60–1300 μm. Between alginate and alginate-starch formulations, there was no statistically significant difference in the yield of viable cells that could be encapsulated; however, during the encapsulation process, some substantial losses in bacterial viability occurred. In the *in vitro* wash-off mucoadhesion test, chitosan-containing microparticles exhibited good mucoadhesive behavior, but in alginate-resistant starch-containing microparticles, low mucoadhesive strength on gastric mucosa was observed (Ta et al., 2021).

Chen et al. formulated an innovative delivery system of chitosan-coated emodin-entrapped nanomicelles incorporated in gastroretentive beads. The *in vitro* cytotoxicity test indicated that blank nanomicelles had no cytotoxicity and emodin-loaded nanomicelles showed better antitumor activity than emodin suspensions. Mucoadhesion, *in vitro* floating behavior, and swelling of nanomicelles incorporated floating mucoadhesive beads were investigated. The formation of nanomicelles and nanomicelles incorporated mucoadhesive floating beads were identified by FT-IR spectroscopy. The drug release of beads was followed by an anomalous transport mechanism. X-ray imaging study revealed that these beads stayed at least 8 hours in the stomach of rabbits for *in vivo* gastroretention. The beads can prolong the duration that antitumor drugs are retained in the stomach and can release the drugs locally. They can also significantly inhibit gastric cancer cells by releasing nanomicelles (Chen et al., 2019).

Kristó et al. formulated chitosan-based ascorbic acid-containing mucoadhesive films using the solvent casting method. Ascorbic acid was used in various concentrations (2–5%), providing an acidic pH to the formulation. Various parameters such as mucoadhesion force, surface free energy, and breaking strength were measured. These films were further analyzed using atomic force microscopy and free volume. XPRD and FTIR studies were also conducted to characterize the structure of the films. Ascorbic acid works well in chitosan films because it is biodegradable and acts as a permeability enhancer when combined with chitosan (Kristó et al., 2022).

Hussain et al. (2020) formulated chitosan-based milnacipran HCl-loaded mucoadhesive matrix tablets combined with polycaprolactone using the wet granulation method. *In vitro* release, DSC, and FTIR analysis were used to characterize the formulations. The intestinal mucosa of rabbits was used for *in vitro* mucoadhesive analysis. *In vivo*, analysis of drug release was performed on dogs. DSC and FTIR studies confirmed that there were no significant interactions between the polymers and milnacipran HCl in these matrix formulations. Adhesion time was improved by approximately 12 hours in the optimized formulations on the gastrointestinal mucosa. The *in vivo* investigation resulted in a longer milnacipran HCl elimination half-life, demonstrating that the matrix formulation is a sustained-release drug delivery system. The T_{max} was increased, and the C_{max} of the milnacipran HCl matrix decreased from 149.22 ± 9.942 ng/ml in the milnacipran HCl solution to 121.60 ± 9.496 ng/ml. Comparing the matrix formulation to the milnacipran HCl solution, there was a 272.20% ± 48.11% improvement in the bioavailability of the matrix formulation (Hussain et al., 2020).

Agarwal et al. formulated chitosan-containing tranexamic acid-based gastroretentive mucoadhesive tablets in combination with Carbopol 934 and HPMC K4M using the wet granulation technique. Weight variation, shape, size, appearance, hardness, friability, swelling, mucoadhesiveness, and *in vitro* release of the drug of the formulated tablets were performed.

Different batches of tablet formulations were developed using various ratios of chitosan and hydroxypropyl methylcellulose K4M. Tablets containing a higher concentration of chitosan resulted in faster drug release, higher swelling, and less mucoadhesion (Agarwal et al., 2015).

Samprasit et al. fabricated chitosan and thiolated chitosan-loaded α-mangostin mucoadhesive nanoparticles for colonic drug delivery for colorectal cancer. These nanoparticles were characterized by surface morphology study and particle size. Drug release patterns were observed. The wash-off method was used to examine *in vitro* mucoadhesive study. Antitumor properties were tested using colorectal cancer cells. The shape of the nanoparticles was a little bit oblong, with particle sizes ranging from 300 to 900 nm. The thiolated chitosan-containing nanoparticles using genipin and Eudragit® L100 displayed stronger mucoadhesion to colon mucosa than uncoated nanoparticles and chitosan-containing nanoparticles. Nanoparticles also showed antitumor action. For a controlled-release drug delivery strategy of α-mangostin to the colon, the mucoadhesive thiolated chitosan-coated nanoparticles may be a potential candidate (Samprasit et al., 2021).

Abruzzo et al. formulated chitosan-gelatin microparticles-based propranolol hydrochloride mucoadhesive tablets for buccal delivery. These microparticles were fabricated with various ratios of chitosan-gelatin by the spray drying technique. In order to prepare buccal tablets, microparticles were compressed. Various studies were performed such as mucoadhesion, drug release, *in vitro* water uptake, and drug permeation for the characterization of tablets. Tablets containing the maximum amounts of chitosan revealed the best mucoadhesive characteristics, permitted the maximum amount of drug permeation among all formulations, and the delivery of propranolol hydrochloride via the buccal route must be very promising (Abruzzo et al., 2015).

Jiang et al. developed chitosan-based benzydamine hydrochloride-loaded microspheres by the spray drying technique for oromucosal delivery. High and low-molecular-weight chitosan are used for this formulation. All chitosan-containing formulations showed less moisture content and enhanced bioavailability. Chitosan significantly extended the release of the drug from the formulations. Mucoadhesive interaction of the chitosan-modified microparticles was also significantly improved, and it can be strengthened by increasing the quantity of chitosan in the formulations. These studies clearly showed that the addition of chitosan improves the mucoadhesion force of formulations to the mucous membrane as well as extends the drug release which may be beneficial for local actions (Jiang et al., 2017).

9.4.3 CARRAGEENAN-BASED MUCOADHESIVE DRUG DELIVERY

Yermak et al. developed echinochrome A-incorporated gelled beads containing carrageenan and chitosan. The prepared mucoadhesive beads were found to maintain good stability. Chitosan was employed for the layer-wise coating of carrageenan beads. The swelling study indicated pH-dependent swelling behavior. The *in vitro* dissolution study confirms the sustained release of echinochrome A from the polymer matrix (Yermak et al., 2021).

Leong et al. developed a lecithin-functionalized carboxymethylated kappa-carrageenan-based microparticulate system using the ionotropic gelation method for oral insulin delivery. The *in vivo* and *in vitro* effects of the drug were investigated. These prepared microparticles were surface-functionalized with lecithin to enhance intestinal muco-adhesiveness. These oral formulations of insulin reported a hypoglycemic effect for a longer duration of up to 12–24 hours in diabetic rats. These microparticles showed sustained drug-release behavior due to the existence of sulfate groups in carboxymethylated kappa-carrageenan. This oral insulin delivery approach was very useful because of the release behavior and lower toxicity (Leong et al., 2011).

Sahoo et al. aimed to develop carboxymethylated iota-carrageenan and chitosan-loaded insulin-containing nanoparticles for oral delivery through response surface methodology to improve the delivery efficiency of the drug. In these formulations, zeta potential, drug loading capacity, mean particle size, and entrapment efficiency were 52.5 ± 0.5 mV, 10.7 ± 0.6%, 613 ± 41 nm, and 86.9 ± 2.6%, respectively. In an artificial gastric acidic environment, the release of the drug from the nanoparticles was low; however, in an intestinal fluid-like environment, it was high within 12 hours of the study. In simulated intestinal fluid, the release of the drug showed pH-responsive release behavior (Sahoo et al., 2017).

In another study, Sahoo et al. investigated muco-adhesiveness, *in vivo* drug release, stability, and transport mechanism of chitosan and carboxymethylated iota-carrageenan polymer complex of insulin nanoparticles. In a simulated gastrointestinal environment, the formulations of insulin were fully stable and biologically active. The formulations reported good muco-adhesiveness. Chitosan gradually opens the tight junctions of Caco-2 cells, allowing released insulin from intestinal lumen particles to pass through and enter the systemic circulation. In contrast to the estimated bioavailability of insulin from CS/CMCi particles in humans, which was estimated to be 44.7–46.9% and increased roughly threefold when compared to rats, the *in vivo* study also demonstrated an extended blood glucose-lowering effect with a relative bioavailability (%) of approximately 16.1% over 36 h (insulin 100 IU/kg). From the above findings, these formulations showed good potential in oral insulin delivery (Sahoo et al., 2019).

Vigani et al. formulated κ-carrageenan and hydroxypropyl cellulose incorporated in situ mucoadhesive gel formulations for therapeutic use in oral mucositis and esophagitis. κ-carrageenan was used as a polymer to maintain the gelling capacity, and hydroxypropyl cellulose was used to enhance the mucoadhesive characteristics. To improve the interaction between κ-carrageenan and saliva ions, $CaCl_2$ was used. The washability study was performed using pig esophageal mucosa, which validated the formulations' capacity to create a gel with saliva ion interaction and biological substrate adhesion. According to the overall findings, these formulations are prospective therapeutic agents for the management of oral mucositis and esophagitis induced by radiotherapy and chemotherapy (Vigani et al., 2019).

In another study, Vigani et al. developed κ-carrageenan and hydroxypropyl cellulose loaded in situ mucoadhesive gel containing *Hibiscus sabdariffa* hydroalcoholic extract for oral mucositis and esophagitis treatment. *Hibiscus sabdariffa* extract consists of various phytochemicals such as polysaccharides, polyphenols, and organic acids. These phytochemicals are responsible for anti-inflammatory and antioxidant characteristics. Rheological and mucoadhesive characteristics of *Hibiscus sabdariffa* extract-loaded formulations were studied. Such formulations had pronounced elastic and mucoadhesive properties at 37°C, which were useful for a protective function towards the damaged mucosa, as well as low viscosity at 25°C, which ensured convenient administration. Finally, *in vitro* testing on human skin fibroblasts demonstrated the biocompatibility, proliferative impact, antioxidant, and anti-inflammatory capabilities of *Hibiscus sabdariffa* extract-containing formulations (Vigani et al., 2019).

Kianfar et al. prepared a buccal wafer containing κ-carrageenan, polyethylene glycol 600, and poloxamer by lyophilization of gels. A freeze-annealing cycle at 55°C for 7 hours was determined with the help of the eutectic temperature (12.8°C) of the system that contained both pure solids and a liquid solution under a fixed pressure. Mechanical resistance to compression, hydration, and mucoadhesion experiments revealed that optimal wafers were produced from aqueous gels containing 2% w/w κ-carrageenan, 4% w/w poloxamer, and 4.4% w/w polyethylene glycol 600. A porous polymeric network that enabled simple hydration was visible in the SEM and TGA, which both indicated residual water of about 1%. The wafers demonstrated optimal swelling and mucoadhesion in settings that replicated saliva compared to saline, and the wafers were stable throughout 6 months of storage. They may be used for buccal mucosal delivery (Kianfar et al., 2013).

9.4.4 AGAROSE-BASED MUCOADHESIVE DRUG DELIVERY

Syed et al. investigated the mucoadhesive efficacy of agarose in gel formulations of benzocaine and tibezonium iodide. Agarose used in combination with carbopol 934P and hydrocolloid gel was formulated through a homogenization method. The *ex-vivo* mucoadhesive strength was found to vary depending on agarose concentration and the presence of carbopol 934P in the formulations. A 0.3% w/v agarose concentration provided the highest mucoadhesion time, facilitating the sustained release of drugs. The formulation indicated excellent stability without the presence of any inflammation (Syed et al., 2022).

Jillani et al. developed zolmitriptan succinate-loaded agarose and guar gum-containing buccal films. Glycerol was used as a plasticizer. The mucoadhesive strength and swelling behavior were determined using different polymer ratios. The buccal mucosal layer was employed to investigate mucoadhesion time along with the drug release pattern. The drug release report indicated a steady-state release of zolmitriptan succinate. The morphology study revealed the presence of small crystals on the surface of the buccal film, and an increased flux was observed in the ex-vivo permeation study. Histopathological investigation suggested the absence of a necrotic effect on the buccal mucosa (Jillani et al., 2022).

9.5 CONCLUSION

Marine polysaccharides are widely used in several pharmaceutical areas since they are abundant in nature and have beneficial characteristics. These polysaccharides possess huge therapeutic value such as anticancer, anti-inflammatory, and antiviral characteristics. In the pharmaceutical research and industry, they are extensively studied as basic biomaterials. Marine polysaccharides are a potential biopolymer for pharmaceutical applications due to their stimuli-responsive nature as well as their biodegradability and biocompatibility. In this chapter, the classification of polysaccharides from marine sources, their applications, and the mucoadhesive properties of these marine polysaccharides are discussed. Excellent mucoadhesive performance is observed with marine polymers. The manifestation of mucoadhesion depends on the mucoadhesive polymer and the environment in which it is adhered. For polymers containing non-ionic functional groups or charged groups capable of electrostatic interactions, along with hydrogen bonds or hydrophobic forces with mucosal surfaces, excellent mucoadhesive activity is seen. By interacting with mucus through diffusion as well as secondary chemical bonds, polymers including chitosan, alginate, carrageenan, and other marine polymers create weakly crosslinked networks. These polysaccharides have a high capacity to adsorb water, which can enhance drug dissolution and increase the oral bioavailability of drugs that have low water solubility. In this review, the ability of marine polysaccharides to interact with mucin and mucosal epithelia, which has been studied in several systems, is discussed. These marine-sourced polysaccharides can be processed into various forms, such as hydrogels, microspheres, and films, and are frequently used in the biomedical industry due to their physical and chemical characteristics, such as the ability to form gels and increase the viscosity of solutions. Such marine biopolymer-based mucoadhesive drug delivery systems have the ability to significantly improve the efficacy of the drug employed and are characterized by their ease of application.

KEYWORDS

- **carrageenans**
- **gastrointestinal tract**
- **glycosaminoglycans**
- **molecular weight**
- **mucoadhesion**

REFERENCES

Abd El-Hack, M. E., El-Saadony, M. T., Shafi, M. E., Zabermawi, N. M., Arif, M., Batiha, G. E., Khafaga, A. F., Abd El-Hakim, Y. M., & Al-Sagheer, A. A. (2020). Antimicrobial and antioxidant properties of chitosan and its derivatives and their applications: A review. *International Journal of Biological Macromolecules, 164*, 2726–2744.

Abodinar, A., Tømmeraas, K., Ronander, E., Smith, A. M., & Morris, G. A. (2016). The physicochemical characterization of pepsin degraded pig gastric mucin. *International Journal of Biological Macromolecules, 87*, 281–286.

Abruzzo, A., Cerchiara, T., Bigucci, F., Gallucci, M. C., & Luppi, B. (2015). Mucoadhesive buccal tablets based on chitosan/gelatin microparticles for delivery of propranolol hydrochloride. *Journal of Pharmaceutical Sciences, 104*(12), 4365–4372.

Agarwal, S., & Murthy, R. S. (2015). Effect of different polymer concentration on drug release rate and physicochemical properties of mucoadhesive gastroretentive tablets. *Indian Journal of Pharmaceutical Sciences, 77*(6), 705–714.

Albarkah, Y. A., Green, R. J., & Khutoryanskiy, V. V. (2015). Probing the mucoadhesive interactions between porcine gastric mucin and some water-soluble polymers. *Macromolecular Bioscience, 15*(11), 1546–1553.

Andrews, G. P., Laverty, T. P., & Jones, D. S. (2009). Mucoadhesive polymeric platforms for controlled drug delivery. *European Journal of Pharmaceutical Sciences, 71*(5), 505–518.

Araújo, F., Pedro, J., Granja, P. L., Santos, H. A., & Sarmento, B. (2017). Functionalized materials for multistage platforms in the oral delivery of biopharmaceuticals. *Progress in Materials Science, 89*, 306–344.

Banerjee, A., Qi, J., Gogoi, R., Wong, J., & Mitragotri, S. (2016). Role of nanoparticle size, shape and surface chemistry in oral drug delivery. *Journal of Controlled Release, 238*, 176–185.

Bassi da Silva, J., de Ferreira, S. B. S., de Freitas, O., & Bruschi, M. L. (2017). A Critical Review about methodologies for the analysis of mucoadhesive properties of drug delivery systems. *Drug Development and Industrial Pharmacy, 43*(7), 1053–1070.

Bassi da Silva, J., Ferreira, S., Reis, A., Cook, M., & Bruschi, M. (2018). Assessing Mucoadhesion in polymer gels: The effect of method type and instrument variables. *Polymers, 10*(3), 254.

Becker, T. A., & Brandon, T. (2001). Calcium alginate gel: a biocompatible and mechanically stable polymer for endovascular embolization. *Journal of Biomedical Materials Research, 54*(1), 76–86.

Bera, H., Kandukuri, S. G., Nayak, A. K., & Boddupalli, S. (2015). Alginate-sterculia gum gel-coated oil-entrapped alginate beads for gastroretentive risperidone delivery. *Carbohydrate Polymers, 120*, 74–84.

Biswas, N., & Sahoo, R. K. (2016). Tapioca starch blended alginate mucoadhesive-floating beads for intragastric delivery of Metoprolol Tartrate. *International Journal of Biological Macromolecules, 83*, 61–70.

Brito, A. S., Cavalcante, R. S., Palhares, L. C., Hughes, A. J., Andrade, G. P., Yates, E. A., Nader, H. B., Lima, M. A., & Chavante, S. F. (2014). A non-hemorrhagic hybrid heparin/heparan sulfate with anticoagulant potential. *Carbohydrate Polymers, 99*, 372–378.

Brown, E. M., Allsopp, P. J., Magee, P. J., Gill, C. I., Nitecki, S., Strain, C. R., & McSorley, E. M. (2014). Seaweed and human health. *Nutrition Reviews, 72*(3), 205–216.

Butnarasu, C., Barbero, N., Pacheco, D., Petrini, P., & Visentin, S. (2019). Mucin binding to therapeutic molecules: the case of antimicrobial agents used in cystic fibrosis. *International Journal of Pharmaceutics, 564*, 136–144.

Cardoso, M. J., Costa, R. R., & Mano, J. F. (2016). Marine origin polysaccharides in drug delivery systems. *Marine Drugs, 14*(2), 34.

Carvalho, F. C., Bruschi, M. L., Evangelista, R. C., & Gremião, M. P. D. (2010). Mucoadhesive drug delivery systems. *Brazilian Journal of Pharmaceutical Sciences, 46*(1), 1–17.

Caterson, B., & Melrose, J. (2018). Keratan sulfate, a complex glycosaminoglycan with unique functional capability. *Glycobiology, 28*(3), 182–206.

Chaudhari, V. A., Sarode, S. M., Sathe, B. S., & Vadnere, G. P. (2014). Mucoadhesive buccal drug delivery system: A review. *Pharmaceutical Sciences Monitor, 5*(4), 142–162.

Chen, N., Li, Q., Li, J., Ren, Y., Wu, G., Liu, Y., & Shi, Y. (2019). Development and evaluation of a new gastroretentive drug delivery system: Nanomicelles-loaded floating mucoadhesive beads. *Journal of Drug Delivery Science and Technology, 51*, 485–492.

Cheung, R. C., Ng, T. B., Wong, J. H., & Chan, W. Y. (2015). Chitosan: An update on potential biomedical and pharmaceutical applications. *Marine Drugs, 13*(8), 5156–5186.

Choi, H. J., Kim, M. C., Kang, S. M., & Montemagno, C. D. (2014). The osmotic stress response of split influenza vaccine particles in an acidic environment. *Archives of Pharmacal Research, 37*(12), 1607–1616.

Claverie, M., McReynolds, C., Petitpas, A., Thomas, M., & Fernandes, S. C. M. (2020). Marine-derived polymeric materials and biomimetics: An overview. *Polymers, 12*(5), 1002.

Cook, S. L., Bull, S. P., Methven, L., Parker, J. K., & Khutoryanskiy, V. V. (2017). Mucoadhesion: A food perspective. *Food Hydrocolloids, 72*, 281–296.

Cook, S. L., Woods, S., Methven, L., Parker, J. K., & Khutoryanskiy, V. V. (2018). Mucoadhesive polysaccharides modulate sodium retention, release and taste perception. *Food Chemistry, 240*, 482–489.

Demitri, C., Lamanna, L., De Benedetto, E., Damiano, F., Cappello, M. S., Siculella, L., & Sannino, A. (2017). Encapsulation of Lactobacillus kefiri in alginate microbeads using a double novel aerosol technique. *Materials Science and Engineering: C, 77*, 548–555.

Diós, P., Nagy, S., Pál, S., Pernecker, T., Kocsis, B., Budán, F., Horváth, I., Szigeti, K., Máthé, D., & Dévay, A. (2015). Preformulation studies and optimization of sodium alginate based floating drug delivery system for eradication of Helicobacter pylori. *European Journal of Pharmaceutical Sciences, 96*, 196–206.

Djerbal, L., Lortat-Jacob, H., & Kwok, J. (2017). Chondroitin sulfates and their binding molecules in the central nervous system. *Glycoconjugate Journal, 34*(3), 363–376.

Dodou, D., Breedveld, P., & Wieringa, P. (2005). Mucoadhesives in the gastrointestinal tract: Revisiting the literature for novel applications. *European Journal of Pharmaceutical Sciences, 60*(1), 1–16.

Esim, O., Savaser, A., Ozkan, C. K., Tas, C., & Ozkan, Y. (2020). Investigation of the mucoadhesivity, swelling, and drug release mechanisms of indomethacin buccal tablets: Effect of formulation variables. *Drug Development and Industrial Pharmacy, 46*(12), 1979–1987.

Fei, J., Zhe, C., & Yuanyuan, D. (2021). Wound dressing hydrogel of *Enteromorpha prolifera* polysaccharide–polyacrylamide composite: A facile transformation of marine blooming into biomedical material. *ACS Applied Materials & Interfaces, 13*(12), 14530–14542.

Fitton, J. H., Stringer, D. N., & Karpiniec, S. S. (2015). Therapies from fucoidan: An update. *Marine Drugs, 13*(10), 5920–5946.

Ghiselli, G. (2017). Drug-mediated regulation of glycosaminoglycan biosynthesis. *Medicinal Research Reviews, 37*(5), 1051–1094.

Ghumman, S. A., Noreen, S., & Tul Muntaha, S. (2020). *Linum usitatissimum* seed mucilage-alginate mucoadhesive microspheres of metformin HCl: Fabrication, characterization and evaluation. *International Journal of Biological Macromolecules, 155*, 358–368.

Gómez-Ordóñez, E., Jiménez-Escrig, A., & Rupérez, P. (2014). Bioactivity of sulfated polysaccharides from the edible red seaweed *Mastocarpus stellatus*. *Bioactive Carbohydrates and Dietary Fibre, 3*(1), 29–40.

Gupta, R. C., Lall, R., Srivastava, A., & Sinha, A. (2019). Hyaluronic acid: molecular mechanisms and therapeutic trajectory. *Frontiers in Veterinary Science, 6*, 192.

Hägerström, H., Edsman, K., & Strømme, M. (2003). Low-frequency dielectric spectroscopy as a tool for studying the compatibility between pharmaceutical gels and mucus tissue. *Journal of Pharmaceutical Sciences, 92*(9), 1869–1881.

Hamdan, M. A., Ramli, N. A., Othman, N. A., Amin, K. N. M., & Adam, F. (2020). Characterization and property investigation of Microcrystalline Cellulose (MCC) and Carboxymethyl Cellulose (CMC) filler on the carrageenan-based bio composite film. *Materials Today: Proceedings, 42*, 56–62.

Hecht, H., & Srebnik, S. (2016). Structural characterization of sodium alginate and calcium alginate. *Biomacromolecules, 17*(6), 2160–2167.

Henrotin, Y., Mathy, M., Sanchez, C., & Lambert, C. (2010). Chondroitin sulfate in the treatment of osteoarthritis: from in vitro studies to clinical recommendations. *Therapeutic Advances in Musculoskeletal Disease, 2*(6), 335–348.

Hu, Q., & Luo, Y. (2018). Recent advances of polysaccharide-based nanoparticles for oral insulin delivery. *International Journal of Biological Macromolecules, 120*, 775–782.

Hussain, T., Ijaz, M., Shamim, R., Hussain, K., Abbas, N., Hussain, A., Griessinger, J. A., & Bukhari, N. I. (2020). In vivo evaluation of a novel chitosan-polycaprolactone based mucoadhesive gastro-retentive sustained release drug delivery system for milnacipran HCl. *AAPS Pharm SciTech, 21*(2), 58.

Inic-Kanada, A., Stein, E., Stojanovic, M., Schuerer, N., Ghasemian, E., Filipovic, A., Marinkovic, E., Kosanovic, D., & Barisani-Asenbauer, T. (2018). Effects of iota-carrageenan on ocular *Chlamydia trachomatis* infection in vitro and in vivo. *Journal of Applied Phycology, 30*(5), 2601–2610.

Jain, S. K., Prajapati, N., Rajpoot, K., & Kumar, A. (2016). A novel sustained release drug–resin complex-based microbeads of ciprofloxacin HCl. *Artificial Cells, Nanomedicine, and Biotechnology, 44*(7), 1891–1900.

Jana, S., & Sen, K. K. (2016). Alginate based nanocarriers for drug delivery applications. *Current Pharmaceutical Design, 22*(22), 3399–3410.

Jana, S., & Sen, K. K. (2017). Chitosan—Locust bean gum interpenetrating polymeric network nanocomposites for delivery of aceclofenac. *International Journal of Biological Macromolecules, 102*, 878–884.

Jana, S., Das, A., Nayak, A. K., Sen, K. K., & Basu, S. K. (2013a). Aceclofenac-loaded unsaturated esterified alginate/gellan gum microspheres: in vitro and in vivo assessment. *International Journal of Biological Macromolecules, 57*, 129–137.

Jana, S., Maji, N., Nayak, A. K., Sen, K. K., & Basu, S. K. (2013b). Development of chitosan-based nanoparticles through inter-polymeric complexation for oral drug delivery. *Carbohydrate Polymers, 98*, 870–876.

Jiang, W. Z., Cai, Y., & Li, H. Y. (2017). Chitosan-based spray-dried mucoadhesive microspheres for sustained oromucosal drug delivery. *Powder Technology, 312*, 124–132.

Jiao, G., Yu, G., Zhang, J., & Ewart, H. (2011). Chemical Structures and Bioactivities of Sulfated Polysaccharides from Marine Algae. *Marine Drugs, 9*(2), 196–223.

Jillani, U., Mudassir, J., Arshad, M. S., Mehta, P., Alyassin, Y., Nazari, K., Yousef, B., Patel, M., Zaman, A., & Sayed, E. (2022). Design and evaluation of agarose based buccal films containing zolmitriptan succinate: Application of physical and chemical enhancement approaches. *Journal of Drug Delivery Science and Technology, 69*, 103041.

Karmakar, S., Manna, S., Kabiraj, S., & Jana, S. (2022). Recent progress in alginate-based carriers for ocular targeting of therapeutics. *Food Hydrocolloids Health, 20*, 100071.

Khutoryanskiy, V. V. (2011). Advances in mucoadhesion and mucoadhesive polymers. *Macromolecular Bioscience, 11*(6), 748–764.

Kianfar, F., Ayensu, I., & Boateng, J. S. (2014). Development and physico-mechanical characterization of carrageenan and poloxamer-based lyophilized matrix as a potential buccal drug delivery system. *Drug Development and Industrial Pharmacy, 40*(3), 361–369.

Kim, I. Y., Seo, S. J., Moon, H. S., Yoo, M. K., Park, I. Y., Kim, B. C., & Cho, C. S. (2008). Chitosan and its derivatives for tissue engineering applications. *Biotechnology Advances, 26*(1), 1–21.

Kristó, K., Módra, S., Hornok, V., Süvegh, K., Ludasi, K., Aigner, Z., Kelemen, A., Sovány, T., Pintye-Hódi, K., & Regdon, G. (2022). Investigation of Surface properties and free volumes of chitosan-based buccal mucoadhesive drug delivery films containing ascorbic acid. *Pharmaceutics, 14*(3), 345.

Kwon, K., & Daniell, H. (2016). Oral delivery of protein drugs bioencapsulated in plant cells. *Molecular Therapy, 24*(8), 1342–1350.

Leal, J., Smyth, H. D. C., & Ghosh, D. (2017). Physicochemical properties of mucus and their impact on transmucosal drug delivery. *International Journal of Pharmaceutics, 532*, 555–572.

Leong, K. H., Chung, L. Y., Noordin, M. I., Onuki, Y., Morishita, M., & Takayama, K. (2011). Lectin-functionalized carboxymethylated kappa-carrageenan microparticles for oral insulin delivery. *Carbohydrate Polymers, 86*(2), 555–565.

Li, B., Lu, F., & Wei, X. (2008). Fucoidan: Structure and bioactivity. *Molecules, 13*(7), 1671–1695.

Li, J. P., & Kusche-Gullberg, M. (2016). Heparan sulfate: Biosynthesis, structure, and function. *International Review of Cell and Molecular Biology, 325*, 215–273.

Liu, J., Zhan, X., Wan, J., Wang, Y., & Wang, C. (2015). Review for carrageenan-based pharmaceutical biomaterials: Favorable physical features versus adverse biological effects. *Carbohydrate Polymers, 121*, 27–36.

Luthuli, S., Wu, S., Cheng, Y., Zheng, X., Wu, M., & Tong, H. (2019). Therapeutic effects of fucoidan: A review on recent studies. *Marine Drugs, 17*(9), 487.

Ma, S., Wang, L., Huang, X., Wang, X., Chen, S., Shi, W., Qiao, X., & Jiang, Y. (2018). Oral recombinant Lactobacillus vaccine targeting the intestinal microfold cells and dendritic cells for delivering the core neutralizing epitope of porcine epidemic diarrhea virus. *Microbial Cell Factories, 17*(1), 1–12.

Mackie, A. R., Goycoolea, F. M., Menchicchi, B., Caramella, C. M., Saporito, F., Lee, S., Stephansen, K., Chronakis, I. S., Hiorth, M., Adamczak, M., Waldner, M., Nielsen, H. M., & Marcelloni, L. (2017). Innovative Methods and Applications in Mucoadhesion Research. *Macromolecular Bioscience, 17*(4), 1600534.

Maharjan, S., Singh, B., Jiang, T., Yoon, S., Li, H., Kim, G., Jeong, M., Ji, S., Park, O., Hyun, S., et al. (2016). Systemic administration of RANKL overcomes the bottleneck of oral vaccine delivery through microfold cells in ileum. *Biomaterials, 84*, 286–300.

Manivasagan, P., & Oh, J. (2016). Marine polysaccharide-based nanomaterials as a novel source of nanobiotechnological applications. *International Journal of Biological Macromolecules, 82*, 315–327.

Manna, S., & Jana, S. (2021). Carrageenan-based nanomaterials in drug delivery applications. In *Biopolymer-Based Nanomaterials in Drug Delivery and Biomedical Applications* (pp. 365). Academic Press.

Manna, S., Banerjee, A., Jana, S., & Bhowmik, M. (2022a). Chitosan-based nanoengineered drug delivery system. In *Micro-and Nanoengineered Gum-Based Biomaterials for Drug Delivery and Biomedical Applications* (p. 77). Elsevier.

Manna, S., Jana, S. (2022b). Marine polysaccharides in tailor-made drug delivery. *Current Pharmaceutical Design, 28*(10), 1046–1066.

Manna, S., Jana, S., Mandal, D., & Bhowmik, M. (2022c). Marine biopolymer-based oral in situ gel: drug delivery application. In *Marine Biomaterials* (pp. 139). Springer: Singapore.

Manna, S., Nandi, G., & Jana, S. (2022d). Alginate-based Carriers for Transdermal Drug Delivery. In *Polysaccharide-Based Biomaterials: Delivery of Therapeutics and Biomedical Applications* (pp. 69). RSC Publishing.

Marco, B., Remy, T., Grace, V., Dennis, N., Aurelie, R., Ronan, P. V., & Prasad, S. (2021). Hydrogel-forming algae polysaccharides: From seaweed to biomedical applications. *Biomacromolecules, 22*(3), 1027–1052.

Melcher, R. L., Neumann, M., Fuenzalida, W. J. P., Gröhn, F., & AlMoerschbacher, B. M. (2017). Revised domain structure of ulvan lyase and characterization of the first ulvan binding domain. *Scientific Reports, 7*, 1–9.

Menchicchi, B., Fuenzalida, J. P., Hensel, A., Swamy, M. J., David, L., Rochas, C., & Goycoolea, F. M. (2015). Biophysical analysis of the molecular interactions between polysaccharides and Mucin. *Biomacromolecules, 16*(3), 924–935.

Merry, C. L. R. (2021). Exciting new developments and emerging themes in glycosaminoglycan research. *Journal of Histochemistry & Cytochemistry, 69*(1), 9–11.

Mishra, S., & Ganguli, M. (2021). Functions of, and replenishment strategies for, chondroitin sulfate in the human body. *Drug Discovery Today, 26*(5), 1185–1199.

Najdahmadi, A., Lakey, J. R., & Botvinick, E. (2018). Structural characteristics and diffusion coefficient of alginate hydrogels used for cell-based drug delivery. *MRS Advances, 3*(38), 2399–2408.

Nayak, A. K., Pal, D., & Santra, K. (2013). Plantago ovata F. Mucilage-alginate mucoadhesive beads for controlled release of glibenclamide: Development, optimization, and in vitro-in vivo evaluation. *Journal of Pharmaceutics, 2013*, 151035.

O'Brien, F. J. (2011). Biomaterials & scaffolds for tissue engineering. *Materials Today, 14*(3), 88–95.

Orivea, G., Tamb, S. K., Pedraza, J. L., & Jean-P, Halle. (2006). Biocompatibility of alginate–poly-l-lysine microcapsules for cell therapy. *Biomaterials, 27*(17), 3691–3700.

Palem, C. R., Dudhipala, N. R., Battu, S. K., Repka, M. A., & Yamsani, M. R. (2016). Development, optimization and in vivo characterization of domperidone-controlled release hot-melt-extruded films for buccal delivery. *Drug Development and Industrial Pharmacy, 42*(3), 473–484.

Papakonstantinou, E., Roth, M., & Karakiulakis, G. (2012). Hyaluronic acid: A key molecule in skin aging. *Dermato-Endocrinology, 4*(3), 253–258.

Paul, R., Kabiraj, S., Manna, S., & Jana, S. (Eds.). (2022). Marine polysaccharides in pharmaceutical applications. In *Marine Biochemistry* (pp. 111). CRC Press.

Pham, Q. D., Nöjd, S., Edman, M., Lindell, K., Topgaard, D., & Wahlgren, M. (2021). Mucoadhesion: mucin-polymer molecular interactions. *International Journal of Pharmaceutics, 610*, 121245.

Prydz, K. (2015). Determinants of glycosaminoglycan (GAG) structure. *Biomolecules, 5*(3), 2003–2022.

Punitha, S., & Girish, Y. (2010). Polymers in Mucoadhesive Buccal Drug Delivery System – A Review. *International Journal of Pharmaceutical Sciences, 1*(1), 170–186.

Rhein-Knudsen, N., Ale, M., & Meyer, A. (2015). Seaweed Hydrocolloid Production: An Update on Enzyme Assisted Extraction and Modification Technologies. *Marine Drugs, 13*(6), 3340–3359.

Rossi, S., Vigani, B., Bonferoni, M. C., Sandri, G., Caramella, C., & Ferrari, F. (2018). Rheological analysis and mucoadhesion: A 30-year-old and still active combination. *Journal of Pharmaceutical and Biomedical Analysis, 156*, 232–238.

Ruocco, N., Costantini, S., Guariniello, S., & Costantini, M. (2016). Polysaccharides from the marine environment with pharmacological, cosmeceutical and nutraceutical potential. *Molecules, 21*(5), 551.

Russo, E., Selmin, F., Baldassari, S., Gennari, C. G. M., Caviglioli, G., Cilurzo, F., Minghetti, P., & Parodi, B. (2016). A focus on mucoadhesive polymers and their application in buccal dosage forms. *Journal of Drug Delivery Science and Technology, 32*, 113–125.

Sahoo, P., Leong, K. H., Nyamathulla, S., Onuki, Y., Takayama, K., & Chung, L. Y. (2017). Optimization of pH-responsive carboxymethylated iota-carrageenan/chitosan nanoparticles for oral insulin delivery using response surface methodology. *Reactive and Functional Polymers, 119*, 145–155.

Sahoo, P., Leong, K. H., Nyamathulla, S., Onuki, Y., Takayama, K., & Chung, L. Y. (2019). Chitosan complexed carboxymethylated iota-carrageenan oral insulin particles: Stability, permeability and in vivo evaluation. *Materials Today Communications, 20*, 100557.

Samprasit, W., Opanasopit, P., & Chamsai, B. (2021). Mucoadhesive chitosan and thiolated chitosan nanoparticles containing alpha mangostin for possible Colon-targeted delivery. *Pharmaceutical Development and Technology, 26*(3), 362–372.

Shaikh, R., Raj Singh, T. R., Garland, M. J., Woolfson, A. D., & Donnelly, R. F. (2011). Mucoadhesive drug delivery systems. *Journal of Pharmacy & Bioallied Sciences, 3*(1), 89–100.

Siddhesh, N., Kevin, P., & Edgar, J. (2012). Alginate derivatization: a review of chemistry properties and applications. *Biomaterials, 33*(13), 3279–3305.

Singh, I., & Rana, V. (2012). Techniques for the Assessment of Mucoadhesion in Drug Delivery Systems: An Overview. *Journal of Adhesion Science and Technology, 26*(14–15), 2251–2267.

Smart, J. D. (2005). The basics and underlying mechanisms of mucoadhesion. *Advanced Drug Delivery Reviews, 57*(11), 1556–1568.

Sosnik, A., das Neves, J., & Sarmento, B. (2014). Mucoadhesive polymers in the design of nano-drug delivery systems for administration by non-parenteral routes: A review. *Progress in Polymer Science, 39*, 2030–2075.

Syed, M. A., Hanif, S., Ain, N. U., Syed, H. K., Zahoor, A. F., Khan, I. U., Abualsunun, W. A., Jali, A. M., Qahl, S. H., Sultan, M. H., Madkhali, O. A., Ahmed, R. A., Abbas, N., Hussain, A., & Qayyum, M. A. (2022). Assessment of binary agarose-carbopol buccal gels for mucoadhesive drug delivery: Ex vivo and in vivo characterization. *Molecules, 27*(24), 7004.

Ta, L. P., Bujna, E., Kun, S., Charalampopoulos, D., & Khutoryanskiy, V. V. (2021). Electrosprayed mucoadhesive alginate-chitosan microcapsules for gastrointestinal delivery of probiotics. *International Journal of Pharmaceutics, 597*, 120342.

Tai, G. H., Nieduszynski, I. A., Fullwood, N. J., & Huckerby, T. N. (1997). Human corneal keratan sulfates. *Journal of Biological Chemistry, 272*, 28227–28231.

van Weelden, G., Bobiński, M., Okła, K., van Weelden, W. J., Romano, A., & Pijnenborg, J. M. (2019). Fucoidan structure and activity in relation to anti-cancer mechanisms. *Marine Drugs, 17*(1), 32.

Venkatesan, J., Lowe, B., Anil, S., Manivasagan, P., Kheraif, A. A., Kang, K. H., & Kim, S. K. (2015). Seaweed polysaccharides and their potential biomedical applications. *Starch-Stärke, 67*(5–6), 381–390.

Vigani, B., Faccendini, A., Rossi, S., Sandri, G., Bonferoni, M. C., Gentile, M., & Ferrari, F. (2019). Development of a mucoadhesive and in situ gelling formulation based on κ-Carrageenan for application on oral mucosa and esophagus walls. I. A functional in vitro characterization. *Marine Drugs, 17*(2), 112.

Vigani, B., Rossi, S., Gentile, M., Sandri, G., Bonferoni, M. C., Cavalloro, V., Martino, E., Collina, S., & Ferrari, F. (2019). Development of a mucoadhesive and an in-situ gelling formulation based on κ-carrageenan for application on oral mucosa and esophagus walls. II. Loading of a bioactive hydroalcoholic extract. *Marine Drugs, 17*(3), 153.

Wang, Z., Dhurandhare, V. M., Mahung, C. A., Arnold, K., Li, J., Su, G., Xu, D., Maile, R., & Liu, J. (2021). Improving the sensitivity for quantifying heparan sulfate from biological samples. *Analytical Chemistry, 93*(28), 11191–11199.

Westergren-Thorsson, G., Onnervik, P. O., Fransson, L. A., & Malmstrom, A. (1991). Proliferation of cultured fibroblasts is inhibited by L-iduronate-containing glycosaminoglycans. *Journal of Cellular Physiology, 147*(3), 523–530.

Woertz, C., Preis, M., Breitkreutz, J., & Kleinebudde, P. (2013). Assessment of test methods evaluating mucoadhesive polymers and dosage forms: An overview. *European Journal of Pharmaceutical Sciences, 85*, 843–853.

Wu, Y., Zhang, W., Huang, J., Luo, Z., Li, J., Wang, L., & Di, L. (2020). Mucoadhesive improvement of alginate microspheres as potential gastroretentive delivery carrier by blending with Bletilla striata polysaccharide. *International Journal of Biological Macromolecules, 156*, 1191–1201.

Yazdi, M. K., Taghizadeh, A., Taghizadeh, M., Stadler, F. J., Farokhi, M., Mottaghitalab, F., Zarrintaj, P., Ramsey, J. D., Seidi, F., Saeb, M. R., & Mozafari, M. (2020). Agarose-based biomaterials for advanced drug delivery. *Journal of Controlled Release, 326*, 523–543.

Yermak, I. M., Gorbach, V. I., Karnakov, I. A., Davydova, V. N., Pimenova, E. A., Chistyulin, D. A., Isakov, V. V., & Glazunov, V. P. (2021). Carrageenan gel beads for echinochrome inclusion: Influence of structural features of carrageenan. *Carbohydrate Polymers, 272*, 118479.

Younes, I., & Rinaudo, M. (2015). Chitin and chitosan preparation from marine sources. Structure, properties and applications. *Marine Drugs, 13*(2), 1133–1174.

Zia, K. M., Tabasum, S., Nasif, M., Sultan, N., Aslam, N., Noreen, A., & Zuber, M. (2017). A review on synthesis, properties and applications of natural polymer-based carrageenan blends and composites. *International Journal of Biological Macromolecules, 96*, 282–301.

CHAPTER 10

Algae-Based 3D Printed Hydrogel Patch in Diabetic Wound Healing

MD. JASIM UDDIN,[1] MD. FAIYAZUDDIN,[2] JASMIN HASSAN,[3] TASLIMA SULTANA POLY,[3] SAFIYA MOHAMMED SAEED,[3] IFFAT ARA BRISHTI,[3] ADWIZA CHAKRABORTY BISHAKHA,[3] and FOUZIA TASNIM,[3]

[1]*Department of Pharmaceutical Technology, Faculty of Pharmacy, Universiti Malaya, 50603 Kuala Lumpur, Malaysia*

[2]*School of Pharmacy, Al-Karim University, Katihar, Bihar, India*

[3]*Drug Delivery and Therapeutics Lab, London IG2 6TT, United Kingdom*

ABSTRACT

Marine-derived algae-based hydrogel is used in the effective delivery of drugs. Utilizing the hydrogels in the delivery of drugs to the CNS, ocular, oral, cancer, cardiovascular, stomach, and bones is definitely possible. Recently, hydrogel has been used in arthritis and bone-related diseases. Hydrogels can help supply sustained medication release which occurs for a prolonged period of time and can even span several days. There is also a subclass of hydrogel called nanogel which is highly efficient in delivering drugs. In this chapter, the author discusses the current trend in the management of diabetic wound healing using 3D-printed hydrogel patches derived from marine origin. However, it is an emerging and rising field of research that can provide potential opportunities to treat wound healing using cutting-edge 3D printing technology with biocompatible and biodegradable marine-derived algae hydrogel patches.

Handbook of Research in Marine Pharmaceutics: Exploring Oceanic Microbial Diversity for Human Health and Wellness. Md. Faiyazuddin & Khalid Rehman Hakeem (Eds.)

10.1 INTRODUCTION

10.1.1 DIABETIC WOUND HEALING

Diabetes mellitus (DM) is a disease that is becoming increasingly prevalent in the current world. The major implication associated with this disease condition involves complications in prime organs such as the retina, kidney, and skin of diabetic patients, which tend to injure frequently. It involves chronic as well as systemic conditions, which are categorized by nephropathy, retinopathy, and neuropathy (Quondamatteo, 2014). It is known as the leading cause of reduced longevity and increased morbidity and is known for its multifaceted metabolic disease, affecting as many as 340 million people throughout the globe. Among this population, 20% of individuals develop diabetic wounds. Foot ulcers or leg ulcers are one of the common conditions for diabetic wounds in these patients (Patel et al., 2019). It has been estimated in a study that one in three individuals, compared to one in five with diabetes mellitus, ends up developing a chronic non-healing wound in their lifespan. Diabetic foot ulcer (DFU) is the aggressive form that arises from a chronic non-healing wound and has an increased rate of recurrence, seen to be 40% within one year and surprisingly developing to 60% within five years, with no reliable method for predicting the condition beforehand (Burgess et al., 2021). Diabetes has been the largest public health concern; however, the incidence of it has subsided in many regions of the world. The number of diabetes patients was seen to be 451 million in the year 2017 and in the future, it is estimated to reach 693 million by the year 2045 if the proposed prevention methods are not imposed. In Figure 10.1, the global cases of diabetes can be seen to increase from the year 1990 to 2017 when the incidence was initially 11.3 million and it has seen a dramatic increase in the following two decades, estimated to be 211.2 million and reached 476.0 million in the year 2017, showing an increase of 129.7%. The events of death associated with diabetes have also increased from 0.61 million in the year 1990 to 1.37 million in the year 2017. The incidence and prevalence of diabetes are shown in Figure 10.1 (Saeedi et al., 2019).

Diabetic foot ulcer (DFU) is the leading cause for those admitted to the hospital for severe cases of diabetes. While undergoing treatment, DFU is the most common cause of long-term complications, leading to a noticeable amount of mortality and morbidity (Mariadoss et al., 2022). The wounds resulting from diabetes are topical wounds that fall into different categories such as arterial ulcers, foot ulcers, and common ulcers, among others. These topical wounds are categorized into two major types: acute and chronic

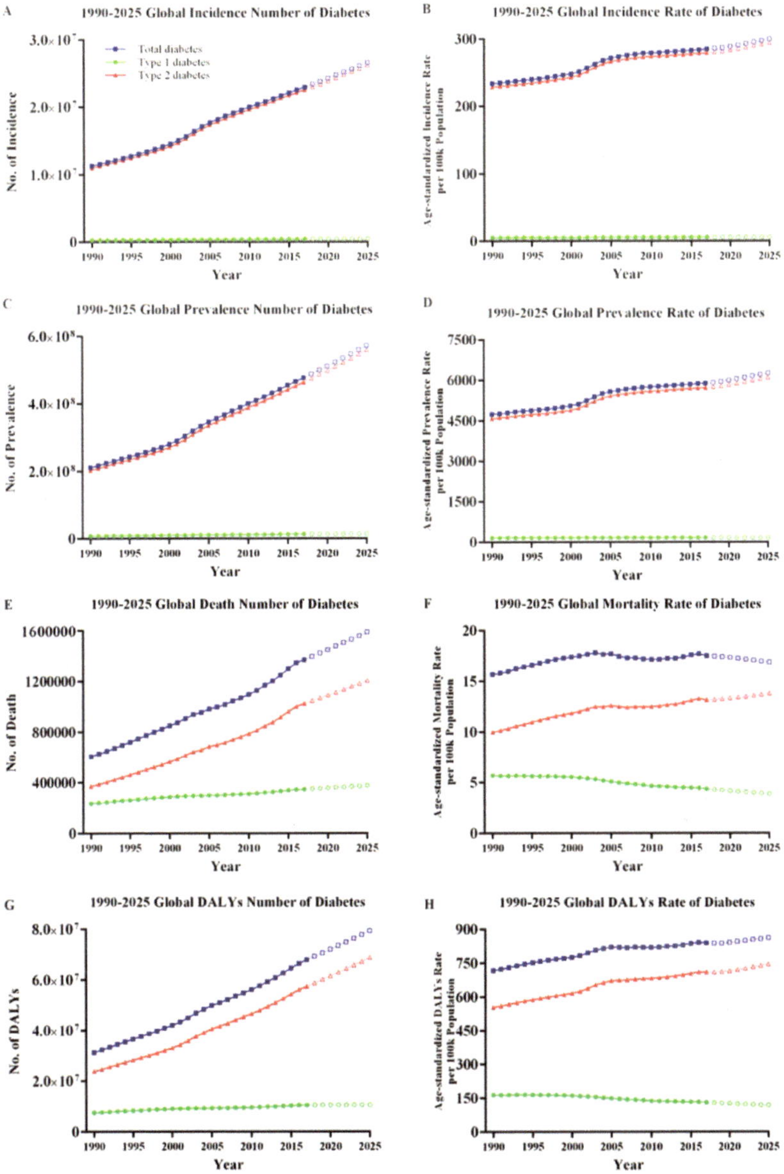

FIGURE 10.1 The worldwide trend of the diabetic burden from the year 1990 to 2025 (from Global, regional, and national burden and trend of diabetes in 195 countries and territories: an analysis from 1990 to 2025). Global burden of diabetes mellitus from 1990 to 2025. (A) Incidence number; (B) Age-standardized incidence rate; (C) Prevalence number; (D) Age-standardized prevalence rate; (E) Death number; (F) Age-standardized mortality rate; (G) DALYs number; and (H) Age-standardized DALYs rate [DALYs: disability-adjusted life-years].

Source: Reprinted from Lin et al., 2020. http://creativecommons.org/licenses/by/4.0/.

wounds. Acute wounds are associated with surgical and trauma wounds, utilizing the body's normal wound healing cascade to recover. Acute wounds can often result from scratching or deep wounds that cause damage to blood vessels, tissues, nerves, and internal organs. On the other hand, chronic wounds do not progress through the normal healing process, fail to initiate a normal healing response, and do not produce a sustainable functional and anatomical result. As a result, they do not restore the functions of healthy skin. This type of wound occurs in individuals with a history of diabetes, the elderly, and obese individuals, and falls into three categories: diabetic ulcers, venous ulcers, and pressure ulcers (Saghazadeh et al., 2018).

Diabetes is one of the myriad factors that delay the wound healing cascade and also cause non-healing ulcers resulting as a post-impact of the disease (Jia et al., 2019). Diabetes especially affects the wound healing cascade through obstruction of the balance of vessel growth and the proliferation phase (Okonkwo & Dipietro, 2017). This is how the wound healing process becomes prolonged and results in fatal, chronic wounds, and non-healing ulcers (Ridiandries et al., 2018). The healing process in the wound gets stuck in the inflammation stage (Nuutila et al., 2014). The non-healing ulcers often show post-complications which are accompanied by pain, infections, loss of function of the organ, amputation, and then sepsis (Okonkwo & Dipietro, 2017). The diabetic wounds become severe and chronic non-healing wounds, and they also get stuck in a self-perpetuating inflammatory stage. The wounds tend to lock in an extended and heightened inflammatory state through the damage of the subsequent tissue, and perpetuation of the cycle occurs through neutrophil infiltration with the reactive oxygen species and noxious enzymes. They become intractable although they possess sufficient wound management. If the destructive factors can be separated, many of the chronic wounds can be treated with success (Volpe et al., 2018). Both understanding and addressing the challenges in the treatment of chronic wounds would still lead to better clinical outcomes, which involve faster and more durable closure, and this may help result in improved life quality in patients and reduced healthcare-associated costs. Currently, diabetic wounds are often treated through controlling inflammation and promoting cell proliferation. It is, however, sad that the existing strategies used are not satisfactory and very few alternatives exist as treatment options (Frykberg & Banks, 2015).

Skin works as the first protective barrier which deals with a number of threats that can either be physical or chemical and acts as a barrier against the stress that occurs due to radiation or temperature. Most importantly, in case of the entry of a foreign pathogen into the body (Nuutila et al., 2014).

The other important functions involve protection against the loss of fluid, helping to maintain the homeostasis of the body, calling for the presence of defending cells such as keratinocytes, melanocytes, dendritic cells, mast cells, Langerhans cells, appendages of the skin, and many more which helps with the immunity of the human body (Cañedo-Dorantes & Cañedo-Ayala, 2019). The presence of a wound can greatly reduce the functionality of these barriers, making the wound vulnerable to microbes and pathogens. Therefore, it is important to consider the depth and size of the wound along with the nearby environment as significant factors for wound infection. The recuperation of the wound involves complex processes, and it is crucial to restore the normal functions of the skin as a barrier. However, there are instances when the wound healing process may result in a fibrotic scar, while some wounds can be chronic in nature, leading to several impairments in the individual that may require amputations. These chronic wounds are often found in individuals with diabetes, vascular diseases, and aged patients (Rodrigues et al., 2019).

In the case of a healthy individual, the wound is rectified *via* a wound-healing cascade, and the complex process of wound healing is arbitrarily divided into 4 stages: hemostasis, inflammation, proliferation, and remodeling (Velnar et al., 2009). These stages are discussed below:

1. **Hemostasis Phase and Coagulation:** Right after the injury to the tissues, the coagulation and hemostasis phase begins at the wound site (Jespersen, 1988). This stage aims to prevent exsanguination and protect the vascular system to keep the cells intact and the major organs functioning despite the injuries. Another aim is to provide a matrix for invading cells needed for later phases in wound healing (Pool, 1977). A dynamic balance occurs between endothelial cells, thrombocytes, fibrinolysis, coagulation, and the determination of the fibrin amount required to be deposited at the wound site, influencing the progress of reparative processes (Lawrence, 1998). The extracellular matrix component helps bind platelets, activating them when the skin is injured. The damaged tissues and aggregated platelets trigger the extrinsic and intrinsic pathways. The clot of fibrin and platelets stabilizes the combined works of pathways, resulting in the formation of a scaffold for the wound healing cascade to migrate and proliferate. It also serves as a reservoir for growth factors and cytokines (Bergmeier & Hynes, 2012).
2. **Inflammation:** The next stage of the wound healing cascade involves the inflammation stages which involve the lysis of the cell

as well as the removal of the debris including cellular, extracellular as well as pathogenic factors (Holzer-Geissler et al., 2022). An influx in the case of inflammatory cells occurs in these stages which boosts the pro-angiogenesis of the molecules in the wound site (Costa et al., 2007). The chemokines are involved in the process of inflammation and help remove the dead cells, foreign materials, and debris from the wound through the recruitment of inflammatory cells. They also ensure the closure of the wound by facilitating the migration, proliferation, and differentiation of the endothelial cells with the progenitor cells and the keratinocytes (Piipponen et al., 2020).

3. **Proliferation:** There are a number of endothelial cells, keratinocytes, and fibroblasts which increase in the wound area. They result in the formation of the extracellular matrix which is temporary. While the inflammatory stage is characterized by hemostasis, chemotaxis, and increased vascular permeability and helps in limiting further damage by closing the wound and removing cellular debris, and bacteria while fostering cellular migration, the proliferative phase is characterized by the formation of granulation tissue, reepithelialization, and neovascularization. This phase is said to last for several weeks (Wallace et al., 2023). Many of the pro-angiogenic chemokines support the robust proliferation and migration of the cells and involve CXCL1 and CXCL2 in the wounds (Moldobaeva et al., 2010).
4. **Remodeling:** In the last stages of the wound healing cascade, the expression of alpha-smooth muscle actin gets a boost from the cytokines which include TGF-beta, fibroblasts, and help in transforming the myofibroblasts. The wound is contracted by the myofibroblasts which use their receptors. When the expression of type 1 collagen increases, the lysis of the MMPs causes the type 3 collagen to become disorganized. Thus, a normal scar is formed upon the balance of the formation of collagen and the breakdown of the old ones. Apoptosis of the vascular cells and myofibroblast marks the ending of the entire healing cascade (Klingberg et al., 2013).

10.1.2 CURRENT THERAPY AND LIMITATIONS FOR DIABETIC WOUND HEALING

The major aim of the treatment involving chronic wounds includes prevention of the occurrence of infection and keeping the area clean from non-viable tissue material. Infections may arise easily because of the loss of innate barriers, which are provided by the skin layers, and the pathogens that accumulate in the wound area, further interfering with the process of wound healing (Burgess et al., 2021). Similarly, the elimination of both non-viable tissue through debridement allows the exposure of healthy tissue, enabling cells to migrate and proliferate to repair the wounded site. Growth factors are involved in every phase of the wound healing process, and due to general deregulation in chronic wound healing, growth factors like platelet-derived growth factor (PDGF), epidermal growth factor (EGF), and basic fibroblast growth factor (FGFb) have been considered as potential strategies for diabetic wound healing, showing promise in small animal models of diabetic wound healing. Likewise, the NGF supplementation demonstrates the potential to promote healing after 5–14 weeks of treatment; however, these results came from a very minimal sample size of the conducted study. Currently, there has been an increase in the realization that ulceration in ischemic feet is a more common form of DFU than ulceration in purely neuropathic feet. Ulceration in the case of ischemic feet and neuro-ischemic ulcers develops in mildly or moderately ischemic feet. Neuro-ischemic feet ulceration in the case of lesser degree ischemia occurs due to the coexistence of the neuropathy itself. Neuro-ischemic foot ulcers with ischemic ulcers are more difficult to cure compared to those in non-ischemic neuropathic ulcers and are associated with a higher rate of amputation and mortality (Boulton et al., 2022).

There are many effective therapies available for diabetic wounds, such as neuropathic and antibiotic drugs, bioengineered skin substitutes (amniotic membrane, bi-layered bioengineered skin substitutes, autologous stem cell therapy, human fibroblast-derived dermis), debridement (hydrogels, MDT maggot debridement therapy), inflammatory modulators, growth factors, and so on (Karri et al., 2016).

Hydrogels are known as three-dimensional (3D) networks made of molecular chains embedded in a water-rich environment. They are prepared through several cross-linking mechanisms, exhibit tunable physiochemical properties, and mimic the native tissue's extracellular matrix. Additionally, hydrogels can be chemically modified and incorporate biologically active cues for recognition, including stimuli-responsive molecules and growth factors (GFs), enhancing their biofunctionality (Drury & Mooney, 2003).

Over the last decade, the preference for natural materials has increased, leading to the engineering of the current generation of hydrogels. These hydrogels are tailored to meet specific tissue requirements and are produced using state-of-the-art techniques. Natural-based hydrogels are made with blood derivatives and decellularized tissues as the source of bioinstructive and biomimetic natural-based hydrogel (Gomez-Florit et al., 2020). Thus, hydrogel patches show no adverse effects in patients and provide improved, integrated treatment of wounds, which is the main focus of our study to overcome the drawbacks of other available therapies. Hydrogels help in the rehydration of necrotic tissue. In order to prepare the porous structure of the material of the hydrogel, there are several methods that are followed, which involve gas forming, porogen templating, cryogelation, bicontinuous emulsion templating, electrospinning, 3D printing, freeze-drying, and lastly, the inverse opal hydrogelation (Insperger & Stépán, 2011). The three-dimensional hydrophilic polymer network present in the hydrogels is capable of absorbing a large amount of water without the loss of structure, and there is a vast selection of source materials (Zhang & Khademhosseini, 2017). In the case of tissue engineering, the use of hydrogels using synthetic polymers is still a progressive research area. The advantages of hydrogels involve growth promotion, proliferation, and differentiation of chondrocytes and support cartilage regeneration (Wang et al., 2018). There are also hydrogel dressings with integrated sensors which have provided a new dimension in the treatment of diabetic wound recovery. With the help of hydrogel dressings with integrated sensors, much important information such as general healing status, bacterial density, amount of oxygenation, level of inflammation, temperature, and pH are understood. Thus, it has been made possible with hydrogel patch therapy to get information on the condition of the wound. Moreover, these types of hydrogel patches deliver advantages like more integrated and improved treatment of the wounds, less time required to heal the wound, and thus a reduction in the time of hospitalization and the cost of healthcare. They also have similarities with living tissue and thus make them more compatible for providing more integrated applications in biomedical care (Tavakoli & Klar, 2020).

10.1.3 EFFECTIVENESS OF ALGAE IN WOUND HEALING

Algae belong to the group of eukaryotic or prokaryotic organisms that utilize sunlight and inorganic carbons to produce organic metabolites, also termed biomass, thus called photoautotrophs. They can also live as mixotrophs as they are able to use organic carbon or carbon dioxide as their carbon source (Shea et al., 2019). Microalgae, specifically, are a very cost-effective means of producing many pharmaceutical preparations, such as vaccines, proteins, and nutrients, which would be more expensive if other plant or animal sources were used. Furthermore, they have antiviral, anti-inflammatory, and anticancer properties as well (Poonam & Nivedita, 2017).

Microalgae are key natural resources that possess a large number of physiological, biochemical, and molecular strategies to cope with environmental stresses and produce several active phyto-constituents. Microalgae are living photosynthetic entities containing singular cells. They need carbon dioxide, sunlight, water, and various inorganic substances to grow. For maximum growth rate and vast CO_2 fixation capacity, microalgae provide topmost rearing. They can also be cultivated in wastewater, where these organisms can dissolve organic adulterants, nutrients, and metallic ions, and mitigate atmospheric carbon dioxide. This helps maximize the efficiency of resources, and the fluent growth of microalgal communities has significant pharmacological and medicinal value (Hussein et al., 2019).

In the case of chronic wound dressing, microalgae hydrogel has shown some of the remarkable features that make it advantageous. Adequate oxygen delivery, high capacity of absorbing exudate, antibacterial effect, maintaining a moist environment for accelerating the healing process, exchanging gas, thermal insulation, harmless to the host, easily removable, good handleability, as well as cost-effectiveness and source availability (Op 't Veld et al., 2020) are some of the characteristics of microalgae hydrogel. We hypothesized that *Synechococcus elongatus* could be an effective alga for wound healing (Chen et al., 2020).

Different ways of wound dressing using hydrogel from several sources have been with manifolds of courses. By deduction, we can mention that marine-derived polysaccharides give alginate hydrogel, agarose hydrogel, chitosan hydrogel, gelatin, and hyaluronic acid. These are widely renowned for being abundant in nature and their applications in the field of tissue engineering (Neamtu et al., 2022; Gong et al., 2022). The synthetic polymers usually engaged in wound healing comprehend polyvinyl alcohol, polyacrylamide as well and polyethylene glucose which provide hydrogel wound dressing (Zhang et al., 2020). After that, carrageenan-based hydrogel provides a vast resource of hydrogel in chronic wound healing. Photo-crosslinking hydrogels, gradient hydrogels, floating hydrogels, micro-patterned hydrogels. Nanogels and hydrogel scaffolds are examples of carrageenan-based hydrogel which are individually a way to heal chronic wounds (Neamtu et al., 2022).

10.1.4 ALGAE-BASED HYDROGEL IN DIABETIC WOUND HEALING

Hydrogels are a useful treatment option for wound healing, as they have several advantages affordable (Pan et al., 2021). They can keep wounds moist and allow for proper gas exchange, which helps prevent drying out and worsening of infections. Additionally, they have antibacterial properties and are easy to apply and remove, making them both effective and affordable (Gao et al., 2021). Furthermore, hydrogels are fantastic tools for delivering medicinal molecules that aid in the healing of wounds. They also help in the creation of fresh blood vessels and the regrowth of the skin's outer layer, aided by the activation of skin stem cells. Moreover, hydrogel patches assist in accelerating the creation and maturation of the extracellular matrix, which helps in the healing process (Fan et al., 2021).

Several types of healing abilities for wounds can be offered by two different hydrogel dressing types. Amorphous hydrogel dressings are one of the two varieties; because of their high moisture content, they are unable to absorb much exudate from the wound. Instead, they can create a moist environment in dry wounds, improving the wounds' capacity to absorb, stretchability, moisture retention, and swelling profile. Sheet dressings might be helpful when phlebitis and the epithelial stage are present during the final stage of wound healing (Kibungu et al., 2021). All hydrogels must meet the requirements to be biocompatible, to serve as a vehicle for the transport of biomolecules required for wound healing, and to encourage the expansion of the vessels affected by the wound as well as the proliferation of cells (Tavakoli & Klar, 2020).

An environment of hyperbaric oxygen is created where an alginate patch protects the wound in order to deliver dissolved oxygen. This technique has been used to treat wounds by applying both topical dissolved oxygen and topical gaseous oxygen. Just 300–700 μm of the dermis layer is penetrated, indicating a limited potential for penetration. Moreover, the chronic diabetic wound must first be healed with topical gaseous oxygen (Burgess et al., 2021; Liu et al., 2018). On the contrary, various hydrogel wound dressings with antioxidant properties have emerged to treat chronic wounds as quickly as possible. Its antioxidant property can eliminate residual reactive oxygen species from chronic wounds to reduce oxidative stress and improve the wound environment. This process leads to ultimate wound healing. The use of antioxidant hydrogels as a novel approach to treating chronic wounds is a promising development that demands our attention, as it has the potential to expedite the healing process (Xu et al., 2020).

Hydrogels have the capability to provide the same physicochemical and biological features as human tissues in terms of wound healing. They also offer networks of hydrophilic cross-linked polymer chains that are water-insoluble, similar to biomaterials. Hence, hydrogels have the potential to create a flexible and relaxing 3D environment for stem cell regulation due to their increased biocompatibility and adaptable chemical and mechanical properties. Moreover, it has been demonstrated that stem cells can self-renew and differentiate on hydrogels as a substrate (Schilrreff & Alexiev, 2022).

Conversely, a variety of hydrogel wound dressings have emerged that are enriched with antioxidant characteristics to cure chronic wounds as rapidly as possible. The presence of antioxidants can help eliminate any remaining reactive oxygen species in chronic wounds. As a result, oxidative stress is decreased, and the overall healing conditions of the wound are enhanced. The ultimate result of this process is wound healing. Antioxidant hydrogels' innovative approach to treating chronic wounds will offer new promise for accelerating wound healing, and it is deserving of our consideration (Cui et al., 2022b; Demir et al., 2020).

Previously, wound healing was involved in dressing using cotton, cotton bandage, and gauze. With modern science, many biomedical preparations like hydrocolloids, foams, and hydrogels have been developed for better wound treatment. However such dressing systems are unable to cover and absorb exudates properly. Using multidrug for better wound care, such as adding antimicrobial effect and fast proliferation rate with conventional dressing process, was quite complex to manage. Moreover, traditional wound care systems do not allow moist healing as it was believed that wounds must be kept dry for rapid healing. So, when we think about an open wound in the case of a diabetic patient, there is a high risk of amputation and disability due to wound infection. Among the conventional wound treatment systems, we can use hydrogel for its multipurpose activity, e.g., medical purposes for cellulosic fabric to impart pH sensitivity and application in antimicrobial surgery and transdermal therapy (Benltoufa et al., 2020). A study found that the bioinspired potential application of hydrogel patches using the 3D printing technique has the opportunity to open a new era for wound management with microalgae as they have the capability to release dissolved oxygen to the wound site. Algal polysaccharides can provide strong mechanical support to work as a hydrogel patch. Besides, their biodegradability along with biocompatibility makes them excellent biomedical preparations for wound dressing where hyperbaric oxygen (HBO) and topical gaseous oxygen (TGO) have less skin-penetrating capacity (Chen et al., 2020b). Therefore, algae-based hydrogel patches can be a better choice for diabetic wound treatment due to their exclusive efficacy in solving problems in other wound care methods.

To improve diabetic chronic wounds, the most effective way is to provide oxygen, which can be achieved using microalgae hydrogel patches where other treatments cannot provide this particular advantage. They help improve the growth of fibroblasts, and oxygen delivery in the wound site, and facilitate new vessel growth. Furthermore, they do not produce unwanted immune responses and are nontoxic in nature (Chen et al., 2017). Among the various kinds of hydrogel wound dressings in natural and synthetic sources, chitosan-based hydrogel, algae-based hydrogel, and synthetic hydrogel are much more effective, reliable, and cost-efficient. Among them, hydrogel-based algae patches have better productivity. In alginate hydrogel, there is an additional antibacterial function that provides additional protection with faster wound healing. The development of bio-polymer-based natural algae hydrogel increases their authentic physiochemical and functional properties like biocompatibility, biodegradability, high absorption and moisture retention, antimicrobial activity, and non-cytotoxic properties as well as adequate oxygen supply (Jaiswal et al., 2019; Porto, 2019; Jaiswal et al., 2019), and all of these qualities make microalgae hydrogel patch a safe and effective solution for chronic wound problems. On the other hand, hydrogel transdermal patch provides painless self-drug administration, unlike injections, along with sustained release of the drug for a specified and longer period of time. Its adhesive nature allows it to stick to the skin surface with a slow release of the loaded drug in the hydrogel bed (Nafo, 2022). Not only that, but this patch can also remove the dead skin when it needs to be peeled off. Being transparent, it facilitates easy monitoring and using secondary dressing above the patch. Its soft texture with a cooling effect and effortless removal can comfort diabetic patients. Moreover, this form of drug eases the packaging and transportation, which creates appeal to the manufacturer as well (Jung et al., 2020; Augustine et al., 2021).

Chitosan-based hydrogel is limited by its insolubility at a pH of >6.5, which is a drawback in terms of biomedical applications, especially when the requirement is for drug delivery inside the body (Hanna & Saad, 2019). Additionally, due to a dense scale-mesh-like network, it contains weak mechanical properties and uncontrolled dissolution, allowing passive diffusion of nutrients and metabolic wastes, which is not suitable for biomedical applications (Pellá et al., 2018). Although the cytotoxicity of anchored materials is limited to the surrounding implant devices and imposes no adverse effects on non-contacting tissues, it is still not suitable for *in vivo* applications requiring favorable biocompatibility (Zhao et al., 2020). To address synthetic hydrogel limitations, it retains water from the atmosphere. Therefore, the volume expansion in air is much smaller, and the response time is higher in water. Another limitation is that synthetic hydrogels are partial. If the patch size is several centimeters long, it can take several hours for the response time (Ionov, 2014).

10.2 OVERVIEW OF HYDROGEL AND 3D PRINTING IN DIABETIC WOUND HEALING

10.2.1 PROPERTIES OF HYDROGEL

Usually, the conventional drug delivery approach is plagued with inevitable problems of systemic toxicity and repeated administration (Narayanaswamy & Torchilin, 2019). As a result, novel drug delivery strategies are targeted at ensuring the controlled release of drugs in local administration by means of nanostructured systems, such as liposomes, nanoparticles, membranes, hydrogels, etc., (Dreiss, 2020), and these structures can be fabricated using techniques like thermal inkjet 3D printing, binder jet 3D printing, TIZ-SFD (spray freeze drying incorporated thermal inkjet 3D printing), etc. (Uddin et al., 2022). Hydrogels are three-dimensional hydrophilic networks composed of cross-linked polymers, proteins, small molecules, or colloids. Cross-linking between ionic interactions and hydrogen bonding provides adequate mechanical strength and physical integrity for polymeric hydrogels (Sharma & Tiwari, 2020). Therefore, it is an effective form of carrier for targeted drug delivery because of its capacity to protect the drug by providing conditions resembling natural tissue, which in turn results in the rate of drug degradation being reduced due to prolonged duration of action along with sustained drug release controlled by diffusion from the degradation of the matrix or solely by endogenous or exogenous stimuli (Dreiss, 2020).

The three-dimensional (3D), hydrophilic, polymeric network that is useful in terms of absorption of a great quantity of water or biological fluids is called a hydrogel. The potential features of the hydrogel are high water content, soft consistency, porosity, ability to simulate living tissue, low tensile strength, robustness, and steadiness, biodegradability without generating toxic species, colorless, odorless, biocompatible and non-toxic, photostability, re-wetting competency, pH-neutrality, required particle size, and absorption, etc., (Garg & Garg, 2016). The preparation of hydrogel is achieved through the crosslinking of polymers. Both synthetic and natural polymers are combined with hydrogels to prepare hydrogels and accelerate the wound-healing cascade. Thus, some synthetic and natural polymeric elements make hydrogels an effective healing agent (Stoica et al., 2020).

These polymers are the agents that make the hydrogel an effective healing agent. Some of these agents and their functions are mentioned in Tables 10.1 and 10.2.

TABLE 10.1 Name and Role of Natural Polymers Used in the Preparation of Hydrogel in Wound Healing

Polymer	Function in Wound Healing
Alginate	Promotes the healing process, controls the infection rate, and also provides a hemostatic effect
Cellulose	Provides support to the cell and promotes healing.
Heparin	Ensures active delivery of the drug molecules to the site of injury.
Dextran	Provides support to cells and ensures delivery of drug molecules
Gelatine	Reinforcement of tissues, recruitment of tissues, and delivery of drug substances

TABLE 10.2 Synthetic Polymers Employed in Hydrogel Production and Their Functions in Healing Wounds

Polymer	Function in Wound Healing
PEG (polyethylene glycol)	Supports the tissue, aids in infection control, ensures medicine molecules are delivered, and encourages cell attachment
Polyvinyl alcohol	Ensures distribution of therapeutic molecules, regulates the progression of healing and controls infection
Polyvinyl pyrrolidone (PVP)	Antiseptic action encourages microcirculation to speed wound healing
Acrylic polymers (Carbopol)	Encourages the administration of antibiotics and cell adhesion

10.2.2 HYDROGEL IN WOUND HEALING

The effective management of wound healing relies on the precise and regulated administration of therapeutic substances to the specific site of the injury. Lately, hydrogels are used not only in drug delivery but also in wound dressing materials, tissue regeneration materials, etc. According to the composition of hydrogel and its mechanical aspect from the polymers of biomaterial, it is apparent that hydrogels are like extracellular matrix (Kurakula et al., 2020). Hydrogels possess high potential characteristics such as high water or moisture content, flexibility, and biocompatibility which enables them to play a major role in treating wounds. Hydrogels are considered to be suitable for all the stages of wound healing cascades: homeostasis, inflammation, cell proliferation, and maturation. They are compatible with a great number of polymers and thus fabrication of hydrogels of different applications can be formed. They are actively employed as carriers of active drug substances as they constitute characteristics such as fluid affinity, water uptake, and rheological properties. It is also suitable for maintaining the moisture of wound surfaces while ensuring the delivery of active substances. It delivers bioactive molecules such as di-butyryl cyclic adenosine monophosphate, nitrofurazone, antibiotics, polyphenols, antimicrobial peptides, etc. The delivery of these active compounds enhances the healing process, re-epithelialization, control of infection, antimicrobial and antioxidant effects, as well as the generation of skin, proliferation of cells, and acceleration of the closure of wounds. Moreover, it also absorbs the exudate of the wound and keeps it away from the wound site, while also soothing the wound through its high-water content. Additionally, due to controlled cross-linking, polymer hydrogels enable the active therapeutics to remain longer and increase the local concentration of active substances, making it an excellent medium for delivering drugs to accelerate the closure of wounds. The network possesses a 3D configuration and allows cellular attachment, migration, and deposition in the extracellular matrix while effectively delivering active drug molecules to the wound site. Furthermore, it obstructs the entry of bacteria in a swollen state and enables the delivery of functional molecules to the site of interest (Stoica et al., 2020; Kalantari et al., 2020; Lim et al., 2017).

In diabetic patients, the difficult-to-heal factor is tremendously high because of persistent hyperglycemia. Hyperglycemia triggers bacterial colonization and sustained inflammation due to the abundance of Reactive Oxygen Species produced from advanced glycation end products (AGEs). Moreover, vascular damage from microvascular complications results in a deficit of oxygen and nutrients at the wound site. Primarily, wound dressing is vital to protect, cover, and heal the wound. However, there is no alternative to keeping the wound site moist so that the area has the proper environment for tissue regeneration in the presence of balanced moisture. Hence, in this situation, the inherent ability of hydrogel to retain water content gives an additional advantage compared to conventional wound dressing. Usually, hydrogel loaded with antimicrobial and glucose control drugs is recommended (Gao et al., 2021). Therefore, the compromised healing capacity of diabetic patients requires hydrocolloid dressing for its exudate absorption property. The fundamental roles of hydrogel in wound management are summarized in Figure 10.2.

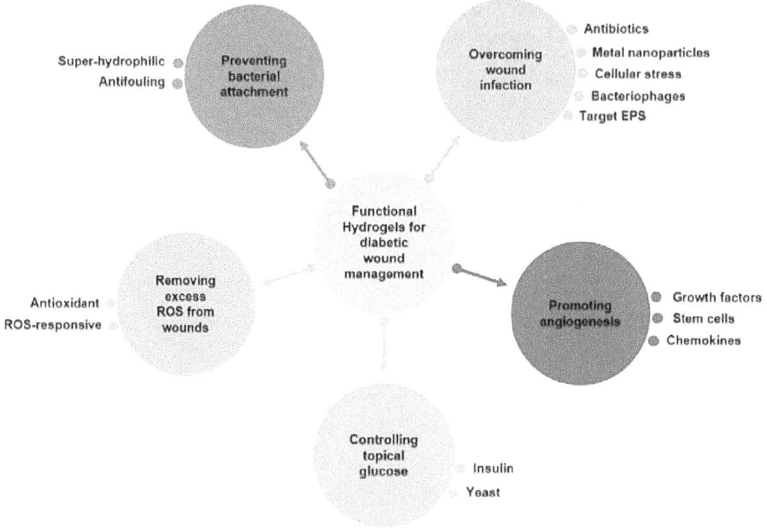

FIGURE 10.2 Functional hydrogels for diabetic wound management.

Source: Reprinted from Gao et al., 2021. Copyright © 2021 Author(s). http://creativecommons.org/licenses/by/4.0/

The polymer chitosan in combination with marine peptides is used to prepare a Chitosan-Marine Peptides hydrogel or CSMP, which is developed to work as a burn dressing. CSMP showed antibacterial activity, proliferation of pro-cells, and enhanced healing of burns. CSMP demonstrates pronounced wound healing efficiencies when it was pathologically examined by HE staining. Moreover, when treated with CSMP for burn wounds, there was the occurrence of re-epithelialization and deposition of fiber in the first week. Then in the next week, the regeneration of epithelium was achieved. Within 21 days of treatment, the wounds were shown to be healed completely. Furthermore, the expression of FGF2 and VEGF is also regulated by CSMP. Thus, cell migration is enhanced by CSMP along with the enhancement of the regeneration of skin, which shows the promising benefits of CSMP in the treatment of chronic burn wounds (Ouyang et al., 2018).

When treated with a hydrogen sponge loaded with platelet-rich plasma exosomes (PRP-Exos) and one homogeneous polysaccharide (ZWP), effective acceleration of wound healing is observed in diabetic rats *in vivo* in comparison with treating the wound with a blank hydrogen dressing. In the *in vivo* study of wound healing in rats, experiments were carried out involving using only the PRP-Exos or ZWP alone, and another experiment involved using a combination of PRP-Exos and ZWP to treat wounds in diabetic rats. All the experiments showed a contraction of wounds in diabetic rats, but a more effective result was observed when the combined therapy of PRP-Exos-ZWP was used rather than using a single administration. The synthesis of collagen and the deposition of collagen were regulated along with the enhancement of angiogenesis at the site of the wound. These results provide evidence that the hydrogel patch works efficiently by accelerating the wound healing process, and the study done *in vivo* with diabetic rats provides evidence that hydrogel patches work efficiently by accelerating the wound healing process and enhancing angiogenesis (Xu et al., 2018).

In diabetic patients, firstly, topical dressings with a moist environment are necessary for granulation tissue production and epithelialization, while reducing the risk of further infection and accelerating wound healing. There are different types of dressings that can provide special characteristics to accelerate wound healing: passive, interactive, and bioactive. Passive dressings cover the wound only; interactive dressings provide a barrier to the entry site for pathogens, and bioactive dressings create a moist environment along with growth factors for granulation tissue formation. Secondly, the use of anti-diabetic drugs like metformin, insulin, DPP-4 inhibitors, and thiazolidinediones can be used during the treatment of chronic wounds since they can reduce MMPs, elevate keratinocytes, increase granulation tissue formation, or promote fibroblast proliferation. Thirdly, the introduction of growth factors that stimulate ECM formation helps modulate an inflammatory response and promote re-epithelialization and angiogenesis. Such factors include EGF, VEGF, PDGF, TGFβ1, etc. Lastly, the use of various drugs like statins, phenytoin, metformin, and DPP-4 inhibitors shows great improvement in the wound healing capabilities of diabetic patients (Xu et al., 2018; Hartmeier et al., 2021). Hydrogels are the preparations that can deliver all these properties in the healing of wounds. There are synthetic and natural polymers that are being used in the preparation of hydrogels. These polymeric hydrogels enable the acceleration of the wound healing process, delivery of active molecules, providing an inflammatory response, re-epithelialization, proliferation, promoting adhesion of cells, growth factors, tissues, and other important factors that help in the healing of the wounds (Hartmeier et al., 2021). That is why hydrogel patches are the most effective therapy for diabetic wound healing.

10.2.3 3D PRINTING IN WOUND HEALING

Three-dimensional (3D) printing of organic material, or 3D bioprinting, is a rapidly expanding field with intriguing applications in tissue engineering and regenerative medicine. Bioprinters use cells and biocompatible materials as ink (bio-ink) to construct 3D structures resembling organs and tissues, in a controlled manner and with micrometric precision (Salaris & Rosa, 2019). 3D bioprinting is an emerging technology where scaffold materials and cell-laden hydrogels can be deposited in a predetermined pattern to create 3D porous constructs. Currently, the materials used for 3D printing scaffolds and hydrogel systems are mainly PCL, PLA, gelatin, sodium alginate, etc., (Gong et al., 2022; Uddin et al., 2022).

Normal biopolymer-based hydrogels are generally utilized in biomedical as well as tissue engineering applications. However, printing hydrogels on 3D platforms is still challenging. Pioneers in the field of cell engineering and biomaterials have validated that altering material properties, such as stiffness, bioactive moieties, and degradation, significantly impacts cell behavior and tissue growth. 3D bioprinting techniques have been developed for 3D cell cultures, either with cells directly or by incorporating them into hydrogels that act as an ECM. Additionally, cells have been encapsulated within micro-porous or tunable nanofibrillar structures for 3D bioprinting. 3D systems promote the rapid growth of adherent or encapsulated cells. Nano-fibrous platforms provide a 3D architecture that supports the attachment of ECM proteins during *in vitro* cell cultures. Hydrogels have the unique ability to mimic the structure and mechanics of the native cell microenvironment. Due to their ability to replicate the nature of the softest tissues, hydrogels are highly desirable materials for developing synthetic ECM analogs (Shin et al., 2020).

Currently, 3D printing in wound healing is gaining traction and could be a potential game changer. Different materials, including hydrogels, are being used as potential wound dressings to promote wound healing. Ilhan et al., 2020, studied the effectiveness of *Satureja cuneifolia*-loaded sodium alginate/polyethylene glycol scaffolds to promote wound healing by applying 3D printing technology (Ilhan et al., 2020). Another study by X. L. Xu et al., 2019, consisted of using nanocellulose assisted with gelatin methacrylate as bio-ink for wound healing via 3D printing technology. The creation of self-healing polymers, such as hydrogels and thermosets, using 3D printing has received a lot of attention up to this point, but all of these studies have relied on extrusion-based 3D printing (Heidarian et al., 2019; Kim et al., 2019; Nadgorny et al., 2018).

3D printers come in a variety of designs, each with unique benefits and cons. When buying a printer, it's crucial to take institutional objectives and anticipated model usage into account. The two main categories of printers are additive and reductive processes (Mardis, 2018). The model is gradually transformed into a 3-dimensional shape using additive machines, which deposit layer after layer of material. In a reductive procedure, the printer starts with a block of material and gradually removes pieces until the desired shape is revealed. Additive methods are used by the majority of commercially viable printers for medical purposes. Mardis (2018) described a number of commercially available 3D printers which are detailed in Table 10.3. Tsegay et al. (2018) illustrated the materials compatible with the 3D printer and the accuracy rate for each one (Tsegay et al., 2022).

TABLE 10.3 Categories of Commercially Accessible 3D Printers

Additive Method Printers	Fused deposition modeling (FDM)
	Digital light processing (DLP)
	Stereolithography
	Polyjet
	Selective laser sintering (SLS)
	Selective laser melting (SLM)
	Electron beam melting (EBM)
Reductive Method Printers	Laminated object manufacturing (LOM)

Extrusion-based bioprinters, inkjet-based bioprinters, laser-assisted, and stereolithography are the available 3D printers in the current market. But in recent years, extrusion-based 3D bioprinting has the facilities of hydrogel-based fabrication with advanced techniques and coaxial bio-printing for more complex structures. Furthermore, the multi-material multi-nozzle method in 3D bioprinting has reduced the printing time with uniform and controlled construction. Nowadays, the FRESH method (the freedom reversible embedding of suspended hydrogel) is also used in 3D printing because of its advantages for volumetric patterns. On the other hand, open-source 3D bioprinters have become popular as modified facilities can be added to them according to the requirements of a researcher (Liu et al., 2019b). As hydrogels are polymeric compounds constructed with cross-linked chains, both physical and chemical crosslinking mechanisms can modify the mechanical strength of hydrogels. Physical crosslinking involves the gelation process using its intrinsic properties by adopting thermal condensation methods, self-assembly, and ionic interaction methods. On the other hand, chemical crosslinking techniques have been used for their long-term stability and other facilities. For hydrogel formation, chemical crosslinking is controllable and more beneficial than physical crosslinking, providing a more precise and structurally complex microenvironment (Unagolla & Jayasuriya, 2020).

Where an alginate patch covers up the wound, an atmosphere of hyperbaric oxygen is established for delivering dissolved oxygen, and using this technique algae-based hydrogel has been used for wound treatment as they have the capability to release dissolved oxygen to the wound site. Algal polysaccharides can provide strong mechanical support to work as a hydrogel patch. Besides, their biodegradability along with biocompatibility make them excellent biomedical preparations for wound dressing where hyperbaric oxygen (HBO) and topical gaseous oxygen (TGO) have less skin-penetrating capacity (Chen et al., 2020a). Not only that, but various hydrogels with antioxidant properties have also emerged as wound dressings to treat chronic wounds as quickly as possible. Their antioxidant property can help eliminate residual reactive oxygen species from chronic wounds, reducing oxidative stress and improving the wound environment. The utilization of antioxidant hydrogels in the treatment of chronic and diabetic wounds is a promising strategy that can expedite the healing process, leading to the ultimate closure of the wound. This innovative approach offers a new ray of hope for individuals suffering from chronic wounds (Xu et al., 2020). Another study has highlighted that the hydrogel patch can retain a large amount of wound exudate and facilitate fast healing. Self-drug administration of a hydrogel transdermal patch provides sustained release of the drug for a specified and longer period of time, and its adhesive nature allows it to stick to the skin surface with a slow release of the loaded drug in the hydrogel bed. Its soft texture with a cooling effect and effortless removal can comfort a diabetic patient (Nafo, 2022; Jung et al., 2020).

10.2.4 COMBINATIONAL USE OF HYDROGEL & 3D PRINTING

The integration of hydrogel as a potential material for 3D printing has gained major interest in recent years. A polymer hydrogel is a 3D crosslinked network of stretchable polymer chains that has a significant amount of water content yet still has solid-like qualities. Because of the 3D network's ability to retain liquid, a swelling gel phase may be formed, and the liquid within can keep the polymer network from collapsing, giving it properties resembling those of soft tissues. A polymer sol or solution can quickly shift from a viscous liquid state to a solid phase hydrogel at the gel point (sol-gel transition) (Li et al., 2020). An example of this includes the design and production of voxelated soft matter using multi-material multi-nozzle 3D

(MM3D) printing, which can program the material component, structure, and function at voxel size, was recently described by Skylar-Scott et al. (2019) (Persad & Rocke, 2022). The MM3D printheads enable smooth high-frequency switching between several viscoelastic materials based on diode-like behavior, producing voxels with a volume that is almost equal to the cube of the nozzle diameter. Through the use of MM3D printing, this study reveals a potent technique for producing voxelated soft matter that would bypass current printhead design restrictions on periodicity while simultaneously enhancing feature resolution and speeding up manufacturing time (Li et al., 2020; Persad & Rocke, 2022).

This combination is particularly useful for diabetic patients. To generate unique release profiles for each medicine, Seoane-Viaño et al. (2021) used semi-solid Extrusion Three-Dimensional Printing (SSE 3DP) has been used during the creation of polypills that contain various medications in separate compartments (Seoane-Viaño et al., 2021). This is true of a polypill that contains glipizide, nifedipine, and captopril, all of which are used to treat arterial hypertension in type II diabetes (treatment for type II diabetes). Additionally, a polypill containing three medications for the treatment of type II diabetes was created using SSE 3DP (metformin hydrochloride, glyburide, and acarbose) (Seoane-Viaño et al., 2021). Another study by Bai et al. (2020), stated the importance of nanoparticles/hydrogels, by including both bioactive compounds (such as growth factors, genes, proteins/peptides, stem cells/exosomes, etc.) and non-bioactive components (metal ions, oxygen, nitric oxide, etc.), nanoparticles/hydrogels aid in the healing of diabetic wounds (Khaled et al., 2015). Researchers now choose smart hydrogels among them because they offer a highly promising approach for loading a variety of bioactive components. Additionally, several technologies (such as PTT, LBL self-assembly method, and 3D printing technology) can be used with nanoparticles and hydrogels to treat diabetic wound healing (Bai et al., 2020).

Viscosity and rheological qualities of hydrogel materials are sometimes used interchangeably; however, viscosity alone cannot capture the complicated rheological behavior of hydrogel materials (Johan Göhl & Kajsa Markstedt, 2018; Gao et al., 2018). Two dynamic moduli, such as the storage modulus (G') and loss modulus (G"), are crucial in addition to viscosity. The hydrogel material's solid-like and liquid-like properties are indicated, respectively, by the storage modulus (G') and loss modulus (G") values. Identification of the proper bio-ink components can be aided by knowledge of the gelation (G'=G") point and viscosity. Determining the recovery rate of the released hydrogel is crucial for ensuring the form accuracy of the constructed scaffold at the post-printing stage (Jiang et al., n.d.). It has always been a wonder to researchers whether hydrogel and 3D printing support proper rheological behaviors. To test this theory, Hernández-Sosa et al. (2022), investigated the optimization of rheological properties of self-assembled tripeptide/alginate/cellulose hydrogels for 3D printing (Alejandro et al., 2022). They stipulated that the rheological parameters investigated include (i) minimum viscosity; (ii) crosslinking capability; (iii) shear-thinning characteristics; (iv) yield stress; (v) ratio between the storage modulus (G') and the loss modulus (G"); (vi) retention of solid-like behavior of hydrogels. It was observed that the hydrogel combination proved to have favorable rheological characteristics for 3D printing. Another study by Ma et al., 2020, observed the rheological characteristics and printability of cellular nanocrystal (CNC) hydrogels at various concentrations (0.5–25 wt%) and measured CNC's self-assembly behavior in response to shear during printing. It was concluded that CNC was an ink that showed promise for 3D printing and exhibited the right rheological properties (printing resolution, concentration selection capability) (Ma et al., 2021).

3D printing has significant benefits over conventional processing techniques, including an easy operating process, accurate structural control, and great cost-effectiveness (Costantino Creton, 2017). 3D-structured hydrogels for various functional purposes may be creatively and independently created by layer-by-layer printing (Zhao et al., 2021; Wang et al., 2022; Song et al., 2018). Gel material applications in a variety of industries are anticipated to increase with the widespread usage of 3D-printed hydrogels. Their capability of releasing dissolved oxygen to the wound site and strong mechanical support make them suitable to heal the wounds of diabetes. Besides, their biodegradability along with biocompatibility makes them excellent biomedical preparation for wound dressing where hyperbaric oxygen (HBO) and topical gaseous oxygen (TGO) have less skin-penetrating capacity (Chen et al., 2020a).

Moreover, painless self-drug administration, unlike injections, along with sustained release of drug by a hydrogel transdermal patch for a specified and longer period of time, is beneficial to the diabetic patient, especially those who need help from others to deal with their injuries and have the risk of infection. Its adhesive nature allows it to stick to the skin surface with a slow release of the loaded drug in the hydrogel bed (Ahsan et al., 2021). On the other hand, a study done by Jung highlighted that a hydrogel patch can retain a large amount of wound exudate, which facilitates fast healing. This patch not only removes dead skin when needed but also allows easy monitoring and application of secondary dressing due to its transparent nature. Its soft texture, cooling effect, and easy removal provide comfort to diabetic patients (Jung et al., 2020). Mainly, the rheological behavior of polymers like hydrogel is measured by analysis through a rational rheometer. Shear rate, storage modulus (G), and loss modulus (G') also need to be assessed during the experiment to understand the material's rheology. Besides that, as both intrinsic and extrinsic factors greatly influence the sol-gel behavior, this study also focused on them.

Intrinsic factors such as chemical nature, molecular weight, materials viscosity, as well as the flowing property or elasticity are important. For example, a study by Alcantara showed that the ratio between different compounds and their amount in the hydrogel composition influences changes in elasticity. In his investigation, he described how hydrogel with carrageenan exhibited stronger hydrogen bonds than just gelatin molecules, resulting in the highest G value and excellent elastic behavior when Gelatin: carrageenan = 1:2 was achieved (de Alcântara et al., 2020).

On the other hand, extrinsic factors also have a greater influence on the rheological behavior of polymeric materials to bring modifications during hydrogel preparation. For instance, temperature controls the primary gelation procedure. With an increase in temperature, the viscosity of alginate with different compositions can be decreased. The resolution of alginate also depends on temperature. Therefore, the consideration of the number of materials, elasticity, viscosity, shear rate, temperature, shear thinning, and shear thickening behavior control the sol-gel transition on which the printing procedure depends. To achieve a desirable hydrogel bed using a 3D bioprinter with desirable resolution, these rheological studies need to be assessed before starting printing with different shapes (Liu et al., 2019).

Previously, wound healing involved using cotton, cotton bandage, and gauze for dressing. With advancements in modern science, many biomedical preparations like hydrocolloids, foams, and hydrogels have been developed for improved wound treatment. However, these dressing systems may not effectively cover and absorb exudates. Utilizing multidrug approaches for enhanced wound care, such as incorporating antimicrobial effects and promoting fast proliferation rates with conventional dressing processes, can be quite complex to manage. Additionally, traditional wound care systems that do not support moist healing were believed to be necessary for rapid healing (Zhang & Zhao, 2020). When considering open wounds in diabetic patients, there is a high risk of amputation and disability due to wound infections. Among conventional wound treatment systems, hydrogels can be utilized for their multipurpose activities, such as medical purposes for cellulosic fabric to provide pH sensitivity and applications in antimicrobial surgical and transdermal therapy (Benltoufa et al., 2020). Not only that, painless self-drug administration with sustained release of drug by this hydrogel patch is also convenient for diabetic patients (Ahsan et al., 2021). Therefore, an algal-based hydrogel patch can be a better choice for diabetic wound treatment due to its exclusive efficacy in solving the problems compared to other conventional wound care methods.

10.3 FABRICATION & PHARMACOLOGICAL MECHANISM

10.3.1 ALGAE SELECTION AND HYDROGEL PREPARATION

One type of algae that could potentially be used for a hydrogel patch is *Chlorella vulgaris*, as it has been shown to possess properties that make it suitable for use in hydrogel-based products. This alga has a high content of polysaccharides, which can be extracted and used to create a hydrogel matrix. Additionally, *C. vulgaris* has been shown to have good biocompatibility and biodegradability (Li et al., 2020).

Another potential candidate for a hydrogel patch is the algae species *Porphyridium cruentum*. This alga has been shown to produce a highly hydrated and elastic hydrogel when its exopolysaccharides are extracted and crosslinked (Beaumont et al., 2021). *P. cruentum* has also been shown to have good biocompatibility, making it a promising candidate for use in medical devices such as hydrogel patches (Tsegay et al., 2022). Seven hydrogel-forming algae-extracted polysaccharides are alginate, λ-carrageenan, ulvan, starch, agarose, ι-carrageenan, κ-carrageenan, porphyran, and (nano)cellulose. Figure 10.3 shows the types of algae used in hydrogel patches.

According to a study published in 2020 (Mariadoss et al., 2020), the hydrogel preparation/patch demonstrated high cell viability, with over 90% of cells remaining viable after being cultured on the patch for three days. Another study published in 2022 (Jang et al., 2022) also found that the hydrogel patch had a high cell viability rate, with over 85% of cells remaining viable after being cultured on the patch for five days. The isolation process (Figure 10.4 for details) involves the following methods;

a. **Enzymatic Extraction:** This method involves the use of enzymes to break down the cell walls of the algae, allowing the hydrogel to be extracted. According to a 2020 article published in the Journal of Biotechnology, the first step in this process is to homogenize the algae cells using a blender or food processor. Next, the cells are incubated with enzymes such as pectinase or cellulase to break down the cell walls. The resulting mixture is then filtered to separate the hydrogel from the other cell debris (Yang et al., 2021).

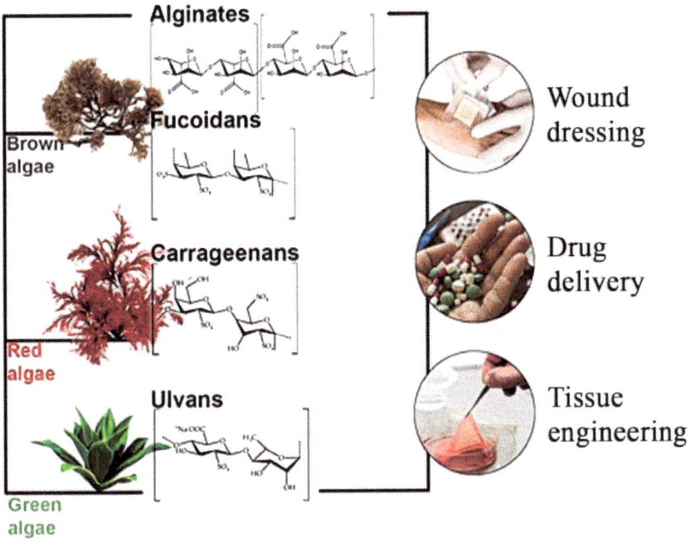

FIGURE 10.3 The types of algae Used in the hydrogel patch.

Source: Reprinted from Kuznetsova et al., 2020. Copyright © 2019 by the authors. Licensee MDPI, Basel, Switzerland. http://creativecommons.org/licenses/by/4.0

b. **Chemical Extraction:** This method involves the use of chemicals to dissolve the hydrogel from the algae cells. According to a 2021 article published in the Journal of Materials Science and Engineering, the first step in this process is to grind the algae into a fine powder using a blender or food processor. Next, the powder is mixed with a solvent such as water or ethanol to dissolve the hydrogel. The mixture is then filtered to separate the hydrogel from the other cell debris (Beaumont et al., 2021).

c. **Mechanical Extraction:** This method involves mechanically separating the hydrogel from the algae cells using techniques such as centrifugation or filtration. According to a 2022 article published in the Journal of Environmental Science and Technology, the first step in this process is to homogenize the algae cells using a blender or food processor. Next, the cells are subjected to centrifugation or filtration to separate the hydrogel from the other cell debris (Lin et al., 2022).

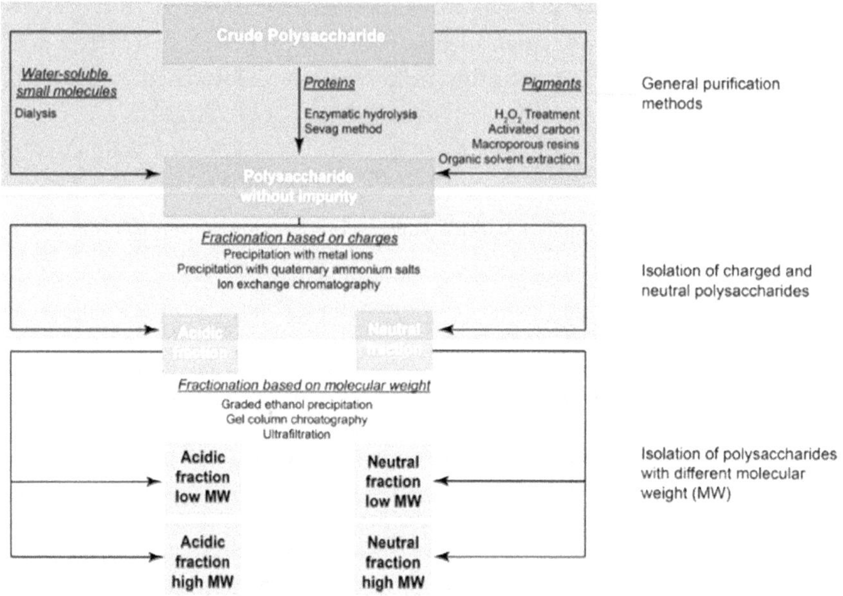

FIGURE 10.4 Routes of purification and extraction can be employed to isolate a polysaccharide according to its chemical structure.

Source: Reprinted from Beaumont et al., 2021. Copyright © 2021 The Authors. Published by American Chemical Society. http://creativecommons.org/licenses/by/4.0

We hypothesized that *Synechococcus elongatus* could be an effective alga for wound healing (Chen et al., 2020). Where an alginate patch covers the wound, an atmosphere of hyperbaric oxygen is established for delivering dissolved oxygen. *Synechococcus elongatus*, a unicellular cyanobacterium, is filled in the gel beads of the scaffolds containing active *Synechococcus elongatus* PCC7942. PCC7942 can make use of inorganic carbon such as CO_2, CO_3^{2-}, and HCO_3^- to release O_2 by photosynthesis. So, these additional carbonates in PCC7942 are responsible for the O_2 production, and in a particular range of concentration of PCC7942, the production rate of oxygen was positively correlated, and the optimum concentration was recorded as 1×10^9/ml. By using this oxygen delivery method, it is notable that the topical dissolved oxygen, as well as topical gaseous oxygen, passes through about 300–700μm through the dermis layer. Furthermore, topical gaseous oxygen is prior to healing the chronic diabetic wound (Chen et al., 2020). On the contrary, various hydrogel wound dressings flourished with antioxidant properties have arisen to treat chronic wounds as quickly as possible. A unique property of antioxidants is that they can eliminate residual reactive oxygen species from chronic wounds to reduce oxidative stress and improve the wound environment.

This process leads to ultimate wound healing. Among the various kinds of natural and synthetic sourced hydrogels used for wound dressing, chitosan-based hydrogel, algae-based hydrogel, and synthetic hydrogel have much more reliability from the view of effectiveness and cost-efficiency. Hydrogel-based algae patches have better productivity among these. In alginate hydrogel, additional protection against health-threatening bacterial infection is provided by its additional antibacterial functions. The bio-polymer-based natural algae hydrogel development increases their authentic physicochemical and functional properties like biocompatibility, biodegradability, high absorption, and moisture retention, antimicrobial activity, and non-cytotoxic properties as well as adequate oxygen supply (Porto, 2019; Jacob et al., 2021; Jaiswal et al., 2019). Hydrogel preparation by the coordinate cross-linking of multi-arm thiolated polyethylene glycol provides a tight mesh structure. This structure ultimately prevents microbial infection at the wound site (Wang et al., 2018; Kamoun et al., 2017). Hydrogel is abundant with the potential to biomimic the topical microenvironment of the skin because of its unique structure with porous and hydrated anatomy (Saghazadeh et al., 2018).

To be applied in wound dressing, hydrogels need to meet the requirements of biocompatibility and also need to be suitable for wound application. After fulfilling the prerequisite conditions, hydrogel can be used as a permanent or temporary wound dressing that can help reinforce and heal the injured area (Tavakoli & Klar, 2020). An important point to note is that the 3D polymeric structures of the hydrogel can retain water in large quantities, much more than the hydrogel's dry weight, and maintain a high moisture level at the wound site, which helps accelerate the healing process (Op 't Veld et al., 2020). Therefore, the hydrogel can provide a platform to load cells, growth factors, antibacterial agents, as well as various supplementary and biomacromolecules, achieving several important aspects for wound healing such as preventing bacterial infection, absorbing moisture or excess exudates, allowing oxygen permeability, minimizing tissue trauma, providing higher adhesiveness, no toxicity, preventing exuding necrotic wounds, offering thermal insulation, and being abundant and cost-effective (Tavakoli & Klar, 2020). Additionally, features like softness, malleability, and biocompatibility make hydrogel more reliable to use as a patch, and the healing time is shorter compared to the normal wound healing process. The hostile environment of the wound, enriched with degrading enzymes, elevated pH, and deflective ionic strength, helps the hydrogel to expand and release water molecules, indicating a specific environmental response to temperature, pH, ionic strength, and hydrophobic forces (Kamoun et al., 2017; Saghazadeh et al., 2018; Yahia, 2015).

8% Sodium Alginate (8%) was added to the PCC7942 solution (with 500μM Na_2CO_3), and using a syringe (1 ml), the solution was dripped into $CaCl_2$ (3%). A magnetic stirrer was used to stir them so that uniform-sized alga-gel particles could be formed. To transform the algal gel into an Alga gel patch, a hydrogel bed (1 mm in diameter) was sealed between a 0.22-micrometer porous hydrophilic polytetrafluoroethylene (PTFE) membrane, which was used as the inner liner, and a clear polyurethane film as the back liner (Chen et al., 2020).

10.3.2 DRUG CHOICE & LOADING

The synthesis of a hydrogel loaded with the drugs can be achieved in situ and pre-loading methods. The in situ method is efficient for the hydrophilic type of drug. It is achieved by dissolving the drug into the water along with the powder of the polymers. The pre-loading method refers to the immersion of drug solution into the films of dry hydrogel for some time. After the incorporation of the drug in both methods, the state of the hydrogel achieved is dry. (Wong & Dodou, 2017) have prepared polyhydrogels through cross-linking with pentaerythritol tetra-acrylate. Different types of drugs were loaded in the hydrogels including hydrochloride, diclofenac sodium, and ibuprofen. Moreover, the hydrogel formulations showed a positive outcome *in vitro* and pre-clinical studies. As a result, these are being marketed for various therapeutic applications (Narayanaswamy & Torchilin, 2019). In Table 10.4, some of the hydrogel formulations and what drugs they are delivering are mentioned:

Loading of the Dexamethasone into hydrogel was efficiently done, along with an enhanced loading capacity (5 mg/g). The adequate colon-targeting attribute of the pH-sensitive P(LE–IA–MEG) hydrogel (T_{max} = 1.0 h, C_{max} = 2.16 μg/ml of dexamethasone; T_{max} = 3.9 h, C_{max} = 0.43 μg/ml of dexamethasone hydrogel) was seen in the *in vivo* pharmacokinetic study as well as in the *in vitro* release study (Vigata et al., 2020). In order to dissolve both dexamethasone and hydrogel monomer materials (poly (ethylene glycol) methyl ether (MPEG)–poly (lactide acid)–acryloyl chloride macromonomer, itaconic acid (IA) and MPEG–methacrylate), different types of solvents (ethanol and 1,2-propanediol) were used. Afterward, by means of the heat-initiated free radical polymerization method, the hydrogel is prepared by mixing these together (Dong et al., 2014).

According to a review by (Zhang et al., 2023), the choice of drugs to be incorporated into algae hydrogels is largely based on their potential therapeutic effects and the ability to be easily released from the hydrogel matrix. Algae hydrogels have the ability to retain a wide range of drugs due to their high water content and ability to form strong physical crosslinks (Chen et al., 2020a). Additionally, algae hydrogels have been shown to be biocompatible and biodegradable, making them a desirable delivery system for a variety of drugs. This versatility in drug choice and the potential for widespread therapeutic effects make the use of algae hydrogels a universal approach to drug delivery. One hypothesis is that the choice of drugs is influenced by their physical and chemical properties, such as solubility, stability, and bioavailability (Hai, 2022). For example, drugs with low solubility may be more difficult to incorporate into the hydrogel matrix, leading to reduced drug release and bioavailability. Another hypothesis is that the choice of drugs is influenced by the specific application of the hydrogel, such as the intended use and target tissue or organ (He et al., 2018). For example, drugs with anti-inflammatory properties may be preferred for the treatment of inflammatory diseases, while drugs with anti-cancer properties may be preferred for the treatment of cancer. Overall, the choice of drugs in algae hydrogel formulations appears to be universal, as many different drugs have been successfully incorporated into these formulations for a variety of applications.

TABLE 10.4 Formulations of Hydrogels and the Drug Delivered (Narayanaswamy & Torchilin, 2019)

Name of the Product	Type of Hydrogel	Therapeutic Application	Drug Delivered
Sericin	Dextran	Delivery of drug for malignant melanoma that is trackable optically	Doxorubicin
Hyalofemme	Carbomer propylene glycol, a derivative of hyaluronic acid	Vaginal dryness, an alternative to estrogen	Derivative of hyaluronic acid
muGard	Mucoadhesive 2% Poloxamer	Oral licher planus Recurrence of cervical cancer	Carboplatin
Regranex	Carboxymethyl cellulose	Diabetic foot ulcer	Recombinant human platelet-derived growth factor

Hydrogels, which are chitosan-based, are capable of enhancing the process of wound healing at different stages as well as mitigating factors associated with slowing down the wound healing process. A bee venom-infused chitosan–PVA hydrogel was developed, producing an anti-inflammatory effect similar to that of the ideal anti-inflammatory drug diclofenac gel. When bee venom was combined with chitosan, it significantly accelerated the wound-healing process in diabetic rats (Guan et al., 2021). Chitosan hydrogels have good biocompatibility and antimicrobial properties, making them suitable for use in wound healing and drug delivery applications (Zhang et al., 2014). The combination of a drug and hydrogel can be effective in wound healing due to the unique properties of both substances. Hydrogels, composed of water and crosslinked polymers, can provide a moist environment for wound healing and serve as a delivery system for drugs (Gao et al., 2018). This allows for the sustained release of the drug over time, increasing the chances of the drug being absorbed by the body and having a therapeutic effect (Gao et al., 2018).

One example of this is the use of a hydrogel containing the drug pentoxifylline in the treatment of diabetic foot ulcers. A study by (Abbas et al., 2019) found that the use of pentoxifylline-containing hydrogel helped to shrink the size of the ulcers significantly and caused an increase in the rate of wound healing compared to the control group. The authors attribute these positive effects to the anti-inflammatory and vasodilatory properties of pentoxifylline, which can improve blood flow to the wound site and promote healing (Heidarian et al., 2019). Overall, the combination of drugs and hydrogel can be an effective approach to wound healing due to the ability of hydrogels to deliver drugs in a sustained manner and the potential therapeutic effects of the drugs themselves.

Algae hydrogel has been shown to effectively carry drug molecules through a process called encapsulation. Encapsulation involves enclosing the drug molecules within the hydrogel matrix, which can then be administered to the desired location in the body. One study found that alginate, a type of algae hydrogel, was able to effectively encapsulate and release drugs such as ibuprofen and paracetamol (Bashir et al., 2018). Another study demonstrated the potential of using algae hydrogel for drug delivery in cancer treatment, as it was able to effectively carry and release the chemotherapy drug doxorubicin (Zhao et al., 2022). Overall, the ability of algae hydrogel to effectively carry and release drug molecules makes it a promising material for drug delivery applications.

The active dressings of a hydrogel patch with heparin show promising results in the healing of chronic wounds compared to using a hydrogel patch alone. Hydrogel dressing that is encapsulated with heparin and basic fibroblast growth factor speeds up or enhances the healing of chronic wounds. The benefits of the theory of wet healing and the exogenous growth factor supplement are combined with the heparin-hydrogel dressing. Moreover, inflammation is decreased, and the wound-healing process is enhanced as a result of the encapsulation of heparin into hydrogel patches. It also helps in the stabilization of the basic fibroblast growth factors. *In vitro* tests provide evidence that, within 10 days, the hydrogel patch network releases basic growth factors in a sustainable manner due to the regulation of heparin. Fibroblast proliferation is promoted by the released growth factor *in vitro*. Furthermore, when treated with a heparin-hydrogel patch, effective acceleration of wound healing is observed in diabetic rats *in vivo* compared to treating the wound with a blank hydrogel dressing (Bai et al., 2020). Thus, a hydrogel patch combined with a polymer or a polymeric drug is more effective in wound treatment than only using a blank hydrogel patch.

The conventional methods of drug administration need several administrations to exert the potential therapeutic effect. As a result, the efficacy is affected, and patient compliance is also reduced. Moreover, this results in serious complications including systemic toxicities, poor targeting, and short duration of action (Sánchez-Cid et al., 2022). Thus, researchers have focused on controlled drug delivery systems through hydrogels, nanoparticles, liposomes, etc. These systems provide controlled drug delivery and also regulate the time of availability in tissues or cells. Especially hydrogels are drawing more attention in drug delivery and have spread to several branches of healthcare including cardiology, immunology, wound healing, and the management of pain, etc. The different sizes, architectures, functions, and overall functions determine the drug delivery mechanism of hydrogels (Jacob et al., 2021). Hydrogels can be formulated in the required size and shape as per the requirements of delivery routes of the human body and to achieve controlled release. The controlled release by hydrogels can be achieved by regulation of the degradation of the network or by swelling or through deformation (Li & Mooney, 2016).

- **Controlled Release by Degradation of Mesh Network:** Upon degradation of the mesh network, the size of the mesh increases which permits more drug diffusion from the hydrogel. The backbone of the polymer can be degraded through hydrolysis or by enzyme degradation. PEG hydrogels are formed by using the ester bonds that slowly hydrolyze the network to ensure the slow release of proteins. Matrix metalloproteinases (MMPs) cleave the oligopeptide bonds, and this is useful in the incorporation of MMP-cleavable peptide (GGRMSMPV) into a hyaluronic acid hydrogel. Recombinant tissue inhibitor of MMPs was delivered through this MMP-responsive hydrogel. There are similar cleavage systems that are being used to respond to biomolecules of the body to ensure the delivery of drugs. For instance, a glucose-responsive hydrogel is used to deliver insulin in diabetic patients, and thrombin-responsive hydrogel for the regulation of blood coagulation (J. Li & Mooney, 2016).

- **Controlled Release Via Hydrogel Swelling:** The release of drugs to the site of interest can also be achieved by hydrogel swelling to release the entrapped drugs. Upon swelling of hydrogels, the size of the mesh also increases and aids in the delivery. Hydrogels have a swelling nature that may be sensitive to several external factors which include temperature, pH, electric fields, and light. The hydrogels that are pH-responsive are useful in the delivery of oral and cancer therapeutics. The hydrogel swelling in the acidic condition of the stomach is minimal, which ultimately protects the drug and entraps it physically. Then the hydrogel moves down the intestinal tract where the hydrogels swell dramatically as designed to permit rapid diffusion of drugs (Drury & Mooney, 2003). The polymer coating has been used to control the release of drugs, but polymeric encapsulation limits environmental exposure. Many possible pathways such as solution-processing by means of ink-jet printing, spraying, or drop-casting might be used, but there are some limitations due to which unwanted drug/regulatory interactions like (re)crystallization or restriction in solvent selection may occur. The aim of a study done by Werzer et al. was to show the impact of different variables like pH, and drug solubility on the encapsulated drugs which are dependent on temperature, where a hydrogel layer of p (NIPAAM-co-DEGDVE) is used for the release of the drugs that are deposited by techniques like initiated chemical vapor deposition (iCVD). This technique is a solvent-free technique that helps in loading drugs by ignoring hampers in the drug layer. The conformation of the polymer changes under the low critical solution temperature (LCST) which is selected around 22°C or 29°C when a drug is present. As a result, exposure of polar groups to the interface with the environment occurs and water adsorption by the polymer in its network occurs. Above the LCST, the contracted state of hydrogel causes the polar isopropyl groups exposure towards the interface and faster release of all model drugs occurs at low pH. It must be noted that the formation of the H-bonds that take place between the drugs and the polymers is important in this case. So bonding between polymers and the hydrophobic drug molecules can trap the drugs which lead to hindering their release. But in the presence of phenytoin in the buffer solution, swelling capacities decrease. Again, reversed release behavior was noticed for indomethacin at neutral pH as in this state deprotonated indomethacin is less likely to make H-bonding with the polymer and result in faster release when the freer volume was available i.e., below the LCST. All these cases explain how the molecular species and the polymer (the polymer-drug interaction) have an immense influence on the drug release characteristics (Werzer et al., 2019).

The one characteristic of the majority of drugs that is posed as a disadvantage in hydrogel-based drug delivery systems is hydrophobicity. Since a hydrophilic polymeric core is not favorable for holding incompatible hydrophobic drugs, a problem arises in the efficacy of hydrophobic therapy. The weakness in the tensile strength of hydrogels often leads to early drug release before the desired time (Narayanaswamy & Torchilin, 2019). Another study has described the limitations of protein therapeutics, for example, the encapsulating formulation process of proteins has a deficiency in precise control over the loaded drug, and unacceptable interactions between proteins and organic co-solvents, cross-linking agents, or the materials themselves cause denaturation and/or immunogenicity. Synthetic or natural polymer-based hydrogels with aqueous interiors are compatible with protein encapsulation in the long term. However, stronger interactions lead to slower rates of release. Moreover, although electrostatic interactions are appealing, charged protein encapsulation into material meshes of opposite charge leads to protein-matrix interactions and becomes responsible for the aggregation of proteins or protein denaturation (Miller et al., 2019).

10.3.3 D PRINTING OF HYDROGEL PATCH

The most preferable hydrogel depends on the specific application and desired properties of the scaffold (Table 10.5). One type of hydrogel that has been widely used for 3D printed scaffold construction is polyethylene glycol (PEG) hydrogel, due to its tunable mechanical properties, good biocompatibility, and ability to mimic the extracellular matrix (ECM). PEG hydrogels have been used in various tissue engineering applications, including bone and cartilage repair (Li et al., 2018), and the ability to be functionalized with various biomolecules. In addition, PEG hydrogels can be crosslinked using a variety of methods, including UV irradiation and chemical crosslinking agents, allowing for precise control over the mechanical and structural properties of the final scaffold (Song et al., 2021). To bio-fabricate a 3D-printed hydrogel patch, the following steps can be followed:

a. **Choose a Suitable Hydrogel Material:** The choice of hydrogel material depends on the specific application of the patch. For example–polyethylene glycol (PEG), polyacrylamide (PAAm), and alginate.
b. **Prepare the Hydrogel Ink:** The hydrogel ink can be prepared by mixing the chosen hydrogel material with a crosslinking agent and a solvent to form a homogeneous solution. The concentration of the hydrogel material, crosslinking agent, and solvent can be optimized based on the desired properties of the patch.
c. **Design the 3D Model of the Patch:** The 3D model of the patch can be designed using computer-aided design (CAD) software such as AutoCAD or SolidWorks. The design should consider the shape, size, and thickness of the patch, as well as the intended application.
d. **3D Printing the Patch:** The 3D printed patch can be fabricated using a 3D printer with a nozzle that can dispense the hydrogel ink. The printer should be calibrated to ensure that the patch is printed with the desired resolution and accuracy.
e. **Crosslink the Hydrogel Patch:** After printing, the patch should be subjected to a crosslinking process to improve its mechanical strength and stability. This can be achieved by exposing the patch to a crosslinking agent, heat, or UV light (Chaudhary & Chakraborty, 2022).

3D printing has been shown to be effective in the production of hydrogel patches for a variety of applications. According to a review by (Zhang et al., 2022), 3D printing technology has the potential to "provide precise control over the structure and composition of hydrogel patches, leading to improved mechanical properties and functionality" (p. 187). In particular, 3D printing has been used to produce hydrogel patches for drug delivery and tissue engineering (Wei et al., 2022). One example of the use of 3D printing in hydrogel patch production is the development of a 3D-printed hydrogel patch for the delivery of insulin to diabetic patients (Gao et al., 2021). The patch was able to effectively release insulin over a period of time and showed promising results in both *in vivo* as well as *in vitro* tests. Another example is the use of 3D printing to produce hydrogel patches for wound healing (Cui et al., 2022a). The patches were able to effectively promote the healing of skin

TABLE 10.5 Categories of Hydrogel for Construction Of 3D Printed Scaffolds (Shi et al., 2015)

Scale	Material	Advantages	Disadvantages	Application
Undegradable	Hydroxyethyl methacrylate (HEMA)	Cheap, high ratio of functional group	Not preferable for *in vivo* use	Guided cell growth, cell engineering
Degradable	Polyethylene glycol diacrylate acid (PEGDA)	Cost-effective, mechanical strength, biocompatible	Low rate of degradation, no cell adhesion motifs	Extracellular microenvironment
	Gelatin	Contains RGD group, thermo-reversible, preferable for *in vivo* and vitro use	Extrusion-based printing, low resolution	Multi-nozzle co-printing
	GelMA	Contains RGD group, photocurable, preferable for *in vivo* and vitro use	Hard to produce, UV is harmful to pre-seeded cells	Vasculature networks
	Alginate	Cheap, easy jellification, reliable, tunable properties, high strength	Low cell adhesion motifs, need Ca^{2+} to keep gel-like	Tubing, wound patch, boost cell growth

wounds in a mouse model, demonstrating the potential for 3D printing to produce hydrogel patches with therapeutic properties.

Alginate, a linear polysaccharide that is water-soluble, is derived from brown algae. It is biocompatible, easily degradable, and can readily hydrolyze. It can convert into a gel under certain conditions, and these properties make it an alternative for cell encapsulation (Figure 10.5). We can fabricate gel from different mechanisms. At a pH below 3, alginate self-assembles into acidic gels by forming intermolecular hydrogen bonds. Alternatively, alginate forms a physical hydrogel by cooperative binding with divalent or trivalent cations such as Ca^{2+}, which can interact with carboxylic acid groups in the sugars. An alginate chain consists of mannuronic acid (M unit) and guluronic acid (G unit) and is arranged in blocks that are rich in G units, blocks that are rich in M units, and blocks that comprise alternating G and M units. Alginate hydrogels have a density of crosslinking which is a function of the G blocks and molecular weight. The viscosity of alginate solution and its overall stiffness once gelled depend on the polymer concentration and its molecular weight distribution (Li et al., 2020). To optimize the composition of alginate hydrogel, fish gelatin can also be introduced in the gelation mechanism. It gives hydrogel mechanical strength. As F-GelMA contains RDG motifs, it is easily biodegradable, and photocurable, and gives a positive response in *in-vivo* studies with alginate (Kamoun et al., 2017; Zhang et al., 2018).

Biofabrication of hydrogel is basically done with laser-induced forward transfer, inkjet 3D bioprinting, and robotic dispensing methods (*see* Figure 10.6). 3D printing improves the capacity of the produced matrix to promote cell proliferation as opposed to 2D scaffolds, which are essential for rapid wound healing (Smandri et al., 2020).

10.3.4 DRUG RELEASE & WOUND HEALING MECHANISM

Healthy intact skin has a pH range from 4.0 to 6.0. But when a wound happens, the pH range of the site increases underlying cells of the wound site. When a drug is loaded in hydrogel, its pH declines to acidic and it becomes anionic hydrogel. Also, algae's natural pH accelerates this process. So, when it is exposed to the basic environment of the wound site, it swells and releases all the drugs at the action site (*see* Figure 10.7) (Rizwan et al., 2017; Jones et al., 2015).

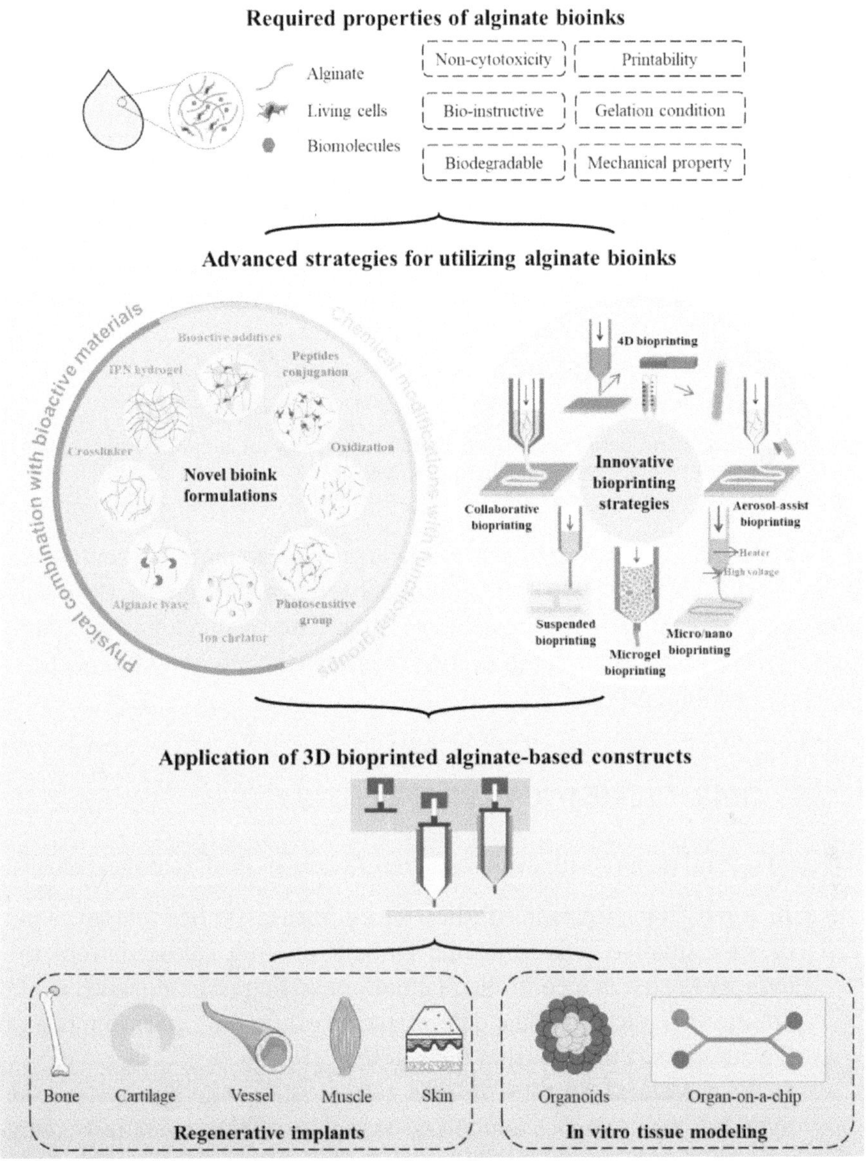

FIGURE 10.5 Process of bio fabrication of alginate-based bioinks and their applications (Q. Gao et al., 2021).

Source: Reprinted from Gao et al., 2021. Copyright © 2019 by the authors. Licensee MDPI, Basel, Switzerland. http://creativecommons.org/licenses/by/4.0

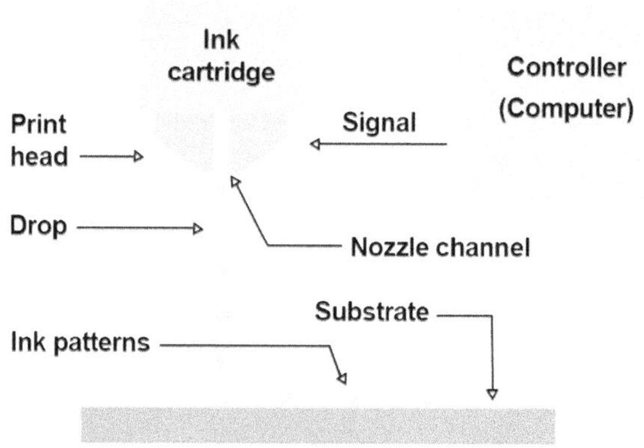

FIGURE 10.6 Underlying principle of inkjet-based 3D bioprinting.

Source: Reprinted from Uddin et al., 2022. Copyright © 2022 by the authors. Licensee MDPI, Basel, Switzerland. http://creativecommons.org/licenses/by/4.0/

As in the patch, there are live microalgae present, so there is a continuous supply of dissolved oxygen. This dissolved oxygen delivery is >100-fold better than topical gaseous oxygen, so it initiates wound healing more quickly (Chen et al., 2020a). Figure 10.8 mentions the effects of hydrogel on the wound-healing process.

10.3.5 CHARACTERIZATION AND EVALUATION

There are several methods for evaluating the effectiveness of hydrogel patches.

- ***In Vitro* Drug Release Studies:** Drug release studies that are done *in vitro* involve measuring the amount of drug released from the hydrogel patch in a controlled laboratory setting (Abasalizadeh et al., 2020). This allows for the determination of the drug release rate and the duration of release from the patch.
- ***In Vivo* Animal Studies:** *In vivo* animal studies involve testing the hydrogel patch on live animals to assess its effectiveness in a living system (Jain et al., 2016). This can include measuring drug levels in the blood or tissue after administration of the patch to determine its bioavailability.
- **Clinical Trials:** Clinical trials involve testing the hydrogel patch on human subjects to assess its effectiveness in a clinical setting (Niemeyer et al., 2022). This can include measuring drug levels in the

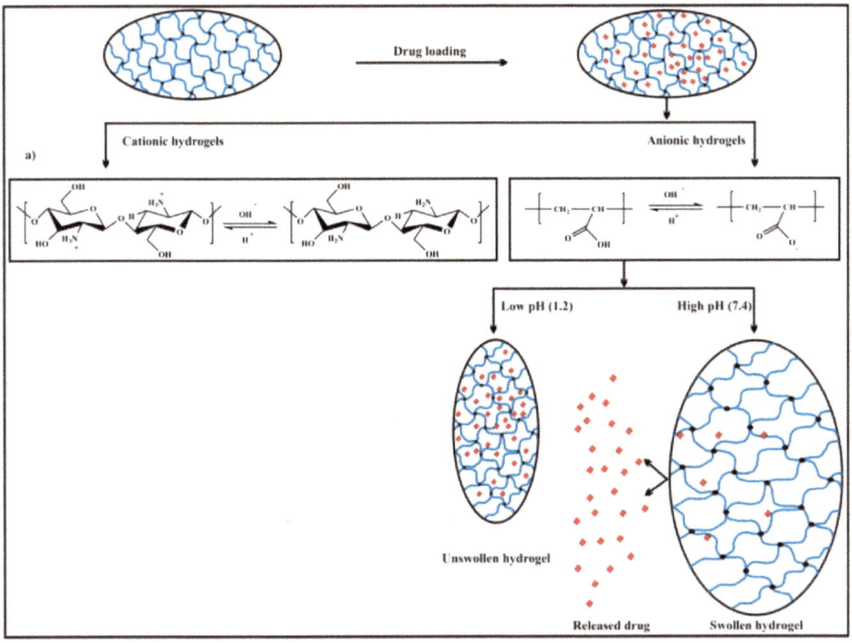

FIGURE 10.7 Release of the drug from the hydrogel patch with pH sensitivity.

Source: Reprinted from Rizwan et al., 2017. Copyright © 2017 by the authors. Licensee MDPI, Basel, Switzerland. http://creativecommons.org/licenses/by/4.0/

blood or tissue after administration of the patch, as well as subjective measures of effectiveness such as patient-reported outcomes. One study by (Li et al., 2020) investigated the use of a chitosan-based hydrogel patch for wound healing in rats. The authors found that the patch effectively promoted the proliferation of fibroblasts and collagen synthesis, leading to significant improvements in wound healing (Li et al., 2020). Another study (Abate & Salini, 2012) examined the use of a hyaluronic acid-based hydrogel patch for the treatment of knee osteoarthritis in a rabbit model. The authors found that the patch significantly reduced inflammation and cartilage degradation, suggesting its potential as a therapeutic option for osteoarthritis. To evaluate the healing area after applying a hydrogel patch, the following steps can be taken (Liu et al., 2021).

- **Assess the Size and Shape of the Wound:** Measure the length and width of the wound and note any changes in shape. Observe the color

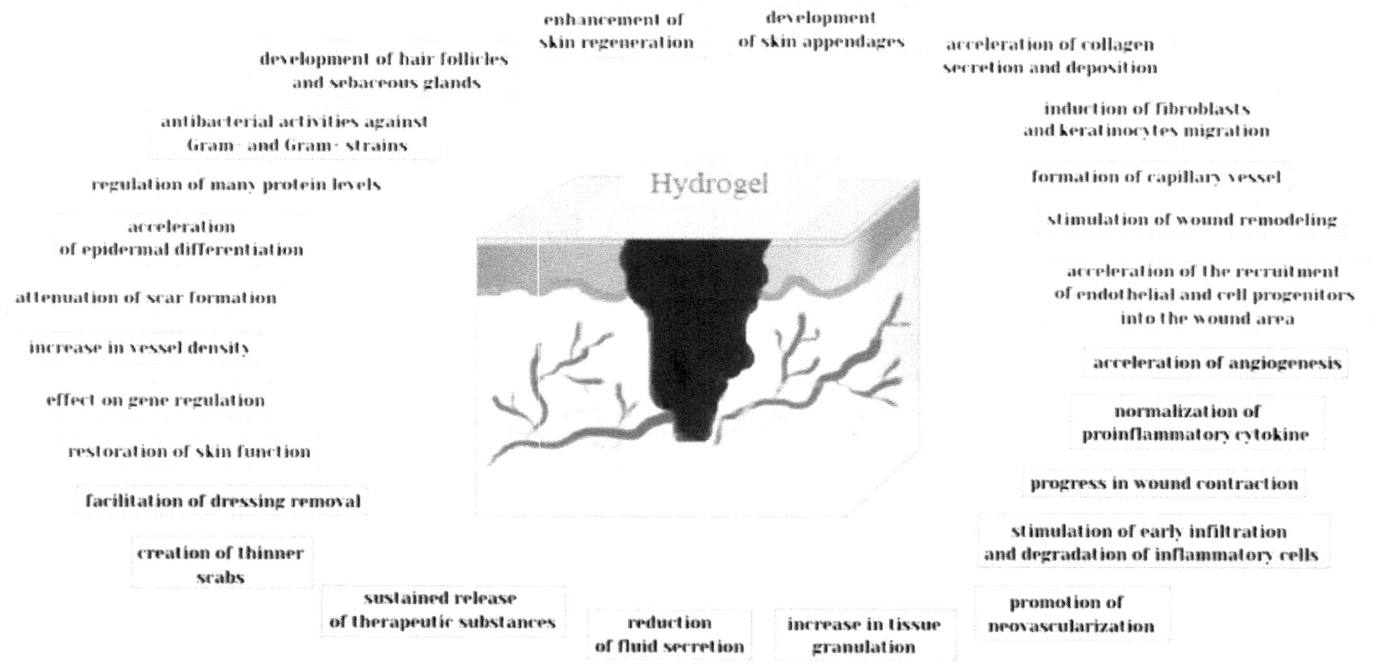

FIGURE 10.8 Effect of hydrogel on wound healing process.

Source: Reprinted from Zagórska-Dziok & Sobczak, 2020. https://creativecommons.org/licenses/by/

and appearance of the wound: Look for signs of inflammation, such as redness or swelling, and note any changes in the color of the wound.

- **Check for Signs of Infection:** Look for signs of infection, such as fever, redness, or pus, and seek medical attention if necessary.
- **Assess the Amount of Exudate:** Exudate is the fluid that is produced by the body to clean the wound. If the wound is producing too much exudate, it may be a sign that the hydrogel patch is not working as effectively as it should.
- **Check for Signs of Granulation Tissue:** Granulation tissue is a type of tissue that forms during the healing process. If the hydrogel patch is working effectively, the wound should be covered in granulation tissue.
- **Monitor the Healing Progress:** Keep track of the healing progress by taking pictures of the wound at regular intervals.

Sustainability evaluation after examining the wound of a diabetic patient involves assessing the patient's overall health and the effectiveness of their wound care treatment. This includes evaluating factors such as the patient's blood sugar control, nutrition status, and overall medical management of their diabetes (Jacob et al., 2021). According to a study conducted by (Gao et al., 2021), the use of a hydrogel patch for wound management in diabetic patients resulted in a significant decrease in healing time and improvement in wound closure. In addition, the hydrogel patch had a low incidence of adverse effects and was found to be cost-effective compared to traditional wound dressings. Another study by (Zhang et al., 2023) also found that the use of a hydrogel patch in diabetic wound care resulted in a decrease in pain and inflammation, as well as an improvement in overall wound healing.

In terms of sustainability, the use of a hydrogel patch for wound management in diabetic patients can also have environmental benefits. Traditional wound dressings often require frequent changes, leading to increased waste and potential negative impacts on the environment (Wu et al., 2023). In contrast, the hydrogel patch can be left in place for extended periods of time, reducing the need for frequent dressing changes and potentially reducing waste (Zhang et al., 2023). Overall, the use of a hydrogel patch for wound management in diabetic patients has been shown to be effective in improving healing outcomes and reducing the incidence of adverse effects, while also having potential environmental benefits through reduced waste.

10.4 DISCUSSION AND OPINION

10.4.1 RESPONSES AND RELATION

Unbalanced immune-based reactions in the skin tissue are known to be linked to some chronic inflammatory diseases, including atopic dermatitis, and medications like steroids result in a variety of undesirable side effects. There was a connection between the etiology of atopic dermatitis pathogenesis from high levels of oxidative stress from reactive oxygen species (ROS), which exhibit high ROS levels and encourage the production of T-cell differentiation and pro-inflammatory cytokines (Peng et al., 2021). This can further contribute to the beginning and worsening of the disease's course (Bertino et al., 2020). The therapeutic hydrogel patches reduce the high levels of oxidative stress caused by atopic dermatitis, and a hydrogel that has ROS-scavenging ceria nanoparticles embedded in it helps to reduce intracellular as well as extracellular ROS while exhibiting cytoprotective effects under highly oxidative conditions. Diabetic wounds pose difficulties for traditional wound dressing, which consists of bandages, gauze, films, and other simple materials. Bacterial cells may form within the biofilm because of hyperglycemia in the wound bed, and the wound would have strong medication resistance (Zhang et al., 2023).

There has been a significant decrease in epidermal thickness and levels of AD-associated biomarkers as well (Kim et al., 2022). Infants with premature skin or diabetic patients with extra-fragile skin are at risk for adhesive-induced injury because adhesion causes inflammation and pain while worsening the harm when the patch is removed. The adhesion of the hydrogel can be smartly activated after being in contact with warm skin, and the detachment can be easily done by adding a bag on the surface of the hydrogel patch. The hydrogel exhibits immunomodulatory performance, which prevents allergic reactions and irritation with the skin. The bioelectronics in the hydrogel patch were seen to be comfortable and non-irritating to the skin health of infants, creating ideal conditions for managing diabetic wounds that speed up the recovery process (Figure 10.9) (Jiang et al., 2022; Yuk et al., 2019).

10.4.2 COMPATIBILITY & COMPARATIVE STUDIES

Hydrogels are three-dimensional structures that have the potential to absorb and retain water within the mesh through their porous structure. The field of hydrogels for biomedical purposes prefers natural biopolymers more

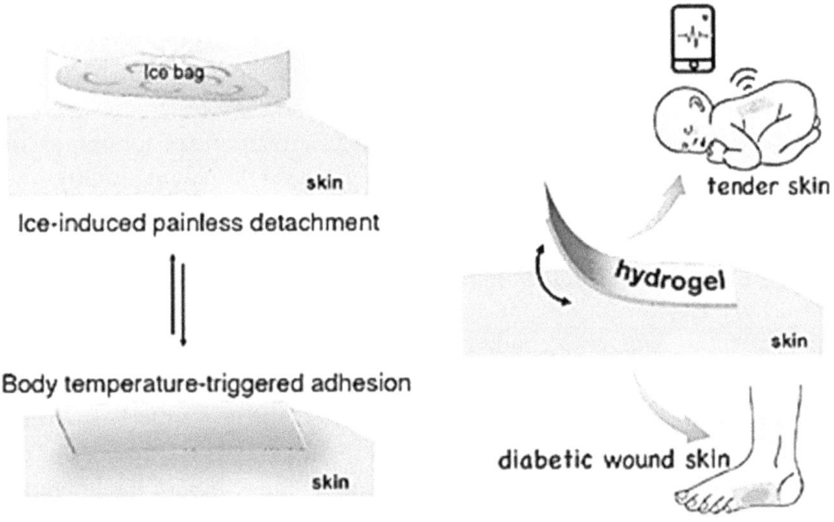

FIGURE 10.9 Use of bioelectronic hydrogel patch in infants and diabetic wound healing.

Source: Reprinted with permission from Jiang et al., 2022. Copyright © 2022, American Chemical Society

because of their blood suitability, ability to bind with proteins, and platelet adhesion, as well as other advantages including simplicity of administration of components to the action site. Different combinations of gelatin, dialyzed SF, curcumin, and N, N methylene bis-acrylamide (MBA) are used, and these mixtures are subjected to a stringent inflammation test, biodegradation analysis, Fourier-transform infrared spectroscopy, scanning electron microscopy, various *in vitro* coagulation tests, the determination of cholesterol level, and total serum cholesterol analysis (Alula et al., 2022).

It is necessary to incorporate varied appropriate compounds and hence include functional chemicals because not doing so could sometimes cause the hydrogel's desired qualities to be lost to achieve the desired properties of the hydrogel. This may aid in protecting the device from enzymatic, mechanical, and physical deterioration (Yahia, 2015; Buwalda et al., 2014). When hydrogels partially degrade, they may release oligomers, monomers, nanoparticles, or molecules that may not be properly eliminated by the kidney and may subsequently build up in the body (Zagórska-Dziok & Sobczak, 2020). The traditional hydrogel patches are more improved after the development of printable hydrogel ink, for extrusion-based 3D printing has introduced new possibilities for improved pharmaceutical application.

The incorporation of starch in the alginate-based 3D hydrogel for topical delivery is one of the quality designs that had more biocompatibility, low toxicity, biodegradability, and relatively low cost compared to the traditional patches. More focus has been given to printing accuracy, viscoelastic properties, gelation time, release rates, and microstructures for improving drug delivery through 3D-printed hydrogel patches (Bom et al., 2020).

10.4.3 CURRENT TRENDS AND CHALLENGES

Hyperglycemia, a diverse condition, characterizes diabetes mellitus (DM). Type 1 and type 2 diabetes are often distinct from one another, and they are generally the two categories of diabetes. Type 1 diabetes, formerly known as teenage onset or insulin-dependent diabetes mellitus (IDDM), makes up only 5% to 10% of all cases of diabetes (depending on the region of the universe). Scleredema diabeticorum is a rare, chronic connective tissue condition connected to DM. The affected skin is unpleasant, indurated, thick, and occasionally erythematous. It has been estimated that between 2.5% and 14% of DM patients have it. The condition known as diabetic skin microangiopathy can affect any part of the body. This painful, superficial ulcer has a complicated structure and is an acknowledged composition. Because of this, diabetic individuals struggle to quickly heal their wounds (Yan et al., 2019).

To prevent veins from being exposed to the air, lessen patient suffering, and aid in the delivery of medication, hydrogels are used as the medication load material in skin ulcers. These gels are characterized by adequate biocompatibility and water retention. The past few decades have seen a tremendous amount of interest in hydrogels, which are sensitive and contain a substantial amount of water. Because of their resemblance to extracellular grids, exceptional biocompatibility, and innate ability to collaborate with other cells, they have been widely used for biomedical applications comprising tissue creation and drug delivery. Chitosan is a copolymer made from D-glucosamine and N-acetylglucosamine, which are obtained from chitin. Chitosan should be interconnected due to its hydrophilic properties in delivery applications. Moreover, chitosan typically needs an acidic environment to break down, and some researchers have found that the incorporation of polysaccharides like carboxymethyl chitosan (CMC) into the hydrogel structure surprisingly altered the characteristics of the gel. Carboxymethyl chitosan, a water-soluble chitosan derivative created by the reaction of chitosan with monochloroacetic acid, is an attractive biocompatible, benign, and biodegradable polymer that has been investigated for use in a variety of products, including medications, food, and cosmetics in an acidic state. In this manner, hydrogel aids in the healing of diabetic patients' wounds (Yan et al., 2019).

In the case of diabetic patients, first, topical dressings with a moist environment are crucial for granulation tissue development and epithelialization, while lowering the risk of contracting a further infection and speeding up the healing process. There are various types of dressings that can offer unique properties to speed up wound healing, including passive, interactive, and bioactive ones. Passive dressings only cover the wound; interactive dressings offer a barrier to the pathogens' point of entry; and bioactive dressings provide a moist environment as well as some growth factors for the development of granulation tissue. Second, since anti-diabetic medications including metformin, insulin, DPP-4 inhibitors, and thiazolidinediones can lower MMPs, enhance keratinocytes, and promote the growth of granulation tissue or fibroblast, they can be used to treat chronic wounds. Thirdly, the introduction of growth factors, such as EGF, VEGF, PDGF, TGFβ1, etc., can encourage the development of ECM, help to control an inflammatory response, perform re-epithelization, and also promote angiogenesis. Finally, the ability of diabetic individuals to heal wounds has been shown to significantly improve with the use of several medications like statin, phenytoin, metformin, and DPP-4 inhibitors (Spampinato et al., 2020). While it resembles dermal tissues and can partially promote or expedite skin generation, hydrogels offer a better therapeutic possibility for healing chronic wounds. However, the focus of advanced wound care is on the creation and application of hydrogels in conjunction with elements that will improve the wound-healing properties of hydrogels known as active dressings. Hydrogels are coupled with natural as well as synthetic polymers to speed up the wound-healing process. Their cross-linked structure prevents the material from dissolving quickly, increasing its longevity and mechanical stability. For instance, to improve wound management, hydrogel patches are coupled with alginate, chitosan, gelatin, dextran, heparin, polyethylene glycol, polyvinyl alcohol, and carbopol. These various polymer dressings support and encourage wound healing, infection control, hemostasis, anti-inflammatory action, active molecule delivery, tissue recruitment, cell support, wound closure, cell adhesion, and other processes (Catoira et al., 2019).

Researchers are of the view that Iodozyme is a dressing that should be taken into consideration for treating chronic wounds based on the favorable patient and physician comments and the encouraging clinical results. The preliminary results are positive, and a randomized controlled trial will now be conducted to further explore them (Wood et al., 2010). And another treatment used is MET-HCL. When applied to the stomach region, MET-HCL hydrogel was successfully used to reduce intrabdominal collected fats and aid in weight loss. To treat women with problems with fertilization, sildenafil citrate was made as a thermo-responsive gel for vaginal organization (Tawfeek et al., 2020).

- Another type of antimicrobial hydrogel patch for use on polluted shallow wounds is iodozyme (Archimed). Iodine is produced inside the dressing by its normal protein structure. Newer products, that have overcome these issues, will typically use lower fixations or perhaps slow delivery specialists, such as povidone-iodine and cadexomer iodine. Modern hydrogel injury prevention products are now practical, acceptable, easy to use, and cost-effective. By absorbing extra exudate or delivering moisture, they effectively control the quantity of hydration on the wound surface. To speed up recovery, hydrogels have also been shown to adhere to the elimination of necrotic or dead tissue. Iodozyme is meant to saddle iodine's biocidal activity in a superior hydrogel formulation. When compared to various dressings, the supported antimicrobial sufficiency of Iodozyme has been demonstrated *in vitro*. The sole clinical trial for Iodozyme is a previously unreported open, non-near trial that was accepted for enrollment and to gather data supporting its safety and effectiveness. This was a single-site, non-randomized, open study with 32 participants who had superficial injuries with variable etiologies and clinical signs of increased bioburden (swelling, warmth, malodor, and distress). Immunosuppressants and basic anti-infection medications were not given to any of the patients. Unpublished outcomes for 30 patients were available; three (10%) of them completely recovered, 21 (70%) significantly improved, three (10%) showed no change, and three (10%) disintegrated. An open, non-randomized case-arrangement study was conducted to gain field experience with dressing a larger number of patients who had various kinds of difficult-to-recuperate wounds. A case-arrangement structure was used since it provided the opportunity for quick enrollment at a variety of master centers as well as organized feedback from their experienced staff. Also, it provided valuable information about the suitability of the improved dressing when used in practical settings by attendants (Wood et al., 2010).
- It was shown that MET-HCL significantly contributes to wound healing by directing certain signal routes that affect the polarization of macrophages. For example, nano liposomal description

and nanofiber-eluting layers for wound-healing applications were created using MET-HCL. However, the researched plans are very expensive, and complex, and require special tools and materials to be made. So, the purpose of our investigation was to study the potential injury-recovering effect of specifically applied MET-HCL following incorporation into a simple hydrogel. The medication's dosage, pH, homogeneity, rheology, width, epidermal perturbation implementation, and *in vitro* outflow were all measured after it was stacked into different hydrogel bases. Furthermore, rodents were the primary target of the injury-healing effect. Also, patients with lower appendage wounds or ulcers were subjected to human clinical examination. To show how the applied formula affected the healing of the injury, histological and immunohistochemical alterations following gel therapy were also evaluated.

By using these treatments, many chronic wounds like arterial leg ulcers, ulcers in the foot from diabetes, pressure ulcers, surgical wounds, venous leg ulcers, and miscellaneous wounds were healed. So, it can also heal inner wounds like osteochondritis defects as well (Wood et al., 2010). Scientists concocted the 3D-printed stage lodging a cell phone as the light hotspot for a working OEW-based microfluidic chip. This 3D-printed stage can be utilized as a possible apparatus for on-site monitoring of HABs in a waterway with aptamer-based identification (Dias Meirelles et al., 2022; Li et al., 2016; Li & Mooney, 2016). Algae cells could be recognized within a couple of moments, allowing for early detection and screening of microbiological pollutants, which would prevent the spread of harmful microorganisms in aquatic environments. For the most part, the regular checking strategies, for example, culture-based colony-forming units (CFU), flow cytometry (FCM), and using a hemocytometer require expensive equipment, time, expertise, and large volumes of reagents to conduct the measures (Zahavy et al., 2018). Hence, the traditional methods are generally inconvenient and unsuitable for in situ and continuous microbial water quality monitoring in the field (Lee et al., 2019).

Despite the initial etiology, the healing mechanisms are advanced in the diabetes patients as a group of subjects. Longer healing times for wounds frequently increase the risk of consequences, including repeated infections, sepsis, and in the worst instance, amputation. Due to the negative implications of extended wound healing, the importance of wound regeneration and closure is paramount (Mieczkowski et al., 2022). Endotheliopathy raises the risk of damage and slow wound healing by causing micro and macrovascular abnormalities that result in neuropathic changes (Richard et al., 2012). The most typical osteoarthritis symptom, which affects millions of individuals worldwide, is cartilage deformities. In terms of cartilage regeneration, hydrogels—a class of biomaterials with elastic properties, a smooth surface, and a high water content—seem to hold promise. In this instance, the osteochondral disease site is injected with hydrogel scaffolds comprised of synthetic polymers (polyethylene glycol, polyvinyl alcohol, and polyacrylamide) and proteins, polysaccharides, and peptides (Wei et al., 2021). According to studies (Feng et al., 2019; Mellati et al., 2017), hydrogels can repair cartilage defects in two ways: first, by encapsulating autologous cells in the hydrogel, which are then implanted in the defect; second, by using stem cells to help and encourage the participation of the repair process (Guo et al., 2018; Sharma et al., 2013). The most desired hydrogels are those with both characteristics. The 3D hydrogel scaffolding has both significant benefits and downsides. The disadvantages of fabrication methods, for instance, solvent casting and particulate leaching include limited pore connections, few solvent options, and severe chemical use, while the advantages include tunable pore size and control of the face process (Sola et al., 2019). Although the advantage is desirable to structure control, phase separation-based manufacturing processes have few solvent options and are not very user-friendly (Ganesan et al., 2018). The advantages of the gas foaming-based fabrication technology include a high level of interconnected pores, while the disadvantages entail unsustainable processing and low kinetic stability (Gunathilake et al., 2017).

10.4.4 CLINICAL & COMMERCIAL PROSPECT

The hydrogels are viable clinically due to the fact that there are currently 30 injectable hydrogel-based products that are approved by the FDA and/or EMA. The examples include SpaceOAR by Augmenix, Inc. and Vantas by Endo Pharmaceuticals, and in the case of facial correction, which includes Radiesse by Merz Pharmaceuticals and Radiesse by Bioform Medical, Inc. (Mandal et al., 2020). Hydrogels are appealing in different types of drug delivery systems and are clinically accepted in many branches of medicine, which include oncology, cardiology, wound healing, immunology, and pain management. It is still possible to commercialize the hydrogel if fabrication techniques, cost, and regulatory complexity are overcome. The high water content can pose difficulty in comparison to biodegradable polymers, which include PLGA microspheres. The hydrogels are required to be dehydrated after fabrication in order to prevent premature degradation during storage and prevent changes in hydrogel structure and drug bioactivity (Wurm et al., 2014). Overall, the application of hydrogel includes dye removal, biomedical applications, heavy metal ion removal, sanitary diapers, agriculture, biosensors, pH sensors, etc. In the case of wound dressing-based hydrogel, the market offers amorphous gels, as well as gauze saturated with hydrogel and hydrogel films. The amorphous gel has been used in cases of dry wounds with low exudates where moist healing is promoted through autolytic debridement. This also promotes the growth of epithelial cells from various wounds including post-surgical wounds, diabetic ulcers, cavity wounds, leg ulcers, pressure ulcers, and abrasions (Aswathy et al., 2020). 3D printing technology has been popularized through fused deposition modeling (FDM), polyjet process, and the use of selective laser sintering (SLS), digital light processing (DLP), 3D inkjet, etc. Technologies, even though they have a high capacity for processing different materials with accuracy and better resolution, with the low generation of waste and less manual labor, and increased risk mitigation scopes, the whole process is rather complex and requires significant cost in order to control and design the sub-micro components of the patches and bandages (Tsegay et al., 2022). Recently, attempts have been made to commercialize 3D-printed stem cell-laden hydrogel patches for patient use (Melhem et al., 2017). While the modern commercialization of 3D hydrogel patches remains a horizon for new hope, the commercialization of such patches for reaching general patients would still take time until the pharmaceutical industries consider all the mentioned variables for mass production.

10.4.5 LIMITATIONS

Interfaces between stripes are a problem area in 3D-printed intense hydrogels, making them a significant challenge to produce. Even though these fragile patches were strengthened by a hydrogel with two distinct organizational structures—firm calcium alginate linkages and delicate polyacrylamide chains—the breaking zone was still clearly visible at the intersection of adjacent stripes following elastic testing. Intense hydrogels were produced using needles with different checks as was observed. Although the elasticity of 3D printed stripes increased linearly with expanding check, the 3D printed stripes' width decreased, and their breaking strain and Young's modulus remained comparable at ~1.7 mm/mm and ~25 kPa, respectively. The hydrogels created by measure 18 via 3D printing had an impressive breaking strength, which was 423.04 kPa. Due to the same process used to create these 3D printed hydrogels, the comparable breaking strain, as well as Young's modulus, were acquired. Little stripes' improved interface characteristics and fairly elevated contact territory proportion serve as indicators of their improved breaking strength. The ink solution's Young's modulus was 38.14 kPa, somewhat greater than that of 3D printed extreme hydrogel via gauge 18, exhibiting a Young's modulus of 28.40 kPa. This was compared to their previous result of 3D printing hydrogel by ink arrangement of alginate, agar, and acrylamide monomer with poor form consistency and printing objective, but the forward section of both pressure strain bends was equivalent (28.40 kPa). Moreover, the breaking strength of 3D-printed robust hydrogel via gauge 18 (423.04 kPa) was significantly weaker than that of mixing ink (488.75 kPa). These findings came about because of the weak spot between neighboring stripes. The study revealed that during fracture propagation, the agar chain pullout mechanism caused an extended breaking strain, exhibiting the effectiveness of 3D printed hydrogel ink solution (Wang et al., 2021).

The F127-BUM hydrogel living materials were fit for advancing magnificent cells by means of ability for different realms of microorganisms and can fill in as a defensive framework for the immobilized cells. The utilization of a multimineral 3D printer empowered the affidavit of hydrogel inks embedded with microbial animal types into a solitary printed part, while microscopy affirms that the cells remain bound to their individual printed districts. While cells were seen to escape into the surrounding watery medium following a few days of brooding, we noticed no evidence of the penetration of cells in liquid suspension into the hydrogel. Besides, cells didn't move across boundaries of neighboring printed locales of cell-loaded hydrogels. This 3D printing compatibility of the F127-BUM hydrogels, combined with the exact isolation of the different microbial species will allow for future investigations into cell-cell interactions and behavior of microbial consortia, or the utilization of these consortia for distributed metabolic cycles in biomanufacturing. The ease of designing with F127-BUM hydrogels, combined with its cell-friendly attributes detailed here and in our previous work, offers flexibility for consortium studies and applications that have not yet been possible with other existing materials or liquid culture designs (Johnston et al., 2020).

Despite the fact that one-component hydrogels, such as alginate- and gelatin-based ink formulations for 3D printing purposes, have been designed with notable success (Klotz et al., 2016; J. Jia et al., 2014), a key obstacle to the field's quick development has been identified as the scarcity of a wider range of printable hydrogel systems (Chimene et al., 2020). Additionally, the cell-loaded bioink formulations will be more rigorous compared to biomaterial inks for 3D printing of cell-loaded hydrogels, also known as 3D bioprinting (Abdulghani & Mitchell, 2019; Gungor-Ozkerim et al., 2018; Lee & Yeong, 2016). Due to their higher water content, excellent biocompatibility, and structural resemblance to the extracellular matrix (ECM), hydrogels offer a damp environment that promotes wound healing. Along with being highly permeable to oxygen, nutrients, and other water-soluble metabolites, hydrogels may also absorb wound exudates (Jang et al., 2021; Tabriz & Douroumis, 2022). However, with any innovative idea, there are bound to be limitations. Such is the case with preparing a hydrogel patch using a 3D printer. According to Jang et al., 2021, they are materially characteristic-based and broadly consist of (i) decreased vascularization capacity; (ii) poor wound closure rates; (iii) weak mechanical characteristics; and (iv) higher manufacturing costs, despite their benefits in promoting wound healing (Jang et al., 2021). Liu et al., 2021, stated that it was challenging to utilize standard hydrogel as a material for 3D printing because of its weak intrinsic structural rigidity and poor printability due to its unpredictable expansion, as not all materials required for hydrogel preparation can be used as the ink for 3D printing (Liu et al., 2021). The quick spread of the application of bioprinting technology is severely constrained by the necessity for pricey professional-level biological printers and expert technical people. Bioprinting technology is still in its infancy; if used in clinical settings, bioprinted skin constructions might pose certain safety issues (He et al., 2018).

In order to correct these materials' shortcomings, the addition of other compounds or employing their derivatives often overcomes these limitations (Liu et al., 2021). The modeling and progress of printable hydrogels have emerged as an important study focus in this discipline due to the discrepancy between quickly advancing 3D printing and constrained printable hydrogel approaches. Research is being focused on creating functional hydrogels by modifying composites, using them, and conjugating bioactive compounds (Jang et al., 2021). For hydrogels to have better mechanical characteristics, complicated processing is necessary, including crosslinking (Tabriz & Douroumis, 2022).

10.4.6 FUTURE ASPECTS

The need for custom-made materials for certain wounds, such as burn and chronic wounds, drives ongoing efforts in the extremely difficult and challenging field of wound treatment (Kamoun et al., 2017). Foams and hydrogels can be used in these situations to absorb exudates, supply moisture, lessen discomfort, and restrict bacterial development (Espinosa et al., 2019). The wound healing process has been demonstrated to be accelerated by medicated wound dressings, which are widely accessible. This improvement can be attributed to either the direct enhancement of wound healing phases or the indirect removal of necrotic tissue (Mohseni et al., 2019). These medicinal dressings use alginates, hydrogels, hydrocolloids, polyurethane films, silicone gels, etc., to administer these agents to wounds (Saghazadeh et al., 2018). This study is beneficial as algae are proven by several researchers to be an effective hydrogel material for promoting wound healing. The future applications of wound management studies could expand to further testing in clinical trials with algae hydrogel via 3D printing technology, with the application of varied algae to experiment with their wound healing potential, as well as carry out their effectiveness on minor injured patients in hospitals and healthcare institutions for further testing. To understand physical properties, further research into the potential and applications via rheological studies is required.

The mechanical strength and regulated release of bioactive substances are two features of scaffolds that may be highly controlled by nanoparticles (NPs). Additionally, the NPs are among the best choices for bioactive agent administration and monitoring due to restrictions and limiting factors including limited solubility, unstable bioactivity, and short circulation half-life of bioactive molecules (growth factors, cytokines, inhibitors, genes, medicines, etc.) and contrast agents (Mi et al., 2016; Bahal et al., 2016). Fadilah et al. (2022) stipulated that when compared to analog composites without nanoparticles, polymer composites containing nanoparticles have been shown to improve cellular interactions, scaffolding performance, and physical, chemical, and biological characteristics. Another researcher, Jiang et al. (2020), coined this combination as novel superstructures that are growing in popularity, particularly in biomedical research. These nanoparticle-hydrogel superstructures have easily modifiable individual parts, making them extremely adaptable and naturally multipurpose platforms. The advantages of the hydrogel and nanoparticle components may be coupled synergistically, allowing these superstructures to perform better in situations when using each component alone would not be the best option.

The use of 3D-printed hydrogel scaffolds as drug carriers in biological applications has considerable promise (Cárcamo-Martínez et al., 2021; Wei et al., 2020). Liu et al. (2021) designed a 3D-printed double network with chitosan and polyvinyl alcohol combination as a framework matrix. It was observed to have improved mechanical characteristics and decreased swelling behavior. This study also showed antibacterial activity due to the double network which could be the answer to tissue regeneration (Liu et al., 2021). Another study by Fujioka-Kobayashi et al., 2016, used varied hyaluronic acid gel-based scaffolds in combination with recombinant human bone morphogenetic protein 9 as a prospective carrier to improve growth factors (Fujioka-Kobayashi et al., 2016). Not only did growth factors improve with increased bone regeneration due to the hydrogel carrier used, but also the involved bone protein immensely improved its osteogenic potential. Hence, hydrogel scaffolds could be used as a universal carrier in various diseases. Hydrogel is perhaps the most predominant dressing for injuries. The 3D organizational nature of hydrogel confers flexibility like delicate tissues to help tissue recovery. It keeps moisture on the injury surface, upholds support with discomfort, and provides an ideal environment for wound healing. In addition, bioactive particles can also be encapsulated in hydrogels and delivered through dispersion, achieving prolonged retention at the injury site. These advantages demonstrate that the combination of hydrogel with PRP might become a potential therapeutic method to enhance the healing effects of PRP. Sodium alginate (SA), a regularly obtained polysaccharide separated from the cell walls of brown algae, is a polysaccharide that has been approved by the Food and Drug Administration (FDA) and is suitable for aiding PRP. It is screened from clinically authorized materials. SA has recently been evolved for every hydrogel dressing that is commercially available, including Flaminal Hydro Alginate Gel (Flen Wellbeing UK Ltd.), Nu-Gel Hydrogel (Systa-genix Wound Administration Ltd.), and Purilon Gel (Coloplast Ltd.). Moreover, these advantages helped realize that SA can structure hydrogels in the company of divalent cations like Ca2+, which facilitates the formation of an "egg-box" cross-connecting model between the guluronate squares of nearby alginate chains. SA-PRP twofold organization hydrogel production likelihood is increased due to this, using a basic "one-venture" implementation measure. For the purpose of producing homogenous entering networks, calcium/thrombin can be employed specifically as one gelation regulator in a pre-blended PRP and alginate solution. The hydrogel-PRP composite may be given the potential results to completely cover the wounded area and achieve a reinforced emergence of substantial particles obtained from PRP thanks to the enhanced main approach and adaptability (Zhang et al., 2021).

Nanoparticles have offered an interesting array of properties for drug delivery, including high drug loading capacity, combinatorial delivery, controlled and sustained drug release, prolonged stability and lifespan, and targeted delivery. Nanoparticles and hydrogels have gradually been combined to create a hybrid biomaterial framework for regulated drug delivery in order to further enhance therapeutic efficacy, particularly for specific applications. Nanoparticle technology offers a range of benefits for drug delivery, including efficient loading, combination therapy, controlled release, extended circulation, and targeted delivery. Advances in biotechnology and biomedicine continue to accelerate the development of novel therapeutic nanoparticles aimed at further enhancing therapeutic efficacy. For example, recent advancements in genomics and proteomics have led to novel nanoparticle designs tailored for rational drug combinations for personalized therapy. Nanoparticle formulations have been developed to precisely modulate the timing and spatial distribution of drug molecules to target the internal state of potentially harmful cells for bioactivity, influenced by biological insights into the organization of interconnected pathways within a cell. A deeper understanding of disease pathogenesis, particularly that of malignant growth and infection, has sparked the development of highly responsive nanoparticles that, when exposed to external stimuli, cause direct physicochemical changes that enable the release of specific medications at the target site. The development of more sophisticated, biomimetic nanoparticles built of common biomaterials has exhibited prevalent biointerfacing qualities, inspiring novel therapies, particularly for drug delivery, detoxification, and immunization. In this light, stacking nanoparticles inside hydrogels has drawn a lot of interest.

The 3D polymer networks, which are hydrophilic and known as hydrogels, have several applications in medication delivery and tissue engineering. By blending in with the monomer arrangement, followed by gelation, nanoparticles may be included in the hydrogel network. Yet, following gel creation, they may be connected into a grid by allowing the gel organization to grow and "breathe in" nanoparticles for ensnarement. This method is especially useful when nanoparticles interfere with the gelation process. Moreover, inorganic nanoparticles are routinely filled in situ inside gel lattices by first stacking nanoparticle precursors into the gel and then decreasing responses for nanoparticle synthesis. With suitable syntheses, hydrogels not only preserve the primary integrity and functionalities of the contained nanoparticles but also provide additional design flexibility to enhance the overall therapeutic effectiveness. Another method for incorporating nanoparticles to impart hydrogel-like properties (nanoparticle colloidal hydrogel), in addition to hydrogel entanglement, is to directly utilize nanoparticles as cross-linkers to form 3D hydrogel networks. In this process, nanoparticles can be linked by mixing oppositely surface-charged particles or by forming strong hydrophobic interactions. However, nanoparticles may also establish covalent bonds with each other, serving as nodes to construct a hydrogel framework (Gao et al., 2016).

There are various techniques for creating frameworks, including pre-made porous platforms, decellularized extracellular scaffolds, cell sheets with secreted extracellular matrices, and cells with self-assembled hydrogels (Chan & Leong, 2008). Due to their similarity to the native extracellular matrix, hydrogels have garnered significant attention in tissue engineering. Several methods are available for fabricating hydrogel platforms, such as (i) lyophilization; (ii) emulsification; (iii) solvent casting-particulate leaching; (iv) gas foaming-solvent casting; (v) photolithography; (vi) microfluidics; (vii) micromolding; and (viii) 3D printing. Hydrogel-based frameworks are utilized for mending bone, cartilage, and cardiovascular tissues. According to research by Ayesha et al. creating ligament tissue may be accomplished using a hydrogel made of sodium alginate and nanocellulose that has been crosslinked with calcium chloride. According to a second analysis, gelatin-based hydrogel made with eggshell particles aids during the division of fully developed cells to form osteoblasts (Wu et al., 2019). Because injectable hydrogels may be used to easily fill up random flaws and don't require complicated operations, they are deemed to be more valuable compared to typical hydrogel scaffolds. Hyaluronan hydrogel that can be cross-linked in situ has been shown to be useful for tissue design by Xiao et al. The component of the extracellular grid known as hyaluronan aids in cell bonding, augmentation, and division (Zheng Shu et al., 2004). Polyethylene glycol (PEG)-adjusted with RGD themes (in situ photopolymerized injectable hydrogels) hydrogels were researched as osteoblast encapsulation matrices to determine their suitability for assisting bone tissue engineering (Burdick & Anseth, 2002).

10.5 CONCLUSION

Hydrogels are essentially hydrophilic polymers, which consist of 3D visco-elastic networks that can store water, often at a rate of once or double their dry weight, and expand in a physiological setting. Actual communications, as well as substance cross-connecting of hydrogels, could supply in the direction of underlying along with actual uprightness equally. The materialness of hydrogels as bio-materials remains in the distinctiveness of their properties, for example, water content, delicate and flexible texture, and depleted glue power with water or natural liquids. Because of particular actual qualities, hydrogels can provide controlled disintegration, shield the labile drug from corruption as well as regulate the entry of other active ingredients, such as little atom medications, macromolecules, and cells. Accordingly, hydrogels possess the extraordinary capacity to address several formulations – as well as medication-associated challenges, along these lines proving their usage in different medication conveyance frameworks, wound dressings, hygiene products, and regenerative medication. In contrast to earlier published writing, the current chapter provides a thorough overview of the most recent developments in hydrogels for use in various medication conveyance frameworks. Besides, the appropriateness of hydrogels in injury management as well as tissue engineering is widely understood and discussed. Hydrogels can be extensively characterized depending on some of their attributes, including side groups (ionic or non-ionic), primary structures (homopolymer or co-polymer), physical nature (crystalline, amorphous, supramolecular), and responsiveness to various external stimuli, such as: (i) light; (ii) ultrasound; (iii) temperature; (iv) pH; (v) electromagnetic radiation; (vi) ionic strength; (vii) proteins; and (viii) glucose. Several physical and chemical cross-linking methods are already in use to provide hydrogels with sufficient mechanical strength as well as biodegradability (Jacob et al., 2021).

Hydrogels used as biomaterials receive recognition due to their versatility, water-retaining capacity, potential to mimic tiny atoms along with macromolecular medications, and regulated degradability. In this study, there is a lot of discussion regarding functionalization methods that address the shortcomings of customary hydrogels and increase the functionality of cutting-edge hydrogels, including DNA hydrogels. Hydrogels are categorized under many cross-connecting methods, resources employed, methodology, merits, and challenges. The use of hydrogels is illustrated, particularly in buccal, oral, vaginal, and transdermal medication delivery systems. This chapter also focuses on composite hydrogels, exhibiting better characteristics and being developed to meet various demands for abrasion dressing equipment. The innovative merits of hydrogel nanoparticles in targeted and intracellular delivery of various therapeutic agents are elucidated. Moreover, various types of hydrogel-based materials are used for tissue engineering purposes, and the creation of interaction focal points is discussed. This chapter also provides an overview of selected examples of commercial products launched, especially in the field of oral and ocular drug delivery systems and wound dressing materials.

KEYWORDS

- **3D printing**
- **carboxymethyl chitosan**
- **computer-aided design**
- **diabetes mellitus**
- **marine algae**
- **nanoparticles**
- **wound healing**

REFERENCES

Abasalizadeh, F., Moghaddam, S. V., Alizadeh, E., Akbari, E., Kashani, E., Fazljou, S. M. B., Torbati, M., & Akbarzadeh, A. (2020). Alginate-based hydrogels as drug delivery vehicles in cancer treatment and their applications in wound dressing and 3D bioprinting. *Journal of Biological Engineering, 14*, 8. https://doi.org/10.1186/s13036-020-0227-7.

Abate, M., & Salini, V. (2012). Hyaluronic acid in the treatment of osteoarthritis: What is new. In *Osteoarthritis-Diagnosis, Treatment and Surgery. InTech eBooks,* 101–121. https://doi.org/10.5772/26818.

Abbas, N., Parveen, K., Hussain, A., Latif, S., Zaman, S. U., Shah, P. A., & Ahsan, M. (2019). Nanosponge-based hydrogel preparation of fluconazole for improved topical delivery. *Tropical Journal of Pharmaceutical Research, 18*(2), 215–222. https://doi.org/10.4314/tjpr.v18i2.1.

Abdulghani, S., & Mitchell, G. R. (2019). Biomaterials for in situ tissue regeneration: A review. *Biomolecules, 9*(11), 750. https://doi.org/10.3390/biom9110750.

Ahsan, A., Tian, W. X., Farooq, M. A., & Khan, D. H. (2021). An overview of hydrogels and their role in transdermal drug delivery. *International Journal of Polymeric Materials and Polymeric Biomaterials, 70*(8), 574–584. https://doi.org/10.1080/00914037.2020.1740989.

Alejandro Hernández-Sosa, Rosa Ana Ramírez-Jiménez, Luis Rojo, Fouzia Boulmedais, María Rosa Aguilar, & Miryam Criado-Gonzalez, R. H. (2022). Optimization of the Rheological Properties of Self-Assembled Tripeptide/Alginate/Cellulose Hydrogels for 3D Printing. *Polymers* 2022, *14*(11), 2229. https://doi.org/10.3390/polym14112229.

Alula, K., Adali, T., & Han Ebedal, O. (2022). Preparation characterization and blood compatibility studies of silk fibroin/gelatin/curcumin injectable hydrogels. *Bio-Medical Materials and Engineering*, *34*(1), 77–93. https://doi.org/10.3233/BME-221407.

Aswathy, S. H., Narendrakumar, U., & Manjubala, I. (2020). Commercial hydrogels for biomedical applications. *Heliyon, 6*(4), e03719. https://doi.org/10.1016/j.heliyon.2020.e03719.

Augustine, R., Zahid, A. A., Hasan, A., Dalvi, Y. B., & Jacob, J. (2021). Cerium Oxide Nanoparticle-loaded gelatin methacryloyl hydrogel wound-healing patch with free radical scavenging activity. *ACS Biomaterials Science & Engineering, 7*(1), 279–290. https://doi.org/10.1021/acsbiomaterials.0c01138.

Babich, O., Sukhikh, S., Larina, V., Kalashnikova, O., Kashirskikh, E., Prosekov, A., Noskova, S., Ivanova, S., Fendri, I., Smaoui, S., et al. (2022). *Algae: Study of Edible and Biologically Active Fractions, Their Properties and Applications. Plants. 11*(6), 780. https://doi.org/10.3390/plants11060780.

Bahal, R., Ali McNeer, N., Quijano, E., Liu, Y., Sulkowski, P., Turchick, A., Lu, Y.-C., Bhunia, D. C., Manna, A., Greiner, D. L., Brehm, M. A., Cheng, C. J., López-Giráldez, F., Ricciardi, A., Beloor, J., Krause, D. S., Kumar, P., Gallagher, P. G., Braddock, D. T., & Glazer, P. M. (2016). In vivo correction of anaemia in β-thalassemic mice by γPNA-mediated gene editing with nanoparticle delivery. *Nature Communications, 7*, 13304. https://doi.org/10.1038/ncomms13304.

Bai, H., Wang, Z., Li, M., Liu, Y., Wang, W., Sun, P., Wei, S., Wang, Z., Li, J., & Dardik, A. (2020). Hyaluronic acid–heparin conjugated decellularized human great saphenous vein patches decrease neointimal thickness. *Journal of Biomedical Materials Research – Part B Applied Biomaterials, 108*(6), 2417–2425. https://doi.org/10.1002/jbm.b.34574.

Bai, Q., Han, K., Dong, K., Zheng, C., Zhang, Y., Long, Q., & Lu, T. (2020). Potential applications of nanomaterials and technology for diabetic wound healing. *International Journal of Nanomedicine, 15*, 9717–9743. https://doi.org/10.2147/IJN.S276001.

Bashir, S., Teo, Y. Y., Ramesh, S., & Ramesh, K. (2018). Synthesis and characterization of karaya gum-g-poly (acrylic acid) hydrogels and in vitro release of hydrophobic quercetin. *Polymer, 147*, 108–120. https://doi.org/10.1016/j.polymer.2018.05.071.

Beaumont, M., Tran, R., Vera, G., Niedrist, D., Rousset, A., Pierre, R., Shastri, V. P., & Forget, A. (2021). Hydrogel-forming algae polysaccharides: From seaweed to biomedical applications. *Biomacromolecules, 22*(3), 1027–1052. https://doi.org/10.1021/acs.biomac.0c01406.

Benltoufa, S., Miled, W., Trad, M., Slama, R. Ben, & Fayala, F. (2020). Chitosan hydrogel-coated cellulosic fabric for medical end-use: Antibacterial properties, basic mechanical and comfort properties. *Carbohydrate Polymers, 227*, 115352. https://doi.org/10.1016/j.carbpol.2019.115352.

Bergmeier, W., & Hynes, R. O. (2012). Extracellular matrix proteins in hemostasis and thrombosis. *Cold Spring Harbor Perspectives in Biology, 4*(2), a005132. https://doi.org/10.1101/cshperspect.a005132.

Bertino, L., Guarneri, F., Cannavò, S. P., Casciaro, M., Pioggia, G., & Gangemi, S. (2020). Oxidative Stress and Atopic Dermatitis. *Antioxidants (Basel, Switzerland), 9*(3), 196. https://doi.org/10.3390/antiox9030196.

Bom, S., Santos, C., Barros, R., Martins, A., Paradiso, P., Cláudio, R., Pinto, P., Ribeiro, H., & Marto, J. (2020). Effects of starch incorporation on the physicochemical properties and release kinetics of alginate-based 3d hydrogel patches for topical delivery. *Pharmaceutics, 12*(1), 719. https://doi.org/10.3390/pharmaceutics12080719.

Boulton, A., Armstrong, D., Londehal, M., Frykberg, R., Game, F., Edmonds, M., Orgill, D., Kramer, K., Gurtner, G., Januszyk, M., & Vileikyte, L. (2022). New evidence-based therapies for complex diabetic foot wounds. *ADA Clinical Compendia, 2022*(2), 1–23. https://doi.org/10.2337/db2022-02.

Burdick, J. A., & Anseth, K. S. (2002). Photoencapsulation of osteoblasts in injectable RGD-modified PEG hydrogels for bone tissue engineering. *Biomaterials, 23*(22), 4315–4323. https://doi.org/10.1016/s0142-9612(02)00176-x.

Burgess, J. L., Wyant, W. A., Abujamra, B. A., Kirsner, R. S., & Jozic, I. (2021). Diabetic Wound-Healing Science. pp. 1–15.

Buwalda, S. J., Boere, K. W. M., Dijkstra, P. J., Feijen, J., Vermonden, T., & Hennink, W. E. (2014). Hydrogels in a historical perspective: From simple networks to smart materials. *Journal of Controlled Release, 190*, 254–273. https://doi.org/10.1016/j.jconrel.2014.03.052.

Cañedo-Dorantes, L., & Cañedo-Ayala, M. (2019). Skin acute wound healing: A comprehensive review. *International Journal of Inflammation, 2019*, 115. https://doi.org/10.1155/2019/3706315.

Cárcamo-Martínez, Á., Mallon, B., Domínguez-Robles, J., Vora, L. K., Anjani, Q. K., & Donnelly, R. F. (2021). Hollow microneedles: A perspective in biomedical applications. *International Journal of Pharmaceutics, 599*, 120455. https://doi.org/10.1016/j.ijpharm.2021.120455.

Catoira, M. C., Fusaro, L., Di Francesco, D., Ramella, M., & Boccafoschi, F. (2019). Overview of natural hydrogels for regenerative medicine applications. *Journal of Materials Science: Materials in Medicine, 30*(10), 1–10. https://doi.org/10.1007/s10856-019-6318-7.

Chan, B. P., & Leong, K. W. (2008). Scaffolding in tissue engineering: General approaches and tissue-specific considerations. *European Spine Journal, 17*(Suppl 4), 467–479. https://doi.org/10.1007/s00586-008-0745-3.

Chaudhary, S., & Chakraborty, E. (2022). Hydrogel-based tissue engineering and its future applications in personalized disease modeling and regenerative therapy. *Beni-Suef University Journal of Basic and Applied Sciences, 11*(1), 3. https://doi.org/10.1186/s43088-021-00172-1.

Chen, H., Cheng, Y., Tian, J., Yang, P., Zhang, X., Chen, Y., Hu, Y., & Wu, J. (2020). Dissolved oxygen from microalgae-gel patch promotes chronic wound healing in diabetes. *Science Advances, 6*(20), 8297–8316. https://doi.org/10.1126/sciadv.aba4311.

Chen, Y., Huang, L., Dai, X., Tian, Q., Yu, M., Agheb, M., Chan, H. N., Poon, E., Guo, Z., Boheler, K. R., & Wu, H. (2017). Facile formation of a microporous chitosan hydrogel based on self-crosslinking. *Journal of Materials Chemistry B, 5*(47), 9291–9299. https://doi.org/10.1039/C7TB02736B.

Chimene, D., Kaunas, R., & Gaharwar, A. K. (2020). Hydrogel bioink reinforcement for additive manufacturing: A focused review of emerging strategies. *Advanced Materials, 32*(1), 1–22. https://doi.org/10.1002/adma.201902026.

Costa, C., Incio, J., & Soares, R. (2007). Angiogenesis and chronic inflammation: Cause or consequence? *Angiogenesis, 10*(3), 149–166. https://doi.org/10.1007/s10456-007-9074-0.

Creton, C. (2017). 50th anniversary perspective: Networks and gels: soft but dynamic and tough. *Macromolecules, 50*(21), 8297–8316. https://doi.org/10.1021/acs.macromol.7b01698.

Cui, H., Su, Y., Wei, W., Xu, F., Gao, J., & Zhang, W. (2022a). How microalgae is effective in oxygen deficiency aggravated diseases? A comprehensive review of literature. *International Journal of Nanomedicine, 17*, 3101–3122. https://doi.org/10.2147/IJN.S368763.

Cui, H., Su, Y., Wei, W., Xu, F., Gao, J., & Zhang, W. (2022b). How microalgae is effective in oxygen deficiency aggravated diseases? A comprehensive review of literature. *International Journal of Nanomedicine, 17*, 3101–3122. https://doi.org/10.2147/IJN.S368763.

de Alcântara, M. G., de Freitas Ortega, N., Souza, C. J. F., & Garcia-Rojas, E. E. (2020). Electrostatic hydrogels formed by gelatin and carrageenan induced by acidification: Rheological and structural characterization. *Food Structure, 24*, 100137. https://doi.org/10.1016/j.foostr.2020.100137.

Demir, I., Besson, A., Guiraud, P., & Formosa-Dague, C. (2020). Towards a better understanding of microalgae natural flocculation mechanisms to enhance flotation harvesting efficiency. *Water Science and Technology, 82*(6), 1009–1024. https://doi.org/10.2166/wst.2020.177.

Dias Meirelles, A. A., Rodrigues Costa, A. L., Michelon, M., Viganó, J., Carvalho, M. S., & Cunha, R. L. (2022). Microfluidic approach to produce emulsion-filled alginate microgels. *Journal of Food Engineering, 315*, 110812. https://doi.org/10.1016/j.jfoodeng.2021.110812.

Dong, K., Dong, Y., You, C., Xu, W., Huang, X., Yan, Y., Zhang, L., Wang, K., & Xing, J. (2014). Assessment of the drug loading, in vitro and in vivo release behavior of novel pH-sensitive hydrogel. *Drug Delivery, 23*(1), 174–184. https://doi.org/10.3109/10717544.2014.908329.

Dreiss, C. A. (2020). Hydrogel design strategies for drug delivery. *Current Opinion in Colloid & Interface Science, 48*, 1–17. https://doi.org/10.1016/j.cocis.2020.02.001.

Drury, J. L., & Mooney, D. J. (2003). Hydrogels for tissue engineering: Scaffold design variables and applications. *Biomaterials, 24*(24), 4337–4351. https://doi.org/10.1016/s0142-9612(03)00340-5.

Espinosa, E., Filgueira, D., Rodríguez, A., & Chinga-Carrasco, G. (2019). Nanocellulose-based inks—effect of alginate content on the water absorption of 3D printed constructs. *Bioengineering, 6*(3), 65. https://doi.org/10.3390/bioengineering6030065.

Fadilah, N. I. M., Isa, I. L. M., Zaman, W. S. W. K., Tabata, Y., & Fauzi, M. B. (2022). The effect of nanoparticle-incorporated natural-based biomaterials towards cells on activated pathways: A systematic review. *Polymers, 14*(3), 476. https://doi.org/10.3390/polym14030476.

Fan, F., Saha, S., & Hanjaya-Putra, D. (2021). Biomimetic hydrogels to promote wound healing. *Frontiers in Bioengineering and Biotechnology, 9*, 1–24. https://doi.org/10.3389/fbioe.2021.718377.

Feng, Q., Li, Q., Wen, H., Chen, J., Liang, M., Huang, H., Lan, D., Dong, H., & Cao, X. (2019). Injection and self-assembly of bioinspired stem cell-laden gelatin/hyaluronic acid hybrid microgels promote cartilage repair in vivo. *Advanced Functional Materials, 29*(50), 1–11. https://doi.org/10.1002/adfm.201906690.

Frykberg, R. G., & Banks, J. (2015). Challenges in the treatment of chronic wounds. *Advances in Wound Care, 4*(9), 560–582. https://doi.org/10.1089/wound.2015.0635.

Fujioka-Kobayashi, M., Schaller, B., Kobayashi, E., Hernandez, M., Zhang, Y., & Miron, R. J. (2016). Hyaluronic acid gel-based scaffolds as potential carrier for growth factors: An in vitro bioassay on its osteogenic potential. *Journal of Clinical Medicine, 5*(12), 112. https://doi.org/10.3390/jcm5120112.

Ganesan, K., Budtova, T., Ratke, L., Gurikov, P., Baudron, V., Preibisch, I., Niemeyer, P., Smirnova, I., & Milow, B. (2018). Review on the production of polysaccharide aerogel particles. *Materials, 11*(11), 1–37. https://doi.org/10.3390/ma11112144.

Gao, D., Zhang, Y., Bowers, D. T., Liu, W., & Ma, M. (2021). Functional hydrogels for diabetic wound management. *APL Bioengineering, 5*(3), 031503. https://doi.org/10.1063/5.0046682.

Gao, F., Jiao, C., Yu, B., Cong, H., & Shen, Y. (2021). Preparation and biomedical application of injectable hydrogels. *Materials Chemistry Frontiers, 5*(13), 4912–4936. https://doi.org/10.1039/D1QM00489A.

Gao, Q., Kim, B. S., & Gao, G. (2021). Advanced strategies for 3D bioprinting of tissue and organ analogs using alginate hydrogel bioinks. *Marine Drugs, 19*(12), 708. https://doi.org/10.3390/md19120708.

Gao, T., Gillispie, G. J., Copus, J. S., Pr, A. K., Seol, Y.-J., Atala, A., Yoo, J. J., & Lee, S. J. (2018). Optimization of gelatin-alginate composite bioink printability using rheological parameters: A systematic approach. *Biofabrication, 10*(3), 034106. https://doi.org/10.1088/1758-5090/aacdc7.

Gao, W., Zhang, Y., Zhang, Q., & Zhang, L. (2016). Nanoparticle-hydrogel: A hybrid biomaterial system for localized drug delivery. *Annals of Biomedical Engineering, 44*(6), 2049–2061. https://doi.org/10.1007/s10439-016-1583-9.

Garg, S., & Garg, A. (2016). Hydrogel: Classification, properties, preparation and technical features. *Asian Journal of Biomaterial Research, 2*, 163–170.

Göhl, J., Markstedt, K., Mark, A., Håkansson, K., Gatenholm, P., & Edelvik, F. (2018). Simulations of 3D bioprinting: predicting bioprintability of nanofibrillar inks. *Biofabrication, 10*(3), 034105. https://doi.org/10.1088/1758–5090/aac872.

Gomez-Florit, M., Pardo, A., Domingues, R. M. A., Graça, A. L., Babo, P. S., Reis, R. L., & Gomes, M. E. (2020). Natural-based hydrogels for tissue engineering applications. *Molecules, 25*(24), 5858. https://doi.org/10.3390/molecules25245858.

Gong, H., Li, W., Sun, J., Jia, L., Guan, Q., Guo, Y., & Wang, Y. (2022). A review on plant polysaccharide-based drug delivery systems for construction and application, with emphasis on traditional Chinese medicine polysaccharide. *International Journal of Biological Macromolecules, 211*, 711–728. https://doi.org/10.1016/j.ijbiomac.2022.05.087.

Guan, Y., Niu, H., Liu, Z., Dang, Y., Shen, J., Zayed, M., Ma, L., & Guan, J. (2021). Sustained oxygenation accelerates diabetic wound healing by promoting epithelialization and angiogenesis and decreasing inflammation. *Science Advances, 7*(35), 915–946. https://doi.org/10.1126/sciadv.abj0153.

Gunathilake, T. M. S. U., Ching, Y. C., & Chuah, C. H. (2017). Enhancement of curcumin bioavailability using nanocellulose reinforced chitosan hydrogel. *Polymers, 9*(2), 64. https://doi.org/10.3390/polym9020064.

Gungor-Ozkerim, P. S., Inci, I., Zhang, Y. S., Khademhosseini, A., & Dokmeci, M. R. (2018). Bioinks for 3D bioprinting: An overview. *Biomaterials Science, 6*(5), 915–946. https://doi.org/10.1039/C7BM00765E.

Guo, T., Noshin, M., Baker, H. B., Taskoy, E., Meredith, S. J., Tang, Q., Ringel, J. P., Lerman, M. J., Chen, Y., Packer, J. D., & Fisher, J. P. (2018). 3D printed biofunctionalized scaffolds for microfracture repair of cartilage defects. *Biomaterials, 185*, 219–231. https://doi.org/10.1016/j.biomaterials.2018.09.022.

Hai, N. C. (2022). Factors affecting the development of cultural tourism in the Mekong Delta, Vietnam. *Journal of Asian Finance, Economics and Business, 9*(3), 123–133. https://doi.org/10.13106/jafeb.2022.vol9.no3.0123.

Hanna, D. H., & Saad, G. R. (2019). Encapsulation of ciprofloxacin within modified xanthan gum-chitosan based hydrogel for drug delivery. *Bioorganic Chemistry, 84*, 115–124. https://doi.org/10.1016/j.bioorg.2018.11.036.

Hartmeier, P. R., Pham, N. B., Velankar, K. Y., Issa, F., Giannoukakis, N., & Meng, W. S. (2021). Hydrogel Dressings for Chronic wound Healing in Diabetes: Beyond Hydration. *J Pharm Drug Deliv Res, 10*(1), 1000197. https://pubmed.ncbi.nlm.nih.gov/36110983/.

He, P., Zhao, J., Zhang, J., Li, B., Gou, Z., Gou, M., & Li, X. (2018). Bioprinting of skin constructs for wound healing. *Burns & Trauma, 6*, 1–10. https://doi.org/10.1186/s41038-017-0104-x.

Heidarian, P., Kouzani, A. Z., Kaynak, A., Paulino, M., & Nasri-Nasrabadi, B. (2019). Dynamic hydrogels and polymers as inks for three-dimensional printing. *ACS Biomaterials Science & Engineering, 5*(6), 2688–2707. https://doi.org/10.1021/acsbiomaterials.9b00047.

Holzer-Geissler, J. C. J., Schwingenschuh, S., Zacharias, M., Einsiedler, J., Kainz, S., Reisenegger, P., Holecek, C., Hofmann, E., Wolff-Winiski, B., Fahrngruber, H., Birngruber, T., Kamolz, L. P., & Kotzbeck, P. (2022). The impact of prolonged inflammation on wound healing. *Biomedicines, 10*(4), 856. https://doi.org/10.3390/biomedicines10040856.

Hussein, R. A., Salama, A. A. A., El Naggar, M. E., & Ali, G. H. (2019). Medicinal impact of microalgae collected from high-rate algal ponds; phytochemical and pharmacological studies of microalgae and its application in medicated bandages. *Biocatalysis and Agricultural Biotechnology, 20*, 101237. https://doi.org/10.1016/j.bcab.2019.101237.

Ilhan, E., Cesur, S., Guler, E., Topal, F., Albayrak, D., Guncu, M. M., Cam, M. E., Taskin, T., Sasmazel, H. T., Aksu, B., Oktar, F. N., & Gunduz, O. (2020). Development of *Satureja cuneifolia*-loaded sodium alginate/polyethylene glycol scaffolds produced by 3D-printing technology as a diabetic wound dressing material. *International Journal of Biological Macromolecules, 161*, 1040–1054. https://doi.org/10.1016/j.ijbiomac.2020.06.086.

Insperger, T., & Stépán, G. (2011). Engineering applications. In *Applied Mathematical Sciences (Switzerland)* (Vol. 178, pp. 93–149). Springer. https://doi.org/10.1007/978-1-4614-0335-7_5.

Ionov, L. (2014). Hydrogel-based actuators: Possibilities and limitations. *Materials Today, 17*(10), 494–503. https://doi.org/10.1016/j.mattod.2014.07.002.

Jacob, S., Nair, A. B., Shah, J., Sreeharsha, N., Gupta, S., & Shinu, P. (2021). Emerging role of hydrogels in drug delivery systems, tissue engineering and wound management. *Pharmaceutics, 13*(3), 357. https://doi.org/10.3390/pharmaceutics13030357.

Jain, S., Patel, N., Madan, P., & Lin, S. (2016). Formulation and rheological evaluation of ethosome-loaded carbopol hydrogel for transdermal application. *Drug Development and Industrial Pharmacy, 42*(8), 1315–1324. https://doi.org/10.3109/03639045.2015.1132227.

Jaiswal, L., Shankar, S., & Rhim, J.-W. (2019). Carrageenan-based functional hydrogel film reinforced with sulfur nanoparticles and grapefruit seed extract for wound healing application. *Carbohydrate Polymers, 224*, 115191. https://doi.org/10.1016/j.carbpol.2019.115191.

Jang, M. J., Bae, S. K., Jung, Y. S., Kim, J. C., Kim, J. S., Park, S. K., Suh, J. S., Yi, S. J., Ahn, S. H., & Lim, J. O. (2021). Enhanced wound healing using a 3D printed VEGF-mimicking peptide incorporated hydrogel patch in a pig model. *Biomedical Materials (Bristol, England), 16*(4), 045013. https://doi.org/10.1088/1748-605X/abf1a8.

Jespersen, J. (1988). Pathophysiology and clinical aspects of fibrinolysis and inhibition of coagulation. *Danish Medical Bulletin, 35*(1), 1–33.

Jia, J., Richards, D. J., Pollard, S., Tan, Y., Rodriguez, J., Visconti, R. P., Trusk, T. C., Yost, M. J., Yao, H., Markwald, R. R., & Mei, Y. (2014). Engineering alginate as bioink for bioprinting. *Acta Biomaterialia, 10*(10), 4323–4331. https://doi.org/10.1016/j.actbio.2014.06.034.

Jia, W., Weng, J., Zhu, D., Ji, L., Lu, J., Zhou, Z., Zou, D., Guo, L., Ji, Q., Chen, L., Chen, L., Dou, J., Guo, X., Kuang, H., Li, L., Li, Q., Li, X., Liu, J., Ran, X., & Zhao, Z. (2019). Standards of medical care for type 2 diabetes in China 2019. *Diabetes/Metabolism Research and Reviews, 35*(6), 1–26. https://doi.org/10.1002/dmrr.3158.

Jiang, Y., Zhang, X., Zhang, W., Wang, M., Yan, L., Wang, K., Han, L., & Lu, X. (2022). Infant skin-friendly adhesive hydrogel patch activated at body temperature for bioelectronics securing and diabetic wound healing. *ACS Nano, 16*(6), 8662–8676. https://doi.org/10.1021/acsnano.2c00662.

Jiang, Y., Zhou, J., Feng, C., Shi, H., Zhao, G., & Bian, Y. (2020). Rheological behavior, 3D printability and the formation of scaffolds with cellulose nanocrystals/gelatin hydrogels. Journal of Materials Science, 55(33), 15709–15725. https://doi.org/10.1007/s10853–020–05128-x.

Johnston, T. G., Fillman, J. P., Priks, H., Butelmann, T., Tamm, T., Kumar, R., Lahtvee, P. J., & Nelson, A. (2020). Cell-laden hydrogels for multikingdom 3D printing. *Macromolecular Bioscience, 20*(8), 1–7. https://doi.org/10.1002/mabi.202000121.

Jones, E. M., Cochrane, C. A., & Percival, S. L. (2015). The effect of pH on the extracellular matrix and biofilms. *Advances in Wound Care, 4*(7), 431–439. https://doi.org/10.1089/wound.2014.0538.

Jung, H., Kim, M., Lee, J., Choi, S. W., & Kim, J. (2020). Hydrogel patch: Adhesive hydrogel patch with enhanced strength and adhesiveness to skin for transdermal drug delivery (Adv. Funct. Mater. 42/2020). *Advanced Functional Materials, 30*, 2070280. https://doi.org/10.1002/adfm.202070280.

Kalantari, K., Mostafavi, E., Saleh, B., Soltantabar, P., & Webster, T. J. (2020). Chitosan/PVA hydrogels incorporated with green synthesized cerium oxide nanoparticles for wound healing applications. *European Polymer Journal, 134*, 109853.

Kamoun, E. A., Kenawy, E.-R. S., & Chen, X. (2017). A review on polymeric hydrogel membranes for wound dressing applications: PVA-based hydrogel dressings. *Journal of Advanced Research, 8*(3), 217–233. https://doi.org/10.1016/j.jare.2017.01.005.

Karri, V. V. S. R., Kuppusamy, G., Talluri, S. V., Yamjala, K., Mannemala, S. S., & Malayandi, R. (2016). Current and emerging therapies in the management of diabetic foot ulcers. *Current Medical Research and Opinion, 32*(3), 519–542. https://doi.org/10.1185/03007995.2015.1128888.

Khaled, S. A., Burley, J. C., Alexander, M. R., Yang, J., & Roberts, C. J. (2015). 3D printing of tablets containing multiple drugs with defined release profiles. *International Journal of Pharmaceutics, 494*(2), 643–650. https://doi.org/10.1016/j.ijpharm.2015.07.067.

Kibungu, C., Kondiah, P. P. D., Kumar, P., & Choonara, Y. E. (2021). Recent advances in chitosan and alginate-based hydrogels for wound healing application. *Frontiers in Materials, 8*(July), 1–15. https://doi.org/10.3389/fmats.2021.681960.

Kim, S. W., Kim, D. Y., Roh, H. H., Kim, H. S., Lee, J. W., & Lee, K. Y. (2019). Three-dimensional bioprinting of cell-laden constructs using polysaccharide-based self-healing hydrogels. *Biomacromolecules, 20*(5), 1860–1866. https://doi.org/10.1021/acs.biomac.8b01589.

Kim, Y. E., Choi, S. W., Kim, M. K., Nguyen, T. L., & Kim, J. (2022). Therapeutic hydrogel patch to treat atopic dermatitis by regulating oxidative stress. *Nano Letters, 22*(5), 2038–2047. https://doi.org/10.1021/acs.nanolett.1c04899.

Klingberg, F., Hinz, B., & White, E. S. (2013). The myofibroblast matrix: Implications for tissue repair and fibrosis. *The Journal of Pathology, 229*(2), 298–309. https://doi.org/10.1002/path.4104.

Klotz, B. J., Gawlitta, D., Rosenberg, A. J. W. P., Malda, J., & Melchels, F. P. W. (2016). Gelatin-methacryloyl hydrogels: Towards biofabrication-based tissue repair. *Trends in Biotechnology, 34*(5), 394–407. https://doi.org/10.1016/j.tibtech.2016.01.002.

Kurakula, M., Rao, G. K., Kiran, V., Hasnain, M. S., & Nayak, A. K. (2020). Alginate-based hydrogel systems for drug releasing in wound healing. In Elsevier eBooks (pp. 323–358). https://doi.org/10.1016/b978–0-12–817640–5.00013–3.

Kuznetsova, T. A., Andryukov, B. G., Besednova, N. N., Zaporozhets, T. S., & Kalinin, A. V. (2020). Marine algae polysaccharides as basis for wound dressings, drug delivery, and tissue engineering: A review. *Journal of Marine Science and Engineering, 8*(7), 481. https://doi.org/10.3390/jmse8070481.

Lawrence, W. T. (1998). Physiology of the acute wound. *Clinics in Plastic Surgery, 25*(3), 321–340.

Lee, J. M., & Yeong, W. Y. (2016). Design and printing strategies in 3D bioprinting of cell-hydrogels: A review. *Advanced Healthcare Materials, 5*(22), 2856–2865. https://doi.org/10.1002/adhm.201600435.

Lee, S., Thio, S. K., Park, S. Y., & Bae, S. (2019). An automated 3D-printed smartphone platform integrated with optoelectrowetting (OEW) microfluidic chip for on-site monitoring of viable algae in water. *Harmful Algae, 88*, 101638. https://doi.org/10.1016/j.hal.2019.101638.

Li, J., & Mooney, D. J. (2016). Designing hydrogels for controlled drug delivery. *Nature Reviews Materials, 1*(12), 16071. https://doi.org/10.1038/natrevmats.2016.71.

Li, J., Wu, C., Chu, P. K., & Gelinsky, M. (2020). 3D printing of hydrogels: Rational design strategies and emerging biomedical applications. *Materials Science and Engineering: R: Reports, 140*, 100543. https://doi.org/10.1016/j.mser.2020.100543.

Li, S., Dong, S., Xu, W., Tu, S., Yan, L., Zhao, C., Ding, J., & Chen, X. (2018). Antibacterial hydrogels. *Advanced Science, 5*(5), 1700527. https://doi.org/10.1002/advs.201700527.

Lim, J. Z. M., Ng, N. S. L., & Thomas, C. (2017). Prevention and treatment of diabetic foot ulcers. *Journal of the Royal Society of Medicine, 110*(3), 104–109. https://doi.org/10.1177/0141076816688346.

Lin, J., Jiao, G., & Kermanshahi-Pour, A. (2022). Algal polysaccharides-based hydrogels: Extraction, synthesis, characterization, and applications. *Marine Drugs, 20*(5), 306. https://doi.org/10.3390/md20050306.

Lin, X., Xu, Y., Pan, X., Xu, J., Ding, Y., Sun, X., Song, X., Ren, Y., & Shan, P.-F. (2020). Global, regional, and national burden and trend of diabetes in 195 countries and territories: An analysis from 1990 to 2025. *Scientific Reports, 10*(1), 14790. https://doi.org/10.1038/s41598-020-71908-9.

Liu, F., Li, W., Liu, H., Yuan, T., Yang, Y., Zhou, W., Hu, Y., & Yang, Z. (2021). Preparation of 3D printed chitosan/polyvinyl alcohol double network hydrogel scaffolds. *Macromolecular Bioscience, 21*(4), 1–10. https://doi.org/10.1002/mabi.202000398.

Liu, H., Wang, C., Li, C., Qin, Y., Wang, Z., Yang, F., Li, Z., & Wang, J. (2018). A functional chitosan-based hydrogel as a wound dressing and drug delivery system in the treatment of wound healing. *RSC Advances, 8*(14), 7533–7549. https://doi.org/10.1039/c7ra13510f.

Liu, P., Shen, H., Zhi, Y., Si, J., Shi, J., Guo, L., & Shen, S. G. (2019). 3D bioprinting and in vitro study of bilayered membranous construct with human cells-laden alginate/gelatin composite hydrogels. *Colloids and Surfaces B: Biointerfaces, 181*(January), 1026–1034. https://doi.org/10.1016/j.colsurfb.2019.06.069.

Ma, T., Lv, L., Ouyang, C., Hu, X., Liao, X., Song, Y., & Hu, X. (2021). Rheological behavior and particle alignment of cellulose nanocrystal and its composite hydrogels during 3D printing. *Carbohydrate Polymers, 253*, 117217. https://doi.org/10.1016/j.carbpol.2020.117217.

Mandal, A., Clegg, J. R., Anselmo, A. C., & Mitragotri, S. (2020). Hydrogels in the clinic. *Bioengineering and Translational Medicine, 5*(2), 1–12. https://doi.org/10.1002/btm2.10158.

Mardis, N. J. (2018). Emerging technology and applications of 3D printing in the medical field. *Missouri Medicine, 115*(4), 368–373.

Mariadoss, A. V. A., Sivakumar, A. S., Lee, C. H., & Kim, S. J. (2022). Diabetes mellitus and diabetic foot ulcer: Etiology, biochemical and molecular based treatment strategies via gene and nanotherapy. *Biomedicine and Pharmacotherapy, 151*(March), 113134. https://doi.org/10.1016/j.biopha.2022.113134.

Melhem, M. R., Park, J., Knapp, L., Reinkensmeyer, L., Cvetkovic, C., Flewellyn, J., Lee, M. K., Jensen, T. W., Bashir, R., Kong, H., & Schook, L. B. (2017). 3D printed stem-cell-laden, microchanneled hydrogel patch for the enhanced release of cell-secreting factors and treatment of myocardial infarctions. *ACS Biomaterials Science and Engineering, 3*(9), 1980–1987. https://doi.org/10.1021/acsbiomaterials.6b00176.

Mellati, A., Fan, C.-M., Tamayol, A., Annabi, N., Dai, S., Bi, J., Jin, B., Xian, C., Khademhosseini, A., & Zhang, H. (2017). Microengineered 3D cell-laden thermoresponsive hydrogels for mimicking cell morphology and orientation in cartilage tissue engineering. *Biotechnology and Bioengineering, 114*(1), 217–231. https://doi.org/10.1002/bit.26061.

Mi, P., Kokuryo, D., Cabral, H., Wu, H., Terada, Y., Saga, T., Aoki, I., Nishiyama, N., & Kataoka, K. (2016). A pH-activatable nanoparticle with signal-amplification capabilities for non-invasive imaging of tumour malignancy. *Nature Nanotechnology, 11*(8), 724–730. https://doi.org/10.1038/nnano.2016.72.

Mieczkowski, M., Mrozikiewicz-Rakowska, B., Kowara, M., Kleibert, M., & Czupryniak, L. (2022). The problem of wound healing in diabetes—From molecular pathways to the design of an animal model. *International Journal of Molecular Sciences, 23*(14), 7930. https://doi.org/10.3390/ijms23147930.

Miller, S. E., Yamada, Y., Patel, N., Suárez, E., Andrews, C., Tau, S., Luke, B. T., Cachau, R. E., & Schneider, J. P. (2019). Electrostatically driven guanidinium interaction domains that control hydrogel-mediated protein delivery in vivo. *ACS Central Science, 5*(11), 1750–1759. https://doi.org/10.1021/acscentsci.9b00501.

Mohseni, M., Shamloo, A., Aghababaie, Z., Afjoul, H., Abdi, S., Moravvej, H., & Vossoughi, M. (2019). A comparative study of wound dressings loaded with silver sulfadiazine and silver nanoparticles: In vitro and in vivo evaluation. *International Journal of Pharmaceutics, 564*, 350–358. https://doi.org/10.1016/j.ijpharm.2019.04.068.

Moldobaeva, A., Baek, A., Eldridge, L., & Wagner, E. M. (2010). Differential activity of pro-angiogenic CXC chemokines. *Microvascular Research, 80*(1), 18–22. https://doi.org/10.1016/j.mvr.2010.01.011.

Nadgorny, M., Collins, J., Xiao, Z., Scales, P. J., & Connal, L. A. (2018). 3D-printing of dynamic self-healing cryogels with tuneable properties. *Polymer Chemistry, 9*(13), 1684–1692. https://doi.org/10.1039/C7PY01945A.

Nafo, W. (2022). Hydrogel biomaterials for drug delivery: Mechanisms, design, and drugs. In L. Popa, M. V. Ghica, & C. E. Dinu-Pîrvu (Eds.), *IntechOpen*, 386–204.https://doi.org/10.5772/intechopen.103156.

Narayanaswamy, R., & Torchilin, V. P. (2019). Hydrogels and Their Applications in Targeted Drug Delivery. *Molecules (Basel, Switzerland), 24*(3), 603, https://doi.org/10.3390/molecules 24030603.

Neamtu, B., Barbu, A., Negrea, M. O., Berghea-Neamțu, C. Ștefan, Popescu, D., Zăhan, M., & Mireșan, V. (2022). Carrageenan-Based Compounds as Wound Healing Materials. *International Journal of Molecular Sciences, 23*(16), 9117. https://doi.org/10.3390/ijms23169117.

Niemeyer, P., Hanus, M., Belickas, J., László, T., Gudas, R., Fiodorovas, M., Cebatorius, A., Pastucha, M., Hoza, P., Magos, K., Izadpanah, K., Paša, L., Vásárhelyi, G., Sisák, K., Mohyla, M., Farkas, C., Kessler, O., Kybal, S., Spiro, R., & Gaissmaier, C. (2022). Treatment of large cartilage defects in the knee by hydrogel-based autologous chondrocyte implantation: Two-year results of a prospective, multicenter, single-arm phase iii trial. *Cartilage, 13*(1), 19476035221085144. https://doi.org/10.1177/19476035221085146.

Nuutila, K., Katayama, S., Vuola, J., & Kankuri, E. (2014). Human Wound-Healing Research: Issues and Perspectives for Studies Using Wide-Scale Analytic Platforms. In *Advances in wound care (Vol. 3, Issue 3, pp. 264–271)*, Mary Ann Liebert Publishers. https://doi.org/10.1089/wound.2013.0502.

Okonkwo, U. A., & Dipietro, L. A. (2017). Diabetes and wound angiogenesis. *International Journal of Molecular Sciences, 18*(7), 1–15. https://doi.org/10.3390/ijms18071419.

Op 't Veld, R. C., Walboomers, X. F., Jansen, J. A., & Wagener, F. A. D. T. G. (2020). Design considerations for hydrogel wound dressings: Strategic and molecular advances. *Tissue Engineering. Part B, Reviews, 26*(3), 230–248. https://doi.org/10.1089/ten.TEB.2019.0281.

Ouyang, Q. Q., Hu, Z., Lin, Z. P., Quan, W. Y., Deng, Y. F., Li, S. D., Li, P. W., & Chen, Y. (2018). Chitosan hydrogel in combination with marine peptides from tilapia for burns healing. *International Journal of Biological Macromolecules, 112*, 1191–1198. https://doi.org/10.1016/j.ijbiomac.2018.01.217.

Pan, Z., Ye, H., & Wu, D. (2021). Recent advances on polymeric hydrogels as wound dressings. *APL Bioengineering, 5*(1), 11504. https://doi.org/10.1063/5.0038364.

Patel, S., Srivastava, S., Singh, M. R., & Singh, D. (2019). Mechanistic insight into diabetic wounds: Pathogenesis, molecular targets and treatment strategies to pace wound healing. *Biomedicine & Pharmacotherapy, 112*, 108615. https://doi.org/10.1016/j.biopha.2019.108615.

Richard, J. L., Lavigne, J. P., & Sotto, A. (2012). Diabetes and foot infection: more than double trouble. *Diabetes/Metabolism Research and Reviews, 28*(Suppl 1), 46–53. https://doi.org/10.1002/dmrr.2234.

Ridiandries, A., Tan, J. T. M., & Bursill, C. A. (2018). The role of chemokines in wound healing. *International Journal of Molecular Sciences, 19*(10), 3217. https://doi.org/10.3390/ijms19103217.

Rizwan, M., Yahya, R., Hassan, A., Yar, M., Azzahari, A. D., Selvanathan, V., Sonsudin, F., & Abouloula, C. N. (2017). pH sensitive hydrogels in drug delivery: Brief history, properties, swelling, and release mechanism, material selection and applications. *Polymers, 9*(4), 137. https://doi.org/10.3390/polym9040137.

Rodrigues, M., Kosaric, N., Bonham, C. A., & Gurtner, G. C. (2019). Wound healing: A Cellular perspective. *Physiological Reviews, 99*(1), 665–706. https://doi.org/10.1152/physrev.00067.2017.

Saeedi, P., Petersohn, I., Salpea, P., Malanda, B., Karuranga, S., Unwin, N., Colagiuri, S., Guariguata, L., Motala, A., Ogurtsova, K., Shaw, J., Bright, D., & Williams, R. (2019). Global and regional diabetes prevalence estimates for 2019 and projections for 2030 and 2045: Results from the International Diabetes Federation Diabetes Atlas, 9th edition. *Diabetes Research and Clinical Practice, 157*, 107843. https://doi.org/10.1016/j.diabres.2019.107843.

Saghazadeh, S., Rinoldi, C., Schot, M., Kashaf, S. S., Sharifi, F., Jalilian, E., Nuutila, K., Giatsidis, G., Mostafalu, P., Derakhshandeh, H., Yue, K., Swieszkowski, W., Memic, A., Tamayol, A., & Khademhosseini, A. (2018). Drug delivery systems and materials for wound healing applications. *Advanced Drug Delivery Reviews, 127*, 138–166. https://doi.org/10.1016/j.addr.2018.04.008.

Salaris, F., & Rosa, A. (2019). Construction of 3D in vitro models by bioprinting human pluripotent stem cells: Challenges and opportunities. *Brain Research, 1723*, 146393. https://doi.org/10.1016/j.brainres.2019.146393.

Sánchez-Cid, P., Jiménez-Rosado, M., Romero, A., & Pérez-Puyana, V. (2022). Novel trends in hydrogel development for biomedical applications: A review. *Polymers, 14*(15), 3023. https://doi.org/10.3390/polym14153023.

Schilrreff, P., & Alexiev, U. (2022). Chronic inflammation in non-healing skin wounds and promising natural bioactive compounds treatment. *International Journal of Molecular Sciences, 23*(9), 4928. https://doi.org/10.3390/ijms23094928.

Seoane-Viaño, I., Januskaite, P., Alvarez-Lorenzo, C., Basit, A. W., & Goyanes, A. (2021). Semi-solid extrusion 3D printing in drug delivery and biomedicine: Personalized solutions for healthcare challenges. *Journal of Controlled Release: Official Journal of the Controlled Release Society, 332*, 367–389. https://doi.org/10.1016/j.jconrel.2021.02.027.

Sharma, B., Fermanian, S., Gibson, M., Unterman, S., Herzka, D. A., Cascio, B., Coburn, J., Hui, A. Y., Marcus, N., Gold, G. E., & Elisseeff, J. H. (2013). Human cartilage repair with a photoreactive adhesive-hydrogel composite. *Science Translational Medicine, 5*(167), 167ra6. https://doi.org/10.1126/scitranslmed.3004838.

Sharma, S., & Tiwari, S. (2020). A review on bio macromolecular hydrogel classification and its applications. *International Journal of Biological Macromolecules*, 737–747.

Shi, W., He, R., & Liu, Y. (2015). 3D printing scaffolds with hydrogel materials for biomedical applications. *European Journal of BioMedical Research, 1*(3), 3. https://doi.org/10.18088/EJBMR.1.3.2015.PP3-8.

Shin, W. C., Dutta, S. D., Lim, K. T., & Patel, D. K. (2020). Naturally derived 3D printed hydrogel scaffolds for agroecosystem and biomedical applications. *Multifunctional Hybrid Nanomaterials for Sustainable Agri-Food and Ecosystems, 1*, 537–554. https://doi.org/10.1016/B978-0-12-821354-4.00022-4.

Smandri, A., Nordin, A., Hwei, N. M., Chin, K. Y., Abd Aziz, I., & Fauzi, M. B. (2020). Natural 3D-printed bioinks for skin regeneration and wound healing: A systematic review. *Polymers, 12*(8), 1782. https://doi.org/10.3390/polym12081782.

Sola, A., Bertacchini, J., D'Avella, D., Anselmi, L., Maraldi, T., Marmiroli, S., & Messori, M. (2019). Development of solvent-casting particulate leaching (SCPL) polymer scaffolds as improved three-dimensional supports to mimic the bone marrow niche. *Materials Science & Engineering. C, Materials for Biological Applications, 96*, 153–165. https://doi.org/10.1016/j.msec.2018.10.086.

Song, J., Zhang, Y., Chan, S. Y., Du, Z., Yan, Y., Wang, T., Li, P., & Huang, W. (2021). Hydrogel-based flexible materials for diabetes diagnosis, treatment, and management. *Npj Flexible Electronics, 5*(1), 26. https://doi.org/10.1038/s41528-021-00122-y.

Song, P., Dai, J., Chen, G., Yu, Y., Fang, Z., Lei, W., Fu, S., Wang, H., & Chen, Z. G. (2018). Bioinspired design of strong, tough, and thermally stable polymeric materials via nanoconfinement. *ACS Nano, 12*(9), 9266–9278. https://doi.org/10.1021/acsnano.8b04002.

Spampinato, S. F., Caruso, G. I., De Pasquale, R., Sortino, M. A., & Merlo, S. (2020). The treatment of impaired wound healing in diabetes: Looking among old drugs. *Pharmaceuticals, 13*(4), 1–17. https://doi.org/10.3390/ph13040060.

Stoica, A. E., Chircov, C., & Grumezescu, A. M. (2020). Hydrogel dressings for the treatment of burn wounds: An up-to-date overview. *Materials (Basel, Switzerland), 13*(12), 2853. https://doi.org/10.3390/ma13122853.

Tabriz, A. G., & Douroumis, D. (2022). Recent advances in 3D printing for wound healing: A systematic review. *Journal of Drug Delivery Science and Technology, 74*, 103564. https://doi.org/10.1016/j.jddst.2022.103564.

Tavakoli, S., & Klar, A. S. (2020). Advanced hydrogels as wound dressings. *Biomolecules, 10*(8), 1–20. https://doi.org/10.3390/biom10081169.

Tawfeek, H. M., Abou-Taleb, D. A. E., Badary, D. M., Ibrahim, M., & Abdellatif, A. A. H. (2020). Pharmaceutical, clinical, and immunohistochemical studies of metformin hydrochloride topical hydrogel for wound healing application. *Archives of Dermatological Research, 312*(2), 113–121. https://doi.org/10.1007/s00403-019-01982-1.

Tsegay, F., Elsherif, M., & Butt, H. (2022). Smart 3D printed hydrogel skin wound bandages: A review. *Polymers, 14*(5), 1012. https://doi.org/10.3390/polym14051012.

Uddin, M. J., Hassan, J., & Douroumis, D. (2022). Thermal inkjet printing: Prospects and applications in the development of medicine. *Technologies, 10*(5), 108. https://doi.org/10.3390/technologies10050108.

Unagolla, J. M., & Jayasuriya, A. C. (2020). Hydrogel-based 3D bioprinting: A comprehensive review on cell-laden hydrogels, bioink formulations, and future perspectives. *Applied Materials Today, 18*. https://doi.org/10.1016/j.apmt.2019.100479.

Velnar, T., Bailey, T., & Smrkolj, V. (2009). The wound healing process: An overview of the cellular and molecular mechanisms. *Journal of International Medical Research, 37*(5), 1528–1542. https://doi.org/10.1177/147323000903700531.

Vigata, M., Meinert, C., Hutmacher, D. W., & Bock, N. (2020). Hydrogels as Drug Delivery Systems: A Review of Current Characterization and Evaluation Techniques. In *Pharmaceutics (Vol. 12, Issue 12)*, MDPI. https://doi.org/10.3390/pharmaceutics12121188.

Volpe, C. M. O., Villar-Delfino, P. H., Dos Anjos, P. M. F., & Nogueira-Machado, J. A. (2018). Cellular death, reactive oxygen species (ROS) and diabetic complications. *Cell Death & Disease, 9*(2), 119. https://doi.org/10.1038/s41419-017-0135-z.

Wallace, H. A., Basehore, B. M., & Zito, P. M. (2023). Wound Healing Phases, PubMed. https://pubmed.ncbi.nlm.nih.gov/29262065/.

Wang, J., Liu, Y., Zhang, X., Rahman, S. E., Su, S., Wei, J., Ning, F., Hu, Z., Martínez-Zaguilán, R., Sennoune, S. R., Cong, W., Christopher, G., Zhang, K., & Qiu, J. (2021). 3D printed agar/ calcium alginate hydrogels with high shape fidelity and tailorable mechanical properties. *Polymer, 214*, 123238. https://doi.org/10.1016/j.polymer.2020.123238.

Wang, L., Deng, F., Wang, W., Li, A., Lu, C., Chen, H., Wu, G., Nan, K., & Li, L. (2018). Construction of injectable self-healing macroporous hydrogels via a template-free method for tissue engineering and drug delivery. *ACS Applied Materials & Interfaces, 10*(43), 36721–36732. https://doi.org/10.1021/acsami.8b13077.

Wang, M., Zhang, P., Shamsi, M., Thelen, J. L., Qian, W., Truong, V. K., Ma, J., Hu, J., & Dickey, M. D. (2022). Tough and stretchable ionogels by in situ phase separation. *Nature Materials, 21*(3), 359–365. https://doi.org/10.1038/s41563-022-01195-4.

Wang, X., Gao, P., Yang, Y., Guo, H., & Wu, D. (2018). Dynamic and programmable morphology and size evolution via a living hierarchical self-assembly strategy. *Nature Communications, 9*(1), 2772. https://doi.org/10.1038/s41467-018-05142-3.

Wei, S., Li, J., He, H., Shu, C., Dardik, A., & Bai, H. (2022). A three-layered hydrogel patch with hierarchy releasing of PLGA nanoparticle drugs decrease neointimal hyperplasia. *Smart Materials in Medicine, 3*, 139–147. https://doi.org/10.1016/j.smaim.2021.12.005.

Wei, W., Ma, Y., Yao, X., Zhou, W., Wang, X., Li, C., Lin, J., He, Q., Leptihn, S., & Ouyang, H. (2021). Advanced hydrogels for the repair of cartilage defects and regeneration. *Bioactive Materials, 6*(4), 998–1011. https://doi.org/10.1016/j.bioactmat.2020.09.030.

Wei, X., Liu, C., Wang, Z., & Luo, Y. (2020). 3D printed core-shell hydrogel fiber scaffolds with NIR-triggered drug release for localized therapy of breast cancer. *International Journal of Pharmaceutics, 580*, 119219. https://doi.org/10.1016/j.ijpharm.2020.119219.

Werzer, O., Tumphart, S., Keimel, R., Christian, P., & Coclite, A. M. (2019). Drug release from thin films encapsulated by a temperature-responsive hydrogel. *Soft Matter, 15*(8), 1853–1859. https://doi.org/10.1039/C8SM02529K.

Wong, R. S. H., & Dodou, K. (2017). Effect of drug loading method and drug physicochemical properties on the material and drug release properties of poly (ethylene oxide) hydrogels for transdermal delivery. *Polymers, 9*(7), 286. https://doi.org/10.3390/polym9070286.

Wood, L., Wood, Z., Davis, P., & Wilkins, J. (2010). Clinical experience with an antimicrobial hydrogel dressing on recalcitrant wounds. *Journal of Wound Care, 19*(7), 287–293. https://doi.org/10.12968/jowc.2010.19.7.48902.

Wu, T., Zhang, X., Liu, Y., Cui, C., Sun, Y., & Liu, W. (2023). Wet adhesive hydrogel cardiac patch loaded with anti-oxidative, autophagy-regulating molecule capsules and MSCs for restoring infarcted myocardium. *Bioactive Materials, 21*, 20–31. https://doi.org/10.1016/j.bioactmat.2022.07.029.

Xu, N., Wang, L., Guan, J., Tang, C., He, N., Zhang, W., & Fu, S. (2018). Wound healing effects of a *Curcuma zedoaria* polysaccharide with platelet-rich plasma exosomes assembled on chitosan/silk hydrogel sponge in a diabetic rat model. *International Journal of Biological Macromolecules, 117*, 102–107. https://doi.org/10.1016/j.ijbiomac.2018.05.066.

Xu, X. L., Chen, H. H., Li, Y., & Chen, Q. X. (2019). The role of equity balance and executive stock ownership in the innovation efficiency of renewable energy enterprises. *Journal of Renewable and Sustainable Energy, 11*(5), 055901. https://doi.org/10.1063/1.5116849.

Xu, Z., Han, S., Gu, Z., & Wu, J. (2020). Advances and impact of antioxidant hydrogel in chronic wound healing. *Advanced Healthcare Materials, 9*(5), e1901502. https://doi.org/10.1002/adhm.201901502.

Yahia, Lh. (2015). History and Applications of Hydrogels. *Journal of Biomedical Sciencies, 04*(02), 1–23. https://doi.org/10.4172/2254-609x.100013.

Yan, J., Wang, Y., Zhang, X., Zhao, X., Ma, J., Pu, X., Wang, Y., Ran, F., Wang, Y., Leng, F., & Zhang, W. (2019). Snake gourd root/Astragalus polysaccharide hydrogel preparation and application in 3D printing. *International Journal of Biological Macromolecules, 121*, 309–316. https://doi.org/10.1016/j.ijbiomac.2018.10.008.

Yang, Q., Zhong, W., Xu, L., Li, H., Yan, Q., She, Y., & Yang, G. (2021). Recent progress of 3D-printed microneedles for transdermal drug delivery. *International Journal of Pharmaceutics, 593*, 120106. https://doi.org/10.1016/j.ijpharm.2020.120106.

Zagórska-Dziok, M., & Sobczak, M. (2020). Hydrogel-based active substance release systems for cosmetology and dermatology application: A review. *Pharmaceutics, 12*(5), 396. https://doi.org/10.3390/pharmaceutics12050396.

Zahavy, E., Rotem, S., Gur, D., Aloni-Grinstein, R., Aftalion, M., & Ber, R. (2018). Rapid antibiotic susceptibility determination for yersinia pestis using flow cytometry Spectral Intensity Ratio (SIR) fluorescence analysis. *Journal of Fluorescence, 28*(5), 1151–1161. https://doi.org/10.1007/s10895-018-2279-3.

Zhang, A., Liu, Y., Qin, D., Sun, M., Wang, T., & Chen, X. (2020). Research status of self-healing hydrogel for wound management: A review. *International Journal of Biological Macromolecules, 164*, 2108–2123. https://doi.org/10.1016/j.ijbiomac.2020.08.109.

Zhang, J., Liu, X. F., Liu, Y., Xu, L. Z., Zhou, L. L., Tang, L. L., Zhuang, J., Li, T. T., Guo, W. Q., Hu, R., Qiu, D. S., & Han, D. W. (2014). Environmental risk factors for women with polycystic ovary syndrome in china: a population-based case-control study. *Journal of Biological Regulators and Homeostatic Agents, 28*(2), 203–211. http://europepmc.org/abstract/MED/25001653.

Zhang, M., & Zhao, X. (2020). Alginate hydrogel dressings for advanced wound management. *International Journal of Biological Macromolecules, 162*, 1414–1428. https://doi.org/10.1016/j.ijbiomac.2020.07.311.

Zhang, S., Ge, G., Qin, Y., Li, W., Dong, J., Mei, J., Ma, R., Zhang, X., Bai, J., Zhu, C., Zhang, W., & Geng, D. (2023). Recent advances in responsive hydrogels for diabetic wound healing. *Materials Today Bio, 18*, 100508. https://doi.org/10.1016/j.mtbio.2022.100508.

Zhang, X. N., Zheng, Q., & Wu, Z. L. (2022). Recent advances in 3D printing of tough hydrogels: A review. *Composites Part B: Engineering, 238*, 109895. https://doi.org/10.1016/j.compositesb.2022.109895.

Zhang, X., Kim, G. J., Kang, M. G., Lee, J. K., Seo, J. W., Do, J. T., Hong, K., Cha, J. M., Shin, S. R., & Bae, H. (2018). Marine biomaterial-based bioinks for generating 3d printed tissue constructs. In *Marine Drugs (Vol. 16, Issue 12)*, MDPI. https://doi.org/10.3390/md16120484.

Zhang, X., Yao, D., Zhao, W., Zhang, R., Yu, B., Ma, G., Li, Y., Hao, D., & Xu, F. J. (2021). Engineering platelet-rich plasma based dual-network hydrogel as a bioactive wound dressing with potential clinical translational value. *Advanced Functional Materials, 31*(8), 1–14. https://doi.org/10.1002/adfm.202009258.

Zhang, Y. S., & Khademhosseini, A., (2017). Advances in engineering hydrogels. *Science. 356*(6337), eaaf3627. doi: 10.1126/science.aaf3627. PMID: 28473537; PMCID: PMC5841082.

Zhao, C., Zhou, L., Chiao, M., & Yang, W. (2020). Antibacterial hydrogel coating: Strategies in surface chemistry. *Advances in Colloid and Interface Science, 285*, 102280. https://doi.org/10.1016/j.cis.2020.102280.

Zhao, X., Chen, X., Yuk, H., Lin, S., Liu, X., & Parada, G. (2021). Soft Materials by design: unconventional polymer networks give extreme properties. *Chemical Reviews, 121*(8), 4309–4372. https://doi.org/10.1021/acs.chemrev.0c01088.

Zhao, Y., Ran, B., Xie, X., Gu, W., Ye, X., & Liao, J. (2022). Developments on the smart hydrogel-based drug delivery system for oral tumor therapy. In *Gels (Vol. 8, Issue 11)*, MDPI. https://doi.org/10.3390/gels8110741.

Zheng Shu, X., Liu, Y., Palumbo, F. S., Luo, Y., & Prestwich, G. D. (2004). *In situ* crosslinkable hyaluronan hydrogels for tissue engineering. *Biomaterials, 25*(7–8), 1339–1348. https://doi.org/10.1016/j.biomaterials.2003.08.014.

CHAPTER 11

Recent Trends in Marine Nutraceuticals and Potential Therapeutic Applications

BELA RANI BHAGAT,[1] JENNY SIVAKUMAR,[2] GOWRI SUNDARAM,[3] and MD. FAIYAZUDDIN[4,5]

[1]*Department of Agriculture, Usha Martin University, Ranchi, Jharkhand, India*

[2]*Department of Microbiology, Cauvery College for Women (Autonomous), Tiruchirappalli, Tamil Nadu, India*

[3]*PG & Research Department of Physics, Cauvery College for Women (Autonomous), Tiruchirappalli, Tamil Nadu, India*

[4]*School of Pharmacy, Al-Karim University, Katihar, Bihar, India*

[5]*Nano Drug Delivery, Raleigh-Durham, NC, USA*

ABSTRACT

The use of functional foods and supplements in recent years has shown that bioactive chemicals have a significant therapeutic impact in treating human disease. Researchers in the fields of biomedicine, food science, and nutrition are collaborating to identify novel bioactive compounds with improved therapeutic potential. Many marine species, including seaweed, contain a variety of chemicals, many of which have biological properties known as bioactive compounds. The marine surroundings are a gigantic basis of nutraceuticals. These substances can be isolated and added to various food matrices, resulting in the creation of new functional foods. Marine lipids (including both animal and microalgal origin), macroalgal polysaccharides, probiotics, natural marine colors, chitin and other related products, bioactive

Handbook of Research in Marine Pharmaceutics: Exploring Oceanic Microbial Diversity for Human Health and Wellness. Md. Faiyazuddin & Khalid Rehman Hakeem (Eds.)

marine peptides/enzymes, and vitamins. The current study summarizes commercially accessible marine-based nutraceuticals as well as recent investigations into the separation, characterization, and therapeutic potential of marine-derived bioactive chemicals converging on their prospective as nutraceuticals with positive effects on health.

11.1 INTRODUCTION

Earth is mostly surrounded by water, and the richness of the oceans is unparalleled. Researchers now have access to previously unreachable oceanic depths, increasing their ability to study marine biota. Along with offering a fantastic chance for the discovery of several marine bioactive chemicals, the Indian coastline also boasts a wide variety of marine algae, especially in Tamil Nadu in the south. Since then, other new bioactive secondary metabolites, especially from these marine microbes, have been discovered and studied (Figure 11.1). Even though the algal flora in India is very varied, there aren't many publications on the screening of marine algicolous endophytic fungi for a variety of biological activities. However, although making up nearly 70% of the Earth's surface, marine ecosystems have not been studied from an industrial perspective (Gerwick & Moore, 2012; Wang et al., 2020). Over time, biomedical research has paid more and more attention to marine resources. So far, 28,000 novel chemicals with marine origins have been found. Because of their diverse biological activity, natural substances derived from marine resources have been suggested as potential replacements for traditional medicines in the treatment of illness. These natural marine compounds have demonstrated an extremely important role in cancer research primarily due to their less lethal approach than conventional chemotherapy agents. The wide range of bioactivities evinced by marine bioactives, including antioxidant, antimicrobial, antithrombotic, antiproliferative, anti-inflammatory, anti-diabetic, anti-hypertensive, and cholesterol-lowering effects, can be attributed to their prospective health aids.

In most instances, they are also affordably priced, functional, and readily available. The efficiency is demonstrated by their potential to stop the spread and progression of cancer. Several marine-derived metabolites as well as cancer clinical trials conducted *in vivo* and *in vitro* show this inhibition of tumor cell growth (Centella et al., 2017). Due to their biological and significant characteristics that contribute to their potential for use in nutraceuticals and pharmaceuticals, it appears that marine-derived bioactive compounds can be well-considered to be safe replacements for some existing synthetic drugs. To create functional foods and nutraceuticals, new bioprocessing technologies are being developed to separate certain bioactive substances with antioxidant properties from marine food products. This chapter focuses on a variety of marine-based valuable complexes with a wide range of potential applications for nutraceuticals, potential health benefits, and commercial interests.

Species Diversity of Marine Algae along Indian Coastline

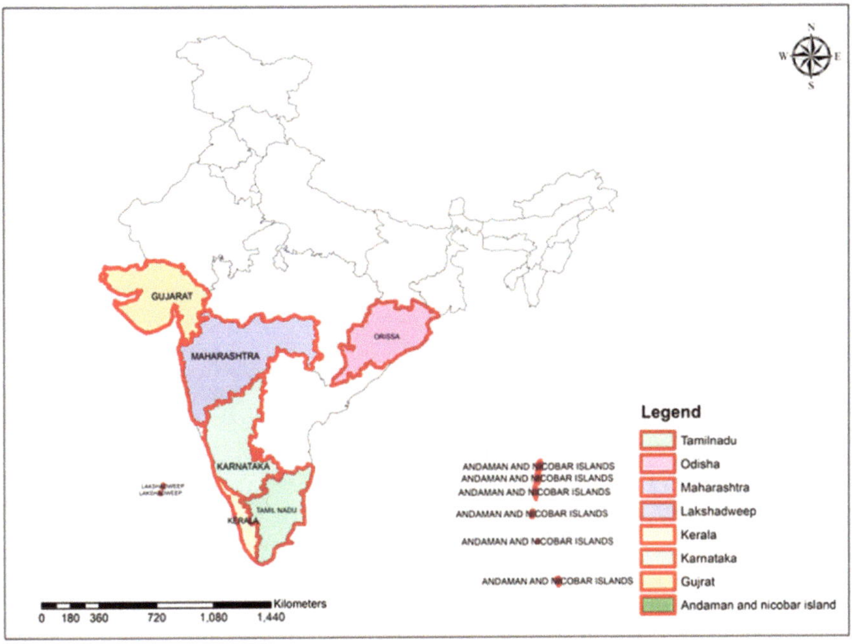

FIGURE 11.1 Marine algae species diversity along Indian coastline.

11.2 MARINE-DERIVED CHEMICALS AS FUNCTIONAL FOOD INGREDIENTS

Large bioactive compounds can be delivered by marine organisms in a variety of amounts, which is a requirement for the food industry. In recent years, the importance of marine products, marine-derived nutraceuticals, and their use has grown significantly. Due to their widespread use in the food and health industries, marine compounds include fish oil, gelatin, chitosan, chitin, enzymes, proteins and peptides, EPA and DHA, EAA, polysaccharides, omega-3 polyunsaturated fatty acids (PUFA), polyphenolic compounds, pigments, vitamins, and minerals. Fish, shellfish, mollusks (such as mussels, oysters, scallops, abalones, and snails), cephalopods (such as squid, cuttlefish, and octopus), crustaceans (such as crayfish, crab, shrimp, and lobster), echinoderms, seaweeds, and microalgae are all sources of these compounds (Figures 11.2 and 11.3) (Nover et al., 2011; Ruiz-Ruiz et al., 2017). Because of their special capacity to form foam and gel, marine proteins are used in food. In this sense, seaweeds have been used as a source of protein for many years (Pooja et al., 2020). Muscles from fish are rich in water-soluble amino acids such as taurine, Pro, Gln, Ala, Gly, and Arg. Therefore, it is assumed that eating fish muscle proteins may increase human nutrition since they may be rich in certain amino acids, which will boost nourishment (Dewapriya & Kim, 2014; Romano et al., 2017).

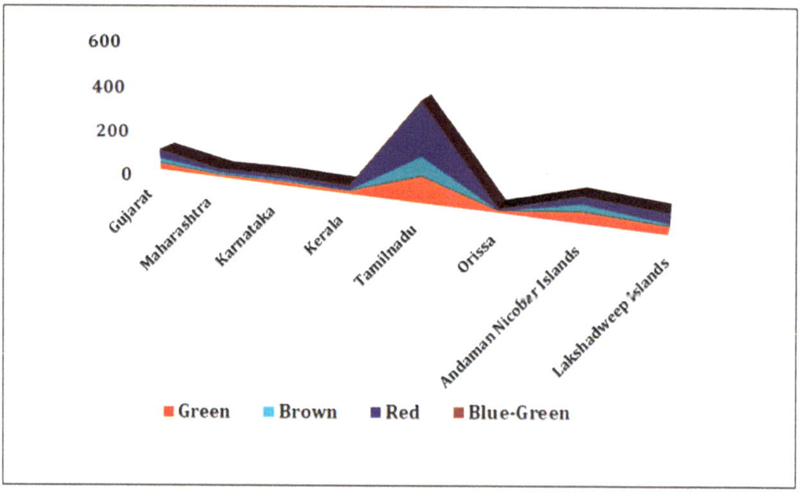

FIGURE 11.2 Marine algae species composition during different surveys along the Indian coastline.

Source: Adapted from Sharma et al., (2019).

Eicosapentaenoic acid (EPA) and docosahexaenoic acid are the most prevalent omega-3 PUFAs produced from aquatic bases (DHA). The only ways to get these PUFAs are from seafood and supplements (Figure 11.4).

Due to its ability to form gels, improve consistency, hold onto water, and prolong the shelf life of food products, marine gelatin is used as a food additive following the fractional hydrolysis of collagen. Due to their many qualities, marine polysaccharides are excellent for use in food. The three most common

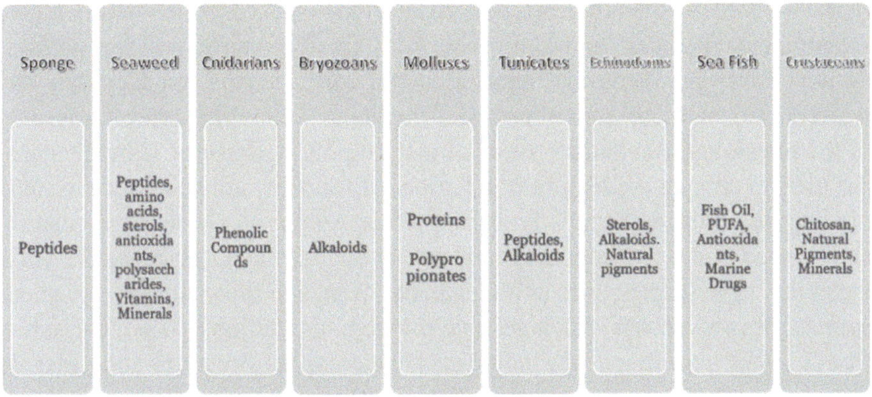

FIGURE 11.3 Marine-originated bioactive compounds as functional food supplements.

Source: Adapted from Ghosh et al., 2022

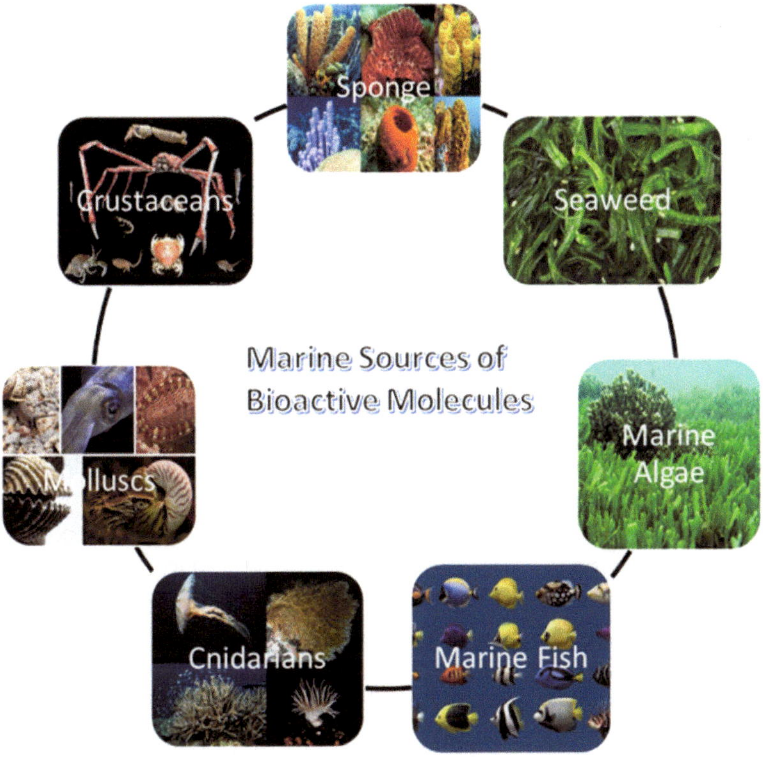

FIGURE 11.4 Marine sources for bioactive molecules.

Source: Adapted from Ghosh et al. 2022.

types of soluble dietary fiber polysaccharides derived from brown algae are alginate, fucoidan, and laminarin. Additionally, marine polysaccharides derived from seaweeds like agar, alginate, carrageenan, and fucoidan are widely used in food, agriculture, and medicine. The ability to bind is one of the physiological characteristics of marine polysaccharides. Carrageenans are used to change the texture of a variety of foods by adjusting their water-binding, foaming, and emulsifying properties. Agar with a high sugar content can be used in the confectionery industry. Other marine polysaccharides with the ability to form gels, such as alginate, chitosan, and fucoidan, are also good sources of raw material for ingestible, biodegradable films. When extracted from various marine species, fucoidans have complex structures and differ in terms of saccharide composition, sulfate content, positions of sulfate groups, molecular weight, linkage mode, and sequence of saccharide residues (Li et al., 2008). Sulfate-based organizations advance the biological characteristics of fucoidans, enabling the dairy industry to request them as nutraceuticals (Vidanarachchi et al., 2011). Chitin, chitosan, and their derivatives are rarely used in food except as additives, nutraceuticals, edible films, and antimicrobials (Suleria et al., 2016). Chitosan, chitosan oligosaccharides, and glucosamine are all products of deacetylated chitin. These chitosan derivatives are more complementary as nutraceutical ingredients than the completely acetylated, mysterious variety of chitin because of their improved water solubility. The diversity of biological features of chitosan and its derivatives have flickered commercial interest in chitosan in several fields due to its biodegradability, biocompatibility, and non-toxicity (Gurpilhares et al., 2019) (Nagaoka et al., 2011; Zhang et al., 2012). According to Shibata et al. (2008), flavonoids and phenolic compounds, typically phlorotannins made up of phloroglucinol, Eckol, and dieckol, are the most prevalent polyphenols found in marine brown algae. Nutraceuticals use enzymes extensively. Due to their distinguishing activity, enzymes can be predominantly valuable in the food industry and are generally thought to be a substantial profitable resource for biotechnological applications. Lipase, chitinolytic enzymes, polyphenol oxidase (Catecholase, tyrosinase, cresolase, polyphenolase, catechol oxidase, phendase), and transglutaminase are some of the enzymes derived from marine sources. Lipase, transglutaminase, chitinolytic enzymes, polyphenol oxidase, and red algal enzymes are utilized in the degradation of starch (such as -1,4-glucanase) and used as culinary additives. Vitamins and minerals function in intracellular transportation and serve as cofactors in metabolic processes. Vitamins and minerals, such as iron, are abundant in seaweeds (Suleria et al., 2015). Numerous species of seaweed have been found to have implications for food diets and conventional treatments in Eastern nations, and more recently in Europe and America. Seaweeds, therefore, have a lot of potential for use in neuroprotection (Zarros, 2009).

According to a recent study of the worldwide nutraceutical market (Research & Markets, 2012), the competition, market drivers, and industry structure are all carefully described. Due to their greater product knowledge, higher income levels, and preference for nutraceutical items, defensive medicine, and self-treatment, the USA, Europe, and Japan control 85% of the world's market. These three regional markets are anticipated to remain the leaders in the nutraceutical sector, both as manufacturers and consumers. According to estimates, the Asia Pacific area, in particular China and India, will see significant growth over the long term because of increased attention, earnings, and wealth.

11.3 PROSPECTIVE THERAPEUTIC APPLICATIONS OF MARINE BIOACTIVE

In the current situation with mental awareness, lifestyle changes with longer lifespans, and nutrition intake have become important factors. The term "Nutraceutical" in this case refers to a term created by fusing the meanings of the terms "nutrition" and "pharmaceutical," and it has a lively character in enhancing human health. The term "nutraceuticals" refers to compounds that may be regarded as food or a component of food and have medical or human health advantages, such as the ability to prevent and cure illness (Rajasekaran et al., 2008). The use of active chemicals obtained from marine sources as nutraceuticals is a growing area in the field of natural nutraceuticals. The active ingredients from natural sources are widely accepted in the nutraceutical industry because they are known to be safe for human consumption and because populations who eat a lot of plant-based foods and seafood are known to have a lower frequency of diseases like cardiovascular disease (CVD) and various types of cancers (Voutilainen et al., 2006). Active substances obtained from the ocean may one day be employed to improve human health thanks to their significant biological features (Table 11.1). While recent advancements in marine research have increased the applicability of identified marine-derived compounds in the industrial scenario, including the nutraceutical industry, many marine-derived biologically active compounds and their significant biological properties remain underutilized. Globally, the market for marine habitats is growing significantly. According to statistics, the global market for marine-derived compounds has increased fourfold since 2018 and is projected to reach $22 billion by 2025 at a compound annual growth rate of 11.3% from 2019 to 2025. Additionally, due to their distinctive biological properties, including anti-oxidative, anti-inflammatory, anticancer, anti-angiogenesis, antimicrobial, anticoagulant, prebiotic, and probiotic activities, marine-derived nutraceuticals such as polysaccharides, polyunsaturated fatty acids, polyphenols, and carotenoids have been identified as health-beneficial compounds.

11.3.1 ANTIOXIDANT THERAPEUTIC ACTIVITY

Diverse metabolites with unique chemical compositions and physiologically active characteristics were created by marine species. Antioxidants, which can be either natural or synthetic, can help prevent or delay cell damage. These substances shield cells from the damaging effects of free radicals, which are unstable molecules. Bioactive compounds are produced by seaweeds, marine sponges, and other similar bacteria, and they may promote health and ward off disease. These therapeutic benefits are derived from a complex mixture of phytochemicals having antioxidant, antibacterial, anticancer, and antiviral characteristics. These actions are carried out by sulfated polysaccharides, phenolic chemicals, and organic acids. Polyphenolic substances like flavonoids, benzoic acid, gallic acid, phlorotannins, and quercetin are produced in the marine environment (Dominguez, 2013).

Several of these antioxidant compounds can stop tissue damage in aging, atherosclerosis, and cancer by neutralizing reactive oxygen species (ROS). The industry is paying a lot of attention to marine-derived antioxidants as nutraceuticals due to their capacity to shield tissues from oxidative damage. Additionally, the anti-inflammatory actions of several marine antioxidants (Guerin et al., 2003). According to Kochkina & Chirkov (2000), marine-derived chemicals may aid in the body's defense against harm in a variety of ways, including scavenging free radicals, preventing erythrocyte hemolysis, and binding to metals. Some compounds derived from marine sources have strong antioxidant properties, which means they can help shield your body from harm. Molecules called reactive oxygen species (ROS) have the potential to harm cells. When oxygen and other chemicals are combined, ROS are produced. Among ROS, hydrogen peroxide is one type (H_2O_2). ROS can alter the chemical composition of cells when they are present. These alterations can cause molecules to oxidize, which can harm cells more severely. Cells contain proteins as vital components. Some proteins are capable of killing cells (Jing et al., 2011).

TABLE 11.1 Biological Activity of Marine Nutraceutical Compounds (Anake & Pichan, 2004; Maria et al., 2013; Hafiz et al., 2015)

Compounds	Chemical Structure	Biological Activity
Fucoidan	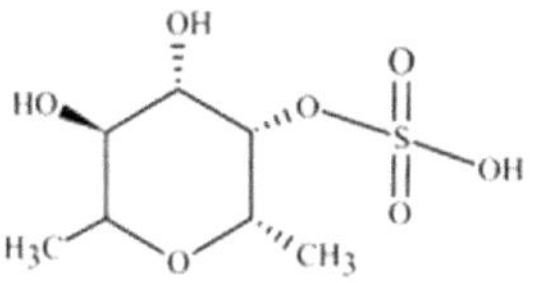	Antioxidant, antiangiogenic and antitumor activities, anticoagulant, immunomodulating and Hypolipidemic, anti-inflammatory
Alginate	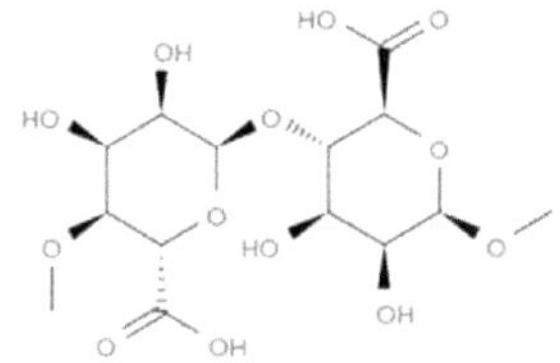	Stimulates the immune system, reduces intestinal absorption, modulates colonic microflora, and elevates colonal barrier function
Carrageenan	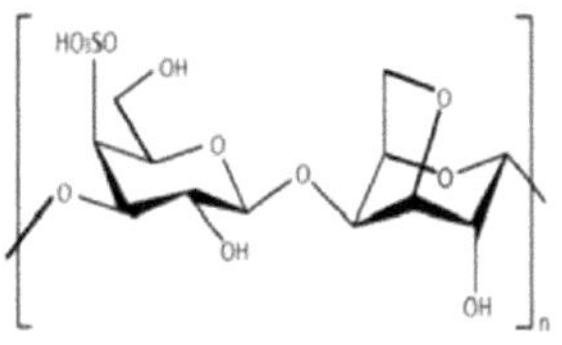	Antibacterial, anti-tumor, antiviral and anti-inflammatory activities
Caulerpicin	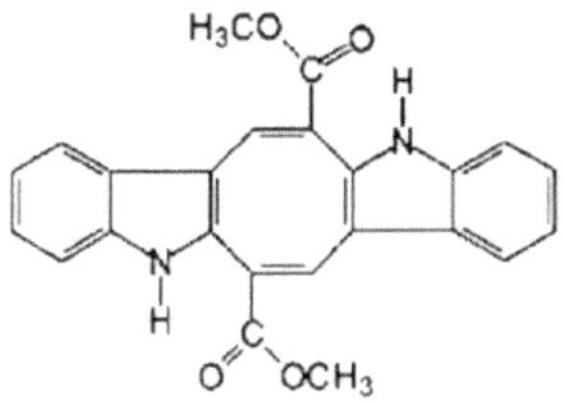	Antiviral activity
Sulfated polysaccharides	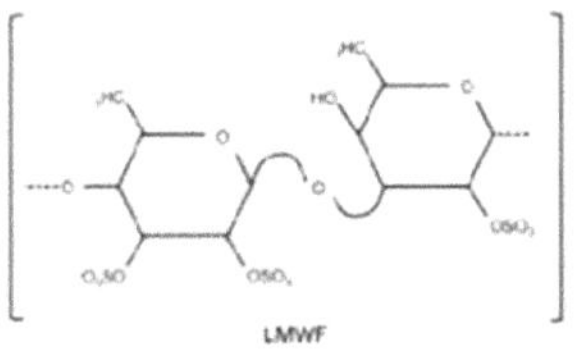	Antioxidant, antithrombin activity, antitumor, cell recognition, and cell adhesion or regulation of receptor functions
Xylofucans	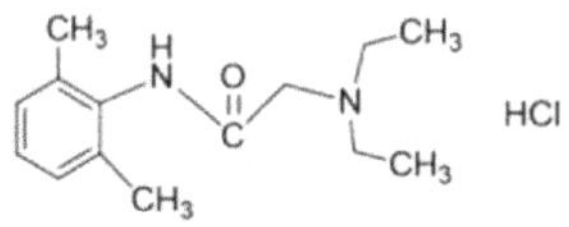	Antiviral activity
Alginic acid	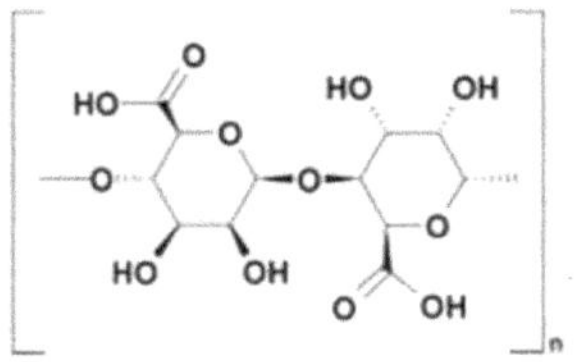	Antiviral activity

TABLE 11.1 *(Continued)*

Compounds	Chemical Structure	Biological Activity
Chitin	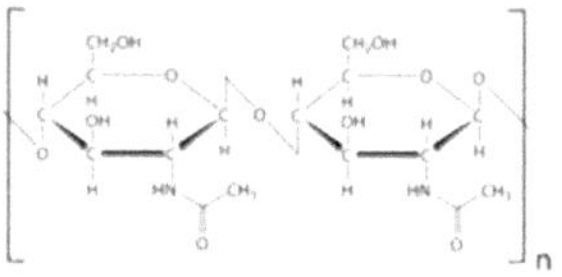	Antioxidant properties
Eicosapentaenoic acid	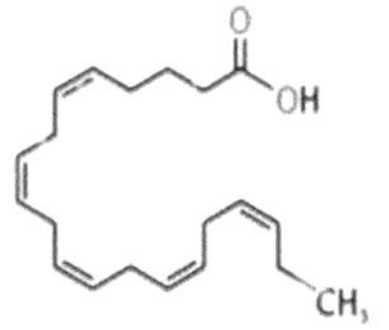	Nutraceutical, anti-inflammatory

11.3.2 ANTIMICROBIAL THERAPEUTIC ACTIVITY

Some organism defense mechanisms include antibacterial peptides. These peptides, which have a sequence of 12 to 50 amino acids and are small, cationic, amphipathic proteins, act as the body's initial line of defense against infections (2002; Cunha & Pintado, 2021). It has been shown that some bacteriocin, chitin derivatives, chitosan, COS, and extracts from marine algae are significant antibacterial substances with the ability to reduce the risk of illness in people (Vidanarachchi et al., 2010).

11.3.3 ANTICANCER THERAPEUTIC ACTIVITY

Cells that develop cancer cease to respond to signals from other cells. As a result, the cells might eventually perish. Researchers look for substances that can trigger apoptosis, or "programmed cell death." If this occurs, treating cancer might be simpler. It's possible that eating fish increases the signals that can cause cancer. According to research, several signaling mechanisms might cause cancer, and one of these mechanisms involves fish. Eating fish can increase the amount of apoptosis, which is an important anticancer target and a process that aids in the death of cancer cells. Low concentrations of salt and lipids, as well as a variety of molecules with weights ranging from 1 to 7 kDa, all have an impact on cell growth (Kim et al., 2013). Recently, many scientists have concluded that marine organisms are rich sources of novel molecules and anticancer drugs. A lot of marine-derived nutraceuticals are employed as anticancer medications, including beta-carotene, fucoidans, chitosan, chitosan oligosaccharides, astaxanthin, and phlorotannin. Marine oligosaccharides are of particular interest to researchers because they have demonstrated fewer negative effects than the other agents. These marine oligosaccharides target the tumor's blood vessel network, stimulate the immune system and other mechanisms to kill tumor cells, and damage the tumor's nutritional support system.

The anti-proliferative gene P53 is activated, cell apoptosis is induced via a variety of pathways, angiogenesis is inhibited, and matrix metalloproteinases (MMPs) are inhibited as some of the cancer-inhibiting pathways of these marine-derived anticancer agents (Vidanarachchi et al., 2011). Numerous cytotoxic substances have been reported over the past 10 years; some of them, like dehydrodidemnin B and bryostatins, are currently undergoing anticancer clinical trials. Resmi et al. (2020) used two sequential processes, such as demineralization and deproteinization, to separate chitosan nanoparticles from waste shrimp shells. When compared to chitosan produced chemically, chitosan NP demonstrated decreased cytotoxicity toward healthy L929 fibroblast cells and an inhibitory effect on the proliferation of MCF-7 breast cancer cells. *Procambarus clarkii* freshwater crayfish waste was utilized by El-Naggar et al. (2020) to produce chitin, which was later deacetylated to produce chitosan. The chitosan was converted into Schiff bases and Chitosan NP. When tested against three cell lines, chitosan Schiff bases and chitosan NP both demonstrated anti-tumor efficacy (HepG-2, HCT-116, and MCF-7). Chitosan showed the least anti-tumor activity of all the substances tested. Sedghi et al. (2020) created nanofibers from a chitosan derivative. The results of the MMT test indicated that the substance was effective against 4T1 breast cancer cells and had no negative effects on healthy cells, suggesting that nanofibers might be used to prevent the recurrence of breast cancer. Chemoresistance has emerged as the primary issue in the management of cancer, stressing the urgent need for improved therapies. There are now innovative molecules that can be utilized to develop new cancer treatments that are affected by marine natural chemicals. Polysaccharides, polyunsaturated fatty acids (PUFAs), vitamins, minerals, and proteins from the marine diet all makeup marine products, which also comprise both essential and non-essential amino acids.

11.4 CONCLUSION

Marine creatures develop and become extraordinary sources of chemicals that have uses other than feeding humans. Due to their biological characteristics or health status, several of these chemicals have been used as functional food ingredients or for treating or preventing a variety of disorders. The identification of new chemical compounds with physiological activity and the development of their potential applications in the manufacture of foods, functional foods, and supplements are receiving more attention. Chitosan, chitin, fish oil, EPA, DHA, EAA, peptides, gelatin, polysaccharides, polyphenols, pigments, vitamins, and minerals have all been found to have antibacterial, antioxidant, and anticancer properties. These substances are also used in the food industry to increase the nutritional value of food.

11.5 FUTURE PROSPECTIVES

How far marine-origin components can be used in the future will depend on their bioavailability and the advancement of environmentally friendly technology for their high-yielding production. The sustainability of marine resources must be taken into account when they are used industrially. This entails extraction, preservation against deterioration, and the discovery of individual elements that can be purified for use in a particular form. The difficult task of farming, growing organisms for a particular component, and figuring out the conditions needed for its high-yield recovery calls for the creation of new effective technologies. To prevent the deterioration of these beneficial substances, marine-based bioactive compounds may be used to create nanosystems using nanotechnology techniques. Clinical trials are also necessary to confirm the therapeutic effect of recently discovered chemicals, as well as to ascertain their role in promoting health and the proper dosages for daily use. These would help the food industry produce appealing, functional foods for consumers. Nano-based delivery systems can be created to modify the bioactive component's metabolism and bioactivity within the body by changing the pathway of uptake. Food bioactive ingredients can be protected from harmful environmental factors, have their water dispersibility and food matrix compatibility improved, have off flavors reduced, have their release profiles controlled, have improved stability in the gastrointestinal tract (GIT), and have their bioavailability increased using nanoscaled structures. There are many potential medical applications for nanoparticles, according to recent research trends. Additionally, because it produces no poisonous by-products, the process of generating nanoparticles from a variety of natural resources is frequently used as an eco-friendly technique. Numerous techniques have been developed for the production of various inorganic nanoparticle types, including gold, zinc, titanium, magnesium, and silver. Marine Organisms and Substances for the Synthesis of Green NP Current marine science research and innovation are facilitating the exponential growth of numerous industries, including pharmaceuticals, environmental trends, nanomedicine, and food. A major accomplishment has been reached in the development of nanoscale delivery systems to increase the bioavailability of chemicals derived from aquatic sources. Nanofibers are used in fish meat to improve protein and amino acid stability. Chitosan-based nanofibers and electrospun chitosan-based nanofibers loaded with thymol may be used to successfully deliver nicotinamide acid, pyridoxal, and other vitamins, while fish fillets contain pyridoxine and pyridoxamine. EPA, DHA, and EAA are important marine-based substances for human health, like tryptophan or the B group of vitamins. Probiotic microorganisms are essential for the human immune and digestive systems. Consuming nanoprobiotic-coated fish meat (*L. reuteri* & *L. rhamnosus*) could be an alternative method. Because of these nanoprobiotics, the fatty acids in fish meat were more consistently stable. Algae, on the other hand, are excellent sources of a wide range of essential fatty acids. Algae are also good sources of a variety of essential fatty acids, including omega-3 and omega-6, which are important for the immune system. Furthermore, gold nanoparticles derived from brown algae have anti-diabetic and anti-inflammatory properties.

Nanotechnology-enabled marine-based goods might be blended with other probiotic-rich dietary ingredients. Nanoemulsion technology has improved the physicochemical properties of fish oil-fortified yogurt. Nano-based materials for biological applications may potentially be produced using medications derived from microalgae. Obtained nanoparticles from seaweeds like *Cymodocea* sp. have maximum diameters between 29 and 60 nm, and marine fungi, for instance, Aspergillus flavus, showed cytotoxic and anticancer properties. A marine algal product called fucoxanthin is safe for humans, and its nano-suspension delivery may enhance the efficacy of supplements. Nanostructures often provide a larger contact surface area and may provide an effectively controlled release profile. To improve health, researchers are also searching for novel sources of nanobioactive chemicals to include in functional food items (Table 11.2).

A major milestone has been reached in the development of nanoscale delivery systems to increase the bioavailability of chemicals derived from aquatic sources. Nanofibers, for instance, enhance the stability of fatty acids and amino acids in fish meat. The consumption and stability of vitamins are also crucial, and nicotinamide acid, pyridoxal, pyridoxine, and pyridoxamine can be successfully added to fish fillets using chitosan-based nanofibers and electrospun chitosan-based nanofibers that have been loaded with thymol. Similar to tryptophan or the B group of vitamins, EPA, DHA, and EAA are

TABLE 11.2 Marine-Bioinspired Metallic Nanoparticles Treating Infectious Diseases

Name of Marine-Derived Compound/Product	Name of NPs	Size Range of MNPs	Shape/ Morphology	Sources/Organism	Antimicrobial Types	References
Extracts	AgNPs	12 nm	Spherical	*Ulva rigida* *Cystoseiramyrica* *Gracilariafoliifera*	Antibacterial	Algotiml et al. (2022)
Extracts	SNPs	200–450 nm	Spherical	*Spirulina platensis*	Antibacterial	Raja et al. (2022)
Extracts	SeNP	85 nm	Spherical	*U. lactuca*	Antibacterial	Vikneshan et al. (2020)
Extracts	AuNPs	Less than 20 nm	Spherical	*Acanthophoraspicifera*	Antibacterial	Babu et al. (2020)
Extracts	MgONPs	68.06 nm	Flower	*S. wighitii*	Antibacterial	Pugazhendhi et al. (2019)
Chitosan	Chitosan-AgNPs	4.5 ± 20.0–50.2 ± 74.0 nm	Spherical	*Aspergillus* sp.	Antibacterial	Mostafa et al. (2022)
Chitosan	Chitosan-AuNPs	3.47 ± 2.00–35.50 ± 2.00 nm	Spherical	*Alternaria* sp.	Antibacterial	
Chitosan	AgNPs	5–20 nm	Spherical	Marine Seafood	Antibacterial	Alharthi et al. (2022)
Extracts	ZnONPs	23–200 nm	Rectangular	*C. crinita*	Antibacterial	Elrefaey et al. (2022)
Extracts	AgNPs	12 nm	Spherical	*U. rigida*	Antifungal	Algotiml et al. (2022)
Extracts	AgNPs	5.52–35.00 nm	Spherical	*Streptomyces* sp. *R. rhodochrous*	Antiviral	Alam et al. (2021)

TABLE 11.2 *(Continued)*

Name of Marine-Derived Compound/Product	Name of NPs	Size Range of MNPs	Shape/ Morphology	Sources/Organism	Antimicrobial Types	References
Extracts	AgNPs	30–80 nm	Circular	*Nocardiopsis dassonvillei-DS013*	Antibacterial	Dhanaraj et al. (2020)
Extracts	AgNPs	11–21 nm	Cubic	*Streptomyces* sp. *Al-Dhabi-87*	Antibacterial	Al-Dhabi et al. (2022)

Source: Adapted from Šimat et al., 2020.

crucial marine-based substances for human health. Probiotic bacteria are crucial for the human immune system and food systems. Consuming fish meat that has been coated with nanoprobiotics (*L. reuteri* & *L. rhamnosus*) might offer an alternative method. The fatty acids in fish meat were more consistently stable thanks to these nanoprobiotics. Algae, on the other hand, are excellent sources of a variety of essential fatty acids, including -3 and -6, which are crucial for the immune system. Additionally, the anti-diabetic and anti-inflammatory properties of gold nanoparticles made from brown algae seaweed are used in medical applications.

Applications of marine-based goods in nanotechnology might be combined with other probiotic-rich dietary ingredients. For instance, yogurt supplemented with fish oil now has better physicochemical qualities thanks to nanoemulsion technology. The synthesis of nano-based materials might be accomplished using microalgae-based medications and their medicinal uses. For instance, nanoparticles with maximum sizes of 29 and 60 nm isolated from seaweeds like *Cymodocea* sp. and marine fungi like *Aspergillus flavus* showed strong anticancer and cytotoxic action. Fucoxanthin, a marine algal product, has been determined to be safe for people, and its distribution by nanosuspension may improve the effectiveness of supplements. Nanostructures often provide a greater surface area of contact and may offer an excellent regulated release profile. As a result, scientists are searching for fresh sources of nanobioactive compounds that can be added to functional food items to promote health.

KEYWORDS

- **bioactive compounds**
- **functional foods**
- **gastrointestinal tract**
- **marine bio-actives**
- **marine drug**
- **marine resources**
- **nutraceuticals**
- **reactive oxygen species**

REFERENCES

Alam, A., Tanveer, F., Khalil, A. T., Zohra, T., Khamlich, S., Alam, M. M., Salman, M., Ali, M., Ikram, A., Shinwari, Z. K., et al. (2021). Silver nanoparticles biosynthesized from secondary metabolite producing marine actinobacteria and evaluation of their biomedical potential. *Antonie Van Leeuwenhoek, 114*(10), 1497–1516.

Al-Dhabi, N. A., Ghilan, A. K. M., Arasu, M. V., & Duraipandiyan, V. (2018). Green biosynthesis of silver nanoparticles produced from marine *Streptomyces* sp. Al-Dhabi-89 and their potential applications against wound infection and drug resistant clinical pathogens. *Journal of Photochemistry and Photobiology. B, Biology, 189*, 176–184.

Algotiml, R., Gab-Alla, A., Seoudi, R., Abulreesh, H. H., El-Readi, M. Z., & Elbanna, K. (2022). Anticancer and antimicrobial activity of biosynthesized Red Sea marine algal silver nanoparticles. *Scientific Reports, 12*(1), 2421.

Alharthi, S. S., Gomathi, T., Joseph, J. J., Rakshavi, J., Florence, J. A. K., Sudha, P. N., Rajakumar, G., & Thiruvengadam, M. (2022). Biological activities of chitosan-salicylaldehyde Schiff base assisted silver nanoparticles. *Journal of King Saud University – Science, 34*(1), 102177.

AnakeKijjoa, & PichanSawangwong. (2004). Drugs and cosmetics from the sea. *Marine Drugs, 2*, 73–82.

Babu, B., Palanisamy, S., Vinosha, M., Anjali, R., Kumar, P., Pandi, B., Tabarsa, M., You, S., & Prabhu, N. M. (2020). Bioengineered gold nanoparticles from marine seaweed *Acanthophoraspicifera* for pharmaceutical uses: Antioxidant, antibacterial, and anticancer activities. *Bioprocess and Biosystems Engineering, 43*(12), 2231–2242.

Centella, M. H., Arévalo-Gallegos, A., Parra-Saldivar, R., & Iqbal, H. M. N. (2017). Marine-derived bioactive compounds for value-added applications in bio- and non-bio sectors. *Journal of Cleaner Production, 168*, 1559–1565.

Cunha, S. A., & Pintado, M. E. (2021). Bioactive peptides derived from marine sources: Biological and functional properties. *Trends in Food Science & Technology, 119*, 348–370.

Dewapriya, P., & Kim, S. K. (2014). Marine microorganisms: An emerging avenue in modern nutraceuticals and functional foods. *Food Research International, 56*, 115–125.

Dhanaraj, S., Thirunavukkarasu, S., Allen John, H., Pandian, S., Salmen, S. H., Chinnathambi, A., & Alharbi, S. A. (2020). Novel marine *Nocardiopsis dassonvillei*-DS013 mediated silver nanoparticles characterization and its bactericidal potential against clinical isolates. *Saudi Journal of Biological Sciences, 27*(4), 991–995.

Domínguez, H. (2013). Algae as a source of biologically active ingredients for the formulation of functional foods and nutraceuticals. In H. Dominguez (Ed.), *Functional Ingredients from Algae Foods Nutraceuticals*, 1–19. Amsterdam: Elsevier.

El-Naggar, M. M., Haneen, D. S. A., Mehany, A. B. M., & Khalil, M. T. (2020). New synthetic chitosan hybrids bearing some heterocyclic moieties with potential activity as anticancer and apoptosis inducers. *International Journal of Biological Macromolecules, 150*, 1323–1330.

Elrefaey, A. A. K., El-Gamal, A. D., Hamed, S. M., & El-belely, E. F. (2022). Algae-mediated biosynthesis of zinc oxide nanoparticles from *Cystoseira crinite* (*Fucales; Sargassaceae*) and its antimicrobial and antioxidant activities. *Egyptian Journal of Chemistry, 65*, 231–240.

Gerwick, W. H., & Moore, B. S. (2012). Lessons from the past and charting the future of marine natural products drug discovery and chemical biology. *Chemistry & Biology, 19*(1), 85–98.

Guerin, M., Huntley, M. E., & Olaizola, M. (2003). *Haematococcus astaxanthin*: Applications for human health and nutrition. *Trends in Biotechnology, 21*(5), 210–216.

Gurpilhares, D. D. B., Cinelli, L. P., Simas, N. K., Pessoa, A., & Sette, L. D. (2019). Marine prebiotics: Polysaccharides and oligosaccharides obtained by using microbial enzymes. *Food Chemistry, 280*, 175–186.

Jing, J., Wang, L., & Hao, W. (2011). Bio-function summary of marine oligosaccharides. *International Journal of Biology, 3*, 74–86.

Kim, Y. K., Lim, C. W., & Yeun, S. (2013). Dipeptide (Tyr-Ile) acting as an inhibitor of angiotensin-I-converting enzyme (ACE) from the hydrolysate of jellyfish *Nemopilema nomurai*. *Fisheries and Aquatic Sciences, 14*(4), 283–288.

Kochkina, Z. M., & Chirkov, S. N. (2000). Effect of chitosan derivatives on the reproduction of coliphages T2 and T7. *Microbiology, 69*(2), 208–211.

Li, L. Y., Sattler, I., Deng, Z. W., Groth, I., Walther, G., Menzel, K. D., Peschel, G., Grabley, S., & Lin, W. H. (2008). A Seco-oleane-type triterpenes from Phomopsis sp. (strain HK10458) isolated from the mangrove plant *Hibiscus tiliaceus*. *Phytochemistry, 69*, 511–517.

Marine-Derived Drugs Market Growing at a CAGR of 11.20% and Expected to Reach $21,955.6 Million by 2025—Exclusive Report by Infinium Global Research. (2019). Infinium Global Research. https://www.infiniumglobalresearch.com/reports/global-marine-derived-drugs-market (accessed on 15 October 2020).

Mostafa, E. M., Abdelgawad, M. A., Musa, A., Alotaibi, N. H., Elkomy, M. H., Ghoneim, M. M., Badawy, M. S. E. M., Taha, M. N., Hassan, H. M., & Hamed, A. A. (2022). Chitosan silver and gold nanoparticle formation using endophytic fungi as powerful antimicrobial and anti-biofilm potentialities. *Antibiotics, 11*, 668.

Nagaoka, I., Igarashi, M., Hua, J., Ju, Y., Yomogida, S., & Sakamoto, K. (2011). Recent aspects of the anti-inflammatory actions of glucosamine. *Carbohydrate Polymers, 84*, 825–830.

Noyer, C., Thomas, O. P., & Becerro, M. A. (2011). Patterns of chemical diversity in the Mediterranean sponge *Spongia lamella*. *PLoS One, 6*(4), e20844.

Pooja, K., Rani, S., Rana, V., & Pal, G. K. (2020). Aquatic plants as a natural source of antimicrobial and functional ingredients. Functional and Preservative Properties of Phytochemicals, edited by Bhanu Prakash 2020, pp. 93–118 Amsterdam: Elsevier. https://doi.org/10.1016/B978–0-12–818593–3.00003–8.

Pugazhendhi, A., Prabhu, R., Muruganantham, K., Shanmuganathan, R., & Natarajan, S. (2019). Anticancer, antimicrobial, and photocatalytic activities of green synthesized magnesium oxide nanoparticles (MgONPs) using aqueous extract of *Sargassum wightii*. *Journal of Photochemistry and Photobiology B: Biology, 190*, 86–97.

Raja, S., Anjali, G., Anitha, J., Ephsy, D., Sahana, K., & Mohan, K. (2022). Optimization of silver nanocrystals reduced from the functional molecules enriched spirulina—A potent antibiotic against human and marine pathogen vibrio parahaemolyticus. Research Square: Durham, NC, USA., pp. 1–26. https://doi.org/10.21203/rs.3.rs-1611077/v2.

Rajasekaran, A., Sivagnanam, G., & Xavier, R. (2008). Nutraceuticals as therapeutic agents: A review. *Research Journal of Pharmacy and Technology, 1*, 328–340.

Raposo, M. F. J., Santos, R. M., Costa de Morais, A. M., & Miranda Bernardo de Morais, A. (2013). Health applications of bioactive compounds from marine microalgae. *Life Sciences, 93*(15), 479–486.

Research and Markets: Nutraceuticals—2012, Global strategic business report annual estimates and forecasts for 2010–2018. https://www.businesswire.com/news/home/20120920005987/en/Research-and-Markets-Nutraceuticals---2012-Global-Strategic-Business-Report---Annual-Estimates-and-Forecasts-for-2010–2018.

Resmi, R., Yoonus, J., & Beena, B. (2020). Anticancer and antibacterial activity of chitosan extracted from shrimp shell waste. *Materials Today Proceedings*, *41*(3), 2021, 570–576.

Romano, G., Costantini, M., Sansone, C., Lauritano, C., Ruocco, N., & Ianora, A. (2017). Marine microorganisms as a promising and sustainable source of bioactive molecules. *Marine Environmental Research, 128*, 58–69.

Ruiz-Ruiz, F., Mancera-Andrade, E. I., & Iqbal, H. M. N. (2017). Marine-derived bioactive peptides for biomedical sectors: A review. *Protein and Peptide Letters, 24*, 109–117.

Sedghi, R., Gholami, M., Shaabani, A., Saber, M., & Niknejad, H. (2020). Preparation of novel chitosan derivative nanofibers for prevention of breast cancer recurrence. *European Polymer Journal, 123*, 109421.

Shibata, T., Ishimaru, K., Kawaguchi, S., Yoshikawa, H., & Hama, Y. (2008). Antioxidant activities of phlorotannins isolated from Japanese Laminariaceae. *Journal of Applied Phycology, 20*, 705–711.

Suleria, H. A. R., Gobe, G., Masci, P., & Osborne, S. A. (2016). Marine bioactive compounds and health promoting perspectives; innovation pathways for drug discovery. *Trends in Food Science & Technology, 50*, 44–55.

Suleria, H. A. R., Osborne, S., Masci, P., & Gobe, G. (2015). Marine-based nutraceuticals: An innovative trend in the food and supplement industries. *Marine Drugs, 13*(10), 6336–6351.

Vidanarachchi, J. K., & Kurukulasuriya, M. (2011). Industrial applications of marine cosmeceuticals. In *Marine Cosmeceuticals: Latest Trends and Prospects*, 1–36. Taylor & Francis.

Vidanarachchi, M. S., Kurukulasuriya, W. M. N. M. W., Eaton, P., Fernandes, J. C., & Pereira, E. (2008). Atomic force microscopy study of the antibacterial effects of chitosans on *Escherichia coli* and *Staphylococcus aureus*. *Ultramicroscopy, 108*(10), 1128–1134.

Vikneshan, M., Saravanakumar, R., Mangaiyarkarasi, R., Rajeshkumar, S., Samuel, S. R., Suganya, M., & Baskar, G. (2020). Algal biomass as a source for novel oral nano-antimicrobial agent. *Saudi Journal of Biological Sciences, 27*(12), 3753–3758.

Voutilainen, S., Nurmi, T., & Mursu, J. (2006). Carotenoids and cardiovascular health. *American Journal of Clinical Nutrition, 83*(6), 1265–1271.

Wang, X., Yuen, K. F., Wong, Y. D., & Li, K. X. (2020). How can the maritime industry meet sustainable development goals? An analysis of sustainability reports from the social entrepreneurship perspective. *Transportation Research Part D: Transport and Environment, 78*, 102173.

Zhang, C., & Kim, S. K. (2012). Application of marine microbial enzymes in the food and pharmaceutical industries. *65*, 423–435. doi: 10.1016/B978–0-12–416003–3.00028–7. PMID: 22361204.

CHAPTER 12

Insights on Phytochemicals in Macroalgae (Seaweeds) in the Treatment of Cardiovascular Diseases

KAMRAN ASHRAF,[1] HANNIS FADZILLAH MOHSIN,[1] ZOLKAPLI ESHAK,[1] WASIM AHMAD,[2] and MD. FAIYAZUDDIN[3,4]

[1]*Faculty of Pharmacy, Universiti Teknologi MARA (UiTM), Cawangan Selangor, Kampus Puncak Alam, Bandar Puncak Alam, Selangor Darul Ehsan, Malaysia*

[2]*Department of Pharmacy, Mohammed Al-Mana College for Medical Sciences, Dammam, Saudi Arabia*

[3]*School of Pharmacy, Al-Karim University, Katihar, Bihar, India*

[4]*Nano Drug Delivery, Raleigh-Durham, NC, USA*

ABSTRACT

In Southeast Asia, seaweed is used as part of a traditional diet, and it could prevent several chronic diseases including cardiovascular diseases. Cardiovascular disease (CVD), which entails the emergence and aggravation of several circumstances, including dyslipidemia, renin-angiotensin system activation, vascular endothelial cell damage, and oxidative stress, is a significant factor in high mortality rates, accounting for one-third of all fatalities globally. As a result, numerous studies have been conducted in the past to explore the potential utility of foods, especially seaweeds, and their bioactive components for treating and preventing CVD. This review highlights research that focused on the therapeutic effects of seaweed and its natural compounds for treating and preventing CVD. The results of this

Handbook of Research in Marine Pharmaceutics: Exploring Oceanic Microbial Diversity for Human Health and Wellness. Md. Faiyazuddin & Khalid Rehman Hakeem (Eds.)

review analysis can be used as a foundation for future research in the field of marine pharmaceuticals.

12.1 INTRODUCTION

Cardiovascular diseases (CVDs) are conditions that affect the heart and vascular system, such as ischemic heart issues like myocardial infarction. Conditions affecting the heart or blood vessels are collectively referred to as cardiovascular disease (CVD). Atherosclerosis, a buildup of fatty deposits inside the arteries, and a higher risk of blood clots are typically connected with it. It may also be linked to artery damage in several organs, including the kidneys, eyes, brain, and heart (Roth et al., 2015). Diabetes and obesity both increase the risk of cardiovascular disease and cause arteriosclerosis, which may result in CVD (Murray et al., 1997).

One of the risk factors for cardiovascular disorders like atherosclerosis and coronary artery disease is hypercholesterolemia, which is characterized as a high level of LDL cholesterol in blood plasma. By damaging the artery endothelium and inducing further vascular tissue alterations that result in cholesterol buildup, hypertension causes atherosclerosis (Mayer et al., 2019; Trajkovska et al., 2017).

According to the National Health and Nutrition Examination recent Survey, the prevalence of CVD including coronary heart disease (CHD), heart failure (HF), and hypertension among those over 20 years of age is approximately 50%. According to the World Health Organization, CVD accounts for 32% of all global deaths. This might be the leading cause of global mortality as well as a significant contributor to disability in the future. Because of this, there is growing worry that CVD may raise the risk of COVID-19 severe symptoms (Ryu et al., 2021).

Sea macroalgae has long been recognized for its health and nutritional benefits. This serves a remarkable significance as a key functional food in diets all over the world as they are abundant and potentially renewable resources (Barbosa et al., 2014).

Due to its positive benefits in extending life expectancy and preventing CVDs, seaweed has recently been one of the most well-liked foods in various Western diets. Seaweed is traditionally eaten as food in Southeast Asian regions including China, Japan, and South Korea (Seca et al., 2018; Brown et al., 2014).

A healthy eating profile like seaweed has been associated with a lower risk of cardiovascular disease and its risk factors (Déléris et al., 2016). Seaweeds contain many nutrients, ions, antioxidants, polyphenols, and proteins, all of which are important components for avoiding CVD. These are important in the reduction of the incidence of numerous CVD risk factors such as obesity, hyperlipidemia, diabetes, hypertension, and metabolic syndrome (Peñalver et al., 2020).

Seaweeds also contain lipids, polyunsaturated fatty acids, and a significant amount of the omega-3 and omega-6 fatty acids eicosapentaenoic acid, and arachidonic acid, which are helpful in lowering cholesterol levels in the body (Holdt et al., 2011).

There is scope for further diversification of the natural substances obtained from marine seaweeds, which have made major contributions to modern medicine (Sigwart et al., 2021). Still, macroalgae therapeutic properties are limited to conventional therapies, and their bioactive potential remains untapped (Paiva et al., 2017).

This review showcases studies that looked into the therapeutic effects of seaweed and its natural compounds for treating and preventing CVD. The findings of this review analysis could serve as a basis for future research in the marine pharmaceuticals field.

12.2 BIOACTIVE CONSTITUENTS OF SEAWEEDS USEFUL IN THE TREATMENT OF CVD

Many bioactive compounds obtained from algae have shown a number of significant biological effects, such as antihypertensive and antioxidant, which are closely related to the chemical structure of the peptides or carbohydrates and others. A wide range of biological activity has been shown by a number of bioactive compounds made from seaweed. The structurally and physiologically distinct compounds produced by sea microorganisms are thought to have a wide range of pharmaceutical applications (Zhao et al., 2018; Jesumani et al., 2021). Below is an overview of several phytochemicals found in seaweeds that may be useful in the treatment of CVD.

12.2.1 POLYPHENOLS

Seaweeds particularly green and red algae were reported rich in polyphenols, such as flavonols, fluorotannins, and catechins. Epidemiological, clinical, and nutritional studies support the hypothesis that dietary polyphenols play crucial roles in human health and that regular intake can decrease the risk of various chronic diseases, including cancer, metabolic and neurodegenerative disorders, and CVDs (Gómez-Guzmán et al., 2018).

Seaweed contains valuable resources of polyphenolic compounds such as flavonols, catechins, and phlorotannins. Phenolic acids, flavonoids, and bromophenols constitute the primary phenolic compounds found in green and red algae. Phlorotannins, the primary polyphenolic compounds that consist of a series of complex polymers of phloroglucinol (1,3,5-trihydroxybenzene), are identified only in marine brown algae (Eom et al., 2012; Sivagnanam et al., 2015; Tibbetts et al., 2016). Phlorotannins can be classified into six subclasses: eckols, fuhalols, fucophlorethols, phlorethols, fucols, and ishofuhalols (Figure 12.1). Phlorotannins have drawn the attention of pharmaceutical companies as the main targets of phlorotannin supplements for cardiovascular diseases. The focus is generally on arteriosclerosis prevention and the increase of protective high-density lipoprotein cholesterol (HDL-C) (Cotas et al., 2020).

A compound known as (+)-3-(2,3-dibromo-4,5-dihydroxy-phenyl)-4-bromo-5,6-dihydroxy-1,3-dihydroiso-benzofuran isolated from *L. marina* exhibited inhibitory action against thrombin through *in vitro* and *in vivo* studies. The compound has the potential to treat cardiovascular diseases, such as thrombosis and thromboembolism, in which thrombin plays a key role in disease development. Thrombin is a proteinase that plays a key role in procoagulant factors (Figure 12.2).

12.2.2 POLYSACCHARIDES

Carrageenans are a class of biomolecules made up of linear polysaccharide chains with sulfate half-esters linked to the sugar unit. There are three general forms (kappa, lambda, and iota), each with its gelling property (Figure 12.3). Because of the biomolecules group, which consists of linear polysaccharide chains with sulfate half esters linked to the sugar unit, it dissolves readily in water. Carrageenan is more commonly used than agar as an emulsifier and stabilizer in many food products. It could potentially be used as a promising pharmaceutical for its anti-tumor, anti-viral, anti-coagulant, and immunomodulation activities (Pal et al., 2014). It has also been used in pharmaceutical products as suspension agents and stabilizers in drugs, lotions, and creams.

Brown seaweed contains fucoidan, a polysaccharide composed of L-fucose. The type of seaweed and the methods of extraction have a significant impact on the chemical structure. The most studied and commercially available form of fucoidan is prepared from *Fucus vesiculosus*, which comprises 44% fucose and 26% sulfate. This compound contains a linear backbone of (1→3)-α-L-Fucose or alternating (1→3)-α-L-Fucose and

(1→4)-α-L-Fucose, whereas the sulfate groups dominate the C-2, C-3, and C-4 of fucose (Deniaud-Bouët et al., 2017). It was reported that fucoidan is responsible for atherosclerosis by alteration in lipid uptake and metabolism (Patil et al., 2018).

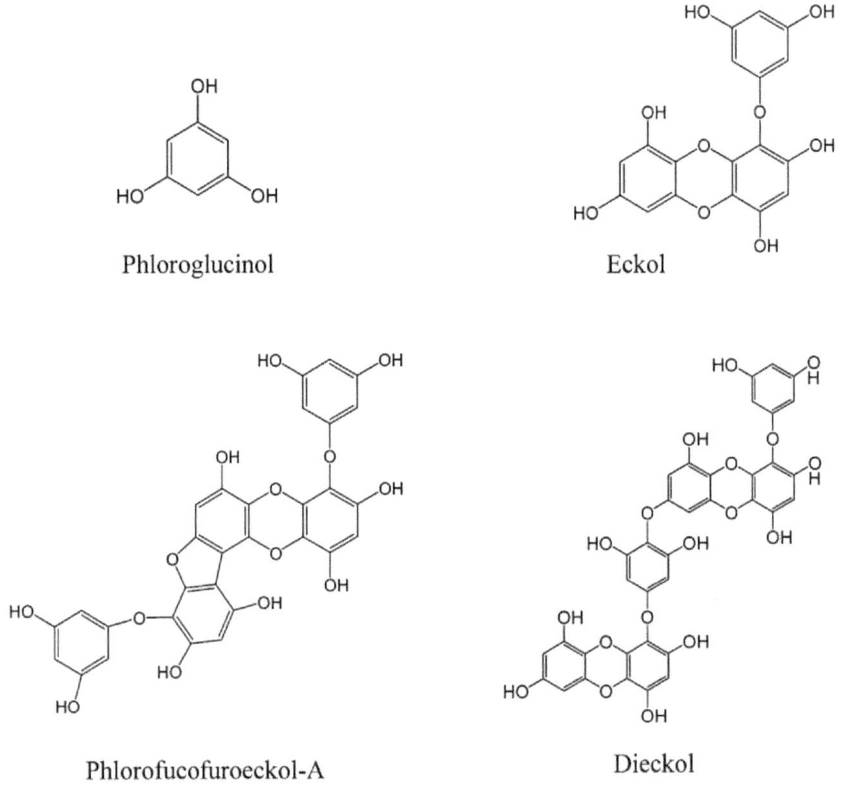

FIGURE 12.1 Chemical structures of some phlorotannins; phloroglucinol, Eckol, phlorofucofuroeckol-A, and dieckol.

FIGURE 12.2 Chemical structures of basic bromophenols.

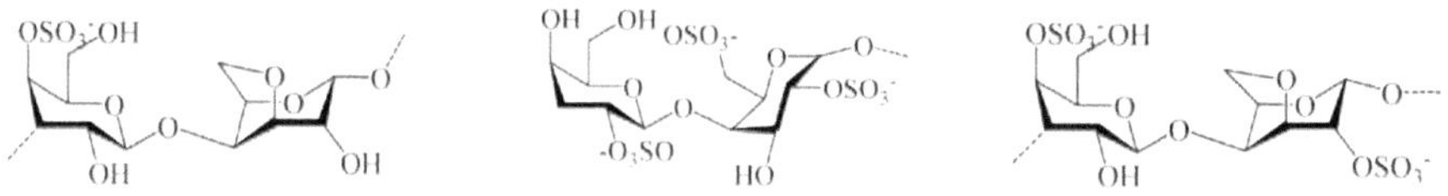

FIGURE 12.3 Basic chemical structures of different types of carrageenan.

Alginate is a polymer derived from brown seaweed that has been widely utilized in biotechnology. Alginate has been used for many years in the food and beverage industry as a thickener, gelling agent, and colloidal stabilizer. Alginates are linear block copolymers composed of 1,4-linked β-D-mannuronic acid (M) with 4C1 ring conformation and α-L-guluronic acid (G) with 4C1 conformation, both in the pyranosic conformation and present in varying amounts (Patil et al., 2018). Numerous *in vivo* studies have demonstrated that different types of alginates induce hypo-cholesterolemic and hypolipidemic responses. It was demonstrated that the alginates extracted from the brown seaweed from Yogyakarta, known as *Sargassum duplicatum* and *Turbinaria* sp., could reduce total blood cholesterol levels in hypercholesterolemic rats that had been pretreated.

Another polysaccharide, namely rhamnan sulfate, was found to be the primary bioactive compound of *Monostroma Nitidum*. Structurally, L-rhamnose sugars are linked through the α-1,3 carbons. The anticoagulant effect of the chemical has been explored to some extent, although there has been limited research on its applicability in cardiovascular disorders.

Laminarin is a polysaccharide from brown algae composed of D-glucopyranose connected through 1,3-glycosidic linkages with an average length of 20–25 disaccharides (Nelson & Lewis, 1974). Laminarin was reported to be abundant in the seaweed from the genus Laminaria.

The sulfated type of laminarin was found to have a more significant effect on atherosclerosis risk (Patil et al., 2018) (Figure 12.4).

12.2.3 CAROTENOIDS

Balasubramaniam et al. (2020) investigated the carotenoid pattern of seaweed by employing LC-MS/MS approach. Based on their fragmentation pattern from collision-induced dissociation on LC-MS/MS analysis, various carotenoids from different seaweeds were explored. Around 12 different forms of carotenoids were discovered from the ethanolic extract

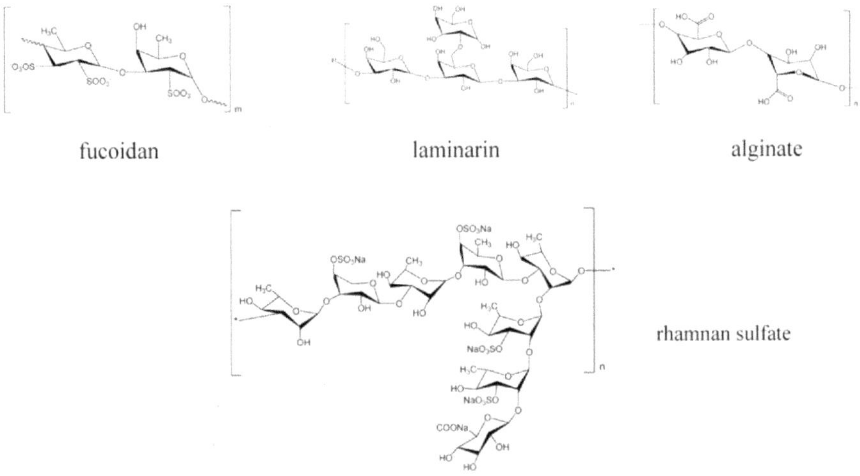

FIGURE 12.4 Polysaccharides derived from various seaweeds.

of various seaweeds based on the molecular mass (m/z). As shown in the study, *C. lentillifera* has high concentrations of β-carotene and canthaxanthin, which were reported to be 4.6 mg/100 g and 19.5 mg/100 g of its dried weight, respectively (Figure 12.5). Zeaxanthin, another prevalent carotenoid, is present in *C. lentillifera*, as well as other varieties of green seaweed and higher plants. As found in *C. lentillifera*, green seaweed has an abundance of xanthophylls such as canthaxanthin and astaxanthin (Balasubramaniam et al., 2020). Another research on *C. lentillifera* discovered that the growth rate and mineral content increased for indoor cultivation, but its carotenoid and fiber content were higher when grown in the sea (Syamsuddin et al., 2019).

Seaweed such as *E. cottonii* and *C. lentifera* can be a source of vitamin C and α-tocopherol, which is able to inhibit LDL oxidation and thromboxin formation (Matanjun et al., 2009). On the other hand, beta-carotene content varies in different species of seaweed. Brown seaweeds contain high levels of carotenoids, including fucoxanthin, beta-carotene, pheophytins, and violaxanthin, however, green and red seaweeds merely have α-tocopherol. Fucoxanthin from the methanol extract of *S. subrepandum* shows the improvement of lipid profile with the oral administration of the extract in a dose (Cardoso et al., 2015). Foods high in antioxidative carotenoids lower the risk of cardiovascular disease, cancer, and other eye-related diseases (Mohamed et al., 2012).

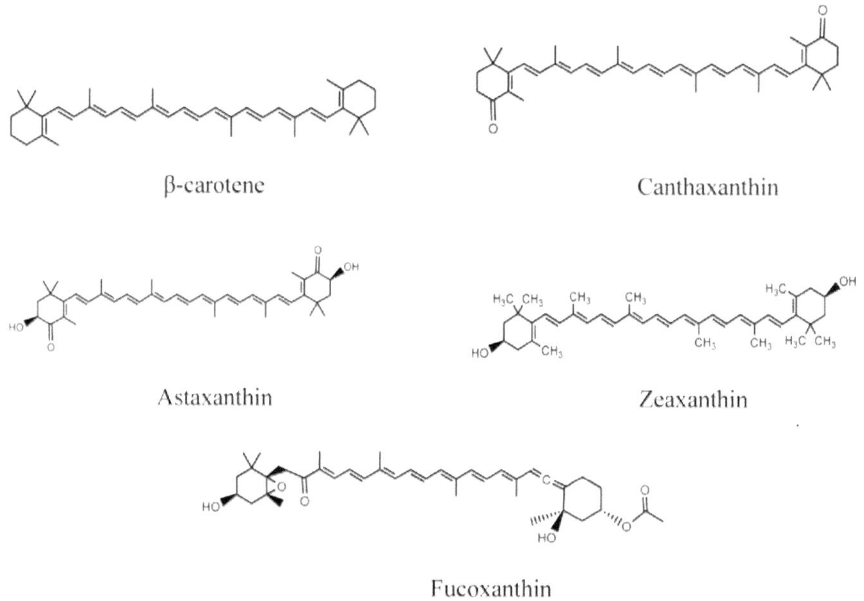

FIGURE 12.5 Various carotenoid chemical structures were identified from seaweed.

12.2.4 *FATTY ACIDS*

It has been established that, from a dietary perspective, a balanced ratio of omega-3 to omega-6 PUFA is essential for the prevention of chronic diseases, including cardiovascular disease. Although seaweeds are low in total lipids, they may still contain polyunsaturated fatty acids. Eicosapentaenoic acid, an omega-3 fatty acid, is the most abundant fatty acid in *K. alvarezii*, making up 24.98% of all fatty acids (Matanjun et al., 2010). It is stated that *S. polycystum* has the highest cardioprotective effect and effectively reduces body weight gain, followed by *C. lentillifera* and lastly *K. alvarezii*. However, *K. alvarezii* and *C. lentillifera* showed better hypolipidemic effects than *S. polycystum* and increased plasma HDL-C to normal levels. For total hypocholesterolemic effects, *C. lentillifera* exhibited the highest effect compared to *K. alvarezii* and *S. polycystum*.

Among the fatty acid contents, there were 15 different forms of saturated fatty acids, 9 different types of monounsaturated fatty acids, and 11 different types of polyunsaturated fatty acids, which varied from C8 to C24 (PUFAs). Polyunsaturated fatty acids like omega-3 and omega-6, which are powerful antioxidants, are found in much higher concentrations in *C. lentillifera* than in terrestrial plants (Mendis & Kim, 2011). Both environmental and genetic variations are believed to have contributed to the observed disparities in the levels of fatty acid content.

The methods for lowering cholesterol-lowering effects of increasing bile acid or other digestive enzyme excretion is one of the benefits of seaweeds. Bound substances delay the digestion and absorption of carbs, and lipids, and result in a rise in probiotic microorganisms' mass, in addition to the creation of short-chain fatty acids. Dietary cholesterol and polyionic polysaccharides form a binding complex, facilitating their fecal output, and leading to lower blood cholesterol levels. Some seaweeds like alginic acid (e.g., funoran and carrageenan) enhance fecal cholesterol excretion by creating an unpalatable ionic colloid.

12.2.5 STEROLS

Sterols are an essential chemical constituent of microalgae and the most significant nutritional component of seaweed. Sterols predominate in plant, animal, and fungal cells, with cholesterol being the most well-known animal sterol. Cholesterol is essential for cellular activity and regulates the fluidity of animal cell membranes; it also functions as a secondary signaling agent during development. Cholesterol functions as a precursor for fat-soluble vitamins and steroid hormones (Pal et al., 2014). Different types of seaweed have different sterols, including green seaweeds having 28 isomers of ucocholesterol, cholesterol, 24 methylene cholesterol, and β-sitosterol, and brown seaweeds containing fucosterol, cholesterol, and brassicasterol. Red marine algae possess sterols such as desmosterol, cholesterol, sitosterol, fucosterol, and chalinasterol (Sánchez-Machado et al., 2004). The red seaweed *M. afaqhusainii*, which contains 0.46–0.01% sterols, also exerted lipid-lowering effects. It was reported that cholesterol is the most commonly occurring sterol in red seaweed and can reduce blood cholesterol levels (Bhakuni & Silva, 1974). In addition, another study also mentions that the non-toxic sterols of red algae can lower blood cholesterol and fat accumulation in the heart and liver (Ryu et al., 2021). According to reports, plant sterols such as β-sitosterol and fucosterol (Figure 12.6) reduce the levels of cholesterol in animal and human serum experimentally.

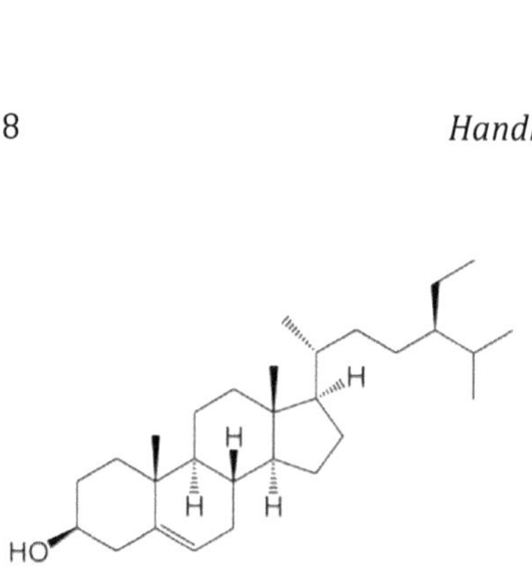

β-sitosterol

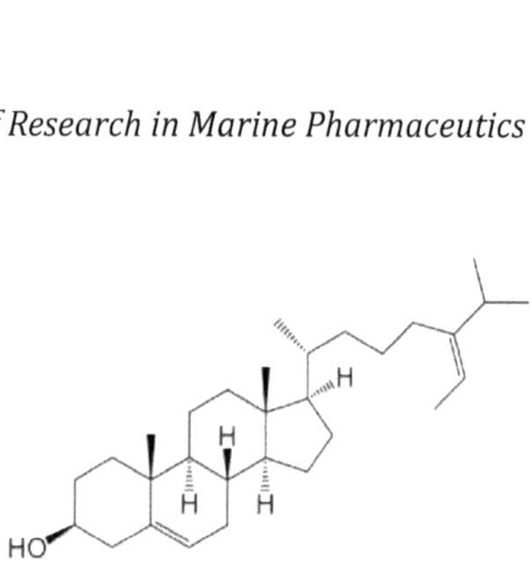

fucosterol

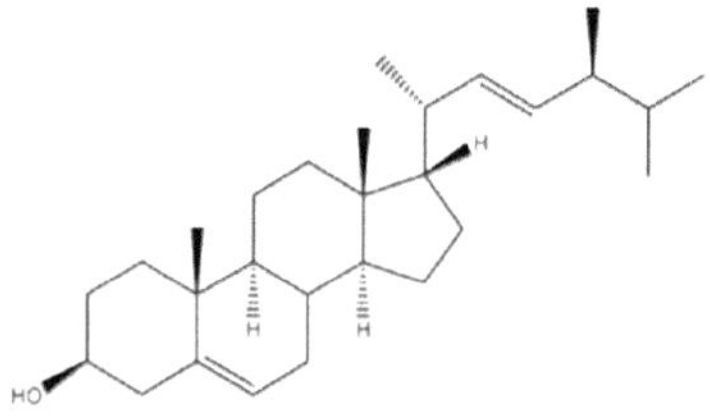

Ergosterol

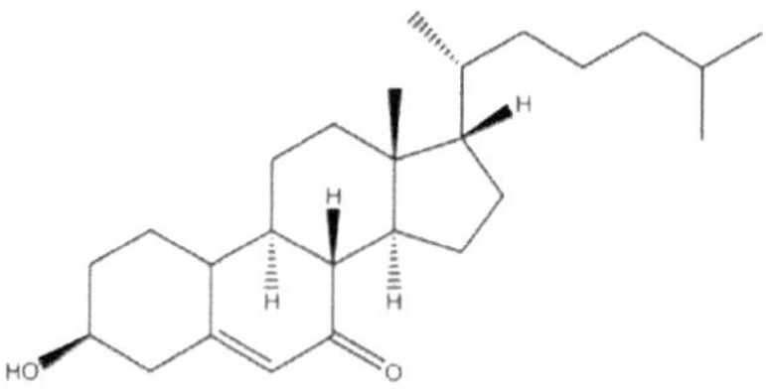

7-oxocholestrol

FIGURE 12.6 Structures of some sterol identified from seaweed.

12.2.6 ALKALOIDS

Alkaloids are nitrogenous compounds with a variety of biological activities (Fernando et al., 2016; Souza et al., 2020). Compared to other compound classes, the chemical properties of alkaloids, such as their ionization ability and capability to form stable salts, may suggest that they have a higher potential as drug candidates for many diseases. Caulerpin, an indolic alkaloid with a reported anti-inflammatory effect, is prevalent in the algae extracts of the genus *Caulerpa* (Lucena et al., 2018). Caulerpin has been discovered in many species of the genus *Caulerpa*, such as *C. peltata, C. racemosa, C. cupressoides, C. paspaloides, C. prolifera, C. sertularioides, C. mexicana,* as well as *C. lentillifera* (Lunagariya et al., 2019). Racemosin A, B, and C, as well as caulersin (Figure 12.4), are additional Caulerpa indolic alkaloids that have been discovered in algae (Liu et al., 2013; Lucena et al., 2018; Lunagariya et al., 2019; Souza et al., 2020) (Figure 12.7).

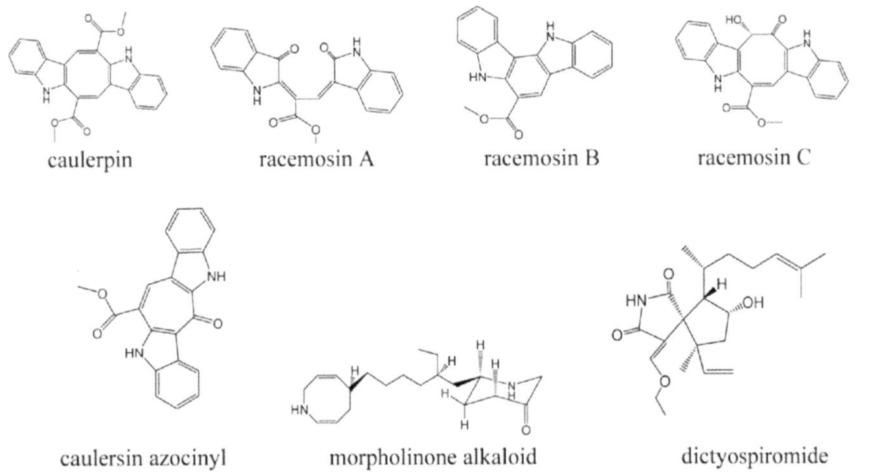

FIGURE 12.7 Marine indole alkaloids are derived from algae.

12.2.7 PROTEIN AND AMINO ACID

Individually, seaweed proteins are high in aspartic and glutamic amino acids, whereas their cysteine concentration is moderate (Fernando et al., 2016). The protein content of different kinds of seaweed varies. The levels of proteins are higher in *Rhodophyta* (red) compared to *Chlorophyta* (green) and *Ochrophyta* (brown) (Mohamed et al., 2012). The protein content of *C. lentillifera* ranged from 0.43 to 19.38% and could be affected by various external factors, such as water temperature, season, geography, weather, and other factors (Rajapakse & Kim, 2011). It was reported that the seaweed's protein content was higher in the winter than in the autumn and summer. The free amino acid fragment of seaweed consists primarily of taurine, alanine, amino butyric acid, ornithine, citrulline, and hydroxyproline (Holdt & Kraan, 2011). Taurine, an amino acid abundant in red algae that can function as an antioxidant, has been demonstrated to reduce the release of serum lipids and apolipoprotein B100, a structural component of low-density lipoproteins, hence decreasing the risk of atherosclerosis and coronary heart disease (Mendis & Kim, 2011).

Almost all essential amino acids present in *C. lentillifera* are deemed to have high-quality proteins and the major ones are aspartic and glutamic acid, which contribute to the umami flavor. The highest essential amino acid was phenylalanine, while the amino acid profile consisted of aspartic acid, glutamic acid, serine, glycine, histidine, arginine, threonine, alanine, proline, tyrosine, valine, methionine, isoleucine, leucine, phenylalanine, and lysine (Matanjun et al., 2009).

12.3 CONCLUSION

Seaweeds show various positive benefits to human health because they contain a variety of bioactive chemicals. In this review, the prevention of CVDs was discussed, and it was observed that numerous elements of seaweed have a significant impact on lowering blood cholesterol. However, longer-term human study is required to assess the potency of these seaweed components against CVD.

ACKNOWLEDGMENTS

The authors would also like to thank Universiti Teknologi MARA for the financial support under reference no 600-UiTM/SEL (PI.5/4) (019/2022).

KEYWORDS

- **bioactive compounds**
- **cardiovascular disease**
- **macroalgae**
- **polyunsaturated fatty acids**
- **seaweeds**

REFERENCES

Balasubramaniam, V., June Chelyn, L., Vimala, S., Mohd Fairulnizal, M. N., Brownlee, I. A., & Amin, I. (2020). Carotenoid composition and antioxidant potential of *Eucheuma denticulatum*, *Sargassum polycystum* and *Caulerpa lentillifera*. *Heliyon, 6*(8), e04668.

Barbosa, M., Valentão, P., & Andrade, P. B. (2014). Bioactive compounds from macroalgae in the new millennium: Implications for neurodegenerative diseases. *Marine Drugs, 12*(9), 4934–4972.

Bhakuni, D. S., & Silva, M. (1974). Biodynamic substances from marine flora. *Botanica Marina. XVII*, 40–51.

Brown, E. S., Allsopp, P. J., Magee, P. J., Gill, C. I., Nitecki, S., Strain, C. R., & McSorley, E. M. (2014). Seaweed and human health. *Nutrition Reviews, 72*, 205–216.

Cotas, J., Leandro, A., Monteiro, P., Pacheco, D., Figueirinha, A., Gonçalves, A. M. M., Da Silva, G. J., & Pereira, L. (2020). Seaweed phenolics: From extraction to applications. *Marine Drugs, 18*(8), 384.

Déléris, P., Nazih, H., & Bard, J. M. (2016). Seaweeds in human health. In J. Fleurence & I. Levine (Eds.), *Seaweed in Health and Disease Prevention* (pp. 319–367), 1st ed. Academic Press; London.

Deniaud-Bouët, E., Hardouin, K., Potin, P., Kloareg, B., & Hervé, C. (2017). A review about brown algal cell walls and fucose-containing sulfated polysaccharides: Cell wall context, biomedical properties and key research challenges. *Carbohydrate Polymers, 175*, 395–408.

Eom, S. H., Kim, Y. M., & Kim, S. K. (2012). Antimicrobial effect of phlorotannins from marine brown algae. *Food and Chemical Toxicology, 50*(9), 3251–3255.

Fernando, I. P. S., Nah, J. W., & Jeon, Y. J. (2016). Potential anti-inflammatory natural products from marine algae. *Environmental Toxicology and Pharmacology, 48*, 22–30. https://doi.org/10.1016/j.etap.2016.09.023.

Gómez-Guzmán, M., Rodríguez-Nogales, A., Algieri, F., & Gálvez, J. (2018). Potential role of seaweed polyphenols in cardiovascular-associated disorders. *Marine Drugs, 16*(8), 1–21. https://doi.org/10.3390/md16080250.

Holdt, S. L., & Kraan, S. (2011). Bioactive compounds in seaweed: Functional food applications and legislation. *Journal of Applied Phycology, 23*, 543–597.

Holdt, S. L., & Kraan, S. (2011). Bioactive compounds in seaweed: Functional food applications and legislation. *Journal of Applied Phycology, 23*(3), 543–597. https://doi.org/10.1007/s10811- 010-9632-5.

Jesumani, V., Du, H., Aslam, M., Pei, P., & Huang, N. (2019). Potential use of seaweed bioactive compounds in skincare—A review. *Marine Drugs, 17*, 688.

Liu, D., Mao, S., Zhang, H., Yu, X., Feng, M., Wang, B., Feng, L., & Guo, Y. (2013). Racemosins A and B, two novel bisindole alkaloids from the green alga *Caulerpa racemosa*. *Fitoterapia, 91*, 15–20. https://doi.org/10.1016/j.fitote.2013.08.014.

Lucena, A. M. M., Souza, C. R. M., Jales, J. T., Guedes, P. M. M., De Miranda, G. E. C., de Moura, A. M. A., Araújo-Júnior, J. X., Nascimento, G. J., Scortecci, K. C., Santos, B. V. O., & Souto, J. T. (2018). The bisindole alkaloid caulerpin, from seaweeds of the genus Caulerpa, attenuated colon damage in murine colitis model. *Marine Drugs, 16*(9), 1–18. https://doi.org/10.3390/md16090318.

Lunagariya, J., Bhadja, P., Zhong, S., Vekariya, R., & Xu, S. (2019). Marine natural product bis-indole alkaloid caulerpin: Chemistry and biology. *Mini Reviews in Medicinal Chemistry, 19*(9), 751–761.

Matanjun, P., Mohamed, S., Muhammad, K., & Mustapha, N. M. (2010). Comparison of cardiovascular protective effects of tropical seaweeds, *Kappaphycus alvarezii, Caulerpa lentillifera*, and *Sargassum polycystum*, on high-cholesterol/high-fat diet in rats. *Journal of Medicinal Food, 13*(4), 792–800. https://doi.org/10.1089/jmf.2008.1212.

Matanjun, P., Mohamed, S., Mustapha, N. M., & Muhammad, K. (2010). Nutrient content of tropical edible seaweeds, *Eucheuma cottonii, Caulerpa lentillifera* and *Sargassum polycystum*. *Journal of Applied Phycology, 21*(1), 75–80. https://doi.org/10.1007/s10811-008-9326-4.

Mayer, F. J., & Binder, C. J. (2019). Atherosclerosis. In M. Geiger (Ed.), *Fundamentals of Vascular Biology. Learning Materials in Biosciences*, pp. 195–233. Springer, Cham. https://doi.org/10.1007/978-3-030-12270-6_10.

Mendis, E., & Kim, S. K. (2011). Present and future prospects of seaweeds in developing functional foods. In *Advances in Food and Nutrition Research* (Vol. 64). Elsevier Inc. https://doi.org/10.1016/B978-0-12-387669-0.00001-6.

Mohamed, S., Hashim, S. N., & Rahman, H. A. (2012). Seaweeds: A sustainable functional food for complementary and alternative therapy. *Trends in Food Science and Technology, 23*(2), 83–96. https://doi.org/10.1016/j.tifs.2011.09.001.

Murray, C. J., & Lopez, A. D. (1997). Global mortality, disability, and the contribution of risk factors: Global burden of disease study. *The Lancet, 349*(9063), 1436–1442.

Nelson, T. E., & Lewis, B. A. (1974). Separation and characterization of the soluble and insoluble components of insoluble laminaran. *Carbohydrate Research, 33*(1), 63–74. https://doi.org/10.1016/S0008-6215(00)82940-7.

Paiva, L., Lima, E., Neto, A. I., Marcone, M., & Baptista, J. (2017). Nutritional and functional bioactivity value of selected Azorean macroalgae: *Ulva compressa*, *Ulva rigida*, *Gelidium microdon*, and *Pterocladiella capillacea*. *Journal of Food Science, 82*(7), 1757–1764.

Pal, A., Kamthania, M. C., & Kumar, A. (2014). Bioactive compounds and properties of seaweeds—A review. *OALib, 01*(04), 1–17. https://doi.org/10.4236/oalib.1100752.

Patil, N. P., Le, V., Sligar, A. D., Mei, L., Chavarria, D., Yang, E. Y., & Baker, A. B. (2018). Algal polysaccharides as therapeutic agents for atherosclerosis. *Frontiers in Cardiovascular Medicine, 5*(October), 1–18. https://doi.org/10.3389/fcvm.2018.00153.

Peñalver, R., José, M., Lorenzo, Gaspar Ros, L. G., Amarowicz, R., Pateiro, M., & Gema Nieto, G. (2020). Seaweeds as a functional ingredient for a healthy diet. *Marine Drugs, 18*(6), 301.

Rajapakse, N., & Kim, S. K. (2011). Nutritional and digestive health benefits of seaweed. In *Advances in Food and Nutrition Research* (Vol. 64). Elsevier Inc. https://doi.org/10.1016/B978-0-12-387669-0.00002-8.

Roth, G. A., Nguyen, G., Forouzanfar, M. H., Mokdad, A. H., Naghavi, M., & Murray, C. J. (2015). Estimates of global and regional premature cardiovascular mortality in 2025. *Circulation, 132*(14), 1270–1282.

Ryu, B., Kim, Y. S., & Jeon, Y. J. (2021). Seaweeds and their natural products for preventing cardiovascular associated dysfunction. *Marine Drugs, 19*(9), 507. https://doi.org/10.3390/md19090507.

Sánchez-Machado, D. I., López-Hernández, J., Paseiro-Losada, P., & López-Cervantes, J. (2004). An HPLC method for the quantification of sterols in edible seaweeds. *Biomedical Chromatography, 18*(3), 183–190. https://doi.org/10.1002/bmc.316.

Seca, A. M. L., & Pinto, D. (2018). Overview on the antihypertensive and anti-obesity effects of secondary metabolites from seaweeds. *Marine Drugs, 16*, 237.

Sigwart, J. D., Blasiak, R., Jaspars, M., Jouffray, J. B., & Tasdemir, D. (2021). Unlocking the potential of marine biodiscovery. *Natural Product Reports, 38*(8), 1235–1242.

Sivagnanam, S. P., Yin, S., Choi, J. H., Park, Y. B., Woo, H. C., & Chun, B. S. (2015). Biological properties of fucoxanthin in oil recovered from two brown seaweeds using supercritical CO_2 extraction. *Marine Drugs, 13*(6), 3422–3442. https://doi.org/10.3390/md13063422.

Souza, C. R. M., Bezerra, W. P., & Souto, J. T. (2020). Marine alkaloids with anti-inflammatory activity: Current knowledge and future perspectives, *Mar Drug 2020 Mar 2;18*(3), 147.

Syamsuddin, R., Azis, H. Y., Badraeni, Rustam. (2019). Comparative study on the growth, carotenoid, fiber and mineral content of the seaweed *Caulerpa lentillifera* cultivated indoors and in the sea. *IOP Conference Series: Earth and Environmental Science, 370*(1), 012019. https://doi.org/10.1088/1755-1315/370/1/012019.

Tibbetts, S. M., Milley, J. E., & Lall, S. P. (2016). Nutritional quality of some wild and cultivated seaweeds: Nutrient composition, total phenolic content and in vitro digestibility. *Journal of Applied Phycology, 28*(6), 3575–3585. https://doi.org/10.1007/s10811-016-0863-y.

Trajkovska, K. T., & Topuzovska, S. (2017). High-density lipoprotein metabolism and reverse cholesterol transport: Strategies for raising HDL cholesterol. *Anatolian Journal of Cardiology, 18*, 149–154.

Zhao, C., Yang, C., Liu, B., Lin, L., Sarker, S. D., Nahar, L., et al. (2018). Bioactive compounds from marine macroalgae and their hypoglycemic benefits. *Trends in Food Science & Technology, 72*, 1–12.

CHAPTER 13

Cosmeceuticals from the Sea

SATISH POLSHETTIWAR,[1] SHILPA S. CHAPADGAONKAR,[2] DAYANAND KANNUR,[3] VIKAS MUSALE,[4] and SEJAL KABRA[1]

[1]*School of Pharmacy, Dr. Vishwanath Karad MIT World Peace University, Kothrud, Pune, Maharashtra, India*

[2]*School of Biology, Dr. Vishwanath Karad MIT World Peace University, Kothrud, Pune, Maharashtra, India*

[3]*SCES's, Indira College of Pharmacy, Pune, Maharashtra, India*

[4]*Department of Pharmaceutics, R. C. Patel Institute of Pharmaceutical Education and Research, Shirpur, Maharashtra, India*

ABSTRACT

Over the past two decades, the cosmetics industry has experienced significant growth and transformation. Beauty products are no longer limited to those used for enhancing and protecting the appearance of the human body. This has led to the creation of new terminologies such as "cosmeceuticals" and "nutricosmetics" to describe the health benefits of beauty products that work from within. Natural-based cosmetics are perceived to be safe and have gained popularity in recent years. The demand for cosmeceuticals made from marine resources has also increased due to their unique chemical and biological features. This chapter discusses the importance of marine-derived compounds, focusing on novel chemical entities with cosmeceutical potential and the mechanisms through which they affect biological functions, providing associated health benefits to humans. Marine settings are considered the most significant source of biologically active chemicals with untapped potential for use in medicines, nutraceuticals, and cosmeceuticals. Marine organisms

Handbook of Research in Marine Pharmaceutics: Exploring Oceanic Microbial Diversity for Human Health and Wellness. Md. Faiyazuddin & Khalid Rehman Hakeem (Eds.)

such as sea algae, seaweed, and corals offer a rich and sustainable source of small molecules and bulk compounds that have important applications in the cosmetics industry. For example, compounds like ectoine, trichodin A, and mytiloxanthin derived from marine organisms have properties that promote skin hydration, prevent microbial contamination, and provide color to cosmetic products. In addition, marine-derived chemicals like kojic acid, sargafuran, chrysophanol, scytonemin, mycosporine-like amino acids, antioxidants, astaxanthin, and PUFAs are key constituents in cosmeceuticals that provide anti-tyrosinase, anti-acne, whitening, UV protection, and anti-wrinkle effects. In recent years, the beauty and skincare industry in India has seen a significant surge in demand for anti-aging products. With an increasing aging population and rising disposable income, more and more Indian consumers are willing to invest in skincare products that can help them maintain a youthful appearance.

Moreover, the Indian climate and lifestyle factors, such as pollution and sun exposure, can cause premature aging and skin damage. This has further driven the demand for anti-aging skincare products.

13.1 THE UNIQUE METABOLIC POTENTIAL OF MARINE MICROORGANISMS

Life, as we all know, began in the oceans and hence marine microorganisms are the closest relatives of the early cells. Microbes inhabit diverse oceanic niches from hydrothermal vents to the extremely cold regions of the Arctic and Antarctica. The variety of habitat niches in oceans is the cradle of microbial diversity. Microorganisms inhabiting these extreme environments have developed metabolic responses that help them to survive and thrive in the powerful environmental influences in terms of geographical location, turbulence, light, temperature, nutrients, salinity, pH, and other macro and microorganisms. They are the mainstays of biogeochemical cycles of carbon, nitrogen, silica, iron, and many trace elements. They produce oxygen and organic matter which is vital for sustaining life on Earth. The metabolic potential of the oceanic organisms is waiting to be discovered. Several compounds of oceanic microbial origin have already proven to be of use in pharmaceuticals, cosmetics, nutraceuticals, agriculture, and other fine chemicals (Yang, Wang, & Proksch, 2020).

Cosmetology is a highly evolving field where novel products are being introduced each day. Several successful cosmetics/cosmeceuticals of marine origin have been discovered and commercialized, and scientists have begun to acknowledge the potential for the discovery of novel compounds with unique properties. Some of the novel and interesting compounds of the major cosmeceutical categories have been discussed based on their indications, such as moisturizing agents, anti-aging compounds, skin lightening, scar reduction, sunscreens, antioxidants, hair strengthening, treatment of acne, rosacea, etc.

13.2 MOISTURIZERS

The osmolarity of marine habitats compels marine organisms to develop resistance against osmotic fluid loss. Marine organisms accumulate large amounts of specific small organic compounds called compatible solutes, which do not interfere with normal cellular functions but balance intra and extracellular osmotic pressure. These compounds prevent moisture loss and have high hydrating potential. Moisturizing cosmetic preparations can be classified as emollients, humectants, or occlusives. Emollients penetrate and deposit between the corneocytes, smoothing the skin texture. On the other hand, humectants, owing to their hygroscopic property, hydrate the skin. In contrast, occlusive compounds provide strong barriers against water loss. A variety of marine natural compounds impart these properties and are therefore renowned ingredients in many high-value moisturizer formulations.

The beneficial ingredients found in brown and red macroalgae have already been recognized for their valuable properties in the cosmetics industry. These compounds are beneficial as emollients, skin conditioners, viscosity controllers, and stabilizers in various cosmetic formulations (Alves, Sousa, Kijjoa, & Pinto, n.d.). Brown algae constitute the major groups of macroalgae that are cultured on the seashores. They can be cultured easily, and have a high growth rate, and the bioactive components' yields can be manipulated by controlling the growth media. Brown marine algae *Laminaria japonica* and *Laminaria ochroleuca* extracts have excellent moisturizing potential. Alginic acid and other ingredients such as polysaccharides, vitamins, and amino acids provide nutrition, detoxification, and anti-inflammatory activity. This extract is hypoallergenic and hence considered superior to its synthetic alternatives (Pereira, 2018; Choi, 2013). Cosmetic formulations that contain laminaria extract have already been commercialized.

Fucoidan, a sulfated polysaccharide, is a structural component of the cell walls of brown algae. This component acts as a barrier against cellular desiccation in their natural habitat, which explains their excellent bioactivity as moisturizers (Kim, Lee, Kim, & Kang, n.d,). Apart from polysaccharides, lipids from marine microalgae possess emollient and skin-softening properties. This property is attributed to the capacity to deposit between corneocytes and smoothen the skin texture. This also makes the skin more radiant since it depends on the amount of light reflected by the skin's surface. It is widely acknowledged that omega-6 polyunsaturated fatty acids, particularly C-18 fatty acids like linoleic acid (i); and γ-linolenic acid (ii), have the ability to replenish moisture in the skin and prevent water loss. Heterotroph microalgae, including genera *Thraustochytrium* and *Schizochytrium*, and dinoflagellate *Crypthecodinium cohnii* accumulate EPA (eicosapentaenoic acid) and DHA (docosahexaenoic acid). These species are particularly of interest from a commercial point of view as they are non-pathogenic and non-toxigenic due to DHA/EPA (Barta et al., 2021). Apart from algae, novel bioactives from marine bacteria have also been identified. Ectoine and hydroxyectoine are classes of compatible solutes that act as emollients, stabilizers, and osmotic protective agents (Liu et al., 2021).

13.3 ANTI-AGING COMPOUNDS

The aging of the skin is a multifaceted process that is affected by internal and external factors. Internal factors include genetics and metabolic conditions, while external factors include exposure to pollution, radiation, chemicals, and toxins. The aging process also involves significant alterations to the components of the extracellular matrix. Collagen, elastin, and glucosaminoglycans are the primary structural components of the epidermis and validated targets for anti-aging therapy (Ganceviciene et al., 2012). During aging, the overall concentration of collagen reduces, structurally it gets degraded and unevenly distributed. This is brought about by higher activity of matrix metalloproteases and disruption of transforming growth factor-β signaling due to reactive oxygen species production while aging. Recent research in cosmeceutical science has recognized three important anti-aging strategies: (i) maintaining a healthy skin barrier, (ii) replenishing and inhibiting degradation of structural components of the skin especially collagen and other extracellular matrix components, and (iii) preventing oxidative skin damage.

Dermatologists are strong advocates of skincare routines that maintain a healthy and functional skin barrier, which protects against dehydration, infection, reactive oxygen species, radiation damage, allergy, and irritation.

This is imperative to prevent extrinsic aging factors. However, at the same time, it is important to inhibit the degradation of structural components of the skin, such as collagen and elastin, which are known to be responsible for skin wrinkles. Another principle is to use topical and systemic antioxidants to reduce inflammation. These are used in combination with sunscreen agents as effective anti-aging formulations.

13.4 REPLENISHING EXTRACELLULAR MATRIX (ECM) COMPONENTS AND PREVENTING THEIR DEGRADATION

Marine collagen: Collagen, the famed constituent of the extracellular matrix, has been proven as an important anti-aging agent having excellent properties such as water solubility, compatibility, and bioavailability, which can be formulated easily in different types of cosmeceutical formulations. Collagen has been shown to be effective in diminishing skin wrinkles, enhancing skin elasticity, and improving the overall appearance of the skin. It stimulates the migration of fibroblasts and keratinocytes and promotes epidermal cell growth and attachment, vascularization, and wound healing. Marine organisms such as fish, jellyfish, sponges, and other invertebrates are an important source of collagen (Geahchan et al., 2022). Hyaluronic acid is isolated from terrestrial and marine sources. Researchers have isolated hyaluronic acid from marine bivalves and fishes (Giji & Arumugam, 2014; Volpi & Maccari, 2003).

13.5 PROMOTING BIOSYNTHESIS OF EXTRACELLULAR MATRIX COMPONENTS

Kim et al. (2022) investigated the activation of the ECM by a mixture of amino acids, copper, and hyaluronic acid and its effect on skin moisture, inner elasticity, thickness, density, as well as wrinkles around the eyes. They claimed that dermal fibroblasts had an increase in collagen and elastin expression of over two and three folds, respectively (Kim et al., 2022). Chajra et al. (2016) showed that applying a cosmetic composition containing N-acetyl-d-glucosamine-6-phosphate, d-glucuronic acid, and magnesium sulfate stimulates the de novo synthesis of sulfated glucosaminoglycans and proteoglycans by stimulating the expression of enzymes involved in their biosynthetic pathways. The results were also confirmed in human volunteer studies (Kim, Chun, Lee, & Song, 2022).

13.6 INHIBITING COLLAGENASE

A research study was carried out on 22 halophytic plants found along the coast of South Korea. The study investigated the effectiveness of ethanol extracts from these plants in various assays, including DPPH and hydrogen peroxide scavenging, anti-elastase, anti-collagenase, anti-hyaluronidase, and anti-tyrosinase. The results revealed that the extract of Salicornia europaea, a specific type of halophytic plant, demonstrated the most potent anti-wrinkle potential with anti-elastase and anti-hyaluronidase effects. On the other hand, the extract of the Rosa rugosa plant exhibited anti-collagenase effects (Jiratchayamaethasakul, Ding, Hwang, Im, Jang, Myung et al., 2020).

13.7 PREVENTING OXIDATIVE SKIN DAMAGE

Carotenoids have superior antioxidant potential which act by removing free radicals. Several carotenoids are widely known in the cosmetic industry, including β-carotene from the halotolerant microalga *Dunaliella salina*, astaxanthin from *Haematococcus pluvialis,* and canthaxanthin from *Coelastrella striolata.* (Hosseini Tafreshi, & Shariati, 2009; Grama, Liu, & Li, 2022). *Dunaliella salina* is also capable of producing 9-cis-β-carotene, which is a notable antioxidant with tissue repair and anti-infective properties. This carotenoid has also been identified for its effectiveness in treating psoriasis (Harvey & Ben-Amotz, 2020). Besides the carotenes, the *Dunaliela salina* extract is a rich source of minerals that stimulate skin rejuvenation. Therefore, *Dunaliela salina* extract is an important component of cosmetic formulations. Astaxanthin is another important carotenoid pigment obtained from marine algae, crustaceans, and salmon. Astaxanthin is known for its oxidative stress inhibition, photoprotection, anti-aging, and anti-inflammatory activities.

13.8 SKIN REJUVENATING POLYSACCHARIDES

Microalgae produce alguronic acid, a potent regenerative substance that helps them survive and grow in difficult environmental conditions. *Alteromonas macleodii,* isolated from a polychaete found at a depth of 2600 m in the East Pacific Rise, was discovered to secrete the exopolysaccharide known as deepsane. Deepsane consists of uncommon trisubstituted galacturonic carbohydrate structures (Le Costaouec et al., 2012). Deepsane is commercially available under the brand name Abyssine® by Lucas Meyers in cosmetics to soothe the irritation of sensitive skin caused by chemical, mechanical, and UV-B exposure.

13.9 SKIN WHITENING

Skin lightening or depigmentation is in high demand as a cosmetic therapy or for the prevention of disorders like hyperpigmentation. The majority of depigmentation techniques involve blocking one or more steps of the melanogenic pathway to lower melanin levels. To reduce the level of melanin in the skin, it is necessary to suppress the activity of the tyrosinase enzyme, inhibit microphthalmia-associated transcription factor (MITF), decrease the functionality of the melanocortin 1 receptor, and prevent melanosome maturation and transfer.

Some natural tyrosinase inhibitors like (i) hydroquinone; (ii) Kojic acid; and (iii) arbutin have been popularly used in cosmetic preparations. However, these ingredients have limitations such as toxicity and low efficacy (Qian et al., 2020). Recently, hydroquinone use has been banned in cosmetic formulations as it has been associated with irritation, dermatitis, ochronosis, and carcinogenesis. Kojic acid isolated from the fungus *Aspergillus oryzae* and arbutin, a glycosylated natural derivative of hydroquinone, have been widely used as skin-whitening agents with tyrosinase activity (Phasha, Senabe, Ndzotoyi, Okole, Fouche, & Chuturgoon, 2022). Arbutin is distributed widely, especially in plants of the Ericaceae family, for example, in bearberry, marjoram (*Origanum majorana*), and pear (Sasaki, Ichitani, Kunimoto, Asada, & Nakamura, 2014).

Cha et al. (2011) investigated 43 marine algae species for their tyrosinase activity and identified two promising extracts from Ecklonia cava and *Sargassum silquastrum*. Shimoda et al. (2010) reported the anti-tyrosinase activity of Fucoxanthin from Laminaria japonica, which decreased melanogenesis in mice and guinea pig models. Ishige okamurae extracts from brown algae were also reported to have a skin-whitening effect attributed to diphlorethohydroxycarmalol that inhibits tyrosinase and reduces the expression of MITF. Phloroglucinol derivatives from brown algae have anti-tyrosinase activity because of the chelation of available copper required for tyrosinase activity. Phlorotannins such as phloroeckol and dioxinohydroeckol from red algae such as Ecklonia cava are potent tyrosinase inhibitors. Additionally, marine Pseudomonas species' methylene chloride extract showed marked anti-tyrosinase activity, reducing melanin pigment in both melan-a cells and zebrafish models (Kang, Yoon, & Lee, 2011).

13.10 PHOTOPROTECTION

Exposure to UVA (400–320 nm) and UVB (320–290 nm) radiations are major contributors to pigmentation, radiation damage, skin aging, as well as some forms of skin cancer (Amaro-Ortiz, Yan, & D'Orazio, 2014). A good photoprotective formulation is generally advised by dermatologists for regular skin care (Amaro-Ortiz, Yan, & D'Orazio, 2014). Many countries have recently prohibited the use of widely used sunscreen ingredients like Oxybenzone (BP-3) and octinoxate (OMC) due to concerns about their toxicity to both humans and marine ecosystems (Suh, Pham, Smith, & Mesinkovska, 2020). The search for effective and safe photoprotective agents has led to the discovery of novel marine compounds with photoprotective activity. Mycosporine-like amino acids (MAAs) are a class of compounds that have a cyclohexanone or cyclohexenimine ring structure with different functional groups attached to them. These compounds have UV radiation absorbing properties due to the presence of a conjugated double bond in the structure, and their absorbance maxima are 310 nm for cyclohexanones and 360 nm for cyclohexenimines. Some drugs currently available in the market contain extracts from *Porphyra umbilicalis*, which is a type of red alga. Palythine, another MAA found in the red algae *Chondrus yendoi,* has been found to protect human skin from UVA and UVB rays (Alparslan, Sekeroglu, & Kijjoa, 2014) (Table 13.1).

KEYWORDS

- **antiaging**
- **anti-tyrosinase**
- **anti-wrinkle**
- **cosmeceuticals**
- **marine-derived compounds**
- **nutricosmetics**
- **UV protection**

TABLE 13.1 Cosmeceuticals from Marine Organisms

Compound	Source	Application
Brown marine algal extracts	*Laminaria japonica* *Laminaria ochroleuca*	Moisturization
Fucoidan	*Brown marine algae*	
ω-6 polyunsaturated fatty acids	*Thraustochytrium Schizochytrium Crypthecodinium cohnii (dinoflagellate)*	Emollients
Ectoine hydroxyectoine	*Marine bacteria (Halomonas salina, Halomonas elongata*	Emollient, stabilizer
Collagen	*Fish, Jellyfish, Sponges*	Extracellular matrix component Anti-wrinkle, antiaging
Hyaluronic acid	Vitreous humor(eyeballs)	Fish
Chondroitin sulfate	Skeleton, cartilage	Fish, shark, sea cucumber
A cosmetic product has been developed with a combination of N-acetyl-d-glucosamine-6-phosphate, d-glucuronic acid, and magnesium sulfate as key ingredients.	Several	Stimulation of production of ECM components
Halophytic plant extract *Salicornia europea*	*Salicornia europea*	Anti-elastase, Anti-hyaluronidase
Halophytic plant *Rosa rugosa* extract	*Rosa rugosa*	An anti-collagenase
β Carotene	*Dunaliella salina*	Anti-oxidative skin damage, Photoprotection, Antiaging
Haematococcus pluvialis	Astaxanthin	Anti-oxidative skin damage
Dunaliella salina	9-cis-β-carotene	High antioxidant, tissue repair as well as anti-infective activities, anti-psoriasis activity
Alteromonas macleodii	Deepsane (exopolysaccharide)	Compound that may have chemical, mechanical, and UV-B aggression

TABLE 13.1 *(Continued)*

Compound	Source	Application
Alguronic acid	Microalgae	Regenerative activity
Marine algae	*Ecklonia cava* and *Sargassum silquastrum*	Anti-tyrosinase depigmentation activity
Fucoxanthin	*Laminaria japonica*	Anti-tyrosinase (skin whitening)
Extract from *Ecklonia cava*	*Marine algae*	Anti-tyrosinase (skin whitening)
Sargassum silquastrum		
diphlorethohydroxycarmalol	*Ishige okamurae*	Anti-tyrosinase (skin whitening)
Phloroglucinol derivatives	Brown algae	Anti-tyrosinase (skin whitening)
Phloroeckol and dioxinohydroeckol	Red alga Ecklonia cava	Anti-tyrosinase (skin whitening)
Mycosporine-like amino acids (MAAs)	*Pophyra umbicalis.* Palythine, red alga *Chondrus yendoi*	UV Protection

REFERENCES

Alparslan, L., Şekeroğlu, N., & Kıjjoa, A. (2018). The Potential of Marine Resources in Cosmetics. *Current Perspectives on Medicinal and Aromatic Plants, 1*(2), 53–66. https://doi.org/10.38093/cupmap.488904.

Alves, A., Sousa, E., Kijjoa, A., & Pinto, M. (2020). Marine-derived compounds with potential use as cosmeceuticals and nutricosmetics. Molecules, 25(11), 1–38 https://doi.org/10.3390/molecules25112536.

Amaro-Ortiz, A., Yan, B., & D'Orazio, J. A. (2014). Ultraviolet radiation, aging and the skin: Prevention of damage by topical cAMP manipulation. *Molecules, 19*(5), 6202–6219. doi: 10.3390/molecules19056202.

Barta, D. G., Coman, V., & Vodnar, D. C. (2021). Microalgae as sources of omega-3 polyunsaturated fatty acids: Biotechnological aspects. *Algal Research, 58*, 1–13. https://doi.org/10.1016/j.algal.2021.102410.

Cha, S. H., Ko, S. C., Kim, D., & Jeon, Y. J. (2011). Screening of marine algae for potential tyrosinase inhibitor: Those inhibitors reduced tyrosinase activity and melanin synthesis in zebrafish. *Journal of Dermatology, 38*(4), 354–363. doi: 10.1111/j.1346-8138.2010.00983.x.

Chajra, H., Auriol, D., Joly, F., Pagnon, A., Rodrigues, M., Allart, S., et al. (2016). Reactivating the extracellular matrix synthesis of sulfated glycosaminoglycans and proteoglycans to improve the human skin aspect and its mechanical properties. *Clinical, Cosmetic and Investigational Dermatology, 9*, 461–472. https://doi.org/10.2147/CCID.S116548.

Choi, J. S., Moon, W. S., Choi, J. N., Do, K. H., Moon, S. H., Cho, K. K., Han, C. J., & Choi, I. S. (2013). Effects of seaweed Laminaria japonica extracts on skin moisturizing activity in vivo. Journal of cosmetic science, 64(3), 193–205.

Ding, Y., Jiang, Y., Im, S. T., Myung, S., Kim, H. S., & Lee, S. H. (2021). Diphlorethohydroxycarmalol inhibits melanogenesis via protein kinase A/cAMP response element-binding protein and extracellular signal-regulated kinase-mediated microphthalmia-associated transcription factor downregulation in α-melanocyte stimulating hormone-stimulated B16F10 melanoma cells and zebrafish. *Cell Biochem. Funct., 39*(4), 546–554. doi: 10.1002/cbf.3620.

Ganceviciene, R., Liakou, A. I., Theodoridis, A., Makrantonaki, E., & Zouboulis, C. C. (2012). Skin anti-aging strategies. *Dermato-endocrinology, 4*(3), 308–319. https://doi.org/10.4161/derm.22804.

Geahchan, S., Baharlouei, P., & Rahman, M. A. (2022). Potential role of seaweed polyphenols in cardiovascular-associated disorders. *Marine Drugs, 16*(8), 1–21. https://doi.org/10.3390/md16080250.

Giji, S., & Arumugam, M. (2014). Isolation and characterization of hyaluronic acid from marine organisms. In: *Advances in Food and Nutrition Research* (Vol. 72, pp. 61–77). Academic Press Inc. https://doi.org/10.1016/B978-0-12-800269-8.00004-X.

Grama, S. B., Liu, Z., & Li, J. (2022). Emerging trends in genetic engineering of microalgae for commercial applications. *Marine Drugs, 20*(5), 1–24. https://doi.org/10.3390/md20050285.

Harvey, P. J., & Ben-Amotz, A. (2020). Towards a sustainable *Dunaliella salina* microalgal biorefinery for 9-cis β-carotene production. *Algal Research, 50*, 1–9. https://doi.org/10.1016/j.algal.2020.102002.

Hosseini Tafreshi, A., & Shariati, M. (2009). Dunaliella biotechnology: Methods and applications. *Journal of Applied Microbiology, 107*(1), 14–35. https://doi.org/10.1111/j.1365-2672.2009.04153.x.

Jiratchayamaethasakul, C., Ding, Y., Hwang, O., Im, S. T., Jang, Y., Myung, S. W., et al. (2020). In vitro screening of elastase, collagenase, hyaluronidase, and tyrosinase inhibitory and antioxidant activities of 22 halophyte plant extracts for novel cosmeceuticals. *Fish & Shellfish Immunology, 23*(1), 1–9. https://doi.org/10.1186/s41240-020-00149-8.

Kang, H. Y., Yoon, T. J., & Lee, G. J. (2011). Whitening effects of marine pseudomonas extract. *Annals of Dermatology, 23*(2), 144–149. doi: 10.5021/ad.2011.23.2.144.

Kim, J. H., Lee, J. E., Kim, K. H., & Kang, N. J. (2018). Beneficial effects of marine algae-derived carbohydrates for skin health. *Marine Drugs, 16*(11), 1–20. https://doi.org/10.3390/md16110459.

Kim, M. S., Chun, K. E., Lee, D. K., & Song, S. H. (2022). Evaluation of the efficacy of an elastin-inducing composition containing amino acids, copper, and hyaluronic acid: Results of an open single-center clinical trial study. *Cosmetics, 9*(3), 1–10. https://doi.org/10.3390/cosmetics9030051.

Liu, M., Liu, H., Shi, M., Jiang, M., Li, L., & Zheng, Y. (2021). Microbial production of ectoine and hydroxyectoine as high-value chemicals. *Microbial Cell Factories, 20*(1), 1–11. https://doi.org/10.1186/s12934-021-01567-6.

Maheswari, V., & Babu, P. A. S. (2022). Comprehensive reviews on phenolic compounds from Phaeophyceae as potential therapeutic agent. *Journal of Applied Biology & Biotechnology, 10*(5), 14–21. doi: 10.7324/JABB.2022.100502.

Pereira, L. (2018). Seaweeds as source of bioactive substances and skin care therapy-Cosmeceuticals, algotheraphy, and thalassotherapy. *Cosmetics, 5*(4), 1–41. https://doi.org/10.3390/cosmetics5040068.

Phasha, V., Senabe, J., Ndzotoyi, P., Okole, B., Fouche, G., & Chuturgoon, A. (2022). Review on the Use of Kojic Acid—A Skin-Lightening Ingredient. *Cosmetics, 9*(3), 1–11. https://doi.org/10.3390/cosmetics9030064.

Qian, W., Liu, W., Zhu, D., Cao, Y., Tang, A., Gong, G., et al. (2020). Natural skin whitening compounds for the treatment of melanogenesis (Review). *Experimental and Therapeutic Medicine, 20*(1), 173–185. https://doi.org/10.3892/etm.2020.8687.

Sasaki, C., Ichitani, M., Kunimoto, K. K., Asada, C., & Nakamura, Y. (2014). Extraction of arbutin and its comparative content in branches, leaves, stems, and fruits of Japanese pear *Pyrus pyrifolia* cv. Kousui. *Bioscience, Biotechnology, and Biochemistry, 78*(5), 874–877. https://doi.org/10.1080/09168451.2014.893185.

Shimoda, H., Tanaka, J., Shan, S. J., & Maoka, T. (2010). Anti-pigmentary activity of fucoxanthin and its influence on skin mRNA expression of melanogenic molecules. *Journal of Pharmacy and Pharmacology, 62*(9), 1137–1145. doi: 10.1111/j.2042-7158.2010.01139.x.

Suh, S., Pham, C., Smith, J., & Mesinkovska, N. A. (2020). The banned sunscreen ingredients and their impact on human health: a systematic review. *International Journal of Dermatology, 59*(9), 1033–1042. doi: 10.1111/ijd.14824.

Volpi, N., & Maccari, F. (2003). Purification and characterization of hyaluronic acid from the mollusc bivalve *Mytilus galloprovincialis*. *Biochimie, 85*(6), 619–625. https://doi.org/10.1016/S0300-9084(03)00083-X.

Yang, X. W., Wang, B. G., & Proksch, P. (Eds.). (2020). Natural Products from Marine Microorganisms. Frontiers Media, SA. https://doi.org/10.3389/978-2-88963-358-6.

CHAPTER 14

Marine Therapeutics Featuring Anti-Cancer, Anti-Viral, Antimalarial, and Anti-HIV Properties

ISHRAT ZAHOOR[1] and SHAHID NAZIR WANI[2,3]

[1]*Department of Pharmaceutics, MM College of Pharmacy, Maharishi Markandeshwar (Deemed to be University), Mullana–Ambala, 133207, Haryana, India*

[2]*Chitkara College of Pharmacy, Chitkara University, Rajpura–140401, Punjab, India*

[3]*Aman Pharmacy College, Dholakhera, Udaipurwati Jhunjhunu, Rajasthan, India*

ABSTRACT

The marine microenvironment is inhabited by a large number of distinct kinds of organisms, each of which is accountable for producing a huge number of substances that are very beneficial to human beings in the treatment of various disorders. The search for innovative medications is becoming more urgent, and over the course of the past few centuries, there have been significant advancements made in the development of marine-based drugs. As a result, we now have numerous natural products of marine origin that are approved to treat pain and cancer, as well as a number of promising leads currently being tested in clinical settings. The range of products that may be extracted from the ocean floor functions both as potentially therapeutic ingredients and as inert carriers that carry the drug of interest to the desired target region. The phytopharmaceuticals that are native to marine habitats have a diverse range of structural characteristics as well as

Handbook of Research in Marine Pharmaceutics: Exploring Oceanic Microbial Diversity for Human Health and Wellness. Md. Faiyazuddin & Khalid Rehman Hakeem (Eds.)

different mechanisms of action. There has been a considerable exponential rise in the percentage of marine drugs that have been granted approval for usage in clinical settings over the period of the last few decades. This rise may be linked to the increased amount of research that has been conducted on marine chemicals in laboratories all around the world. There are still a number of composites that are unknown under the blue canopy that have the potential to provide a solution for untreated diseases that continue to occasionally haunt the earth. These marine-derived active compounds have been used extensively for a broad variety of biological applications. Because of their ease of availability, cheap production costs, biocompatibility, and minimal cytotoxicity toward eukaryotic cells, materials obtained from marine sources are used for therapeutic purposes in various infectious diseases caused by fungi and viruses, as well as non-infectious diseases like cancer, inflammatory diseases, etc. These materials may be used either as entire extracts or as pure components. The secondary metabolites obtained from marine-derived species are structurally novel and biologically active, making them appealing and significant resources for drug development. Among more than 30,000 marine-derived chemicals, numerous metabolites from diverse chemical classes have shown potential antimalarial activity. The marine-derived species also produce various compounds that are biologically active; peptides and macrolides obtained from marine species have antiviral properties as well as antiretroviral activities against HIV. The chapter summarizes the recent developments regarding bioactive compounds obtained from various marine organisms that act as anti-cancer, anti-viral, antimalarial, and anti-HIV.

14.1 INTRODUCTION

Around two-thirds of the Earth's surface is occupied by water, and this water contains a great deal of valuable things, including food and medicines. Since the beginning of human civilization, people have been using marine products in their daily routines, such as the salt that comes from the evaporation of seawater, and other foods that come from the ocean. The oceans are the most amazing ecosystem on Earth and a substantial reserve of aquatic habitats with greater levels of both natural and biochemical biodiversity (Kiuru, D'Auria, Muller, Tammela, Vuorela, & Yli-Kauhaluoma, 2014). In the search for powerful inhibitors of many molecular targets, natural compounds derived from the marine environment have been an important reservoir of novel chemical substances. Marine life has a wide variety of bioactive chemicals, some of which have shown tremendous promise as potential ingredients in functional foods and pharmaceuticals (Qi & Ma, 2011; Kabir, Uddin, Jeandet, Emran, Mitra, Albadrani, Sayed, Abdel-Daim, & Simal-Gandara, 2021). Nearly 28,000 novel metabolites have been identified in marine species such as sponges, algae, starfish, and seaweed. This diversity of species produces a broad spectrum of secondary active substances and metabolic products (Montaser & Luesch, 2011). Numerous bioactive chemicals originating from marine species, including chitosan, chitin, vitamins, minerals, polyunsaturated fatty acids, bioactive peptides, carotenoids, and many more, may have the potential to improve one's health. They are also famous for providing anti-inflammatory and anti-carcinogenic actions, in addition to decreasing cardiovascular diseases (Haefner, 2003). Following the scientific breakthrough due to the development of arabinosyl thymidine (spongothymidine), a cytotoxic molecule in the 1950s, a growing number of chemists and biomedical researchers became interested in and hopeful about the prospect of discovering marine-derived natural products that have therapeutic applications (Begum & Hemalatha, 2022). After the first discovery, scientists and researchers from all over the globe worked tirelessly to identify other marine chemicals as well as synthetic analogs of marine compounds (Mayer, Glaser, Cuevas, Jacobs et al., 2010). Before the year 1970, protamine sulfate and cytarabine (arabinosyl cytosine, ara-C), both therapeutically significant chemicals of marine origin, were in clinical use, with vidarabine (ara-A, arabinosyl adenine) being introduced in the year 1976 (Rangel & Falkenberg, 2015; Miklosz, Kalaska, Kaminski, Rusak, Szczubialka, Nowakowska, Pawlak, & Mogielnicki, 2019; Sagar, Kaur, & Minneman, 2010). Another marine chemical, called ziconotide, was finally given approval for usage in clinical settings after almost three decades. Since that time, there has been rapidly increasing curiosity among researchers in the investigation of marine life for the purpose of drug development (Malve, 2016). The chapter summarizes the recent developments regarding bioactive molecules obtained from different marine species that act as anti-cancer, anti-viral, antimalarial, and anti-HIV.

14.2 ANTI-CANCER AGENTS OF MARINE ORIGIN

Cancer continues to be one of the illnesses that pose the greatest risk to human life around the globe (Vasarri, Barletta, & Degl'Innocenti, 2022). There is a significant opportunity to identify fresh substances in marine environments that may have an involvement in either the treatment or prophylaxis of cancer. There is a growing interest in utilizing the diversity and complexity of marine natural product scaffolds for the purpose of rational drug discovery (Newman & Cragg, 2016). This is due to the fact that natural product scaffolds have a significant amount of potential. In general, natural products have been an important reservoir of molecules used for the prophylaxis or treatment of a wide range of malignancies. They also serve as a potential alternative for the assessment of not only novel chemical classes of anticancer medicines, but also distinctive and possibly important mechanisms of action (Bhanot, Sharma, & Noolvi, 2011). The following anticancer therapeutic agents obtained from marine sources have been approved for human use in various countries around the world.

14.2.1 CYTARABINE

The accidental identification of the nucleosides spongouridine and spongothymidine, which were produced from the *Tethya crypta*, a Caribbean sponge, resulted in the development of marine-based cytostatic drugs, which were a significant step towards the fight against cancer (Pereira, Evdokimov, Lefranc, Valentão et al., 2019). Based on this discovery, the synthetic reference cytosine arabinoside cytarabine has been developed as an efficient antileukemic substance and was therefore approved in 1969 by the U.S. Food and Drug Administration (FDA) (Jimenez, Wilke, & Costa-Lotufo, 2018). The prodrug of cytarabine is the pyrimidine nucleoside, and when it enters the cells, it converts into cytosine arabinoside triphosphate and damages Deoxyribonucleic acid (DNA) by competing with deoxycytidine, thereby inhibiting DNA polymerase and DNA synthesis (Jimeno, Faircloth, Sousa-Faro, Scheuer, & Rinehart, 2004; Lindequist, 2016).

14.2.2 TRABECTEDIN

The *Ecteinascidia turbinata* extract's anticancer actions have been recognized since the 1960s, but its separation and structural characterization did not occur until more than two decades later. The chemical known as trabectedin belongs to the family of alkaloids known as tetrahydroisoquinolines (Carter & Keam, 2007). Trabectedin was the first marine-derived anticancer chemical to be approved in 2007 by the European Union (EU) for the therapeutic purpose of platinum-sensitive ovarian cancer as well as sarcoma recurrences. The FDA approved this in the year 2015. This chemical was identified as a DNA alkylator having a mechanism distinct from that of the majority of common alkylating agents, based on examinations of its mode of action (Ganjoo & Patel, 2009; Fajarningsih, 2012). Trabectedin attaches to guanine residues placed in precise sequences in the minor grooves of the double helix, effectively bending DNA strands away from the alkylation site. These adducts, which are produced by trabectedin, inhibit the action of Ribonucleic acid (RNA) polymerase II, which in turn stops DNA transcription (D'incalci, Badri, Galmarini, & Allavena, 2014). One of the most important functions is the suppression of the MDR1 gene transcription since this gene is in charge of the formation of P-glycoprotein, which is a component of the cellular detoxification mechanism (Gajdos & Elias, 2011).

14.2.3 ERIBULIN MESYLATE

Halichondrin B is a synthetic analog of halichondrin B that has been obtained from marine sponges, such as *Lyssodendoryx* sp. and *Halichondria okadai*. Biochemically, it is a nontaxane derivative of polyketides. Hirata and Uemura isolated halichondrins from the *Halichondria okadai*, a sponge. These halichondrins exhibit substantial anticancer actions both *in vitro* and *in vivo*. Among these, halichondrin B, a molecule with a large molecular weight (about 1,110 g/mol), was used as a prototype for eribulin, an analog in the form of a mesylate salt (Jimenez, Wilke, & Costa-Lotufo, 2018; Shakya & Naik, 2022; Radjasa, Vaske, Navarro, Vervoort, Tenney, Linington, & Crews, 2011). In 2010, the FDA granted approval for the drug to be used as an anticancer therapy, and later the EU granted approval for the drug in 2011 (Shetty & Gupta, 2014). The drug is intended for patients who have locally advanced or metastatic breast cancer. Eribulin mesylate is a powerful chemical that inhibits the dynamics of microtubules and acts as an antimitotic agent. It does this by slowing the advancement phase of microtubules and, further, by segregating tubulin into masses that are not productive, which ultimately leads to an inhibition in the G2/M phase of the cell cycle as well as apoptosis (Smith, Wilson, Azarenko, Zhu, Lewis, Littlefield, & Jordan, 2010).

14.2.4 BRYOSTATIN 1

Bryostatin 1, the most well-known anticancer agent, was obtained from *Bugula neritina* and is capable of inducing the ubiquitination of Bcl-2 in lymphoblastic leukemia as well as the proteasome's degradation of Bcl-2. Additionally, it enables the development of progenitor cells that originate from bone marrow (Sekar, Poomalai, Gunasekaran, Mani, & Krishnamurthy, 2015). Cellular activation, proliferation, and differentiation are all controlled by bryostatins, which are powerful activators of the protein kinase C enzyme (Wall, Mohammad, Reddy, & Al-Katib, 2000). A few other proposed mechanisms of action were revealed which involve the arresting of the cell cycle, suppression of protein production, and antiangiogenic activities that correspond to didemnin B and aplidine (Nobili, Lippi, Witort, Donnini, Bausi, Mini, & Capaccioli, 2009).

14.2.5 BRENTUXIMAB VEDOTIN

It was discovered in 1972 that extracts extracted from the mollusk *Dolabella auricularia* (gastropod), which is native to the Indian Ocean, have potent anticancer properties (Cheung, Ng, & Wong, 2015). The anticancer drug brentuximab vedotin is made up of the cytotoxic substance known as monomethyl auristatin, as well as an antibody that is directed against CD30, which may be found on the tumor cell surfaces like those seen in systemic anaplastic large cell lymphoma or Hodgkin lymphoma. Both of the constituents are connected by an enzyme-cleavable valine-citrulline dipeptide (Barreca, Spanò, Montalbano, Cueto et al., 2020; Ansell, 2011; Doronina, Mendelsohn, Bovee, Cerveny et al., 2006). Auristatin is a synthetic derivative of dolastatin 10, a marine naturally occurring substance. The cytotoxic peptides known as dolastatins were first extracted from the mollusk wedge sea hare (*Dolabella auricularia*), which lives in the Indian Ocean. Dolastatin 10 is a depsipeptide molecule with a linear structure. Due to its severe toxicity, it is impossible to deliver it alone, but with the aid of a conjugated antibody, it is possible to target it specifically to the tumor cells that carry the specified antigen (Nigam, Suleria, Farzaei, & Mishra, 2019). It was also shown that tubulin is the primary target of these peptides, suggesting that dolastatins suppress tumor cell proliferation by preventing microtubule polymerization and, thus, cell division (Jimenez, Wilke, & Costa-Lotufo, 2018). Some of the common anti-cancer agents of marine origin and their chemical structures are depicted in Figure 14.1.

14.3 ANTIVIRAL AGENTS OF MARINE ORIGIN

Viral diseases, such as influenza, herpes simplex virus, and human immunodeficiency virus (HIV), have killed millions of people globally throughout

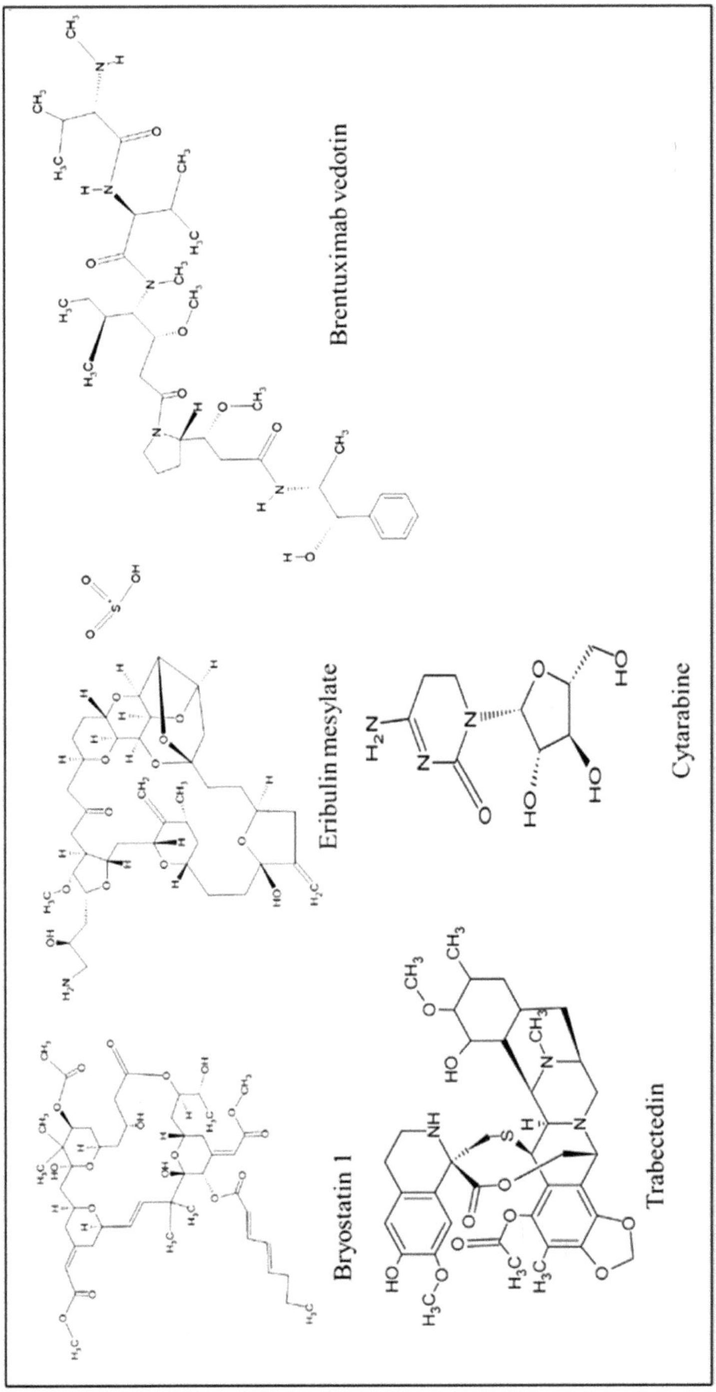

FIGURE 14.1 Chemical structures of some common anti-cancer agents of marine origin.

the decades. However, the choice of appropriate therapies for infectious disorders remains restricted since the majority of viral infections have developed resistance to a variety of medical treatments (Yi, Lin, Zhang, Jin, & Ding, 2020). Nowadays, marine microorganisms are one of the most abundant reservoirs of pharmacologically active natural compounds, which have consistently yielded promising antiviral agents. This is due to the fact that marine microbes are rich in a broad spectrum of natural products. Natural products obtained from marine species have also played an essential role in the development of antiviral medication. The antiviral properties of a vast variety of secondary metabolites formed by marine microorganisms have been extensively reported up to the present time (Yasuhara-Bell & Lu, 2010).

14.3.1 VIDARABINE OR ARA-A

The nucleoside spongouridine was the first chemical compound to be isolated from *Tethya crypta*, a marine sponge, leading to the production of the first antiviral agent commonly known as Ara-A (Sigwart, Blasiak, Jaspars, Jouffray, & Tasdemir, 2021). Vidarabine suppresses the DNA production of viruses by converting it into adenine arabinoside triphosphate, which in turn prevents DNA polymerase of viruses and the DNA formation of herpes, varicella zoster, and vaccinia viruses. Clinically, it has been used to treat infections caused by the herpes virus. Vidarabine was the first antiviral molecule obtained from a sponge to be approved by the U.S. FDA; however, its marketing was subsequently discontinued since it was discovered to be less effective and have more adverse effects compared to the more modern drug acyclovir (Jaspars, De Pascale, Andersen, Reyes, Crawford, & Ianora, 2016; Siddiqui, Iram, Siddiqui, & Sahu, 2014).

14.3.2 FUCOIDAN

Fucoidan is a sulfated polysaccharide that is extracted from the *Fucus vesiculosus*, a brown seaweed. Fucoidan is a polysaccharide that is negatively charged and very hygroscopic (Cui, Tai, Shan, Hao, Li, & Yu, 2018). It has been demonstrated to have an inhibitory action on the DNA replication of viruses, namely herpes virus (HSV-1, HSV-2), as well as Human Cytomegalovirus. It has been demonstrated that these chemicals are effective against RNA viruses such as Vesicular stomatitis virus, HIV-1, and Sindbis virus. However, this substance does not show any antiviral activity against the poliovirus, coxsackievirus, and parainfluenza virus (Luthuli, Wu, Cheng, Zheng, Wu, & Tong, 2019). In an investigation, it was found that fucoidan from *F. vesiculosus* has the ability to prevent hepatitis B virus (HBV) replication both *in vivo* and *in vitro* (Li, Li, Tang, Lin et al., 2017). Fucoidan inhibited the replication of the hepatitis B virus by activating the ERK signaling pathway and enhancing the generation of type I interferon by stimulating the host immune system. This recently identified pathway suggests an alternative strategy that might successfully restrict HBV replication (Luthuli, Wu, Cheng, Zheng, Wu, & Tong, 2019).

14.3.3 GALACTAN SULFATE

Galactan sulfate is a polysaccharide that was obtained from *Agardhiella tenera*, a red alga and has been proven to have antiviral action against human immunodeficiency virus-1 and human immunodeficiency virus-2 (Xu, Huang, & Cheong, 2017). It prevents the adherence of the virus to the cell and also prevents gp120 on $CD4^+$ T-cell receptors from attaching to HIV-1 gp120. *Schizymenia binderi*, an additional sulfated galactan that was isolated from red seaweed, has shown highly specific antiviral actions against HSV-1 and HSV-2 by inhibiting the adherence of the virus to host cells (Delattre, Fenoradosoa, & Michaud, 2011; Sansone, Brunet, Noonan, & Albini, 2020). Sulfated galactan of various structural types has been demonstrated to have strong antiviral action against a broad spectrum of enveloped viruses, like dengue virus-2, HSV-1, HSV-2, HIV-1, and HIV-2 (Salih, Thissera, Yaseen, Hassane, El-Seedi, Sayed, & Rateb, 2021).

14.3.4 ALGINATES

The term "alginates" refers to a collection of naturally produced anionic polysaccharides that are generated from the cell walls of brown algae, like *Laminaria hyperborea, Macrocystis pyrifera*, and *Ascophyllum nodosum,* as well as several strains of bacteria (*Azotobacter, Pseudomonas*). This term is most frequently utilized to refer to alginic acid and the salts of alginic acid; however, it may also be utilized to refer to all alginic acid derivatives (Sachan, Pushkar, Jha, & Bhattacharya, 2009; Remminghorst & Rehm, 2006). It was shown that alginic acid-containing components and sulfated polysaccharides isolated from algae showed antiviral actions against a broad spectrum of virus families, such as *Rhabdoviridae, Flaviviridae,*

Herpesviridae, and *Togaviridae*. These compounds have the potential to be used for treating virus infections (Lee, Takeshita, Hayashi, & Hayashi, 2011). The mechanism of action is associated with the sulfated alginates' potent anionic charge, which is able to interact with the host cell's positively charged surfaces. As a consequence of this interaction, the viruses are unable to make contact with the positively charged surfaces of the host cell, which means that the viral material is unable to enter the cells and cause infection (Wang, Wang, & Guan, 2012; Szekalska, Puciłowska, Szymańska, Ciosek, & Winnicka, 2016).

14.3.5 MYCALAMIDE

Mycalamides are very effective antiviral and anticancer chemicals, and they were obtained from marine sponges. Mycalamide A and Mycalamide B were first discovered and extracted from the marine sponge of the genus *Mycale* that was found in New Zealand (Devkar, Thakur, & Kaur, 2022). Mycalamide A reduced both Polio type I and Herpes simplex type I viruses at 5 ng/disc concentration. Mycalamide B was shown to have a higher potency compared to mycalamide A, which only required 1–2 ng/disc in order to be active (Hassan, Masarčíková, & Berchová, 2015; Gogineni, Schinazi, & Hamann, 2015). The mycalamides exhibit immunosuppressive effects through suppression of $CD4^+$ T-cell activation and are considerably more effective than FK 506 or cyclosporin A (Roush & Pfeifer, 2000).

14.3.6 NAVICULAN

The sulfated polysaccharide Naviculan was extracted from the diatom *Navicula directa*, which was obtained from the deep-sea water of Toyama Bay in Japan. This chemical suppresses HSV-1 and HSV-2 by interfering with the early phases of viral replication (IC_{50} = 7–14 μg/ml), most probably through influencing viral adherence and infiltration into host cells (Lee, Hayashi, Hirata, Kuroda et al., 2006; Tounsi, Hentati, Hlima, Barkallah et al., 2022). In addition, it was demonstrated that naviculan has a strong inhibitory action on the fusion that occurred between cells expressing the CD4 receptor and an HIV gp160 that expresses the HeLa cell line. The HeLa cell line served as a model for the HIV infection process (Ahmadi, Zorofchian, Abubakar, & Zandi, 2015).

14.3.7 CARRAGEENAN

Carrageenans are anionic sulfated polysaccharides that occur naturally and are extracted in large amounts from some red algae (Rhodophyta), such as Gigartina, Chondrus, and Hypnea, Eucheuma. They have a structural function analogous to that of cellulose present in plants (Jiao, Yu, Zhang, & Ewart, 2011). Carrageenans are a type of selective inhibitor that can inhibit the replication of various viruses, both nonenveloped and enveloped. They primarily achieve this by preventing viruses from attaching to or entering host cells, where they can then replicate (Grassauer, Weinmuellner, Meier, Pretsch, Prieschl-Grassauer, & Unger, 2008). *In vitro,* carrageenans suppress the initial stage of Human papillomavirus (HPV) infection, making them a highly effective HPV inhibitor (Buck, Thompson, Roberts, Müller, Lowy, & Schiller, 2006). Figure 14.2 illustrates some common antiviral agents of marine origin and their chemical structures.

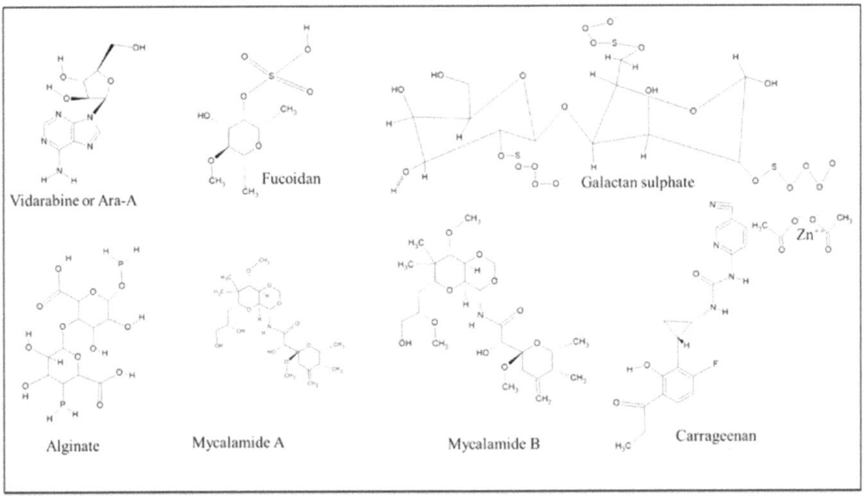

FIGURE 14.2 Chemical structures of some common antiviral agents of marine origin.

14.4 ANTIMALARIAL AGENTS OF MARINE ORIGIN

Malaria is a life-threatening infectious disease that is accountable for the mortalities of at least 1 million people every year and is caused by the protozoans of the genus Plasmodium (*P. ovale, P. vivax, P. falciparum,* and

P. malariae). Unfortunately, there are not enough chemical compounds available currently for its treatment, which also adds more to the highly destructive and dangerous nature of the disease (Sharma, Saini, Sheikh et al., 2022). Natural products derived from species of marine origin are becoming an increasingly essential source for the generation of novel antimalarial drugs, and some of the commonly used marine antimalarial drugs are as follows.

14.4.1 MANZAMINES

The manzamine alkaloids, which may be found in a variety of marine sponge species, belong to the family of β-carboline alkaloids. These alkaloids have intricate structures and have shown potent antimalarial action. They are highly intricate polycyclic alkaloids with seven to eight rings or more, and these compounds were first isolated in 1986 from an Okinawan sponge of the genus *Haliclona* by Higa and coworkers (Murti & Agrawal, 2010). The development of *P. falciparum* was shown to be potentially inhibited by manzamine A and 8-hydroxy manzamine, its derivative, both *in vitro* and *in vivo* (Fattorusso & Taglialatela-Scafati, 2009).

14.4.2 AXISONITRILE-3

Axisonitrile-3 was extracted from the *Acanthella klethra* Pulitzer-Finali sponge first in 1992. It was discovered that it exhibits greater potential antimalarial actions on both the chloroquine-resistant (W2, 17 ng/mL) and chloroquine-sensitive (D6, 142 ng/mL) strains of P. falciparum. It was also found that it does not show any cytotoxicity toward KB cells (Murti & Agrawal, 2010; Fattorusso & Taglialatela-Scafati, 2009).

14.4.3 BROMOPHYCOLIDES

Bromophycolides J-Q were obtained from extracts of the Fijian red algae *Callophycus serratus* (Lane, Stout, Lin, Prudhomme et al., 2009). The bromophycolides showed a wide spectrum of biological actions, including being antiviral (effective against HIV-1), antimicrobial (effective against vancomycin-resistant *Enterococcus faecium* and methicillin-resistant *Staphylococcus aureus*), antimalarial, and anticancer (Stout, Cervantes, Prudhomme, France,

La Clair, Le Roch, & Kubanek, 2011). These bromophycolides demonstrated strong antimalarial action against *P. falciparum* with IC50 values ranging from 0.5–2.9 μM (Lane, Stout, Lin, Prudhomme et al., 2009). Bromophycolides A has been shown to have sub-micromolar blood stage antimalarial action. This activity targets the human malarial parasite *Plasmodium falciparum's* heme crystallization process. *In vivo*, experiments performed on a mouse model of malaria showed a minimal level of toxicity and satisfactory bioavailability. The molecule, however, is susceptible to fast metabolism in the liver, which results in a short *in vivo* half-life (Teasdale, Prudhomme, Torres, Braley et al., 2013; Chhetri, Tedbury, Sweeney-Jones, Mani et al., 2022). In addition, Lin et al. have isolated further bromophycolides R, S, and U from the same alga. These bromophycolides exhibit significant action against *P. falciparum* (IC_{50} values range from 0.9–2.1 M) (Lin, Stout, Prudhomme, Roch et al., 2010).

14.4.4 BASTIMOLIDE A

Bastimolide A is a polyhydroxy macrolide that has a 40-membered ring, and it was obtained from a novel genus of the cyanobacterium *Okeania hirsuta*, which is found in tropical marine habitats (Shao, Linington, Balunas, Centeno et al., 2015). It demonstrated strong antimalarial efficacy against four resistant strains of *P. falciparum*, such as TM90-C2B, TM90-C2A, W2, and TM91-C235, having IC50 values of 90, 80, 140, and 270 nM, respectively. Bastimolide A is a candidate for use as a lead chemical in the development of an antimalarial medication. It has antimalarial action against parasites that are resistant to other medications (Shao, Mou, Cao, Spadafora et al., 2018).

14.4.5 PALSTIMOLIDE A

Palstimolide A, a complex polyhydroxy macrolide containing a 40-membered macrolactone ring, was produced by a tropical cyanobacterium on Palmyra Atoll (Das, Rauf, Mitra, Emran et al., 2022). Palstimolide A has strong antimalarial action and also has preliminary results that are encouraging in terms of its potential as an antileishmanial agent (Keller, Siqueira-Neto, Souza, Eribez, LaMonte, Smith, & Gerwick, 2020). Palstimolide A had high antimalarial potential against the blood stage of the *P. falciparum* Dd2 (IC50 = 172.5 nM), and it also exhibited modest toxicity to HepG2 liver cells (IC_{50} = 5040 nM), resulting in a selectivity index of 29.2 for the compound (Das, Rauf, Mitra,

Emran et al., 2022; Keller, Siqueira-Neto, Souza, Eribez, LaMonte, Smith, & Gerwick, 2020; Gammone & D'Orazio, 2020). This newly discovered natural product is a member of a select group of polyhydroxy macrolides that are closely connected to one another. The constituents of this group of naturally occurring substances were extracted from marine cyanobacteria obtained from various tropical locations worldwide (Keller, Siqueira-Neto, Souza, Eribez, LaMonte, Smith, & Gerwick, 2020). Some common antimalarial agents of marine origin and their chemical structures are depicted in Figure 14.3.

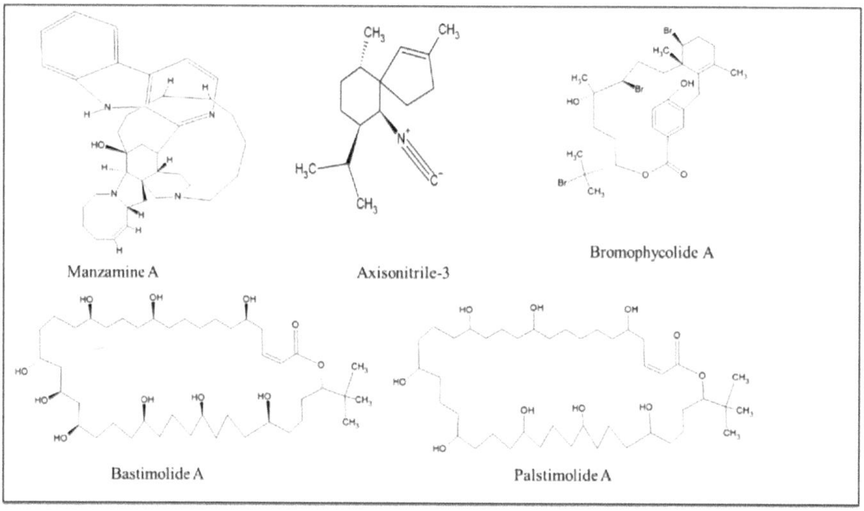

FIGURE 14.3 Chemical structures of some common antimalarial agents of marine origin.

14.5 ANTI-HIV AGENTS OF MARINE ORIGIN

Acquired immune deficiency syndrome, also termed AIDS, is a condition that may be caused by an infection with HIV, which is a concern for public health all over the world (Ojewole, Mackraj, Naidoo, & Govender, 2008). The anti-HIV treatment that involves the use of chemical medications has resulted in an improvement in the patient's quality of life. The most common causes of anti-HIV treatment failure are, however, the development of HIV medication resistance, the occurrence of adverse effects, and the need for continuing treatment for an extended period of time. As a result, the extraction of innovative anti-HIV therapies from natural resources is an absolute need (Vo & Kim, 2010). Natural products that are produced from marine species are becoming an increasingly valuable resource for the formation of innovative anti-HIV medications, and some of the marine anti-HIV compounds that are most widely utilized are described below:

14.5.1 PHLOROTANNINS

Brown algae have been reported to contain phlorotannins, which are synthesized through the acetate-malonate pathway from the polymerization of phloroglucinol (1,3,5-trihydroxybenzene) monomer units. Phlorotannins are extremely hydrophilic components that may vary in weight from 126 daltons to 650 kilodaltons (Shibata, Kawaguchi, Hama, Inagaki, Yamaguchi, & Nakamura, 2004; Steevensz, MacKinnon, Hankinson, Craft et al., 2012). Brown marine algae are known to store a diverse range of phloroglucinol-based polyphenols, which are known as phlorotannins. These phlorotannins may vary in molecular weight from low to high and include both phenoxy and phenyl units (Singh & Bharate, 2006). In addition to this, phlorotannins are widely distributed throughout marine species, particularly brown and red algae, and are made up of phloroglucinol units that are joined to one another in a variety of different ways. Phlorotannins can be divided into four different subclasses, each of which is determined by the type of linkage they contain: phlorethols and fuhalols (phlorotannins that contain an ether linkage), fucophloroethols (contain both phenyl and an ether linkage), Eckols (having a dibenzodioxin linkage), and fucols (contain a phenyl linkage). It has been observed that brown algae with a greater concentration of phlorotannins have anti-HIV action (Vo & Kim, 2010). Ahn et al. isolated 8,4‴dieckol and 8,8′bieckol phlorotannins from *Ecklonia cava* (brown algae) and then evaluated the anti-HIV activities of the obtained compounds, and it was discovered that 8,8′bieckol (IC_{50} = 0.5 μm) inhibits HIV-1 reverse transcriptase (RT) 10 times more effectively than 8,4‴dieckol (IC_{50} = 5.3 μm). Moreover, the phlorotannin 8,8′-bieckol preferentially suppresses the RT enzyme than the protease enzyme, and its inhibiting action was similar to that of nevirapine (IC_{50} = 0.28 μm). 8,8′-bieckol acts as a non-competitive inhibitor of the RNA-dependent DNA synthesis activities of the HIV-1 reverse transcriptase enzyme (Ahn, Yoon, Min, Lee et al., 2004).

14.5.2 POLYSACCHARIDES

Polysaccharides obtained from algal species of marine origin are intricate macromolecules that demonstrate a broad variety of therapeutic characteristics, including anticancer, antiviral, anti-inflammatory, and anti-HIV actions, among other activities. The antiviral properties of these polysaccharides, which are structurally sulfated derivatives, vary with their degree of sulfation, component sugars, conformation, molecular weight, dynamic stereochemistry, and other factors. It has been shown that polysaccharides may suppress the enveloped virus' replication. Moreover, it was discovered that they prevent the adherence of the viral peptide, making them candidates for use in anti-HIV therapies (Nagarajan & Mathaiyan, 2015; Zhou, Liu, Yang, Lin, Yang, & Liu, 2013). Lee et al. investigated the HIV-1 inhibiting properties of sulfated polysaccharides derived from the *Navicula directa*, a marine algae. The obtained naviculan polysaccharide exhibits strong HIV-1 inhibitory action by preventing the fusion of cells with an IC50 value of 53 ± 6.2μg ml^{-1} (Lee, Hayashi, Hirata, Kuroda, Suzuki, Kubo, & Hayashi, 2006). Wang et al. investigated the HIV-1 inhibiting properties of sulfated galactans GLP and GFP isolated from the red algae *Grateloupia longifolia* and *Grateloupia filicina,* respectively. It was discovered that these extracted sulfated glycans have significant anti-HIV action, with an EC50 value that ranged from 0.010 to 0.003μm (Wang, Bligh, Shi, Wang et al., 2007).

14.5.3 MANZAMINE

Alkaloids that are related to manzamine and have been identified from an Indo-Pacific sponge include manzamine A, manzamine E, 8-hydroxymanzamine A, manzamine F, 6-deoxymanzamine X, and neokauluamine (Laport, Santos, & Muricy, 2009). The EC50 values for the effectiveness of manzamines A and E against HIV-1 were 12.9 and 18.7 μM, respectively. Despite this, the effectiveness of these two manzamines was much higher than that of either (–)-8-hydroxymanzamine or (–)-manzamine F (Gochfeld, El Sayed, Yousaf, Hu et al., 2003). Ircinal A, which is derived from the Okinawan sponge *Ircinia* sp. and is also a β-carboline alkaloid, is thought to be a possible biogenic precursor of the manzamines. Ircinal A was effective against HIV-1 at a concentration that was comparable to that of compound (–)-8-hydroxymanzamine A, despite the fact that it had a far lower cytotoxic potential (Wei, Fu, Wang, Yang, Zhao, Wang, Ji, & Song, 2022).

14.5.4 FUCOIDAN

The antiviral properties of fucoidan, which may be extracted from brown algae, have been shown to be efficacious against HIV, HSV, human cytomegalovirus,

SARS-CoV-2, and many other DNA and RNA viruses. The antiviral activities of fucoidan are thought to result from the compound's capacity to prevent the adherence of virus particles to the host cell (Wang, Wang, & Guan, 2012; Wei, Fu, Wang, Yang, Zhao, Wang, Ji, & Song, 2022). Dinesh et al. isolated three different types of fucoidan, namely FF1, FF2, and CFF, from the seaweed *Sargassum swartzii* and revealed that FF2 was useful against HIV-1 at concentrations of 1.56–6.25 μg/mL. This was demonstrated by a considerable decrease in the levels of p24 antigen (95.6 ± 1.1%) and reverse transcriptase (78.9 ± 1.43%), which occurred at a concentration of 25 μg/mL (Dinesh, Menon, Hanna, Suresh, Sathuvan, & Manikannan, 2016). Lee et al. revealed that fucoidan isolated from *Sargassum trichophyllum* and Mekabu substantially reduced viral infections caused by HSV-1, HSV-2, influenza A, and H5N3 while also boosting immune function (Lee, Hayashi, Hashimoto, Nakano, & Hayashi, 2004).

14.5.5 POLYMANNUROGULURONATE

Polymannuronate, commonly known as PMG, is a form of alginate that has a low molecular weight (Wang, Lu, Wang, Liu et al., 2020). The sulfated polymannuronate (SPMG) works by inhibiting the interaction between $CD4^+$ T lymphocyte receptor and HIV-1 gp120 protein, which prevents the virus from entering the lymphocyte (Meiyu, Fuchuan, Xianliang, Jing, Zuowei, & Huashi, 2003). Furthermore, Miao et al. proposed that the interaction of SPMG with $CD4^+$ T-lymphocytes may provide a mechanistic explanation for SPMG's immunoenhancement and anti-AIDS effectiveness in HIV-infected people. This suggests that PMGS is a potentially beneficial new approach for preventing HIV infection and needs more research (Miao, Geng, Li, Li, Chen, Guan, & Ding, 2004).

14.5.6 BROMOPHYCOLIDES

Bromophycolides are diterpenes extracted from the Fijian red algae *Callophycus serratus* that have powerful anti-HIV action. Bromophycolide A, B, and C are obtained, and both bromophycolides A and B exhibited only modest levels of antibacterial and antifungal action, while bromophycolides A demonstrated significant antiviral activity against two different HIV strains, with IC50 values of 9.1 and 9.8 μM (Kubanek, Prusak, Snell, Giese et al., 2005). Some of the common anti-HIV agents of marine origin and their chemical structures are depicted in Figure 14.4.

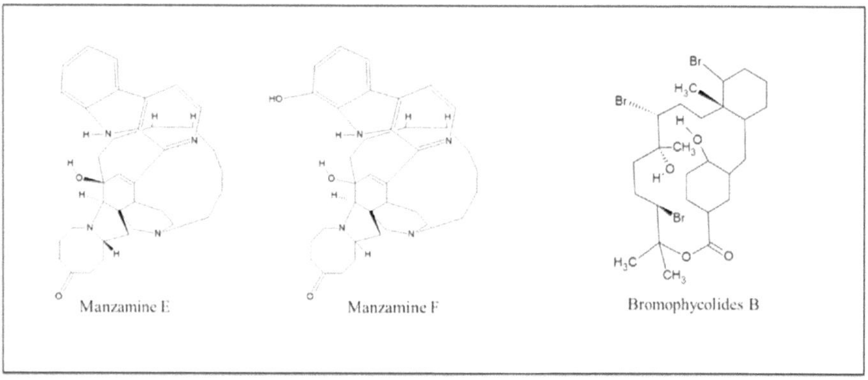

FIGURE 14.4 Chemical structures of some common anti-HIV agents of marine origin.

Some of the common compounds derived from marine organisms with their source and therapeutic effects which are discussed in this chapter are summarized in Table 14.1.

14.6 CONCLUSION

Recent studies have shown evidence that medicinal compounds produced from marine sources may play a significant function in the battle against a wide range of diseases. There is reason to be optimistic about the prospects for developing novel therapeutic candidates and medicines with the goal of supporting the reduction or regulation of diseases associated with chronic malfunctions. In addition, the information provided here suggests that marine-derived therapeutic agents have the potential to be used as active components in the development of innovative pharmaceutical products since marine-derived therapeutic agents perform significant biological functions that have positive impacts on human health.

14.7 FUTURE PERSPECTIVES

Recent studies highlight the potential benefits of using a diverse range of marine organisms as a novel source of naturally occurring medicinal

TABLE 14.1 Marine Drugs from Natural Sources

Compound	Source	Effects
Cytarabine	*Tethya crypta*	Antileukemic
Trabectedin	*Ecteinascidia turbinata*	Platinum-sensitive ovarian cancer as well as sarcoma recurrences.
Eribulin Mesylate	*Lissodendoryx sp*	Antineoplastic
Bryostatin-1	*Bugula neritina*	Antineoplastic
Brentuximab Vedotin	*Dolabella auricularia*	Anticancer
Vidarabine	*Tethya crypta*	Active against herpes, varicella zoster, and vaccinica viruses.
Fucoidan	*Fucus vesiculosus*	Active against HSV-1, HSV-2, Human Cytomegalovirus, Vesicular stomatitis virus, HIV-1, and Sinbis virus.
Galactan sulfate	*Agardhiella tenera*	HSV 1, HSV 2, HIV-1, HIV-2, and dengue virus-2.
Alginates	*Laminaria hyperborea, Macrocystis pyrifera,* and *Ascophyllum nodosum as well as several strains of bacteria (Azotobacter, Pseudomonas).*	Active against Rhabdoviridae, Flaviviridae, Herpesviridae, and Togaviridae families.
Mycalamides	Genus Mycale	Polio type I and Herpes simplex type I viruses
Naviculan	*Navicula directa*	HSV-1 and HSV-2
Carrageenans	Red algae (Rhodophyta), like as Gigartina, Chondrus, and Hypnea, Eucheuma	Human papillomavirus
Manzamines	Genus Haliclona	Inhibit the development of *P. falciparum.*
Axisonitrile-3	*Acanthella klethra*	Active against chloroquine-resistant (W2, 17 ng/mL) and chloroquine-sensitive (D6, 142 ng/mL) strains of *P. falciparum.*
Bromophycolides	*Callophycus serratus*	Antimalarial action against P. falciparum, Anti-HIV

TABLE 14.1 *(Continued)*

Compound	Source	Effects
Bastimolide A	*Okeania hirsuta*	Active against four resistant strains of *P. falciparum*, such as TM90-C2B, TM90-C2A, W2, and TM91-C235.
Palstimolide A	*Cyanobacteria*	Anti-malarial and antileishmanial agent
Phlorotannins	Brown algae (*Ecklonia cava*)	Anti-HIV inhibits HIV-1 reverse transcriptase
Polymannuroguluronate	Brown algae	Anti-AIDS drug

compounds. Now we need to consider whether or not it is possible to produce medicinal compounds from marine species in a sustainable manner. This might entail the cultivation of marine species as well as the operations of a biorefinery, which could include numerous stages of marine species extraction, as well as the utilization of a greater amount of biomass. Extensive studies must be carried out on marine therapeutics in order to ensure that they have the potential of achieving success not only in *in-vitro* models of animals but also in clinical trials of humans. This research is also necessary in order to make use of this resource to overcome the challenges that are currently encountered when developing novel drugs.

KEYWORDS

- **anti-cancer**
- **anti-HIV**
- **antimalarial**
- **anti-viral**
- **marine drugs**
- **marine natural products**
- **phytopharmaceuticals**

REFERENCES

Ahmadi, A., Zorofchian M. S., Abubakar, S., & Zandi, K. (2015). Antiviral potential of algae polysaccharides isolated from marine sources: A review. *Biomed Res. Int, 2015*, 1–10.

Ahn, M. J., Yoon, K. D., Min, S. Y., Lee, J. S., Kim, J. H., Kim, T. G., Kim, S. H., Kim, N. G., Huh, H., & Kim, J. (2004). Inhibition of HIV-1 reverse transcriptase and protease by phlorotannins from the brown alga *Ecklonia cava. Biol. Pharm. Bull.*, *27*, 544–547.

Ansell, S. M. (2011). Brentuximab vedotin: delivering an antimitotic drug to activated lymphoma cells. *Expert Opin. Investig. Drugs, 20*, 99–105.

Barreca, M., Spanò, V., Montalbano, A., Cueto, M., Díaz Marrero, A. R., Deniz, I., Erdoğan, A., Lukić Bilela, L., Moulin, C., Taffin-de-Givenchy, E., & Spriano, F. (2020). Marine anticancer agents: An overview with a particular focus on their chemical classes. *Mar. Drugs, 18*, 619.

Begum, S. M., & Hemalatha, S. (2022). Marine natural products—A vital source of novel biotherapeutics. *Curr. Pharmacol. Rep.*, *8*, 1–11.

Bhanot, A., Sharma, R., & Noolvi, M. N. (2011). Natural sources as potential anti-cancer agents: A review. *Int. J. Phytomedicine, 3*, 09–26.

Buck, C. B., Thompson, C. D., Roberts, J. N., Müller, M., Lowy, D. R., & Schiller, J. T. (2006). Carrageenan is a potent inhibitor of papillomavirus infection. *PLoS Pathog.*, *2*(7), e69.

Carter, N. J., & Keam, S. J. (2007). Trabectedin. *Drugs, 67*, 2257–2276.

Cheung, R. C., Ng, T. B., & Wong, J. H. (2015). Marine peptides: Bioactivities and applications. *Mar. Drugs, 13*, 4006–4043.

Chhetri, B. K., Tedbury, P. R., Sweeney-Jones, A. M., Mani, L., Soapi, K., Manfredi, C., Sorscher, E., Sarafianos, S. G., & Kubanek, J. (2022). Marine Natural Products as Leads against SARS-CoV-2 Infection. *J. Nat. Prod.*, *85*(3), 657–665.

Cui, K., Tai, W., Shan, X., Hao, J., Li, G., & Yu, G. (2018). Structural characterization and anti-thrombotic properties of fucoidan from Nemacystus decipiens. *Int. J. Biol. Macromol, 120*, 1817–1822.

D'incalci, M., Badri, N., Galmarini, C. M., & Allavena, P. (2014). Trabectedin, a drug acting on both cancer cells and the tumor microenvironment. *Br. J. Cancer, 111*, 646–650.

Das, R., Rauf, A., Mitra, S., Emran, T. B., Hossain, M. J., Khan, Z., Naz, S., Ahmad, B., Meyyazhagan, A., Pushparaj, K., & Wan, C. C. (2022). Therapeutic potential of marine macrolides: An overview from 1990 to 2022. *Chem.-Biol. Interact., 365,* 110072.

Delattre, C., Fenoradosoa, T. A., & Michaud, P. (2011). Galactans: an overview of their most important sourcing and applications as natural polysaccharides. *Braz. Arch. Biol. Technol, 54*, 1075–1092.

Devkar, H., Thakur, N. L., & Kaur, P. (2023). Marine-derived antimicrobial molecules from the sponges and their associated bacteria. *Can. J. Microbiol.*, *69*, 1–16.

Dinesh, S., Menon, T., Hanna, L. E., Suresh, V., & Sathuvan, M. (2016). In vitro anti-HIV-1 activity of fucoidan from *Sargassum swartzii. Int. J. Biol. Macromol.*, *82*, 83–88.

Doronina, S. O., Mendelsohn, B. A., Bovee, T. D., Cerveny, C. G., Alley, S. C., Meyer, D. L., Oflazoglu, E., Toki, B. E., Sanderson, R. J., Zabinski, R. F., & Wahl, A. F. (2006). Enhanced activity of monomethyl auristatin F through monoclonal antibody delivery: effects of linker technology on efficacy and toxicity. *Bioconjugate Chem, 17*, 114–124.

Fajarningsih, N. D. (2012). An emerging marine biotechnology: marine drug discovery. *Squalen Bull. Mar. Fish. Postharvest Biotechnol, 7*, 89–96.

Fattorusso, E., & Taglialatela-Scafati, O. (2009). Marine antimalarials. *Mar. Drugs.*, *7*(2), 130–152.

Gajdos, C., & Elias, A. (2011). Trabectedin: safety and efficacy in the treatment of advanced sarcoma. *Clin. Med. Insights Oncol, 5*, CMO–S4907.

Gammone, M. A., & D'Orazio, N. (2020). Potential applications of marine macrolides: New drugs from the sea? *Int. Aquat. Res.*, *12*(2), 151–159.

Ganjoo, K. N., & Patel, S. R. (2009). Trabectedin: an anticancer drug from the sea. *Expert Opin. Pharmacother., 10*, 2735–2743.

Gochfeld, D. J., El Sayed, K. A., Yousaf, M., Hu, J., Bartyzel, P., Dunbar, D. C., Wilkins, S. P., Zjawiony, J. K., Schinazi, R. F., Schlueter Wirtz, S., & Tharnish, P. M. (2003). Marine natural products as lead anti-HIV agents. *Mini Rev. Med. Chem.*, *3*, 401–424.

Gogineni, V., Schinazi, R. F., & Hamann, M. T. (2015). Role of marine natural products in the genesis of antiviral agents. *Chem. Rev, 115*, 9655–9706.

Grassauer, A., Weinmuellner, R., Meier, C., Pretsch, A., Prieschl-Grassauer, E., & Unger, H. (2008). Iota-Carrageenan is a potent inhibitor of rhinovirus infection. *Virol. J.*, *5*, 1–3.

Haefner, B. (2003). Drugs from the deep: marine natural products as drug candidates. *Drug Discov. Today, 8*, 536–544.

Hassan, S. T., Masarčíková, R., & Berchová, K. (2015). Bioactive natural products with anti-herpes simplex virus properties. *J. Pharm. Pharmacol, 67*, 1325–1336.

Jaspars, M., De Pascale, D., Andersen, J. H., Reyes, F., Crawford, A. D., & Ianora, A. (2016). The marine biodiscovery pipeline and ocean medicines of tomorrow. *J. Mar. Biol. Assoc. U. K, 96*, 151–158.

Jiao, G., Yu, G., Zhang, J., & Ewart, H. S. (2011). Chemical structures and bioactivities of sulfated polysaccharides from marine algae. *Mar. Drugs, 9*, 196–223.

Jimenez, P. C., Wilke, D. V., & Costa-Lotufo, L. V. (2018). Marine drugs for cancer: Surfacing biotechnological innovations from the oceans. *Clinics, 73*, e482s.

Jimeno, J., Faircloth, G., Sousa-Faro, J. F., Scheuer, P., & Rinehart, K. (2004). New marine derived anticancer therapeutics—A journey from the sea to clinical trials. *Mar. Drugs, 2*, 14–29.

Kabir, M. T., Uddin, M. S., Jeandet, P., Emran, T. B., Mitra, S., Albadrani, G. M., Sayed, A. A., Abdel-Daim, M. M., & Simal-Gandara, J. (2021). Anti-Alzheimer's molecules derived from marine life: Understanding molecular mechanisms and therapeutic potential. *Mar. Drugs, 19*, 251.

Keller, L., Siqueira-Neto, J. L., Souza, J. M., Eribez, K., LaMonte, G. M., Smith, J. E., & Gerwick, W. H. (2020). Palstimolide A: A complex polyhydroxy macrolide with antiparasitic activity. *Molecules, 25*(7), 1604.

Kiuru, P., D'Auria, M. V., Muller, C. D., Tammela, P., Vuorela, H., & Yli-Kauhaluoma, J. (2014). Exploring marine resources for bioactive compounds. *Planta Med, 80*, 1234–1246.

Kubanek, J., Prusak, A. C., Snell, T. W., Giese, R. A., Hardcastle, K. I., Fairchild, C. R., Aalbersberg, W., Raventos-Suarez, C., & Hay, M. E. (2005). Antineoplastic Diterpene–Benzoate Macrolides from the Fijian Red Alga Callophycus serratus. *Org. Lett., 7*, 5261–5264.

Lane, A. L., Stout, E. P., Lin, A. S., Prudhomme, J., Le Roch, K., Fairchild, C. R., Franzblau, S. G., Hay, M. E., Aalbersberg, W., & Kubanek, J. (2009). Antimalarial bromophycolides J–Q from the Fijian red alga Callophycus serratus. *J. Org. Chem., 74*(7), 2736–2742.

Laport, M. S., Santos, O. C., & Muricy, G. (2009). Marine sponges: potential sources of new antimicrobial drugs. *Curr. Pharm. Biotechnol., 10*, 86–105.

Lee, J. B., Hayashi, K., Hashimoto, M., Nakano, T., & Hayashi, T. (2004). Novel antiviral fucoidan from sporophyll of *Undaria pinnatifida* (Mekabu). *Chem. Pharm. Bull., 52*, 1091–1094.

Lee, J. B., Hayashi, K., Hirata, M., Kuroda, E., Suzuki, E., Kubo, Y., & Hayashi, T. (2006). Antiviral sulfated polysaccharide from *Navicula directa*, a diatom collected from deep-sea water in Toyama Bay. *Biol. Pharm. Bull, 29*, 2135–2139.

Lee, J. B., Takeshita, A., Hayashi, K., & Hayashi, T. (2011). Structures and antiviral activities of polysaccharides from Sargassum trichophyllum. *Carbohydr. Polym., 86*, 995–999.

Li, H., Li, J., Tang, Y., Lin, L., Xie, Z., Zhou, J., Zhang, L., Zhang, X., Zhao, X., Chen, Z., & Zuo, D. (2017). Fucoidan from Fucus vesiculosus suppresses hepatitis B virus replication by enhancing extracellular signal-regulated kinase activation. *Virol. J, 14*, 1–8.

Lin, A. S., Stout, E. P., Prudhomme, J., Roch, K. L., Fairchild, C. R., Franzblau, S. G., Aalbersberg, W., Hay, M. E., & Kubanek, J. (2010). Bioactive Bromophycolides R– U from the Fijian red alga *Callophycus serratus*. *J. Nat. Prod., 73*(2), 275–278.

Lindequist, U. (2016). Marine-derived pharmaceuticals–challenges and opportunities. *Biomol. Ther., 24*, 561.

Luthuli, S., Wu, S., Cheng, Y., Zheng, X., Wu, M., & Tong, H. (2019). Therapeutic effects of fucoidan: A review on recent studies. *Mar. Drugs, 17*, 487.

Malve, H. (2016). Exploring the ocean for new drug developments: Marine pharmacology. *J. Pharm. Bioallied Sci, 8*, 83.

Mayer, A. M., Glaser, K. B., Cuevas, C., Jacobs, R. S., Kem, W., Little, R. D., McIntosh, J. M., Newman, D. J., Potts, B. C., & Shuster, D. E. (2010). The odyssey of marine pharmaceuticals: A current pipeline perspective. *Trends Pharmacol. Sci, 31*, 255–265.

Meiyu, G., Fuchuan, L., Xianliang, X., Jing, L., Zuowei, Y., & Huashi, G. (2003). The potential molecular targets of marine sulfated Polymannuroguluronate interfering with HIV-1 entry: interaction between SPMG and HIV-1 rgp120 and CD4 molecule. *Antivir. Res., 59*, 127–135.

Miao, B., Geng, M., Li, J., Li, F., Chen, H., Guan, H., & Ding, J. (2004). Sulfated polymannuroguluronate, a novel anti-acquired immune deficiency syndrome (AIDS) drug candidate, targeting CD4 in lymphocytes. *Biochem. Pharmacol., 68*, 641–649.

Miklosz, J., Kalaska, B., Kaminski, K., Rusak, M., Szczubialka, K., Nowakowska, M., Pawlak, D., & Mogielnicki, A. (2019). The inhibitory effect of protamine on platelets is attenuated by heparin without inducing thrombocytopenia in rodents. *Mar. Drugs, 17*, 539.

Montaser, R., & Luesch, H. (2011). Marine natural products: a new wave of drugs? *Future Med. Chem, 3*, 1475–1489.

Murti, Y., & Agrawal, T. (2010). Marine derived pharmaceuticals-development of natural health products from marine biodiversity. *Int. J. Chemtech Res., 2*(4), 2198–2217.

Nagarajan, S., & Mathaiyan, M. (2015). Emerging novel anti-HIV biomolecules from marine algae: An overview. *J. Appl. Pharm. Sci., 5*, 153–158.

Newman, D. J., & Cragg, G. M. (2016). Drugs and drug candidates from marine sources: An assessment of the current "state of play." *Planta Med, 82*, 775–789.

Nigam, M., Suleria, H. A., Farzaei, M. H., & Mishra, A. P. (2019). Marine anticancer drugs and their relevant targets: a treasure from the ocean. *DARU J. Pharm. Sci, 27*, 491–515.

Nobili, S., Lippi, D., Witort, E., Donnini, M., Bausi, L., Mini, E., & Capaccioli, S. (2009). Natural compounds for cancer treatment and prevention. *Pharmacol. Res, 59*, 365–378.

Ojewole, E., Mackraj, I., Naidoo, P., & Govender, T. (2008). Exploring the use of novel drug delivery systems for antiretroviral drugs. *Eur. J. Pharm. Biopharm., 70*(3), 697–710.

Pereira, R. B., Evdokimov, N. M., Lefranc, F., Valentão, P., Kornienko, A., Pereira, D. M., Andrade, P. B., & Gomes, N. G. (2019). Marine derived anticancer agents: Clinical benefits, innovative mechanisms, and new targets. *Mar. Drugs, 17*, 329.

Qi, Y., & Ma, S. (2011). The medicinal potential of promising marine macrolides with anticancer activity. *ChemMedChem, 6*, 399–409.

Radjasa, O. K., Vaske, Y. M., Navarro, G., Vervoort, H. C., Tenney, K., Linington, R. G., & Crews, P. (2011). Highlights of marine invertebrate-derived biosynthetic products: Their biomedical potential and possible production by microbial associants. *Bioorg. Med. Chem., 19*, 6658–6674.

Rangel, M., & Falkenberg, M. (2015). An overview of the marine natural products in clinical trials and on the market. *J. Coast. Life Med., 3*, 421–428.

Remminghorst, U., & Rehm, B. H. (2006). Bacterial alginates: from biosynthesis to applications. *Biotechnol. Lett, 28*, 1701–1712.

Roush, W. R., & Pfeifer, L. A. (2000). Total Synthesis of Mycalamide A and 7-e pi-Mycalamide A. *Org. Lett, 2*, 859–862.

Sachan, N. K., Pushkar, S., Jha, A., & Bhattcharya, A. (2009). Sodium alginate: the wonder polymer for controlled drug delivery. *J. Pharm. Res, 2*(8), 1191–1199.

Sagar, S., Kaur, M., & Minneman, K. P. (2010). Antiviral lead compounds from marine sponges. *Mar. Drugs, 8*, 2619–2638.

Salih, A. E., Thissera, B., Yaseen, M., Hassane, A. S., El-Seedi, H. R., Sayed, A. M., & Rateb, M. E. (2021). Marine sulfated polysaccharides as promising antiviral agents: A comprehensive report and modeling study focusing on SARS-CoV-2. *Mar. Drugs, 19*, 406.

Sansone, C., Brunet, C., Noonan, D. M., & Albini, A. (2020). Marine algal antioxidants as potential vectors for controlling viral diseases. *Antioxidants, 9*, 392.

Sekar, M., Poomalai, S., Gunasekaran, M., Mani, P., & Krishnamurthy, A. (2015). Bioactive compounds from marine yeast inhibits lung cancer. *J. Appl. Pharm. Sci, 5*, 007–015.

Shakya, A. K., & Naik, R. R. (2022). The Chemotherapeutic Potentials of Compounds Isolated from the Plant, Marine, Fungus, and Microorganism: Their Mechanism of Action and Prospects. *J. Trop. Med., 2022*, 5919453.

Shao, C. L., Linington, R. G., Balunas, M. J., Centeno, A., Boudreau, P., Zhang, C., Engene, N., Spadafora, C., Mutka, T. S., Kyle, D. E., & Gerwick, L. (2015). Bastimolide A, a potent antimalarial polyhydroxy macrolide from the marine cyanobacterium *Okeania hirsuta. J. Org. Chem.*, *80*(16), 7849–7855.

Shao, C. L., Mou, X. F., Cao, F., Spadafora, C., Glukhov, E., Gerwick, L., Wang, C. Y., & Gerwick, W. H. (2018). Bastimolide B, an antimalarial 24-membered marine macrolide possessing a tert-butyl group. *J. Nat. Prod.*, *81*(1), 211–215.

Sharma, V., Saini, P., Sheikh, I., Upadhyay, S. K., Sharma, U., Sharma, B., & Sharma, A. K. (2022). Role of plant secondary metabolites as potential antimalarial drugs. *Int. J. Mosq. Res.*, *9*, 13–22.

Shetty, N., & Gupta, S. (2014). Eribulin drug review. *South Asian J. Cancer, 3*, 057–059.

Shibata, T., Kawaguchi, S., Hama, Y., Inagaki, M., Yamaguchi, K., & Nakamura, T. (2004). Local and chemical distribution of phlorotannins in brown algae. *J. Appl. Phycol.*, *16*(4), 291–296.

Siddiqui, A. A., Iram, F., Siddiqui, S., & Sahu, K. (2014). Role of natural products in drug discovery process. *Int. J. Drug Dev. Res*, *6*(2), 172–204.

Sigwart, J. D., Blasiak, R., Jaspars, M., Jouffray, J. B., & Tasdemir, D. (2021). Unlocking the potential of marine biodiscovery. *Nat. Prod. Rep, 38*, 1235–1242.

Singh, I. P., & Bharate, S. B. (2006). Phloroglucinol compounds of natural origin. *J. Nat. Prod., 23*, 558–591.

Smith, J. A., Wilson, L., Azarenko, O., Zhu, X., Lewis, B. M., Littlefield, B. A., & Jordan, M. A. (2010). Eribulin binds at microtubule ends to a single site on tubulin to suppress dynamic instability. *Biochemistry, 49*, 1331–1337.

Steevensz, A. J., MacKinnon, S. L., Hankinson, R., Craft, C., Connan, S., Stengel, D. B., & Melanson, J. E. (2012). Profiling phlorotannins in brown macroalgae by liquid chromatography–high resolution mass spectrometry. *Phytochem Anal., 23*, 547–553.

Stout, E. P., Cervantes, S., Prudhomme, J., France, S., La Clair, J. J., Le Roch, K., & Kubanek, J. (2011). Bromophycolide A targets heme crystallization in the human malaria parasite Plasmodium falciparum. *ChemMedChem*, *6*(9), 1572–1577.

Szekalska, M., Puciłowska, A., Szymańska, E., Ciosek, P., & Winnicka, K. (2016). Alginate: current use and future perspectives in pharmaceutical and biomedical applications. *Int. J. Polym. Sci., 2016*. 1–17.

Teasdale, M. E., Prudhomme, J., Torres, M., Braley, M., Cervantes, S., Bhatia, S. C., La Clair, J. J., Le Roch, K., & Kubanek, J. (2013). Pharmacokinetics, metabolism, and in vivo efficacy of the antimalarial natural product bromophycolide A. *ACS Med. Chem. Lett.*, *4*(11), 989–993.

Tounsi, L., Hentati, F., Hlima, H. B., Barkallah, M., Smaoui, S., Fendri, I., Michaud, P., & Abdelkafi, S. (2022). Microalgae as feedstock for bioactive polysaccharides. *Int. J. Biol. Macromol., 221*, 1238–1250.

Vasarri, M., Barletta, E., & Degl'Innocenti, D. (2022). Marine Migrastatics: A Comprehensive 2022 Update. *Mar. Drugs, 20*, 273.

Vo, T. S., & Kim, S. K. (2010). Potential anti-HIV agents from marine resources: An overview. *Mar. Drugs.*, *8*(10), 2871–2892.

Wall, N. R., Mohammad, R. M., Reddy, K. B., & Al-Katib, A. M. (2000). Bryostatin 1 induces ubiquitination and proteasome degradation of Bcl-2 in the human acute lymphoblastic leukemia cell line, Reh. *Int. J. Mol. Med.*, *5*, 165–236.

Wang, S. C., Bligh, S. W., Shi, S. S., Wang, Z. T., Hu, Z. B., Crowder, J., Branford-White, C., & Vella, C. (2007). Structural features and anti-HIV-1 activity of novel polysaccharides from red algae *Grateloupia longifolia* and *Grateloupia filicina*. *Int. J. Biol. Macromol.*, *41*, 369–375.

Wang, S., Lu, Z., Wang, S., Liu, W., Gao, J., Tian, L., Wang, L., Zhang, X., Zhao, X., Wang, W., & Li, C. (2020). The inhibitory effects and mechanisms of polymannuroguluronate sulfate against human papillomavirus infection in vitro and in vivo. *Carbohydr. Polym.*, *241*, 116365.

Wang, W., Wang, S. X., & Guan, H. S. (2012). The antiviral activities and mechanisms of marine polysaccharides: an overview. *Mar. Drugs*, *10*, 2795–2816.

Wei, Q., Fu, G., Wang, K., Yang, Q., Zhao, J., Wang, Y., Ji, K., & Song, S. (2022). Advances in research on antiviral activities of sulfated polysaccharides from seaweeds. *Pharmaceuticals*, *15*, 581.

Xu, S. Y., Huang, X., & Cheong, K. L. (2017). Recent advances in marine algae polysaccharides: Isolation, structure, and activities. *Mar. Drugs*, *15*, 388.

Yasuhara-Bell, J., & Lu, Y. (2010). Marine compounds and their antiviral activities. *Antivir. Res*, *86*, 231–240.

Yi, M., Lin, S., Zhang, B., Jin, H., & Ding, L. (2020). Antiviral potential of natural products from marine microbes. *Eur. J. Med. Chem.*, *207*, 112790.

Zhou, X., Liu, J., Yang, B., Lin, X., Yang, X. W., & Liu, Y. (2013). Marine natural products with anti-HIV activities in the last decade. *Curr. Med. Chem.*, *20*, 953–973.

CHAPTER 15

Carrageenan in Drug Delivery Therapeutics

TITILOPE JOHN JAYEOYE,[1] FREDRICK NWUDE EZE,[2] NONGNUJ MUANGSIN,[1] and SUDARSHAN SINGH[3]

[1]*Department of Chemistry, Faculty of Science, Chulalongkorn University, Bangkok, Thailand*

[2]*Faculty of Pharmaceutical Sciences, Prince of Songkla University, Hat Yai, Songkhla, Thailand*

[3]*Department of Pharmaceutical Sciences, Faculty of Pharmacy, Chiang Mai University, Chiang Mai, Thailand*

ABSTRACT

With environmental sustainability consistently garnering more popularity in the scientific community, the use of environmentally benign materials for diverse human activities has been adduced as a pragmatic strategy to stem the tide of negative anthropogenic-mediated consequences. Carrageenan is a class of biopolymers derived from marine red seaweed, with wide applications in foods, pharmacy, and biomedical fields. They are anionic, sulfated oligosaccharides with high water solubility, which imbue them with enormous advantages for their wide-ranging applications. In this write-up, sources and preparation of carrageenan, and physico-chemical/biological properties of carrageenan are discussed, after which their use in drug delivery therapeutics is elucidated. We have showcased carrageenan-based structures including hydrogels, microparticles, tablets, blends, etc., formulation, carrageenan applications in liposomes and emulsion formulation for the delivery of bioactive compounds, carrageenan-based nanoparticles fabrication delivery

Handbook of Research in Marine Pharmaceutics: Exploring Oceanic Microbial Diversity for Human Health and Wellness. Md. Faiyazuddin & Khalid Rehman Hakeem (Eds.)

of drugs, and carrageenan nanocomposites for therapeutic drug delivery. The latest information on the subject matter is aggregated to present a detailed and holistic range of knowledge on carrageenan application in drug delivery, while also highlighting the challenges and prospects of the subject matter.

15.1 INTRODUCTION

Among the numerous biopolymers available nowadays, special interest has been devoted to the Carrageenan class, based on their special and interesting physicochemical properties. The term biopolymers simply denotes nature-sourced macromolecules with origins traceable to micro-organisms, animals, and plants (Priyadarshi, Kumar, Deeba, Kulshreshtha, & Negi, 2019). They are negatively charged (anionic), sulfated polysaccharides, commonly extracted from reddish-colored seaweeds (Irish moss) of the Rhodophyceae family, such as *Eucheuma cottonii, Chondrus crispus, Gigartina stellata, Gigartina acicularis, uma spinosum, Chondrus ocellatus,* and *Gigartina pistillata*, etc. (Hasizah, Mahendradatta, Laga, Metusalach, & Salengke, 2021; Kalsoom Khan, Saba, Nawazish et al., 2017). With rich sources readily available in the biosphere, the global market value of Carrageenan in the year 2020 was valued at about 780.5 US dollars, which is projected to a global growth of 6% from 2020 to 2028. Moreover, the standing market demand for Carrageenan in 2020 stood at 54.08 kilotons, expected to hit about 68.45 kilotons in 2028, at an estimated demand growth rate of 3.4% (Grandviewresearch.com). In terms of food texture demand, Carrageenan is rated fourth behind starches, gelatin, and pectin. Carrageenan is in high demand cutting across several human fields of endeavors, not limited to food-related applications, pharmaceuticals, medical, and industrial applications. Their ease of preparation and a perfect blend with multiple food products imbue carrageenan with its wide-ranging popularity as a notable constituent of foods.

15.1.1 PHYSICO-CHEMICAL PROPERTIES OF CARRAGEENAN

Carrageenan is a naturally occurring, viscous, highly hydrophilic biopolymer with perfect water solubility, but insoluble in ethanol and many other organic solvents. It is composed of repeating galactose monomeric units linked perfectly together through glycosidic linkages of α-1,3 and β-1,4 configurations (Necas & Bartosikova, 2013). Since it is highly sulfated (pKa of about 2), with about 15–40% of ester sulfate groups, hence, it is negatively charged, and this dominates its main chemical properties (Blakemore & Harpell, 2010). Quite notably, carrageenan has a molecular weight of about 100 kilo-Daltons.

15.1.2 CLASSIFICATION OF CARRAGEENAN

Carrageenan is generally classified into Lambda (λ), kappa (k), Mu (μ), Theta (θ), Iota (τ), and Nu (ν). The three most important among these, from the commercial viewpoint, are kappa (k), Lambda (λ), and Iota (τ) carrageenan. The main difference between the carrageenan forms is in the number/position of their ester sulfate groups and the composition of their 3,6 anhydro-galactose groups or (3,6-AG). The level of the sulfated ester groups in carrageenan could influence the solubility temperature and the gel properties in the solution (Li, Ni, Shao, & Mao, 2014). Table 15.1 shows the summary of the most important commercially available carrageenan.

TABLE 15.1 Summary of Notable Properties of the Commonly Available Carrageenan Type (Li et al., 2014)

Carrageenan Type	3,6-anhydro Bridges	Sulphate Esters Content	Calculated Sulfate Content (%)	Commercial Type
Kappa (k)	Present	1	20	25% ester sulfate, 34% 3,6-AG unit.
Iota (τ)	Present	2	33	32% ester sulfate, 30% 3,6-AG unit
Lambda (λ)	Absent	3	41	35% ester sulfate, no 3,6-AG unit

15.1.3 PREPARATION OF CARRAGEENAN

Carrageenan is prepared from generally similar extraction processes involving collection from the marine environment, drying, washing, alkali extraction, the release of bound carrageenan cells, and the final powder collected, as detailed in Figure 15.1.

The final property of carrageenan obtained from the scheme is greatly affected by the solution pH and temperature, as well as the alkali pre-treatment exposure duration (Zia, Tabasum, Nasif, Sultan, Aslam, Noreen, & Zuber, 2017). The most popular methods of releasing carrageenan from the solution include freeze drying (Awalludin, Mingu, Yaser et al., 2021) and the alcohol precipitation method (Shrgawi, Shamsudin, Hanibah, Mohd Noor, & Kasim, 2021). With these final steps, carrageenan powder of peculiar properties is obtained. Moreover, under enzymatic treatment and alkaline hydrolysis, the less common carrageenan types, Mu (μ), Lambda (λ), and Nu (ν), may be transformed into the commonly available and commercially important carrageenan classes kappa (k), Lambda (λ), and Iota (τ) carrageenan, by forming the 3,6-anhydro bridge (Khotimchenko, Tiasto, Kalitnik, Begun et al., 2020).

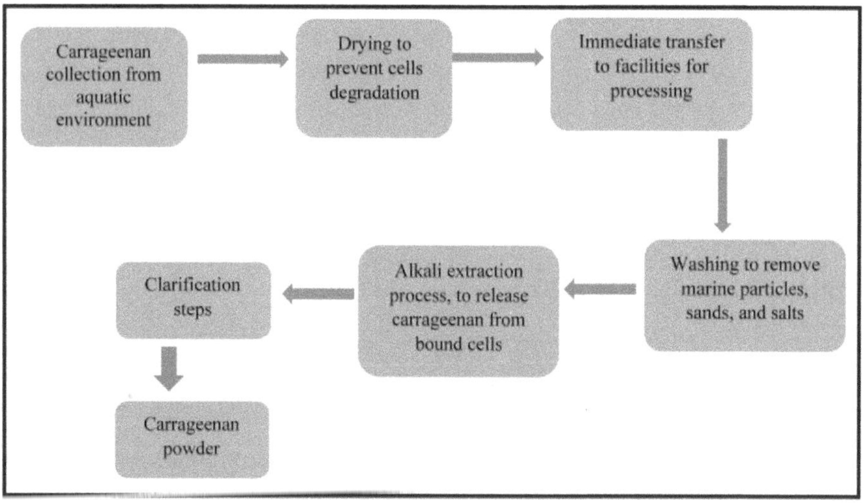

FIGURE 15.1 General extraction protocol of carrageenan from marine red seaweeds.

Source: Adapted from Rowe et al., 2009

15.1.4 USES OF CARRAGEENAN

Common uses of carrageenan, especially in food-related applications, include acting as a thickener for ice cream, and yogurt, as a stabilizer for milk and dairy products, and as a gelling agent for cookies and confectioneries, owing to its heat-reversible gel property. Carrageenan can stabilize the α-casein component of evaporated milk and further prevent creaming of the product (BeMiller, 2018). Its stabilizing properties have also been extended to soft drinks, fruit juices, etc. Carrageenan can also act as a substitute for fats in meat preservation and food processing. The addition of carrageenan to low-fat meats such as pork and hamburger can improve product tenderness and moisture content retention (Dong, Wei, & Xue, 2021).

15.1.5 APPLICATIONS OF CARRAGEENAN IN DRUG DELIVERY

The delivery of therapeutics to their specific action sites is of utmost importance in drug design and biomedicine. The excellent biological properties of carrageenan confer numerous positives on it as a material of interest for various drug delivery activities. Generally, carrageenan has antiherpetic and anticoagulant potentiality (Carlucci, Pujol, Ciancia, Noseda et al., 1997), immunomodulatory activity (Yuan, Song, Li, Li, & Dai, 2006), anticancer or antitumor activity (Liu, Gao, Yang, Meng, Zhan, Jiang, & Sun, 2019; Luo, Shao, Nie, Wei et al., 2015), antihyperglycemic and antihyperlipidemic activity (Qiu & Zhong, 2018), antibacterial activity (Wang, Yao, Wu, Zhang, Zhu, & Gai, 2012; Zhu, Ge, Lyu, Zi, Li, Li, & Mu, 2017). Resulting of these properties, carrageenan-based drug delivery platforms have garnered huge research interest from scientists. Central to the improved drug delivery efficacy of carrageenan is the various structural interplay which in no small measure facilitate the ability of carrageenan to successfully aid drug therapeutics delivery. Aside from the single carrageenan materials, hybrid-type carrageenan materials are also popular.

15.2 VARIOUS CARRAGEENAN STRUCTURES IN DRUG DELIVERY SYSTEM

Drug delivery is an important research area that is explored through various design strategies to maximize drug loading and its eventual discharge or delivery to the medium of interest. This is summarized in Figure 15.2, where nanomaterials-based delivery strategies offer wide-ranging opportunities to achieve effective drug loading and delivery. Notable biopolymer-based drug delivery systems, including carrageenan, are integrated into various nanomaterial designs and fabrications. In a simple definition, nanomaterials are materials with a size distribution within the nanoscale (1–100 nm) (Kreyling, Semmler-Behnke, & Chaudhry, 2010). This does not imply that materials with sizes greater than 100 nm are not nanomaterials. The definition is used for reference and simplification of the basics of nanomaterial classification. At the nano level, materials exhibit distinct properties that are significantly different from their bulk or parent materials. Their small size provides them with a high surface area to volume ratio, allowing easy surface tunability and drug loading for important therapeutic delivery.

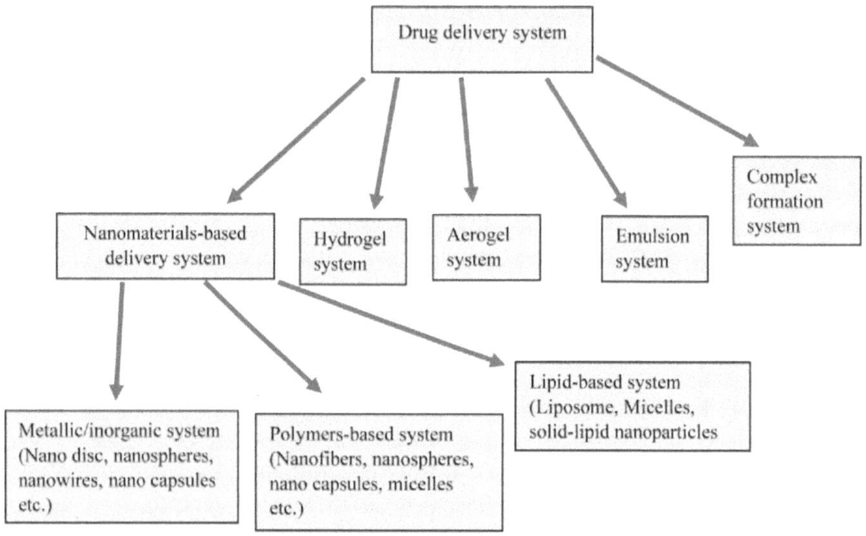

FIGURE 15.2 Overview of drug therapeutics delivery systems.

Source: Adapted from Volpedo et al., 2019

15.2.1 HYDROGEL-BASED DRUG DELIVERY SYSTEM

Hydrogels are three-dimensional (3D), highly cross-linked, hydrophilic polymer networks that absorb a huge amount of biological fluids and water (90–99%), then swell significantly while still retaining their 3D network, without solubilizing (Hoare & Kohane, 2008). Arising from their starting materials, hydrogels are biocompatible, with excellent drug pharmacokinetic features, while also demonstrating good chemically tailorable properties, porosity, and impressive viscoelasticity (He, Dai, Wei, Zhu, Li, Chen, & Ni, 2022). Hydrogels are fabricated through chemical interactions between the biocompatible polymer supports and the other components. Consequently, their constituting materials are firmly joined or cross-linked, either by covalent or non-covalent cross-linking methods. Covalent cross-linking methods rely on ordered chemical interactions between the substrate and the cross-linkers, using chemical reactions such as acetalization (organic chemical interaction between an alcohol and an aldehyde or a ketone), oftentimes using glutaraldehyde as a cross-linking agent. Imine formation reaction is another form of covalent cross-linking reaction, where an aldehyde reacts with an amine group. It is also a form of Schiff-based interaction. For instance, Ooi et al. designed a hydrogel system through imine formation between numerous hydroxyl groups of cellulose nanocrystals and gelatin, using glutaraldehyde (Ooi, Ahmad, & Amin, 2016). The glutaraldehyde reacted with the amine group of gelatins to form a Schiff base. Polymerization is another form of covalent cross-linking strategy to develop hydrogels for drug delivery. In the presence of appropriate monomers and chemical initiators, polymerization of monomers can be realized to obtain hydrogels with 3D architecture, which can provide high surface binding for the drugs of interest and responsive releasing profiles. The popular free-radical polymerization is a popular polymerization strategy in this regard. Non-covalent cross-linking methods explore the not-too-strong intermolecular forces/interactions to design supramolecular hydrogels for drug release. Such forces include electrostatic interaction, hydrogen bonding, and ionic interactions. Though this interaction shows less mechanical strength for the hydrogel realized, they compensate for their injectable and self-healing property (Yan, Chen, Song, Ye et al., 2020). Table 15.2 shows a summarized carrageenan-based hydrogel construct, with varying applications in drug delivery activity.

15.2.2 AEROGEL-BASED DRUG DELIVERY SYSTEM

Similar to hydrogels, aerogels are 3D, ultra-light, interconnected solid structures filled with air (99.9%). The presence of air-filled porous nanostructure enhances the physicochemical properties, as well as the structural peculiarities at the nanoscale level of aerogel. As a result, aerogels are efficient drug delivery substrates (Noman, Amor, Ali, Petrik, Coufal, Adach, & Fijalkowski, 2021). Mostly, aerogels are classified into hybrid, inorganic, and organic, based on the precursor materials used for their fabrication. Furthermore, based on their surface chemical materials, aerogels have also been classified as hydrophilic and hydrophobic. Polysaccharides-derived aerogels, owing to their biocompatibility, are of significant interest in the biomedical fields for the delivery of diverse drug regimens. Carrageenan-derived aerogels generally follow the following steps, including dissolving carrageenan in water, then forming carrageenan hydrogel through emulsion formation by potassium carbonate addition (known as the gelation stage), and finally, drying the carrageenan hydrogel through the freeze-drying method (Ganesan & Ratke, 2014). The gel drying process is a critical stage in aerogel fabrication steps. This is because if the drying process is not properly executed, it may affect the air composition and the nanoporous properties of aerogels. Common drying processes include supercritical drying, which may be either low-temperature or high-temperature drying methods (Wang, Su, Wang, Fang, Riffat, & Jiang, 2019). Others include microwave drying

TABLE 15.2 Summarized Carrageenan-Based Hydrogel System Drug Delivery Network

Hydrogel Compositions	Other Constituents	Carrageenan Type	Encapsulated Drugs/ Active Materials	Comments	References
k-carrageenan/chitosan stabilized sulfur nanoparticles (SNPs)/ grape seed extract (GSE) hydrogel film	Chitosan capped SNPs and GSE	k-carrageenan	SNPs and GSE	Encapsulated SNPs and GSE could improve the antibacterial property of the hydrogel against common wound bacteria, Staphylococcus epidermis and *Escherichia coli* while chitosan and k-carrageenan provided good biocompatibility when tested on mouse fibroblast or (L929 cell lines).	(Jaiswal, Shankar, & Rhim, 2019)
k-carrageenan/PVA	PVA	k-carrageenan	β-carotene	Hydrogel encapsulating β-carotene as a model drug was designed using Genipin, as a cross-linking agent. pH-modulated release of the drug was studied and the release was tested using *in-vitro* assays.	(Hezaveh & Muhamad, 2013)
Chitosan/k-carrageenan hydrogel beads	Chitosan	k-carrageenan	5-Amino salicylic acid (5-ASA)	The ionotropic gelation method was used for hydrogel fabrication, for 5-ASA release within the colon for the treatment of intestinal bowel disease (IBD).	(Stavarache, Ghebaur, Dinescu, Samoilă, Vasile, Vlasceanu, Iovu, & Gârea, 2021)
Agar/k-carrageenan	Agar	k-carrageenan	Metformin hydrochloride	Electrostatic interaction between polysaccharide mixes and drug (MET), enhanced sustained release of drug from the matrix.	(Zhao, Sun, Li, Dong, & Zhang, 2020)

TABLE 15.2 *(Continued)*

Hydrogel Compositions	Other Constituents	Carrageenan Type	Encapsulated Drugs/ Active Materials	Comments	References
Sodium alginate/k-carrageenan	Sodium alginate (SA)	k-carrageenan	Chloramphenicol	K^+-mediated cross-linking between SA and k-carrageenan improved the sustained release of the drug at pH 4.1.	(Fahmy, Aly, Sayed, Abou-Okeil, 2021)
PHBV/k-carrageenan/ locust bean gum	PHBV microparticles	k-carrageenan	Ketoprofen and Mupirocin hydrophobic drugs	PHBV microparticles were embedded in the hydrophilic polymer of k-carrageenan and locust bean gum to study the controlled release of hydrophobic drugs for drug healing applications.	(Pettinelli, Rodríguez-Llamazares, Farrag, Bouza et al., 2020)
Magnetic Fe_3O_4/k-carrageenan/ chitosan	Magnetic Fe_3O_4	k-carrageenan	Methotrexate anticancer drug	Magnetic Fe_3O_4 was synthesized in the k-carrageenan matrix and then cross-linked with chitosan, to form hydrogel for anti-cancer drug encapsulation.	(Mahdavinia, Mosallanezhad, Soleymani, & Sabzi, 2017)

PVA = Polyvinyl alcohol; PHBV = Poly (hydroxybutyrate-co hydroxy valerate).

(Zuo, Zhang, Zhang, Miao, Fan, & Liu, 2015) and vacuum drying processes (Schwan & Ratke, 2013). Table 15.3 showcases some reported carrageenan-derived aerogel delivery systems and their encapsulated drugs.

15.2.3 EMULSION-BASED DRUG DELIVERY SYSTEM

Emulsion systems contain two immiscible liquids (e.g., oil and water), in which one of the liquids is dispersed as small droplets in the other (McClements, Decker, & Weiss, 2007), supplemented with a surface-active molecule called surfactants. The average diameter of the droplets can range between 100 nm to 100 μm. The organization of the oil and water phases is used in the specific classification of emulsions as it relates to drug delivery applications. An *oil-in-water* (O/W) emulsion is made up of oil droplets perfectly dispersed in an aqueous phase, while a *water-in-oil* (W/O) emulsion system is composed of water droplets dispersed in an oil phase. The droplet component of the substance is called the dispersed phase and is of importance in influencing the major properties of the emulsion system. The constituents outside of the dispersed phase are called the continuous phase. Aside from the two-phase emulsion components, triple systems or multiple phases are also feasible. For example, *oil-in-water-in-oil* (O/W/O) emulsions and *water-in-oil-water* (W/O/W) components. Emulsions are generally thermodynamically unstable or unfavorable systems that are easily broken down owing to the influence of multiple physical and chemical factors such as flocculation, Ostwald ripening, and gravitational influence which induces separation (McClements, 2004). Thus, conventional emulsion systems can be further stabilized using stabilizers or surface-active molecules, which can absorb on the surface of the common emulsion interphase, i.e., between the water and the oil phases (Yang, Fang, Chen, Zhang et al., 2017). This class of emulsion has huge biomedical and food applications. They are mostly more stable and resistant against coalescence in comparison to the common emulsion phase system, relying only on the use of surfactants. The stabilizers (mostly solids) form a steric protection or barrier through adsorption at the oil-water interphase (Du Sorbier, Aimable, & Pagnoux, 2015), which consequently confers additional stability on Pickering-based emulsion systems. Solid stabilizing agents used for Pickering emulsion fabrication include silica, hydroxyapatite, chitosan, cyclodextrin, carrageenan, zein, starch, etc. Table 15.4 shows some reported drug delivery systems based on conventional emulsion and Pickering emulsion-derived systems.

TABLE 15.3 Summarized Carrageenan-Based Aerogel System Drug Delivery Network

Aerogel Compositions	Other Constituents	Encapsulated Drugs	Comments	References
k-carrageenan/ionic liquids (ILs)	ILs and other	Tetracycline (Tc)	Potassium salts mediated cross-linking was adopted and the aerogel mixed with ILs was investigated for the delivery of Tc as a model drug.	(Agostinho, Paninho, Cordeiro, Nunes et al., 2020)
Carrageenan microparticles aerogel	-	meloxicam and atorvastatin	The emulsion gelation method was used for aerogel fabrication, using a CO_2-based super-critical drying approach, and was tested for the delivery of type II drugs.	(Alnaief, Obaidat, & Mashaqbeh, 2019)
Carrageenan microparticles aerogel	-	Ibuprofen	Emulsion gelation with supercritical fluid drying process and assessment of ibuprofen loading.	(Obaidat, Alnaief, & Mashaqbeh, 2018)
Monolithic k-carrageenan hydrogel	Potassium thiosulfate is a source of ions.	-	Fibrillar structure aerogel was realized, with the impressive property as a host material for drugs.	(Ganesan & Ratke, 2014)
Carboxymethyl cellulose/k-carrageenan aerogel	ECH, epichloro-hydrin as chemical cross-linker for the starting polymers.	Adsorption of cipro-floxacin (pharma-ceutical drug) from contaminated water	Ciprofloxacin adsorption from simulated polluted water was investigated using the carefully fabricated 3D semi-interpenetrating aerogel. Electrostatic interaction between amine groups of drugs and the carboxyl groups of polymers was adduced as the basis of interaction.	(Li, Gao, Yang, & Yang, 2022)
Lanthanum modified-k-carrageenan/sodium alginate-based aerogel	Sol-gel synthesis approach on the starting materials	Ciprofloxacin (CIP) adsorption	Aerogel was investigated for the adsorption of CIP as a simulated contaminant from water. Electrostatic interaction, hydrogen bonding, and π-π interaction between support aerogel and the drug were explained as the mode of interaction.	(Yu, Yang, Zhang, Yang, & Ma, 2021)

TABLE 15.4 Summarized Carrageenan-Based Conventional Emulsion and Pickering Emulsion System Drug Delivery Network

Emulsion Compositions	Emulsion Type	Encapsulated Drugs	Comments	References
k-carrageenan based oil-in-gel emulsion	Oil-in-gel (O/g)	β-carotene	In-situ ionic gelation method was applied for system fabrication for the delivery of the drug. Hence ionic gelation increased the lipid droplet size.	(Soukoulis, Tsevdou, Andre, Cambier, Yonekura, Taoukis, & Hoffmann, 2017)
Pickering emulsion of alginate/κ-carrageenan hydrogel beads	Pickering emulsion	resveratrol	The drug-loaded Pickering emulsion was fabricated using the high-pressure homogenization technique, and the role of cations K^+ and Ca^{2+} were tested on the system capacity for drug loading.	(Wu, Li, Li, Li, & Xia, 2022)
Porcine plasma protein (PPP) nanoparticles supported by carrageenan	High internal phase emulsion (HIPEs)	-	PPP nanoparticles interacted with carrageenan through non-covalent interaction and formed a hybrid structure while acting as an emulsifier for the system.	(Yan, Yuan, Kang, Lv, & Liu, 2022)
Pickering emulsions supported in zein/carboxymethyl dextrin nanoparticles and k-carrageenan	Pickering emulsion gel	Curcumin	The emulsion system was realized based on the conjugation chemistry with EDC (N-Ethyl-N′-(3-dimethylaminopropyl) carbodiimide hydrochloride), which facilitated the bioaccessibility of the drug.	(Zhang, Meng, Li, Liu, Cheng, & Wang, 2021)
K-carrageenan stabilized emulsion	Nanoemulsion	Fucoxanthin-rich seaweed oil	Ultrasound-assisted emulsification method was used to fabricate the system, and the role of k-carrageenan on the emulsion droplet size and its overall influence on drug encapsulation was investigated.	(Saravana, Shanmugapriya, Gereniu, Chae, Kang, Woo, & Chun, 2019)

TABLE 15.4 *(Continued)*

Emulsion Compositions	Emulsion Type	Encapsulated Drugs	Comments	References
k-carrageenan system	Oil-in-gel emulsion	β-carotene	The role of k-carrageenan was investigated, with support from 20 mM Ca^{2+}, to offer protection against oxidative degradation of the drug. k-carrageenan (34–38%) was found to reduce drug bioaccessibility.	(Soukoulis, Tsevdou, Yonekura, Cambier, Taoukis, & Hoffmann, 2017)
k-carrageenan/polydextrose	Emulsion	caffeine	The role of potassium chloride on some important properties of the emulsion was investigated especially on the glass transition property of the system. Molecular transport of the drug in the matrix was equally investigated, with higher KCl concentration providing a counter-ion effect on k-carrageenan.	(Ikasari, Paramita, & Kasapis, 2022)

15.2.4 COMPLEX-BASED DRUG DELIVERY SYSTEM

Drug-carrier matrix interaction presents an important challenge in drug delivery processes; hence, appropriate interactions must be understood in order to achieve controlled and efficacious release of encapsulated drugs. Complexation between drugs and the intended host carrier is a common type of complex formed in drug delivery systems. Molecular complexes are formed between a ligand and its substrate through non-covalent interactions (hydrogen bonding, van der Waals forces, electrostatic interaction, and hydrophobic interactions). Inclusion complex formation is peculiar to the cyclodextrin families (host material) and their substrates (guest) through special binding interactions involving hydrophobic forces. The host dimensions and chemical peculiarities especially match with the substrates, forming a strong inclusion complex accordingly (Chen & Jiang, 2011). Polyelectrolyte complexes (PECs) are another form of a complex between oppositely charged molecules, including polymer-to-polymer type, polymer-to-drug type, and polymer-to-drug-to-polymer type complex (Lankalapalli & Kolapalli, 2009). Simply put, polyelectrolytes are polymers with net positive or negative charges at almost neutral pH. Hence, they have very good water solubility, arising from their ability to form electrostatic interactions between water molecules and their charged monomeric parts. Carrageenan, DNA, proteins, and some nanocellulose species are good examples. Table 15.5 shows a summarized carrageenan-based complex formation type drug delivery system.

15.2.5 NANOMATERIALS-BASED DRUG DELIVERY SYSTEM

This approach relies on the formation of nanoparticles between polymers on which the drug of interest is encapsulated. Due to the high surface area to volume ratio and their small sizes, a nanoparticle-based delivery strategy holds positive and exciting prospects in drug delivery across diverse media. The optical, physical, and chemical properties of the nanoparticles may be significantly modulated by the presence of carrageenan, owing to its good water dispersity, negative charges, biocompatibility, and aqueous stability. In a polymer-based delivery method, nanomaterials are employed as the drug carrier, resulting in an increase in the drug solubility and facilitating controlled release of the drug to specific target points, especially in cancer-based research where specific delivery to cancer sites is of utmost interest (Samprasit, Akkaramongkolporn, Jaewjira, & Opanasopit, 2018; Pridgen, Alexis, & Farokhzad, 2014). Based on the size of the carrier, microparticles

TABLE 15.5 Summarized Carrageenan-Based Complex Formation Drug Delivery System

Complex Compositions	Type of Complex	Encapsulated Drugs	Comments	References
Chitosan/k-carrageenan system	Polyelectrolyte complexation method	Gemcitabine	Pre-ionic gelation of chitosan, followed by polyelectrolyte complexation in the presence of $CaCl_2$, was investigated for the delivery of anti-cancer drugs.	Yilmaz, (2022)
Gelatin/l-carrageenan system	Polyelectrolyte complexation method	Polyphenol rich extract	The interaction between biopolymers was investigated for the delivery of polyphenol-rich extract toward the intestinal-type delivery system. Stable interaction between gelatin and polyphenol was instrumental in drug-based controlled release.	Gómez-Mascaraque, Llavata-Cabrero, Martínez-Sanz, Fabra, & López-Rubio (2018)
chitosan-alginate or chitosan–carrageenan	Polyelectrolyte complexes	Diltiazem clorhydrate	Comparative investigation of the different polymer matrices was investigated for the controlled release study of the drug. Chitosan-alginate matrix's good swelling behavior was found to support better drug release.	Tapia, Escobar, Costa, Sapag-Hagar, Valenzuela, Basualto, Gai, & Yazdani-Pedram (2004)
soy protein isolate (SPI) and κ-carrageenan complex	-	Quercetagetin	The addition of the polymer matrix for pH-mediated protection (pH 2.3 & 6.5) of the drug was studied. It was noted that the addition of the polymers performed better than the single polymer addition towards offering light protection and stability to the drug.	Wang, Liu, & Gao (2016)
Carboxymethyl cellulose/k-carrageenan	-	Probiotic bacteria (*L. plantarum*)	CMC and k-Carr were ionically cross-linked using K^+ and Ca^{2+}. The polymer blend complex protected the encapsulated bacterial cells against gastrointestinal tract adverse effects and thus facilitated specific drug delivery to the colon site	Dafe, Etemadi, Zarredar, & Mahdavinia (2017)

TABLE 15.5 *(Continued)*

Complex Compositions	Type of Complex	Encapsulated Drugs	Comments	References
soy protein isolate/κ-carrageenan/quercetagetin	Noncovalent complex	β-carotene	The protective effect of the complex was studied for drug stabilization. Small particle sizes and good water dispersity enhanced the delivery of the tested drug β-carotene.	Mao, Wang, Tai, Yuan, & Gao, (2017)
Chitosan/kappa (k), Lambda (λ), and Iota (τ) carrageenan	Polyelectrolyte complexation	Glucose oxidase (GOD)	The capacity of different carrageenan types and chitosan to form a controlled-release substrate or system was studied. K carrageenan/chitosan system showed the best encapsulation system for GOD.	Briones, & Sato (2010)

with larger sizes than nanoparticles are equally efficient as drug delivery carriers. Other shapes such as microbeads, nanotubes, and microcapsules are also well endowed with excellent delivery properties for varying classes of drugs. This is summarized in Table 15.6.

15.2.6 CARRAGEENAN-BASED METAL/METAL OXIDE NANOPARTICLES SYNTHESIS AND ITS DRUG DELIVERY POTENTIALITY

Metal nanoparticles (MNPs) and metal oxide nanoparticles (MONPs), particularly of copper, palladium, platinum, zinc, iron, gold, and silver, are of huge interest to scientists due to their unique optical and electrical properties. They are often prepared in an aqueous environment through the reduction of their salts by reducing agents. Major synthesis regimens include chemical, physical, and biological synthesis (Slavin, Asnis, Häfeli, & Bach, 2017). Biological synthesis has been significantly promoted as the cornerstone, especially from the perspective of environmental sustainability and protection. Biopolymer-based synthesis is one of the popular synthesis methods, characteristic of the well-exploited biological synthesis system (Jayeoye, Sirimahachai, Wattanasin, & Rujiralai, 2022; Jayeoye, Eze, Olatunji, & Tyopine, 2022; Eze, Ovatlarnporn, Jayeoye, Nalinbenjapun, & Sripetthong, 2022; Jayeoye, Eze, Singh, Olatunde, Benjakul, & Rujiralai, 2021; Jayeoye & Rujiralai, 2020). Carrageenan is an important biopolymer and has been well exploited in MNPs/MONPs fabrication for multifunctional applications. For example, Vikayahumar et al. reported the synthesis of κ-carrageenan-wrapped zinc oxide nanoparticles (ZnONPs) (Vijayakumar, Saravanakumar, Malaikozhundan, Divya, Vaseeharan, Durán-Lara, & Wang, 2020) for the delivery of antimicrobial drugs against common Methicillin-resistant *Staphylococcus aureus* (MRSA) bacterial strains. The synthesis was achieved by blending zinc acetate solution with κ-carrageenan solution and then subjected to pH alteration using sodium hydroxide solution under stirring. The white precipitate obtained after a 2-hour reaction was washed and dried, then calcined in a muffled furnace to realize ZnONPs. Moreover, Chen et al. prepared ι-carrageenan-reduced and stabilized AgNPs and Au@Ag core-shell nanoparticles gel for wound healing application (Chen, Tao, Zhang, Lu et al., 2022). The gel was realized by blending the synthesized biopolymer reducer/stabilized nanoparticles with poloxamer 407 (Pluronic F127) and boric acid. The synthesized gel was applied in wound healing-based research, where the gel was found to demonstrate photothermal-directed hyperthermia and cell death on immunogenic cells under a near-infrared

TABLE 15.6 Summarized Carrageenan-Based Nanomaterials Drug Delivery System

Nanomaterial Compositions	Type of Nanomaterial/ Diameter	Encapsulated Drugs	Comments	References
Lectin modified carboxymethylated kappa-carrageenan (CMkCar)	Microparticles/1273 µm.	Oral insulin	The protective effect of Lectin capped (CMkCar) was investigated against the gastrointestinal tract, acidic and proteolytic effect, and degradation, on oral insulin and thus enhanced its delivery to the target site.	Leong, Chung, Noordin, Onuki, Morishita, & Takayama (2011)
Carboxymethylated iota-carrageenan (CMiCar)/chitosan nanoparticles	Nanoparticles/613 nm.	Oral insulin	The protective effect of (CMiCar) on oral insulin was investigated in the simulated gastric fluid was studied. Nanoparticles' good stability and small size improved the pH-controlled release of the drug.	Sahoo, Leong, Nyamathulla, Onuki, Takayama, & Chung (2017)
Gelatin/tannic acid/κ-carrageenan (GTC)	Microparticles/46 µm.	–	GTC microparticles were formed from polyelectrolyte interaction occurring between gelatin and k-carrageenan and hydrogen bonding with Tannic acid, which was used and applied for hemostatic materials to manage hemorrhage.	Arunagiri, Tsai, Darge, Chiang, Thankachan, Mei, & Lai (2021)
Zein nanoparticles/i-carrageenan	Nanoparticles/500 nm.	–	i-carrageenan stabilization property on zein nanoparticles was investigated. I-carrageenan was efficient in offering protection within slightly acidic to neutral pH.	Cheng, & Jones (2017)
zein-carrageenan CSNPs. core-shell	Nanoparticles/	curcumin and piperine	A nanoparticles delivery system was formed using hydrophobic zein nanoparticles as core and anionic ι-carrageenan, as the shell material. Curcumin and piperin drugs could load on nanoparticle systems through a multitude of binding interactions ranging from hydrogen bonding, hydrophobic interaction, and electrostatic attraction.	Chen, Li, McClements, Han, Dai, Mao, & Gao (2020)

TABLE 15.6 *(Continued)*

Nanomaterial Compositions	Type of Nanomaterial/ Diameter	Encapsulated Drugs	Comments	References
Superparamagnetic/κ-carrageenan	Nanoparticles	Anti-cancer drug 5-fluorouracil	DMA was firstly grafted on the k-carrageenan surface and then coated on Fe_3O_4 nanoparticles, to achieve a magnetic and pH-sensitive delivery system for anti-cancer drugs.	Geyik, & Işıklan, (2022)
Chitosan/k-carrageenan	Spherical Nanoparticles/200–400 nm.	α-mangostin	Polymeric nanoparticles of chitosan and k-carrageenan improved drug solubility and selectivity against cancer cells. Nanoparticles were prepared through ionic gelation, coupled with the spray pyrolysis method.	Wathoni, Meylina, Rusdin, Mohammed et al. (2021)
Bovine serum albumin (BSA)/k-carrageenan	Nanotubes/400–800 nm.	Curcumin	Efficient electrostatic interaction between BSA and k-carrageenan allowed for the formation of bilayers, and subsequent polyelectrolyte complex system. The system was used for the delivery of curcumin as a model drug.	Maldonado, Chough, Bonilla, Kim, & Kokini (2019)
Sodium Caseinate/κ-Carrageenan	Microbeads/25.3–176.1 µm.	β-carotene	Ionotropic gelation between the polymer's substrates at pH 3.5 to 7 affected the property of microbeads realized. It was used for β-carotene delivery in food products.	(Perrechil, Maximo, Sato, & Cunha (2020)
Carrageenan Coated Magnetic Nanoparticles	Nanoparticles	-	Carrageenan stabilization protects nanoparticles from aggregation and is thus helpful in anti-hemolytic activity.	(Shanmuga, Singhal, & Sen (2015)
Ovalbumin/k-carrageenan complex	Nanoparticles	Curcumin	Polymer mix enhanced curcumin stability in an acidic environment conferred biocompatibility, and ensured the controlled release of the drug.	(Xie, Xiang, Li, Wang et al. (2019)

CSNPs = Core-shell nanoparticles; DMA = Dimethyl aminoethyl methacrylate.

(NIR) laser source. One major difficulty often encountered during MNPs and MONPs synthesis is the need for heating and sometimes catalyst application using sodium hydroxide solution. This is owing to the poor reductive capacity of biopolymers (carrageenan inclusive) to effect the reduction of precursor metal salts at ambient conditions. To circumvent this challenge, sometimes prior hydrolysis/oxidation of biopolymers is conducted to imbue the biopolymer with more reductive functional groups/functionality so that MNPs/MONPs synthesis could be achieved at a less demanding condition. In this light, carrageenan oligosaccharide (CarOLI) mediated synthesis of stable and dispersed gold nanoparticles (AuNPs), destined for photothermal therapy and drug delivery application, was fabricated (Chen, Zhao, Gao, Yin, Bai, & Wang, 2018). The CarOLI was used as the reductant and stabilizing agent for the synthesized AuNPs. CarOLI was obtained from mild acid hydrolysis of k-Carr, which was substantiated by assessing the degree of polymerization of carrageenan before and after hydrolysis. Through the microwave synthesis method, k-carrageenan stabilized silver nanoparticles (AgNPs) were synthesized and applied to bacterial biofilm eradication (Goel, Meher, Gupta, Gulati, Pruthi, & Poluri, 2019). The reductive efficiency of the biopolymer was enhanced by the energy supplied using the common kitchen microwave. AgNPs with well-ordered morphology were realized with an average size distribution of 50 ± 10 nm. The AgNPs demonstrated reliable antibacterial properties since AgNPs could serve as nanodrug material against bacterial cells (Jayeoye, Olatunde, Benjakul, & Rujiralai, 2020). Besides the application of carrageenan in modulating the MNP/MONPs synthesis process, the drug delivery system comprising carrageenan, initially synthesized MNP/MONPs has also been well studied. This approach leads to a composite system where different base materials are allowed to react together through intermolecular interactions to form a carefully architected delivery system.

15.2.7 CARRAGEENAN-BASED COMPOSITES FOR DRUG DELIVERY APPLICATIONS

Carrageenan-based composite drug delivery system is another reliable approach to fabricating a drug carrier matrix for a productive and conserved drug delivery regimen. The individual compositions are made to react through various chemical interactions to yield a highly mixed or cross-linked network in which the drug of interest is encapsulated and protected from facile destruction. Composites-based drug delivery systems may combine some of the mentioned methods (gels, hydrogels, films, beads, etc.), in conjunction with nanomaterials, all interacting through intermolecular forces. MNP/MONPs are also mixed or polymer nanoparticles such as chitosan nanoparticles or protein-based nanoparticles. Bantang et al. fabricated a carrageenan-based composite gels and film, through in-situ generated AgNPs and applied the composites for fungi and bacterial inactivation (Bantang, Bigol, & Camacho, 2021). Applying the reductive strength of carrageenan (κ-, ι-, and λ), then supplemented with heating, silver nitrate was reduced into its nanoparticles form and then cast in a plate to form a film. The film was found to be capable of holding in their polymeric structures both AgNPs/Ag^+, which can be delivered into fungi and bacterial cells environment for inactivation. Exploration of protein-based nanoparticles, such as zein nanoparticles (obtained from maize), for composite fabrication, has been well investigated. Zein nanoparticles have demonstrated impressive biological attributes such as antioxidant, antibacterial, and high encapsulation potentials for bioactive compounds (Luque-Alcaraz, Velazquez-Antillón, Hernández-Téllez, Graciano-Verdugo et al., 2022). Composites of zein nanoparticles with carrageenan would provide materials with inherent multifunctionalities, which can be attributed to the synergistic effects of the various components of the composites. Zou et al. reported a multifunctional zein nanoparticles/carrageenan composite nanoparticle as a drug carrier material for the delivery of hydrophobic nutraceuticals (quercetin) (Zou, Qian, Rong, Cao, McClements, & Hu, 2021). The Quercetin-loaded carrageenan-based nanoparticles composite exhibited a particle size of about 150 nm. The composite nanoparticle strategy enhanced quercetin bioaccessibility, which resulted in improved *in-vitro* antioxidant ability and intracellular reactive oxygen species (ROS) generation. Table 15.7 shows some more carrageenan composite-based delivery systems for drug delivery applications.

15.2.8 CARRAGEENAN-BASED ELECTRO-SPUN NANOFIBERS FOR DRUG DELIVERY APPLICATIONS

The electrospinning technique is another very prominent method for fabricating solid platforms with porous and nanofibrous morphology, with an inherent capacity for drug encapsulation and sustained release. Electrospinning involves the use of high voltage loads of electric field applied to a mixed polymeric material (in a syringe) to generate thin nanofibers, with porous three-dimensional (3D) structures connected through a metal needle. This technique has become the darling of nanotechnology as it allows for the fabrication of carrier systems with huge potential in food, pharmaceuticals, biomedicine, industrial applications (water purification and

TABLE 15.7 Summarized Carrageenan-Based Composites Drug Delivery System

Carrageenan Based Composites	Type of Material/ Diameter	Encapsulated Drugs	Comments	References
k-carrageenan stimuli-responsive composite	Nanogel/hydrogel	Levodopa	A radical polymerization method was adopted to synthesize the nanogel-yielding poly(N-isopropyl acrylamide). The formed nanogel was then introduced in hydrogel during synthesis to obtain nanogel-hydrogel soft nanocomposite, which can be used for enhanced release of LDOPA, as a model drug.	Bardajee, Khamooshi, Nasri, & Vancaeyzeele (2020)
κ-carrageenan/ cinnamon essential oil/hydroxypropyl-β-cyclodextrin composite	Gel type composite	Cinnamon essential oil.	The hardness of the carrageenan-based gel containing the essential oil was affected by the concentration of the composites. K-carrageenan of about 5% offered excellent protection of the essentials which was adopted for food preservation.	Wang, Yuan, Liu, & Cui (2021)
k-carrageenan (Kc)/halloysite nanotube (HNT) nanocomposite	Hydrogel type	Rhodamine B, (cationic) and Orange G (anionic) model drugs.	Halloysite nanotube incorporation improved the thermal stability of the composites. The composite was investigated to show its promising features in oral drug delivery systems.	Sharifzadeh, Wahit, Soheilmoghaddam, Whye, & Pasbakhsh (2016)
Ultrasonically derived AgNPs on cotton fabric using i-carrageenan as biopolymer substrate.	Colloidal bio nanocomposite	-	AgNPs-iota carrageenan-cotton fabric was fabricated using ultrasonic waves generated from an ultrasonic system. Strong interaction between AgNPs and i-carrageenan enhanced facile deposition on cotton fabric with strong potential for biomedical applications, including drug delivery.	Noralian, Gashti, Moghaddam, Tayyeb, & Erfanian (2021)
Polyvinyl alcohol (PVA)-k-carrageenan-magnetite Fe_3O_4 nanoparticles	Nanocomposite hydrogel system	Sodium diclofenac.	Interpenetrating network (IPN), nanocomposite hydrogel was fabricated for drug delivery. The magnetite nanoparticles were synthesized in situ inside the PVA/k-carrageenan matrix. The drug was investigated on Gram-positive bacterial strain (*S aureus*), while k-carrageenan was revealed to induce muco-adhesiveness on the nanocomposites.	Mahdavinia, & Etemadi (2014)

TABLE 15.7 *(Continued)*

Carrageenan Based Composites	Type of Material/ Diameter	Encapsulated Drugs	Comments	References
k-carrageenan/ magnetic nanofiller	Hydrogel nanocomposite	Methylene Blue	Magnetic iron nanoparticles were prepared in the k-carrageenan matrix. The nanofiller content or concentration plays a significant role in the physical properties of the composite including its gel swelling and strength. Sustained release of drug was performed.	Daniel-da-Silva, Moreira, Neto, Estrada, Gil, & Trindade (2012)
Alginate/k-carrageenan or alginate/gelatin composite	Hydrogel beads	Curcumin	Comparative encapsulation efficiency and drug release profiling for hydrophobic drug curcumin were investigated. The platform provided good surfaces for drug encapsulation for enhanced drug release.	PM, Joy, Pradeep, John, & George, (2019)
Poly (vinyl alcohol)/ Carrageenan composite	Film system	Minocycline	Film was fabricated through ionic cross-linking of polymers with borax and K^+ addition, in an aqueous environment. The drug was loaded for improved anti-microbial properties.	Bajpai, Daheriya, Ahuja, & Gupta (2016)
Starch/κ-carrageenan composite	Hydrogel system	Zaltoprofen	Release pellets for the drug were developed through a cross-linked hydrogel composite arrangement between starch and k-Carr. The composite showed good *in-vitro* drug release and muco-adhesion properties, which facilitated enhanced release of the drug.	Sonawane & Patil, (2018)

pollution management), energy storage systems (super-capacitance devices), analytical sensing, air purification, etc. (Xue, Wu, Dai, & Xia, 2019). The generated ultra-fine nanofibers are collected on a conductive surface called the collector placed a few millimeters away from the needle. The material to be spun, contained inside the syringe, can be polymeric, ceramic material, and composite materials. The morphology of the resulting fiber depends on several factors that can affect the electrospinning process. These parameters include the viscosity of the solution inside the syringe, the distance between the needle tip and the collector (tip-to-collector distance), and the applied voltage (whether alternating or direct current). Other factors are referred to as environmental factors, i.e., the conditions of the environment where the electrospinning process is carried out, which mainly depend on temperature and humidity (Liu, Gough, Deng, Gu, Wang, & Hu, 2020). The electrospinning method has gained wide scientific popularity over the past few decades, owing to its robustness, cost-effectiveness, reproducibility, stability, and high encapsulation potentiality, attributed to its nanofibrous morphology and its high surface area to volume ratio (Ray, Chen, Li, Nguyen, & Nguyen, 2016). For drug encapsulation, the drug of interest may be spun with the polymer solution in the syringe through mixing. This is known as the in-situ process. On the other hand, the nanofiber may be prepared and; afterward, treated with drugs to be encapsulated. This approach is known as *ex-situ* modification (e.g., covalent modification) (Haider, Haider, & Kang, 2018). The two processes differ in their drug encapsulation efficiency; hence the property of the intended drug would determine which treatment regimen to be applied for drug loading on nanofibers. Table 15.8 shows a summarized list of some carrageenan-based nanofiber nets for drug delivery purposes, realized through the electrospinning method.

15.2.9 LIPOSOMES/MICELLES CARRAGEENAN-BASED DRUG DELIVERY SYSTEMS

Nanoliposomes, liposome nanoparticles, or simply liposomes are lipid-derived nanoparticles with different compositions, each providing distinct functionality. They contain a lipid bilayer with membranes that are spherical, providing both a hydrophilic and hydrophobic environment. Their ease of functionalization, surface layers/sizes tunability, adjustability, and biodegradability have positioned liposomes as an attractive material for biomedicine, especially in drug delivery (Trucillo, Campardelli, & Reverchon, 2020; Kashapov, Ibragimova, Pavlov, Gabdrakhmanov et al., 2021). Liposomes are classified based on their synthesis strategy; hence they include microfluidization-type

TABLE 15.8 Summarized Carrageenan-Based Nanofibrous Drug Delivery System Through Electrospinning Technique

Carrageenan Based Electro Spun Fibers	Type of Material/ Diameter	Encapsulated Drugs	Comments	References
Hyaluronic Acid/Oxidized K-Carrageenan/polyvinyl alcohol	-	Profiled for Biomedical application	Sodium periodate ($NaIO_4$), oxidation of k-carrageenan was carried out to reduce its hydrophilicity and gel property. The nanofiber demonstrated improved anti-bacterial properties against Gram-positive (*S aureus*) and Gram-negative (*E coli*).	Abou-Okeil, Fahmy, Fouda, Aly, & Ibrahim (2021)
ZnONPs/rosemary essential oil incorporated zein nanoparticles/κ-carrageenan	Nanofibrous mat/640 ± 240 nm	ZnONPs/rosemary essential oil	ZnONPs/rosemary essential oil perfectly dispersed in the nanofiber matrix and could improve the biological properties of the system (cytotoxicity, biocompatibility) and thus, may serve as an active layer for food packaging.	Amjadi, Almasi, Ghorbani, & Ramazani (2020)
Carboxymethyl-kappa-carrageenan (CMKC) in polyvinyl alcohol (PVA) system	Nanofiber	-	CMKC-PVA fiber reduced the viability and adhesion of bacterial strains *S. aureus* and *P. aeruginosa* and also helped in improving blood coagulation activity, all pointing to good dressing materials support for wound healing.	(Madruga, Popat, Balaban, & Kipper (2021)
Quaternized chitosan/k-carrageenan/caffeic acid-coated poly (3-hydroxy-butyrate)	Nanofibrous material/	Caffeic acid	The material was prepared by electrospinning, dip-coating, and complex formation (polyelectrolyte complex). The drug-coated mat was efficient against Gram-positive (*S. aureus*), and Gram-negative (*E. coli*), coupled with antioxidant activity enhancement.	Ignatova, Manolova, Rashkov, & Markova (2016)
k-carrageenan-PHB and k-carr-PHBV Fibers	Nanofiber	-	K-carrageenan blending led to a much smaller fiber dimension, while also promoting osteogenic differentiation property and bioactivity, as scaffold materials for bone tissue regeneration.	Goonoo, Khanbabaee, Steuber, Bhaw-Luximon et al. (2017)

TABLE 15.8 *(Continued)*

Carrageenan Based Electro Spun Fibers	Type of Material/ Diameter	Encapsulated Drugs	Comments	References
PVA/k-carr/Au/ camptothecin/PEG-PU/ paclitaxel	Core-shell nanofiber system	Camptothecin (CMPT) and paclitaxel (PTX) anticancer drugs	A composite nanofibrous membrane was prepared through a two-nozzle/co-axial electrospinning system for the delivery of anti-cancer drugs. Results confirmed that the loading of anti-cancer drugs with gold nanoparticles improved drug treatment for lung cancer.	Irani & Nodeh (2022)
Polycaprolactone/ chitosan/κ-carr	Nanofiber	-	k-carr improved the blending/stability among polymers as a result of electrostatic interaction between k-Carr and chitosan. Improved cell viability on ST-2 cells which accentuates the potential of the mat for wound healing related applications.	Vargas-Osorio, Ruther, Chen, Sengupta et al. (2022)
Silk fibroin/k-carr	Hybrid nanofiber	-	Silk fibroin-k-Carr nanofiber was prepared through a genipen cross-linking strategy. The nanofiber mat provided good mechanical properties, which may act as a support for bone regeneration.	Roshanfar, Hesaraki, & Dolatshahi-Pirouz (2022)

ZnONPs = Zinc oxide nanoparticles; PHB = Poly hydroxybutyrate; PHBV = Poly hydroxybutyrate Valerate; k-carr = k-carrageenan; PVA = Polyvinyl alcohol; PEG-PU = Pegylated polyurethane.

liposomes, extrusion method-based liposomes, heating-derived liposomes, lyophilization-based liposomes, solvent dispersion-derived liposomes, and reverse phase evaporation (REV) method-based liposomes (Nakhaei, Margiana, Bokov, Abdelbasset et al., 2021). The major characteristic of the above-realized liposome is dependent on the lipid composition. For example, liposomes fabricated using unsaturated phosphatidylcholine groups result in lipid bilayers with low stability; however, those realized with saturated phospholipids-mediated synthesis (dipalmitoyl phosphatidylcholine) result in bilayers with a rigid structure and impermeable bilayer arrangements (Sahoo & Labhasetwar, 2003). Carrageenan-based liposomes have been investigated and exploited for drug delivery-purposes. Yermak et al. studied an Echinochrome drug loaded on Liposomal Carrageenan Complex for sustained drug release application (Yermak, Gorbach, Glazunov, Kravchenko, Mishchenko, Pimenova, & Davydova, 2018). The Liposomal complex was prepared through the thin film hydration method coupled with sonication. It was noted that the drug was not oxidized after encapsulation, and it retained its property after liposome formation and subsequent freeze-drying. The carrageenan-protected drug improved the mucoadhesive properties of the drug. The effect of chitosan-coated oleic acid liposome on carbohydrate-derived gels (iota-carrageenan, carboxymethyl cellulose) has been reported (Tan & Misni, 2014). It was noted that the rheological properties of the gel mix confirmed that chitosan-coated oleic acid liposome impregnation in the gel influenced the viscoelastic properties of the mix. This property might be beneficial for the design of liposomal matrix with reliable flow properties. Furthermore, the stability and stability of multi-layered liposomes were investigated using biopolymer deposition (Chun, Choi, Min, & Weiss, 2013). In the work, small-sized liposomes were prepared from lecithin based on the ultrasonic homogenization method. The effects of biopolymers (high methoxy pectin, chitosan, λ-carrageenan) on phospholipid membranes of liposomes were investigated. A liposome system of about 80 nm was realized, which was built by a layer-by-layer approach. The system was found to be suitable for the design and obtention of a highly stable emulsion system for food-related compound encapsulation.

Micelles or Micellar drug delivery systems using (polymeric micelles) are spherical assemblies comprising aggregated colloids possessing amphiphilic polymer components. The polymers, below a certain concentration (critical micellar concentration, CMC), are present as a single entity; on the other hand, above the CMC, they self-assemble into micelles containing an ordered architecture with a lipophilic core (internal) and a hydrophilic shell (external) (Ghezzi, Pescina, Padula, Santi et al., 2021). The typical micelle shape and drug loading are displayed in Figure 15.3. Polymeric micelles are described as being amphiphilic owing to the combination of hydrophilic and hydrophobic sections in their structure (Figure 15.3). These sections confer diverse functional advantages on polymeric micelles, which is manifested in huge research interests devoted to such studies in the last decade.

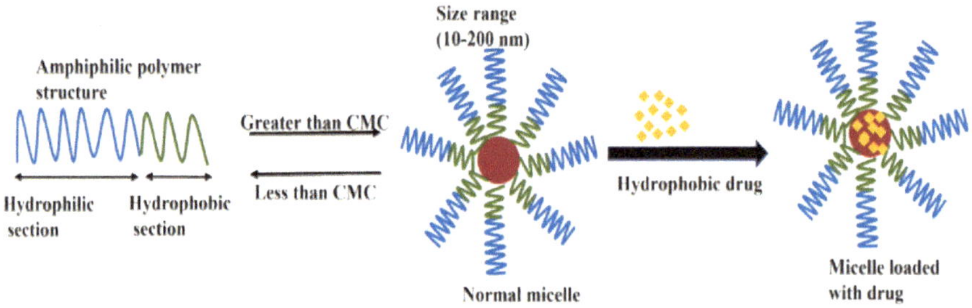

FIGURE 15.3 Detailed representation of a typical polymeric micelle before and after drug loading.

Source: Adapted from Ghezzi et al.,2021

Polymeric micelles, owing to their unique structure, stability, and ease of fabrication, provide a versatile drug carrier system for the delivery of hydrophilic and hydrophobic drugs. Their amphiphilic polymer structure can interact with biological membranes and consequently demonstrate important biological activities. Youssouf et al. reported the encapsulation of curcumin as a model drug on polycaprolactone grafted oligo carrageenan nano micelle system (Youssouf, Bhaw-Luximon, Diotel, Catan et al., 2019). The amphiphilic nano micelle system of polycaprolactone grafted oligo carrageenan showed an average particle size of 187 ± 21 nm and provided a platform for the encapsulation of the hydrophobic drug curcumin, with sustained release property to different organs such as lungs, liver, and brain in a mouse model. Moreover, protein-based micellar systems with polysaccharides are also very prominent. Finally, Carrageenan-based delivery systems provide surfaces for the efficient delivery of drugs, using the inherent biological properties of Carrageenans. They are biocompatible, non-toxic, stable, and highly water-soluble. They provide good stability for food products and nanomaterials. The above discussion is not exhaustive but has been carefully discussed to capture some of the common areas of carrageenan exploitation in drug delivery platforms.

15.3 CONCLUSION

In this chapter, the roles of carrageenan in drug-delivery therapeutics have been discussed. Carrageenan is a noble polysaccharide, extracted from marine seaweeds, with a huge quantity in nature. This is indeed exciting, as the regular availability of carrageenans is guaranteed by virtue of the amount in the marine environment. The popularity of carrageenan as a food additive, gelling agent, and stabilizing agent for diverse formulations is attributed to the impressive biological properties of this class of biopolymer. From the discussion above, the challenges still encountered in carrageenan-based drug therapeutics may be due to the high viscosity of native carrageenan. The hydrolysis step and oxidation of carrageenan may provide a practical approach to circumvent this challenge. Another area where carrageenan has been less explored is in the fabrication of metal/metal oxide nanoparticles. This is indeed worth paying attention to. Benefiting from the high surface area to volume ratio, metal nanoparticles (MNPs) and metal oxide nanoparticles (MONPs) can be well surface-tuned for drug encapsulation and onward delivery, especially for cancer-related research works. Their toxicity reservation may be overlooked. This is emphasized because during chemotherapy, especially at the late cancer treatment stage, the toxicity profile may be of less worry, but rather, the potency of the drug may be of central concern. Furthermore, carrageenan polymeric micelles were equally found to be less reported. This, we presume, may be owing to the high hydrophilicity of carrageenan, which might disrupt the amphiphilic properties of micelles. Thus, these research gaps can be focused on for the design of novel research works to fill up the identified gaps.

KEYWORDS

- **bioactive compounds**
- **carrageenan**
- **drug delivery**
- **metal nanoparticles**
- **metal oxide nanoparticles**
- **sulfated oligosaccharides**
- **therapeutics**

REFERENCES

Abou-Okeil, A., Fahmy, H. M., Fouda, M. M., Aly, A. A., & Ibrahim, H. M. (2021). Hyaluronic acid/oxidized Κ-carrageenan electrospun nanofibers synthesis and antibacterial properties. *BioNanoScience, 11*(3), 687–695. https://doi.org/10.1007/s12668-021-00884-9.

Agostinho, D. A., Paninho, A. I., Cordeiro, T., Nunes, A. V., Fonseca, I. M., Pereira, C., & Ventura, M. G. (2020). Properties of κ-carrageenan aerogels prepared by using different dissolution media and its application as drug delivery systems. *Materials Chemistry and Physics, 253*, 123290. https://doi.org/10.1016/j.matchemphys.2020.123290.

Alnaief, M., Obaidat, R., & Mashaqbeh, H. (2019). Loading and evaluation of meloxicam and atorvastatin in carrageenan microspherical aerogels particles. *Journal of Applied Pharmaceutical Science, 9*(1), 83–88. https://doi.org/10.7324/JAPS.2019.90112.

Amjadi, S., Almasi, H., Ghorbani, M., & Ramazani, S. (2020). Reinforced ZnONPs/rosemary essential oil-incorporated zein electrospun nanofibers by κ-carrageenan. *Carbohydrate Polymers, 232,* 115800. https://doi.org/10.1016/j.carbpol.2019.115800.

Arunagiri, V., Tsai, H. C., Darge, H. F., Chiang, H. W., Thankachan, D., Mei, C. J., & Lai, J. Y. (2021). Preparation of physically crosslinked polyelectrolyte gelatin-tannic acid-κ-carrageenan (GTC) microparticles as hemostatic agents. *International Journal of Biological Macromolecules, 191*, 324–334. https://doi.org/10.1016/j.ijbiomac.2021.09.008.

Awalludin, A., Mingu, N., Yaser, A. Z., Mamat, H., Kamaruzaman, K. A., Jamin, Z., Rahman, M. L., Abd Majid, M. H., & Sarjadi, M. S. (2021). Extraction and Physicochemical properties of refined kappa-carrageenan from *Kappaphycus alvarezii* originated from Semporna, Sabah. *Journal of Applied Science and Engineering, 4*, 11–16. https://doi.org/10.6180/jase.202206_25(3).0006.

Bajpai, S. K., Daheriya, P., Ahuja, S., & Gupta, K. (2016). Water absorption and antimicrobial behavior of physically cross-linked poly (vinyl alcohol)/carrageenan films loaded with minocycline. *Designed Monomers and Polymers, 19*(7), 630–642. http://dx.doi.org/10.1080/15685551.2016.1187441.

Bantang, J. P., Bigol, U. G., & Camacho, D. H. (2021). Gel and film composites of silver nanoparticles in κ-, ι-, and λ-carrageenans: One-pot synthesis, characterization, and bioactivities. *Bionanoscience, 11*(1), 53–56. https://doi.org/10.1007/s12668-020-00806-1.

Bardajee, G. R., Khamooshi, N., Nasri, S., & Vancaeyzeele, C. (2020). Multi-stimuli responsive nanogel/hydrogel nanocomposites based on κ-carrageenan for prolonged release of levodopa as model drug. *International Journal of Biological Macromolecules, 153,* 180–189. https://doi.org/10.1016/j.ijbiomac.2020.02.329.

BeMiller, J. N. (2018). Carrageenan chapter. In *Carbohydrate Chemistry for Food Scientists, 3rd Edition* (pp. 279–291). Elsevier Inc. https://doi.org/10.1016/B978-0-12-812069-9.00013-3.

Blakemore, W. R., & Harpell, A. R. (2010). Carrageenan. In *Food Stabilisers, Thickeners and Gelling Agents* (pp. 73–94), Wiley.

Briones, A. V., & Sato, T. (2010). Encapsulation of glucose oxidase (GOD) in polyelectrolyte complexes of chitosan–carrageenan. *Reactive and Functional Polymers, 70*(1), 19–27. https://doi.org/10.1016/j.reactfunctpolym.2009.09.009.

Carlucci, M. J., Pujol, C. A., Ciancia, M., Noseda, M. D., Matulewicz, M. C., Damonte, E. B., & Cerezo, A. S. (1997). Antiherpetic and anticoagulant properties of carrageenans from the red seaweed *Gigartina skottsbergii* and their cyclized derivatives: Correlation between structure and biological activity. *International Journal of Biological Macromolecules, 20*(2), *97–105.* https://doi.org/10.1016/S0141-8130(96)01145-2.

Chen, G., & Jiang, M. (2011). Cyclodextrin-based inclusion complexation bridging supramolecular chemistry and macromolecular self-assembly. *Chemical Society Reviews, 40*(5), 2254–2266. https://doi.org/10.1039/C0CS00153H.

Chen, S., Li, Q., McClements, D. J., Han, Y., Dai, L., Mao, L., & Gao, Y. (2020). Co-delivery of curcumin and piperine in zein-carrageenan core-shell nanoparticles: Formation, structure, stability and in vitro gastrointestinal digestion. *Food Hydrocolloids, 99*, 105334. https://doi.org/10.1016/j.foodhyd.2019.105334.

Chen, X., Tao, J., Zhang, M., Lu, Z., Yu, Y., Song, P., Wang, T., Jiang, T., & Zhao, X. (2022). Iota carrageenan gold-silver NPs photothermal hydrogel for tumor postsurgical anti-recurrence and wound healing. *Carbohydrate Polymers, 298,* 120123. https://doi.org/10.1016/j.carbpol.2022.120123.

Chen, X., Zhao, X., Gao, Y., Yin, J., Bai, M., & Wang, F. (2018). Green synthesis of gold nanoparticles using carrageenan oligosaccharide and their in vitro antitumor activity. *Marine Drugs, 16*(8), 277. https://doi.org/10.3390/md16080277.

Cheng, C. J., & Jones, O. G. (2017). Stabilizing zein nanoparticle dispersions with ι-carrageenan. *Food Hydrocolloids, 69*, 28–35. http://dx.doi.org/10.1016/j.foodhyd.2017.01.022.

Chun, J. Y., Choi, M. J., Min, S. G., & Weiss, J. (2013). Formation and stability of multiple-layered liposomes by layer-by-layer electrostatic deposition of biopolymers. *Food Hydrocolloids, 30*(1), 249–257. http://dx.doi.org/10.1016/j.foodhyd.2012.05.024.

Dafe, A., Etemadi, H., Zarredar, H., & Mahdavinia, G. R. (2017). Development of novel carboxymethyl cellulose/κ-carrageenan blends as an enteric delivery vehicle for probiotic bacteria. *International Journal of Biological Macromolecules, 97*, 299–307. http://dx.doi.org/10.1016/j.ijbiomac.2017.01.016.

Daniel-da-Silva, A. L., Moreira, J., Neto, R., Estrada, A. C., Gil, A. M., & Trindade, T. (2012). Impact of magnetic nanofillers in the swelling and release properties of κ-carrageenan hydrogel nanocomposites. *Carbohydrate Polymers, 87*(1), 328–335. http://dx.doi.org/10.1016/j.carbpol.2011.07.051.

Dong, Y., Wei, Z., & Xue, C. (2021). Recent advances in carrageenan-based delivery systems for bioactive ingredients: A review. *Trends in Food Science & Technology, 112*, 348–361. https://doi.org/10.1016/j.tifs.2021.04.012.

Du Sorbier, Q. M., Aimable, A., & Pagnoux, C. (2015). Influence of the electrostatic interactions in a Pickering emulsion polymerization for the synthesis of silica–polystyrene hybrid nanoparticles. *Journal of Colloid and Interface Science, 448*, 306–314. https://doi.org/10.1016/j.jcis.2015.02.017.

Eze, F. N., Ovatlarnporn, C., Jayeoye, T. J., Nalinbenjapun, S., & Sripetthong, S. (2022). One-pot biofabrication and characterization of Tara gum/Riceberry phenolics–silver nanogel: A cytocompatible and green nanoplatform with multifaceted biological applications. *International Journal of Biological Macromolecules, 206,* 521–533. https://doi.org/10.1016/j.ijbiomac.2022.02.140.

Fahmy, H. M., Aly, A. A., Sayed, S. M., & Abou-Okeil, A. (2021). Κ-carrageenan/Na-alginate wound dressing with sustainable drug delivery properties. *Polymers for Advanced Technologies, 32*(4), 1793–1801. https://doi.org/10.1002/pat.5218.

Ganesan, K., & Ratke, L. (2014). Facile preparation of monolithic κ-carrageenan aerogels. *Soft Matter, 10*(18), 3218–3224. https://doi.org/10.1039/C3SM52862F.

Ganesan, K., & Ratke, L. (2014). Facile preparation of monolithic κ-carrageenan aerogels. *Soft Matter, 10*(18), 3218–3224. https://doi.org/10.1039/C3SM52862F.

Geyik, G., & Işıklan, N. (2022). Multi-stimuli-sensitive superparamagnetic κ-carrageenan based nanoparticles for controlled 5-fluorouracil delivery. *Colloids and Surfaces A: Physicochemical and Engineering Aspects*, *634*, 127960. https://doi.org/10.1016/j.colsurfa.2021.127960.

Ghezzi, M., Pescina, S., Padula, C., Santi, P., Del Favero, E., Cantù, L., & Nicoli, S. (2021). Polymeric micelles in drug delivery: An insight of the techniques for their characterization and assessment in biorelevant conditions. *Journal of Controlled Release, 332,* 312–336. https://doi.org/10.1016/j.jconrel.2021.02.031.

Goel, A., Meher, M. K., Gupta, P., Gulati, K., Pruthi, V., & Poluri, K. M. (2019). Microwave assisted κ-carrageenan capped silver nanocomposites for eradication of bacterial biofilms. *Carbohydrate Polymers, 206,* 854–862. https://doi.org/10.1016/j.carbpol.2018.11.033.

Gómez-Mascaraque, L. G., Llavata-Cabrero, B., Martínez-Sanz, M., Fabra, M. J., & López-Rubio, A. (2018). Self-assembled gelatin-ι-carrageenan encapsulation structures for intestinal-targeted release applications. *Journal of Colloid and Interface Science, 517*, 113–123. https://doi.org/10.1016/j.jcis.2018.01.101.

Goonoo, N., Khanbabaee, B., Steuber, M., Bhaw-Luximon, A., Jonas, U., Pietsch, U., Jhurry, D., & Schönherr, H. (2017). κ-Carrageenan enhances the biomineralization and osteogenic differentiation of electrospun polyhydroxybutyrate and polyhydroxybutyrate valerate fibers. *Biomacromolecules, 18*(5), 1563–1573. http://dx.doi.org/10.1021/acs.biomac.7b00150.

Grand View Research. (n.d.). Carrageenan Market. Size, Share and Trend analysis report. Retrieved from: https://www.grandviewresearch.com/industry-analysis/carrageenan-market (accessed on 7 June 2024).

Haider, A., Haider, S., & Kang, I. K. (2018). A comprehensive review summarizing the effect of electrospinning parameters and potential applications of nanofibers in biomedical and biotechnology. *Arabian Journal of Chemistry, 11*(8), 1165–1188. http://dx.doi.org/10.1016/j.arabjc.2015.11.015.

Hasizah, A., Mahendradatta, M., Laga, A., Metusalach, M., & Salengke, S. (2021). Extraction of carrageenan from Eucheuma spinosum using ohmic heating: Optimization of extraction conditions using response surface methodology. *Food Science and Technology, 41*, 928–937. https://doi.org/10.1590/fst.26220.

He, P., Dai, L., Wei, J., Zhu, X., Li, J., Chen, Z., & Ni, Y. (2022). Nanocellulose-based hydrogels as versatile drug delivery vehicles: A review. *International Journal of Biological Macromolecules, 222,* 830–843. https://doi.org/10.1016/j.ijbiomac.2022.09.214.

Hezaveh, H., & Muhamad, I. I. (2013). Controlled drug release via minimization of burst release in pH-response kappa-carrageenan/polyvinyl alcohol hydrogels. *Chemical Engineering Research and Design, 91*(3), 508–519. https://doi.org/10.1016/j.cherd.2012.08.014.

Hoare, T. R., & Kohane, D. S. (2008). Hydrogels in drug delivery: Progress and challenges. *Polymer, 49*(8), 1993–2007. https://doi.org/10.1016/j.polymer.2008.01.027.

Ignatova, M., Manolova, N., Rashkov, I., & Markova, N. (2016). Quaternized chitosan/κ-carrageenan/caffeic acid–coated poly (3-hydroxybutyrate) fibrous materials: Preparation, antibacterial and antioxidant activity. *International Journal of Pharmaceutics, 513*(1–2), 528–537. http://dx.doi.org/10.1016/j.ijpharm.2016.09.062.

Ikasari, D., Paramita, V. D., & Kasapis, S. (2022). Glass transition effects on the molecular transport of caffeine from condensed k-carrageenan/polydextrose systems. *Food Hydrocolloids, 126*, 107401. https://doi.org/10.1016/j.foodhyd.2021.107401.

Irani, M., & Nodeh, S. M. (2022). PVA/κ-carrageenan/Au/camptothecin/pegylated-polyurethane/paclitaxel nanofibers against lung cancer treatment. *RSC Advances, 12*(25), 16310–16318. https://doi.org/10.1039/D2RA02150A.

Jaiswal, L., Shankar, S., & Rhim, J. W. (2019). Carrageenan-based functional hydrogel film reinforced with sulfur nanoparticles and grapefruit seed extract for wound healing application. *Carbohydrate Polymers, 224*, 115191. https://doi.org/10.1016/j.carbpol.2019.115191.

Jayeoye, T. J., & Rujiralai, T. (2020). Green, in situ fabrication of silver/poly (3-aminophenyl boronic acid)/sodium alginate nanogel and hydrogen peroxide sensing capacity. *Carbohydrate Polymers, 246,* 116657. https://doi.org/10.1016/j.carbpol.2020.116657.

Jayeoye, T. J., Eze, F. N., Olatunji, O. J., & Tyopine, A. A. (2022). Synthesis of biocompatible Konjac glucomannan stabilized silver nanoparticles, with Asystasia gangetica phenolic extract for colorimetric detection of mercury (II) ion. *Scientific Reports, 12*(1), 1–5. https://doi.org/10.1038/s41598-022-13384-x.

Jayeoye, T. J., Eze, F. N., Singh, S., Olatunde, O. O., Benjakul, S., & Rujiralai, T. (2021). Synthesis of gold nanoparticles/polyaniline boronic acid/sodium alginate aqueous nanocomposite based on chemical oxidative polymerization for biological applications. *International Journal of Biological Macromolecules, 179,* 196–205. https://doi.org/10.1016/j.ijbiomac.2021.02.199.

Jayeoye, T. J., Olatunde, O. O., Benjakul, S., & Rujiralai, T. (2020). Synthesis and characterization of novel poly (3-aminophenyl boronic acid-co-vinyl alcohol) nanocomposite polymer stabilized silver nanoparticles with antibacterial and antioxidant applications. *Colloids and Surfaces B: Biointerfaces, 193,* 111112. https://doi.org/10.1016/j.colsurfb.2020.111112.

Jayeoye, T. J., Sirimahachai, U., Wattanasin, P., & Rujiralai, T. (2022). Eco-friendly poly (aniline boronic acid)/gum tragacanth stabilized silver nanoparticles nanocomposite for selective sensing of Hg2+. *Microchemical Journal, 182*, 107949. https://doi.org/10.1016/j.microc.2022.107949.

Kalsoom Khan, A., Saba, A. U., Nawazish, S., Akhtar, F., Rashid, R., Mir, S., Nasir, B., Iqbal, F., Afzal, S., & Pervaiz, F., & Murtaza, G. (2017). Carrageenan based bionanocomposites as drug delivery tool with special emphasis on the influence of ferromagnetic nanoparticles. *Oxidative Medicine and Cellular Longevity, 2017*(1), p81583015. https://doi.org/10.1155/2017/8158315.

Kashapov, R., Ibragimova, A., Pavlov, R., Gabdrakhmanov, D., Kashapova, N., Burilova, E., Zakharova, L., & Sinyashin, O. (2021). Nanocarriers for biomedicine: From lipid formulations to inorganic and hybrid nanoparticles. *International Journal of Molecular Sciences, 22*(13), 7055. https://doi.org/10.3390/ijms22137055.

Khotimchenko, M., Tiasto, V., Kalitnik, A., Begun, M., Khotimchenko, R., Leonteva, E., Bryukhovetskiy, I., & Khotimchenko, Y. (2020). Antitumor potential of carrageenans from marine red algae. *Carbohydrate Polymers, 246*, 116568. https://doi.org/10.1016/j.carbpol.2020.116568.

Kreyling, W. G., Semmler-Behnke, M., & Chaudhry, Q. (2010). A complementary definition of nanomaterial. *Nano Today, 5*, 165–168. https://doi.org/10.1016/j.nantod.2010.03.004.

Lankalapalli, S., & Kolapalli, V. R. (2009). Polyelectrolyte complexes: A review of their applicability in drug delivery technology. *Indian Journal of Pharmaceutical Sciences, 71*(5), 481. https://doi.org/10.4103%2F0250-474X.58165.

Leong, K. H., Chung, L. Y., Noordin, M. I., Onuki, Y., Morishita, M., & Takayama, K. (2011). Lectin-functionalized carboxymethylated kappa-carrageenan microparticles for oral insulin delivery. *Carbohydrate Polymers, 86*(2), 555–565. https://doi.org/10.1016/j.carbpol.2011.04.070.

Li, L., Ni, R., Shao, Y., & Mao, S. (2014). Carrageenan and its applications in drug delivery. *Carbohydrate Polymers, 103*, 1–1. https://doi.org/10.1016/j.carbpol.2013.12.008.

Li, N., Gao, B., Yang, R., & Yang, H. (2022). Simple fabrication of carboxymethyl cellulose and κ-carrageenan composite aerogel with efficient performance in removal of fluoroquinolone antibiotics from water. *Frontiers of Environmental Science & Engineering, 16*(10), 1–4. https://doi.org/10.1007/s11783-022-1568-x.

Liu, H., Gough, C. R., Deng, Q., Gu, Z., Wang, F., & Hu, X. (2020). Recent advances in electrospun sustainable composites for biomedical, environmental, energy, and packaging applications. *International Journal of Molecular Sciences, 21*(11), 4019. https://doi.org/10.3390/ijms21114019.

Liu, Z., Gao, T., Yang, Y., Meng, F., Zhan, F., Jiang, Q., & Sun, X. (2019). Anti-cancer activity of porphyran and carrageenan from red seaweeds. *Molecules, 24*(23), 4286. https://doi.org/10.3390/molecules24234286.

Luo, M., Shao, B., Nie, W., Wei, X. W., Li, Y. L., Wang, B. L., & Wei, Y. Q. (2015). Antitumor and adjuvant activity of λ-carrageenan by stimulating immune response in cancer immunotherapy. *Scientific Reports, 5*(1), 1–2. https://doi.org/10.1038/srep11062.

Luque-Alcaraz, A. G., Velazquez-Antillón, M., Hernández-Téllez, C. N., Graciano-Verdugo, A. Z., García-Flores, N., Iriqui-Razcón, J. L., Silvas-García, M. I., Zazueta-Raynaud, A., Moreno-Vásquez, M. J., & Hernández-Abril, P. A. (2022). Antioxidant effect of nanoparticles composed of zein and orange (*Citrus sinensis*) extract obtained by ultrasound-assisted extraction. *Materials, 15*(14), 4838. https://doi.org/10.3390/ma15144838.

Madruga, L. Y., Popat, K. C., Balaban, R. C., & Kipper, M. J. (2021). Enhanced blood coagulation and antibacterial activities of carboxymethyl-kappa-carrageenan-containing nanofibers. *Carbohydrate Polymers, 273,* 118541. https://doi.org/10.1016/j.carbpol.2021.118541.

Mahdavinia, G. R., & Etemadi, H. (2014). In situ synthesis of magnetic CaraPVA IPN nanocomposite hydrogels and controlled drug release. *Materials Science and Engineering: C, 45,* 250–260. http://dx.doi.org/10.1016/j.msec.2014.09.023.

Mahdavinia, G. R., Mosallanezhad, A., Soleymani, M., & Sabzi, M. (2017). Magnetic-and pH-responsive κ-carrageenan/chitosan complexes for controlled release of methotrexate anticancer drug. *International journal of biological macromolecules, 97*, 209–217. https://doi.org/10.1016/j.ijbiomac.2017.01.012.

Maldonado, L., Chough, S., Bonilla, J., Kim, K. H., & Kokini, J. (2019). Mechanism of fabrication and nano-mechanical properties of α-lactalbumin/chitosan and BSA/κ-carrageenan nanotubes through layer-by-layer assembly for curcumin encapsulation and determination of in vitro cytotoxicity. *Food Hydrocolloids*, *93*, 293–307. https://doi.org/10.1016/j.foodhyd.2019.02.040.

Mao, L., Wang, W., Tai, K., Yuan, F., & Gao, Y. (2017). Development of a soy protein isolate–carrageenan–quercetagetin non-covalent complex for the stabilization of β-carotene emulsions. *Food & Function*, *8*(12), 4356–4363. https://doi.org/10.1039/C7FO01238A.

McClements, D. J. (2004). Food Emulsions: Principles, Practices, and Techniques. CRC Press. pp. 632. https://doi.org/10.1201/9781420039436.

McClements, D. J., Decker, E. A., & Weiss, J. (2007). Emulsion-based delivery systems for lipophilic bioactive components. *Journal of Food Science, 72*(8), R109–R124. https://doi.org/10.1111/j.1750-3841.2007.00507.x.

Nakhaei, P., Margiana, R., Bokov, D. O., Abdelbasset, W. K., Jadidi Kouhbanani, M. A., Varma, R. S., Marofi, F., Jarahian, M., & Beheshtkhoo, N. (2021). Liposomes: structure, biomedical applications, and stability parameters with emphasis on cholesterol. *Frontiers in Bioengineering and Biotechnology, 58*(4), 187–205. https://doi.org/10.3389/fbioe.2021. 705886.

Necas, J., & Bartosikova, L. (2013). Carrageenan: A review. *Veterinarni medicina, 58*(6), 264.

Noman, M. T., Amor, N., Ali, A., Petrik, S., Coufal, R., & Adach, K. (2021). Aerogels for biomedical, energy and sensing applications. *Gels, 7*(4), 264. https://doi.org/10.3390/gels 7040264.

Noralian, Z., Gashti, M. P., Moghaddam, M. R., Tayyeb, H., & Erfanian, I. (2021). Ultrasonically developed silver/iota-carrageenan/cotton bionanocomposite as an efficient material for biomedical applications. *International Journal of Biological Macromolecules, 180,* 439–457. https://doi.org/10.1016/j.ijbiomac.2021.02.204.

Obaidat, R. M., Alnaief, M., & Mashaqbeh, H. (2018). Investigation of carrageenan aerogel microparticles as a potential drug carrier. *AAPS PharmSciTech, 19*(5), 2226–2236. https://doi.org/10.1208/s12249-018-1021-4.

Ooi, S. Y., Ahmad, I., & Amin, M. C. (2016). Cellulose nanocrystals extracted from rice husks as a reinforcing material in gelatin hydrogels for use in controlled drug delivery systems. *Industrial Crops and Products, 93,* 227–234. https://doi.org/10.1016/j.indcrop.2015.11.082.

Perrechil, F. A., Maximo, G. J., Sato, A. C., & Cunha, R. L. (2020). Microbeads of sodium caseinate and κ-carrageenan as a β-carotene carrier in aqueous systems. *Food and Bioprocess Technology*, *13*(4), 661–669. https://doi.org/10.1007/s11947-020-02426-9.

Pettinelli, N., Rodríguez-Llamazares, S., Farrag, Y., Bouza, R., Barral, L., Feijoo-Bandín, S., & Lago, F. (2020). Poly (hydroxybutyrate-co-hydroxyvalerate) microparticles embedded in κ-carrageenan/locust bean gum hydrogel as a dual drug delivery carrier. *International Journal of Biological Macromolecules, 146,* 110–118. https://doi.org/10.1016/j.ijbiomac.2019.12.193.

PM, S., Joy, R., Pradeep, N., John, F., & George, J. (2019). Alginate/k-carrageenan and alginate/gelatin composite hydrogel beads for controlled drug release of curcumin. *Advanced Materials Letters, 10*(7), 508–514. http://dx.doi.org/10.5185/amlett.2019.2149.

Pridgen, E. M., Alexis, F., & Farokhzad, O. C. (2014). Polymeric nanoparticle technologies for oral drug delivery. *Clinical Gastroenterology and Hepatology*, *12*(10), 1605–1610. https://doi.org/10.1016/j.cgh.2014.06.018.

Priyadarshi, R., Kumar, B., Deeba, F., Kulshreshtha, A., & Negi, Y. S. (2019). Food packaging: natural and synthetic biopolymers. In *Encyclopedia of Polymer Applications* (pp. 1325–1342). Taylor and Francis Publishers.

Qiu, X., & Zhong, W. (2018). Antihyperglycemic and antihyperlipidemic effects of low-molecular-weight carrageenan in rats. *Open Life Sciences, 13*(1), 379–384. https://doi.org/10.1515/biol-2018-0046.

Ray, S. S., Chen, S. S., Li, C. W., Nguyen, N. C., & Nguyen, H. T. (2016). A comprehensive review: Electrospinning technique for fabrication and surface modification of membranes for water treatment application. *RSC Advances, 6*(88), 85495–85514. https://doi.org/10.1039/c6ra14952a.

Roshanfar, F., Hesaraki, S., & Dolatshahi-Pirouz, A. (2022). Electrospun Silk Fibroin/kappa-Carrageenan Hybrid Nanofibers with Enhanced Osteogenic Properties for Bone Regeneration Applications. *Biology, 11*(5), 751. https://doi.org/10.3390/biology11050751.

Rowe, R. C., Sheskey, P., & Quinn, M. (2009). *Handbook of Pharmaceutical Excipients*. Libros Digitales-Pharmaceutical Press. http://repositorio.ub.edu.ar/handle/123456789/5143.

Sahoo, P., Leong, K. H., Nyamathulla, S., Onuki, Y., Takayama, K., & Chung, L. Y. (2017). Optimization of pH-responsive carboxymethylated iota-carrageenan/chitosan nanoparticles for oral insulin delivery using response surface methodology. *Reactive and Functional Polymers*, *119*, 145–155. http://dx.doi.org/10.1016/j.reactfunctpolym.2017.08.014.

Sahoo, S. K., & Labhasetwar, V. (2003). Nanotech approaches to drug delivery and imaging. *Drug Discovery Today, 8*(24), 1112–1120. https://doi.org/10.1016/S1359-6446(03)02903-9.

Samprasit, W., Akkaramongkolporn, P., Jaewjira, S., & Opanasopit, P. (2018). Design of alpha mangostin-loaded chitosan/alginate controlled-release nanoparticles using genipin as crosslinker. *Journal of Drug Delivery Science and Technology*, *46*, 312–321. https://doi.org/10.1016/j.jddst.2018.05.029.

Saravana, P. S., Shanmugapriya, K., Gereniu, C. R., Chae, S. J., Kang, H. W., Woo, H. C., & Chun, B. S. (2019). Ultrasound-mediated fucoxanthin rich oil nanoemulsions stabilized by κ-carrageenan: Process optimization, bio-accessibility and cytotoxicity. *Ultrasonics Sonochemistry*, *55*, 105–116. https://doi.org/10.1016/j.ultsonch.2019.03.014.

Schwan, M., & Ratke, L. (2013). Flexibilisation of resorcinol–formaldehyde aerogels. *Journal of Materials Chemistry A*, *1*(43), 13462–13468. https://doi.org/10.1039/C3TA13172F.

Shanmuga, S. I., Singhal, M., & Sen, S. (2015). Synthesis and characterization of carrageenan coated magnetic nanoparticles for drug delivery applications. *Translational Biomedicine*, *6*(3), 1–5 https://doi.org/10.21767/2172-0479.100019.

Sharifzadeh, G., Wahit, M. U., Soheilmoghaddam, M., Whye, W. T., & Pasbakhsh, P. (2016). Kappa-carrageenan/halloysite nanocomposite hydrogels as potential drug delivery systems. *Journal of the Taiwan Institute of Chemical Engineers, 67*, 426–434. http://dx.doi.org/10.1016/j.jtice.2016.07.027.

Shrgawi, N., Shamsudin, I. J., Hanibah, H., Mohd Noor, S. A., & Kasim, N. (2021). Purification on kappa carrageenan by re-precipitation technique. In *Solid State Phenomena* (Vol. 317, pp. 327–332). Trans Tech Publications Ltd.

Slavin, Y. N., Asnis, J., Häfeli, U. O., & Bach, H. (2017). Metal nanoparticles: Understanding the mechanisms behind antibacterial activity. *Journal of Nanobiotechnology*, *15*(1), 1–20. https://doi.org/10.1186/s12951-017-0308-z.

Sonawane, R. O., & Patil, S. D. (2018). Fabrication and statistical optimization of starch-κ-carrageenan cross-linked hydrogel composite for extended-release pellets of zaltoprofen. *International Journal of Biological Macromolecules, 120*, 2324–2334. https://doi.org/10.1016/j.ijbiomac.2018.08.177.

Soukoulis, C., Tsevdou, M., Andre, C. M., Cambier, S., Yonekura, L., Taoukis, P. S., & Hoffmann, L. (2017). Modulation of chemical stability and in vitro bioaccessibility of beta-carotene loaded in kappa-carrageenan oil-in-gel emulsions. *Food Chemistry, 220*, 208–218. https://doi.org/10.1016/j.foodchem.2016.09.175.

Soukoulis, C., Tsevdou, M., Yonekura, L., Cambier, S., Taoukis, P. S., & Hoffmann, L. (2017). Does kappa-carrageenan thermoreversible gelation affect β-carotene oxidative degradation and bioaccessibility in o/w emulsions? *Carbohydrate Polymers, 167*, 259–269. https://doi.org/10.1016/j.carbpol.2017.03.007.

Stavarache, C. E., Ghebaur, A., Dinescu, S., Samoilă, I., Vasile, E., Vlasceanu, G. M., & Gârea, S. A. (2021). 5-aminosalicylic acid loaded chitosan-carrageenan hydrogel beads with potential application for the treatment of inflammatory bowel disease. *Polymers, 13*(15), 2463. https://doi.org/10.3390/polym13152463.

Tan, H. W., & Misni, M. (2014). Effect of chitosan-modified fatty acid liposomes on the rheological properties of the carbohydrate-based gel. *Applied Rheology, 24*(3), 16–24. https://doi.org/10.3933/ApplRheol-24-34839.

Tapia, C., Escobar, Z., Costa, E., Sapag-Hagar, J., Valenzuela, F., Basualto, C., Gai, M. N., & Yazdani-Pedram, M. (2004). Comparative studies on polyelectrolyte complexes and mixtures of chitosan–alginate and chitosan–carrageenan as prolonged diltiazem clorhydrate release systems. *European Journal of Pharmaceutics and Biopharmaceutics*, *57*(1), 65–75. https://doi.org/10.1016/S0939-6411(03)00153-X.

Trucillo, P., Campardelli, R., & Reverchon, E. (2020). Liposomes: From bangham to supercritical fluids. *Processes, 8*(9), 1022. https://doi.org/10.3390/pr8091022.

Vargas-Osorio, Z., Ruther, F., Chen, S., Sengupta, S., Liverani, L., Michálek, M., Galusek, D., & Boccaccini, A. R. (2022). Environmentally friendly fabrication of electrospun nanofibers made of polycaprolactone, chitosan and κ-carrageenan (PCL/CS/κ-C). *Biomedical Materials, 17*(4), 045019. https://doi.org/10.1088/1748-605X/ac6eaa.

Vijayakumar, S., Saravanakumar, K., Malaikozhundan, B., Divya, M., Vaseeharan, B., Durán-Lara, E. F., & Wang, M. H. (2020). Biopolymer K-carrageenan wrapped ZnO nanoparticles as drug delivery vehicles for anti MRSA therapy. *International Journal of Biological Macromolecules, 144,* 9–18. https://doi.org/10.1016/j.ijbiomac.2019.12.030.

Volpedo, G., Costa, L., Ryan, N., Halsey, G., Satoskar, A., & Oghumu, S. (2019). Nanoparticulate drug delivery systems for the treatment of neglected tropical protozoan diseases. *Journal of Venomous Animals and Toxins Including Tropical Diseases, 25,* e144118. https://doi.org/10.1590/1678-9199-JVATITD-1441-18.

Wang, F. F., Yao, Z., Wu, H. G., Zhang, S. X., Zhu, N. N., & Gai, X. (2012). Antibacterial activities of kappa-carrageenan oligosaccharides. In *Applied Mechanics and Materials* (Vol. 108, pp. 194–199). Trans Tech Publications Ltd.

Wang, W., Liu, F., & Gao, Y. (2016). Quercetagetin loaded in soy protein isolate–κ-carrageenan complex: Fabrication mechanism and protective effect. *Food Research International, 83,* 31–40. http://dx.doi.org/10.1016/j.foodres.2016.02.012.

Wang, Y., Su, Y., Wang, W., Fang, Y., Riffat, S. B., & Jiang, F. (2019). The advances of polysaccharide-based aerogels: Preparation and potential application. *Carbohydrate polymers, 226,* 115242. https://doi.org/10.1016/j.carbpol.2019.115242.

Wang, Y., Yuan, C., Liu, Y., & Cui, B. (2021). Fabrication of kappa–carrageenan hydrogels with cinnamon essential oil/hydroxypropyl–β–cyclodextrin composite: Evaluation of physicochemical properties, release kinetics and antimicrobial activity. *International Journal of Biological Macromolecules, 170,* 593–601. https://doi.org/10.1016/j.ijbiomac. 2020.12.176.

Wathoni, N., Meylina, L., Rusdin, A., Mohammed, A. F., Tirtamie, D., Herdiana, Y., & Muchtaridi, M. (2021). The potential cytotoxic activity enhancement of α-mangostin in chitosan-kappa carrageenan-loaded nanoparticle against MCF-7 cell line. *Polymers, 13*(11), 1681. https://doi.org/10.3390/polym13111681.

Wu, B., Li, Y., Li, Y., Li, H., & Xia, Q. (2022). The influence of Ca2+/K+ weight ratio on the physicochemical properties and in vitro digestion behavior of resveratrol-loaded Pickering emulsions encapsulated in alginate/κ-carrageenan hydrogel beads. *Reactive and Functional Polymers, 181,* 105414. https://doi.org/10.1016/j.reactfunctpolym.2022.105414.

Xie, H., Xiang, C., Li, Y., Wang, L., Zhang, Y., Song, Z., & Fang, W. (2019). Fabrication of ovalbumin/κ-carrageenan complex nanoparticles as a novel carrier for curcumin delivery. *Food Hydrocolloids, 89,* 111–121. https://doi.org/10.1016/j.foodhyd.2018.10.027.

Xue, J., Wu, T., Dai, Y., & Xia, Y. (2019). Electrospinning and electrospun nanofibers: Methods, materials, and applications. *Chemical Reviews, 119*(8), 5298–5415. http://dx.doi.org/10.1021/acs.chemrev.8b00593.

Yan, P., Yuan, J. L., Kang, X., Lv, J. Y., & Liu, X. C. (2022). Characteristics, formation mechanism and stability of high internal phase emulsions stabilized by porcine plasma protein (PPP)/carrageenan (CG) hybrid particles. *Food Bioscience, 47, 101751.* https://doi.org/10.1016/j.fbio.2022.101751.

Yan, X., Chen, Y. R., Song, Y. F., Ye, J., Yang, M., Xu, B. B., & Yu, J. K. (2020). Advances in the application of supramolecular hydrogels for stem cell delivery and cartilage tissue engineering. *Frontiers in Bioengineering and Biotechnology, 8*, 847. https://doi.org/10.3389/fbioe.2020.00847.

Yang, Y., Fang, Z., Chen, X., Zhang, W., Xie, Y., Chen, Y., Liu, Z., & Yuan, W. (2017). An overview of Pickering emulsions: Solid-particle materials, classification, morphology, and applications. *Frontiers in Pharmacology, 8*, 287. https://doi.org/10.3389/fphar.2017.00287.

Yermak, I. M., Gorbach, V. I., Glazunov, V. P., Kravchenko, A. O., Mishchenko, N. P., Pimenova, E. A., & Davydova, V. N. (2018). Liposomal form of the echinochrome-carrageenan complex. *Marine Drugs, 16*(9), 324. https://doi.org/10.3390/md16090324.

Yılmaz, H. (2022). Preparation and characterization of chitosan/κ-carrageenan based polymeric nanoparticles for gemcitabine delivery. *Erzincan University Journal of Science and Technology, 15*(2), 636–648. https://doi.org/10.18185/erzifbed.1134093.

Youssouf, L., Bhaw-Luximon, A., Diotel, N., Catan, A., Giraud, P., Gimié, F., Koshel, D., Casale, S., Bénard, S., Meneyrol, V., & Lallemand, L. (2019). Enhanced effects of curcumin encapsulated in polycaprolactone-grafted oligocarrageenan nanomicelles, a novel nanoparticle drug delivery system. *Carbohydrate Polymers, 217,* 35–45. https://doi.org/10.1016/j.carbpol.2019.04.014.

Yu, F., Yang, Z., Zhang, X., Yang, P., & Ma, J. (2021). Lanthanum modification κ-carrageenan/sodium alginate dual-network aerogels for efficient adsorption of ciprofloxacin hydrochloride. *Environmental Technology & Innovation, 24*, 102052. https://doi.org/10.1016/j.eti.2021.102052.

Yuan, H., Song, J., Li, X., Li, N., & Dai, J. (2006). Immunomodulation and antitumor activity of κ-carrageenan oligosaccharides. *Cancer Letters, 243*(2), *228–234.* https://doi.org/10.1016/j.canlet.2005.11.032.

Zhang, B., Meng, R., Li, X. L., Liu, W. J., Cheng, J. S., & Wang, W. (2021). Preparation of Pickering emulsion gels based on κ-carrageenan and covalent crosslinking with EDC: Gelation mechanism and bioaccessibility of curcumin. *Food Chemistry, 357*, 129726. https://doi.org/10.1016/j.foodchem.2021.129726.

Zhao, J., Sun, C., Li, H., Dong, X., & Zhang, X. (2020). Studies on the physicochemical properties, gelling behavior and drug release performance of agar/κ-carrageenan mixed hydrogels. *International Journal of Biological Macromolecules, 154*, 878–887. https://doi.org/10.1016/j.ijbiomac.2020.03.087.

Zhu, M., Ge, L., Lyu, Y., Zi, Y., Li, X., Li, D., & Mu, C. (2017). Preparation, characterization and antibacterial activity of oxidized κ-carrageenan. *Carbohydrate Polymers, 174*, 1051–1058. https://doi.org/10.1016/j.carbpol.2017.07.029.

Zia, K. M., Tabasum, S., Nasif, M., Sultan, N., Aslam, N., Noreen, A., & Zuber, M. (2017). A review on synthesis, properties and applications of natural polymer-based carrageenan blends and composites. *International Journal of Biological Macromolecules, 96*, 282–301. https://doi.org/10.1016/j.ijbiomac.2016.11.095.

Zou, Y., Qian, Y., Rong, X., Cao, K., McClements, D. J., & Hu, K. (2021). Encapsulation of quercetin in biopolymer-coated zein nanoparticles: Formation, stability, antioxidant capacity, and bioaccessibility. *Food Hydrocolloids, 120,* 106980. https://doi.org/10.1016/j.foodhyd.2021.106980.

Zuo, L., Zhang, Y., Zhang, L., Miao, Y. E., Fan, W., & Liu, T. (2015). Polymer/carbon-based hybrid aerogels: preparation, properties and applications. *Materials, 8*(10), 6806–6848. https://doi.org/10.3390/ma8105343.

CHAPTER 16

Lead Role of Cyanobacteria in Drug Discovery Using the Genome Mining Approach

HUDA GHAZANFAR, MUSAWAR HUSSAIN, and
MUHAMMAD TAHIR SHEHZAD

Institute of Soil and Environmental Sciences, University of Agriculture, Faisalabad, Pakistan

ABSTRACTS

Cyanobacteria are an immense source of potential bioactive compounds, so they serve as potential candidates for drug discovery. Cyanobacteria greatly possess an extensive range of cryptic and non-cryptic biosynthetic gene clusters (BGCs) capable of producing natural products. These natural products are not just directly used as drugs but also serve as templates for the synthesis of novel drugs. As natural product-based drug discovery has drawn scientific attention due to the complications associated with synthetic drugs, cyanobacteria have a significant role to play in this regard. Advances in the knowledge of biosynthesis, metabolomics, and predictive tools of bioinformatics have led to the discovery of a genome mining approach that facilitates the prediction and identification of these biosynthetic gene clusters. Genome mining of cyanobacterial genomes has unfolded the diversity of BGCs responsible for the production of various bioactive compounds. These compounds show various bioactivities such as antibacterial, anticancer, antifungal, anti-inflammatory, anti-tuberculosis, antiviral, immunosuppressive, etc. These cyanobacterial secondary metabolites follow different biosynthetic pathways, notably the non-ribosomal peptide synthetases (NRPS) or polyketide synthases (PKS) route. Research shows that the widespread application of genome mining

Handbook of Research in Marine Pharmaceutics: Exploring Oceanic Microbial Diversity for Human Health and Wellness. Md. Faiyazuddin & Khalid Rehman Hakeem (Eds.)

approaches drives the rapid development of marine natural products. Also in the last decade, different web tools and databases have been integrated with the approach of genome mining to enhance its efficiency. Hence, this chapter spotlights the contribution, significance, and pharmaceutical potential of marine cyanobacteria in drug discovery. Additionally, it provides a brief understanding of BGCs identification and biosynthetic pathways of the marine cyanobacteria using genome mining and the associated approaches, and thereby the applications in pharmaceutics.

16.1 INTRODUCTION

Recently, marine pharmacology has been the center of focus across the globe as it offers great scope for research on marine-origin drugs. The occurrence of pharmacological resistance in conventional treatments and difficulty in treating various diseases lead to shifting from terrestrial to marine natural products (Cappello & Nieri, 2021). The products and biologically active compounds of marine origin are given preference over synthetic formulations based on chemicals due to their lesser side effects and ease of accessibility (Centella et al., 2017). Marine cyanobacteria act as a significant and richest source for both novel and known bioactive compounds, having a wide range of applications in the field of pharmaceutics as well as nutraceutics (Nilsson et al., 2020) and cosmeceutics (Mourelle et al., 2017). These cyanobacteria possess different biochemical pathways for the production of bioactive compounds that play a lead role in pharmaceutics based on their antibiotic, anticancer, anticoagulant, antifungal, anti-HIV, anti-inflammatory, antimalarial, antituberculosis, antiprotozoal, antiviral, immunosuppressant, and neuroprotective activities (Tiwari & Tiwari, 2020).

Marine cyanobacteria are oxygenic photoautotrophs, gram-negative bacteria that can live in almost all types of ecosystems (Singh et al., 2011; Tiwari & Tiwari, 2020). There are also heterotrophic species; those use glucose as a carbon source and energy source in the dark, but mostly are autotrophs. They may live as individuals or form colonies, which may be protected by a variety of sheaths depending on the surrounding environment (Singh et al., 2011). The production of secondary metabolites is a conspicuous feature of cyanobacteria, which is why scientists are focusing on it for drug discovery, where the recently emerged field of study associated with cyanobacterial drugs is termed 'cyanotherapeutics.' One of the most useful properties of cyanobacteria for drug discovery is that they can inhabit all kinds of ecosystems on the planet, from bare rock to air, freshwater to oceans, and hot springs to ice sheets. This property makes them one of the most useful organisms because they are easy to culture in multiple media, as well as cyanobacteria are autotrophs which means they need fewer nutrients for growth (Mazard et al., 2016). Another additional feature is that their genome is very simple compared to other organisms with a high diversity of gene clusters (Singh et al., 2011; Tiwari & Tiwari, 2020). This means that it can produce a huge variety of natural products that can be used for pharmaceutical purposes. Furthermore, cyanobacteria have multiple pathways to produce novel natural products with a wide range of applications. The presence of non-ribosomal peptide synthetase and polyketide synthetase genes reveals the potential for drug discovery using cyanobacteria (Mazard et al., 2016; Singh et al., 2011; Tiwari & Tiwari, 2020).

Therefore, the importance of further exploring the resources of marine cyanobacteria for their pharmaceutical application is due to the wide spectrum of bioactivity that the natural products from cyanobacteria provide (Dhakal et al., 2021). Research on cyanobacteria from marine sources has the benefit of a higher rate of discovery (i.e., greater than 95%) of new bioactive compounds compared to conventional sources of microbes (Uzair et al., 2012). Conventionally, the identification of the compounds was done through bioactivity-guided isolation methods, which were time and resource-consuming and also had some other major challenges. However, with the advent of the novel approach of genome mining, natural product drug discovery has been greatly facilitated. Although like all other approaches, this approach also comes with its strengths and weaknesses (Albarano et al., 2020).

Advances in bioinformatics, genome sequencing, metabolomics, and synthetic biology have served as the driving force behind the development of genome-mining techniques for drug discovery (Baltz, 2021). This has accelerated the production of novel drugs from marine cyanobacteria, as this approach is used to identify the new biosynthetic gene clusters (BGCs) that are then involved in the synthesis of pharmaceutically significant bioactive compounds, responsible for the production of natural products (Albarano et al., 2020). The genome mining approach is also said to be a "bottom-up" process in comparison with the conventional natural product discovery technique known as the "top-down" technique, as it leverages the availability and abundance of genome data, analysis of bioinformatics, and tools for genetic modification in order to explore the BGCs that encode new natural products (Malit et al., 2022). The tools and different techniques complementarily used with genome mining have the utmost importance in narrowing down the biosynthetic gene clusters from a broader/larger pool acquired by using various prediction software, one of these being AntiSMASH (Machado et al., 2015).

Many natural product BGCs, for instance, toxins, are thought to be dispersed from the plasmids of cyanobacteria through horizontal gene transfer, and thereby, plasmids may present the potential to produce toxins and play a significant role in the progress of various cyanobacterial natural product BGCs (Popin et al., 2021).

16.2 CONTRIBUTION OF MARINE CYANOBACTERIA IN DRUG DISCOVERY

Microbial sources, especially marine microbes, are becoming a prominent source of the discovery of natural products. Medicines derived from marine sources are gradually gaining more attention along with land-based medicines due to their multiple benefits. Antibiotic resistance is an important factor that pushes us towards natural product drug discovery; moreover, synthetic drugs are now effective for one-third of all diseases due to resistance, so the discovery of more useful natural products is the need of the hour (Singh et al., 2011). Cyanobacteria, particularly marine cyanobacteria, are a major source of drug discovery due to their distinguished properties (Singh et al., 2011; Tiwari & Tiwari, 2020).

Compounds produced by cyanobacteria not only have health benefits for humans, but can also be used in fuels, plastics, drugs, additives, and cosmetics (Mazard et al., 2016; Tiwari & Tiwari, 2020). Some strains of cyanobacteria are rich food sources, containing good amounts of nutrients. *Nostoc* and *Anabaena* are commonly used for food as a source of protein and fiber in many countries. *Spirulina,* scientifically known as *Arthrospira platensis*, is a very prominent type of cyanobacteria that is used in capsule, powder, or tablet form for health and nutritional purposes and is a common source of proteins, vitamins, and β-carotene. Spirulina has been approved by the European Space Agency and NASA as an important source of food, dietary supplements, and nutrition. A few other species of cyanobacteria have also been identified for food, pharmaceutical, and nutraceutical potential. Consumption of cyanobacteria imparts many health benefits and is conventionally known as "food for fitness." Now scientific communities claim it as "food for the future" due to the health and therapeutic benefits attributed to cyanobacterial strains (Khalifa et al., 2021; Niccolai et al., 2019; Tiwari & Tiwari, 2020).

Marine cyanobacteria are more diverse in nature than terrestrial cyanobacteria because they live in different niches from very simple to complex. Several new peptides have been identified in different marine cyanobacterial strains with pharmacological properties (Mi et al., 2017). Studies show that the pharmaceutical, nutraceutical, and therapeutic focus is shifting towards marine natural medicine. These drugs have broad applications and use in the field of pharmacology, and marine cyanobacteria are one of the easily accessible organisms with a wide range of secondary metabolites and macrolides (Mazard et al., 2016; Mi et al., 2017; Wang et al., 2017). The physiological, genetic, and morphological characteristics all together make cyanobacteria a great source of natural drugs. In the future, marine cyanobacteria will be a great source of medicine (Kini et al., 2020).

16.2.1 MARINE CYANOBACTERIA AND ITS SIGNIFICANCE

Marine-based natural products have been a prominent drug source for a long time (Zhang et al., 2016). Marine ecosystems are the world's largest ecosystems, covering nearly 70% of the Earth's surface with a large pool of biological diversity. They contain a heterogeneous range of organisms, from micro to macro-organisms, but microbes are the major contributors to the marine environment (Bajpai et al., 2018; Mondal et al., 2020). These organisms provide us with numerous natural compounds that can be used as medicines for various diseases. Terrestrial organisms have a long history attributed to their use in pharmaceuticals, particularly fauna, but over the past decade, marine pharmaceutics and synthetic biology have emerged as important areas in drug discovery (Mondal et al., 2020). Compounds produced by marine life are more bioactive and effective than terrestrial life sources. This has increased the importance of marine life in drug discovery. Marine life, in contrast to terrestrial life forms, contains structurally diverse and effective secondary metabolites. Therefore, natural medicines obtained from marine sources have more efficacy and treatment than terrestrial medicines (Zhang et al., 2016). The immense potential of marine life for drug discovery drives the scientific community to explore marine life for pharmaceutical purposes. Especially, microbial life that could produce fatty acids, proteins, lipids, vitamins, and other compounds of pharmaceutical importance. Among all other microbes, cyanobacteria have certain advantages such as being easy to grow, handle, and harvest. Additionally, most species are extremophiles that can be cultured on waste and low-nutrient-containing materials, which ultimately reduce production costs. These features make it eco-friendly and cost-effective (Bajpai et al., 2018).

Studies show that marine cyanobacteria are one of the most important organisms in natural drug discovery. They produce secondary metabolites for defense mechanisms in the marine environment, with a wide structural variety that is very useful in drug discovery (Mazard et al., 2016; Mi et al., 2017; Mondal et al., 2020). It is believed that through detailed and systematic studies, cyanobacteria can become a substantial source of drug compounds in the future. Brentuximab vedotin and dolastatin-10 are common examples of drugs extracted from marine cyanobacteria and are very effective for cancer and other diseases. As already discussed, marine cyanobacteria are more diverse in nature. They have a wide spectrum of bioactive compounds with a large structural variety. Mi et al. (2017) claim that the Panama ICGB (International Cooperative Biodiversity Group) program identified approximately 400 natural products in marine cyanobacteria, many of them with therapeutic properties, especially peptide-containing groups. Marine cyanobacteria yielded 126 new compounds by the end of 2016, including peptide-containing groups, primarily comprising the genera Symploca, Langbia, and *Oscillatoria*. In short, marine cyanobacteria are highly sustainable and a potential source for marine natural drug discovery due to cultivable properties, genetic tractability, and rapid growth.

16.2.2 PHARMACEUTICAL POTENTIAL OF MARINE CYANOBACTERIA

A broad spectrum of secondary metabolites defines the potential of marine cyanobacteria. This spectrum of compounds shows huge structural variability which enhances the area of therapeutic applications (Singh et al., 2011; Tiwari & Tiwari, 2020). The spectrum of natural products contains 4% macrolides, 4.2% fatty acids, 5.6% amino acids, 9.4% amides, 40% lipopeptides, and others (Singh et al., 2011). These compounds include some toxins such as saxitoxins, microcystins, and cytotoxins (Jafari Porzani et al., 2021; Zahra et al., 2020). Marine cyanobacteria produce these compounds through biochemical pathways with therapeutic applications based on their anticancer, anti-HIV, anti-inflammatory, antimicrobial, antifungal, antibacterial, anticoagulant, antituberculosis, antimalarial, and antiprotozoal characteristics (Bajpai et al., 2018; Khalifa et al., 2021; Mazard et al., 2016; Mondal et al., 2020; Zahra et al., 2020; Zhang et al., 2016). Marine cyanobacteria symbiotically interact with other organisms and lead to the production of unique compounds. Studies showed that macro-organisms like ascidians and sponges produce natural bioactive compounds by cyano-bacterium associations (Mazard et al., 2016). Cyanobacteria have notable genetic plasticity and variability, making many strains easy to transform. Many cyanobacterial genomes have been fully sequenced and are available in the databases. Bioactivity along with other properties of cyanobacteria indicate the potential of marine cyanobacteria in drug discovery (Khalifa et al., 2021; Singh et al., 2011; Zahra et al., 2020). Additionally, planktonic marine cyanobacteria contain some allelopathic products that inhibit the growth of other microbes. For instance, a specific *Synechococcus* produces some metabolites that inhibit the growth of other microbes (Mazard et al., 2016). These characteristics combine to enhance the potential of marine cyanobacteria for drug discovery.

As the chemical structure of marine compounds extracted from marine cyanobacteria is very broad, it includes polyketides, polysaccharides, lipids, terpenes, lactones, peptides, depsipeptides, alkaloids, and lipopeptides. Khalifa et al. (2021) reported that 450 natural products were identified in different marine cyanobacteria species, especially *Symploca* (7%), *Lyngbya* (35%), and *Oscillatoria* (58%). Apratoxin D is a common example, which is a strong cytotoxin for human cancer and is identified in *Lyngbya* sp. Moreover, symplocamide A is very helpful in the treatment of lung cancer and is extracted from *Symploca* species (Khalifa et al., 2021). Sulfated polysaccharides metabolites show anti-tumor and anti-viral activities, particularly produced by the *Arthrospira* species. Furthermore, *Arthrospira* species also form GLA (gamma-linolenic acid), which is helpful in controlling cholesterol, blood pressure levels, and heart diseases. The presence of amino acids, some minerals, chlorophylls, carotenoids, and phycocyanin results in pharmaceutical properties (Tiwari & Tiwari, 2020).

A number of drugs, sourced from marine cyanobacteria, are prominent anti-microtubule agents which are in clinical and pre-clinical trials (Keller et al., 2020; Tan & Phyo, 2020). Dolastatin and Curacin (A) are potential anticancer drugs which are now in the clinical and preclinical trials. These compounds are also used in the production of synthetic drugs, such as soblidotin (TZT-1027), LU-103793, and tasidotin (ILX-651), compounds with improved therapeutical characteristics. Soblidotin (TZT-1027) is a very potential anti-tumor agent and is derived from dolastatin 10. Tasidotin (ILX-651) is also the derivative of 3rd generation dolastatin 10 and used as an anti-cancer agent.

Neurotoxins can also be produced through the marine cyanobacteria either as blockers or activators. Common blockers are jamaicamide and kalkitoxin, additionally antillatoxins (use sodium-gated channels) are good activators derived from marine cyanobacteria. Studies reveal the novelty of compounds in different fields like antillotoxin has mechanistic novelty, which is useful in the modification of sodium channels (Tan, 2007). Tutuilamides are cyclic peptides and very potent treatments for several diseases. These act as inhibitors for the elastase and have an intermediate type of potency for lung tumors (Keller et al., 2020). Marine cyanobacteria have valuable structural and molecular tools that are helpful in producing a number of novel natural products with high variability of structure and application (Figure 16.1) (Tan, 2007). Marine cyanobacteria are a great source of natural drug discovery.

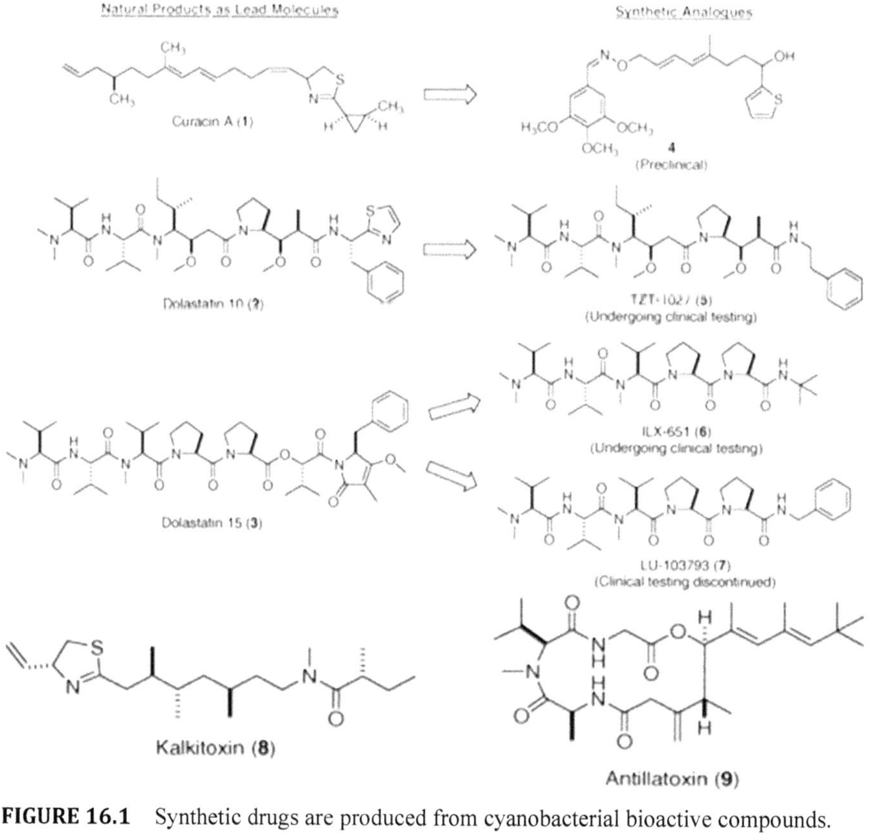

FIGURE 16.1 Synthetic drugs are produced from cyanobacterial bioactive compounds.
Source: Adapted from Tan, 2007.

16.3 NATURAL PRODUCT DRUG DISCOVERY USING MARINE CYANOBACTERIA

Cyanobacteria produce two kinds of metabolites: primary metabolites and secondary metabolites. The former are essential for survival, while the latter are used for defense mechanisms. Secondary metabolites have a range of applications, with pharmaceutical applications being one of the most important. In the later section of this chapter, we will describe the detailed pharmaceutical applications of the natural products produced by marine cyanobacteria. Compounds included in secondary metabolites are non-ribosomal peptides (NRPs), terpenes, polyketides (PKs), post-translationally modified peptides (RiPPs), and alkaloids. These are produced through biological gene clusters (BGCs) (Sweeney-Jones et al., 2020). Table 16.1. will indicate a few compounds extracted from cyanobacteria.

16.4 GENOME MINING APPROACH

Before the natural product discovery used to be done through the bioactivity-guided approach for isolation, but in the early 2000s, this field undertook a major shift when the sequencing of the very first bacterial genomes of *Streptomyces* took place. These genomes showed that immense microbially produced small molecules were yet to be revealed, hence paving the way for future discovery attempts through complementing approaches focused on the genome (Bauman et al., 2021). The novel approach of genome mining was first started in the year 2000 (Challis & Ravel, 2000). Thereby, the first-ever natural product derived from the process of genome mining, which in 2003 went for human clinical trials, was farnesylated benzodiazepine (ECO4601) (Bachmann et al., 2014).

16.4.1 THE PROCESS OF GENOME MINING

Genome mining is said to be an integrated approach that is used for discovering novel natural products (NPs) based on the sequences of the gene clusters and the pathways followed for biosynthesis (Yang et al., 2020). The metagenomic and bacterial genomic large sequence depository offers a basis for the process of genome mining in order to unveil the potential of novel secondary metabolites with great structural diversity (Aleti et al., 2015).

TABLE 16.1 Natural Product Drug Discovery Using Marine Cyanobacteria

Product	**Source/Strain**	**Uses**	**Reference**
Gallinamide	*Schizothrix*	Inhibits cathepsin L	Zaporozhets & Besednova (2020)
Grassistatin	*Schizothrix*	Inhibits cathepsins D	Hampson et al. (2006)
Glycogen	*Synechococcus sp*	Hepatoprotective,	Badary et al. (2018) Hampson et al. (2006)
Phycocyanin	All marine cyanobacteria	Hepatoprotective, Antioxidant, neuroprotective, anti-inflammatory	Sweeney-Jones et al. (2020)
Largazole	*Symploca* sp.	Cytotoxic Anti-tumor	Carson & Clarke (2018); Seiser et al. (2008)
Eucapsitrione	*Eucapsis* sp.	Antibacterial Cytotoxic	Demay et al. (2019b)
Saxitoxin	*Cylindrospermopsis, Anabaena, Lyngbya, Aphanizomenon, Planktothrix,*	Neurotoxin	Sweeney-Jones et al. (2020)
Macrolide	*Lyngbya* sp.	Antioxidant, neuroprotective, anti-inflammatory, anti-cancer	Carson & Clarke (2018)
Abietic acid	*Nostoc sp, Plectonema radiosum, Chroococcidiopsis* sp., *Synechocystis salina, Synechocystis* sp., *Nodosilinea* sp., *Leptolyngbya ectocarpi,*	Antibacterial	Demay et al. (2019b)
Anatoxin-a	*Oscillatoria, Anabaena, Aphanizomenon, Cylindrospermum, Planktothrix*	Neurotoxin, anti-inflammatory	Sweeney-Jones et al. (2020)
Hapalindole	*Fischerella musicola, Hapalosiphon Westiellopsis* sp., *Fontinalis, Welwitschii, Hapalosiphon, Westiella intricata,*	Antibacterial Antifungal Antialgal Cytotoxic Insecticidal Lethal activity Reverse multidrug resistance (MDR) VGSC modulator	Demay et al. (2019b)

Various kinds of sequenced data such as epigenomics, genomics, metabolomics, multi-omics, proteomics, and transcriptomics, and many tools of bioinformatics are used for genome mining (Rosic, 2022a). The process starts by accessing the omics data from publicly available databases such as NCBI (National Center for Biotechnology Information) (Sayers et al., 2010) and other combined database resources like INSDC (International Nucleotide Sequence Database Collaboration) (Karsch-Mizrachi et al., 2018). This process of genome mining encompasses the whole spectrum of the up-to-date central dogma of molecular biology involving gene prediction and pathway functions. It is carried out through bioinformatics, gene expression control and translation, and identifying and structurally elucidating novel metabolites from the producing organism's metabolome (Bachmann et al., 2014). The various approaches involved in the process of genome mining are illustrated in (Figure 16.2).

Various strategies can be used to perform genome mining. If the cyanobacterial genome sequence is known, it can be done with the help of web-based tools for genome mining (https://www.ncbi.nlm.nih.gov/). One such tool is antiSMASH, while there are other alternatives such as EvoMining, BAGEL (BActeriocin GEnome Mining tool), and RODEO (Rapid ORF Description and Evaluation Online), where https://secondarymetabolites.org contributes to a higher variety based on a planned analysis and offers a good outline for various tools used for different strategies to analyze the biosynthetic gene clusters or the associated secondary metabolites (Weber, 2020). On the contrary, conserved biosynthetic sequences of genes (from LanC or NRPS) could also be utilized for searching highly similar sequences of genomes from NCBI through the Basic Local Alignment Search Tool (BLAST) (Sandiford, 2017). In this way, screening of cyanobacteria with respect to their genome sequence possession for the specific bioactive compound production from the genome database takes place. A similar example, Micallef et al. (2015), conducted the genome mining methodology with the help of antiSMASH to detect the BGCs in subsection V cyanobacteria.

16.4.2 GENOME MINING SIGNIFICANCE

The significance of the method of genome mining lies in the urgent need to discover novel entities of drugs because of the increased occurrence of serious diseases (cancer) and the decreased efficiency of the present drugs (Soumya Haider, 2015). Such an approach allows us to overcome the over-exploitation of aquatic resources and the utilization of destructive practices for the collection process in this regard, and more importantly, it is a very eco-friendly approach compared to the conventional methods used for drug discovery (Costantini, 2020). Furthermore, genome mining avoids problems regarding isolation, production, and scalability (Bauman et al., 2021).

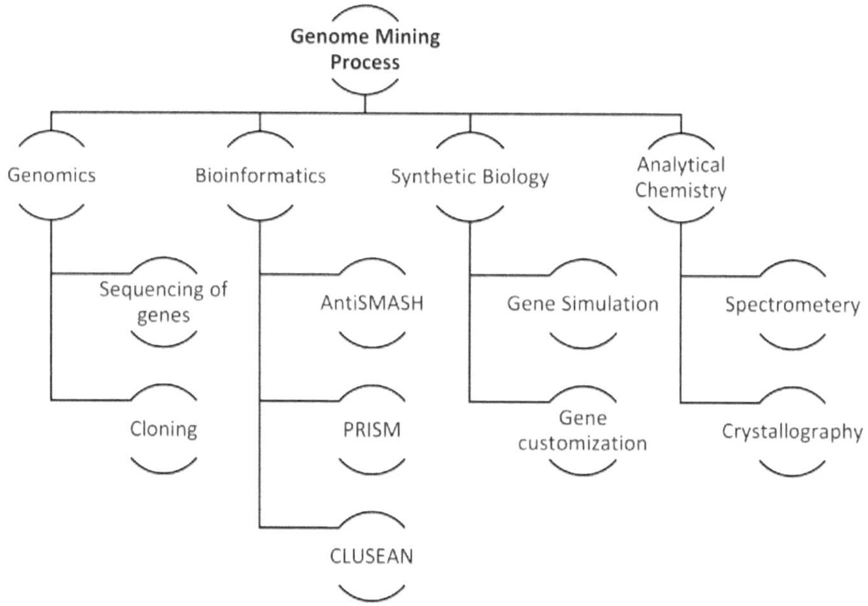

FIGURE 16.2 Procedures involved in genome mining.

The approach of genome mining is comparatively cheaper than the conventional ones and is easier to practice in the laboratory, and it is not requisite for the operator to have specific skills or experience (Albarano et al., 2020). Also, this approach is useful in detecting a huge amount of biologically active natural products driven by the rapid reduction of cost and enhanced throughput of technologies for DNA sequencing.

16.4.3 GENOME MINING ASSOCIATED TOOLS AND TECHNIQUES

The bioinformatics-associated tools and techniques, such as the Antibiotics and Secondary Metabolites Analysis Shell abbreviated as AntiSMASH, are used for identifying and annotating the BGCs. AntiSMASH 5 has additional detection guidelines for clusters that encode for the biosynthesis of poly-brominated diphenyl ethers, acyl-amino acids, β-lactones, C-nucleosides, lipolanthines, and PPY-like ketones. It also provides further in-detail predictions for gene clusters that encode for type II polyketide synthase (Blin et al., 2019). Prediction Informatics for Secondary Metabolism (PRISM) is a tool for identifying the type I and II polyketides, non-ribosomal peptides, and RiPPs (Skinnider et al., 2017). BAGEL 4 is used for identifying the clusters of genes that encode RiPPs and bacteriocins (van Heel et al., 2018). CLUSEAN, i.e., CLUster SEuence ANalyzer, is a cluster mining tool that provides automated searches for homology, detection in the NRPS and PKS clusters of genes for the conserved protein domains, enzyme classification, and specific predictions for A-domains of NRPS (Weber et al., 2009). The bioinformatics tools associated with the process of genome mining are listed in Table 16.2.

TABLE 16.2 Genome Mining Tools and Techniques

Genome Mining Associated Tool and Techniques	Key Feature	References
RIPPER (RiPP Precursor Peptide Enhanced Recognition)	Detection of BGCs and precursor peptides of-RIPPs	Moffat et al. (2021)
BIG-SCAPE (Biosynthetic Gene Similarity Clustering and Prospecting Engine)	Using data from metabolomics, it creates a report of sequence similarity for the novel BGCs.	Navarro-Muñoz et al. (2020)
Evo-Mining	Identifies the expansions of recruitments of gene families, into BGCs.	Sélem-Mojica et al. (2019)
ARTS (Antibiotic Resistant Target Seeker)	Allows targeted directed genome mining for antibiotic resistance.	Mungan et al. (2020)
BiG-SLiCE (Biosynthetic Gene clusters – Super Linear Clustering Engine)	Analyzes the BGCs data on a large scale and identifies patterns of cross-species BGCs.	Kautsar et al. (2021)

There are many approaches to synthetic biology that work in association with genome mining. The simulation and gene customization techniques, such as gene knockouts, genetic overexpression of transcription factors, or other promoter exchange experiments, pleiotropic regulators, etc., have eased the eventual fate by triggering the production of bioactive compounds (Bhattarai et al., 2020).

The process of genome mining has not only turned out to be a promising approach to detecting natural compounds but also considers the cryptic biosynthetic gene clusters due to the immense data available through genomic arrangement examination information (Sarsaiya et al., 2021). Over time, the progress of natural product discovery from cyanobacterial genomes has been slowed down as the corresponding BGCs were silent (i.e., cryptic) (Dhakal et al., 2021). But now, many different combinatorial biosynthesis approaches associated with genome mining are used for the activation of these silent BGCs, which involve techniques of gene activation/inactivation and mutasynthesis (Albarano et al., 2020). Other approaches include direct cloning approaches, ribosome engineering, OSMAC (one strain/many compounds) method, and heterologous expression, etc., (Baral et al., 2018; Dhakal et al., 2021; Nguyen et al., 2020).

16.5 GENOME MINING OF CYANOBACTERIA

In this regard, a meta-analysis of 670 papers showed that more than 90 cyanobacterial genera were observed/found producing potent beneficial compounds, among which mostly belonged to orders Chroococcales, Nostocales, Oscillatoriales, and Synechococcales. While the rest of the orders of cyanobacteria are explored poorly with respect to their relative bioactivities and structural diversity i.e., Chroococcidiopsales, Gloeobacterales, and Pleurocapsales (Demay et al., 2019a).

16.5.1 CYANOBACTERIAL GENOME AND BIOSYNTHETIC GENE CLUSTERS

Recent studies in genome mining revealed that the Cyanobacterial genome dedicated 5–6% of it to the production of secondary metabolites, encompassing a massive variety of cryptic BGCs, whereby 33 are inactive i.e., silent, thus their products are unknown by conventional conditions of culture (Gómez-Espinoza et al., 2022). A phylum-wide study was conducted to analyze 1923 genomes of cyanobacteria; these were then scanned using the antiSMASH (v 5.1.2), which resulted in the identification of 12,323 BGCs in total (Leão et al., 2021a).

The biosynthetic gene clusters mainly encode peptides and polyketides (PKs), but these also encode compounds of some other pharmaceutically crucial families such as alkaloids, lipids, and terpenes. Out of the known NPs of cyanobacteria, 66% of these are members of the peptide family. Where molecules of peptide are biosynthesized through ribosomal and non-ribosomal origins by actions of the biosynthetic machinery of ribosomally synthesized and post-translationally modified peptides (RiPPs) and non-ribosomal peptide synthetases (NRPSs) (Dhakal et al., 2021).

The major biosynthetic gene clusters in cyanobacteria that encode for RiPPs are lantionines, cyanobactins, and microviridines. It is interesting that prochlorosin is the only lantionine isolated to date. Cyanobactins are abundant in the *Arthrospira, Microcystis,* and *Oscillatoria* genera but are scarcely present in the *Synechococcus* and *Prochlorococcus* genera. Microviridins hold the potential for serine protease inhibitory activity (Ferreira Leao, 2019). The number of BGCs identified from various cyanobacterial genomes is given in Figure 16.3.

Characterization of many cyanobacterial bioactive compounds, which are potentially novel, is yet to be done. For instance, the comparative genome examination of *Moorea* species showed that each genome had 30–45 biosynthetic gene clusters, while only a few of the compounds were isolated from each strain (Leao et al., 2017). To facilitate functionalities regarding cyanobacterial genome mining, a web interface presents both the application of Palantir and the first SQL database dedicated to secondary metabolites of cyanobacteria. This database comprises 1616 curated groups of cyanobacterial genomes in relation to annotations of antiSMASH and Palantir. Out of these, 1488 genomes were expected to encode at least one biosynthetic gene cluster (Meunier et al., 2020).

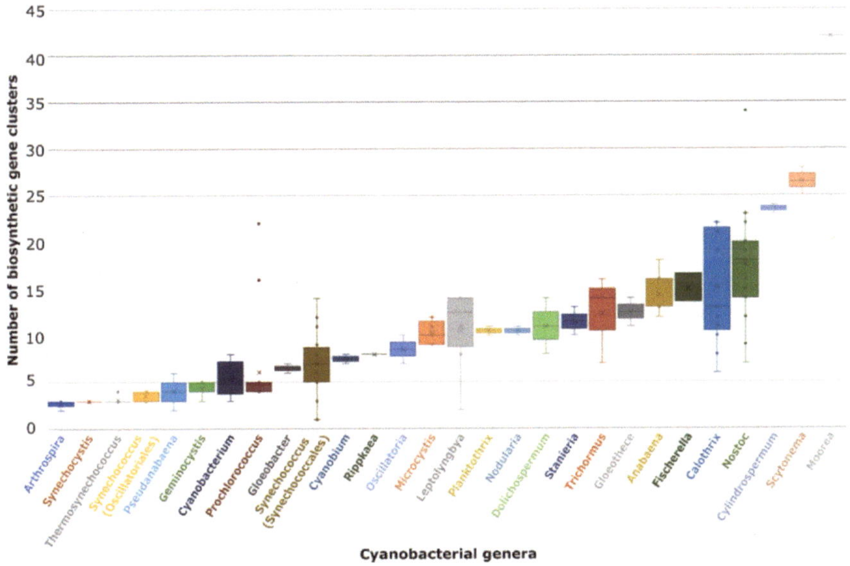

FIGURE 16.3 Number of BGC from different cyanobacteria.

Source: Adapted from Clark et al., 2016

16.5.2 THE BIOSYNTHETIC PATHWAYS

If NRPS and PKS-I (type-I) genes are used as a standard/benchmark for each microbe to identify the production capacity of bioactive compounds, then it is evident that microbes with larger genomes will provide higher production. For instance, 60% of all important biological compounds discovered in the last half century are attributed to NRPS, PKS-I, or both NRPS-PKS. Popin et al. (2021) showed that 424 plasmids were identified using 185 genomes, and 17% of these plasmids contained natural drug-producing BGCs. Some plasmids contained only one natural BGC, but there were 99 identified BGCs. 26 BGCs fall into the hybrid category (NRPS-PKS), while 20 represent only NRPS. Another study showed that 1649 cyanobacterial genomes were used to identify the BGCs, which indicated that marine cyanobacterial genomes contain high beta diversity. The main classes of clusters found in 24 marine cyanobacteria lie under NRPS (peptides), T1PKS-NRPS hybrid (lipopeptides), and terpenes (Leão et al., 2021b).

Notably, paracyclophanes and hierridins are the only two type-III PKS products that have been isolated from cyanobacteria. Larsen et al. (2021) reported 517 genomes of cyanobacteria containing type-III PKS BGCs through genome mining. Of these, 17% of the examined genomes encoded one or more type-III PKS. Using the genome mining approach, the synthesis of Mycosporine-like Amino Acids (MAA) through the pentose phosphate pathway and/or the shikimate pathway was examined, and the genes were isolated through the Mys cluster (known as the MAA pathway) using the cyanobacterium Anabaena variabilis. This cluster comprised four genes encoding enzymes: adenosine triphosphate (ATP) grasp, dehydroquinate synthase (DHQS), nonribosomal peptide synthetase (NRPS), and O-methyl-transferase (O-MT). The diversity in these genes, crucial for the biosynthesis pathways of MAA, was validated through genome data involving about 300 cyanobacteria (Rosic, 2022a).

16.6 APPLICATIONS IN PHARMACEUTICS

Pharmaceutical applications are very broad and are attributed to natural drugs with a long history. The genetics involved in these processes are very complex and make drug discovery a challenge for current needs. However, genome mining is a great novel technique to resolve these complexities. For example, MAA produced by marine cyanobacteria are skin protectants that act to protect humans against UV-A and UV-B rays. These produce four types of enzymes which are very complex in nature, but genome mining allows us to explore these through the exploration of the most suitable BGCs. Genome mining allows us to discover countless drugs through unconventional ways (Rosic, 2022b).

Conventionally, we follow biological pathways for the production of novel bioactive products which involve extraction, isolation, production, and other processes. All these processes make it a more complex and time-consuming process, but we urgently need to produce numerous products that will serve as a cure for many diseases. This requires a short time and simple processes for the production of products. Genome mining is one of the ways that enables us to extract products with minimum time and resource exploitation (Ghareeb et al., 2020; Tewari et al., 2021).

Applications of secondary metabolites of cyanobacteria are not only restricted to pharmaceuticals but also used in food, food additives, nutraceuticals, cosmetics, and others (Hassan et al., 2022; Khalifa et al., 2021; Mazard et al., 2016; Tiwari & Tiwari, 2020). *Spirulina* is a marine cyanobacterium that provides omega-6 fatty acids, which function as nutraceuticals for growth (Hassan et al., 2022). In previous sections of this chapter, we discussed cyanobacteria, metabolites of cyanobacteria, and the applications of genome mining in extracting beneficial compounds from marine cyanobacteria. In this section, we focus on the pharmaceutical/therapeutic applications of secondary metabolites of cyanobacteria. Secondary metabolites produced through different channels in cyanobacteria are helpful in defense mechanisms. Due to their protective properties, these compounds are also useful for human medication as antiviral, antibacterial, anti-inflammatory, anti-HIV, anti-cancer, anti-malaria, skin protectors, and more (Bajpai et al., 2018; Demay et al., 2019b; Hassan et al., 2022; Mazard et al., 2016; Sweeney-Jones et al., 2020).

Antibiotic resistance is an emerging issue in the pharmaceutical field, and we need to identify more antibiotic compounds (Hassan et al., 2022; Saeed et al., 2022). The cyanobacterial compounds serve as potential antibiotics; for example, noscomin is an extracellular compound extracted from the *Nostoc cyanobacterium*. Commonly, noscomin acts as an antibiotic for *E. coli, Staphylococcus epidermidis*, and *Bacillus cereus* (Hassan et al., 2022). These antibiotics are produced through bioactive gene clusters that are highly conserved by gene sequences. These gene sequences are highly complex, and genome mining is capable of solving these complex structures and helps in identifying novel products (Strieth et al., 2022).

Along with the antibacterial characteristics, the bioactive compounds of the cyanobacteria also act against viruses (Pradhan & Ki, 2022). Sulfated polysaccharides such as spirulan show bioactivity against viruses. Calcium spirulan extracted from *Arthrospira platensis* is very effective against HSV-I (herpes simplex virus type-I) and HIV (human immunodeficiency virus type-I) (Hassan et al., 2022). These compounds also contribute as antiparasitic and antifungal agents (Hassan et al., 2022; Saeed et al., 2022). Macrolides including glycolipopeptide hassallidin and scytophycin function as antifungal agents and are found in the *Nostoc* and *Anabaena* species. *Viridamide A* acts against *P. falciparum, T. cruzi*, and *L. Mexicana* and can be extracted from *Oscillatoria nigro virdis*. Viridamide is an antiparasitic agent isolated from cyanobacteria (Hassan et al., 2022). *Amphidinium* sp. acts as both antiprotozoal and antifungal (Saeed et al., 2022).

In addition to the above characteristics, these compounds show cytotoxicity. There are several compounds that contribute to the cure of cancer. The metabolic compounds extracted from marine cyanobacteria, including Brentuximab, Dolafexin, Auristatin derivative, Lurbinectedin, and Enfortumab Vedotin, are potential anti-cancer agents (Kamat et al., 2022).

16.7 CONCLUSION

Marine cyanobacteria possess great pharmaceutical potential, as evidenced by their antibiotic, anticancer, anticoagulant, antifungal, anti-HIV, anti-inflammatory, antituberculosis, antiprotozoal, antiviral, immunosuppressant, and neuroprotective bioactivities. Bioactive compounds are synthesized from biosynthetic gene clusters of the marine cyanobacterial genome following a specific biosynthetic pathway, and these biosynthetic gene clusters are now easily identified using the novel and eco-friendly approach of genome mining. Different tools and techniques associated with this approach have further improved and increased the synthesis of natural products from marine cyanobacteria, leading to applications in the fields of pharmaceutics, nutraceutics, and cosmeceutics. This highlights the leading role of marine cyanobacteria in the discovery of natural products via genome mining. Ongoing research on genome mining will provide new insights and opportunities to unveil the pharmaceutical and biotechnological potential of cyanobacteria from all environments (marine, terrestrial, etc.) and contribute to the emerging field of cyanotherapeutics.

KEYWORDS

- **biosynthetic gene clusters**
- **biosynthetic pathways**
- **data mining**
- **drug discovery**
- **genome mining**
- **marine drugs**

REFERENCES

Albarano, L., Esposito, R., Ruocco, N., & Costantini, M. (2020). Genome mining as new challenge in natural products discovery. *Marine Drugs, 18*(4), 199. https://doi.org/10.3390/md18040199.

Aleti, G., Sessitsch, A., & Brader, G. (2015). Genome mining: Prediction of lipopeptides and polyketides from Bacillus and related Firmicutes. *Computational and Structural Biotechnology Journal, 13,* 192–203. https://doi.org/10.1016/j.csbj.2015.03.003.

Bachmann, B. O., Van Lanen, S. G., & Baltz, R. H. (2014). Microbial genome mining for accelerated natural products discovery: Is a renaissance in the making? *Journal of Industrial Microbiology and Biotechnology, 41*(2), 175–184. https://doi.org/10.1007/s10295-013-1389-9.

Badary, A., Takamatsu, S., Nakajima, M., Ferri, S., Lindblad, P., & Sode, K. (2018). Glycogen production in marine cyanobacterial strain *Synechococcus* sp. NKBG 15041c. *Marine Biotechnology (NY), 20*(2), 109–117. https://doi.org/10.1007/s10126-017-9792-2.

Bajpai, V. K., Shukla, S., Kang, S.-M., Hwang, S. K., Song, X., Huh, Y. S., & Han, Y.-K. (2018). Developments of cyanobacteria for nano-marine drugs: Relevance of nanoformulations in cancer therapies. *Marine Drugs, 16*(6), 179.

Baltz, R. H. (2021). Genome mining for drug discovery: Progress at the front end. *Journal of Industrial Microbiology and Biotechnology, 48,* 9–10. https://doi.org/10.1093/jimb/kuab044.

Baral, B., Akhgari, A., & Metsä-Ketelä, M. (2018). Activation of microbial secondary metabolic pathways: Avenues and challenges. *Synthetic and Systems Biotechnology, 3*(3), 163–178. https://doi.org/10.1016/j.synbio.2018.09.001.

Bauman, K. D., Butler, K. S., Moore, B. S., & Chekan, J. R. (2021). Genome mining methods to discover bioactive natural products. *Natural Product Reports, 38*(11), 2100–2129. https://doi.org/10.1039/d1np00032b.

Bhattarai, K., Bastola, R., & Baral, B. (2020). Chapter 6 – Antibiotic drug discovery: Challenges and perspectives in the light of emerging antibiotic resistance. In D. Kumar (Ed.), *Advances in Genetics* (Vol. 105, pp. 229–292). Academic Press.

Blin, K., Shaw, S., Steinke, K., Villebro, R., Ziemert, N., Lee, S. Y., & Weber, T. (2019). Anti-SMASH 5.0: Updates to the secondary metabolite genome mining pipeline. *Nucleic Acids Research, 47*(W1), W81–W87. https://doi.org/10.1093/nar/gkz310.

Cappello, E., & Nieri, P. (2021). From life in the sea to the clinic: The marine drugs approved and under clinical trial. *Life (Basel), 11*(12), 1390. https://doi.org/10.3390/life11121390.

Carson, M. A., & Clarke, S. A. (2018). Bioactive compounds from marine organisms: Potential for bone growth and healing. *Marine Drugs, 16*(9), 340.

Centella, M. H., Arévalo-Gallegos, A., Parra-Saldivar, R., & Iqbal, H. M. N. (2017). Marine-derived bioactive compounds for value-added applications in bio- and non-bio sectors. *Journal of Cleaner Production, 168,* 1559–1565. https://doi.org/10.1016/j.jclepro.2017.05.086.

Challis, G. L., & Ravel, J. (2000). Coelichelin, a new peptide siderophore encoded by the *Streptomyces coelicolor* genome: Structure prediction from the sequence of its non-ribosomal peptide synthetase. *FEMS Microbiology Letters, 187*(2), 111–114. https://doi.org/10.1111/j.1574-6968.2000.tb09145.x.

Clark, K., Karsch-Mizrachi, I., Lipman, D. J., Ostell, J., & Sayers, E. W. (2016). GenBank. *Nucleic Acids Research, 44*(D1), D67–D72.

Costantini, M. (2020). Genome mining and synthetic biology in marine natural product discovery. *Marine Drugs, 18*(12), 615. https://doi.org/10.3390/md18120615.

Demay, J., Bernard, C., Reinhardt, A., & Marie, B. (2019). Natural products from cyanobacteria: Focus on beneficial activities. *Marine Drugs, 17*(6), 320.

Dhakal, D., Chen, M., Luesch, H., & Ding, Y. (2021). Heterologous production of cyanobacterial compounds. *Journal of Industrial Microbiology and Biotechnology, 48*, 3–4. https://doi.org/10.1093/jimb/kuab003.

Meng-Yun Chen, Wen-Kai Teng, Liang Zhao, Chun-Xiang Hu, Yang-Kai Zhou, Bo-Ping Han, Li-Rong Song, & Wen-Sheng Shu, (2021). Comparative genomics reveals insights into cyanobacterial evolution and habitat adaptation, *The ISME Journal, 15*(1), 211–227, https://doi.org/10.1038/s41396–020–00775-z.

Ghareeb, M. A., Tammam, M. A., El-Demerdash, A., & Atanasov, A. G. (2020). Insights about clinically approved and preclinically investigated marine natural products. *Current Research in Biotechnology, 2,* 88–102. https://doi.org/10.1016/j.crbiot.2020.09.001.

Gómez-Espinoza, O., Núñez-Montero, K., & Barrientos Díaz, L. (2022). Chapter 6– Antimicrobial activity of cyanobacteria-derived compounds. In G. Lopes, M. Silva, & V. Vasconcelos (Eds.), *The Pharmacological Potential of Cyanobacteria* (pp. 145–172). Academic Press.

Hampson, L. J., Arden, C., Agius, L., Ganotidis, M., Kosmopoulou, M. N., Tiraidis, C., & Oikonomakos, N. G. (2006). Bioactivity of glycogen phosphorylase inhibitors that bind to the purine nucleoside site. *Bioorganic & Medicinal Chemistry, 14*(23), 7835–7845. https://doi.org/10.1016/j.bmc.2006.07.060.

Hassan, S., Meenatchi, R., Pachillu, K., Bansal, S., Brindangnanam, P., Arockiaraj, J., & Selvin, J. (2022). Identification and characterization of the novel bioactive compounds from microalgae and cyanobacteria for pharmaceutical and nutraceutical applications. *Journal of Biological Macromolecules*, *62*(9), 999–1029.

Jafari Porzani, S., Konur, O., & Nowruzi, B. (2021). Cyanobacterial natural products as sources for antiviral drug discovery against COVID-19. *Journal of Biological Systems and Dynamics*, *40*(16), 1–17.

Kamat, S., Kumari, M., Sajna, K. V., Shruthi, M., & Jayabaskaran, C. (2022). Marine endophytes from the Indian coasts: The untapped sources of sustainable anticancer drug discovery. *Sustainable Chemistry and Pharmacy, 27,* 100675. https://doi.org/10.1016/j.scp.2022.100675.

Karsch-Mizrachi, I., Takagi, T., Cochrane, G., & Collaboration, I. N. S. D. (2018). The international nucleotide sequence database collaboration. *Nucleic Acids Research, 46*(D1), D48–D51.

Kautsar, S. A., van der Hooft, J. J. J., de Ridder, D., & Medema, M. H. (2021). BiG-SLiCE: A highly scalable tool maps the diversity of 1.2 million biosynthetic gene clusters. *GigaScience, 10*(1), 1–17. https://doi.org/10.1093/gigascience/giaa154.

Keller, L., Canuto, K. M., Liu, C., Suzuki, B. M., Almaliti, J., Sikandar, A., & Duggan, B. M. (2020). Tutuilamides A–C: Vinyl-chloride-containing cyclodepsipeptides from marine cyanobacteria with potent elastase inhibitory properties. *ACS Chemical Biology, 15*(3), 751–757.

Khalifa, S. A., Shedid, E. S., Saied, E. M., Jassbi, A. R., Jamebozorgi, F. H., Rateb, M. E., & Al-Hammady, M. A. (2021). Cyanobacteria—From the oceans to the potential biotechnological and biomedical applications. *Marine Drugs, 19*(5), 241.

Kini, S., Divyashree, M., Mani, M. K., & Mamatha, B. S. (2020). Chapter 12 – Algae and cyanobacteria as a source of novel bioactive compounds for biomedical applications. In P. K. Singh, A. Kumar, V. K. Singh, & A. K. Shrivastava (Eds.), *Advances in Cyanobacterial Biology* (pp. 173–194). Academic Press.

Larsen, J. S., Pearson, L. A., & Neilan, B. A. (2021). Genome mining and evolutionary analysis reveal diverse Type III polyketide synthase pathways in cyanobacteria. *Genome Biology and Evolution, 13*(4), 1–15. https://doi.org/10.1093/gbe/evab056.

Leao, T., Castelão, G., Korobeynikov, A., Monroe, E. A., Podell, S., Glukhov, E., & Gerwick, L. (2017). Comparative genomics uncovers the prolific and distinctive metabolic potential of the cyanobacterial genus *Moorea. Proceedings of the National Academy of Sciences, 114*(12), 3198–3203. https://doi.org/10.1073/pnas.1618556114.

Leão, T., Wang, M., Moss, N., da Silva, R., Sanders, J., Nurk, S., & Gerwick, L. (2021). A multi-omics characterization of the natural product potential of tropical filamentous marine cyanobacteria. *Marine Drugs, 19*(1), 20. https://doi.org/10.3390/md19010020.

Machado, H., Sonnenschein, E. C., Melchiorsen, J., & Gram, L. (2015). Genome mining reveals unlocked bioactive potential of marine Gram-negative bacteria. *BMC Genomics, 16*(1), 158. https://doi.org/10.1186/s12864-015-1365-z.

Malit, J. J. L., Leung, H. Y. C., & Qian, P. Y. (2022). Targeted large-scale genome mining and candidate prioritization for natural product discovery. *Marine Drugs, 20*(6), 398. https://doi.org/10.3390/md20060398.

Mazard, S., Penesyan, A., Ostrowski, M., Paulsen, I. T., & Egan, S. (2016). Tiny microbes with a big impact: The role of cyanobacteria and their metabolites in shaping our future. *Marine Drugs, 14*(5), 97. https://doi.org/10.3390/md14050097.

Meunier, L., Tocquin, P., Cornet, L., Sirjacobs, D., Leclère, V., Pupin, M., & Baurain, D. (2020). Palantir: A springboard for the analysis of secondary metabolite gene clusters in large-scale genome mining projects. *Bioinformatics, 36*(15), 4345–4347. https://doi.org/10.1093/bioinformatics/btaa517.

Mi, Y., Zhang, J., He, S., & Yan, X. (2017). New peptides isolated from marine cyanobacteria: An overview over the past decade. *Marine Drugs, 15*(5), 132. https://doi.org/10.3390/md15050132.

Micallef, M. L., D'Agostino, P. M., Al-Sinawi, B., Neilan, B. A., & Moffitt, M. C. (2015). Exploring cyanobacterial genomes for natural product biosynthesis pathways. *Marine Genomics, 21,* 1–12. https://doi.org/10.1016/j.margen.2015.01.001.

Moffat, A. D., Santos-Aberturas, J., Chandra, G., & Truman, A. W. (2021). A user guide for the identification of new RiPP biosynthetic gene clusters using a RiPPER-based workflow. In C. Barreiro & J.-L. Barredo (Eds.), *Antimicrobial therapies: Methods and protocols* (pp. 227–247). New York, NY: Springer US. https://doi.org/10.1007/978-1-0716-1246-0_14.

Mondal, A., Bose, S., Banerjee, S., Patra, J. K., Malik, J., Mandal, S. K., & Bishayee, A. (2020). Marine cyanobacteria and microalgae metabolites—A rich source of potential anticancer drugs. *Marine Drugs, 18*(9), 476. https://doi.org/10.3390/md18090476.

Mourelle, M. L., Gómez, C. P., & Legido, J. L. (2017). The potential use of marine microalgae and cyanobacteria in cosmetics and thalassotherapy. *Cosmetics, 4*(4), 46. https://doi.org/10.3390/cosmetics4040046.

Mungan, M. D., Alanjary, M., Blin, K., Weber, T., & Ziemert, N. (2020). ARTS 2.0: Feature updates and expansion of the antibiotic resistant target seeker for comparative genome mining. *Nucleic Acids Research, 48*(W1), W546–W552. https://doi.org/10.1093/nar/gkaa 374.

Navarro-Muñoz, J. C., Selem-Mojica, N., Mullowney, M. W., Kautsar, S. A., Tryon, J. H., Parkinson, E. I., & Abubucker, S. (2020). A computational framework to explore large-scale biosynthetic diversity. *Nature Chemical Biology, 16*(1), 60–68. https://doi.org/10.1038/s41589-019-0421-1.

Nguyen, C. T., Dhakal, D., Pham, V. T. T., Nguyen, H. T., & Sohng, J. K. (2020). Recent advances in strategies for activation and discovery/characterization of cryptic biosynthetic gene clusters in Streptomyces. *Microorganisms, 8*(4), 616. https://doi.org/10.3390/microorganisms8040616.

Niccolai, A., Chini Zittelli, G., Rodolfi, L., Biondi, N., & Tredici, M. R. (2019). Microalgae of interest as food source: Biochemical composition and digestibility. *Algal Research, 42,* 101617. https://doi.org/10.1016/j.algal.2019.101617.

Nilsson, A. K., Jiménez, C., & Wulff, A. (2020). Nutraceutical fatty acid production in marine microalgae and cyanobacteria. In *Nutraceutical Fatty Acids from Oleaginous Microalgae* (pp. 23–76). https://doi.org/10.1016/b978-0-12-817911-0.00002-0.

Popin, R. V., Alvarenga, D. O., Castelo-Branco, R., Fewer, D. P., & Sivonen, K. (2021). Mining of cyanobacterial genomes indicates natural product biosynthetic gene clusters located in conjugative plasmids. *Frontiers in Microbiology, 12,* 582260. https://doi.org/10.3389/fmicb.2021.582260.

Pradhan, B., & Ki, J.-S. (2022). Phytoplankton toxins and their potential therapeutic applications: A journey toward the quest for potent pharmaceuticals. *Marine Drugs, 20*(4), 271.https://doi.org/10.3390/md20040271.

Rosic, N. (2022). Genome mining as an alternative way for screening the marine organisms for their potential to produce UV-absorbing mycosporine-like amino acid. *Marine Drugs, 20*(8), 478. https://doi.org/10.3390/md20080478.

Saeed, M. U., Hussain, N., Shahbaz, A., Hameed, T., Iqbal, H. M., & Bilal, M. J. J. O. B. M. (2022). Bioprospecting microalgae and cyanobacteria for biopharmaceutical applications. *Journal of Biological Macromolecules, 62*(9), 1110-1124. https://doi.org/10.1002/jbm4.10577.

Sandiford, S. K. (2017). Genome database mining for the discovery of novel lantibiotics. *Expert Opinion on Drug Discovery, 12*(5), 489–495. https://doi.org/10.1080/17460441.2017.1313675.

Sarsaiya, S., Jain, A., Shi, J., & Chen, J. (2021). Fungi endophytes for biofactory of secondary metabolites: Genomics and metabolism. In S. Jogaiah (Ed.), *Biocontrol Agents and Secondary Metabolites* (pp. 1–21). Woodhead Publishing. https://doi.org/10.1016/b978-0-12-822919-8.00001-3.

Sayers, E. W., Barrett, T., Benson, D. A., Bolton, E., Bryant, S. H., Canese, K., & Federhen, S. (2010). Database resources of the national center for biotechnology information. *Nucleic Acids Research, 39*(suppl_1), D38–D51. https://doi.org/10.1093/nar/gkq1170.

Seiser, T., Kamena, F., & Cramer, N. (2008). Synthesis and biological activity of largazole and derivatives. *Angewandte Chemie International Edition, 47*(34), 6483–6485. https://doi.org/10.1002/anie.200801317.

Sélem-Mojica, N., Aguilar, C., Gutiérrez-García, K., Martínez-Guerrero, C. E., & Barona-Gómez, F. (2019). EvoMining reveals the origin and fate of natural product biosynthetic enzymes. *Microbial Genomics, 5*(12), 1–16. https://doi.org/10.1099/mgen.0.000260.

Singh, R. K., Tiwari, S. P., Rai, A. K., & Mohapatra, T. M. (2011). Cyanobacteria: An emerging source for drug discovery. *The Journal of Antibiotics, 64*(6), 401–412. https://doi.org/10.1038/ja.2011.21.

Skinnider, M. A., Merwin, N. J., Johnston, C. W., & Magarvey, N. A. (2017). PRISM 3: Expanded prediction of natural product chemical structures from microbial genomes. *Nucleic Acids Research, 45*(W1), W49–W54. https://doi.org/10.1093/nar/gkx320.

Soumya Haider, K. H. M. (2015). Genome mining for bioactive compounds. In S. K. Kim (Ed.), *Springer Handbook of Marine Biotechnology* (pp. 531–539). Berlin/Heidelberg, Germany: Springer Handbooks. https://doi.org/10.1007/978-3-642-53971-8_33.

Strieth, D., Lenz, S., & Ulber, R. (2022). In vivo and in silico screening for antimicrobial compounds from cyanobacteria. *Microorganisms, 11*(2), e1268. https://doi.org/10.3390/microorganisms110201268.

Sweeney-Jones, A. M., Gagaring, K., Antonova-Koch, J., Zhou, H., Mojib, N., Soapi, K., & Kubanek, J. (2020). Antimalarial peptide and polyketide natural products from the Fijian marine cyanobacterium *Moorea producens*. *Marine Drugs, 18*(3), 167. https://doi.org/10.3390/md18030167.

Tan, L. T. (2007). Bioactive natural products from marine cyanobacteria for drug discovery. *Journal of Pharmacy and Pharmacology, 68*(7), 954–979.

Tan, L. T., & Phyo, M. Y. (2020). Marine cyanobacteria: A source of lead compounds and their clinically-relevant molecular targets. *Molecules, 25*(9), 2197. https://doi.org/10.3390/molecules25092197.

Tewari, D., Atanasov, A. G., Semwal, P., & Wang, D. (2021). Natural products and their applications. *Current Research in Biotechnology, 3, 82–83.*

Tiwari, A. K., & Tiwari, B. S. (2020). Cyanotherapeutics: An emerging field for future drug discovery. *Applied Phycology, 1*(1), 44–57. https://doi.org/10.1080/26388081.2020.1744480.

Uzair, B., Tabassum, S., Rasheed, M., & Rehman, S. F. (2012). Exploring marine cyanobacteria for lead compounds of pharmaceutical importance. *The Scientific World Journal, 2012,* 179782. https://doi.org/10.1100/2012/179782.

van Heel, A. J., de Jong, A., Song, C., Viel, J. H., Kok, J., & Kuipers, O. P. (2018). BAGEL4: A user-friendly web server to thoroughly mine RiPPs and bacteriocins. *Nucleic Acids Research, 46*(W1), W278–W281. https://doi.org/10.1093/nar/gky383.

Wang, M., Zhang, J., He, S., & Yan, X. (2017). A review study on macrolides isolated from cyanobacteria. *Marine Drugs, 15*(5), 126.

Weber, T., & Kim, H. U. (2016). The secondary metabolite bioinformatics portal: computational tools to facilitate synthetic biology of secondary metabolite production. *Synthetic and Systems Biotechnology, 1*(2), 69–79. https://doi.org/10.1016/j.synbio.2015.12.002.

Weber, T., Rausch, C., Lopez, P., Hoof, I., Gaykova, V., Huson, D., & Wohlleben, W. (2009). CLUSEAN: A computer-based framework for the automated analysis of bacterial secondary metabolite biosynthetic gene clusters. *Journal of Biotechnology, 140*(1–2), 13–17.

Yang, Z., He, J., Wei, X., Ju, J., & Ma, J. (2020). Exploration and genome mining of natural products from marine Streptomyces. *Applied Microbiology and Biotechnology, 104*(1), 67–76. https://doi.org/10.1007/s00253-019-10227-0.

Zahra, Z., Choo, D. H., Lee, H., & Parveen, A. (2020). Cyanobacteria: Review of current potentials and applications. *Environment, 7*(2), 13.

Zaporozhets, T. S., & Besednova, N. N. (2020). Biologically active compounds from marine organisms in the strategies for combating coronaviruses. *Acta Medica, 6*(4), 470.

Zhang, G., Li, J., Zhu, T., Gu, Q., & Li, D. (2016). Advanced tools in marine natural drug discovery. *Current Opinion in Biotechnology, 42,* 13–23. https://doi.org/10.1016/j.copbio.2016.02.021.

CHAPTER 17

Pharmacokinetics of Marine-Derived Actives: A Focus on Anticancer Drugs

SHILPA N. SHROTRIYA,[1] SUGANDHA V. MULGUND,[1] and AMELIA M. AVACHAT[2]

[1]*Bharati Vidyapeeth (Deemed to be University), Poona College of Pharmacy, Pune, Maharashtra, India*

[2]*Sinhgad College of Pharmacy, Vadgaon (Bk.), Pune, Maharashtra, India*

ABSTRACT

In recent years, the demand for marine activities has enormously increased due to their unique chemical and biological properties. These properties have paved the way for the development of new drugs from marine resources. The marine milieu is a huge source of species, amongst them, algae, crustaceans, sea cucumber, fungus, sea urchins, sponges, mollusks, tunicate, and bryozoan are major sources of therapeutic active substances. Several marine-derived compounds exhibit numerous applications including anticancer, antimicrobial, antifungal, anti-diabetic, anti-obesity, antihyperlipidemic, anti-Alzheimer, anti-fibrotic, neuroprotective, wound healing, skin protective activities, etc. Many compounds present promising results in *in-vitro* models but show problems in pharmacokinetic studies in animal models. On the other hand, in spite of being effective in preclinical models, more than 90% of new drugs fail in clinical trials. The pharmacokinetics of marine actives have gained growing attention in recent years for the potential and optimal use of a therapeutic arsenal with appropriate dosing. In this chapter, the authors present a summary of pharmacokinetic data (including absorption, distribution, metabolism, and excretion) of marine-derived active

Handbook of Research in Marine Pharmaceutics: Exploring Oceanic Microbial Diversity for Human Health and Wellness. Md. Faiyazuddin & Khalid Rehman Hakeem (Eds.)

compounds with a particular focus on anticancer actives. This chapter also provides an update on the various analytical and bioanalytical techniques used for the evaluation of pharmacokinetic parameters.

17.1 INTRODUCTION

One of the main issues affecting human health today is cancer, which has significant social and financial repercussions. Besides synthetically derived drugs, the discovery of anticancer entities has also come from natural resources. Today, more than 60% of commercially accessible anticancer medications come from natural sources. Significant research has been done for new anticancer medications from natural sources, including plants, microorganisms, and marine organisms because there is an urgent need for drugs with novel mechanisms of action. Many naturally-derived antiproliferative drugs like doxorubicin, daunomycin, bleomycin, mitomycin C, vincristine, and vinblastine play a significant role in curative cancer chemotherapy (Da Rocha Adriana, Rafael M. Lopes, & Gilberto Schwartsmann, 2001). The most significant recent additions to the clinical arsenal in oncology, taxanes, and camptothecins, are also naturally derived substances. Marine organisms are a fantastic source of novel chemicals with enormous promise for the creation of new medicines. Marine sources represent a pool of novel bioactive metabolites with varied sets of chemical structures. Identification of novel marine-derived compounds and the investigation of their pharmacological action have grown rapidly in recent years. In one of the reviews, 68 novel marine-derived anticancer chemical entities were discussed, the majority of which had unclear mechanisms of action (Mayer Alejandro & Kirk R. Gustafson, 2003). A review of the clinical outcomes produced by a number of marine-derived substances, including dolastatins, didemnin B, and bryostatin, was published for their anticancer activity (Amador Maria, Jimeno, Paz-Ares, Cortes-Funes, & Hidalgo, 2003). Jimeno et al. (2004) reviewed three marine-derived anticancer agents—ET-743 (Yondelis), Aplidin, and Kahalalide—and summarized their efficacy as anticancer agents. Alves et al. (2018) discussed the antitumor potential of bioactive molecules derived from algae. This review also summarized the most significant studies exploring the potential of algae as a source of marine pharmaceuticals for cancer therapies. These studies highlight the significance of marine resources as a potential source of anticancer chemicals. The biological activities of numerous kinds of marine-derived substances, such as marine algae, bacteria, sponges, corals, tunicates, and sea cucumbers, were reported in many research articles. Alkaloids, polyketides, terpenes, peptides, polysaccharides, and anthraquinones are some of the chemical classes into which marine natural products are categorized (Ruiz-Torres Verónica, Jose Antonio Encinar, María Herranz-López, Almudena Pérez-Sánchez et al., 2017; Ghareeb Mosad, Mohamed A. Tammam, Amr El-Demerdash, & Atanas G. Atanasov, 2020). Several reviews elaborated on the recent advances in cancer therapy and immune signaling. For better human health, recent developments in the use of nanotechnology-based delivery of anticancer medications and knowledge of cancer immunity are also presented (Saeed, Abdullah Fuh, Jingqian Su, & Songying Ouyang, 2021). Various possible molecular mechanisms along with their chemical classes, therapeutic strategies, and limitations of numerous marine anticancer drugs have been discussed in many papers (Khalifa, Shaden, Nizar Elias, Mohamed A. Farag, Lei Chen, Aamer Saeed, Mohamed-Elamir F. Hegazy, Moustafa S. Moustafa et al., 2019). Many marine-derived substances, including bryostatin, dolastatins, and didemnin B, are currently being clinically tested as potential anticancer medications, and several anti-cancer moieties isolated from marine compounds have been approved for clinical use. These include cytarabine, vidarabine, nelarabine (a prodrug of ara-G), fludarabine phosphate (a pro-drug of ara-A), trabectedin, eribulin mesylate, brentuximab vedotin, polatuzumab vedotin, enfortumab vedotin, belantamab mafodotin, plitidepsin, and lurbinectedin (Barreca, Marilia, Virginia Spanò, Alessandra Montalbano, Mercedes Cueto, Ana R. Díaz Marrero, Irem Deniz, Ayşegül Erdoğan et al., 2020).

The pharmacokinetic study of these marine actives assumes greater importance in clinical studies and has gained growing attention in recent years for the potential and optimal use of a therapeutic arsenal with appropriate dosing. Shikov et al. reviewed and summarized the pharmacokinetic studies and corresponding data from human investigations of various actives obtained from marine sources, with an emphasis on their ADME (Shikov Alexander, Elena V. Flisyuk, Ekaterina D. Obluchinskaya, & Olga N. Pozharitskaya, 2020). In this chapter, for the first time, the authors have discussed and summarized the pharmacokinetic data of some of the anticancer drugs derived from marine sources, including those that have been tested in cancer patients.

17.2 PHARMACOKINETICS OF SELECTIVE MARINE-DERIVED ANTICANCER MOIETIES

Even though the number of studies about various marine-derived moieties has drastically increased, only a few compounds with promising results have been submitted for pharmacokinetic research as only a few marine antineoplastic agents have reached clinical phase trials. For instance, bryostatin 1, ET-743, and dolastatin 10 have recently entered a phase II clinical trial against melanoma, non-Hodgkin's lymphoma, renal cancer, and colorectal cancer. The chemical structures of the marine-derived anticancer moieties discussed in this chapter are depicted in Figure 17.1.

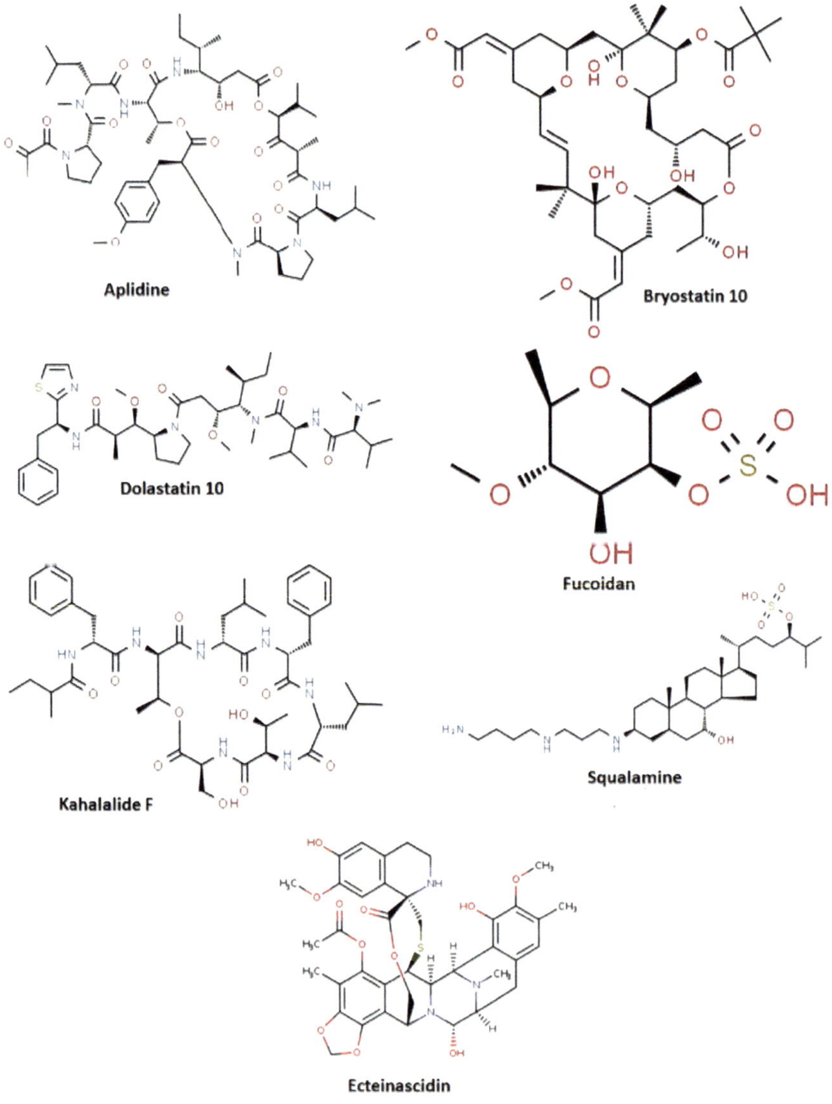

FIGURE 17.1 Chemical structures of marine anticancer moieties.

17.2.1 APLIDINE

Aplidine, a new depsipeptide with strong cytotoxic properties, is generated from the tunicate *Aplidium albicans* (Le Tourneau Christophe, Sandrine Faivre, Eva Ciruelos, María J. Domínguez et al., 2010). Celli et al. (1999) explored its use in rats to study its absorption, distribution, and metabolism. Aplidine concentration (100 ng/mL) in blood was reported two minutes after intravenous administration at a dose of 700 μg/kg. Around 50% of the dose of aplidine was stored in blood cells after intravenous administration. An unmodified form of aplidine was noticed in urine after IV injection in rats. Fast disappearance of aplidine from blood with a $t_{1/2}$ of 9 min was reported after I/V injection. Although it was still visible after 20 minutes, the concentration remained below the quantification threshold (5 ng/mL).

A number of pharmacokinetic studies have been carried out in human volunteers with advanced cancers as a continuation of pre-clinical research on aplidine and its pharmacokinetics *in vivo*. In one of the studies (Faivre Sandrine, Stéphanie Chièze, Catherine Delbaldo, Nora Ady-Vago, Cecilia Guzman, Luis Lopez-Lazaro, Stéphanie Lozahic et al., 2005), 48 patients with non-lymphoma or solid tumors were selected for the phase I clinical trials. Aplidine infusions (0.13–8.00 mg/m^2) by different schemes (1 or 24 h weekly; 3 or 24 h biweekly; 1 h for 5 consecutive days every 3 weeks) were administered to human volunteers to understand its pharmacokinetic data. Substantial prolongation in the individuals' $t_{1/2}$ (up to 43.8 h), prolongation in mean residence time (up to 51 h), elevation in clearance (up to 6.2 L/h), and elevation in blood steady-state volume of distribution V_{ss} were reported after i/v administration of aplidine for 24 hours at doses of 0.2 to 8 mg/m^2. Lower concentration of aplidine in plasma as compared to that in blood and increased AUC and C_{max} with increasing dose were also observed during the study. The unaltered drug was eliminated in urine (with around 15% recovery over 48 h).

Van Andel et al. also reported similar data in one of their research articles where the use of radiolabeled 14C aplidine was reported in six patients (Van Andel Lotte, Hilde Rosing, Jan HM Schellens, & Jos H. Beijnen, 2018). Clinical evidence supported earlier findings made in animals that aplidine accumulates in red blood cells and up to levels of 5 mg/m^2 linearly. Moderate microsomal-mediated metabolism of the aplidine was also reported (Nalda-Molina Ricardo, Belén Valenzuela, Amelia Ramon-Lopez, Bernardo Miguel-Lillo, Arturo Soto-Matos, & Juan Jose Perez-Ruixo, 2009). These findings led to the recommendation of aplidine dosages of 5 and 7 mg/m^2.

A simple LC-ESI-MS/MS was used to understand the pharmacokinetic data of Aplidine in human plasma, whole blood, and urine in cancer human volunteers during the phase I clinical trials. The developed method enabled the demonstration of Aplidine dynamic equilibrium with plasma and blood cells. Additionally, the approach was used to analyze the pharmacokinetics of Aplidin® in cancer patients with success (Celli Nicola, Barbara Mariani, Francesco Di Carlo, Massimo Zucchetti, Luis Lopez-Lazaro, Maurizio D'Incalci, & Domenico Rotilio, 2004).

17.2.2 BRYOSTATIN

Bryostatin is a cyclic macrolide, an anticancer compound isolated from a marine bryozoan, *Bugula neritina*, which acts by modulation of protein kinase C (Kollár Peter, Josef Rajchard, Zuzana Balounová, & Jiří Pazourek, 2014; Raghuvanshi Rinky & Sandip B. Bharate, 2020; Sun Miao-Kun & Daniel L. Alkon, 2006). Zhang et al. used radiolabeled Bryostatin 1 to study the pharmacokinetics of Bryostatin 1 in mice. Bryostatin 1 was administered intravenously to mice at a dose of 40 μg/kg, resulting in a C_{max} and AUC of 92 and 376 ng/mL, respectively. The AUC was better (620 ng/ml.h) but the C_{max} was lower (13.5 ng/mL) with intraperitoneal administration of the same dose. Plasma levels of bryostatin-1 were mostly detected for up to 24 hours, indicating a prolonged half-life. It is widely dispersed throughout the body's organs, but it accumulates most heavily in the adipose tissue, lungs, liver, and GIT. It was discovered that it was mostly removed from the body through urine excretion during the first 12 hours post-administration (Zhang Xueshu, Ruiwen Zhang, Hui Zhao, Hongying Cai et al., 1996). In another study, the pharmacokinetics of Bryostatin 1 was evaluated after intravenous administration in mice at doses of 3.50 and 5.25 μg/kg. The C_{max} value was about 2.26 ng/mL at a high dose. The peak concentration in plasma increased with increasing dose (Nelson, Sen, Alkon, & Sun, 2013). A bioanalytical technique was developed by Zhao et al. (2005) to investigate the pharmacokinetics of Bryostatin 1 in humans. A single patient's intravenous infusion of bryostatin-1 at a dose of 20 μg/m^2 was examined for its single-dose pharmacokinetics. The elimination $t^{1/2}$ was 9 hours, and the C_{max} was determined to be 95 pg/mL. Due to the non-measurability of human PK data, phase I studies' pharmacokinetic analyses, as well as many clinical trial designs, were unsuccessful. So that future clinical trials may be better structured, many attempts were made to create a more sensitive and specific bioanalytical method for the estimation of Bryostatin 1.

17.2.3 DOLASTATINS

Small peptides called Dolastatins 10 and 15 were extracted from the shell-less mollusk sea hare *Dolabella auricularia*, which has anticancer properties (Simmons T. Luke, Eric Andrianasolo, Kerry McPhail, Patricia Flatt, & William H. Gerwick, 2005). The effect of intravenous, subcutaneous, and intraperitoneal administration of radio-labeled dolastatin 10 (0.24 mg/kg), (0.32 mg/kg) on pharmacokinetics data was observed in mice. A sharp decline in dolastatin levels in plasma was seen after IV administration. While in the case of subcutaneous administration, the dolastatin concentration in plasma gradually increased up to a C_{max} of 11 ng/mL. The plasma protein binding for dolastatin in humans, dogs, and mice was estimated to be greater than 81%. A sizable portion of the radioactivity in plasma after dolastatin 10 administration in mice was ascribed to radiolabeled metabolites. Dolastatin 10 is quickly converted to the more polar dihydroxy derivative (*in vitro*) after being incubated with a homogenate of whole rat liver. Elimination of dolastatin 10 was observed from plasma with a $t_{1/2}$ of 5.6 h by IV administration. Faster elimination ($t_{1/2}$ = 3.7 h) was reported after subcutaneous administration. Less than 2% of administered dolastatin 10 was excreted in urine in its unchanged form regardless of the route of administration (Newman, Fuentes, Covey, & Benvenuto, 1994).

Both dolastatins (10 and 15) were found in the plasma five minutes after I.V. administration (1 mg/kg) in mice. The two-exponential plasma concentration-time function was observed with $t_{1/2}$ α of 0.04 h and 0.09 h and $t_{1/2}$ β of 1.6 h and 0.52 h, for dolastatins 10 and 15, respectively. The area under the curve for dolastatin 10 was found to be 1.6-fold higher than for dolastatin 15. The plasma binding of dolastatin 10 was more than 90%. Rapid elimination of both dolastatins (10 and 15) from the plasma after I.V. administration was observed with a shorter elimination speed of dolastatin 15 than dolastatin 10. Traces of dolastatin 10 were noticed in the plasma of all animals at seven hours post-injection, and in the plasma of two of three animals at 24 h (Aherne G. Wynne, Anthea Hardcastle, Melanie Valenti, Alexander Bryant et al., 1996).

HPLC/electrospray ionization mass spectrometry was utilized for quantitation of dolastatin-10 during phase I clinical trials and the pharmacokinetics of dolastatin 10 were assessed in a cancer patient following a bolus I.V. injection at a dose of 65 μg/m^2. Five minutes after injection, dolastatin 10 was found at its highest concentration (29.3 ng/mL). The three-exponential plasma concentration-time function was observed with a $t_{1/2}$ α of 0.087 h, $t_{1/2}$ β of 0.69 h, and $t_{1/2}$ γ of 8 h and V_{ss} of 46.7 L/m^2, and plasma clearance of 3.7

L/h/m^2 were observed. The metabolite (N-dimethyl dolastatin 10) with m/z 772 was detected in the plasma from 0.75 to 1.5 h. Metabolite concentration was almost half that of dolastatin 10 concentration (Garteiz, Madden, Beck, Huie et al., 1998).

17.2.4 ECTEINASCIDINS

The naturally occurring tetrahydroisoquinoline alkaloids known as "ecteinascidins" were isolated from the *Caribbean ascidian* (*Ecteinascidia turbinate*). Although numerous of its analogs showed remarkably potent antitumor activity, the compound designated ET-743 covalently binds to DNA in the minor groove and exerts better antitumor activity (Rinehart, Tom, Nancy, Justin et al., 1990; Wright Amy, Dorilyn A. Forleo, Geewananda P. Gunawardana, Sarath P. Gunasekera, Frank E. Koehn, & Oliver J. McConnell, 1990; Reid Joel, Denise L. Walker, & Matthew M. Ames, 1996). Several preclinical studies revealed the activity of ET-743 against different tumor types. The pharmacokinetic data of ecteinascidin 743 (syn. trabectedin) was estimated in patients with solid tumors after administration of multiple doses (50–900 µg/m^2) by 24 h intravenous infusion. The linear increase in C_{max} (0.06–0.95 ng/mL) and AUC (1.2–36 ng·h/mL) with increased dose were observed suggesting the linear pharmacokinetics of ecteinascidin 743 after 21 days of five repeated courses. Increased $t_{1/2}$ up to 400 µg/m^2 of dose following decrease was reported. After five consecutive courses, an increase in clearance was observed. When taking all of the findings into account, the authors suggested that a 24-hour intravenous infusion every three weeks is an effective and well-tolerated course of treatment (Van Kesteren et al., 2000).

In another study, the pharmacokinetic data of trabectedin (syn. ecteinascidin 743) were estimated in patients after continuous 24-hour intravenous infusion at a dose of 1.3 mg/m^2. The highest concentration (0.6 ng/mL) was recorded at 8 hours and remained constant at this level until 24 hours after injection. An area under the curve of 16.5 ng·h/mL; clearance of 51.3 L/h/m^2 and a half-life of 31.9 hours were reported. The results of this study recommended the use of ecteinascidin 743 treatment outline for liposarcoma patients (Zangarini et al., 2014). In another clinical trial, 44 patients with solid tumors received trabectedin (ecteinascidin 743) by intravenous infusion over 1 or 3 hours every 21 days. A dose-dependent linear increase in the area under the curve was observed in this trial. Over the course of the 1-hour infusion regimen, clearance of trabectedin was not linear. With the 3-hour infusion schedule, it did not greatly depend on the dose and did not change. Higher C_{max} values were substantially correlated with the reported hepatotoxicity. The trial's findings recommended an infusion of 1650 μg/m^2 trabectedin for three hours (Twelves et al., 2003). In one of the phase-I clinical trial studies, the feasibility of administering ecteinascidin-743 at daily intravenous infusion for 5 days every 3 weeks was assessed to recommend the dose for Phase II studies, to characterize its pharmacokinetic behavior, and to gather preliminary evidence of anticancer activity. The study concluded that a dose of 325 μg/m^2/day daily X 5 every 3 weeks of ecteinascidin-743 can be administered repetitively (Villalona-Calero et al., 2002).

17.2.5 FUCOIDAN

Fucoidan is a naturally occurring substance present in several animals and different species of brown algae (*Phaeophyceae*), which has drawn attention to its potential to treat cancer. Algal cell walls are comprised of the sulfated polysaccharide compound fucoidan (Newman David, Gordon M. Cragg, & Kenneth M. Snader, 2003; Synytsya et al., 2010). Very few research papers have discussed the detailed pharmacokinetic parameters (ADME) of fucoidan (Ale Marcel Tutor, & Anne S. Meyer, 2013; Irhimeh, Fitton, Lowenthal, & Kongtawelert. 2005). Fucoidans from various sources have been studied in rabbits, rats, and mice for the estimation of pharmacokinetic parameters including absorption, distribution, metabolism, and elimination using a variety of administration methods. Fluorescein isothiocyanate (FITC)-labeled, commercially available fucoidan (MW = 107.8 kDa) was investigated for pharmacokinetic data after intravenous administration (50 mg/kg) in mice. The report indicates the C_{max} to be 66.37 μg/mg, and AUC as 198.11 μg/g·h in the blood (Bai Xu, Zhang, Bo Hu, Hao Liang, Shuliang Song, & Aiguo Ji, 2020). With a T_{max} of 0.5 hours, fucoidan swiftly traveled to the liver and kidney and achieved its maximum concentration. After intravenous administration in mice, the clearance of FITC-labeled fucoidan was 0.25 mg/(μg/g)/h and was not detected in blood 4 hours after intravenous administration in mice. Another fucoidan (MW = 100 kDa) from L. japonica's pharmacokinetics was studied in rats following its intravenous (6 mg/kg) and oral (20 mg/kg) administration. After an intravenous dose, the C_{max} was 75.59 μg/mL and AUC was 479.07 g/h/mL. Following oral dosing, the C_{max} was 7.33 μg/mL, the T_{max} was 2 h, and the AUC was 42.69 g/h/mL. The author further reports the distribution of fucoidan after intravenous, oral, and transdermal applications. Following intravenous administration and oral administration, respectively, fucoidan with a molecular weight of 100 kDa showed equal values of MRT of 5.48, 1.78 h, and 5.30, 1.65 h (Zhang et al., 2020). In rats, the clearance of fucoidan with a molecular weight of 100 kDa was 0.0037 mg/μg/mL/h following intravenous treatment and 0.138 mg/μg/mL/h following oral administration. Equal $T_{1/2}$ (4.01 and 4.1 h, respectively) following intravenous administration and oral administration in rats was characteristic of fucoidan with a molecular weight of 100 kDa. After intravenous administration in rabbits (50 mg/kg), a low molecular weight fucoidan (7.1 kDa) from *Laminaria japonica* quickly enters the plasma (C_{max} = 110.53 μg/mL; Tmax = 5 min). The two-exponential serum concentration-time function had $t_{1/2}\ \alpha = 11.24 \pm 2.93$ min and $t_{1/2}\ \beta = 98.20 \pm 25.78$ min. Due to the low sensitivity of the analytical method, fucoidan was only detected at 2 h, but not quantified in the serum of rabbits after oral administration (200 mg/kg) (Zhang Wenjing, Delin Sun, Xia Zhao, Weihua Jin, Jing Wang, & Quanbin Zhang, 2016). After being given orally to rats, the high-MW fucoidan displayed a prolonged MRT in blood and accumulation in organs with a filtering function, including the spleen, kidneys, and liver. However, it was not detected in the brain or the heart. Fucoidan (MW = 735 kDa) from Fucus vesiculosus was recently studied for its pharmacokinetics in rats following oral, intravenous, and transdermal administration at doses of 100 mg/kg. Intestinal absorption of fucoidan (737 kDa) was checked in rats. After I/V injection, C_{max} 9.15 ± 0.60 μg/mL was observed (Pozharitskaya Olga, Alexander N. Shikov, Ekaterina D Obluchinskaya, & Heikki Vuorela, 2019). After being administered orally, fucoidan was identified in the plasma 30 minutes later with a C_{max} of 0.125 μg/mL and a T_{max} of 4 hours (Li Shuai, Yuanhong Wang, Tingfu Jiang, Han Wang, Shuang Yang, & Zhihua Lv, 2016), whereas C_{max} was 0.12 μg/mL at 1.2 hours following transdermal ointment administration (Pozharitskaya Olga, Alexander N. Shikov, Ekaterina D. Obluchinskaya, & Heikki Vuorela, 2019). All these results proved that fucoidan rapidly absorbs after different routes of administration. A high quantity of fucoidan was found in the skin of rats 15 minutes after topical treatment, and it began to decline after one hour. This suggests that the medication first saturated the epidermis. Fucoidan concentrations rose in muscles and plasma after an hour, with muscle-plasma concentration ratios at the first- and second-hour post-dose being 1.77 and 1.75, respectively. At dosages of 50–150 mg/kg, linear pharmacokinetics was seen. The authors hypothesized that fucoidan's extended $t_{1/2}$ (20.75 ± 9.43 h) in plasma with low bioavailability (F = 17.7% ± 7.7%) following topical treatment in rats were caused by the quick penetration and retention of substance in the skin. Olga et al. analyzed the pharmacokinetic profile and tissue distribution of Fucoidan in rats after oral administration (Pozharitskaya et al., 2018). From the result of the study, it has been observed that Fucoidan is mainly retained in the kidneys and has an MRT of 6.79 h, indicating a slow absorption rate and prolonged blood circulation. Additionally, research comparing the bioavailability of low molecular weight fucoidan with moderate molecular weight fucoidan revealed that low molecular weight fucoidan has a higher absorption rate and bioavailability, meaning that it has a greater biological potential. Furthermore, recent pharmacokinetic research of *Fucus vesiculosus* indicated that the molecular weight of fucoidan plays a significant role in its absorption and deposition, which may be related to its prolonged half-life after topical treatment. In terms of toxicity, fucoidan has good pharmacokinetics. However, data on its biodistribution in humans is still a challenge to researchers.

After peroral administration, Irhimeh et al. analyzed fucoidan for the first time in human blood. In this study, 3 g of a 713 kDa, 75% sulfated galactofucan isolated from *Undaria pinnatifida* was given three times daily for 12 days to 40 volunteers. Plasma concentration of 12.989 mg/L and around 0.6% bioavailability was reported by authors (Irhimeh, Fitton, Lowenthal, & Kongtawelert. 2005). In another study, 10 healthy male volunteers received 1 g of fucoidan. After 3 h of administration, fucoidan was seen in serum and urine. The time-dependent increase in the concentration of fucoidan in the serum (up to 100 ng/mL) and urine (up to 1000 ng/mL) was reported. The molecular weight (66 kDa) of the administered fucoidan in the serum remained unchanged, whereas the MW (1.8–3.1 kDa) of fucoidan in urine was significantly lower, which evidenced the polysaccharide metabolism (Tokita Yoshiharu, Katsuyuki Nakajima, Hiroshi Mochida, Masahiko Iha, & Takeaki Nagamine, 2010).

17.2.6 KAHALALIDE F

The green algae *Bryopsis* sp. and the Hawaiian mollusk *Elysia rufescens* both contain kahalalide F, a cyclic depsipeptide that has the potential as an anticancer drug (Hamann Mark & Paul J. Scheuer. 1993; Suárez Yajaira, Laura González, Ana Cuadrado, Maite Berciano, Miguel Lafarga, & Alberto Muñoz, 2003; Ciavatta et al., 2017). In one of the preclinical research, mice were used to analyze the pharmacokinetic profile of kahalalide F. The initial plasma concentration of kahalalide F was found to be 1.55M after an intravenous administration of 278 μg/kg. Two exponential plasma concentrations with $t_{1/2}$ α of 15.8 min and $t_{1/2}$ β of 4.4 h were observed as the plasma concentration-time function (Faircloth & Maria del Carmen Cuevas Marchante, 2006). After intravenous administration, the V_{ss} for Kahalalide F was found to be 100 times more than body weight, demonstrating that the drug is broadly disseminated in peripheral tissues. Kahalalide F was not detected in mice's plasma 24 hours after administration, and it did not accumulate after frequent injections at recesses of 24 h. The total body clearance of kahalalide F was found to be 23% of the blood flow in the liver. This demonstrated that kahalalide F has a minimal affinity to extravascular tissues and is rapidly removed from the plasma. In phase I clinical trials intravenous infusion (20–930 μg/m^2) over one hour for five consecutive days every three weeks was given to 32 subjects with advanced androgen-refractory prostate cancer. At the end of infusion, C_{max}, 37.9–183.7 ng/mL, and AUC 28.0–53.4 ng/mL were observed. This was further characterized by dose linearity up to 560 μg/m^2 per day. Rapid drug elimination with $t_{1/2}$ = 0.47–0.88 h, and clearance = 14.5–6.9 L/h were recorded. Further, Kahalalide F did not accumulate during the course. The urinary excretion of the drug was not recorded due to the sensitivity of the analytical method. Finally, by considering all the pharmacokinetic results, 560 μg/m^2/day was considered as a rational dose of Kahalalide F (Faircloth & Maria del Carmen Cuevas Marchante, 2006). Pardo et al. evaluated the pharmacokinetic parameters of Kahalalide F in 35 human subjects with advanced solid tumors after a once-a-week, 1-hour intravenous administration of infusion. The results depicted linear kinetics for C_{max} and AUC, narrow V_{ss} (5.55 L), and a short $t_{1/2}$(0.52 h). In a subsequent study, the pharmacokinetics of Kahalalide F was described from phase II trials in patients with advanced malignant melanoma. Here, patients were infused weekly for one hour intravenously with Kahalalide F at a dose of 650 μg/m^2. The median V_{ss} was 7.0 L, and the median $t_{1/2}$ was 0.46 h (Martín-Algarra et al., 2009). In another study, researchers described the population pharmacokinetics of Kahalalide F in patients with advanced cancer following intravenous dosing. NONMEMTM VII was used to evaluate data from 240 patients who were included in three Phase I and three Phase II trials and received intravenous administration weekly and every three weeks infusion of Kahalalide F at dosages ranging from 266 to 6650 μg/m^2. The pharmacokinetic information from six clinical studies revealed dose-proportional exposure, linear elimination of Kahalalide F from plasma, and time-independent pharmacokinetics (Miguel-Lillo Bernardo, Belén Valenzuela, José Esteban Peris-Ribera, Arturo Soto-Matos, & Juan José Pérez-Ruixo, 2015).

17.2.7 SQUALAMINE

Squalamine is isolated from the tissues of various species of dogfish sharks (*Squalus acanthias*) which has a natural cationic steroid with antiangiogenic properties. Currently, this agent is the subject of clinical trials for a number of cancers (Mammari Nour, Elsa Salles, Audrey Beaussart, Sofiane El-Kirat-Chatel, & Mihayl Varbanov, 2022). Hao et al. assessed the possibility and depicted the pharmacokinetics of squalamine. It was administered as a 5-day continuous intravenous infusion in incremental doses to patients with advanced solid tumors every three weeks. From a starting dose of 6 mg/m^2/day, doses were originally increased by 100% increments, with a single patient being treated at each dose level until substantial toxicity was noticed, at which point more patients were treated. Squalamine was administered in 73 sessions to 33 individuals at 13 different dose levels ranging from 6 to 700 mg/m^2/day. Several pharmacokinetic parameters like the volume of distribution, half-life, and mean clearance at 500 mg/m^2/day were found to be 2.67 L/h/m^2, 9.46 h, and 36.84 L/m^2, respectively. Steady-state concentrations of 20.08 μg/ml were also significantly higher than the levels that inhibit angiogenesis in preclinical models. The results of the studies conclude that squalamine is well tolerated at the Phase II dose of 500 mg/m^2/day and produces plasma concentrations that are at least a factor of two larger than those needed to have noticeable antiangiogenic effects in preclinical trials (Hao Desirée et al., 2003). Bhargava et al. worked on phase I and the pharmacokinetics of squalamine in patients with advanced cancers. The main objective of the study was to find the maximum tolerated dose, dose-limiting toxicity, and pharmacokinetics of squalamine lactate after the administration of continuous intravenous infusion for 120 hours every two weeks; and to acquire data on extended (>120 hours) continuous intravenous infusions in patients who have tolerated 120-hour infusions. During the study, 19 patients received squalamine doses at eight levels. As a result, this study shows that patients can receive squalamine, an antiangiogenic agent, for a 120-hour continuous intravenous infusion up to a dose rate of 192 mg/m^2/day without experiencing any negative side effects. Patients who have never taken squalamine before might be able to handle a dose rate of 384 mg/m^2/day without experiencing any side effects. Pharmacokinetic data showed a linear relationship between the area under the curve and dose rate, with a prolonged terminal squalamine persistence in patient plasma (median $t_{1/2}$ = 18 h; range, 8–48 h) (Bhargava et al., 2001). Roy et al. designed a phase I/IIA study to evaluate the safety, clinical data, and pharmacokinetics of squalamine by administering a 5-day continuous infusion along with carboplatin and paclitaxel every 3 weeks in patients with stage IIIB (pleural effusion) or stage IV non-small cell lung cancer. The result depicts that the combination of squalamine paclitaxel and carboplatin is well tolerated (Herbst Roy S., Lisa A. Hammond, David P. Carbone, Hai T. Tran, Kenneth J. Holroyd, Avinash Desai, Jon I. Williams et al., 2003).

17.3 ANALYTICAL METHODS UTILIZED IN PHARMACOKINETIC STUDIES

Many analytical techniques are reported in the literature to evaluate the pharmacokinetic parameters of marine-derived moieties. The approaches that are most frequently utilized to examine marine-derived chemicals in biomaterials include fluorescein labeling (Bai Xu, Zhang, Bo Hu, Hao Liang, Shuliang Song, & Aiguo Ji, 2020), radioactive labeling and radioimmunoassay, matrix-assisted laser desorption/ionization, matrix-assisted laser desorption/ionization imaging mass spectrometry, gas chromatography, high-performance liquid chromatography (Egorin et al., 1996) or liquid chromatography with evaporative light-scattering detection, fluorescent detection (Zhang et al., 2020; Zhang Wenjing, Delin Sun, Xia Zhao, Weihua Jin, Jing Wang, & Quanbin Zhang, 2016; Zhao et al., 2016), differential mass spectrometry (Celli Nicola, Agata Motos Gallardo, Cosmo Rossi, Massimo Zucchetti, Maurizio D'Incalci, & Domenico Rotilio, 1999), electrospray ionization, and UV detection (Pozharitskaya Olga, Alexander N. Shikov, Ekaterina D. Obluchinskaya, & Heikki Vuorela, 2019; Pozharitskaya et al., 2018; Su Tongwei, Wenhui Wu, Ting Yan, Chaoyan Zhang, Quangang Zhu, & Bin Bao, 2013; Kosman, Faustova, Urakova, Karlina, & Makarov IV, 2020; Shikov et al., 2019). Biomarker-based assays are also utilized in the analysis of several compounds as they provide high sensitivity and broad linearity.

17.4 CONCLUSION

Recently, many novel anti-cancer compounds have been isolated and identified from marine species. The most intriguing and effective candidates were chosen after additional screening and have now entered intensive pre-clinical and clinical trials. New perspectives are constantly emerging in the progress of marine compounds due to constant research in this area to produce treatments derived from marine sources. However, insufficient attention has been paid to pharmacokinetics, and data on the pharmacokinetics of marine medicines are frequently inconsistent. With an emphasis on absorption, distribution, metabolism, and excretion in both preclinical and clinical studies, here we have summarized and analyzed the information on the pharmacokinetics of selective anticancer compounds derived from marine sources. An in-depth understanding of pharmacokinetics can provide data that can be used to determine the right doses and treatment schedules, understand the molecular underpinnings of pharmacological activity, and provide more effective medication applications. It is anticipated that the pharmacokinetic study of substances obtained from marine sources will be taken up by various research groups on a faster track, which will not only help in gaining acceptance for commercialization but will also open doors for newer anticancer drugs derived from marine sources to see the light of day.

KEYWORDS

- **anticancer potential**
- **fluorescein isothiocyanate**
- **marine bioactive molecules**
- **Pharmacokinetics**
- **squalamine**

REFERENCES

Aherne, G. W., Hardcastle, A., Valenti, M., Bryant, A., Rogers, P., Pettit, G. R., & Kelland, L. R. (1996). Antitumor evaluation of dolastatins 10 and 15 and their measurement in plasma by radioimmunoassay. *Cancer Chemotherapy and Pharmacology, 38*(3), 225–232.

Ale, M. T., & Meyer, A. S. (2013). Fucoidans from brown seaweeds: An update on structures, extraction techniques and use of enzymes as tools for structural elucidation. *RSC Advances, 3*(22), 8131–8141.

Alves, C., Silva, J., Pinteus, S., Gaspar, H., Alpoim, M. C., Botana, L. M., & Pedrosa, R. (2018). From marine origin to therapeutics: The antitumor potential of marine algae-derived compounds. *Frontiers in Pharmacology, 9,* 777.

Amador, M. L., Jimeno, J., Paz-Ares, L., Cortes-Funes, H., & Hidalgo, M. (2003). Progress in the development and acquisition of anticancer agents from marine sources. *Annals of Oncology, 14*(11), 1607–1615.

Bai, X., Zhang, E., Hu, B., Liang, H., Song, S., & Ji, A. (2020). Study on absorption mechanism and tissue distribution of fucoidan. *Molecules, 25*(5), 1087.

Barreca, M., Spanò, V., Montalbano, A., Cueto, M., Díaz Marrero, A. R., Deniz, I., & Trombetta, D. (2020). Marine anticancer agents: An overview with a particular focus on their chemical classes. *Marine Drugs, 18*(12), 619.

Bhargava, P., Marshall, J. L., Dahut, W., Rizvi, N., Trocky, N., Williams, J. I., & Hawkins, M. J. (2001). A phase I and pharmacokinetic study of squalamine, a novel antiangiogenic agent, in patients with advanced cancers. *Clinical Cancer Research, 7*(12), 3912–3919.

Celli, N., Gallardo, A. M., Rossi, C., Zucchetti, M., D'Incalci, M., & Rotilio, D. (1999). Analysis of aplidine (dehydrodidemnin B), a new marine-derived depsipeptide, in rat biological fluids by liquid chromatography–tandem mass spectrometry. *Journal of Chromatography B: Biomedical Sciences and Applications, 731*(2), 335–343.

Celli, N., Mariani, B., Di Carlo, F., Zucchetti, M., Lopez-Lazaro, L., D'Incalci, M., & Rotilio, D. (2004). Determination of Aplidin®, a marine-derived anticancer drug, in human plasma, whole blood and urine by liquid chromatography with electrospray ionisation tandem mass spectrometric detection. *Journal of Pharmaceutical and Biomedical Analysis, 34*(3), 619–630.

Ciavatta, M. L., Lefranc, F., Carbone, M., Mollo, E., Gavagnin, M., Betancourt, T., & Kiss, R. (2017). Marine mollusk-derived agents with antiproliferative activity as promising anticancer agents to overcome chemotherapy resistance. *Medicinal Research Reviews, 37*(4), 702–801.

Da Rocha, A. B., Lopes, R. M., & Schwartsmann, G. (2001). Natural products in anticancer therapy. *Current opinion in pharmacology, 1*(4), 364–369.

Egorin, M. J., Sentz, D. L., Rosen, D. M., Ballesteros, M. F., Kearns, C. M., Callery, P. S., & Eiseman, J. L. (1996). Plasma pharmacokinetics, bioavailability, and tissue distribution in CD2F1 mice of halomon, an antitumor halogenated monoterpene isolated from the red algae *Portieria hornemannii. Cancer Chemotherapy and Pharmacology, 39*, 51–60.

Faircloth, G., & Cuevas Marchante, M. del C. (2006). Kahalalide F and ES285: potent anticancer agents from marine molluscs. In *Molluscs. From Chemo-Ecological Study to Biotechnological Application* (pp. 363–379).

Faivre, S., Chièze, S., Delbaldo, C., Ady-Vago, N., Guzman, C., Lopez-Lazaro, L., & Lozahic, S. (2005). Phase I and pharmacokinetic study of aplidine, a new marine cyclodepsipeptide in patients with advanced malignancies. *J. Clin. Oncol., 23*(31), 7871–7880.

Garteiz, D. A., Madden, T., Beck, D. E., Huie, W. R., McManus, K. T., Abbruzzese, J. L., & Newman, R. A. (1998). Quantitation of dolastatin-10 using HPLC/electrospray ionization mass spectrometry: Application in a phase I clinical trial. *Cancer Chemotherapy and Pharmacology, 41*(4), 299–306.

Ghareeb, M. A., Tammam, M. A., El-Demerdash, A., & Atanasov, A. G. (2020). Insights about clinically approved and preclinically investigated marine natural products. *Current Research in Biotechnology, 2*, 88–102.

Hamann, M. T., & Scheuer, P. J. (1993). Kahalalide F: a bioactive depsipeptide from the sacoglossan mollusk *Elysia rufescens* and the green alga Bryopsis sp. *Journal of the American Chemical Society, 115*(13), 5825–5826.

Hao, D., Hammond, L. A., Eckhardt, S. G., Patnaik, A., Takimoto, C. H., Schwartz, G. H., & Goetz, A. D. (2003). A Phase I and pharmacokinetic study of squalamine, an aminosterol angiogenesis inhibitor. *Clinical Cancer Research, 9*(7), 2465–2471.

Herbst, R. S., Hammond, L. A., Carbone, D. P., Tran, H. T., Holroyd, K. J., Desai, A., & Williams, J. I. (2003). A phase I/IIA trial of continuous five-day infusion of squalamine lactate (MSI-1256F) plus carboplatin and paclitaxel in patients with advanced non-small cell lung cancer. *Clinical Cancer Research, 9*(11), 4108–4115.

Irhimeh, M. R., Fitton, J. H., Lowenthal, R. M., & Kongtawelert, P. (2005). A quantitative method to detect fucoidan in human plasma using a novel antibody. *Methods and Findings in Experimental and Clinical Pharmacology, 27*(10), 705–710.

Jimeno, J., Faircloth, G., Fernández Sousa-Faro, J. M., Scheuer, P., & Rinehart, K. (2004). New marine derived anticancer therapeutics—A journey from the sea to clinical trials. *Marine Drugs, 2*(1), 14–29.

Khalifa, S. A. M., Elias, N., Farag, M. A., Chen, L., Saeed, A., Hegazy, M. E. F., & Badria, F. A. (2019). Marine natural products: A source of novel anticancer drugs. *Marine Drugs, 17*(9), 491.

Kollár, P., Rajchard, J., Balounová, Z., & Pazourek, J. (2014). Marine natural products: Bryostatins in preclinical and clinical studies. *Pharmaceutical Biology, 52*(2), 237–242.

Kosman, V. M., Faustova, N. M., Urakova, I. N., Karlina, M. N., & Makarov, V. G. IV. (2020). Dypeptydylpeptidase IV activity inhibition after oral administration to rabbits of Strongylocentrotus droebahiensis gonads extract as possible biomarker of pharmacokinetics. *Drug Development and Registration, 9*, 158–165.

Le Tourneau, C., Faivre, S., Ciruelos, E., Domínguez, M. J., López-Martín, J. A., Izquierdo, M. A., & Raymond, E. (2010). Reports of clinical benefit of plitidepsin (Aplidine), a new marine-derived anticancer agent, in patients with advanced medullary thyroid carcinoma. *American journal of clinical oncology, 33*(2), 132–136.

Li, S., Wang, Y., Jiang, T., Wang, H., Yang, S., & Lv, Z. (2016). Absorption and transport of sea cucumber saponins from *Apostichopus japonicus*. *Marine Drugs, 14*(6), 114.

Mammari, N., Salles, E., Beaussart, A., El-Kirat-Chatel, S., & Varbanov, M. (2022). Squalamine and its aminosterol derivatives: overview of biological effects and mechanisms of action of compounds with multiple therapeutic applications. *Microorganisms, 10*(6), 1205.

Martín-Algarra, S., Espinosa, E., Rubió, J., López López, J. J., Manzano, J. L., Alonso Carrión, L., & Paz-Ares, L. (2009). Phase II study of weekly Kahalalide F in patients with advanced malignant melanoma. *European Journal of Cancer, 45*(5), 732–735.

Mayer, A. M. S., & Gustafson, K. R. (2003). Marine pharmacology in 2000: Antitumor and cytotoxic compounds. *International Journal of Cancer, 105*(3), 291–299.

Miguel-Lillo, B., Valenzuela, B., Peris-Ribera, J. E., Soto-Matos, A., & Pérez-Ruixo, J. J. (2015). Population pharmacokinetics of kahalalide F in advanced cancer patients. *Cancer Chemotherapy and Pharmacology, 76*(2), 365–374.

Nalda-Molina, R., Valenzuela, B., Ramon-Lopez, A., Miguel-Lillo, B., Soto-Matos, A., & Perez-Ruixo, J. J. (2009). Population pharmacokinetics meta-analysis of plitidepsin (Aplidin®) in cancer subjects. *Cancer Chemotherapy and Pharmacology, 64*(1), 97–108.

Nelson, T. J., Sen, A., Alkon, D. L., & Sun, M. K. (2013). Adduct formation in liquid chromatography-triple quadrupole mass spectrometric measurement of bryostatin 1. *Journal of Chromatography B Analytical Technologies in the Biomedical and Life Sciences, 944*, 55–62.

Newman, D. J., Cragg, G. M., & Snader, K. M. (2003). Natural products as sources of new drugs over the period 1981– 2002. *Journal of Natural Products, 66*(7), 1022–1037.

Newman, R. A., Fuentes, A., Covey, J. M., & Benvenuto, J. A. (1994). Preclinical pharmacology of the natural marine product dolastatin 10 (NSC 376128). *Drug Metabolism and Disposition, 22*(3), 428–432.

Pardo, B., Paz-Ares, L., Tabernero, J., Ciruelos, E., García, M., Salazar, R., & Cortés-Funes, H. (2008). Phase I clinical and pharmacokinetic study of kahalalide F administered weekly as a 1-hour infusion to patients with advanced solid tumors. *Clinical Cancer Research, 14*(4), 1116–1123.

Pozharitskaya, O. N., Shikov, A. N., Faustova, N. M., Obluchinskaya, E. D., Kosman, V. M., Vuorela, H., & Makarov, V. G. (2018). Pharmacokinetic and tissue distribution of fucoidan from Fucus vesiculosus after oral administration to rats. *Marine Drugs, 16*(4), 132.

Pozharitskaya, O. N., Shikov, A. N., Obluchinskaya, E. D., & Vuorela, H. (2019). The pharmacokinetics of fucoidan after topical application to rats. *Marine Drugs, 17*(12), 687.

Rademaker-Lakhai, J. M., Horenblas, S., Meinhardt, W., Stokvis, E., de Reijke, T. M., Jimeno, J. M., & Schellens, J. H. (2005). Phase I clinical and pharmacokinetic study of kahalalide F in patients with advanced androgen refractory prostate cancer. *Clinical Cancer Research, 11*(5), 1854–1862.

Raghuvanshi, R., & Bharate, S. B. (2020). Preclinical and clinical studies on bryostatins, a class of marine-derived protein kinase C modulators: A mini-review. *Current Topics in Medicinal Chemistry, 20*(12), 1124-1135.

Reid, J. M., Walker, D. L., & Ames, M. M. (1996). Preclinical pharmacology of ecteinascidin 729, a marine natural product with potent antitumor activity. *Cancer Chemotherapy and Pharmacology, 38*(4), 329–334.

Rinehart, K. L., Holt, T. G., Fregeau, N. L., Stroh, J. G., Keifer, P. A., Sun, F., & Martin, D. G. (1990). Ecteinascidins 729, 743, 745, 759A, 759B, and 770: Potent antitumor agents from the *Caribbean tunicate Ecteinascidia turbinata. The Journal of Organic Chemistry, 55*(15), 4512–4515.

Ruiz-Torres, V., Encinar, J. A., Herranz-López, M., Pérez-Sánchez, A., Galiano, V., Barrajón-Catalán, E., & Micol, V. (2017). An updated review on marine anticancer compounds: The use of virtual screening for the discovery of small-molecule cancer drugs. *Molecules, 22*(7), 1037.

Saeed, A. F., Su, J., & Ouyang, S. (2021). Marine-derived drugs: Recent advances in cancer therapy and immune signaling. *Biomedicine & Pharmacotherapy, 134,* 111091.

Shikov, A. N., Flisyuk, E. V., Obluchinskaya, E. D., & Pozharitskaya, O. N. (2020). Pharmacokinetics of marine-derived drugs. *Marine Drugs, 18*(11), 557.

Shikov, A. N., Pozharitskaya, O. N., Faustova, N. M., Kosman, V. M., Makarov, V. G., Razzazi-Fazeli, E., & Novak, J. (2019). Pharmacokinetic study of bioactive glycopeptide from Strongylocentrotus droebachiensis after intranasal administration to rats using biomarker approach. *Marine Drugs, 17*(10), 577.

Simmons, T. L., Andrianasolo, E., McPhail, K., Flatt, P., & Gerwick, W. H. (2005). Marine natural products as anticancer drugs. *Molecular Cancer Therapeutics, 4*(2), 333–342.

Su, T., Wu, W., Yan, T., Zhang, C., Zhu, Q., & Bao, B. (2013). Pharmacokinetics and tissue distribution of a novel marine fibrinolytic compound in Wistar rat following intravenous administrations. *Journal of Chromatography B, 942*, 77–82.

Suárez, Y., González, L., Cuadrado, A., Berciano, M., Lafarga, M., & Muñoz, A. (2003). Kahalalide F, a new marine-derived compound, induces oncosis in human prostate and breast cancer cells. *Molecular Cancer Therapeutics, 2*(9), 863–872.

Sun, M.-K., & Alkon, D. L. (2006). Bryostatin-1: Pharmacology and therapeutic potential as a CNS drug. *CNS Drug Reviews, 12*(1), 1–8.

Synytsya, A., Kim, W.-J., Kim, S.-M., Pohl, R., Synytsya, A., Kvasnička, F., & Park, Y. I. (2010). Structure and antitumor activity of fucoidan isolated from sporophyll of Korean brown seaweed *Undaria pinnatifida. Carbohydrate Polymers, 81*(1), 41–48.

Tokita, Y., Nakajima, K., Mochida, H., Iha, M., & Nagamine, T. (2010). Development of a fucoidan-specific antibody and measurement of fucoidan in serum and urine by sandwich ELISA. *Bioscience, Biotechnology, and Biochemistry, 4*(2).

Twelves, C., Hoekman, K., Bowman, A., Vermorken, J. B., Anthoney, A., Smyth, J., & Twelves, C. (2003). Phase I and pharmacokinetic study of yondelis™ (ecteinascidin-743; et-743) administered as an infusion over 1 h or 3 h every 21 days in patients with solid tumours. *European Journal of Cancer, 39*(13), 1842–1851.

Van Andel, L., Rosing, H., Schellens, J. H. M., & Beijnen, J. H. (2018). Review of chromatographic bioanalytical assays for the quantitative determination of marine-derived drugs for cancer treatment. *Marine Drugs, 16*(7), 246.

Van Kesteren, C., Cvitkovic, E., Taamma, A., López-Lázaro, L., Jimeno, J. M., Guzman, C., & Beijnen, J. H. (2000). Pharmacokinetics and pharmacodynamics of the novel marine-derived anticancer agent ecteinascidin 743 in a phase I dose-finding study. *Clinical Cancer Research, 6*(12), 4725–4732.

Villalona-Calero, M. A., Eckhardt, S. G., Weiss, G., Hidalgo, M., Beijnen, J. H., van Kesteren, C., & Rosing, H. (2002). A phase I and pharmacokinetic study of ecteinascidin-743 on a daily × 5 schedule in patients with solid malignancies. *Clinical Cancer Research, 8*(1), 75–85.

Wright, A. E., Forleo, D. A., Gunawardana, G. P., Gunasekera, S. P., Koehn, F. E., & McConnell, O. J. (1990). Antitumor tetrahydroisoquinoline alkaloids from the colonial ascidian *Ecteinascidia turbinata. The Journal of Organic Chemistry, 55*(15), 4508–4512.

Zangarini, M., Ceriani, L., Sala, F., Marangon, E., Bagnati, R., D'Incalci, M., & Zucchetti, M. (2014). Quantification of trabectedin in human plasma: Validation of a high-performance liquid chromatography–mass spectrometry method and its application in a clinical pharmacokinetic study. *Journal of Pharmaceutical and Biomedical Analysis, 95,* 107–112.

Zhang, E., Chu, F., Zhao, T., Chai, Y., Liang, H., Song, S., & Ji, A. (2020). Determination of fucoidan in rat plasma by HPLC and its application in pharmacokinetics. *Pakistan Journal of Pharmaceutical Sciences, 33*(1), 1–9.

Zhang, W., Sun, D., Zhao, X., Jin, W., Wang, J., & Zhang, Q. (2016). Microanalysis and preliminary pharmacokinetic studies of a sulfated polysaccharide from *Laminaria japonica. Chinese Journal of Oceanology and Limnology, 34*(1), 177–185.

Zhang, X., Zhang, R., Zhao, H., Cai, H., Gush, K. A., Kerr, R. G., & Kraft, A. S. (1996). Preclinical pharmacology of the natural product anticancer agent bryostatin 1, an activator of protein kinase C. *Cancer Research, 56*(4), 802–808.

Zhao, M., Rudek, M. A., He, P., Smith, B. D., & Baker, S. D. (2005). Validation and implementation of a method for determination of bryostatin 1 in human plasma by using liquid chromatography/tandem mass spectrometry. *Analytical Biochemistry, 337*(1), 143–148.

Zhao, X., Guo, F., Hu, J., Zhang, L., Xue, C., Zhang, Z., & Li, B. (2016). Antithrombotic activity of oral administered low molecular weight fucoidan from *Laminaria japonica. Thrombosis Research, 144*, 46–52.

CHAPTER 18

Marine Bioactives: From Bench to Clinic

UZMA AZEEM[1] and KHALID REHMAN HAKEEM[2,3]

[1]*Government College Malerkotla, Department of Botany, Malerkotla, Punjab, India*

[2]*Department of Biological Sciences, Faculty of Science, King Abdulaziz University, Jeddah, Saudi Arabia*

[3]*University Centre for Research & Development, Chandigarh University, Mohali, Punjab, India.*

ABSTRACT

For the past few decades, more emphasis has been placed on the exploration and utilization of bioactive natural products for clinical drug development. Marine organisms are reservoirs of several constituents with nutritional and medicinal properties. The history of the use of marine bioactive natural products in the development of medicines is very long, particularly in cancer treatment. Over time, more and more marine bioactive natural products are characterized by marine organisms, increasing expectations for a greater number of drugs developed from marine natural products in the future. For further advancement in clinical drug development from marine bioactive natural products, structural elucidation, chemical synthesis elucidation, and/or fermentation production are also needed for adequate and continuous production and supply of these drugs. Additionally, preclinical trials such as *in vitro* studies and *in vivo* experiments with experimental organisms are also required before the production and utilization of drugs from marine bioactive natural products in the treatment of human health disorders.

Handbook of Research in Marine Pharmaceutics: Exploring Oceanic Microbial Diversity for Human Health and Wellness. Md. Faiyazuddin & Khalid Rehman Hakeem (Eds.)

18.1 INTRODUCTION

Marine ecosystems are habitats of a vast diversity of organisms like algae, bacteria, many invertebrates, fungi, etc., rich in various health-benefiting bioactive natural components. The latter includes several biochemicals like polysaccharides, lipids, peptides and proteins, antioxidants, and many other bioactive compounds (Ibañez, Herrero, Mendiola, & Castro-Puyana, 2012; Subramani & Aalbersberg, 2013; Dalmaso, Ferreira, & Vermelho, 2015; Shaik & Sarbon, 2022). From marine bioactive compounds, various drugs have been developed and some are under evaluation. These drugs are used for the treatment of various human disorders such as cancer, inflammation, viral infections, neurodegenerative disorders, etc. (Karthikeyan, Joseph, & Nair, 2022; Al-Khayri, Asghar, Khan, Akhtar et al., 2022). For instance, spongothymidine and spongouridine isolated from the marine sponge *Tethya crypta* led to the development of Ara-C (cytarabine, an antileukemia agent) and Ara-A (vidarabine, an antiviral agent). The clinical use of these drugs has also been approved by the Food and Drug Administration (FDA) in 1969 and 1976, respectively (Bergmann & Feeney, 1951; Bergmann & Burke, 1956; Bergmann & Stempien, 1957). Some other marine natural products like ziconotide (Prialt®), and trabectedin (Yondelis®) are known for the treatment of severe chronic pain, soft tissue sarcomas, and ovarian cancer, respectively (Klotz, 2006). As of October 2020, there are nearly 14 approved marine-derived drugs and various marine-derived compounds are under phases III, II, or I of drug development (http://marinepharmacology.midwestern.edu/). With time, many more marine natural products are being explored and pharmacologically evaluated for novel drug development. This becomes possible with the recent developments in marine natural products research such as improved isolation and extraction methods, and the addition of novel analytical tools and techniques (Duarte, Rocha-Santos, Freitas, & Duarte, 2012; Lafarga, Acién-Fernández, & Garcia-Vaquero, 2020). Moreover, the increasing enthusiasm for the therapeutic use of marine bioactive compounds leads to more advancement in the discovery of novel drugs from the marine ecosystem. The approval of marine natural products for clinical use takes many years of research by academicians and the pharmaceutical industry (Glaser & Mayer, 2009; Rangel & Falkenberg, 2015). This chapter highlights different types of marine bioactive natural compounds and drugs, pharmacological studies (*in vitro* and *in vivo*), and clinical trials with marine bioactive natural products.

18.2 MARINE BIO-ACTIVES

Marine ecosystems inhabit phytoplankton, micro and macroalgae, plants, protozoans, mammals, fishes, fungi, bacteria, viruses, etc. The marine organisms face extreme environmental conditions such as fluctuations in temperature, high pressure, and limited light sources providing ideal conditions for synthesizing secondary metabolites (Das & Mangwani, 2015; Alharbi, 2016). Some of the biochemicals produced by marine organisms are mentioned in Table 18.1. To date, more than 23,000 medicinally important compounds such as alkaloids, phenols, terpenes, peptides, polyketides, enzymes, fatty acids, etc., have been credited to marine microbes (François, François, Travers, Morga et al., 2016; Blunt, Copp, Keyzers, Munro, & Prinsep, 2017). Of these, about 70% are obtained from actinomycetes, 20% from fungi, 7% from *Bacillus* spp., and 1–2% from other microbes (Mehbub, Perkins, Zhang, & Franco, 2016). The bioactive compounds from sponges, bryozoans, algae, cyanobacteria, and soft corals have led to the discovery of various drugs like Pyranonigrin (François, François, Travers, Morga et al., 2016), Rubrumazine (Barka, Vatsa, Sanchez, Gaveau-Vaillant et al., 2016), Echinulin (Lacret, Pérez-Victoria, Oves-Costales, De La Cruz et al., 2016), Dehydroechinulin (Rakotondraibea, Rasolomampianinab, Parka, Lia et al., 2016), Variecolorin (Oliveira Dos Santos, Britta, Bianco, Ueda-Nakamura, Pereira, & Nakamura, 2011), and Cristatumin (Behie, Bonet, Zacharia, Mcclung, & Traxler, 2017). This is because the compounds isolated from marine organisms have antibacterial (Betancur, Naranjo-Gaybor, Vinchira-Villarraga, Moreno-Sarmiento et al., 2017), antifungal (Kimura, 2018), anti-plasmodial (Patin, Schorn, Aguinaldo, Lincecum, Moore, & Jensen, 2017), anti-protozoal (Kannikan & Rebecca, 2018), anti-inflammatory (Niu, Zhou, Xie, Zhang, & Yang, 2017), anticancer (Takahashi & Nakashima, 2018), and antiviral (Iribarren, Serrano, Naranjo, & García-Castillo, 2018) characteristics. Of all the marine organisms, bacteria are the major contributors and the most evaluated species. The bacterial population in a given marine ecosystem is approximately 4-6 × 10^30 (Elbendary, Hessain, El-Hariri, Seida, Moussa, Mubarak, Kabli, Hemeg, & El Jakee, 2018; Kumar, Krishnani, Bhushan, & Brahmane, 2015). Of these, actinomycetes constitute a major portion dwelling in the intertidal zone, salt marshes, lagoons, estuaries, mangroves, coral reefs, and even the sea floors (Hwang, Lee, Hong, Kim, & Kim, 2018). They exhibit symbiosis with several marine macrofauna and flora, such as insects, invertebrates, marine sponges, and cone snails

(Lindequist, 2016). Several taxa of marine fungi occur in association with marine-inhabiting algae, sponges, invertebrates, and also with sediments. Like marine algae, these are also known as a source of secondary metabolites such as alkaloids, polyketides, terpenoids, isoprenoid and non-isoprenoid compounds, quinones, exhibiting antibacterial, antitumor, and anti-inflammatory efficacy (Hasan, Ansari, Ahmad, & Mishra, 2015). From 2000 to 2005, there were about 100 marine fungal metabolites, and from 2006 to 2010, there were approximately 690 natural products isolated from marine fungi (Duarte, Rocha-Santos, Freitas, & Duarte, 2012). Many marine bioactive compounds isolated from fungi and actinobacteria show significant activity against colorectal cancer cell lines. Different strains of *Aspergillus, Penicillium, Westerdykella* sp., *Paradendryphiella* sp., *Dichotomomyces* sp., and *Neosartorya* sp. marine fungi, and different strains of *Streptomyces* and *Nocardiopsis* sp. bacteria exhibit cytotoxic action against various cell lines such as RKO, SW480, HCT15, HCT16, HCT17, HCT18 in MTT, SRB assays with IC_{50} even greater than 277 μM. This cytotoxic effect is credited to the presence of a number of cytotoxic compounds such as 18-oxo-19,20-dihydrophomacin C, 19,20-dihydrophomacin C, Gymnastatin Z, Diorcinol, Dichocerazine A, 3-O-methyldiorcinol, Pityriacitrin, Stellarine A, etc., present in these marine fungi and bacteria (Julianti, Abrian, Wibowo, Azhari et al., 2022). Different compounds exhibit cytotoxicity through different routes. Rosellichalasin and Cytochalasin E are cytotoxic compounds obtained from *Aspergillus* sp. Nov. F1. These compounds inhibit cell division and promote apoptosis by their effect on actin filaments (Haidle & Myers, 2004; Xiao, Liu, Wu, Liu, Wei, Zhang, & Lin, 2013). *Aspergillus versicolor* Ppf48 yielded several compounds, including asperphenin A and asperphenin B. Treatment with the former compound against RKO cells causes cell accumulation toward the sub-G1 apoptotic phase, enhances G2/M phase arrest, stimulates reactive oxygen species (ROS), and inhibits polymerization of tubulin (Bae, Liao, Park, Kim, Shin, & Lee, 2020). The compound androsamide isolated from *Nocardiopsis* sp. CNT-189 blocks Caco-2 cell migration in a dose-dependent manner. The same compound reduces EMT transcription factors corresponding to the expression by Caco-2 cells as well as genes participating in cell motility in Caco-2, e.g., CAPN1 and RAC2 (Lee, Gamage, Kim, Hillman et al., 2020). Several marine bioactive compounds show their potential against Alzheimer's disease. Today, about 50 million people have dementia, and it is foreseen that this number will increase by 60% in 2030 and by 180% in 2050 (Russo, Kisialiou, Lamonaca, Moroni, Prinzi, & Fini, 2016). Alzheimer's disease (AD) is a type of dementia and is becoming common these days. The symptoms of this disease are decreased cholinergic function, loss of memory and intellectual function, neuronal death, and behavioral disorders. Other risk factors for AD are a medical history of the family, aging, the apolipoprotein E"4 allele genotype, cardiac health, lifestyle, and psychosocial factors (Alzheimer's Association, 2019). Alkaloids obtained from marine organisms like fungi, bacteria, sponges, corals, jellyfish, ascidians, and mollusks exhibit therapeutic action against Alzheimer's disorder through different mechanisms such as inhibiting β-amyloid (Aβ) production, suppressing the formation of neurofibrillary tangles, inhibiting pro-inflammatory factors, inhibiting acetylcholinesterase (AChE), and stabilizing nicotinic acetylcholine receptors (nAChR's). Dictyodendrin F, Dictyodendrin H-J, Lamellarin O, Lamellarin O1–O2, and Iianthellidone F inhibit Aβ production exhibiting IC_{50s} from 10 to 1 μM. 6-Bromoindirubin, 8-Hydroxymanzamine A, Manzamine E-F, and Manzamine Y exhibit inhibition of glycogen synthetase kinase-3 beta (GSK3β) with IC_{50s} from 25 to 0.045 μM. Meridianin A-G (IC_{50}~100–0.07 μM) and Variolin B (IC_{50}~0.005 μM) are the compounds causing the inhibition of GSK3β and casein kinase 1 delta (CKlδ). Eleutherobin and Sarcodyctin A-D bring stabilization of microtubules. Caulerpin and Pseudane-VII cause suppression of pro-inflammatory factors. Irene-carboline A-B, Circumdantin C-I, and Quinolactacin A1-A2 suppress AChE. Anabaseine and Fascaplysin cause the stabilization of nAChRs (Lima & Medeiros, 2022). Marine organisms have developed a strong immune system to maintain health under challenging marine environments. Their immune system comprises a wide variety of biomolecules endowed with wide antimicrobial (antibacterial, antifungal, and antiviral) potential. Therefore, these biomolecules are being implemented in biomedical applications to fight against pathogens causing human and animal infectious diseases. Marine invertebrates and vertebrates like fishes possess antimicrobial peptides (AMPs) participating in the innate immune system. AMPs are efficacious against bacteria, fungi, and/or enveloped and non-enveloped viruses. Some AMPs are potent immunomodulators making the development of resistant pathogens more difficult. AMPs target cell membranes of pathogens and envelopes of viruses (Shai, 2002). The possible mode of action of AMPs is explained through pore-formation models to destabilize/permeabilize membranes of the pathogens (Jelinek & Kolusheva, 2005; Hollmann, Martinez, Maturana, Semorile, & Maffia, 2018). A wide range of marine organisms possesses different classes of AMPs (both linear and disulfide cross-linked) such as fishes, crustaceans, chelicerates, cnidarians, and mussels containing piscidins, penaeidins, crustins, tachyplesins, arenicins, and nicomicins, mytilins, and myticins AMPs, respectively (Otero-Gonzáiez, Magalhaes, Garcia-Villarino, López-Abarrategui et al., 2010; Tassanakajon, Rimphanitchayakit, Visetnan, Amparyup et al., 2018; Falco, Ortega-Villaizan, Chico, Brocal, Perez, Coll, & Estepa, 2009; Balseiro, Falcó, Romero, Dios et al., 2011; Yu & Gu, 2015; Riccio, Ruocco, Mutalipassi, Costantini et al., 2020; Macedo, Cunha, Carneiro, Costa et al., 2021; Rey-Campos, Novoa, Pallavicini, Gerdol, & Figueras, 2021). AMPs exhibit a wide range of *in vitro* and *in vivo* antimicrobial potential against animals, and even human pathogens including both gram-positive and –negative and multidrug-resistant bacteria (Lauth, Babon, Stannard, Singh et al., 2005; Huang, Chen, & Kuo, 2007; Noga, Silphaduang, Park, Seo, Stephenson, & Kozlowicz, 2009; Pan, Chen, Lin, & Lin, 2009), viruses (Balseiro, Falcó, Romero, Dios et al., 2011; Masso-Silva & Diamond, 2014; Falco, Chico, Marroqui, Perez, Coll, & Estepa, 2008; Wang, Rajanbabu, & Chen, 2015; Novoa, Romero, Álvarez, Moreiraet al., 2016), fungi (Pan, Chen, Lin, & Lin, 2009; Sung, Lee, & Lee, 2008), and protozoa (Pan, Chen, Lin, & Lin, 2009; Colorni, Ullal, Heinisch, & Noga, 2008). In addition to AMPs, several antimicrobial metabolites have been identified from marine organisms such as microbes, algae, and invertebrates (Habbu, Warad, Shastri, Madagundi, & Kulkarni, 2016; Pietra, 1997) and fishes also (Pereiro, Forn-Cuní, Dios, Coll, Figueras, & Novoa, 2017; Bello-Perez, Falco, Novoa, Perez, & Coll, 2019; Adamek, Davies, Beck, Jordan et al., 2021). Currently, a vast array of antimicrobial bioactive substances are known from marine organisms such as polyethers, polyols, polyphenols, polysaccharides, saponins, steroids, terpenoids, and unsaturated fatty acids, unusual or noncanonical amino acids (i.e., nonproteinogenic amino acids), which can combine, and with other usual amino acids and non-amino acid moieties, to form bioactive peptides (Yu & Gu, 2015; Riccio, Ruocco, Mutalipassi, Costantini et al., 2020; Ruiz-Torres, Encinar, Herranz-López, Pérez-Sánchez et al., 2017; Falco, Adamek, Pereiro, Hoole et al., 2022; Falco, Adamek, Pereiro, Hoole et al., 2022). Dipeptides widely occur in marine ecosystems mainly in microorganisms (gram-negative and gram-positive bacteria, cyanobacteria, and fungi), sponges, eukaryotic algae, invertebrates (such as cnidarians, echinoderms, tunicates), and vertebrates (even mammals) (Yu & Gu, 2015; Huang, Zhou, Xu, Yang, & Liu, 2010; Huang, Yi, Zhou, Su, Peng, & Gao, 2014; Rojas, Rivas, Cárdenas, & Guzmán, 2020). Marine algae are rich sources of carbohydrates (polysaccharides), lipids (cholesterol and clionasterol), proteins, pigments (chlorophyll, carotenoids, and phycobiliproteins), phenolic compounds (catechins, flavonoids, phenolic acids, phlorotannins, tannins, and many other phenolic compounds), vitamins, minerals, etc. The polysaccharides may be in sulfated and non-sulfated form. Sulfated polysaccharides are agars, carrageenans, fucoidans, and galactans while the non-sulfated polysaccharides include alginates and laminaran. These biochemicals have nutraceutical value as well as antiangiogenic, anticancer, antidiabetic, anti-inflammatory, antioxidant, and immunomodulatory properties (Jimenez-Lopez, Pereira, Lourenço-Lopes, García-Oliveira et al., 2021).

Seagrasses are also one of the marine sources of secondary metabolites. These are the true marine angiosperms with 72 species across the world, falling into four families (*Cymodoceaceae*, *Hydrocharitaceae*, *Posidoniaceae*, and *Zosteraceae*) (Short, Short, & Novak, 2016; Danaraj, Saravanakumar, & Mariasingarayan, 2021). They are found in shallow water habitats of temperate, subtropical, and tropical seas. They inhabit shallow coastal areas across the globe, with distribution ranges of approximately 50–90 m in depth (Mehari Ghebretinsae, Awet Adhanom, Henok Ghebremedhin, & Biniam Tsegay, 2019). Seagrasses are known in folk medicine for the treatment of fevers, stomach problems, muscle pain, wounds, skin diseases, and as a remedy for stings of different kinds of rays and as tranquilizers for babies (Grignon-Dubois & Rezzonico, 2015). Secondary metabolites obtained from seagrasses such as polyphenols, terpenoids, and halogenated compounds exhibit anticancer, antifungal, anti-inflammatory, antimicrobial, antiviral, antidiabetic, antimalarial, antioxidant, anti-aging, and cytotoxic characteristics. These metabolites are also found effective in the treatment of human diseases, but no drugs derived from seagrasses have been approved by the FDA or European Medicines Agency (EMA) (Kim, Mahomoodally, Sadeer, Seok et al., 2021; Gono, Ahmadi, Hertiani, Septiana, & Putra, 2022).

18.3 CLINICAL STUDIES

18.3.1 MARINE BIOACTIVES EFFECTIVE AGAINST HUMAN DISORDERS

Several secondary metabolites from marine organisms play a role in the treatment of various human disorders such as tuberculosis, neurodegenerative diseases, Alzheimer's disease, Parkinson's disease, autoimmune diseases, inflammatory diseases, and different types of cancer (Karthikeyan, Joseph, & Nair, 2022). Mitochondrial dysfunction and excessive production of ROS are major causes of neurodegenerative diseases. The high sensitivity of the central nervous system to ROS species and the low content of antioxidant defenses in the brain increase the severity of damage in neurodegenerative

TABLE 18.1 Biochemicals Derived from Major Groups of Marine Organisms

Group of Organisms	Organism	Biochemical	References
Algae	*Avrainvillea longicaulis, A. nigricans, A. rawsonii, Codium fragile, Jania rubense, Lobophora variegata, Meroditerpenoids Sargassum fallax, Odonthalia corymbifera, Padina arborescens, Pterocladia capillacea, Rhodomela confervoides and Sargassum siliquastrum*	Bromophenols	Jimenez-Lopez, Pereira, Lourenço-Lopes, García-Oliveira et al. (2021)
	Callophycus serratus and Laurencia sp.	Terpenoids	
	Prasiola spp. and Porphyra sp.,	Mycosporine-like amino acids	
	Tichocarpus crinitus	Tichocarpol	
	Ecklonia cava, E, kurome, E. stolonifera, Ishige okamurae, Eisenia arborean, and Ei. bicyclis,	Phlorotannins	
	Colpomenia sinuosaq	Colpol	
	Dasycladus vermicularis	Coumarins	
	Cladophora socialis	Vanillic acid	
Bacteria (Actinomycetes)	*Actinomycetospora Chlora, Micromonospora matsumotoense, Microbacterium* sp., *Micromonospora* sp., *Nesterenkonia flava, Nocardiopsis alba, Nocardiopsis* sp., *Pseudonocardia endophytica, Rhodococcus* sp.,	Actinonin, Angucyclinone, Anandins, Ananstreps, Antimycin, Akaeolide, Bohemamine, Bonnevillamides, Borrelidin, Dibenzoxazepinone, Dinactin, Fredericamycin A, 3-(4-Hydroxyphenyl)-N-methyl propanamide, Hormaomycins, Isomethoxyneihumicin,	Kimura (2018); Andrade, Ponting, Gibson, & Bork (2000); Chattopadhyay, Taub, Paul, Weissman, & Sokurenko (2013); Nikolouli & Mossialos (2012); Huang, Hou, Li, Qiu, Ju, & Li (2016); García, Bargiela, Martínez, & Ferrer (2018); Liang, Liu, Low, Ang, & Zhao (2017);

TABLE 18.1 *(Continued)*

Group of Organisms	Organism	Biochemical	References
	Saccharothrix sp., *Salinisporaarenicola, Saccharomonospora* sp., *Salinispora* sp., *Streptomyces anandii, S. chartreusis, S. cheonanensis, S. olivaceus, S. somaliensis, S. spinoverrucosus, S.* sp., *S. xiamenensis and S. zhaozhouensis*	3-Isopropylhexahydro- 1H-pyrido[1,2-a] pyrazine-1,4(6H)-dione, Lagumycin B, Lobosamides, Lobophorin K, Lynamicins, Manzamine A, 2-Methyl butyl propyl phthalate, Marangucycline A, Isoikarugamycin, Isopimara-2-one-3-ol-8,15-diene, Mycemycins C, MDN-0170, 4S-4,11-dihydroxy- 10-methyl-dodec-2-en-1,4-olide, Microindolinone A, N-acetyl-N-demethylmayamycin, Nesterenkoniane, N-(4-minocyclooctyl)-3,5-dinitrobenzamide, Paulomycin, Petrocidin A, Quinomycin G, Retimycin, Rifamycin B, Saccharomonopyrones, Saccharothrixones, Salinipostins Sioxanthin, Somalimycin, Spiroindimicins, Strepchazolins, Streptoanthraquinone A, Succinilenes, Thiasporine A, Xiamenmycin, Violapyrone B and Wailupemycins	Bekiesch, Basitta, & Apel (2016); Gomez-Escribano & Bibb (2014); Nah, Pyeon, Kang, Choi, & Kim (2017); Rosano & Ceccarelli (2014); Martínez-García, Nikel, Aparicio, & Lorenzo (2014); Fu & Macmillan (2015); Zhang, Che, Tan, Qi et al. (2017); Mullowney, Hainmhire, Tanouye, Burdette, Van-Pham, & Murphy (2015); Liu, Song, Shang, & Huang (2015); Hassan, Degen, Jang, Ebright, Fenical, & Salinamide (2015); Cheng, Macintyre, Abdelmohsen, Horn et al. (2015); Bae, Chung, Oh, Shin, & Oh (2015); Schulze, Donia, Siqueira-Neto, Ray et al. (2015); Schulze, Navarro, Ebert, DeRisi, & Linington (2015); Xie, Liu, Xia, Gao et al. (2017); Li, Huang, Hou, Ju, & Li (2017); Kim, Mahomoodally, Sadeer, Seok et al. (2021); Cheng, Othman, Stopper, Ebel, Hentschel, & Abdelmohsen (2017); Braña, Sarmiento-Vizcaíno, Osset, Pérez-Victoria et al. (2017); Bae, Park, Kwon, Lee et al. (2017); Zhang, Li, Hu, Sheng et al. (2016); Yellamanda, Vijayalakshmi, Kavitha, Reddy, & Venkateswarlu (2016); Yang & Sun (2016); Xu, Wang, Bu, Yu

TABLE 18.1 *(Continued)*

Group of Organisms	Organism	Biochemical	References
			et al. (2016); Waters, Peraud, Kasanah, Sims et al. (2014); Mangamuri, Muvva, Poda, Naragani et al. (2016); Mangamuri, Vijayalakshmi, Poda, Manavathi, Chitturi, & Yenamandra (2016); Chen, Hao, Wang, Wang, Kong, & Zhu (2016); Fukuda, Takahashi, Nagai, Harunari, Imada, & Tomoda (2016); Zhou, Komaki, Ichikawa, Hosoyama, Sato, & Igarashi (2015); Zhen, Gong, Liu, Zhang, Zhou, Li, & Zhu (2015); Zhang, Huang, Xu, Wang, Ju, Tan, & Li (2015)
Fungi	*Ascochyta Sali-corniae, Aspergillus fumigatus, A.* sp., *Apiospora montagnei, Cryptosphaeria* sp., *Fusarium* sp., *Halorosellinia* sp., *Leptosphaeria* sp., *Paecilomyces* sp., *Penicillium commune SD-118, P. paxilli (strain MaG) K, P. raistrikii, Penicillium citrinum, Penicillium* sp. *(strain DGM3) 6"C, Montagnei* and *Zygosporium masonii*	Alkaloids, chrodrimanin B, costaclavine, Cryptosphaerolide, (2–3, 3-dimethylprop-1-ene)-costaclavine and (2–3, 3-dimethylprop-1-ene)-epicosta-clavine, (E)-4-methoxy-5-(3-methoxybut-1-enyl)-6-methyl-2H-pyran-2-one, fumgaclavine A and C, indole alkaloids, Leptosins, Norliquexanthone (1,3,6-trihydroxy-8-methyl-9H-xanthen-9-one), polyketide asco-sali pyrrolidinone-A40, 13-desoxy-phomenone, Paeciloxocins A and B, (2-(1-hydroxy-3-methylbutyl) and 2-(1-acetoxy-3-methylbutyl)-11-hydroxy-9-methyl-1-methoxy-5H,7H-dibenzo[b, g]-1,5-dioxocin-5-ones), sesquiterpenoids, xanthocillin X(3) and Zygosporamide	Kossuga, Romminger, Xavier, Milanetto, Valle (2012); Yanagihara, Sasaki-Takahashi, Sugahara, Yamamoto et al. (2005); Zhou, Debbab, Mándi, Wray et al. (2013); Swathi, Narendra, Sowjanya, & Satya (2013); Zhang, Satake, Fukuzawa, Sugahara et al. (2012); Wei, Yang, Shao, Wang et al. (2011); Smetanina, Yurchenko, Kalinovskii, Berdyshev et al. (2011); Mathan, Smith, Kumaran, & Prakash (2011); Joel, Jean-Paul, Michael, & Jonas (2010); Oh, Jensen, Murphy, Fiorilla et al. (2010); Blunt, Copp, Keyzers, Munro, & Prinsep (2012); Wen, Chen, She, Yan, Cai, & Mu (2010); Liu, Yan, Yu, Chen, & Zhang (2010)

disorders. Secondary metabolites isolated from marine organisms, such as *Streptomyces,* aid in novel drug development to prevent oxidative stress-induced neurodegenerative diseases like Alzheimer's disease and Parkinson's disease. Secondary metabolites from *Actinomycetes* species have been found effective against inflammatory diseases. Marine-derived inhibitors have been evaluated for their anticancer efficacy against different types of cancers like human colon cancer, human cervical cancer, human skin cancer, blood cancer, and mammary cancer. Various secondary metabolites obtained from marine organisms are effective against many human disorders as mentioned in Table 18.2.

18.3.2 MARINE-DERIVED DRUGS UNDER CLINICAL TRIALS

Many drugs developed from marine organisms such as bacteria, and invertebrates (corals, sponges, tunicates, etc., have been used for the treatment of many human disorders. These drugs are in different phases of clinical trials, some are in phase III (e.g., Plitidepsin and Tetrodotoxin) and some are in phase II such as Irvalec®, Pseudopterosins, and seco-pseudopterosins as mentioned in Table 18.3.

18.4 CONCLUSION

Marine organisms such as algae, bacteria, fungi, sponges, tunicates, corals, mollusks, seagrasses, etc., are rich in several health-benefiting bioactive metabolites. These bioactive metabolites fall into different categories of biochemicals such as polysaccharides, lipids, peptides and proteins, polyphenols, steroids, antioxidants, alkaloids, etc. There is a great need to explore natural medicinal sources and to develop drugs from them in order to get rid of the high cost and adverse side effects of synthetic drugs usually observed in older adults. Several drugs have been developed from bioactive metabolites isolated from marine organisms, and many are under clinical trials against multiple diseases. For drug development from marine sources, the knowledge of bioactive marine metabolites is very crucial. Genetic technologies and bioinformatics methods, metagenomic approaches, genome mining, heterologous biosynthesis, etc., speed up the exploration, invention, and accessibility of the remaining unexplored marine natural products for use in pharmaceuticals. A big problem in drug development from marine sources arises if the marine bioactive metabolite is difficult to isolate and/

TABLE 18.2 Marine-Derived Secondary Metabolites Useful in the Treatment of Human Diseases

Marine Source	Secondary Metabolite	Activity	Action	References
Streptomyces aureoverticillatus	Aureoverticillactam	Anticancer	Against different cell types of tumors	Mitchell, Nicholson, Teisan, Lam, & Potts (2004)
Streptomyces sp.	Caprolactones		Against cancer cell lines	Stritzke, Schulz, Laatsch, Helmke, & Beil (2004)
	Chinikomycins		Against various cancer cell lines in humans	Li, Maskey, Qin, Sattler, Fiebig et al. (2005)
Salinisporatropica	Salinosporamide A (NPI-0052)		Inhibitory action on the proteosome and activation of NF-κB	Beer & Moore (2007)
Streptomyces hygroscopicus	Himastatin		Contains α-hydroxyisovaleric acid, 5-hydroxypiperazic acid, leucine, threonine, valine, and a dimeric hexahydropyrroloindole	Leet, Schroeder, Golik, Matson et al. (1996)
Marinispora sp. *strain CNQ-140*	Marinomycin		Inhibits proliferation of cancer cells	Kim, Hoang, & Kim, (2006); Kim, Hoang, & Kim (2006)
Streptomyces sp. *M045*	Manumycin		Against different human cancer cell lines	Li, Maskey, Qin, Sattler, Fiebig et al. (2005)
Streptomyces sp.	Marmycin		Cytotoxicity of tumor cells tended to correlate with moderate apoptosis induction and arrest during the G1 cell cycle process	Martin, Tan, Jensen, Dimayuga, Fairchild, Raventos-Suarez, & Fenical (2007)
Streptomyces sp.	Streptochlorin		Effective in the treatment of cholangiocarcinoma	Shin, Jeong, Lee, Park, Kim, & Kwon (2007)
Streptomyces sp.	Marineosins		Strong inhibitory action of human colon carcinoma	Boonlarppradab, Kauffman, Jensen, & Fenical (2008)
Streptomyces variabilis	Ammosamides		Cytotoxic action against the MIA PaCa-2 pancreatic cancer cell lines	Pan, Oswald, Legako, Life, Posner, & Macmillan (2013)
Symploca sp.	Hoiamide D		screening inhibitory activity in contrast to 53/Mdm^2 interaction	Malloy, Choi, Fiorilla, Valeriote, Matainaho, & Gerwick (2012)

TABLE 18.2 *(Continued)*

Marine Source	Secondary Metabolite	Activity	Action	References
Niphates olemda	Niphateolide		p53-Hdm^2/Mdm^2 interaction inhibitor	Kato, Nehira, Matsuo, Kawabata et al. (2015)
Arthrinium sp.	Hexylitaconic acid		Inhibits p53/Mdm^2 binding	Tsukamoto, Yoshida, Hosono, Ohta, & Yokosawa (2006)
Lissoclinum cf. badium	Lissoclinidine B		Killing of altered cells with wild-type p53	Clement, Kitagaki, Yang, Saucedo et al. (2008)
Aspergillus sp.	Himeic acid A		Ubiquitin-activating enzyme (E1) inhibitory action	Tsukamoto, Hirota, Imachi, Fujimuro et al. (2005)
Cymbastela cantharell and Axinella brevistyla	Girolline		Starting G2/M cell cycle arrest in cancer cells	Tsukamoto, Yamashita, Tane, Kizu et al. (2004)
Leucetta aff. microrhaphis	Leucettamol A		Inhibits the ubiquitin E2 enzymes Ubc13 and Uev1A by 50%	Tsukamoto, Takeuchi, Rotinsulu, & Mangindaan (2008)
Dysidea etheria	Dysidiolide		Inhibition of Cdc25 protein phosphatase	Gunasekera, McCarthy, Kelly-Borges, Lobkovsky, & Clardy (1996)
Ircinia sp.	Sulfircin		Inhibition of Cdc25 phosphatase	Nagle, Zhou, Mora, Mohammed, & Kim (2004)
Coscinoderma mathewsi	Coscinosulfate		Inhibition of Cdc25A	Loukaci, Le Saout, Samadi et al. (2001)
Xestospongia carbonaria	Halenaquinone		Inhibition of recombinant human Cdc25B phosphates	Skropeta, Pastro, & Zivanovic (2011)
Penicillium oxalicum	Secalonic acid D		Slows down the course of the cell cycle in human embryonic palatal mesenchymal cells	Cherigo, Lopez, & Martinez-Luis (2015)
Jaspis stellifera	Stellettin B		Reducing CDK and elevating p27 levels	Wang, Zhang, Peng et al. (2016)

TABLE 18.2 *(Continued)*

Marine Source	Secondary Metabolite	Activity	Action	References
Verrucosispora sp.	Abyssomicins		Blocks the pathway between chorismite and para-aminobenzoic acid	Bister, Bischoff, Strobele, Riedlinger et al. (2004)
Streptomyces lateritius	Granaticins		Against P-388 lymphocytic leukemia in mice and against KB cells	Oja, San Martin Galindo, Taguchi, Manner et al. (2015)
Streptomyces griseus	Frigocyclinone		Action against Kaposis Sarcoma Associated Herpesvirus (KSHV)	Bruntner, Binder, Pathom-aree, Goodfellow, Bull et al. (2005)
Streptomyces sp. *FJS31–2*	Zunyimycins		Cytotoxic to the lung cancer cells as it activates apoptosis by an AKT pathway	Lu, Shao, Wang, Qian et al. (2017)
Streptomyces sp. *M-207*	Lobophorin K	Antibacterial	Against gram-positive bacteria	Braña, Sarmiento-Vizcaíno, Osset, Pérez-Victoria et al. (2017)
Streptomyces asenjonii strain KNN 42	Asenjonamide C		Against gram-positive bacteria	Abdelkader, Philippon, Asenjo, Bull et al. (2018)
Streptomyces formicae	Formicamycins A		Against clinical MRSA and VRE isolates	Qin, Munnoch, Devine, Holmes et al. (2017)
Actinoallomurus	Allocyclinones		Against various gram-positive bacteria even the antibiotic-resistant strains	Cruz, Maffioli, Bernasconi, Brunati et al. (2017)
Kocuria sp. *strain MI-67-EC3–038*	Kocurin		Against methicillin-resistant *Staphylococcus aureus* (MRSA)	Mahajan, Thomas, Parab, Patel et al. (2013)
Streptomyces sp.	Fijimycins		Exhibits *in vitro* activity against methicillin-resistant *Staphylococcus aureus* (MRSA) strains	Phan, Jian, Chen, Qiu, Wang, Beach (2004)
Streptomyces chartreusis	Chartreusin		Effective against certain gram-positive bacteria	Butler (2008)
Streptomyces sp.	Himalomycins		Against gram-positive bacteria	Flora, Adeyemi, & George (2015)
Streptomyces fradiae	Tylosin		Effective in respiratory and other infections due to *Mycoplasma* species	Phan, Jian, Chen, Qiu, Wang, Beach (2004)

TABLE 18.2 *(Continued)*

Marine Source	Secondary Metabolite	Activity	Action	References
Micromonospora sp. *GMKU326*	Maklamicin		Against gram-positive bacteria	Igarashi, Sawa, Yamasaki et al. (2017)
Streptomyces fradiae	Urdamycin	Antibacterial and Anticancer	Comprises aminoglycoside and exhibits strong antibacterial and anti-cancer activity	Ganesan, Velsamy, Sivasudha, & Manoharan (2013)
Streptomyces sp. *NTK 937*	Caboxamycin		Inhibitory action against gram-positive bacteria selected human cancer cell lines and phosphodiesterase	Losada, Cano-Prieto, Garcia-Salcedo, Brana, Mendez, Salas, & Olano (2017)
Streptomyces tsukubaensis, S. griseus, S. chrysomallus, and S. werraensi	Nonactin		Strong antineoplastic and antibacterial action	Smith, Xiang, & Shen (2000)
Streptomyces nodosus	Lajollamycin		Effective against drug-sensitive and -resistant gram-positive bacteria and inhibited the growth of B16-F10 tumor cells *in vitro*	Manam, Teisan, White, Nicholson et al. (2005)
Streptomyces variabilis	1-Hydroxy-1-Norresistomycin		Potent cytotoxic activity against cell lines viz. HMO_2 (gastric adenocarcinoma) and HePG2 (hepatic carcinoma) *in vitro*	Ramalingam, Varunkumar, Ravikumar, & Rajaram (2018)
Actinomadura sp.	IB-00208		Various tumor cell lines and antibacterial against gram-positive bacteria	Malet-Cascon, Romero, Espliego-Vazquez, Gravalos, & Fernandez-Puentes (2003)
Streptomyces sp.	Bonactin	Antibacterial and antifungal	Bonactin displayed antimicrobial activity against both gram-positive and gram-negative bacteria as well as antifungal activity	
Streptomyces sp.	Gutingimycin	Antibacterial antifungal and anti-microalgal	Antibacterial antifungal and anti-microalgal activities	Lu, Dong, Liu, & Bie (2009)

TABLE 18.2 *(Continued)*

Marine Source	Secondary Metabolite	Activity	Action	References
Janibacterlimosus	Helquinoline	Antibacterial, antifungal and anti-microalgal	Antibacterial, antifungal, and anti-microalgal activities	Maskey, Helmke, Kayser, Fiebig et al. (2004)
Streptomyces sp.	Buanmycin	Antibiotic	Inhibition of sortase A (a promising target for antibiotic discovery)	Moon, Chung, Shin, Lee et al. (2015)
Streptomyces caelestis	Citreamicin	Antibiotic and anticancer	Cytotoxic against HeLa and Hep62 cells and has antibiotic potential	Liu, Xu, Han, Li et al. (2012)
Streptomyces sioyaensis SA-1758	Altemicidin	Acaricidal and anticancer	Effective against bacteria and tumors	Takahashi, Kurasawa, Ikeda, Okami, & Takeuchi (1989)
Streptomyces strain CNQ-085	Daryamide D	Anticancer and antifungal	Cytotoxicity against HCT-116 cell line of human colon carcinoma and antifungal potential against Candida albicans	Asolkar, Jensen, Kauffman, && Fenical (2006)
Streptomyces nodosus	Amphotericin B	Antifungal	Effective in most systemic infections caused by fungi	Meijer, Thunnissen, White, Garnier et al. (2000)
Micromonosproa sp.	Diazepinomicin (ECO-4601)	Anti-inflammatory	Induces differentiation of neuronal cells in the neuroblastoma cell line, Neuro 2A, and arrests the cell cycle in step G1	Yu, Wu, Lei, Chen et al. (2013)
Agelasidae, Axinellidae, and *Halichondriidae*	Hymenialdisine		Inhibits various pro-inflammatory cytokines (IL-1, IL-2, IL-6, and NO) by suppressing NF-kB signaling pathway	Meijer, Thunnissen, White, Garnier et al. (2000)
Streptomyces sp.	Cyclomarins	Antimicrobial	Against tuberculosis and malaria	Jensen, Williams, Oh, Zeigler, & Fenical (2007)
Streptomyces sp. *HCCB10043*	Arylomycin		Inhibition of an antimicrobial target, type-I signal peptidase (SPase)	Rao, Wei, Ge, Chen, & Sheng (2013)

TABLE 18.2 *(Continued)*

Marine Source	Secondary Metabolite	Activity	Action	References
Streptomyces sp. *SBT345*	Ageloline A	Antioxidant	Decreases oxidative stress and genomic damage due to oxidative stress inducer 4-nitroquinoline-1-oxide (NQO)	Cheng, Othman, Reimer, Grune et al. (2016)
Streptomyces nodosus	Rifampicin	Antituber-culosis	Immunomodulation through the modification of human monocyte production of measured cytokines	Barry, Slayden, Sampson, & Lee (2000)
	Streptomycin		Effective gram-positive and gram-negative bacteria	
	Amikacin		Active against antibiotic-resistant clinical isolates	
	Viomycin		Tuberculostatic agent active against both streptomycin–sensitive and streptomycin-resistant strains	
	Capreomycin		*In vitro* anti-TB drug against nonreplicating tubercle bacilli	
	Kanamycin		A promising effect against staphylococci, pneumococci, *Ulebsiella*, and *Pneumonia* infections	
	Cycloserine		Effective against tuberculous *in vitro* and *in vivo*	
Streptomyces sp. *KS3*	Komodoquinone A	Neurogenic activity	suppresses many pro-inflammatory cytokines (IL-1, IL-2, IL-6, and NO) by inhibition of NF-kB signaling pathway	Itoh, Kinoshita, Aoki, & Kobayashi (2003)

TABLE 18.3 Marine-Derived Drugs Used Against Human Diseases

Source	Drug	Application	References
Aplidium albicans	Plitidepsin	Antiviral against B.1.1.7 variety of SARS-CoV-2	Rinehart, Gloer, Hughes, Renis et al. (1981); Canonico, Pannier, Huggins, & Rienehart (1982); Reuschl, Thorne, Zuliani-Alvarez, Bouhaddou et al. (2021)
Pufferfish and other marine and terrestrial Species	Tetrodotoxin	Analgesic in neuropathies or cancer	Gonzalez-Cano, Ruiz-Cantero, Santos-Caballero, Gomez-Navas, Tejada, & Nieto (2021)
Elysia rufescens	Irvalec® (elisidepsin trifluoroacetate, PM02734)	Anticancer	Teixido, Arguelaguet, Pons, Aracil et al. (2012)
Pseudopterogorgia elisabethae	Pseudopterosins and seco-pseudopterosins	Antimicrobial against (*Staphylococcus aureus*, *Enterococcus faecalis, Pseudomonas aeruginosa,* and *Candida albicans)* and cytotoxic against human cell lines (HeLa, PC-3, HCT116, MCF-7, and BJ)	Correa, Aristizabal, Duque, & Kerr (2011)
Pseudopterogorgia elisabethae	Pseudopterosin H	Anticancer and anti-oxidant	Bowers, Caraballo, & Bentley (2021)
Tethya crypta	Cytarabine (Cytosar-U®, Ara-C, DepoCyt®)	Anticancer (works against acute myelocytic leukemia and non-Hodgkin's lymphoma)	Jaspars, De Pascale, Andersen, Reyes, Crawford, & Ianora (2016); Lowenberg (2013)
Tunicate (*Ecteinascidia turbinata*)	ET-743 (Yondelis®)	Effective to treat tissue sarcomas and ovarian cancer	Menis, & Twelves (2011)
Sponge (*Halichondria okadai)*	Eribulin (Halaven®)	Anticancer against metastatic breast cancer and advanced liposarcoma	
Cone snail *Conus magus*	Ziconotide (Prialt®)	Analgesic against severe and chronic pain	Duarte, Rocha-Santos, Freitas, & Duarte (2012)
sponge *Tethya crypta*	Vidarabine (Ara-A)	Antiviral against herpes simplex	Sagar, Kaur, & Minneman (2010)

or if it is produced in small quantities due to variable marine environment. For preclinical studies, the marine bioactive metabolite is needed in a few milligrams, and for clinical trials in kilograms quantity. Advancements in technological fields e.g., sampling methods, structural elucidation, chemical synthesis, biosynthesis, and genetic engineering give us hope to believe that drugs from marine natural products will flood the market and pharmaceutical industry in the future.

KEYWORDS

- **acetylcholinesterase**
- **bioactive metabolites**
- **clinical trials**
- **Food and Drug Administration**
- **marine organisms**
- **natural products**

REFERENCES

Abdelkader, M. S. A., Philippon, T., Asenjo, J. A., Bull, A. T., Goodfellow, M., Ebel, R., & Rateb, M. E. (2018). Asenjonamides A-C, antibacterial metabolites isolated from *Streptomyces asenjonii* strain KNN 42.f from an extreme-hyper arid Atacama Desert soil. *Journal of Antibiotics (Tokyo), 71*, 425–431.

Adamek, M., Davies, J., Beck, A., Jordan, L., Becker, A. M., Mojzesz, M., & Brogden, G. (2021). Antiviral actions of 25-hydroxycholesterol in fish vary with the virus-host combination. *Frontiers in Immunology, 12*, 3.

Alharbi, N. S. (2016). Novel bioactive molecules from marine actinomycetes. *Bioscience Biotechnology Research Asia, 13*(4), 1905–1927.

Al-Khayri, J. M., Asghar, W., Khan, S., Akhtar, A., Ayub, H., Khalid, N., Alessa, F. M., Al-Mssallem, M. Q., Rezk, A. A. S., & Shehata, W. F. (2022). Therapeutic potential of marine bioactive peptides against human immunodeficiency virus: Recent evidence, challenges, and future trends. *Marine Drugs, 20*(8), 477.

Alzheimer's Association. (2019). Alzheimer's disease facts and figures. *Alzheimer's & Dementia, 15*, 321–387.

Andrade, M. A., Ponting, C. P., Gibson, T. J., & Bork, P. (2000). Homology-based method for identification of protein repeats using statistical significance estimates. *Journal of Molecular Biology, 298*, 521–537.

Asolkar, R. N., Jensen, P. R., Kauffman, C. A., & Fenical, W. (2006). Daryamides A-C, weakly cytotoxic polyketides from a marine-derived actinomycete of the genus Streptomyces strain CNQ-085. *Journal of Natural Products, 69*, 1756–1759.

Bae, M., Chung, B., et al. (2015). Hormaomycins B and C: New antibiotic cyclic depsipeptides from a marine mudflat-derived *Streptomyces sp. Marine Drugs, 13*(8), 5187–5200.

Bae, M., Park, S. H., Kwon, Y., Lee, S. K., Shin, J., Nam, J. W., & Oh, D. C. (2017). QM-HiFSA-aided structure determination of succinilenes A-D, new triene polyols from a marine-derived *Streptomyces* sp. *Marine Drugs, 15*(2), 1–14.

Bae, S. Y., Liao, L., Park, S. H., Kim, W. K., Shin, J., & Lee, S. K. (2020). Antitumor activity of asperphenin A, a lipopeptidyl benzophenone from marine-derived *Aspergillus* sp. fungus, by inhibiting tubulin polymerization in colon cancer cells. *Marine Drugs, 18*, 110.

Balseiro, P., Falcó, A., Romero, A., Dios, S., Martínez-López, A., Figueras, A., Estepa, A., & Novoa, B. (2011). Mytilus galloprovincialis myticin C: A chemotactic molecule with antiviral activity and immunoregulatory properties. *PLoS ONE, 6*, e23140.

Barka, E. A., Vatsa, P., Sanchez, L., Gaveau-Vaillant, N., Jacquard, C., Klenk, H. P., Clément, C., Ouhdouch, Y., & van Wezel, G. P. (2016). Taxonomy, physiology, and natural products of the actinobacteria. *Microbiology and Molecular Biology Reviews, 80*(1), 1–43.

Barry III. C. E., Slayden, R. A., Sampson, A. E., & Lee, R. E. (2000). Use of genomics and combinatorial chemistry in the development of new antimycobacterial drugs. *Biochemical Pharmacology, 59*(3), 221–231.

Beer, L. L., & Moore, B. S. (2007). Biosynthetic convergence of salinosporamides A and B in the marine actinomycete *Salinispora tropica. Organic Letters, 9*, 845–848.

Behie, S. W., Bonet, B., Zacharia, V. M., McClung, D. J., & Traxler, M. F. (2017). Molecules to ecosystems: Actinomycete natural products in situ. *Frontiers in Microbiology, 7*, 2149.

Bekiesch, P., Basitta, P., & Apel, A. K. (2016). Challenges in the heterologous production of antibiotics in Streptomyces. *Archives of Pharmaceutical and Chemistry Life Science, 349*(1–8), 594–601.

Bello-Perez, M., Falco, A., Novoa, B., Perez, L., & Coll, J. (2019). Hydroxycholesterol binds and enhances the anti-viral activities of zebrafish monomeric c-reactive protein isoforms. *PLoS ONE, 14*(10), e0201509.

Bergmann, W., & Burke, D. C. (1956). Contributions to the study of marine products. XL. The nucleosides of sponges. IV. Spongosine. *Journal of Organic Chemistry, 22*, 226–228.

Bergmann, W., & Feeney, R. J. (1951). Contributions to the study of marine products. XXXII. The nucleosides of sponges. I. *Journal of Organic Chemistry, 16*, 981–987.

Bergmann, W., & Stempien, M. F. (1957). Contributions to the study of marine products. XLIII. The nucleosides of sponges. V. The synthesis of spongosine. *Journal of Organic Chemistry, 22*, 1575–657.

Betancur, L. A., Naranjo-Gaybor, S. J., Vinchira-Villarraga, D. M., Moreno-Sarmiento, N. C., Maldonado, L. A., Suarez-Moreno, Z. R., Acosta-González, A., Padilla-Gonzalez, G. F., Puyana, M., Castellanos, L., & Ramos, F. A. (2017). Marine actinobacteria as a source of compounds for phytopathogen control: An integrative metabolic-profiling/bioactivity and taxonomical approach. *PLoS ONE, 12*, e0170148.

Bister, B., Bischoff, D., Strobele, M., Riedlinger, J., Reicke, A., Wolter, F., & Süssmuth, R. D. (2004). Abyssomicin C-A polycyclic antibiotic from a marine *Verrucosispora* strain as an inhibitor of the p-aminobenzoic acid/tetrahydrofolate biosynthesis pathway. *Angewandte Chemie International Edition, 43*(19), 2574–2576.

Blunt, J. W., Copp, B. R., Keyzers, R. A., Munro, M. H., & Prinsep, M. R. (2012). Marine natural products. *Natural Product Reports, 29*(2), 144–222.

Blunt, J. W., Copp, B. R., Keyzers, R. A., Munro, M., & Prinsep, M. R. (2017). Marine natural products. *Natural Product Reports, 34*, 235–294.

Boonlarppradab, C., Kauffman, C. A., Jensen, P. R., & Fenical, W. (2008). Marineosins A and B, cytotoxic spiroaminals from a marine-derived actinomycete. *Organic Letters, 10*, 5505–5508.

Bowers, Z., Caraballo, D., & Bentley, A. (2021). Therapeutic potential of pseudopterosin H on a prostate cancer cell line. *Journal of Cancer Prevention & Current Research, 12*, 82–91.

Braña, A. F., Sarmiento-Vizcaíno, A., Osset, M., Pérez-Victoria, I., Martín, J., De Pedro, N., & García, L. A. (2017). Lobophorin K, a new natural product with cytotoxic activity produced by *Streptomyces* sp. M-207 associated with the deep-sea coral *Lophelia pertusa. Marine Drugs, 15*(5), 144.

Bruntner, C., Binder, T., Pathom-aree, W., Goodfellow, M., Bull, A. T., Potterat, O., & Fiedler, H. P. (2005). Frigocyclinone, a novel angucyclinone antibiotic produced by a *Streptomyces griseus* strain from Antarctica. *Journal of Antibiotics (Tokyo), 58*, 346–349.

Butler, M. S. (2008). Natural products to drugs: natural product-derived compounds in clinical trials. *Natural Product Reports, 25*, 475–516.

Canonico, P. G., Pannier, W. L., Huggins, J. W., & Rienehart, K. L. (1982). Inhibition of RNA viruses in vitro and in Rift Valley fever-infected mice by didemnins A and B. *Antimicrobial Agents and Chemotherapy, 22*(4), 696–697.

Chattopadhyay, S., Taub, F., Paul, S., Weissman, S. J., & Sokurenko, E. V. (2013). Microbial variome database: Point mutations, adaptive or not, in bacterial core genomes. *Molecular Biology and Evolution, 30*, 1465–1470.

Chen, Z., Hao, J., Wang, L., Wang, Y., Kong, F., & Zhu, W. (2016). New α-glucosidase inhibitors from marine algae-derived *Streptomyces* sp. *Scientific Reports, 6*, 1–9.

Cheng, C., Macintyre, L., et al. (2015). Biodiversity, antitrypanosomal activity screening, and metabolomic profiling of actinomycetes isolated from Mediterranean sponges. *PLoS ONE, 10*(9), 1–21.

Cheng, C., Othman, E. M., Reimer, A., Grune, M., Kozjak-Pavlovic, V., Stopper, H., & Abdelmohsen, U. R. (2016). Ageloline A, new antioxidant and antichlamydial quinolone from the marine sponge-derived bacterium Streptomyces sp. SBT345. *Tetrahedron Letters, 57*, 2786–2789.

Cheng, C., Othman, M. E., Stopper, H., Ebel, R., Hentschel, U., & Abdelmohsen, R. U. (2017). Isolation of petrocidin A, a new cytotoxic cyclic dipeptide from the marine sponge-derived bacterium *Streptomyces* sp. SBT348. *Marine Drugs, 15*(12), 383.

Cherigo, L., Lopez, D., & Martinez-Luis, S. (2015). Marine natural products as breast cancer resistance protein inhibitors. *Marine Drugs, 13*, 2010–2029.

Clement, J. A., Kitagaki, J., Yang, Y., Saucedo, C. J., O'Keefe, B. R., Weissman, A. M., & McMahon, J. B. (2008). Discovery of new pyridoacridine alkaloids from *Lissoclinum* cf. *badium* that inhibit the ubiquitin ligase activity of Hdm2 and stabilize p53. *Bioorganic & Medicinal Chemistry, 16*, 10022–10028.

Colorni, A., Ullal, A., Heinisch, G., & Noga, E. J. (2008). Activity of the antimicrobial polypeptide piscidin 2 against fish ectoparasites. *Journal of Fish Diseases, 31*, 423–432.

Correa, H., Aristizabal, F., Duque, C., & Kerr, R. (2011). Cytotoxic and antimicrobial activity of pseudopterosins and seco-pseudopterosins isolated from the octocoral *Pseudopterogorgia elisabethae* of San Andres and Providencia Islands (Southwest Caribbean Sea). *Marine Drugs, 9*, 334–344.

Cruz, J. C., Maffioli, S. I., Bernasconi, A., Brunati, C., Gaspari, E., Sosio, M., & Donadio, S. (2017). Allocyclinones, hyperchlorinated angucyclinones from Actinoallomurus. *Journal of Antibiotics (Tokyo), 70*, 73–78.

Dalmaso, G. Z., Ferreira, D., & Vermelho, A. B. (2015). Marine extremophiles: a source of hydrolases for biotechnological applications. *Marine Drugs, 13*, 1925–1965.

Danaraj, J., Saravanakumar, A., & Mariasingarayan, Y. (2021). Seagrass Metabolomics: A New Insight towards Marine-Based Drug Discovery. In *Metabolomics-Methodology and Applications in Medical Sciences and Life Sciences*, 1–26. IntechOpen: London, UK.

Das, S., & Mangwani, N. (2015). Ocean acidification and marine microorganisms: responses and consequences. *Oceanologia, 57*(4), 349–361.

Duarte, K., Rocha-Santos, T. A., Freitas, A. C., & Duarte, A. C. (2012). Analytical techniques for discovery of bioactive compounds from marine fungi. *Trends in Analytical Chemistry, 34*, 97–110.

Duarte, K., Rocha-Santos, T. A., Freitas, A. C., & Duarte, A. C. (2012). Analytical techniques for discovery of bioactive compounds from marine fungi. *Trends in Analytical Chemistry, 34*, 97–110.

Elbendary, A. A., Hessain, A. M., El-Hariri, M. D., Seida, A. A., Moussa, I. M., Mubarak, A. S., Kabli, S. A., Hemeg, H. A., & El Jakee, J. K. (2018). Isolation of antimicrobial producing actinobacteria from soil samples. *Saudi Journal of Biological Sciences, 25*(1), 44–46.

Falco, A., Adamek, M., Pereiro, P., Hoole, D., Encinar, J. A., Novoa, B., & Mallavia, R. (2022). The immune system of marine organisms as a source for drugs against infectious diseases. *Marine Drugs, 20*(6), 363.

Falco, A., Chico, V., Marroqui, L., Perez, L., Coll, J. M., & Estepa, A. (2008). Expression and antiviral activity of a beta-defensin-like peptide identified in the rainbow trout (*Oncorhynchus mykiss*) EST sequences. *Molecular Immunology, 45*, 757–765.

Falco, A., Ortega-Villaizán, M., Chico, V., Brocal, I., Pérez, L., Coll, J. M., & Estepa, A. (2009). Antimicrobial peptides as model molecules for the development of novel antiviral agents in aquaculture. *Mini Reviews in Medicinal Chemistry, 9*, 1159–1164.

Flora, D. O., Adeyemi, A. I., & George, W. P. (2015). Himalomycin A and cycloheximide-producing marine actinomycete from Lagos Lagoon soil sediment. *Journal of Coastal Life Medicine, 3*, 361–365.

François, C., Travers, M. A., Morga, B., Haure, J., Tremblay, R., Turcotte, F., Pasetto, P., Gastineau, R., Hardivillier, Y., & Leignel, V. (2016). Antimicrobial compounds from eukaryotic microalgae against human pathogens and diseases in aquaculture. *Marine Drugs, 14*, 159.

Fu, P., Macmillan, J. B., et al. (2015). Thiasporines A-C, thiazine and thiazole derivatives from a marine-derived *Actinomycetospora chlora*. *Journal of Natural Products, 78*(3), 548–551.

Fukuda, T., Takahashi, M., Nagai, K., Harunari, E., Imada, C., & Tomoda, H. (2016). Isomethoxyneihumicin, a new cytotoxic agent produced by marine *Nocardiopsis alba* KM6-1. *Journal of Antibiotics, 70*, 590–594.

Ganesan, S., Velsamy, G., Sivasudha, T., & Manoharan, N. (2013). MALDI-TOF mass spectrum profiling, antibacterial and anticancer activity of marine Streptomyces fradiae BDMS1. *World Journal of Pharmacy and Pharmaceutical Sciences, 2*, 5148–5165.

García, C. M., Bargiela, R., Martínez, M., & Ferrer, M. (2018). Metagenomic protocols and strategies. In A. Ferrer (Ed.), *Metagenomics*, 15–34.

Glaser, K. B., & Mayer, A. M. (2009). A renaissance in marine pharmacology: From preclinical curiosity to clinical reality. *Biochemical Pharmacology, 78*(5), 440–448.

Gomez-Escribano, J., & Bibb, M. (2014). Heterologous expression of natural product biosynthetic gene clusters in *Streptomyces coelicolor*: From genome mining to manipulation of biosynthetic pathways. *Journal of Industrial Microbiology and Biotechnology, 41*, 425–431.

Gono, C. M. P., Ahmadi, P., Hertiani, T., Septiana, E., Putra, M. Y., & Chianese, G. (2022). A comprehensive update on the bioactive compounds from seagrasses. *Marine Drugs, 20*(7), 406.

Gonzalez-Cano, R., Ruiz-Cantero, M. C., Santos-Caballero, M., Gomez-Navas, C., Tejada, M. A., & Nieto, F. R. (2021). Tetrodotoxin, a potential drug for neuropathic and cancer pain relief? *Toxins, 13*, 483.

Grignon-Dubois, M., & Rezzonico, B. (2015). Phenolic fingerprint of the seagrass *Posidonia oceanica* from four locations in the Mediterranean Sea: first evidence for the large predominance of chicoric acid. *Botanica Marina, 58*, 379–391.

Gunasekera, S. P., McCarthy, P. J., Kelly-Borges, M., Lobkovsky, E., & Clardy, J. (1996). Dysidiolide: a novel protein phosphatase inhibitor from the Caribbean sponge *Dysidea etheria* de Laubenfels. *Journal of the American Chemical Society, 118*, 8759–8760.

Habbu, P., Warad, V., Shastri, R., Madagundi, S., & Kulkarni, V. H. (2016). Antimicrobial metabolites from marine microorganisms. *Chinese Journal of Natural Medicines, 14*, 101–116.

Haidle, A. M., & Myers, A. G. (2004). An enantioselective, modular, and general route to the cytochalasins: Synthesis of L-696,474 and cytochalasin B. *Proceedings of the National Academy of Sciences of the USA, 101*, 12048–12053.

Hasan, S., Ansari, M. I., Ahmad, A., & Mishra, M. (2015). Major bioactive metabolites from marine fungi: A review. *Bioinformation, 11*(4), 176.

Hassan, H. M., Degen, D., et al. (2015). Salinamide, F., New depsipeptide antibiotic and inhibitor of bacterial RNA polymerase from a marine-derived *Streptomyces sp. Journal of Antibiotics, 68*(3), 206–209.

Hollmann, A., Martinez, M., Maturana, P., Semorile, L. C., & Maffia, P. C. (2018). Antimicrobial peptides: Interaction with model and biological membranes and synergism with chemical antibiotics. *Frontiers in Chemistry, 6*, 204.

http://marinepharmacology.midwestern.edu/ Global marine pharmaceutical website for regular updates on the clinical marine pharmaceutical pipeline.

Huang, H., Hou, L., Li, H., Qiu, Y., Ju, J., & Li, W. (2016). Activation of a plasmid-situated type III PKS gene cluster by deletion of a wbl gene in deepsea-derived Streptomyces somaliensis SCSIO ZH66. *Microbial Cell Factories, 15*(1), 1–9.

Huang, P. H., Chen, J. Y., & Kuo, C. M. (2007). Three different hepcidins from tilapia, Oreochromis mossambicus: Analysis of their expressions and biological functions. *Molecular Immunology, 44*, 1922–1934.

Huang, R. M., Yi, X. X., Zhou, Y., Su, X., Peng, Y., & Gao, C. H. (2014). An update on 2, 5-diketopiperazines from marine organisms. *Marine Drugs, 12*, 6213–6235.

Huang, R., Zhou, X., Xu, T., Yang, X., & Liu, Y. (2010). Diketopiperazines from marine organisms. *Chemistry & Biodiversity, 7*, 2809–2829.

Hwang, C., Lee, H., Hong, M., Kim, J., & Kim, K. (2018). Brain abscess caused by chronic invasive actinomycosis in the nasopharynx: A case report and literature review. *Medicine, 97*(16), e0406.

Ibañez, E., Herrero, M., Mendiola, J. A., & Castro-Puyana, M. (2012). Extraction and characterization of bioactive compounds with health benefits from marine resources: macro and micro algae, cyanobacteria, and invertebrates. In *Marine bioactive compounds* (pp. 55–98). Springer.

Igarashi, M., Sawa, R., Yamasaki, M., Hayashi, C., Umekita, M., Hatano, M., & Nomoto, A. (2017). Kribellosides, novel RNA 5'-triphosphatase inhibitors from the rare actinomycete Kribbella sp. MI481-42F6. *Journal of Antibiotics (Tokyo), 70*, 582–589.

Iribarren, E. M., Serrano, C., Naranjo, R. O., & García-Castillo, E. (2018). Pulmonary actinomycosis in a patient with chronic eosinophilic pneumonia treated with Omalizumab. *Archivos de Bronconeumología, 54*(1), 51–52.

Itoh, T., Kinoshita, M., Aoki, S., & Kobayashi, M. (2003). Komodoquinone A, a novel neuritogenic anthracycline, from marine Streptomyces sp. KS3. *Journal of Natural Products, 66*, 1373–1377.

Jaspars, M., De Pascale, D., Andersen, J. H., Reyes, F., Crawford, A. D., & Ianora, A. (2016). The marine biodiscovery pipeline and ocean medicines of tomorrow. *Journal of Marine Biological Association of the United Kingdom, 96*, 151–158.

Jelinek, R., & Kolusheva, S. (2005). Membrane interactions of host-defense peptides studied in model systems. *Current Protein and Peptide Science, 6*, 103–114.

Jensen, P. R., Williams, P. G., Oh, D. C., Zeigler, L., & Fenical, W. (2007). Species specific secondary metabolite production in marine actinomycetes of the genus Salinispora. *Applied and Environmental Microbiology, 73*, 1146–1152.

Jimenez-Lopez, C., Pereira, A. G., Lourenço-Lopes, C., García-Oliveira, P., Cassani, L., Fraga-Corral, M., & Simal-Gandara, J. (2021). Main bioactive phenolic compounds in marine algae and their mechanisms of action supporting potential health benefits. *Food Chemistry, 341*, 128262.

Joel, Q., Jean-Paul, C., Michael, J. A., & Jonas, C. (2010). Marine Biotechnology. *Advances in Marine Genomics, 1*, 287–313.

Julianti, E., Abrian, I. A., Wibowo, M. S., Azhari, M., Tsurayya, N., Izzati, F., Juanssilfero, A. B., Bayu, A., Rahmawati, S. I., & Putra, M. Y. (2022). Secondary metabolites from marine-derived fungi and actinobacteria as potential sources of novel colorectal cancer drugs. *Marine Drugs, 20*(1), 67.

Kannikan, V., & Rebecca, L. J. (2018). Screening of marine actinomycetes for fibrinolytic enzymes. *Research Journal of Pharmacy and Technology, 11*(10), 43–65.

Karthikeyan, A., Joseph, A., & Nair, B. G. (2022). Promising bioactive compounds from the marine environment and their potential effects on various diseases. *Journal of Genetic Engineering and Biotechnology, 20*(1), 1–38.

Kato, H., Nehira, T., Matsuo, K., Kawabata, T., Kobashigawa, Y., Morioka, H., & Yokosawa, H. (2015). Niphateolide A: isolation from the marine sponge *Niphates olemda* and determination of its absolute configuration by an ECD analysis. *Tetrahedron, 71*, 6956–6960.

Kim, D. H., Mahomoodally, M. F., Sadeer, N. B., Seok, P. G., Zengin, G., Palaniveloo, K., & Rengasamy, K. R. (2021). Nutritional and bioactive potential of seagrasses: A review. *South African Journal of Botany, 137*, 216–227.

Kim, S. K., Hoang, V. L., & Kim, M. M. (2006). Bioactive compounds derived from marine bacteria: anti-cancer activity. *Journal of Marine Biosciences and Biotechnology, 1*, 232–242.

Kimura, N. (2018). Novel biological resources screened from uncultured bacteria by a metagenomic method. In *Metagenomics* (pp. 273–288). Academic Press.

Klotz, U. (2006). Ziconotide-a novel neuron-specific calcium channel blocker for the intrathecal treatment of severe chronic pain-a short review. *International Journal of Clinical Pharmacy and Therapeutics, 44*(10), 478–483.
Kossuga, M. H., Romminger, S., Xavier, C., Milanetto, M. C., Valle, M. Z. D., Pimenta, E. F., & Barroso, V. D. M. (2012). Evaluating methods for the isolation of marine-derived fungal strains and production of bioactive secondary metabolites. *Revista Brasileira de Farmacognosia, 22*, 257–267.
Kumar, S., Krishnani, K. K., Bhushan, B., & Brahmane, M. P. (2015). Metagenomics: Retrospect and prospects in high throughput age. *Biotechnology Research International.*
Lacret, R., Pérez-Victoria, I., Oves-Costales, D., De la Cruz, M., Domingo, E., Martín, J., Díaz, C., Vicente, F., Genilloud, O., & Reyes, F. (2016). MDN-0170, a new napyradiomycin from *Streptomyces* sp. strain CA-271078. *Marine Drugs, 14*(10), 188.
Lafarga, T., Acién-Fernández, F. G., & Garcia-Vaquero, M. (2020). Bioactive peptides and carbohydrates from seaweed for food applications: Natural occurrence, isolation, purification, and identification. *Algal Research, 48*, 101909.
Lauth, X., Babon, J. J., Stannard, J. A., Singh, S., Nizet, V., Carlberg, J. M., Ostland, V. E., Pennington, M. W., Norton, R. S., & Westerman, M. E. (2005). Bass hepcidin synthesis, solution structure, antimicrobial activities and synergism, and in vivo hepatic response to bacterial infections. *Journal of Biological Chemistry, 280*, 9272–9282.
Lee, J., Gamage, C. D. B., Kim, G. J., Hillman, P. F., Lee, C., Lee, E. Y., Choi, H., Kim, H., Nam, S. J., & Fenical, W. (2020). Androsamide, a cyclic tetrapeptide from a marine Nocardiopsis sp., suppresses motility of colorectal cancer cells. *Journal of Natural Products, 83*, 3166–3172.
Leet, J. E., Schroeder, D. R., Golik, J., Matson, J. A., Doyle, T. W., Lam, K. S., & Krishnan, B. S. (1996). Himastatin, a new antitumor antibiotic from *Streptomyces hygroscopicus*. III. Structural elucidation. *Journal of Antibiotics (Tokyo), 49*, 299–311.
Li, F., Maskey, R. P., Qin, S., Sattler, I., Fiebig, H., Maier, A., & Laatsch, H. (2005). Chinikomycins A and B: isolation, structure elucidation, and biological activity of novel antibiotics from a marine *Streptomyces* sp. isolate M045. *Journal of Natural Products, 68*, 349–353.
Li, H., Huang, H., et al. (2017). Discovery of antimycin-type depsipeptides from a wbl gene mutant strain of deep sea-derived *Streptomyces somaliensis* SCSIO ZH66 and their effects on pro-inflammatory cytokine production. *Frontiers in Microbiology, 8*, 1–8.
Liang, J., Liu, Z., Low, X. Z., Ang, E. L., & Zhao, H. (2017). Twin-primer non-enzymatic DNA assembly: An efficient and accurate multipart DNA assembly method. *Nucleic Acids Research, 45*, e94.
Lima, E., & Medeiros, J. (2022). Marine organisms as alkaloid biosynthesizers of potential anti-alzheimer agents. *Marine Drugs, 20*(1), 75.
Lindequist, U. (2016). Marine-derived pharmaceuticals: Challenges and opportunities. *Biomolecules and Therapeutics, 24*, 561–571.
Liu, L. L., Xu, Y., Han, Z., Li, Y. X., Lu, L., Lai, P. Y., & Qian, P. Y. (2012). Four new antibacterial xanthones from the marine-derived actinomycetes *Streptomyces caelestis*. *Marine Drugs, 10*, 2571–2583.
Liu, N., Song, F., et al. (2015). Mycemycins A-E, new dibenzoxazepinones isolated from two different streptomycetes. *Marine Drugs, 13*(10), 6247–6258.
Liu, S., Yan, X., Yu, M., Chen, J., & Zhang, L. (2010). A novel compound from *Penicillium* sp. (M207142). *Chemistry of Natural Compounds, 46*(1), 116–118.
Losada, A. A., Cano-Prieto, C., Garcia-Salcedo, R., Brana, A. F., Mendez, C., Salas, J. A., & Olano, C. (2017). Caboxamycin biosynthesis pathway and identification of novel

benzoxazoles produced by cross-talk in Streptomyces sp. NTK 937. *Microbial Biotechnology, 10*, 873–885.

Loukaci, S., Le Saout, I., Samadi, M., Leclerc, S., Damiens, E., Meijer, L., & Guyot, M. (2001). Coscinosulfate, a CDC25 phosphatase inhibitor from the sponge *Coscinoderma mathewsi*. *Bioorganic & Medicinal Chemistry, 9*, 3049–3054.

Lowenberg, B. (2013). Sense and nonsense of high-dose cytarabine for acute myeloid leukemia. *Journal of the American Society of Hematology, 121*, 26–28.

Lu, Y., Dong, X., Liu, S., & Bie, X. (2009). Characterization and identification of a novel marine *Streptomyces* sp. produced antibacterial substance. *Marine Biotechnology (New York), 11*, 717–724.

Lu, Y., Shao, M., Wang, Y., Qian, S., Wang, M., Wang, Y., & Hu, Y. (2017). Zunyimycins B and C, New chloroanthrabenzoxocinones antibiotics against methicillin-resistant Staphylococcus aureus and Enterococci from Streptomyces sp. FJS31-2. *Molecules, 22*(2), 251.

Macedo, M. W. F. S., Cunha, N. B. D., Carneiro, J. A., Costa, R. A. D., Alencar, S. A. D., Cardoso, M. H., Franco, O. L., & Dias, S. C. (2021). Marine organisms as a rich source of biologically active peptides. *Frontiers in Marine Science, 8*, 667764.

Mahajan, G., Thomas, B., Parab, R., Patel, Z. E., Kuldharan, S., Yemparala, V., & Sivaramkrishnan, H. (2013). In vitro and in vivo activities of antibiotic PM181104. *Antimicrobial Agents and Chemotherapy, 57*(10), 5315–5319.

Malet-Cascon, L., Romero, F., Espliego-Vazquez, F., Gravalos, D., & Fernandez-Puentes, J. L. (2003). IB-00208, a new cytotoxic polycyclic xanthone produced by a marine-derived Actinomadura. I. Isolation of the strain, taxonomy and biological activities. *Journal of Antibiotics (Tokyo), 56*, 219–225.

Malloy, K. L., Choi, H., Fiorilla, C., Valeriote, F. A., Matainaho, T., & Gerwick, W. H. (2012). Hoiamide D, a marine cyanobacteria-derived inhibitor of p53/ MDM2 interaction. *Bioorganic & Medicinal Chemistry Letters, 22*, 683–688.

Manam, R. R., Teisan, S., White, D. J., Nicholson, B., Grodberg, J., Neuteboom, S. T., & Potts, B. C. (2005). Lajollamycin, a nitrotetraene spiro-beta-lactone-gamma-lactam antibiotic from the marine actinomycete Streptomyces nodosus. *Journal of Natural Products, 68*, 240–243.

Mangamuri, U. K., Vijayalakshmi, M., Poda, S., Manavathi, B., Chitturi, B., & Yenamandra, V. (2016). Isolation and biological evaluation of N-(4-aminocyclooctyl)-3, 5-dinitrobenzamide, a new semisynthetic derivative from the mangrove-associated actinomycete *Pseudonocardia endophytica* VUK-10. *Biotechnology, 6*(2), 1–12.

Mangamuri, U., Muvva, V., Poda, S., Naragani, K., Munaganti, R. K., Chitturi, B., & Yenamandra, V. (2016). Bioactive metabolites produced by *Streptomyces Cheonanensis* VUK-A from Coringa mangrove sediments: Isolation, structure elucidation and bioactivity. *Biotechnology, 6*(1), 1–8.

Martin, G. D., Tan, L. T., Jensen, P. R., Dimayuga, R. E., Fairchild, C. R., Raventos-Suarez, C., & Fenical, W. (2007). Marmycins A and B, cytotoxic pentacyclic C-glycosides from a marine sediment-derived actinomycete related to the genus *Streptomyces*. *Journal of Natural Products, 70*, 1406–1409.

Martínez-García, E., Nikel, P. I., Aparicio, T., & Lorenzo, V. (2014). Pseudomonas 2.0: Genetic upgrading of *P. putida* KT2440 as an enhanced host for heterologous gene expression. *Microbial Cell Factories, 13*(1), 1–15.

Maskey, R. P., Helmke, E., Kayser, O., Fiebig, H. H., Maier, A., Busche, A., & Laatsch, H. (2004). Anti-cancer and antibacterial trioxacarcins with high anti-malaria activity from a marine Streptomycete and their absolute stereochemistry. *Journal of Antibiotics, 57*, 771–779.

Masso-Silva, J. A., & Diamond, G. (2014). Antimicrobial peptides from fish. *Pharmaceuticals, 7*, 265–310.

Mathan, S., Smith, A. A., Kumaran, J., & Prakash, S. (2011). Anticancer and antimicrobial activity of *Aspergillus protuberus* SP1 isolated from marine sediments of the South Indian coast. *Chinese Journal of Natural Medicines, 9*(4), 286–292.

Mehari Ghebretinsae, M., Awet Adhanom, B., Henok Ghebremedhin, G., & Biniam Tsegay, B. (2019). A review on bioactive secondary metabolites of seagrass of the southern Red Sea, Eritrea. *International Journal of Advanced Research, 7*, 306–314.

Mehbub, M. F., Perkins, M. V., Zhang, W., & Franco, C. M. (2016). New marine natural products from sponges (porifera) of the order dictyoceratida (2001 to 2012); A promising source for drug discovery, exploration and future prospects. *Biotechnology Advances, 34*, 473–491.

Meijer, L., Thunnissen, A. M., White, A., Garnier, M., Nikolic, M., Tsai, L., & Biernat, J. (2000). Inhibition of cyclin-dependent kinases, GSK-3β and CK1 by hymenialdisine, a marine sponge constituent. *Chemical Biology, 7*, 51–63.

Menis, J., & Twelves, C. (2011). Eribulin (Halaven): a new, effective treatment for women with heavily pretreated metastatic breast cancer. *Breast Cancer: Targets and Therapy, 3*, 101.

Mitchell, S. S., Nicholson, B., Teisan, S., Lam, K. S., & Potts, B. C. (2004). Aureoverticillactam, a novel 22-atom macrocyclic lactam from the marine actinomycete *Streptomyces aureoverticillatus*. *Journal of Natural Products, 67*, 1400–1402.

Moon, K., Chung, B., Shin, Y., Lee, S. K., Oh, K. B., Shin, J., & Oh, D. C. (2015). Discovery of new bioactive secondary metabolites from bacteria in extreme habitats. *Planta Medica, 81*(11), PT24.

Mullowney, M. W., Hainmhire, E., et al. (2015). A pimarane diterpene and cytotoxic angucyclines from a marine-derived *Micromonospora sp.*, Vietnam's East Sea. *Marine Drugs, 13*(9), 5815–5827.

Nagle, D. G., Zhou, Y. D., Mora, F. D., Mohammed, K. A., & Kim, Y. P. (2004). Mechanism targeted discovery of antitumor marine natural products. *Current Medicinal Chemistry, 11*, 1725–1756.

Nah, H. J., Pyeon, H. R., Kang, S. H., Choi, S. S., & Kim, E. S. (2017). Cloning and heterologous expression of a large-sized natural product biosynthetic gene cluster in Streptomyces sp. *Journal of Microbiology and Biotechnology, 27*(3), 394.

Nikolouli, K., & Mossialos, D. (2012). Bioactive compounds synthesized by non-ribosomal peptide synthetases and type-I polyketide synthases discovered through genome mining and metagenomics. *Biotechnology Letters, 34*(8), 1393–1403.

Niu, S., Zhou, T. T., Xie, C. L., Zhang, G. Y., & Yang, X. W. (2017). Microindolinone A, a novel 4,5,6,7-tetrahydroindole, from the deep-sea-derived actinomycete Microbacterium sp. MCCC 1A11207. *Marine Drugs, 15*(7), 4–11.

Noga, E. J., Silphaduang, U., Park, N. G., Seo, J. K., Stephenson, J., & Kozlowicz, S. (2009). Piscidin 4, a novel member of the piscidin family of antimicrobial peptides. *Comparative Biochemistry and Physiology Part B: Biochemistry and Molecular Biology, 152*, 299–305.

Novoa, B., Romero, A., Álvarez, Á. L., Moreira, R., Pereiro, P., Costa, M. M., & Figueras, A. (2016). Antiviral activity of myticin C peptide from mussel: An ancient defense against herpesviruses. *Journal of Virology, 90*, 7692–7702.

Oh, H., Jensen, P. R., Murphy, B. T., Fiorilla, C., Sullivan, J. F., Ramsey, T., & Fenical, W. (2010). Cryptosphaerolide, a cytotoxic Mcl-1 inhibitor from a marine-derived ascomycete related to the genus *Cryptosphaeria*. *Journal of Natural Products, 73*(5), 998–1001.

Oja, T., San Martin Galindo, P., Taguchi, T., Manner, S., Vuorela, P. M., Ichinose, K., & Fallarero, A. (2015). Effective antibiofilm polyketides against *Staphylococcus aureus* from the pyranonaphthoquinone biosynthetic pathways of *Streptomyces* species. *Antimicrobial Agents and Chemotherapy, 59*, 6046–6052.

Oliveira dos Santos, A., Britta, E. A., Bianco, E. M., Ueda-Nakamura, T., Pereira, R. C., & Nakamura, C. V. (2011). 4-Acetoxydolastane diterpene from the Brazilian brown alga Canistrocarpus cervicornis as antileishmanial agent. *Marine Drugs, 9*, 2369–2383.

Otero-González, A. J., Magalhães, B. S., García-Villarino, M., López-Abarrategui, C., Sousa, D. A., Dias, S. C., & Franco, O. L. (2010). Antimicrobial peptides from marine invertebrates as a new frontier for microbial infection control. *FASEB Journal, 24*, 1320–1334.

Pan, C. Y., Chen, J. Y., Lin, T. L., & Lin, C. H. (2009). In vitro activities of three synthetic peptides derived from epinecidin-1 and an antilipopolysaccharide factor against Propionibacterium acnes, Candida albicans and Trichomonas vaginalis. *Peptides, 30*, 1058–1068.

Pan, E., Oswald, N. W., Legako, A. G., Life, J. M., Posner, B. A., & Macmillan, J. B. (2013). Precursor-directed generation of amidine containing ammosamide analogs: ammosamides E-P. *Chemical Science, 4*, 482–488.

Patin, N. V., Schorn, M., Aguinaldo, K., Lincecum, T., Moore, B. S., & Jensen, P. R. (2017). Effects of actinomycete secondary metabolites on sediment microbial communities. *Applied and Environmental Microbiology, 83*, e0267616.

Pereiro, P., Forn-Cuní, G., Dios, S., Coll, J., Figueras, A., Novoa, B., & Mallavia, R. (2017). Interferon-independent antiviral activity of 25-hydroxycholesterol in a teleost fish. *Antiviral Research, 145*, 146–159.

Phan, L. Y., Jian, T., Chen, Z., Qiu, Y. L., Wang, Z., Beach, T., & Or, Y. S. (2004). Synthesis and antibacterial activity of a novel class of 4'-substituted 16-membered ring macrolides derived from tylosin. *Journal of Medicinal Chemistry, 47*(12), 2965–2968.

Pietra, F. (1997). Secondary metabolites from marine microorganisms: Bacteria, protozoa, algae and fungi: Achievements and prospects. *Natural Product Reports, 14*, 453–464.

Qin, Z., Munnoch, J. T., Devine, R., Holmes, N. A., Seipke, R. F., Wilkinson, K. A., & Hutchings, M. I. (2017). Formicamycins, antibacterial polyketides produced by *Streptomyces formicae* isolated from African Tetraponera plant-ants. *Chemical Science, 8*, 3218–3227.

Rakotondraibea, L. H., Rasolomampianina, R., Park, H. Y., Li, J., Slebodnick, C., Brodie, P. J., Blasiak, L. C., Hill, R., TenDyke, K., Shen, Y., & Cassera, M. B. (2016). Antiproliferative and antiplasmodial compounds from selected Streptomyces species L. *Bioorganic & Medicinal Chemistry Letters, 25*(23), 5646–5649.

Ramalingam, V., Varunkumar, K., Ravikumar, V., & Rajaram, R. (2018). p53 mediated transcriptional regulation of long non-coding RNA by 1-hydroxy-1-norresistomycin triggers intrinsic apoptosis in adenocarcinoma lung cancer. *Chemico-Biological Interactions, 287*, 1–12.

Rangel, M., & Falkenberg, M. (2015). An overview of the marine natural products in clinical trials and on the market. *Journal of Coastal Life Medicine, 3*(6), 421–428.

Rao, M., Wei, W., Ge, M., Chen, D., & Sheng, X. (2013). A new antibacterial lipopeptide found by UPLC-MS from an actinomycete Streptomyces sp. HCCB10043. *Natural Product Research, 27*, 2190–2195.

Reuschl, A. K., Thorne, L. G., Zuliani-Alvarez, L., Bouhaddou, M., Obernier, K., Soucheray, M., & Rosales, R. (2021). Host-directed therapies against early-lineage SARS-CoV-2 retain efficacy against B. 1.1. 7 variants. *BioRxiv.*, 1–12.

Rey-Campos, M., Novoa, B., Pallavicini, A., Gerdol, M., & Figueras, A. (2021). Comparative genomics reveals 13 different isoforms of mytimycins (AM) in Mytilus galloprovincialis. *International Journal of Molecular Sciences, 22*, 3235.

Riccio, G., Ruocco, N., Mutalipassi, M., Costantini, M., Zupo, V., Coppola, D., de Pascale, D., & Lauritano, C. (2020). Ten-year research update review: Antiviral activities from marine organisms. *Biomolecules, 10*, 1007.

Rinehart, K. L., Gloer, J. B., Hughes, R. G., Renis, H. E., McGovren, J. P., Swynenberg, E. B., & Kuentzel, S. L. (1981). Didemnins: antiviral and antitumor depsipeptides from a Caribbean tunicate. *Science, 212*(4497), 933–935.

Rojas, V., Rivas, L., Cárdenas, C., & Guzmán, F. (2020). Cyanobacteria and eukaryotic microalgae as emerging sources of antibacterial peptides. *Molecules, 25*, 5804.

Rosano, G. L., & Ceccarelli, E. A. (2014). Recombinant protein expression in Escherichia coli: Advances and challenges. *Frontiers in Microbiology, 5*(172), 1–17.

Ruiz-Torres, V., Encinar, J. A., Herranz-López, M., Pérez-Sánchez, A., Galiano, V., Barrajón-Catalán, E., & Micol, V. (2017). An updated review on marine anticancer compounds: The use of virtual screening for the discovery of small-molecule cancer drugs. *Molecules, 22*, 1037.

Russo, P., Kisialiou, A., Lamonaca, P., Moroni, R., Prinzi, G., & Fini, M. (2016). New drugs from marine organisms in Alzheimer's disease. *Marine Drugs, 14*, 5.

Sagar, S., Kaur, M., & Minneman, K. P. (2010). Antiviral lead compounds from marine sponges. *Marine Drugs, 8*, 2619–2638.

Schulze, C. J., Donia, M. S., et al. (2015). Genome-directed lead discovery: Biosynthesis, structure elucidation, and biological evaluation of two families of polyene macrolactams against *Trypanosoma brucei*. *ACS Chemical Biology, 10*(10), 2373–2381.

Schulze, C. J., Navarro, G., et al. (2015). Salinipostins A-K, long-chain bicyclic phosphotriesters as a potent and selective antimalarial chemotype. *Journal of Organic Chemistry, 80*, 1312–1320.

Shai, Y. (2002). Mode of action of membrane active antimicrobial peptides. *Biopolymers, 66*, 236–248.

Shaik, M. I., & Sarbon, N. M. (2022). A review on purification and characterization of anti-proliferative peptides derived from fish protein hydrolysate. *Food Reviews International, 38*(7), 1389–1409.

Shin, H. J., Jeong, H. S., Lee, H. S., Park, S. K., Kim, H. M., & Kwon, H. J. (2007). Isolation and structure determination of streptochlorin, an antiproliferative agent from a marine-derived *Streptomyces sp.* 04DH110. *Journal of Microbiology and Biotechnology, 17*, 1403–1406.

Short, F., Short, C., & Novak, A. (2016). Seagrasses. In *The Wetland Book: II: Distribution, Description and Conservation*, 1–19. Springer Science: New York, NY, USA.

Skropeta, D., Pastro, N., & Zivanovic, A. (2011). Kinase inhibitors from marine sponges. *Marine Drugs, 9*, 2131–2154.

Smetanina, O. F., Yurchenko, A. N., Kalinovskii, A. I., Berdyshev, D. V., Gerasimenko, A. V., Pivkin, M. V., & Afiyatullov, S. (2011). Biologically active metabolites from the marine isolate of the fungus *Myceliophthora lutea*. *Chemistry of Natural Compounds, 47*(3), 385–390.

Smith, W. C., Xiang, L., & Shen, B. (2000). Genetic localization and molecular characterization of the nonS gene required for macrotetrolide biosynthesis in Streptomyces griseus DSM40695. *Antimicrobial Agents and Chemotherapy, 44*(7), 1809–1817.

Stritzke, K., Schulz, S., Laatsch, H., Helmke, E., & Beil, W. (2004). Novel caprolactones from a marine streptomycete. *Journal of Natural Products, 67*, 395–401.

Subramani, R., & Aalbersberg, W. (2013). Culturable rare Actinomycetes: diversity, isolation and marine natural product discovery. *Applied Microbiology and Biotechnology, 97*, 9291–9321.

Sung, W. S., Lee, J., & Lee, D. G. (2008). Fungicidal effect and the mode of action of piscidin 2 derived from hybrid striped bass. *Biochemical and Biophysical Research Communications, 371*, 551–555.

Swathi, J., Narendra, K., Sowjanya, K. M., & Satya, A. K. (2013). Evaluation of biologically active molecules isolated from obligate marine fungi. *Mintage Journal of Pharmaceutical & Medical Sciences, 2*, 45–47.

Takahashi, A., Kurasawa, S., Ikeda, D., Okami, Y., & Takeuchi, T. (1989). Altemicidin, a new acaricidal and antitumor substance. *Journal of Antibiotics, 42*, 1556–1561.

Takahashi, Y., & Nakashima, T. (2018). Actinomycetes, an inexhaustible source of naturally occurring antibiotics. *Antibiotics, 7*(3), 74.

Tassanakajon, A., Rimphanitchayakit, V., Visetnan, S., Amparyup, P., Somboonwiwat, K., Charoensapsri, W., & Tang, S. (2018). Shrimp humoral responses against pathogens: Antimicrobial peptides and melanization. *Developmental and Comparative Immunology, 80*, 81–93.

Teixido, C., Arguelaguet, E., Pons, B., Aracil, M., Jimeno, J., Somoza, R., & Hernandez-Losa, J. (2012). ErbB3 expression predicts sensitivity to elisidepsin treatment: in vitro synergism with cisplatin, paclitaxel and gemcitabine in lung, breast and colon cancer cell lines. *International Journal of Oncology, 41*, 317–324.

Tsukamoto, S., Hirota, H., Imachi, M., Fujimuro, M., Onuki, H., Ohta, T., & Yokosawa, H. (2005). Himeic acid A: a new ubiquitin-activating enzyme inhibitor isolated from a marine-derived fungus, *Aspergillus* sp. *Bioorganic & Medicinal Chemistry Letters, 15*, 191–194.

Tsukamoto, S., Takeuchi, T., Rotinsulu, H., Mangindaan, R. E., Van Soest, R. W., Ukai, K., & Yokosawa, H. (2008). Leucettamol A: a new inhibitor of Ubc13-Uev1A interaction isolated from a marine sponge, *Leucetta aff. Microrhaphis*. *Bioorganic & Medicinal Chemistry Letters, 18*, 6319–6320.

Tsukamoto, S., Yamashita, K., Tane, K., Kizu, R., Ohta, T., Matsunaga, S., Fusetani, N., Kawahara, H., & Yokosawa, H. (2004). Girolline, an antitumor compound isolated from a sponge, induces G2/M cell cycle arrest and accumulation of polyubiquitinated p53. *Biological and Pharmaceutical Bulletin, 27*, 699–701.

Tsukamoto, S., Yoshida, T., Hosono, H., Ohta, T., & Yokosawa, H. (2006). Hexylitaconic acid: a new inhibitor of p53–HDM2 interaction isolated from a marine-derived fungus, *Arthrinium* sp. *Bioorganic & Medicinal Chemistry Letters, 16*, 69–71.

Wang, R., Zhang, Q., Peng, X., Zhou, C., Zhong, Y., Chen, X., & Gong, M. (2016). Stellettin B induces G1 arrest, apoptosis and autophagy in human non-small cell lung cancer A549 cells via blocking PI3K/Akt/mTOR pathway. *Scientific Reports, 6*, 1–10.

Wang, Y. D., Rajanbabu, V., & Chen, J. Y. (2015). Transcriptome analysis of medaka following epinecidin-1 and TH1-5 treatment of NNV infection. *Fish & Shellfish Immunology, 42*, 121–131.

Waters, A. L., Peraud, O., Kasanah, N., Sims, J. W., Kothalawala, N., Anderson, M. A., & Dass, A. (2014). Abbas, S. H.; Rao, K. V.; Jupally, V. R.; Kelly, M.; Division: An analysis of the sponge *Acanthostrongylophora* igens microbiome yields an actinomycete that produces the natural product manzamine A. *Frontiers in Marine Science, 54*, 1–54.

Wei, M. Y., Yang, R. Y., Shao, C. L., Wang, C. Y., Deng, D. S., She, Z. G., & Lin, Y. C. (2011). Isolation, structure elucidation, crystal structure, and biological activity of a marine natural alkaloid, viridicatol. *Chemistry of Natural Compounds, 47*(2), 322–325.

Wen, L., Chen, G., She, Z., Yan, C., Cai, J., & Mu, L. (2010). Two new paeciloxocins from a mangrove endophytic fungus *Paecilomyces* sp. *Russian Chemical Bulletin, 59*(8), 1656–1659.

Xiao, L., Liu, H., Wu, N., Liu, M., Wei, J., Zhang, Y., & Lin, X. (2013). Characterization of the high cytochalasin E and rosellichalasin producing-*Aspergillus* sp. nov. F1 isolated from marine solar saltern in China. *World Journal of Microbiology and Biotechnology, 29*, 11–17.

Xie, C. L., Liu, Q., et al. (2017). Anti-allergic compounds from the deep-sea-derived *Actino-mycete Nesterenkonia flava MCCC 1K00610. Marine Drugs, 15*(3), 71.

Xu, M. J., Wang, J. H., Bu, X. L., Yu, H. L., Li, P., Ou, H. Y., & Ao, P. (2016). Deciphering the streamlined genome of *Streptomyces xiamenensis* 318 as the producer of the anti-fibrotic drug candidate xiamenmycin. *Scientific Reports, 6*(1), 1–11.

Yanagihara, M., Sasaki-Takahashi, N., Sugahara, T., Yamamoto, S., Shinomi, M., Yamashita, I., & Andoh, T. (2005). Leptosins isolated from marine fungus *Leptoshaeria species* inhibit DNA topoisomerases I and/or II and induce apoptosis by inactivation of Akt/protein kinase B. *Cancer Science, 96*(11), 816–824.

Yang, N., & Sun, C. (2016). The inhibition and resistance mechanisms of actinonin, isolated from marine *Streptomyces sp.* NHF165, against *Vibrio anguillarum. Frontiers in Micro-biology, 7*, 1–11.

Yellamanda, B., Vijayalakshmi, M., Kavitha, A., Reddy, D. K., & Venkateswarlu, Y. (2016). Extraction and bioactive profile of the compounds produced by *Rhodococcus sp.* VLD-10. *3 Biotech, 6*(2), 1–9.

Yu, P., & Gu, H. (2015). Bioactive substances from marine fishes, shrimps, and algae and their functions: Present and future. *Critical Reviews in Food Science and Nutrition, 55*, 1114–1136.

Yu, Y., Wu, J., Lei, F., Chen, L., Wan, W., Hai, L., & Wu, Y. (2013). Design, synthesis and anticancer activity evaluation of diazepinomicin derivatives. *Letters in Drug Design & Discovery, 10*, 369–373.

Zhang, D., Satake, M., Fukuzawa, S., Sugahara, K., Niitsu, A., Shirai, T., & Tachibana, K. (2012). Two new indole alkaloids, 2-(3, 3-dimethylprop-1-ene)-costaclavine and 2-(3, 3-dimethylprop-1-ene)-epicostaclavine, from the marine-derived fungus *Aspergillus fumigatus*. *Journal of Natural Medicines, 66*(1), 222–226.

Zhang, W., Che, Q., et al. (2017). Marine *Streptomyces* sp. derived antimycin analogues suppress HeLa cells via depletion HPV E6/E7 mediated by ROS-dependent ubiquitin-proteasome system. *Scientific Reports, 7*(1), 1–14.

Zhang, Y. M., Li, H. Y., Hu, C., Sheng, H. F., Zhang, Y., Lin, B. R., & Zhou, G. X. (2016). Ergosterols from the culture broth of marine *Streptomyces anandii* H41-59. *Marine Drugs, 14*(5), 1–11.

Zhang, Y., Huang, H., Xu, S., Wang, B., Ju, J., Tan, H., & Li, W. (2015). Activation and enhancement of Fredericamycin A production in deepsea-derived *Streptomyces somaliensis* SCSIO ZH66 by using ribosome engineering and response surface methodology. *Microbial Cell Factories, 14*(1), 1–11.

Zhen, X., Gong, T., Liu, F., Zhang, P. C., Zhou, W. Q., Li, Y., & Zhu, P. (2015). A new analogue of echinomycin and a new cyclic dipeptide from a marine-derived *Streptomyces sp.* LS298. *Marine Drugs, 13*(11), 6947–6961.

Zhou, T., Komaki, H., Ichikawa, N., Hosoyama, A., Sato, S., & Igarashi, Y. (2015). Biosynthesis of akaeolide and lorneic acids and annotation of type I polyketide synthase gene clusters in the genome of *Streptomyces sp.* 2015, *Marine Drugs, 13*(1), 581–596.
Zhou, Y., Debbab, A., Mándi, A., Wray, V., Schulz, B., Müller, W. E., & Aly, A. H. (2013). Alkaloids from the sponge-associated fungus *Aspergillus sp. European Journal of Organic Chemistry, 2013*(5), 894–906.

CHAPTER 19

Exploring the Potential of Marine Drugs: A Patent Perspective

ANAS ISLAM,[1] USAMA AHMAD,[1] MD. FAIYAZUDDIN,[2] ASAD ALI,[1] and MOHD. MUAZZAM KHAN[1]

[1]*Faculty of Pharmacy, Integral University, Lucknow, Uttar Pradesh, India*

[2]*School of Pharmacy, Al-Karim University, Katihar, Bihar, India*

ABSTRACT

The ocean is a potential source of bioactive compounds, but unlike terrestrial plants, it does not have a long history of use in traditional medicine. The marine environment is a one-of-a-kind resource with vast biological diversity that, if properly explored, could lead to breakthrough therapies. The influence of this sector on the pharmaceutical industry is growing as more substances derived from marine sources enter clinical testing. Marine creatures contain a large number of naturally occurring compounds with various structural and pharmacological properties. The recent rise in patent publications amply supports the importance of these compounds as sources of novel therapeutic agents. Marine-derived actives are being researched as prospective sources for anti-cancer, anti-bacterial, antiviral, and anti-inflammatory medications, as well as treatments for a range of other conditions, despite critical supply limitations. As the number of lead marine compounds increases, it becomes essential to explore the ongoing status of patents filed and granted in the field of marine pharmaceuticals. In this chapter, an effort has been made to gather valuable information on the recent patents available on various marine drugs.

Handbook of Research in Marine Pharmaceutics: Exploring Oceanic Microbial Diversity for Human Health and Wellness. Md. Faiyazuddin & Khalid Rehman Hakeem (Eds.)

19.1 INTRODUCTION

Natural compounds from Earth's biodiverse flora and fauna, known as "secondary metabolites," are a crucial, reliable source of promising therapeutic leads (David et al., 2015). With over a million novel chemical entities found to date, this conventional source of medicine is still regarded as the most effective source of prospective medications. Although synthetic pharmaceuticals tend to be stronger than herbal medicines, over 60% of the drugs on the market today have natural origins due to their lower risk of adverse effects and substantially greater absorption, making them a good option for finding lead structures (Dias et al., 2012; Newman et al., 2016; Cheuka et al., 2017; Bushra et al., 2018). But even now, 95% of the estimated 2 million types of fungi, animals, microorganisms, plants, and similar organisms that makeup biodiversity have not been assessed for any biological activity (David et al., 2015). Therefore, the challenge is finding a solution to effectively and efficiently access and value this inherent chemical variety for medicinal purposes.

Up to 70% of the Earth's surface is covered in water bodies, which are home to 32 basic phyla and support the enormous expansion of marine species. They are biologically diverse ocean dwellers who make up over half of all species worldwide and are influenced by the physicochemical characteristics of the ocean (Suleria et al., 2015). Due to their capacity to resist harsh environments including pressure differences, temperature swings, the impacts of mutation, and infection, these marine animals have developed over time by experiencing physical and chemical adaptations. Although immobile organisms do not adjust their physical defense, they are thought to do so chemically (Jimeno et al., 2004). Due to the marine environment's different physical and chemical circumstances, they thus show the existence of a variety of molecules with unique structural characteristics (Skropeta, 2008). The ocean has a wide variety of creatures, making it appropriate to refer to it as a "cure-all." Additionally, it offers an abundance of sources for new bioactive ingredients, a few of which are utilized in foods, cosmetics, and nutraceuticals (Faulkner, 1995). Their requirement to generate effective medicinal entities has expanded due to their investigation into various biological activities (Kim, 2013). These naturally occurring substances are typically produced from secondary metabolites rather than the core metabolic route (Malve, 2016).

Since ancient times, the vast natural resources of the sea have been utilized, including the utilization of fish and other marine animals as well as preparations made from algae as sources of medicine. The most well-known and long-used marine-derived product is fish oils. The study of compounds having active pharmacological qualities found in the marine class of animals and plants is the focus of the pharmaceutical science subfield known as marine pharmacology. Novel bioactive natural products with chemical and structural traits that are frequently missing from land natural products can be found in abundance in the marine environment. Marine creatures are a great source of nutraceuticals and may be used to cure a number of human ailments. Microbes are the primary focus of contemporary marine pharmacology (Kijjoa & Sawangwong, 2004).

Despite the fact that marine pharmacology has been thoroughly studied in the past, both internationally and in India, there is still a need to examine the oceans' potential as a source for the creation of novel medicines, especially given their abundance in nature and large-scale manufacturing. The pharmaceutical industry is now focused on finding novel compounds having previously undisclosed pharmacological features that may be used to create new therapeutic treatments for commercial usage. The development of marine natural substances as medicinal agents is summarized in this chapter.

19.2 PATENTS OF MARINE PRODUCTS HAVING ANTI-CANCER ACTIVITY

One of the worst diseases in the world continues to be cancer. As there is a critical need for treatments with new modes of action, much has been studied on new anti-cancer therapies from natural origin, including plants, microbes, and marine species. Marine ecosystems are a source of new bioactive metabolites with various chemical configurations (Khalifa et al., 2019). The search for novel organisms that can help in the therapy of cancer in marine environments has considerable promise. After 1980, the area of biotechnology arose as one that gave the study of the ocean's direction, focusing on uses like medicine production (Newman et al., 2016). During the preceding 50 years, new research has shown that a variety of natural compounds originating from marine plants and bacteria have beneficial benefits in the fight against cancer. Examples include the marine-based medications eribulin mesylate, cytarabine, trabectedin, and brentuximab vedotin which are used to treat leukemia, ovarian cancer, metastatic breast cancer, and soft tissue sarcoma (Dyshlovoy et al., 2018).

Streptomyces fulvissimus (ZZ40) is a bacterium that has been extracted from the *Haliplanella lineata* seaweed in the marine environment and is known to exist in Wuhan Center, China. This bacterium has the ability to produce two important substances, namely *streptomyces fulvissimus* acid and streptomyces fulvissimus ketone. According to Zhejiang University, these substances have been found to possess anti-tumor properties and can be used for the treatment of gliomas, which are considered to be the most common and deadly types of brain tumors. Researchers conducted an investigation on two enzymes, aconitase 2 of the tricarboxylic acid cycle and cytochrome C of the oxidative phosphorylase, using human glioma U87MG cells. The study found that both enzymes exhibited significant anti-glioma activity by selectively targeting several crucial enzymes involved in glioma glycolysis (Zhang et al., 2018).

According to the South China Sea Institute of Oceanology and the Chinese Academy of Sciences, angucycline compounds have been discovered. These compounds have been derived from a fermented extract obtained from Actinomycetes *Streptomyces rubrogriseus*, a marine bacterium. The three separated angucycline compounds—Angucycline Compound I, Marangucycline A, and Marangucycline B—show cytotoxic activity and have potential for use in the treatment of cancer, including some bacteria may all be prevented by using the chemicals, which also have anti-bacterial action (Jianhua et al., 2017).

According to a Chinese patent, diindolyl diketopiperazine chemicals and phenazine compounds can both be used to make anti-cancer medications. Phenazine compounds, namely Phenaziterpene A and B, have been extracted from the fermented extract of *Streptomyces niveus*, a marine bacterium. These compounds have been found to have a moderate inhibitory effect on the growth of tumor cells, with Phenaziterpene A exhibiting an IC50 value of less than 20 μM for tumor cell lines SF-268 and MCF-7 (Jianhua et al., 2017). The Thia diketopiperazine compounds, including Luteoalbusin A, B, Gliocladine C, and Compound I, have been extracted from the marine fungus *Acrostalagmus luteoalbus*. These compounds exhibit promising growth inhibitory activity against various types of cancers, including human central nervous system cancer cells, human breast cancer cells, non-small cell lung cancer cells, and human liver cancer cells. The compounds have the potential to be used in the development of effective anti-tumor medications for the treatment of multiple types of cancer (Wang et al., 2015).

The separation and synthesis of indole diterpenoid substances from an *Eupenicillium* species strain (*SP. HJ002*) isolated from *Xylocarpus granatum* have been claimed by Hainan Normal University in China. The medicines greatly outperform the positive controls 5-Fu and doxorubicin in their ability to suppress tumor cells such as adenocarcinomic human alveolar basal epithelial cells (A549), HeLa, and HepG2 with an IC50 of 1–25 μg/mL (Zheng et al., 2018). According to Nanjing Agricultural University, substituted triterpenoid saponin chemicals, namely Salbige A and Salbige B, have been extracted from the aerial sections of sea asparagus (Xin et al.,

2015). Shanghai Changzheng Hospital has reportedly discovered hippolide I, a sesterterpenoid-like substance, in sea sponges using ethanol-based isolation techniques. The procedure is considered cost-effective and simple, making it a suitable option for industrial applications. The molecule has been evaluated and found to possess strong anti-tumor, anti-obesity, and anti-diabetic properties. The creators have experimented with a variety of dosage formulations (Shujuan et al., 2016). According to reports, Waseda University in Japan has identified a novel stelliferin compound by utilizing a lipo-soluble solvent to extract stelliferin A, stelliferin B, as well as its diastereomers and enantiomers from the sponge species *Stelleta globostellata* (S12111) (Yoichi et al., 2017). The extract containing the stelliferin component is beneficial for blocking niche formation brought on by cancer cells adhering to supportive tissues and reducing the recurrence of a malignant tumor. According to reports, inventors have developed a pharmaceutical composition using steriferin A in combination with other anti-cancer medications. The composition is intended for the prevention, treatment, and prevention of the repetition of malignant tumors and is available in various dosage forms. The dosage of the mixture is 500 mg applied topically or, more commonly, 10–500 mg taken orally. According to reports, Hangzhou Kexing Biochem Co. Ltd has recently disclosed the discovery of a novel compound containing octahydroperylene with significant Bromodomain. The compound has BRD4 inhibitory activity and is useful for making medications for diseases like blood cancer and breast cancer that are characterized by excessive or abnormal cell proliferation (Ma et al., 2017).

In this research, three patents have been identified that contain assertions about the anti-tumor characteristics of marine microbial peptides. One of these patents, filed by Xiamen University, claims that cyclic dipeptides, specifically cyclo proline-tyrosine (I) dipeptide obtained from marine *Streptomyces* sp. (HNS054), can effectively treat breast cancer, cervical cancer, colon cancer, and regulate hormone and energy metabolism (Zhao et al., 2015). The invention's approach is claimed to be both cost-effective and highly efficient in creating cyclic dipeptides from marine Streptomyces. Another patent, belonging to the Chinese Academy of Fishery Sciences Research Institute, concerns the anti-tumor properties of an active ingredient found in *Bacillus* sp. (N11–8). To investigate the active agent's anti-tumor activity, the human hepatoma cell line BEL-7402 was employed as the target cell, and it was found to have a potent cytotoxic impact on numerous tumor cells, with an inhibition rate exceeding 30% (Zheng et al., 2015). In another patent held by the same assignee, a marine bacillus polypeptide obtained from the metabolites of Antarctic marine *Bacillus* (N11–8) is claimed. The active polypeptide, which has 15 amino acid sequences, is highly cytotoxic to several tumor cells, including human hepatoma carcinoma cells, human ovarian carcinoma cells, human renal clear cell carcinoma cells, and human large cell lung cancer cells. Consequently, the active polypeptide may be beneficial in the treatment and prevention of various malignancies, such as lung cancer, ovarian cancer, hepatocellular carcinoma, and renal cancer (Zheng et al., 2016).

An invention pertains to the discovery of two heteroterpene compounds, referred to as compounds I and II, derived from a marine source. The compounds have shown potential in the treatment of cancer, particularly liver cancer cells HepG-2, breast cancer cells IC50 of MCF-7, glioma cells SF-268, and non-small cell lung carcinoma cells A54950, with a value range of 0.5–17.6 μM. These chemicals, which are identified and extracted from the fermentation culture of the marine fungus *Phomopsis tersa* FS 441, are included in the invention as a preparation technique. The compounds demonstrate significant anti-tumor activity, suggesting their potential for use in the development of anti-tumor drugs (Liu et al., 2020). This invention describes the discovery of two compounds, photeroids A and B, which have potential in the treatment of cancer. Experiments have demonstrated that photeroids A and B have an IC on hepatoma cell HepG-2, breast cancer cell MCF-7, glioma cell SF-268, and non-small cell lung cancer cell A54950, within a value range of 20.0–26.2 μM, indicating their significant anti-tumor activity. These compounds have the potential to be developed as anti-tumor drugs (Liu et al., 2020).

An invention involves the discovery of two compounds, tersaphenatone B and tersaphenatone E, which have potential in the treatment of cancer. Experiments have shown that tersaphenatone B and E have the ability to treat liver cancer cell HepG-2, liver cancer cell MCF-7, glioma SF-268, and non-small cell lung carcinoma A549, within a value range of 5.4–36.7 μM, indicating their significant anti-tumor activity. These compounds can be used in the development of anti-tumor drugs (Chen et al., 2020). This invention pertains to the discovery of six compounds, lithocarols A-F, which have potential in the treatment of cancer. The compounds were extracted and isolated from the fermentation culture medium of the marine fungus *Phomopsis lithocarpus* FS508, and a preparation method for these compounds was provided. Experiments have shown that lithocarols A-F have significant anti-tumor activity, as evidenced by their IC50 on HepG-2 cell, breast cancer cell MCF-7, neuroglial cytoma SF-268, and non-small cell lung cancer cell A549, within a value range of 10.5~87.7 μM. These compounds can be used in the development of anti-tumor drugs (Liu et al., 2019). This invention relates to the discovery of a polyketide compound, Lithocarpin E-G, which has potential in the treatment of cancer. The compound was extracted and prepared from a fermentation culture of the marine fungus *Phomopsis lithocarpus* FS 508, and a preparation method for this compound was provided. Experiments have shown that Lithocarpin E-G has significant anti-tumor activity, as evidenced by its IC50 on liver cancer cell HepG-2, breast cancer cell MCF-7, glioma cell SF-268, and non-small cell lung cancer cell A549, within a value range of 6.3–91.6 μM. The compound can be used in the development of anti-tumor drugs (Chen et al., 2021).

The invention introduces a novel compound named Tenolone K, which has the potential to be developed into anti-tumor medications. The compound's structural formula is provided, and it is derived from the fermentation culture of a marine fungus called *Phomopsis lithocarpus* FS 508 through separation and preparation. The preparation method is also outlined. According to research, Tenolone K has exhibited significant anti-tumor activity by effectively treating liver cancer cells HepG-2, breast cancer cells MCF-7, glioma cells SF-268, and non-small cell lung cancer cells A549, with IC50 values ranging from 11.36–51.62 μM. This compound has potential for use in the production of anti-tumor drugs (Chen et al., 2021).

The current invention is centered around marine natural products and presents a mixed-source terpenoid, its preparation method, and its application in the creation of medicine used to combat breast cancer or dihydroorotate dehydrogenase inhibitors. The mixed-source terpenoid is identified as one of the following: acremochlorin K, acremochlorin L, or acremochlorin M. The compound exhibits significant inhibition of breast cancer cell proliferation and dihydroorotate dehydrogenase, making it a promising candidate for the development of novel and highly effective anti-breast cancer or dihydroorotate dehydrogenase inhibitor medication with a unique structure (Luo et al., 2020). The current invention pertains to the pharmaceutical technology field and discloses an anti-cancer compound called Ananaside A and a highly efficient extraction method. The chemical constitution and spatial configuration of the compound were determined through various modern spectrum analyses, especially through advanced ID NMR spectra and high-resolution mass spectrums. The compound was screened for anti-tumor activity *in vitro* and exhibited significant inhibitory effects on eight types of tumor cell lines. This invention presents a lead compound for developing new anti-tumor drugs and contributes to the development and utilization of marine pharmaceutical organisms in China (Wu et al., 2018).

Weihai Yushanfang Biotechnology Co. Ltd. claims to have developed an herbal marine composition composed of Spirulina, Saccharina japonica, a marine kelp or seaweed, *Sargassum thunbergii, Gloiopeltis furcate,*

Ganoderma, Eleutherococcus senticosus, and selenium-enriched yeast that can effectively stop the spread of cancer and boost immunity. To make this composition, Se-enriched yeast is added to the filtered underflow, and then the raw materials are equally mashed, screened, and combined. Various dose forms of this mixture can be prepared for healthy foods by adding additional auxiliary components (Wang et al., 2015). A marine bioactive compound with great specificity and no adverse effects was the subject of another Chinese patent. Oyster extract, *Hippocampus enzymolysis* extract, seaweed polysaccharide extract, *Gloiopeltis furcata* methanol extract, *Saccharina japonica* (Fucoidan) extract, marine sponge extract, shark liver extract with cartilage, sea cucumber extract, and marine actinomycetes (AFN1007) fermentation broth are all included in the composition in varying amounts. Strong specificity, good tumor-resistance curative effects, and minimal to no adverse effects are all features of the composition. The formulation can be combined with a traditional pharmaceutical vehicle and additives to create tablets, granules, capsules, or oral solutions (Cai, 2017).

19.3 PATENTS OF MARINE PRODUCTS HAVING ANTI-MICROBIAL ACTIVITY

Microbes undergo genetic mutations at a rapid pace, leading to the development of defense mechanisms against antimicrobials. These mechanisms confer an adaptive advantage and become prevalent in microbial populations. Additionally, microbes can transfer these genetic mutations to other species, facilitating the spread of resistance mechanisms (Partridge et al., 2018). The excessive and inappropriate utilization of antimicrobials from the time they were discovered has caused antimicrobial resistance to increase, by functioning as a method of artificial selection that eliminates susceptible microorganisms and favors the survival of resistant ones (McEwen & Collignon, 2018). The decline in the finding of new antimicrobial drugs over the past few decades has exacerbated this issue (Loureiro et al., 2016). The decline in the discovery of new antimicrobial drugs can be attributed to the fact that many environments that have been a source of innovative molecules in the past have been thoroughly researched. Consequently, the search has shifted towards under-explored environments. Organisms found in new ecosystems often have distinct biochemical diversity, and ecosystems that are isolated and extreme are more likely to have organisms with unique and unexplored physiological traits (Katz & Baltz, 2016; Danilovich et al., 2018; Shinde et al., 2019; Donadio et al., 2010).

According to Hainan Normal University, the mangrove fungus *Penicillium citrinum* (HL-5126) strain produces novel antibacterial substances, such as chlorinated anthraquinone or its salt and coumarin compound (Zheng et al., 2017; Zheng et al., 2017). During the production of medications, the chlorinated anthraquinone chemicals are effective in inhibiting the growth of gram-negative bacteria *Vibrio parahemolyticus* and *Staphylococcus aureus*. Meanwhile, the coumarin chemical is utilized in the production of antibacterial medications for suppressing *Staphylococcus epidermis*, *Staphylococcus aureus, Escherichia coli, Bacillus,* or *Vibrio* on the surface of the skin (Zheng et al., 2017). Three novel macrolactin compounds that were obtained from marine Bacillus subtilis and have antibacterial action were revealed by another Korean patent (Shin et al., 2015). A Chinese patent from the Institute of Microbiology Chinese Academy of Sciences describes a process of extracting a component from the supernatant of fermented liquid obtained by inoculating marine *Verrucosispora* sp. conducting liquid fermentation culture and then collecting and centrifuging it (Zhang et al., 2015). The three newly discovered macrolactin compounds – Abyssomicin J, Abyssomicin K, and Abyssomicin L – have been found to possess potent anti-tubercular activity, with minimum inhibitory concentration (MIC) values of 3.125 μg/mL, 25 μg/mL, and more than 100 μg/mL, respectively, against the BCG strain. These compounds can be utilized as the active ingredient in the preparation of anti-tubercular drugs with various dosage forms, using acceptable excipients in the pharmaceutical industry. A marine *Bacillus* sp. strain-isolated chemical with double cyclic depsipeptide is beneficial for avoiding or treating inflammatory and allergy illnesses (Choi et al., 2017).

The current invention concerns a newly discovered anti-bacterial peptide derived from a sea cucumber and its applications. This novel anti-bacterial peptide is composed of a unique amino acid sequence and has shown remarkable efficacy against both gram-positive and gram-negative bacteria. It also possesses potent anti-bacterial activity against bacteria that are resistant to antibiotics, making it a valuable component for a range of applications such as anti-bacterial formulations, antiseptics, cosmetics, and the prevention and treatment of infectious diseases (Kang, 2021). The current invention concerns a newly developed anti-microbial peptide derived from a peptide found in *Styela clava*, a marine organism, and its potential applications. Specifically, the discovery relates to the use of this peptide as an ingredient in anti-bacterial or antiseptic compositions, antibiotic pharmaceutical compositions, pharmaceutical compositions to prevent and treat pathogenic bacterial infections, and antibiotic food additives. The new anti-microbial peptide was developed by replacing isoleucine (I) with lysine (K) at the 5th and 12th amino acids from the N-terminus of the amino acid sequence of the original peptide. This modification significantly reduces the peptide's cytotoxicity while maintaining its potent anti-bacterial activity (Lim, 2021).

The current innovation concerns a chromophoric peptide with antibacterial properties that incorporates the sequence of amino acids Trp-Asp-Cys. The invention also describes a method of preparing the antibacterial chromophoric peptide. The peptide can be obtained from marine shrimps through protease enzymolysis, separation, purification, and freeze-drying. Alternatively, it can be synthesized using peptide synthesis or synthesis and modification methods such as biosynthesis, chemical synthesis, enzymatic synthesis, ligation, addition, coupling, or derivatization. The invention also discloses the antibacterial application of the chromophoric peptide. This natural antibacterial agent has high antimicrobial activity, can replace traditional synthetic antibacterial agents, and has a stronger inhibition capability on pathogenic bacteria of aquatic products. It eliminates the side effects of residue and drug resistance that may be caused by synthetic antibacterial agents and antibiotics. The method can improve the low utilization rate of aquatic products in China and ensure the safety of the antibacterial agent in food for consumers (Lu et al., 2021). The present invention pertains to methods for preparing and utilizing a novel compound of indole alkaloids. The indole alkaloid compound is obtained from the shell mold PRB-2 (*Penicillium crustosum* PRB-2) collected from the deep-sea waters of the Puli Gulf in the South Pole. The indole alkaloid compound has demonstrated strong inhibitory properties against the H1N1 virus (Li et al., 2019).

A substance with outstanding antibacterial action that was isolated from the Antarctic lichen *Stereocaulon alpinum* is the subject of the current invention. The current invention pertains to a composition that has potent antimicrobial and antioxidant properties and includes a particular compound known as Lobaric acid (compound 1), and a pseudodepsidone natural compound named Lobarstin (compound 2) or salt of these compounds extracted from *Stereocaulon alpinum*. This composition is suitable for pharmaceutical products and functional foods as it aids in the prevention and treatment of bacterial infections. Moreover, it can also be utilized in various other applications due to its efficient antioxidative effect (Yim et al., 2014). The present discovery relates to a novel antibiotic peptide derived from polar fish having antimicrobial activity, a food additive, a composition containing the peptide, or a pharmaceutical composition for antibiosis. The antibiotic peptide of the current development displays antimicrobial activity against various pathogenic microorganisms, low cytotoxicity, and maintains antibacterial activity even at a high salt concentration such as *in vivo* (Shin et al., 2018).

The present invention is a method of preparing a degerming eye lotion that contains marine organisms. The composition of the lotion consists of various components, including a detergent, an enzyme preparation, a surfactant, a complexing agent, an adjuvant, and deionized water. These components are present in specific percentages by weight, such as 1.0–3.0% of detergent, 0.002–0.5% of enzyme preparation, 2–15.0% of surfactant, 3.0–8.0% of complexing agent, 1.0–4.0% of adjuvant, and the remaining quantity of deionized water. The degerming eye lotion containing marine organisms has been found to be gentle in cleaning the eyes, capable of rapidly neutralizing the pH of an irritating liquid, and effective in relieving chemical injury to the eyes. It also aids in removing foreign materials from the eyes, is easy to use, and helps relieve discomfort and sudden symptoms caused by eye fatigue. The lotion can provide immediate first aid to an operator on site (Chunyan et al., 2015). This invention pertains to a novel marine organism's spray film that can treat skin injuries and infections, and its method of preparation. It falls under the category of medical materials derived from marine organisms. The raw materials of biology spray film is chitosan, NTx albumen, and Chinese medicine extract, and the raw material of Chinese medicine extract include *Poria, Radix Rehmanniae Preparata, Radix Peoniae Alba, Rhizoma Atractylodis Macrocephalae, Rhizoma Coptidis, Galla Chinensis, Radix Angelicae Dahuricae, Myrrha, Radix Saposhnikoviae, Rhizoma Et Radix Notopterygii, Radix Angelicae Pubescentis, Olibanum, Flos Carthami, Borneolum Syntheticum, Radix Angelicae Sinensis, Radix Glycyrrhizae Preparata.* The biological spray film of the invention has remarkable efficacies such as rapid hemostasis, sterilization, and healing of various types of skin wounds (Shandong Baima Marine Biotechnology Co. Ltd., 2017).

19.4 PATENTS OF MARINE PRODUCTS HAVING ANTI-INFLAMMATORY ACTIVITY

Studies have investigated the occurrence of pharmacologically active compounds such as anti-inflammatory, anti-parasitic, antibiotic, anti-cancer, antiviral, and antithrombotic agents in marine organisms. For instance, a sulfated polysaccharide from the hydrolysates of blacklip abalone (Haliotis rubra) exhibits anticoagulant, antithrombotic, and anti-inflammatory properties (Suleria et al., 2017; Suleia et al., 2017). The non-specific immune response that takes place in response to any form of injury includes inflammation, which is the body's initial line of defense against an injury (Ferrero et al., 2017). In this scenario, mast cells and chemokines are activated and migrate to the site of injury, causing vasodilation through the production of histamine. Chemical mediators such as histamine, prostacyclin, nitric oxide, and neutrophils then penetrate the site of injury via diapedesis. Macrophages play an active role in removing the damaged cells and tissues at the injury site, while B and T cells are released to further fuel the inflammatory response (Kemp et al., 2017). Discovering appropriate substances that can hinder the inflammatory processes of the body is a challenging task due to the intricate regulation of inflammation, and its imbalance can lead to chronic inflammation or, in extreme cases, fatality. However, it is a crucial mechanism for the body's defense against various infections (Gonzalez et al., 2015). Anti-inflammatory drugs are employed to treat inflammatory responses by blocking pro-inflammatory cytokines and other inflammatory mediators, or by promoting the production of anti-inflammatory mediators (Litalien et al., 2011). Numerous natural and synthetic anti-inflammatory drugs are available in the market, but recent studies have highlighted marine organisms as an excellent source of innovative natural anti-inflammatory agents. These novel anti-inflammatory compounds have been identified, and their structures have been determined. Additionally, synthetic chemicals based on these compounds have been developed (Gonzalez et al., 2015).

Emerging acidic lipids and phospholipids from *Nannochloropsis* sp. have an anti-inflammatory action and suppress cytokines (Keisuke et al., 2010). There are a number of patented anti-inflammatory compositions made from microalgae. It is claimed that a composition made of monogalactosyldiacylglycerol (GDG-EPA) and algae from the genera *Chlorella, Chaetoceros, Cyclotella, Ellipsoidon, Isochrysis, Nannochloris,* and *Nannochloropsis* has anti-inflammatory properties (Winget, 1994). A microalgal lipid formulation for topical dermal use is provided as a solution for skin inflammation treatment (Winget, 1997). Patented products include an anti-inflammatory skin spray made from an *Undaria pinnatifida* extract (Yong, 2008).

For an anti-inflammatory action, a manoalide, seco-manoalide, or dehydrosecomanoalide combination is offered (Jacobs & Faulkner, 1984). One of the patents outlines a technique for assessing the anti-inflammatory capabilities of a marine sponge compound based on cell aggregation inhibition (Philip & Gerald, 1986). A skin care product created from peptides extracted from sponges functions as a UV ray blocker and an antiphlogistic (Nakamura & Nakamura, 2004). The anti-cancer, antiviral, and anti-inflammatory properties of a formulation containing the novel bioactive chemical Sorbicillactone A from marine sponges have been studied (Bringmann et al., 2004). Excavatolide B, a diterpene found in the Briareum excavatum coral, and its analog have been found to be effective in treating inflammatory conditions due to their ability to inhibit the overexpression of TNF-α and/or IL-6 (Horngsheu et al., 2013). This coral extract is utilized for skin lightening, moisture retention, increased elasticity, anti-inflammation, and wound closure (Bingzhen et al., 2014). The cembrane diterpenes in *Lobophytum crissum* have an inhibitory effect on the overproduction of nitric oxide (NO) and function as a drug to treat illnesses (inflammation, allergic diseases, etc.). It works as a therapeutic treatment for inflammatory and allergy illnesses because it can stop the cellular production of nitric oxide (Miyamoto & Miyamoto, 2010).

Sesterterpene obtained from the sponge *Dactylopungia elegans* contains abundant active secondary metabolites, and alkaloid compounds are separated from the sponge. Chemical components reported in the literature mainly comprise sesquiterpene quinone/hydroquinone, sesquiterpene acids, macrolides, sesterterpenes, and other types of compounds. The sesterterpene compound has been found to inhibit the release of NO in RAW264.7 cells induced by LPS at a concentration of 10 μM. This effect is not due to cytotoxic activity and demonstrates anti-inflammatory properties. Therefore, the compound may serve as a potential lead compound for the development of anti-inflammatory drugs (Yu et al., 2020). The compound hypoxone A is obtained from a marine fungus *Hypoxylon rubiginosum* FS 5211 by fermentation and purification. In experiments, hypoxone A demonstrated significant inhibition of Nitric Oxide production, with an IC50 of 21.44±0.88 μM in a mouse macrophage RAW264.7 inflammation model induced by lipopolysaccharide (Chen et al., 2020).

The compound trieffusols C-E are obtained and extracted from a marine fungus called *Trichootrys effuse* FS 524 through fermentation. Based on experimental results, these substances have been found to inhibit the production of nitric oxide in a mouse macrophage RAW264.7 inflammation model induced by lipopolysaccharides. They exhibit significant anti-inflammatory activity, as demonstrated by their IC50 values of 51.9±1.4, 54.3±2.2, 55.9±1.6 and 65.5±1.3 μM. As a result, they can be employed in the production of anti-inflammatory medications (Liu et al., 2020). An additional invention discloses derivatives of indole diketopiperazine obtained from deep-sea fungi that inhibit the production of Nitric Oxide (NO) in macrophages RAW264.7 induced by LPS. Compound (1) has an IC50 value of 89 μM and compound (2) has an IC50 value of 20 μM. These derivatives can be utilized to produce anti-inflammatory medications, indicating that the indole diketopiperazine derivative obtained from deep-sea fungi in this invention has the potential to treat inflammation (Chen et al., 2020).

19.5 CONCLUSION

This chapter highlights the application of anti-cancer, anti-microbial, and anti-inflammatory agents from marine organisms and their patents. One, two, or a combination of compounds from the abundance of secondary metabolites present in marine organisms may be the source of the therapeutic function, which needs to be determined. A prospective supply of natural goods, chemicals, and medicines for therapeutic purposes is now the marine environment. Given the vast array of organisms, high biological diversity, and unexplored marine life, the sea has tremendous potential to produce more unique products. Oceans now serve as a possible source for novel medication leads thanks to scientific and industrial curiosity. Scientists have developed drugs of many kinds, but anti-cancer, anti-inflammatory, and anti-microbials are among the most crucial to address. Around the world, these lead compounds are at various preclinical and clinical testing stages. A number of chronic and incurable disorders, like cancer, can be treated with many medications derived from marine sources. They might start a new chapter in the successful and less expensive management of chronic illnesses. The preclinical and clinical data outcomes will determine how promising marine natural products will be marketed and commercially exploited after their identification, extraction, and mass manufacture. Identifying and isolating the bioactive chemical in certain species has been possible, but not in others. Therefore, it is necessary to investigate the region to discover and isolate bioactive compounds from marine species. A chemist can use the chemicals that have been identified as lead molecules to modify their structural properties and create applications that are both safe and effective. Analytical techniques must be developed in order to quantify and standardize the marine-derived chemicals. Consequently, a comprehensive strategy will aid in the development of more effective and secure marine-based pharmaceuticals.

The aim of this chapter on marine drug patents is to investigate the research on new chemical entities from marine sources and their associated biological activities. Developing synthetic analogs of active scaffolds is crucial to create therapeutically effective molecules as marine bioactives have the potential to be drug leads with diverse pharmacological properties. This chapter also provides insights into the rational design of marine scaffolds for preventing and treating various human diseases and aims to contribute to the development of novel chemical compounds with enhanced efficacy and safety for widely available drugs (Table 19.1).

TABLE 19.1 Patents of Compounds Obtained from Marine Sources Having Therapeutic Potential

Sl. No	Name/Type of compounds	Source	Type of Activity	Patent No. (Current Assignee)	Reference
1	Strepto-acid	*Streptomyces fulvissimus* ZZ406	Anti-cancer	CN107619803A (Zhejiang University ZJU)	Zhang et al. (2018)
2	Angucycline compounds	*Streptomyces rubrogriseus*	Anti-cancer	CN104744533A (South China Sea Institute of Oceanology of CAS)	Jianhua et al. (2017)
3	Phenazine compounds	*Streptomyces niveus*	Anti-cancer	CN103360329A (South China Sea Institute of Oceanology of CAS)	Jianhua et al. (2017)
4	Sulfur-containing diindolyl diketopiperazine compounds	*Acrostalagmus luteoalbus*	Anti-cancer	CN102911188A (South China Sea Institute of Oceanology of CAS)	Wang et al. (2015)
5	Indoles diterpene-like compound	*Xylocarpus granatum*	Anti-cancer	CN107827805A (Hainan Normal University)	Zheng et al. (2018)
6	Triterpenoid saponin-reducing compound	*Salicornia bigelovii*	Anti-cancer	CN103304622B (Nanjing Agricultural University)	Xin et al. (2015)
7	Stelliferin compounds	*Sponge (Polifera)*	Anti-cancer	WO2017122736A1 (Waseda University)	Yoichi et al. (2017)
8	Formula I compound	*Alternaria alternata*	Anti-cancer	CN106883113A (Zhejiang Meixin Holding Co., Ltd.)	Ma et al. (2017)
9	Cyclic dipeptides	*Streptomyces* sp.	Anti-cancer	CN105199989A (Xiamen University)	Zhao et al. (2015)
10	Polypeptide	*Bacillus marinus (*N11–8*)*	Anti-cancer	CN104610432A (Yellow Sea Fisheries Research Institute, Chinese Academy of Fishery Sciences)	Zheng et al. (2016)
11	Heteroterpene compounds I and II	*Phomopsis tersa* FS 441	Anti-cancer	CN111004251A (Guangdong Institute of Microbiology)	Liu et al. (2020)

TABLE 19.1 *(Continued)*

Sl. No	Name/Type of compounds	Source	Type of Activity	Patent No. (Current Assignee)	Reference
12	acremochlorin K, acremochlorin L, or acremochlorin M	*Acremonium sclerotiogenum*	Anti-cancer	CN111960937A (Guangxi University of Chinese Medicine)	Luo et al. (2020)
13	Tenellone K	*Phomopsis lithocarpus* FS 508	Anti-cancer	CN112500374A (Guangdong Institute of Microbiology)	Chen et al. (2021)
14	Polyketide lithocarpin E-G	*Phomopsis lithocarpus* FS508	Anti-cancer	CN112592350A (Guangdong Institute of Microbiology)	Chen et al. (2021)
15	Chloroanthraquinone	*Penicillium citrinum* HL-5126	Anti-microbial	CN106631803A (Hainan Normal University)	Zheng et al. (2017)
16	Isocoumarin compound	*Penicillium citrinum* HL-5126	Anti-microbial	CN106432169A (Hainan Normal University)	Zheng et al. (2017)
17	Polycyclic polyketone compounds	*Verrucosispora* sp.	Anti-microbial	CN103319496A (Institute of Microbiology of CAS)	Zhang et al. (2015)
18	Peptides	*Sea cucumber*	Anti-microbial	KR20210007075A (MEPSGEN Co., Ltd.)	Kang (2021)
19	Peptides	*Styela clava*	Anti-microbial	KR20210079616A (Hyunjin Engineering Co., Ltd.)	Lim (2021)
20	Chromophoric peptide	*Metapenaeus ensis*	Anti-microbial	CN112851751A (Jiangsu Ocean University)	Lu et al. (2021)
21	Indole alkaloids	*Penicillium crustosum* PRB-2	Anti-microbial	CN106892854B (Ocean University of China)	Li et al. (2019)
22	Pseudodepsidone compound	*Stereocaulon alpinum*	Anti-microbial	KR101452324B1 (Korea Institute of Ocean Science and Technology)	Yim et al., 2014)
23	Moronecidin like peptides	Polar fish	Anti-microbial	KR101924808B1 (Korea Institute of Ocean Science and Technology)	Shin et al. (2018)

TABLE 19.1 *(Continued)*

Sl. No	Name/Type of compounds	Source	Type of Activity	Patent No. (Current Assignee)	Reference
24	Sesterterpene	*Dactylosospongia elegans*	Anti-inflammatory	CN112159376A (Chinese People Liberation Army Naval Characteristic Medical Center)	Yu et al. (2020)
25	Polyketide Hypoxone A	*Hypoxylon rubiginosum* FS 5211	Anti-inflammatory	CN111217706A (Guangdong Institute of Microbiology)	Chen et al. (2020)
26	Trieffulsols C-E	*Trichobotrys effuse FS524*	Anti-inflammatory	CN111285758A (Guangdong Institute of Microbiology)	Liu et al. (2020)
27	Indole diketopiperazine derivative	*Aspergillus* sp. FS445	Anti-inflammatory	CN112707890A (Guangdong Institute of Microbiology)	Chen et al. (2020)
28	Phospholipid	*Nannochloropsis*	Anti-inflammatory	JP2015174850A (Saga Prefecture)	Keisuke et al. (2010)
29	eicosapentaenoic acid (EPA) bearing monogalactosyldiacylglycerol	Marine algae	Anti-inflammatory	WO1994024984A2 (WINGET, RODNER R)	Winget (1994)
30	Lobophytol acetate, cembranoid C, and dentic claride	*Lobophytum carssum, Lobophytum pauciflorum, Lobophytum denticulatum, Sinularia mayi*	Anti-inflammatory	JP2010222307A (Kyushu University NUC)	Miyamoto & Miyamoto (2010)

KEYWORDS

- **bio-actives**
- ***Escherichia coli***
- **IPR**
- **minimum inhibitory concentration**
- **patents**
- ***Phomopsis lithocarpus***
- ***Verrucosispora* sp**

REFERENCES

Bingzhen, S., Weixian, W., Zhihong, W., Yishan, W., Taiyan, Z., Yinbin, Z., et al. (2014). Coral extract, extraction method and use thereof, and coral extract skin care product. CN 104095883.

Bringmann, G., Gerhard, L., Muchlbacher, J., Muller, W. E. G., Schaumann, K., Steffens, S. (2004). New compounds sorbicillactone A and derivatives, useful as anti-tumor, antiviral and anti-inflammatory agents, obtained, e.g., by culturing Penicillium fungi. DE10238257.

Bushra, N., Aeysha, S., Syeda, L. R. (2018). Comparison of medicinally important natural products versus synthetic drugs-A short commentary. *Natural Product Chemistry Research, 6*(2), 308.

Cai, X. (2017). Shandong Tianwang Pharmaceutical Technology Co. Ltd., Marine bioactive composition and pharmaceutical preparation. CN106668077A.

Chen, Y., Chen, S., Zhang, W. (2020). Compounds tersaphenalone B and tersaphenalone E, preparation method thereof and application thereof in preparation of anti-tumor drugs. CN111808088.

Chen, Y., Li, S., Liu, H., et al. (2021). Compound tenellone K as well as preparation method and application thereof in preparation of anti-tumor drugs. CN112500374.

Chen, Y., Li, S., Liu, H., et al. (2021). Polyketone compound Lithocarpin E-G as well as preparation method and application thereof. CN112592350.

Chen, Y., Li, S., Liu, Z., et al. (2020). Indole diketopiperazine derivative, preparation method thereof and application thereof in preparing anti-inflammatory drugs. CN112707890.

Chen, Y., Zhang, J., Liu, H., et al. (2020). Marine-derived polyketone compound hypoxone A, preparation method thereof, and application of hypoxone A in preparation of anti-inflammatory drugs. CN111217706.

Cheuka, P. M., Mayoka, G., Mutai, P., & Chibale, K. (2017). The role of natural products in drug discovery and development against neglected tropical diseases. *Molecules, 22*(1), E58.

Choi, H. J., Chang, H. W., Kim, G. J., & Lee, X. (2017). Novel marine *Bacillus* sp. compounds isolated from the marine *Bacillus* sp. and pharmaceutical use of the compounds. KR2017003180A.

Danilovich, M. E., et al. (2018). Antarctic bioprospecting: In pursuit of microorganisms producing new anti-microbials and enzymes. *Polar Biology, 41*(7), 1417–1433.
David, B., Jean-Luc Wolfender, J. L., & Dias, D. A. (2015). The pharmaceutical industry and natural products: Historical status and new trends. *Phytochemistry Reviews, 14*(2), 299–315.
Dias, D. A., Urban, S., & Roessner, U. (2012). A historical overview of natural products in drug discovery. *Metabolites, 2*(2), 303–336.
Donadio, S., et al. (2010). Antibiotic discovery in the twenty-first century: Current trends and future perspectives. *The Journal of Antibiotics, 63*(8), 423–430.
Dyshlovoy, S. A., & Honecker, F. (2018). Marine compounds and cancer: 2017 updates. *Marine Drugs, 16*, 41.
Faulkner, D. J. (1995). Chemical riches from the ocean. *Chemistry & Industry, 39*(1), 680–684.
Ferrero-Miliani, L., Nielsen, O., Andersen, P., & Girardin, S. (2007). Chronic inflammation: Importance of NOD2 and NALP3 in interleukin-1 generation. *Clinical and Experimental Immunology, 147*(2), 227–235.
González, Y., Torres-Mendoza, D., Jones, G., & Fernandez, P. (2015). Marine diterpenoids as potential anti-inflammatory agents. *Mediators of Inflammation, 201*, 1–14.
HorngSheu, J., Sun Yang, N., Chi Wei, W., & Yao Huang, C. (2013). Pharmaceutical uses of diterpene excavatolide B from a coral or an analogue thereof. US8530513.
Jacobs, R. S., & Faulkner, D. J. (1984). Manoalide an anti-inflammatory analgesic marine natural product. US4447445.
Jianhua, J., Yongxiang, S., Hongbo, H., & Zhang, Y. (2017). South China Sea Institute of Oceanology Chinese Academy of Sciences. A class phenazine compounds and its application in preparing antineoplastic medicaments. CN103360329A.
Jianhua, J., Yongxiang, S., Jie, L., & Huang, H. (2017). South China Sea Institute of Oceanology Chinese Academy of Sciences. New angucycline compounds useful in preparing drugs for inhibiting bacterial infection and treating cancer. CN104744533A.
Jimeno, J., Faircloth, G., Sousa-Faro, J., Scheuer, P., & Rinehart, K. (2004). New marine derived anti-cancer therapeutics – A journey from the sea to clinical trials. *Marine Drugs, 2*(1), 14–29.
Kang, C. (2021). Novel anti-microbial peptide derived from anti-microbial peptides isolated from Korean sea cucumber and uses thereof. KR20210007075.
Katz, L., & Baltz, R. H. (2016). Natural product discovery: Past, present, and future. *Journal of Industrial Microbiology and Biotechnology, 43*(2–3), 155–176.
Keisuke, T., Takashi, Y., & Hiromi, T. (2010). Anti-inflammatory compounds. JP2015174850A.
Kemp, W., Burns, D., & Brown, T. (2017). From pathology: The big picture. Inflammation and repair. In W. Kemp, D. Burns, & T. Brown (Eds.), *Inflammation and Repair* (pp. 13–22). McGraw Hill Professional.
Khalifa, S. A. M., Elias, N., Farag, M. A., Chen, L., Saeed, A., Hegazy, M.-E. F., Moustafa, M. S., Abd El-Wahed, A., Al-Mousawi, S. M., Musharraf, S. G., & Huang, P. (2019). Marine natural products: A source of novel anti-cancer drugs. *Marine Drugs, 17*(9), 491. https://doi.org/10.3390/md17090491.
Kijjoa, A., & Sawangwong, P. (2004). Drugs and cosmetics from the sea. *Marine Drugs, 2*, 73–82. https://doi.org/10.3390/md202073.
Kim, S. K. (2013). *Marine Pharmacognosy: Trends and Applications* (1st ed.). CRC Press.
Li, C., Wang, X., Li, X., & Qingdao Hifun Marine Biological Technology Co. Ltd. (2015). Degerming eye lotion containing marine organisms (Patent No. CN104548077). China.
Li, D., Yu, G., & Zhu, T. (2019). Indole alkaloid compound as well as preparation method and use thereof (Patent No. CN106892854B). Ocean University of China.

Lim, A. (2021). Novel anti-bacterial peptide using peptide derived from *Styela clava* and the use thereof (Patent No. KR20210079616).

Litalien, C., & Beaulieu, P. (2011). The Pharmacological Basis of Therapeutics. In J. G. Hardman, L. E. Limbird, & A. G. Gilman (Eds.), *New York: McGraw-Hill Education* (pp. 41–72). McGraw-Hill Education.

Liu, H., Chen, S., Zhang, W., et al. (2020). Compound photeoids A and compound photeoids B, and preparation method thereof, and applications in preparation of anti-tumor drugs (Patent No. CN110642827).

Liu, H., Chen, S., Zhang, W., et al. (2020). Marine-derived meroterpenoid compounds I and II, preparation method thereof, and application of marine-derived meroterpenoid compounds I and II in preparation of anti-tumor drugs (Patent No. CN111004251).

Liu, H., Chen, S., Zhang, W., et al. (2020). Preparation method of compound trieffusols C-E and application of compound trieffusols C-E in preparation of anti-inflammatory drugs (Patent No. CN111285758).

Liu, H., Zhang, W., Xu, J., et al. (2019). Compound lithocarols A-F, preparation method and application thereof in preparation of anti-tumor drugs (Patent No. CN109336873).

Loureiro, R. J., & Malve, H. (2016). O uso de antibióticos e as resistências bacterianas: breves notas sobre a sua evolução. *Revista Portuguesa de saúde pública, 34*(1), 77–84.

Lu, M., Liu, X., Wang, S., et al. (2021). Anti-bacterial chromophoric peptide and preparation method and application thereof (Patent No. CN112851751).

Luo, X., Liu, Y., Wang, J., et al. (2020). Mixed source terpenoid, preparation method thereof and application thereof in preparation of anti-breast cancer or dihydroorotate dehydrogenase inhibitor drugs (Patent No. CN111960937).

Ma, Z., Ding, H., Ding, W., Zhang, D., et al. (2017). Compound with BRD4 inhibition activity, and preparation method and application of compound (Patent No. CN106883113A).

Malve, H. (2016). Exploring the ocean for new drug developments: Marine pharmacology. *Journal of Pharmacy & Bioallied Sciences, 8*(2), 83–91.

McEwen, S. A., & Collignon, P. J. (2018). Anti-microbial resistance: A one health perspective. *Microbiology Spectrum, 6*(2), 6–2.

Miyamoto, T., & Miyamoto, S. (2010). Coral-derived therapeutic agent for inflammation and allergic disease (Patent No. JP2010222307).

Nakamura, K., & Nakamura, K. (2004). Skin care preparation for external use (Patent No. JP2004250357).

Newman, D. J., & Cragg, G. M. (2016). Natural Products as Sources of New Drugs from 1981 to 2014. *Journal of Natural Products, 79*(3), 629–661. https://doi.org/10.1021/acs.jnatprod.5b01055.

Newman, D. J., & Gordon, M. C. (2016). Drugs and drug candidates from marine sources: An assessment of the current "state of play." *Planta Medica, 82*, 775–789. https://doi.org/10.1055/s-0036-1572277.

Partridge, S. R., Kwong, S. M., Firth, N., & Jensen, S. O. (2018). Mobile genetic elements associated with anti-microbial resistance. *Clinical Microbiology Reviews, 31*(4), e00088–e00117. https://doi.org/10.1128/CMR.00088-17.

Philip, B. D., & Gerald, W. (1986). Method for measuring anti-inflammatory properties of a composition (Patent No. US4605618).

Shandong Baima Marine Biotechnology Co. Ltd. (2017). Skin wound and infection treating marine organism spraying film and preparation method thereof (Patent No. CN106491568).

Shin, H. J., Lee, H. S., Lee, J. S., & Lee, Y. J. (2015). New compounds having anti-microbial activity and Marine Bacillus subtilis producing the new compounds (Patent No. KR2015000329A).

Shin, S. C., Park, H., & Kim, H. W. (2018). Novel anti-microbial peptides from Antarctic fishes (Patent No. KR101924808B1).
Shinde, P., Banerjee, P., & Mandhare, A. (2019). Marine natural products as source of new drugs: A patent review (2015–2018). *Expert Opinion on Therapeutic Patents, 29*(4), 283–309.
Shujuan, P., Houwen, L., Haiyan, L., & Fan Yang et.al. (2016). Dimeric Sesterterpenoids useful for preventing, diagnosing, detecting, protecting, treating and researching, e.g., tumor, diabetes and obesity (Patent No. CN102617596A).
Skropeta, D. (2008). Deep-sea natural products. *Natural Product Reports, 25*(6), 1131–1166. https://doi.org/10.1039/b800073h.
Suleia, H., Masci, P., Addepalli, R., Gobe, G., & Osborne, S. (2017). In vitro anti-inflammatory activities of blacklip abalone (Haliotis rubra) in RAW 264.7 macrophages. *Food and Agricultural Immunology, 28*(4), 711–724. https://doi.org/10.1080/09540105.2017.1355261.
Suleria, H., Masci, P., Addepalli, R., Chen, W., Gobe, G., & Osborne, S. (2017). In vitro anti-thrombotic and anti-coagulant properties of blacklip abalone (Haliotis rubra) viscera hydrolysatein RAW. *Analytical Bioanalytical Chemistry, 409*(17), 4195–4205. https://doi.org/10.1007/s00216-017-0452-8.
Suleria, H., Osborne, S., Masci, P., & Gobe, G. (2015). Marine-based nutraceutical: An innovative trend in the food and supplement industries. *Marine Drugs, 13*(10), 6336–6351. https://doi.org/10.3390/md13106346.
Wang, F., Zhang, S., Huang, Z., Tian, X., & Li, J. (2015). Sulfur-containing diindolyl diketopiperazine compounds and application thereof in preparing antineoplastic drugs (Patent No. CN102911188A).
Wang, Y., & Weihai Yushanfang Biotechnology Co. Ltd. (2015). Antineoplastic and immunity-enhancing ocean herbal medicine compound preparation and preparation method thereof (Patent No. CN105055473A).
Winget, R. R. (1994). Anti-inflammatory compositions containing eicosapentaenoic acid bearing monogalactosyldiacylglycerol and methods relating thereto (Patent No. WO 1994024984A2).
Winget, R. R. (1997). Anti-inflammatory compositions containing monogalactosyldieicosa-pentaenoyl glycerol and methods of relating thereto (Patent No. US5620962).
Wu, J., Wen, T., & Ding, P. (2018). Novel anti-tumor microtubule stabilizer compound Ananaside A and efficient extraction method thereof (Patent No. CN108676045).
Xin, Z., Zhao, Y., & Wang, H. (2015). Novel sea asparagus triterpenoid saponin reducing compound as well as preparation method and application thereof (Patent No. CN103304622B).
Yim, J. H., Kim, I. C., Han, S. J., Kim, D. G., Lee, H. S., Kim, H. U., Bataral, B. P., & Kim, T. G. (2014). Anti-microbial and antioxidative composition containing depsidone or pseudodepsidone compound originated from *Stereocaulon alpinum* (Patent No. KR 101452324B1).
Yoichi, N., Shigetaka, A., Daisuke, A., Yukari, K. T., & Teppei, S. (2017). Marine organism-derived extract, compound, and medical composition having niche formation suppressing activity of leukemic stem cells (Patent No. WO2017122736A1).
Yong-Ki, H. (2008). An extract of Undaria pinnatifida having anti-inflammatory activity (Patent No. KR100807758).
Yu, H., Lin, H., Yang, F., et al. (2020). Sesterterpene compound and application thereof in preparing anti-inflammatory drugs (Patent No. CN112159376).

Zhang, L., Liu, X., Wang, Q., Song, F., et al. (2015). Polycyclic polyketone compounds from marine verrucosispora sp., and preparation method and application thereof (Patent No. CN103319496).

Zhang, Z., Chen, M., Chai, W., & Lian, X. (2018). Preparation of *Streptomyces fulvissimus* acid and *Streptomyces fulvissimus* ketone and medicinal application thereof (Patent No. CN107619803A).

Zhao, J., Zhao, Y., Chen, J., & Luo, L. (2015). Cyclic dipeptides in marine *Streptomyces* sp. and preparation method thereof (Patent No. CN105199989A).

Zheng, C., Huang, G., Bai, M., Chen, G., et al. (2018). A wood-neem fruit mangrove plant fungal origin indole diterpene compounds, preparation method and application (Patent No. CN107827805A).

Zheng, C., Huang, G., Chen, G., & Bai, M., et.al. (2017). Mangrove fungus derived chloroanthraquinone compound and application of mangrove fungus derived chloroanthraquinone compound serving as anti-bacterial agent (Patent No. CN106631803A).

Zheng, C., Huang, G., Chen, G., & Zheng, C., et.al. (2017). Isocoumarin compound with anti-vibrio activity function and crystal thereof (Patent No. CN106432169A).

Zheng, L., Sun, M., Chen, S., & Yi, Y., et.al. (2015). Bacillus sp. N11-8 and active component thereof having anti-tumor action (Patent No. CN103667097).

Zheng, L., Yang, K., Sun, M., & Lyu, M., et.al. (2016). New peptide from marine Bacillus sp. N11-8, useful for treating and preventing cancer (Patent No. CN104610432A).

CHAPTER 20

Bioactive Compounds from Marine Organisms and Their Potential Applications

SHAHID NAZIR WANI[1,2] and ISHRAT ZAHOOR[3]

[1]*Chitkara College of Pharmacy, Chitkara University, Rajpura, Punjab, India*

[2]*Aman Pharmacy College, Dholakhera, Udaipurwati Jhunjhunu, Rajasthan, India*

[3]*MM College of Pharmacy, Maharishi Markandeshwar (Deemed to be University), Mullana Ambala, Haryana, India*

ABSTRACT

One of the most precious natural resources on the planet is the ocean. More than 70% of the Earth's surface is made up of the ocean, and the variety of the marine environment includes the presence of potentially bioactive compounds. Oceans and seas are also a rich source of new therapeutic molecules and are thought to contain the largest remaining supply of beneficial natural compounds that could be employed as active ingredients in a wide range of dosage forms. Marine species have been shown to have a wide range of physiologically active substances with distinctive structures and pharmacological properties. The natural bioactive compounds found in these species have a wide array of potential uses in different industries, including medicine, nutraceuticals, cosmetics, and food, among others. Extracts and purified compounds from these organisms can be screened for a wide variety of biological activities with therapeutic and industrial significance, including anti-inflammatory, anti-cancer, antiviral, antibacterial, anticoagulant, ion channel/receptor modulation, and plant growth regulation, using different screening techniques. Because

Handbook of Research in Marine Pharmaceutics: Exploring Oceanic Microbial Diversity for Human Health and Wellness. Md. Faiyazuddin & Khalid Rehman Hakeem (Eds.)

of their antioxidant, anticancer, and antibacterial activities, among others, many of these bioactive compounds have been implicated in health benefits. They also play a vital role in the prevention of a number of disorders. The sea organisms or the accompanying microorganisms both create these beneficial compounds. Although there are many natural products derived from marine species, few have a track record in the marketplace. These compounds have also demonstrated importance in pharmaceutical and biotechnological applications. In recent years, the powerful medicinal uses of bioactive compounds, such as steroids, isoprenoids and non-isoprenoid compounds, terpenoids, quinones, nitrogen heterocyclics, brominated compounds, nitrogen-sulfur heterocyclics, antibiotics, etc., have been obtained from marine poriferans, arthropods, cnidarians, mollusks, vertebrates, echinoderms, and seaweeds. Marine research strives to raise public awareness of marine biodiversity and gain a better understanding of how to source marine biotechnology products responsibly. In order to take advantage of this vast chemical variety, substantial work must be done. The chapter discusses a wide spectrum of bioactive molecules that have been extracted from various marine species for use in medicinal drugs and their potential applications.

20.1 INTRODUCTION

The ocean, which is known as the "mother of origin of life," also serves as the reservoir of structurally distinct natural compounds that are mostly collected in living beings. A number of these molecules exhibit therapeutic activity and are beneficial in the development and identification of bioactive molecules, especially for life-threatening conditions such as acquired immunodeficiency syndrome, cancer, arthritis, etc., while other molecules have been produced as analgesics or to treat inflammatory disorders (Jha & Zi-Rong, 2004). In addition, some of these compounds have been found to be advantageous in the treatment of life-threatening conditions. The ocean is believed to be a reservoir of possible pharmaceuticals, and a number of the bioactive molecules and secondary metabolites that may be found there have the potential to be used in medicine. Oceans occupy 70% of the earth's surface, while the marine environment accounts for over 95% of the entire biosphere on the globe. 33–34% of animal phyla occur in the marine environment (Sima & Vetvicka, 2001; Datta, Talapatra, & Swarnaka, 2015). Nearly 80% of the world's flora and fauna are present in the oceans (Datta, Talapatra, & Swarnaka, 2015). The variety of species is astounding, and in tropical areas, there are approximately 1000 distinct species that may be found on just one square meter of land. Environmental pressures such as competition for space, nourishment, and self-defense have resulted in the synthesis of a varied variety of substances that are collectively referred to as secondary metabolites. These species employ secondary metabolites as a means of communicating with one another and with their surroundings. The evolution of these molecules occurred within the context of symbiotic relationships. Since the beginning of medical practice, several diseases that affect humans have been successfully treated using chemical compounds that were obtained from animals, plants, and microbes (Koehn & Carter, 2005). A great number of novel pharmaceuticals derived from microbes, marine creatures, terrestrial plants, invertebrates, and vertebrates have recently been introduced to the market. These medications, along with a number of other natural products or their analogs that are now being tested in clinical settings, continue to show the significance of molecules derived from natural sources in the process of novel drug development (Chin, Balunas, Chai, & Kinghorn, 2006; Kumar & Pal, 2016). Over the course of the last couple of decades, natural compounds extracted from marine sources have captured the interest of chemists and biologists from every corner of the globe. There have been roughly more than 6,800 papers reporting the invention of over 16,000 marine natural products that were extracted from marine species (Datta, Talapatra, & Swarnaka, 2015). In addition to these articles, there are around an additional 9,000 papers that address the topic of marine products (Datta, Talapatra, & Swarnaka, 2015). At this time, over 100 novel products have been investigated in clinical trials, the majority of which are intended to treat cancer or infectious diseases. Even more intriguing is the fact that the use of methods from molecular biology is leading to an increase in the availability of new chemicals that can be easily generated in yeasts or bacteria (Kumar & Pal, 2016). In the process of screening libraries for molecules that are very similar to drugs, combinatorial chemistry techniques that are dependent on natural product-like scaffolds to produce new molecules are very beneficial. The simplicity by which natural products may be integrated into the drug discovery process and data mining is now being addressed by the development of a number of different screening strategies. In addition, the methods of virtual screening were used for databases including natural products (Harvey, 2008). Marine species have only a very recent history of use in the diagnosis, therapy, and prevention of human maladies. An anticancer medicine known as cytosine arabinoside, also known as Ara-C, and an antiviral drug known as adenine arabinoside, sometimes known as Ara-A, were among the initial bioactive chemicals derived from marine organisms and were licensed in 1965 (Harvey, 2008). Because of their discovery, scientists have dedicated a significant amount of time and effort to researching aquatic species during the last three decades. Then, research into the development of pharmaceuticals from marine species is advancing at a rapid rate and currently incorporates a wide range of academic fields, like biochemistry, organic chemistry, biology, ecology, and pharmacology (Capon, 2001; Haefner, 2003). There have been reports of thousands of newly discovered marine natural products. These products fall into the chemical categories of steroids, nonisoprenoids, quinones, terpenoids, isoprenoids, brominated chemicals, nitrogen sulfur heterocyclics, and nitrogen heterocyclics. There have been a lot of significant advancements in this area of study over the last five years, and a lot of noteworthy discoveries of various bioactive chemicals from *in-vitro*, *in-vivo*, as well as in clinical trials that have been reported (Perdicaris, Vlachogianni, & Valavanidis, 2013). Secondary metabolites are a kind of bioactive chemical that is formed mostly by marine species. These might be separated into the categories of isoprenoids, nonisoprenoids, isoprenoids, quinones, steroids, brominated chemicals, terpenoids, nitrogen sulfur heterocyclics, and nitrogen heterocyclics (Datta, Talapatra, & Swarnaka, 2015). A wide variety of these compounds may also be seen as primitive forms of defense. Because of their antiviral, antiprotozoal, antimicrobial, antihelminthic, antifungal, and anticancer activities, bioactive components produced by marine species like protozoans and invertebrates such as poriferans, arthropods, cnidarians, mollusks, annelids, and echinoderms have garnered a lot of attention (Datta, Talapatra, & Swarnaka, 2015). Numerous studies are being conducted on a broad spectrum of metabolites, many of which have peculiar structures and participate in a wide range of biological mechanisms (Sima & Vetvicka, 2001; Martínez-García, Díaz-Valdés, Ramos-Esplá, Salvador, Lopez, Larriba, & Antón, 2007; Alam, Sarkar, & Gomes, 2012; Kiran, Siddiqui, Khan, Ibrar, & Tushar, 2014).

20.2 AVAILABILITY OF MARINE PRODUCTS AND ETHICAL ASPECTS

The ocean is home to a diverse array of living things, each of which has a unique set of physiological characteristics and the capacity to modify its behavior in response to its surroundings. The marine habitat offers a plethora of possibilities for the development of new products from the sea. It is estimated that the marine habitat includes roughly 13,000 compounds, of which only 3,000 compounds are known to possess active characteristics (Singh, Parida, Debbarma, Ray, & Banerjee, 2020). It comprises around

80% of varied plant and animal species. Such genetic variety manifests therapeutic characteristics, which generates an encouraging new drug discovery from the biologically accessible marine living things. A further 12% of pharmaceutical goods are produced from biologically accessible microorganisms found in marine environments, in addition to the approximately 25% of medicinal products that are acquired from plant natural products (Jha & Zi-Rong, 2004). In general, the natural chemicals found in marine habitats function as secondary metabolites. The maritime habitat has a diverse array of temperature ranges, pressure ranges, and nutrient ranges, and it includes vast photic and aphotic regions. This considerable diversity gives novel medicinal agents that may be used in commercial applications. Products derived from natural sources have been utilized for different purposes, including but not limited to foods, perfumes, pigments, pesticides, and medications. Terrestrial plants have long been considered the primary source of medicinally valuable compounds, particularly in the context of traditional or folk medicine, because of the ease with which they may be obtained. Approximately 25% of all pharmaceutical sales are comprised of pharmaceuticals that are obtained from natural products of plant origin, and an additional 12% are comprised of drugs that are based on natural products that are generated by microbes (Jha & Zi-Rong, 2004). The marine environment encompasses a broad range of pressure (1–1000 atm), temperature (from below freezing to around 350°C), and nutrient levels (from oligotrophic to eutrophic), and it comprises significant photic and non-photic regions. The pressure range in the marine habitat ranges from 1 to 1000 atm. This vast variety has made it possible for significant speciation to occur at all phylogenetic levels, ranging from bacteria to mammals. Research into the utilization of natural products obtained from marine sources as pharmacological agents is still in its early stages, even though the marine environment has a biodiversity that considerably surpasses that of the terrestrial environment. This may be attributed to the absence of an ethno-medical history as well as the challenges that are associated with the gathering of aquatic species (Jha & Zi-Rong, 2004). It is now feasible to gather marine samples as a result of the introduction of innovative diving methods, remotely controlled devices, and other such innovations. Over the course of the last 10 years, more than 5000 new chemicals have been obtained from shallow waters all the way down to depths of 900 meters (Haefner, 2003). The existence of several marine species is in jeopardy as a result of human activity, despite the numerous benefits that come from the biological variety of marine species. The main stresses that lead to a loss of biodiversity include uncontrolled use of marine resources, excessive fishing, construction of coastal areas, pollution, the introduction of invasive alien species, and warming of the planet (Kiuru, D'Auria, Muller, Tammela, Vuorela, & Yli-Kauhaluoma, 2014). As more people became aware of the dangers posed by human activity, more politicians and scientists came to the conclusion that drastic measures needed to be done in order to restore ecosystems and natural systems. It is essential to give full thought to the many ethical concerns that may arise throughout the process of marine biodiscovery and product development (Harrell, 2017). In the process of converting marine biodiversity into industrial practicality with the goal of developing breakthrough cosmetic, nutraceutical, and pharmaceutical goods, safety and sustainability are two of the most important principles to keep in mind from an ethical point of view. These fundamental principles are connected to the notion of trust on both a conceptual and experiential level, with the purpose of both constructing and maintaining confidence in the output of the research and industrial operations on all levels. All of the most important processes, such as the various aspects of research, the assessment of industrial feasibility, the development of new products, testing, and the introduction and marketing of those products, are all associated with a perspective of sustainability, which may also include gaining and maintaining trust. Therefore, the idea of trust is predicated on the practice of conveying sustainability, not only in research processes aimed at achieving health but also in relation to concerns pertaining to the environment (Kiuru, D'Auria, Muller, Tammela, Vuorela, & Yli-Kauhaluoma, 2014).

20.3 BIOACTIVE COMPOUNDS FROM MARINE ORGANISMS

20.3.1 ANTI-CANCER COMPOUNDS

Sponges possess a significant quantity of cytotoxic chemicals such as bryostatins. Other chemicals of this kind have been found in sponges. These include the alkaloid-based anti-cancer agent Niphatesine D (Fig. 20.1), which was taken from a *Niphates* species; the *Epinardins,* which were obtained from a species that has not yet been identified; and the globostellatic acids A to D, which are isomalabaracane triterpene components of *Stelletta globostellata* (Harvey, 2008). The macrolides recovered predominantly from the *Bryozoan Clavelina picta* and referred to as bryostatins are perhaps the most well-known of the chemicals that have the potential to be used as anticancer treatments. Some of these macrolides have also been taken from tunicates and sponges. Active chemicals found in tunicates include the cytotoxic alkaloids clavepictine-A (Fig. 20.1) and -B, which were reported for *Clavelina picta* (Agami, Couty, Evano, Darro, & Kiss, 2003), and the cytotoxic dimeric disulfide alkaloid polycarpine (Fig. 20.1), which was recovered from *Polycarpa clavata*. Both of these alkaloids were isolated from *tunicates* (Rudi, Evan, Aknin, & Kashman, 2000). Bacteria found in marine flora have been shown to produce anticancer drugs, such as bryostatins, eleutherobin, discodermolide, and sarcodictyin (Blunden, 2001). Arenicolides is a kind of polyunsaturated macrolactone that has 26 members and is generated by an obligate marine actinomycete called *Salinispora arenicola* strain CNR-005 that was derived from the marine sediment (Olano, Méndez, & Salas, 2009). There are a huge number of triterpenoids, each of which has been shown to demonstrate cytotoxicity against a different kind of tumor cell, and anticancer activity in preclinical animal models, and each of which has been studied (Li, Himaya, & Kim, 2013). It has been reported that various chemicals extracted from seaweeds have been discovered to slow the development of cancer. These observations have led to the conclusion that these compounds have anti-cancer properties. Seaweeds contain bioactive substances that may induce apoptosis, which leads to the death of cancer-causing cells. The anticancer effects of the extracted bioactive substances are achieved by a variety of different modes of action. These mechanisms include the prevention of carcinogenic cell growth, invasion, and metastasis. It is possible for apoptosis to be triggered by either an internal mechanism (which is mediated by mitochondria) or an extrinsic pathway (which is mediated by death receptors) (Singh, Parida, Debbarma, Ray, & Banerjee, 2020). The accidental invention of the nucleosides, spongothymidine and spongouridine, which were extracted from the Caribbean sponge *Tethya crypta*, led to the development of marine-based cytostatic drugs. These nucleosides were used to create spongouridine and spongothymidine. Based on this development, the synthetic reference cytosine arabinoside cytarabine has been developed as an efficient antileukemic medication (Nigam, Suleria, Farzaei, & Mishra, 2019). The chemical known as trabectedin is an alkaloid belonging to the tetrahydroisoquinoline family. The first anticancer chemical compound obtained from marine sources was Trabectedin which got approval in the European Union (EU) in 2007, it was given this permission in order to treat relapsed cases of platinum-sensitive ovarian cancer and sarcoma (soft-tissue) (Mayer, Glaser, Cuevas, Jacobs, Kem, Little et al., 2010).

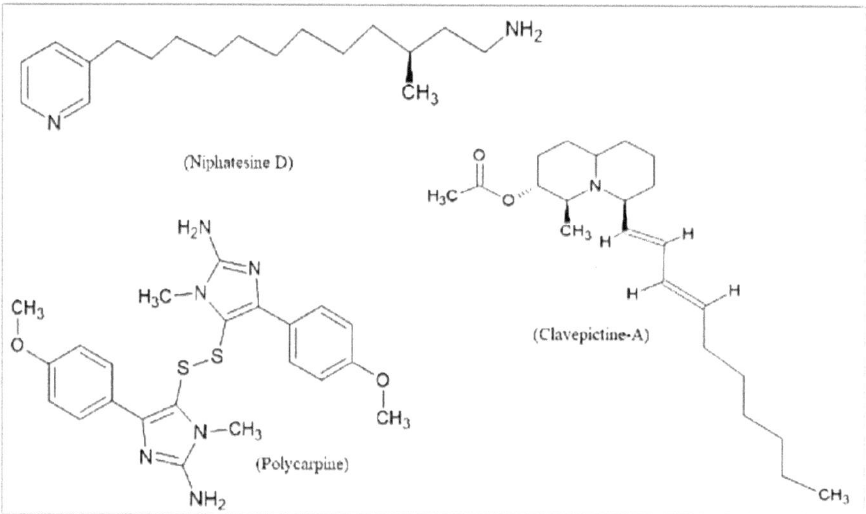

FIGURE 20.1 Anti-Cancer bioactive compounds from marine organisms.

20.3.2 ANTI-BACTERIAL COMPOUNDS

The antimicrobials that were obtained from the sea species belong to the category of fatty acid, cyclic polysulfide, phenol, terpenoid, phthalate, lactone, indole derivative, sterol, sulfated glycoside, protein, acetylene, steroidal glycoside, polysaccharide, terpene, glycerol derivative, glycoproteins, etc. (Kumar & Pal, 2016). Antimicrobial peptides are a "new" family of antimicrobial chemicals that have the potential to be used in the pharmaceutical industry. This is due to the fact that they have a wide spectrum of action against bacteria, fungi, and viruses. Their molecular targets include the membranes of bacteria as well as the molecules found inside cells, such as DNA, RNA, and proteins (Sperstad, Haug, Blencke, Styrvold, Li, & Stensvåg, 2011). Donia and Hamann have conducted an exhaustive study to investigate the possibility of anti-infective compounds originating in marine flora (Donia & Hamann, 2003). Squalamine, the very first aminosterol extracted from the *Squalus acanthias* (Squalidae) commonly known as the dogfish shark, exhibited significant antibacterial action, with a minimal inhibitory concentration (MIC) of 1·0 µg/mL. It was also capable of inhibiting *Staphylococcus aureus* and showing properties that were antiangiogenic and antitumor. It is now being tested in human patients suffering from advanced non-small cell lung cancer as part of clinical studies. Cribrostatins (Fig. 20.2), another chemical derived from the blue sponge *Cribrochalina* sp., showed powerful anticancer and antibacterial properties. Cribrostatin demonstrated significant inhibitory efficacy against *Neisseria gonorrheae*, with a MIC of 0·09 μg/mL (Pettit, Knight, Collins, Herald et al., 2000). Bromosphaerone (Fig. 20.2), a powerful antibacterial diterpene, was discovered in *Sphaerococcus coronopifolius*. It demonstrated antibacterial actions against *Staphylococcus aureus* at a MIC of 0.047 μg/mL (Etahiri, Bultel-Poncé, Caux, & Guyot, 2001). The marine fungus that produced the pest alone was extracted from the surface of the brown alga *Rosenvingea* sp. With a MIC of 0.037 μg/mL against methicillin-resistant *Staphylococcus aureus* and a MIC of 0.078 μg/mL against vancomycin-resistant *Enterococcus faecium*, members of the genus *Pestalotia* exhibited significant antibacterial activities (Cueto, Jense, Kauffman, Fenical, Lobkovsky, & Clardy, 2001). Jorumycin is a dimeric isoquinoline alkaloid that was extracted from the mucus and mantle of the *Jorunna funebris* nudibranch that was found in the Pacific Ocean. At a level of 0.050 μg/mL, it was able to suppress the development of a variety of gram-positive bacteria, such as *Bacillus subtilis* and *Staphylococcus aureus*, and the inhibition zone was 16 mm in diameter. In spite of its great efficacy, its development as a medication was hampered by its cytotoxic actions, which had an IC50 of 0.012 μg/mL, as well as the difficulties of creating a pure and stable form (Fontana, Cavaliere, Wahidulla, Naik, & Cimino, 2000). *Spongosorites* sp. was used to obtain a total of seven monoindole derivatives, five of which were distinct indole pyruvic acid derivatives (Bao, Zhang, Lee, Hong, Lee, & Jung, 2007). Additionally, the bis (indole) alkaloids of the hamacanthin and topsentin classes were extracted from this plant. Antibacterial activities of the hamacanthin class were shown to be much more powerful than that of the topsentin class (Oh, Mar, Kim, Kim, Oh et al., 2005). In the recent past, research has been done to investigate the possibility of using symbiont microbes as an alternate source for the synthesis of host bioactive metabolites. A marine Gram-positive *Bacillus* licheniformis SAB1 bacterium that was characterized by 16S rDNA was obtained from the sponge *Halichondria* sp. and had considerable antimicrobial properties (Devi, Wahidullah, Rodrigues, & Souza, 2010). Johnson and his colleagues investigated the *in vitro* antimicrobial properties of sea sponges of the *Zygomycale* sp. species that were obtained off the coast of Kanyakumari. The sponge extract was evaluated against nine human bacterial pathogens and four human fungal pathogens, and the results indicated that the samples exhibited powerful antibacterial properties against bacteria such as *Klebsiella pneumoniae, Bacillus megaterium*, and *Streptococcus pyogenes,* and it showed antifungal action against all pathogenic test strains. In addition, the extract showed antiviral activity against all pathogenic test strains (Johnson, Citarasu, & Manjusha, 2012).

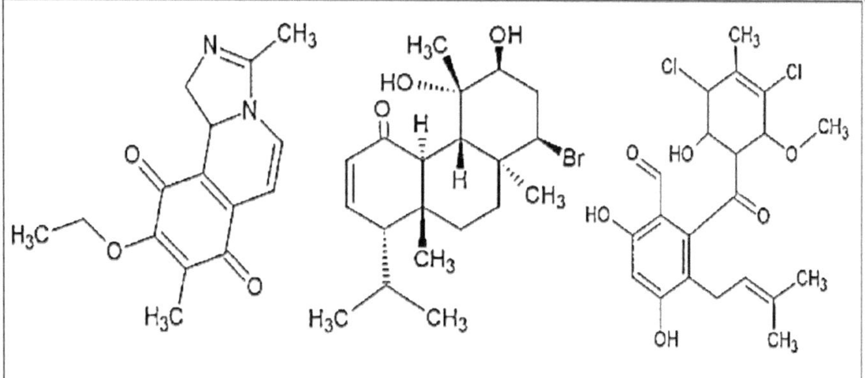

Cribrostatin 6 Bromosphaerone Pestalone

FIGURE 20.2 Anti-bacterial bioactive compounds from marine organisms

20.3.3 ANTI-INFLAMMATORY COMPOUNDS

Even though some compounds have been extracted from cyanophytes (Jha & Zi-Rong, 2004; Beltran & Neilan, 2000), the vast majority of these chemical substances have been extracted from freshwater species. This is because freshwater species are much simpler to cultivate than marine species. Debromoaplysiatoxin (Fig. 20.3) and Lyngbyatoxin-A are two strongly inflammatory metabolites that have structurally distinct differences from each other. Both of these metabolites were separated from toxic strains of Lyngbya mausculata that were collected in Hawaii (Jha & Zi-Rong, 2004). Anatoxin-a was obtained from *Anabaena ciecinalis* (Beltran & Neilan, 2000). It has been shown that the extracellular sheath pigment known as scytonemin, which was first extracted from the *cyanobacterium Stigonema spp.*, had both anti-proliferative and anti-inflammatory characteristics (Stevenson, Capper, Roshak, Marquez et al., 2002; Stevenson, Capper, Roshak, Marquez, Eichman et al., 2002). *Pseudopterocin-E* is a tricyclic diterpene pentoside that is derived from gorgonians belonging to the genus *Pseudopterogorgia.* It has analgesic and anti-inflammatory properties with the same level of effectiveness as the industry standard *indomethacin* (Jha & Zi-Rong, 2004). It has been discovered that the red algae species *Gracilaria textorii* and *Gracilaria* verrucosa both contain elements that have the capacity to reduce inflammation. *Gracilaria tenuistitata* has been shown to be beneficial in reducing inflammation caused by viruses (Singh, Parida, Debbarma, Ray, & Banerjee, 2020). C-phycocyanin is a biliprotein that was isolated from *Spirulina platensis*. It was successful in reducing inflammation by limiting the production of pro-inflammatory cytokines and by inhibiting the expression of cyclooxygenase-2 and inducible nitric oxide synthase. Both of these mechanisms are important in the regulation of inflammation. (Singh, Parida, Debbarma, Ray, & Banerjee, 2020) Research has shown that the brown algae species *Lobophora variegate,* which contains a significant quantity of the fucann compound, has an anti-inflammatory effect. One study's findings revealed that glycyrrhetinic acid has the capability of interacting with a broad variety of xenobiotics or endogenous compounds that are substrates for CYP2C9, CYP2C19, and CYP3A4 (Zhao, Ding, Cao, & Cao, 2012). It not only suppresses the action of mammalian pols, like pol λ, but it also lowers the generation of tumor necrosis factor-α (TNF-α), inhibits the stimulation of nuclear factor Kappa B (NF-κB), and inhibits the inflammation that is caused by tissue plasminogen activator (TPA) in mouse ears. As a result, glycyrrhetinic acid could function as an anti-inflammatory drug due to its ability to inhibit pol λ (Ishida, Mizushina, Yagi, Irino et al., 2012). The gas chromatography and Mass Spectrometry (GC-MS) analysis revealed that the marine sponge *S. halichondriodes* contains trace quantities of glycyrrhetinic acid (Fig. 20.3). Glycyrrhetinic acid is a pentacyclic triterpenoid derivative of the beta-amyrin type that is produced through the hydrolysis of glycyrrhizic acid, which is extracted from the herb liquorice. This is the first report that demonstrates the existence of glycyrrhetinic acid in any marine sponge. The ethyl acetate extract of the sponge demonstrated significant anti-inflammatory action in a study with rats. The rats were given the extract to test for its anti-inflammatory potential (Kumar & Pal, 2016).

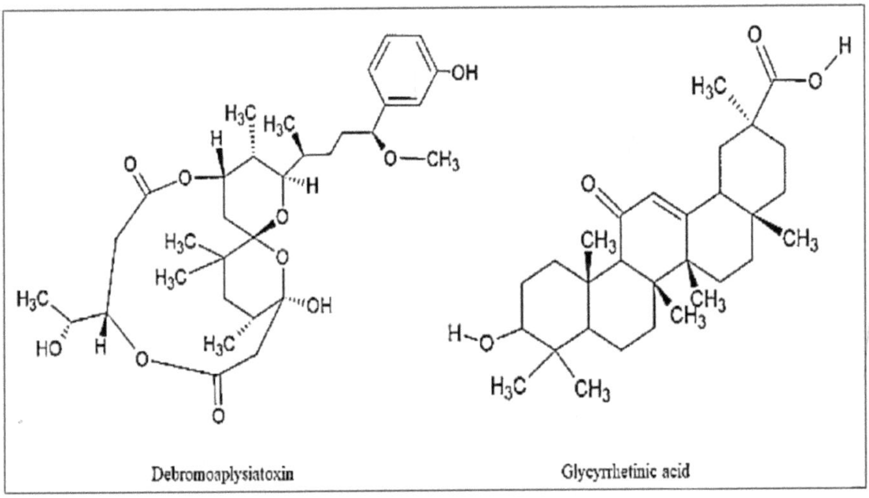

FIGURE 20.3 Anti-inflammatory compounds from marine organisms

20.3.4 ANTI-VIRAL COMPOUNDS

Ara-C, vidarabine (Ara-A or adenine arabinoside) (Fig. 20.4), an antiviral molecule that was licensed by the US FDA in 1976 but is not presently on the market, and zidovudine (or azidothymidine, AZT), one of the most prominent anti-AIDS medications, are all drugs that were derived from nucleoside spongouridine. The so-called pharmacodiversity may be regarded as the most significant among chemicals generated from sponges (Kiuru, D'Auria, Muller, Tammela, Vuorela, & Yli-Kauhaluoma, 2014). A wide variety of antiviral activity may be found in marine-derived polysaccharides, in addition to their oligosaccharide derivatives with lower molecular weights. The replication of enveloped viruses, arenaviruses, rhabdoviruses, orthopoxviruses, and other viruses may be inhibited by sulfated polysaccharides (SPs). The antiviral action of algal SPs is impacted by their conformation, molecular weight, component sugar, degree of sulfation, and chemical composition. The capacity of sulfated polysaccharides to suppress the new HIV virus has been demonstrated and documented (Singh, Parida, Debbarma, Ray, & Banerjee, 2020). There are three main forms of carrageenans: λ-, κ-, and ι-carrageenan. Each type of carrageenan has its unique inhibitory action on a certain virus. Additionally, it aids in the prevention of the beginning stages of human papillomavirus infection (HPV). In mice, low molecular weight carrageenan and its derivatives have been demonstrated to reduce pulmonary edema caused by the influenza virus (FM1). Carrageenan has a highly broad antiviral potential, making it effective against enveloped as well as non-enveloped viruses (Wang, Wang, & Guan, 2012).

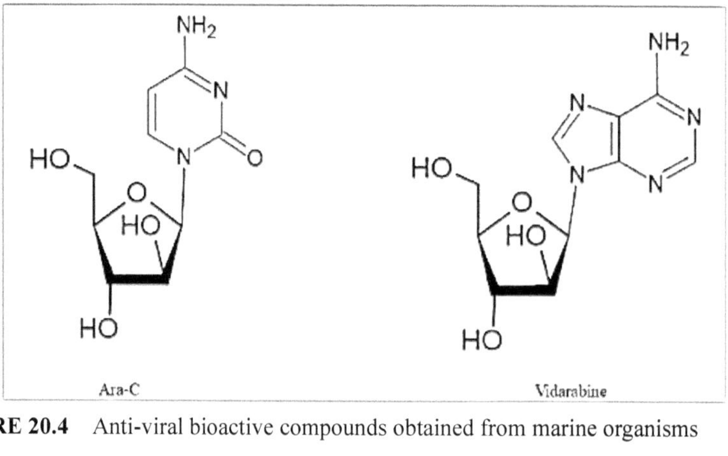

FIGURE 20.4 Anti-viral bioactive compounds obtained from marine organisms

20.3.5 ANTIFUNGAL COMPOUNDS

Amino alcohols, marine macrolides, and related heterocyclic structures such as peptides, oxazoles, and other alkaloids, and even a few representative polyketides and terpenoids are all included in the category of antifungal substances that originate from marine organisms (Kumar & Pal, 2016). The very first example of a cyclodepsipeptide that was obtained from a sponge and discovered as being from a *Jaspis* sp. is called jasplakinolide. Jasplakinolide (Fig. 20.5) was extracted from a sponge. It is a 19-membered macrocyclic depsipeptide having a specific *in-vitro* antimicrobial effect and a MIC of 25 μg/mL against *Candida albicans*. It is also known as jaspamide. It was shown that a 2% solution of jasplakinolide had an *in-vivo* topical efficacy against a *Candida* vaginal infection in mice that was comparable to that of miconazole nitrate (Kumar & Pal, 2016). It has been shown that the majority of antifungal chemicals derived from marine origin are cytotoxic. As a direct result of this, they have not been recognized to have any potential for use in therapeutic applications. Phytosphingosine (Fig. 20.5) is the primary sphinganoid base that may be observed in the sphingolipids of higher plants as well as many other invertebrate organisms. It is the homologous 4-hydroxy analog of D-sphinganine. When it comes to antifungal action, the chain length of sphingoid bases is quite essential. In a comparative study, the antifungal action of sphingolipids against *Candida glabrata* was discovered to be minimal for short-chain 2-amino-3-alkanol homologs, but it enhanced with chain length (C18). This was discovered as a result of the findings that the antifungal action of sphingolipids enhanced with chain length (Nicholas, Li, MacMillan, & Molinski, 2002). Kabiramide C and ulapualides A and B were the initial members of a special family of trisoxazole macrolides in the antifungal class of chemicals to be discovered and reported. In a sponge from Western Australia belonging to the genus *Phorbas*, researchers were able to identify two antifungal macrolides known as phorboxazoles A and B, both of which had new parent ring configurations (Molinski, 2004). Moreover, the recognized plakortides E, F, and H, researchers have isolated three additional antifungal peroxides from two different sponges. These antifungal peroxides were found in the Jamaican sponge *Plakortis halichondriodes* and the Florida sponge *Plakinastrella onkodes*. The compounds such as the cyclic peptide Aciculitin B from Aciculites ciliate, the tetramic acid glycoside aurantoside E from Plakinolopha mirabilis, the 4-isocyanogorgon-11-ene from Axinyssa terpinis, and the biphenyl ether 3,5-dibromo-2-(3,5-dibromo-2-methoxyphenoxy)phenol from Latrunculin B, a candidate molecule for the development of a novel antifungal treatment for the management of common fungal illnesses in aquaculture, were isolated and identified from the marine red-sea sponge *Negombata magnifica* using bioassay-guided isolation and identification (Devi, Ribeiro, & Ciavatta, 2013).

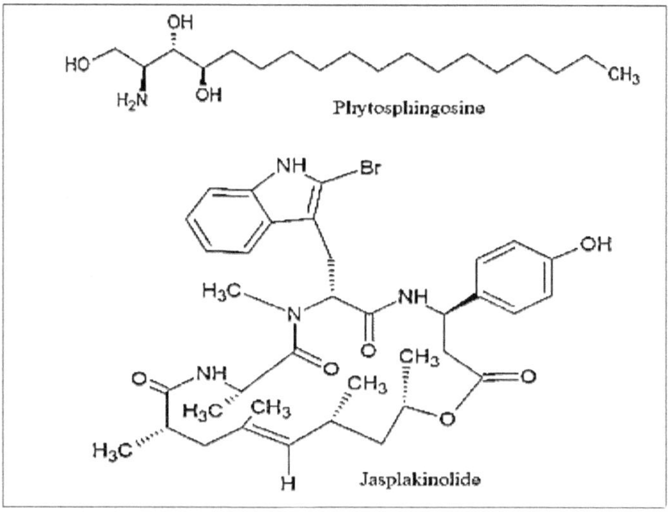

FIGURE 20.5 Antifungal bioactive compounds obtained from marine organisms

20.3.6 IMMUNOSUPPRESSANTS

There has been a relatively small percentage of marine natural compounds that have been shown to have an immunosuppressive effect. A deep-sea marine sponge known as *Agelas flabelliformis* was used to successfully extract two immunosuppressive chemicals for the first time. These compounds are known as 4,5-dibromo-2-pyrrolic acid and 4 alphamethyl-5 alpha-cholest-8-en-3 beta-ol. Both chemicals were very potent in suppressing the response of murine splenocytes in the two-way mixed lymphocyte reaction (MLR), but at lower doses, they demonstrated little to no observable cytotoxicity (Kumar & Pal, 2016). The marine natural products discodermolide (Fig. 20.6), which was extracted from a species of Discodermia, and microcolin A are now considered to be the most attractive prospects for immunosuppressive treatment. In the murine two-way MLR as well as the concanavalin A stimulation of splenocyte cells, the extracted pure natural product discodermolide displayed a significant suppressive effect. Another study demonstrated that the compounds FW635I1 (Daidzein) and FW635I2 (Genistein) (Fig. 20.6) generated from *Micromonospora carbonacea* FIM 02–635 had immunosuppressive and antitumor activity, but no antimicrobial properties at all (Lam, 2007). In another effort, chemicals isolated from the marine sponge *Aurora globostellata* and a total of 10 marine bacterial strains derived from *Callyspongia difusa* were put to the test. Both the ability to combat infections and the ability to produce immunomodulators were shown by the bacterial strains (Chairman, Jeyamala, Sankar, Murugan, & Singh, 2013; Kalirajan, Karpakavalli, Narayanan, Ambiganandham, Ranjitsingh, & Sudhakar, 2013). *Lyngbya majusculata* was used to isolate the immunosuppressive linear peptide microcolin-A, which was able to block the two-way murine mixed lymphocyte response at nanomolar doses (Koehn, Longley, & Reed, 1992). In one of the studies, three different crude extracts of *Halichondria panicea* demonstrated immunostimulative properties at lower doses. On the other hand, they displayed immunosuppressive properties when present in larger doses (Purushottama, Venkateshvaran, Pani Prasad, & Nalini, 2009). Based on the studies conducted on *S. halichondriodes*, it was shown that a methanolic extract obtained from the sponge had the potential to inhibit both cellular and humoral immune responses (Kumar & Pal, 2016). Not only did the extract reduce a nonspecific immunological response, but it was also very efficient in lowering both humoral and cell-mediated immunity. During the course of the study, it was unknown which chemical constituents were responsible for the immunosuppressive action. However, early pharmacognosy chemical testing found evidence of the presence of terpenoids, flavonoids, steroids, and alkaloids. When the presence of sterol molecules in the extract was later confirmed by LC-MS-MS, it was realized that these molecules were the ones responsible for the bioactivity (Kumar, Pandita, & Pal, 2012; Kumar, Desai, & Ak, 2012).

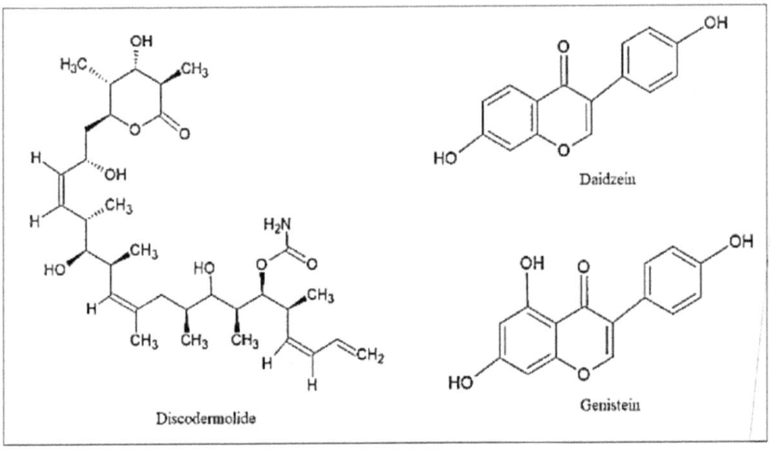

FIGURE 20.6 Immunosuppressants compounds obtained from marine organisms

20.3.7 ANTIOXIDANTS

Antioxidants may safeguard human health from the harm that can be triggered by reactive oxygen species and can reduce the number of macromolecules such as proteins, DNA, and lipids that can lead to conditions like cancer, diabetes, and neurological illnesses. Bioactive peptides generated from marine organisms contain an antioxidant potential that has been controlled by radical scavenging activity. This behavior has been identified by electron spin resonance. A number of the marine algae that contain bioactive chemicals like fucoxanthin and phycoerythrobilin (Fig. 20.7), among other examples, have been shown to have powerful antioxidant properties (Menaa, Wijesinghe, Thiripuranathar, Althobaiti et al., 2021). The polyphenolic chemicals are antioxidants that are created by marine algae. Some examples of these chemicals are flavonoids, benzoic acid, cinnamic acid, quercetin, and gallic acid (Fernando, Kim, Son, Jeong, & Jeon, 2016). Phlorotannins are another kind of antioxidant that may be present in brown macroalgae (Li, Wijesekara, Li, & Kim, 2011) or diatoms (Wijesekara, Yoon, & Kim, 2010). Phlorotannins are a kind of tannin derivative that exhibits a wide variety of bioactivities, including anti-inflammatory and antioxidant properties (Al-Mola, 2009; Malve, 2016; Sansone, Brunet, Noonan, & Albini, 2020). Marine species are a rich reservoir of many different antioxidants, which are potent for the synthesis of medicines for the treatment and prevention of a wide range of acute and chronic human disorders. Indeed, marine species are a rich reservoir of antioxidants, both in terms of the amount and/or variety of the antioxidants they contain. Examples of the richness of marine antioxidants include vitamins B12, C, D, and E, amino acids, carotenoids, peptides, sulfated polysaccharides, phenolic compounds, sterols, phlorotannins, and flavones (Sansone, Brunet, Noonan, & Albini, 2020; Sansone & Brunet, 2019).

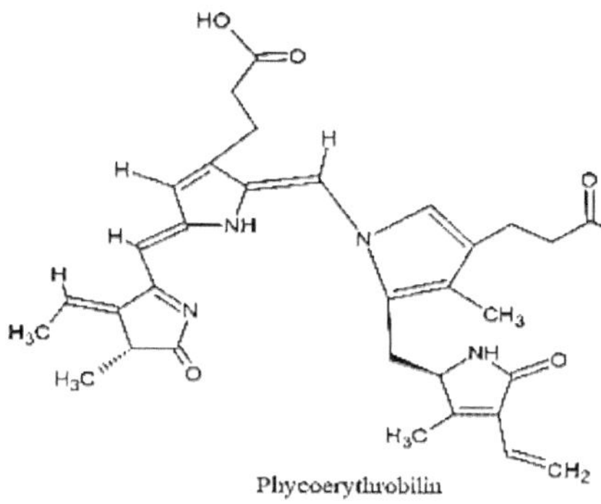

Phycoerythrobilin

FIGURE 20.7 Antioxidant bioactive compound obtained from marine organisms

20.3.8 ANTI-HYPERTENSIVE COMPOUNDS

Hypertension is a global health concern due to its increased incidence and is associated with an increase in morbidity and death. It is the most frequent triggering factor for early cardiovascular disease, being more prevalent than cigarette smoking, dyslipidemia, and diabetes (Lackland & Weber, 2015). It is estimated that hypertension is accountable for 54% of all strokes and 47% of all cases of ischemic heart disease worldwide (Lawes, Vander Hoorn, & Rodgers, 2008). There have been reports of peptides derived from marine sources that exhibit antihypertensive action (Norris, Harnedy, & FitzGerald, 2013). Peptides that lower blood pressure may still be found in rather large quantities in microalgae. Antihypertensive efficacy has been demonstrated by peptides derived from a variety of microalgal species, including *Chlorella vulgaris, Spirulina platensis (Arthrospira platensis),* and *Chlorella ellipsoidea* (Kim & Kang, 2011). A large number of studies have demonstrated that consuming chlorella leads to a reduction in blood pressure. It has been shown that consuming chlorella may bring about a reduction in blood pressure in human beings (Zhao, Wu, Yang, Liu, & Huang, 2015). A hendecapeptide having angiotensin I converting enzyme (ACE) inhibitory action was extracted from the pepsin hydrolysate of proteins from *C. vulgaris* (Sheih, Fang, & Wu, 2009; Samarakoon & Jeon, 2012). The sequence of the hen-decapeptide was as follows: Val-Glu-Cys-Tyr-Gly-Pro-Asn-Arg-Pro-Gln-Phe. The purified hendecapeptide's inhibitory characteristics demonstrated a manner of binding that was not competitive, indicating that it had a significant ACE-inhibitory action. An appealing option for the production of a high-value product for blood pressure (BP) control may be created by using the biochemical features of an isolated hendecapeptide and the low-cost microalgal protein resource (Sheih, Fang, & Wu, 2009). These qualities include: The alcalase-proteolytic hydrolysate of *C. ellipsoidea* protein showed the highest ACE-inhibitory activity as well. This hydrolysate has been fractionated into three ranges according to molecular weight, and the fraction with a molecular weight of less than 5 kDa demonstrated the highest ACE-inhibitory activity (Zhao, Wu, Yang, Liu, & Huang, 2015).

20.3.9 HYPOGLYCEMIC COMPOUNDS

Diabetes mellitus (DM) is a condition that lowers a person's quality of life and puts their life in danger, and its prevalence is rising all over the globe. To date, the majority of naturally occurring diabetes treatments have been extracted from one or more of the 400 different plant species (Kar, Choudhary, & Bandyopadhyay, 2003; Malviya, Jain, & Malviya, 2010). *Chlorella sorokiniana, Chlorella pyrenoidosa,* and *Chlorella vulgaris* have all been shown to have hypoglycemic effects in the research that has been done on them (Noguchi, Konishi, Kumamoto, Maruyama, Ando, & Yanagita, 2013; Jeong, Kwon, & Kim, 2009; Senthilkumar & Ashokkumar, 2012; Chou, Prakash, Huang, Lien et al., 2008). Cherng and Shih demonstrated that *C. vulgaris* increased the hypoglycemic impact of exogenous insulin in streptozotocin-induced mice (Jong-Yuh & Mei-Fen, 2005), while Layam and Reddy demonstrated a similar hypoglycemic response of *Spirulina* in diabetic rats (Layam & Reddy, 2006). Both of these hypoglycemic effects were seen in rats. It was discovered that *Chlorella vulgaris* has the ability to stop the progression of hyperglycemia in obese mice. In obese mice, the prevention of resistance to insulin by *C. vulgaris*, as demonstrated by enhanced glucose and insulin tolerance, was in part because of the increase in the insulin signaling mechanisms at its principal target tissues, which was accomplished by enhancing the protein phosphorylation level. This was demonstrated by the fact that *C. vulgaris* improved glucose and insulin tolerance (Vecina, Oliveira, Araujo, Baggio et al., 2014). The protective role of *C. pyrenoidosa* was investigated by Senthil Kumar and Ashok Kumar using diabetic male albino rats that were subjected to cadmium. *C. pyrenoidosa* has a much higher potential for hypoglycemic and renoprotective effects (Senthilkumar & Ashokkumar, 2012). The chemical aquastatin A (Fig. 20.8), which displayed significant inhibitory action against protein tyrosine phosphatase 1B (PTP1B), was produced by the marine-derived fungus *Cosmospora* sp. SF-5060 (Seo, Sohn, Oh, Kim, & Ahn, 2009). The protein tyrosine phosphatases are the enzymes that are accountable for the control of cellular processes that are reliant on tyrosine phosphorylation (Tonks, 2006). Studies in which aquastatin A was hydrolyzed revealed that the dihydroxypentadecyl benzoic acid moiety contained in the molecule was accountable for the inhibitory action. It has been shown that consuming fucoxanthin may considerably lower both levels of plasma insulin and blood glucose (Maeda, Tsukui, Sashima, Hosokawa, & Miyashita, 2008). Fucoxanthin (Fig. 20.8), is one of the principal marine carotenoids that are prevalent in microalgae. Fucoxanthin may be found in high concentrations in *Phaeodactylum tricornutum* (Kim, Jung, Kwon, Cha et al., 2012). In diabetic and obese mice, Maeda and his colleagues discovered that fucoxanthin significantly reduced both blood glucose levels and plasma insulin concentration, as well as the animals' water consumption (Maeda, Hosokawa, Sashima, & Miyashita,

2007). It has been hypothesized that fucoxanthin may ameliorate resistance to insulin and result in a reduction in the amount of glucose in the blood (Peng, Yuan, Wu, & Wang, 2011). An activity that has not been recorded from any other species of Lyngbya has been revealed to be present in the extract of *Lyngbya confervoides* (Zhao, Wu, Yang, Liu, & Huang, 2015). Lyngbyatoxin A, also known as Teleocidin A-1, was extracted from *Lyngbya majuscula* gomont and was shown to be a powerful activator of protein kinase C (PKC). PKC was found to be involved in the control of a variety of different cellular responses. The epithelium lining the small intestine contains an enzyme called alpha-glucosidase. This enzyme is coupled to a membrane and catalyzes the separation of glucose from disaccharides. It has been shown that inhibiting α-glucosidase is an efficient method for reducing the amount of glucose that is present in the blood Zhao, Wu, Yang, Liu, & Huang, 2015; Zhang, Xue, & Gai, 2001). In addition, β-glucosidase inhibitors could be effective in the treatment as well as the prevention of illnesses including diabetes, obesity, hyperlipoproteinemia, and hyperlipidemia. It has been shown that plants, mushrooms, and fungi all contain significant amounts of these chemicals. On the other hand, relatively few instances have been found of such inhibitors coming from marine microorganisms (Arrieta & Herndl, 2001). The studies conducted by Pandey and his colleagues revealed that the β-glucosidase inhibitors were prevalent in the bacterial associates of the marine sponge known as *Aka coralliphaga*. The bacteria of marine origin were classified as belonging to the *groups Firmicutes, Bacteroidetes, Actinobacteria*, and *Proteobacteria* (Zhao, Wu, Yang, Liu, & Huang, 2015; Pandey, Sree, Dash, Sethi, & Chowdhury, 2013).

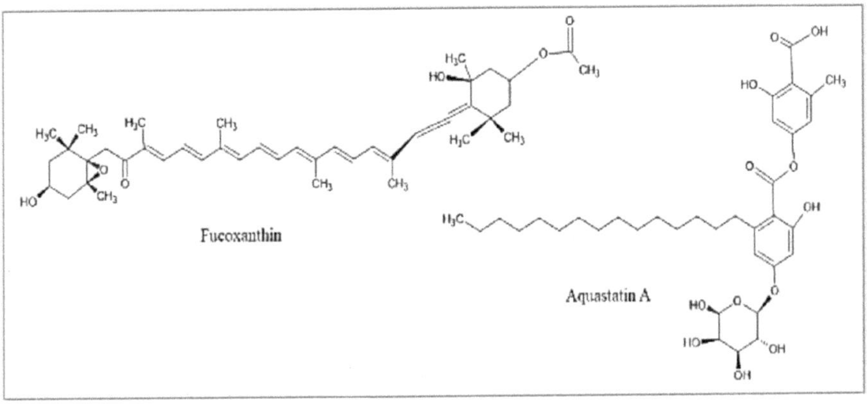

FIGURE 20.8 Hypoglycaemic compounds obtained from marine organisms.

20.3.10 HYPOLIPIDEMIC COMPOUNDS

Farnesylquinone (Fig. 20.9), a chemical produced from marine-sourced *Streptomyces nitrosporeus*, is responsible for the lipid-lowering action. In HepG2 cells, it leads to a significant reduction in the levels of lipids such as triglycerides (TG) and total cholesterol (TC). According to the results, this may lower lipid levels by upregulating the pathway that involves peroxisome proliferator-activated receptor-alpha (Liu, Yang, Wu, Guo, et a., 2014). Mevinolin (Fig. 20.9), a cholesterol-lowering medication generated from fungus, is utilized to decrease blood levels of cholesterol in humans and shows substantial cholesterol-lowering action (Zhao, Wu, Yang, Liu, & Huang, 2015). This cholesterol-lowering substance is utilized to treat cardiovascular disease. In the research that Bose and his colleagues conducted, they discovered that two of the *Salinispora arenicola* strains were able to manufacture mevinolin (Bose, Hodson, Shaw, Fuerst, & Hewavitharana, 2014). The Great Barrier Reef area of Australia was the location where the strains were obtained from marine sponge species (Bose, Hodson, Shaw, Fuerst, & Hewavitharana, 2014). In experimental animals, water-soluble portions of microalgae or extracted algal polysaccharides have been shown to elicit hypocholesterolemic actions, according to the findings of a number of researchers (Zhao, Wu, Yang, Liu, & Huang, 2015; Jong-Yuh & Mei-Fen, 2005). *Spirulina* and its extracts have been shown to have hypolipidemic actions in a variety of animal species, such as hamsters, mice, rats, and rabbits. Dietary treatment of diabetes includes keeping lipid levels and blood glucose levels as close to normal as is practicable in order to minimize the risk of diabetes-related problems arising (Sun, Peng, Liu, Fan, Wang, Chen, 2010). *Spirulina's* ability to lower cholesterol levels has been the subject of many clinical studies involving humans (Kim & Park, 2003). Supplementation with spirulina was shown to be an excellent method for lowering the elevated concentrations of lipids that were present in individuals suffering from hyperlipidemic nephrotic syndrome (Samuels, Mani, Iyer, & Nayak, 2002). *Spirulina* consumption on a daily basis led to a decrease in total blood cholesterol, as well as in the low-density lipids (LDL) fraction and TG levels. Both the ratio of LDL to high-density lipids (HDL) and the ratio of total cholesterol to HDL dropped by a substantial amount. Age is associated with an enhancement in both total and LDL cholesterol levels (Uranga & Keller, 201), which is accompanied by a concurrent rise in the prevalence of cardiovascular disease (Zhao, Wu, Yang, Liu, & Huang, 2015). In rats, a new hypocholesterolemic protein called C-phycocyanin that was produced from *S. platensis* has been shown to have a high hypocholesterolemic action (Nagaoka, Shimizu, Kaneko, Shibayama et al., 2005). This protein may have a dramatic impact on blood cholesterol levels. It has been demonstrated that *Chlorella vulgaris* may affect the levels of lipids in the serum and liver of rats, and it also has anti-atherosclerotic and antilipidemic activities (Shibata, Oda, Onodera-Masuoka, Matsubara et al., 2001). Lee and colleagues found that when given *C. vulgaris*, blood total lipids and liver TG concentrations were considerably lower while fecal total lipids, total cholesterol, and TG excretions were significantly greater, perhaps owing to lipid metabolic regulation and enhanced fecal lipid excretion (Lee, Park, Choi, Huh, & Kim, 2008).

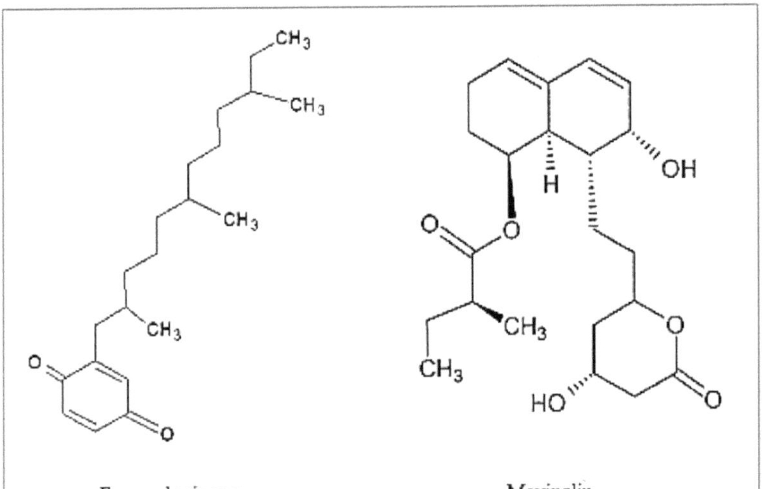

FIGURE 20.9 Hypolipidaemic compounds obtained from marine organisms.

Some of the important compounds which are derived from the marine organisms with their source and therapeutic catagories and discussed in this chapter are summarized in table 20.1.

20.4 APPLICATIONS OF MARINE BIOACTIVE COMPOUNDS

The bioactive substances that have been discovered in marine sponges display the capacity to serve as antiviral, antimalarial, antibacterial, and antifungal agents against human infections as well as fish diseases in the aquaculture sector. These chemical bioactive compounds have been isolated

TABLE 20.1 Bioactive compounds obtained from the marine orgainisms.

Compound	Source	Therapeutic Category
Debromoaplysiatoxin and Lyngbyatoxin-A	Lyngbya mausculata	Anti-Inflammatory
Scytonemin	Cyanobacterium stigonema spp.	anti-proliferative and anti-inflammatory
Pseudopetrocin-E	genus *Pseudopterogorgia*	analgesic and anti-inflammatory
C-phycocyanin	Spirulina plantensis	reducing inflammation
Fucann	Lobophora variegate	anti-inflammatory
Glycyrrhetinic acid	*S. halichondriodes*	anti-inflammatory
Ara-C, viderabine	*Tethya crypta*	Active against herpes, varicella zoster, and vaccinica viruses.
Carrageenans	Red algae (Rhodophyta), like as Gigartina, Chondrus, and Hypnea, Eucheuma	Human papillomavirus
Cyclodepsipeptide	*Jaspis sp*	antimicrobial
Phorboxazoles A and B	genus *Phorbas*	antifungal macrolides
Aciculitin B	*Aciculites ciliate*	antifungal
Aurantoside E	*Plakinolopha mirabilis*	antifungal
4-isocyanogorgon- 11-ene	*Axinyssa terpinis*	antifungal
4, 5-dibromo- 2-pyrrolic acid and 4 alphamethyl- 5 alpha-cholest-8-en-3 beta-ol.	*Agelas flabelliformis*	immunosuppressive
Discodermolide	species of *Discodermia*	immunosuppressive
Daidzein and Genistein	*Micromonospora carbonacea*	immunosuppressive
Microcolin A	*Lyngbya majusculata*	immunosuppressive
Phlorotannins	Brown macroalgae	anti-inflammatory and antioxidant
Hendecapeptide	*Chlorella vulgaris*	Antihypertensive
Aquastatin A	*Cosmospora sp.*	Antidiabetic
Fucoxanthin	*Phaeodactylum tricornutum*	Antidiabetic
Farnesylquinone	*Streptomyces nitrosporeus*	Lipid-lowering action
C-phycocyanin	S. platensis	high hypocholesterolaemic action

from marine sponges. In addition, the use of such substances as antimicrobial agents in many veterinary commodities, like cattle farming, poultry, as well as for domesticated cats, is addressed (Varijakzhan, Loh, Yap, Yusoff et al.,

2021). Natural products obtained from marine organisms offer significant potential for use in antimicrobial applications; nevertheless, the majority of nanoparticle antibiotics now under development come from fungi or actinomycetes (Butler, Blaskovich, Cooper, 2013). Antiviral medications are a highly significant category of antimicrobial therapy; yet, the availability of particular therapeutic therapies is still quite confined (Kiuru, D'Auria, Muller, Tammela, Vuorela, & Yli-Kauhaluoma, 2014). Antiviral drugs are a subcategory of antimicrobial treatment. Despite the fact that they have different structures, many of the compounds extracted from marine sponges have been observed to exhibit antibacterial characteristics against pathogenic bacterial strains in humans. Some examples of these strains include *Escherichia coli, S. aureus,* and *Pseudomonas aeruginosa.* In addition, the bioactive substances have the potential to induce mortality of bacteria by leaking the substances of bacterial cells owing to the breakdown of the cell membrane caused by the production of reactive oxygen species (ROS) (Varijakzhan, Loh, Yap, Yusoff et al., 2021). This may lead to the death of the bacteria. ROS has the ability to oxidize proteins that are located inside bacterial cells by changing the covalent bonds which are quite important for preserving the structure of functional proteins that are necessary for the survival of bacterial cells (Benzaid, Belmadani, Djeribi, & Rouabhia, 2019; Yang, Yusoff, Ajat, Thomas et al., 2019). There have been reports of peptides derived from marine sources that show antihypertensive action (Norris, Harnedy, & FitzGerald, 2013). Peptides that lower blood pressure may still be found in rather large quantities in microalgae. Antihypertensive efficacy has been shown by peptides derived from a variety of microalgal species, including *Chlorella ellipsoidea, Chlorella vulgaris*, and *Spirulina platensis* (*Arthrospira platensis*). Bioactive peptides generated from marine organisms have an antioxidant characteristic that has been controlled by radical scavenging activity. This activity has been identified by electron spin resonance. Some of the marine algae that contain bioactive chemicals like fucoxanthin and phycoerythrobilin, among other examples, have been shown to have powerful antioxidant properties (Menaa, Wijesinghe, Thiripuranathar, Althobaiti et al., 2021). In point of fact, marine species are a very rich reservoir of antioxidants, either in terms of the amount or the variety of their constituents. Examples of marine antioxidant content include vitamins B12, vit. C, vit. D, and vit., amino acids, chitooligosaccharide derivatives, peptides, carotenoids, sulfated polysaccharides, phenolic compounds, phlorotannins, sterols, and flavones (Sansone, Brunet, Noonan, & Albini, 2020; Sansone & Brunet, 2019). Other examples include phenolic compounds and flavones.

The marine bioactive chemicals also have potential in the pharmaceutical and nutritional supplement industries. Oils from fish, omega-fatty acids, and lipids found in marine organisms are most likely the most well-known marine dietary supplements. The Food and Drug Administration has given its stamp of approval to a single omega-3 fish oil product, namely Lovaza®. Out of approximately 40,000 species of algae, only a selective few (such as *Chlorella vulgaris* and *Spirulina pacifera*) have been utilized by the food industry (Kiuru, D'Auria, Muller, Tammela, Vuorela, & Yli-Kauhaluoma, 2014). Sterols, lipids, polyphenols, and carotenoids are all components that may be found in microalgae. In photobioreactors, carotenoid production on an industrial scale has been achieved. Producing the antioxidant and food color astaxanthin in an efficient manner from the biomass of *Haematococcus* is possible (Li, Zhu, Niu, Shen, & Wang, 2011). Extracts of cyanobacteria or macroalgae may have applications in agriculture as agents that regulate plant development, fertilizers, and agents that increase nutrient intake from soil or improve plant resistance. In addition, freshwater cyanobacteria have been investigated for the possibility that they may be used to suppress the development of algae. The vast majority of cyanobacterial herbicides work by inhibiting electron transport in the area around photosystem II. It is worthwhile to do research on the impact that extracts obtained from cyanobacteria have on the photosynthesis process, growth, and development of higher plants and algae. It is helpful to conduct an investigation into the possibility of using the extracts or the separated chemicals as agents that regulate plant development (Kiuru, D'Auria, Muller, Tammela, Vuorela, & Yli-Kauhaluoma, 2014). Extracts of brown algae that are sold commercially have been spread on croplands to boost crop yields and increase plant vigor. Extracts obtained from single-celled green algae, namely chlorella, may boost the production of chlorophyll as well as the development of shoots and roots in grains, vegetables, and fruit trees. In Japan, the use of chlorella extract as a growth-regulating agent has been approved. Cyanobacteria, which are the most prevalent photosynthetic prokaryotic microorganisms, are also known to create substances that have the potential to be used in the agricultural industry as fertilizers and agents that regulate plant development. Nitrogen fixation was first thought to be the cause of their favorable impact on plants. It was later hypothesized that the presence of vitamins, plant hormones, or substances with actions similar to plant hormones, contribute positively to the effect that cyanobacteria have on seed germination and growth. *Hapalosiphon intricatus*, *Calothrix* sp., *Nostoc* sp., and *Arthronema africanum* were among the organisms that were observed to identify effects that were attributable to this sort of activity (Karthikeyan, Prasanna, Nain, & Kaushik, 2007; Stirk, Ördög, Van, Staden, & Jäger, 2002). Marine anti-inflammatory substances will also have a significant potential for use in the cosmetic industry. There is an unfulfilled need for novel UV-absorbing and sustainable substances for sunscreen applications to replace the current sunscreen agents that are allergenic, irritating, or bioaccumulative (Kiuru, D'Auria, Muller, Tammela, Vuorela, & Yli-Kauhaluoma, 2014). The advantages of the various secondary metabolites produced by sponges are not only important in aquaculture but have also been expanded for use in other veterinary products to treat a variety of ailments that may affect animals. It has been shown that extracts obtained from marine sponges in the Southern Brazil region exhibit high antibacterial activity against a condition known as mastitis in cattle, which is caused by Coagulase-negative Staphylococci (Laport, Marinho, da Silva Santos, de Almeida et al., 2012). This illness affects the milk glands of cows. The marine sponges *Haliclona* sp., *Cinachyrella* sp., and *Petromica citrine* were used to prepare aqueous and ethanolic crude extracts, and both types of crude extracts showed antibacterial activity (Laport, Marinho, da Silva Santos, de Almeida et al., 2012).

20.5 CONCLUSION

The aquatic environments contain an infinite number and variety of resources that can be used in the development of new medicines to treat the most serious diseases. It will be of great assistance to us in the process of developing new pharmaceuticals for use in pharmacological as well as biotechnological applications. These marine organisms are widely recognized to have obscured therapeutic effects but are still in the testing stage because marine environments contain immense ecological resources. These ecologically available resources involve a diverse array of organisms, of which some living organisms have been reported while others are still unknown. The marine environment is a profuse source of a wide variety of chemical and biological species. This variety is a huge source of new bioactive compounds that have the possibility for industrial uses such as pharmaceuticals, beauty products, nutraceuticals, fine chemical products, and herbicides and pesticides. Moreover, these novel chemical compounds can be isolated from a broad variety of aquatic organisms. The world's oceans are a vast unexplored resource that could yield the discovery of new compounds with practical applications. Even though the innovation, development, and long-term use of marine-based compounds with health-endorsing applications are considered to be the approaching potential of marine biotechnology, the requirements, approaches, and opportunities are just as applicable to other marine bioproducts. The marine samples that are being collected, including extract as well as pure chemical compound libraries, should be an essential component of the research materials. This is the most important aspect of the study materials. It is necessary to create new extraction procedures using green chemistry as well as biotechnological manufacturing methods. Synthetic chemistry should be employed to supply biologically active compounds and to optimize their performance in biological tests. In the next millennium, the challenge that will be encountered by the marine biotechnology sector is to discover novel supplies of marine bioproducts, establish novel screening techniques, and offer a bearable supply source. The existing marine sample collections, including extracts as well as pure chemical compound libraries, should be an essential component of the research materials. This will help shed light on the effect that these substances have at the gene and protein levels, as well as their effectiveness against pathogenic microorganisms. In order to develop novel drugs that are both effective as well as efficient, which can help mitigate existing problems and provide favorable outcomes for stakeholders, there must first be a string of thorough investigations conducted.

KEYWORDS

- **antibacterial and marine research**
- **anticancer**
- **bioactive compounds**
- **gas chromatography and Mass Spectrometry**
- **human papillomavirus infection**
- **marine biodiversity**
- **protein kinase C**
- **tissue plasminogen activator**
- **tumor necrosis factor-α**

REFERENCES

Agami, C., Couty, F., Evano, G., Darro, F., & Kiss, R. (2003). Enantioselective synthesis of clavepictine analogues and evaluation of their cytotoxic activity. *European Journal of Organic Chemistry, 2003*(11), 2062–2070.

Alam, M. A., Sarkar, S. K., & Gomes, A. (2012). A High Molecular Weight Protein Extract of *Mastobranchus indicus* (Mi-64) Having antiarthritic activity in experimental animals. *Inflammation, 35*, 1223–1231.

Al-Mola, H. F. (2009). Antibacterial activity of crude extracts and phlorotannin isolated from the diatom *Cymbella* spp. *Journal of Pharmacy Research, 2*, 304–308.

Arrieta, J. M., & Herndl, G. J. (2001). Assessing the diversity of marine bacterial β-glucosidases by capillary electrophoresis zymography. *Applied and Environmental Microbiology, 67*, 4896–4900.

Bao, B., Zhang, P., Lee, Y., Hong, J., Lee, C. O., & Jung, J. H. (2007). Monoindole alkaloids from a marine sponge Spongosorites sp. *Marine Drugs, 5*, 31–39.

Beltran, E. C., & Neilan, B. A. (2000). Geographical segregation of the neurotoxin-producing cyanobacterium Anabaena circinalis. *Applied and Environmental Microbiology, 66*, 4468–4474.

Benzaid, C., Belmadani, A., Djeribi, R., & Rouabhia, M. (2019). The effects of *Mentha× piperita* essential oil on *C. albicans* growth, transition, biofilm formation, and the expression of secreted aspartyl proteinases genes. *Antibiotics, 8*, 10.

Blunden, G. (2001). Biologically active compounds from marine organisms. *Phytotherapy Research, 15, 89–94.*

Bose, U., Hodson, M. P., Shaw, P. N., Fuerst, J. A., & Hewavitharana, A. K. (2014). Bacterial production of the fungus-derived cholesterol-lowering agent mevinolin. *Biomedical Chromatography, 28*, 1163–1166.

Butler, M. S., Blaskovich, M. A., & Cooper, M. A. (2013). Antibiotics in the clinical pipeline in 2013. *Journal of Antibiotics, 66*, 571–591.

Capon, R. J. (2001). Marine bioprospecting–Trawling for treasure and pleasure. *European Journal of Organic Chemistry, 2001*(4), 633–645.

Chairman, K., Jeyamala, M., Sankar, S., Murugan, A., & Singh, A. R. (2013). Immunomodulating properties of bioactive compounds present in *Aurora globostellata. International Journal of Marine Science, 3*, 151–157.

Chen, Y., McCarthy, P. J., Harmody, D. K., Schimoler-O'Rourke, R., Chilson, K., Selitrennikoff, C., Pomponi, S. A., & Wright, A. E. (2002). New bioactive peroxides from marine sponges of the family Plakiniidae. *Journal of Natural Products, 65*, 1509–1512.

Chin, Y. W., Balunas, M. J., Chai, H. B., & Kinghorn, A. D. (2006). Drug discovery from natural sources. *The AAPS Journal, 8*, 239–253.

Chou, Y. C., Prakash, E., Huang, C. F., Lien, T. W., Chen, X., Su, I. J., Chao, Y. S., Hsieh, H. P., & Hsu, J. T. (2008). Bioassay-guided purification and identification of PPARα/γ agonists from *Chlorella sorokiniana. Phytotherapy Research, 22*, 605–613.

Cueto, M., Jensen, P. R., Kauffman, C., Fenical, W., Lobkovsky, E., & Clardy, J. (2001). Pestalone, a new antibiotic produced by a marine fungus in response to bacterial challenge. *Journal of Natural Products, 64*, 1444–1446.

Datta, D., Talapatra, S. N., & Swarnaka, S. (2015). Bioactive compounds from marine invertebrates for potential medicines-an overview. *International Letters of Natural Sciences, 7*, 42–61.

Devi, P., Ribeiro, M., & Ciavatta, M. L. (2013). Antifungal potential of marine sponge extract against plant and fish pathogenic fungi. *Oceanography Open Access, 23*, 1–4.

Devi, P., Wahidullah, S., Rodrigues, C., & Souza, L. D. (2010). The sponge-associated bacterium Bacillus licheniformis SAB1: a source of antimicrobial compounds. *Marine Drugs, 8*, 1203–1212.

Donia, M., & Hamann, M. T. (2003). Marine natural products and their potential applications as anti-infective agents. *The Lancet Infectious Diseases, 3*, 338–348.

Etahiri, S., Bultel-Poncé, V., Caux, C., & Guyot, M. (2001). New bromoditerpenes from the red alga *Sphaerococcus coronopifolius*. *Journal of Natural Products, 64*, 1024–1027.

Fernando, I. S., Kim, M., Son, K. T., Jeong, Y., & Jeon, Y. J. (2016). Antioxidant activity of marine algal polyphenolic compounds: a mechanistic approach. *Journal of Medicinal Food, 19*, 615–628.

Fontana, A., Cavaliere, P., Wahidulla, S., Naik, C. G., & Cimino, G. (2000). A new antitumor isoquinoline alkaloid from the marine nudibranch Jorunna funebris. *Tetrahedron, 56*, 7305–7308.

Haefner, B. (2003). Drugs from the deep: marine natural products as drug candidates. *Drug Discovery Today, 8*, 536–544.

Harrell, R. M. (2017). Bioethical considerations of advancing the application of marine biotechnology and aquaculture. *Marine Drugs, 15*, 197.

Harvey, A. L. (2008). Natural products in drug discovery. *Drug Discovery Today, 13*, 894–901.

Ishida, T., Mizushina, Y., Yagi, S., Irino, Y., Nishiumi, S., Miki, I., Kondo, Y., Mizuno, S., Yoshida, H., Azuma, T., & Yoshida, M. (2012). Inhibitory effects of glycyrrhetinic acid on DNA polymerase and inflammatory activities. *Evidence-Based Complementary and Alternative Medicine, 2012*, 1–9.

Jeong, H., Kwon, H. J., & Kim, M. K. (2009). Hypoglycemic effect of *Chlorella vulgaris* intake in type 2 diabetic Goto-Kakizaki and normal Wistar rats. *Nutrition Research and Practice, 3*, 23–30.

Jha, R. K., & Zi-Rong, X. (2004). Biomedical compounds from marine organisms. *Marine Drugs, 2*, 123–146.

Johnson, J. A., Citarasu, T., & Manjusha, W. A. (2012). Antimicrobial screening and identification of bioactive compounds present in marine sponge *Zygomycale* sp. collected from Kanyakumari coast. *Journal of Chemical Biology and Physical Sciences, 2*, 1848.

Jong-Yuh, C., & Mei-Fen, S. (2005). Potential hypoglycemic effects of *Chlorella* in streptozotocin-induced diabetic mice. *Life Sciences, 77*, 980–990.

Jong-Yuh, C., & Mei-Fen, S. (2005). Potential hypoglycemic effects of Chlorella in streptozotocin-induced diabetic mice. *Life Sciences, 77*, 980–990.

Kalirajan, A., Karpakavalli, M., Narayanan, K. R., Ambiganandham, K., Ranjitsingh, A. J., & Sudhakar, S. (2013). Isolation, characterization and phylogeny of sponge-associated bacteria with antimicrobial and immunomodulatory potential. *International Journal of Current Microbiology and Applied Sciences, 2*, 136–151.

Kar, A., Choudhary, B. K., & Bandyopadhyay, N. G. (2003). Comparative evaluation of hypoglycaemic activity of some Indian medicinal plants in alloxan diabetic rats. *Journal of Ethnopharmacology, 84*, 105–108.

Karthikeyan, N., Prasanna, R., Nain, L., & Kaushik, B. D. (2007). Evaluating the potential of plant growth promoting cyanobacteria as inoculants for wheat. *European Journal of Soil Biology, 43*, 23–30.

Kim, S. K., & Kang, K. H. (2011). Medicinal effects of peptides from marine microalgae. *Advances in Food and Nutrition Research, 64*, 313–323.

Kim, S. M., Jung, Y. J., Kwon, O. N., Cha, K. H., Um, B. H., Chung, D., & Pan, C. H. (2012). A potential commercial source of fucoxanthin extracted from the microalga *Phaeodactylum tricornutum*. *Applied Biochemistry and Biotechnology, 166*, 1843–1855.

Kim, W. Y., & Park, J. Y. (2003). The effect of Spirulina on lipid metabolism, antioxidant capacity, and immune function in Korean elderlies. *Korean Journal of Nutrition, 36*, 287–297.

Kiran, N., Siddiqui, G., Khan, A. N., Ibrar, K., & Tushar, P. (2014). Extraction and screening of bioactive compounds with antimicrobial properties from selected species of mollusk and crustacean. *Journal of Clinical & Cellular Immunology, 5*, 1000189.

Kiuru, P., D'Auria, M. V., Muller, C. D., Tammela, P., Vuorela, H., & Yli-Kauhaluoma, J. (2014). Exploring marine resources for bioactive compounds. *Planta Medica, 80*, 1234–1246.

Koehn, F. E., & Carter, G. T. (2005). The evolving role of natural products in drug discovery. *Nature Reviews Drug Discovery, 4*, 206–220.

Koehn, F. E., Longley, R. E., & Reed, J. K. (1992). Microcolins A and B, new immunosuppressive peptides from the blue-green alga *Lyngbya majuscula*. *Journal of Natural Products, 55*, 613–619.

Kumar, M. S., & Pal, A. K. (2016). A review of bioactive compounds from marine organisms with special mention on the potential of marine sponges in pharmacological applications. *Journal of Marine Biology and Assosciation of India, 58*, 84.

Kumar, M. S., & Pandita, N. S., & Pal, A. K. (2012). LC-MS/MS as a tool for identification of bioactive compounds in marine sponge *Spongosorites halichondriodes* (Dendy 1905). *Toxicon, 60*, 1135–1147.

Kumar, M. S., Desai, D., & Pal, A. K. (2012). Immunomodulatory potential of a marine sponge *Spongosorites halichondrioides* (Dendy, 1905). *Turkish Journal of Fisheries and Aquatic Sciences, 12*, 67–71.

Lackland, D. T., & Weber, M. A. (2015). Global burden of cardiovascular disease and stroke: Hypertension at the core. *Canadian Journal of Cardiology, 31*, 569–571.

Lam, K. S. (2007). New aspects of natural products in drug discovery. *Trends in Microbiology, 15*, 279–289.

Laport, M. S., Marinho, P. R., da Silva Santos, O. C., de Almeida, P., Romanos, M. T., Muricy, G., Brito, M. A., & Giambiagi-deMarval, M. (2012). Antimicrobial activity of marine sponges against coagulase-negative staphylococci isolated from bovine mastitis. *Veterinary Microbiology, 155*, 362–368.

Lawes, C. M., Vander Hoorn, S., & Rodgers, A. (2008). Global burden of blood-pressure-related disease, 2001. *The Lancet, 371*, 1513–1518.

Layam, A., & Reddy, C. L. (2006). Antidiabetic property of spirulina. *Diabetologia Croatica, 35*, 29–33.

Lee, E. H., Park, J. E., Choi, Y. J., Huh, K. B., & Kim, W. Y. (2008). A randomized study to establish the effects of Spirulina in type 2 diabetes mellitus patients. *Nutrition Research and Practice, 2*, 295–300.

Li, J., Zhu, D., Niu, J., Shen, S., & Wang, G. (2011). An economic assessment of astaxanthin production by large scale cultivation of *Haematococcus pluvialis*. *Biotechnology Advances, 29*, 568–574.

Li, X. C., Babu, K. S., Jacob, M., Rao, R. R., Agarwal, A. K., Newman, D. J., & Clark, A. M. (2012). Antifungal compounds from four marine sponges. *Planta Medica, 78*, I33.

Li, Y. X., Himaya, S. W., & Kim, S. K. (2013). Triterpenoids of marine origin as anti-cancer agents. *Molecules, 18*, 7886–7909.

Li, Y. X., Wijesekara, I., Li, Y., & Kim, S. K. (2011). Phlorotannins as bioactive agents from brown algae. *Process Biochemistry, 46*, 2219–2224.

Liu, D., Yang, A., Wu, C., Guo, P., Proksch, P., & Lin, W. (2014). Lipid-lowering effects of farnesylquinone and related analogues from the marine-derived *Streptomyces nitrosporeus*. *Bioorganic & Medicinal Chemistry Letters, 24*, 5288–5293.

Maeda, H., Hosokawa, M., Sashima, T., & Miyashita, K. (2007). Dietary combination of fucoxanthin and fish oil attenuates the weight gain of white adipose tissue and decreases blood glucose in obese/diabetic KK-Ay mice. *Journal of Agricultural and Food Chemistry, 55*, 7701–7706.

Maeda, H., Tsukui, T., Sashima, T., Hosokawa, M., & Miyashita, K. (2008). Seaweed carotenoid, fucoxanthin, as a multi-functional nutrient. *Asia Pacific Journal of Clinical Nutrition, 17*, 196–199.

Malve, H. (2016). Exploring the ocean for new drug developments: Marine pharmacology. *Journal of Pharmacy and Bioallied Sciences, 8*, 83.

Malviya, N., Jain, S., & Malviya, S. A. (2010). Antidiabetic potential of medicinal plants. *Acta Poloniae Pharmaceutica, 67*, 113–118.

Martínez-García, M., Díaz-Valdés, M., Ramos-Esplá, A., Salvador, N., Lopez, P., Larriba, E., & Antón, J. (2007). Cytotoxicity of the ascidian *Cystodytes dellechiajei* against tumor cells and study of the involvement of associated microbiota in the production of cytotoxic compounds. *Marine Drugs, 5*, 52–70.

Mayer, A. M., Glaser, K. B., Cuevas, C., Jacobs, R. S., Kem, W., Little, R. D., McIntosh, J. M., Newman, D. J., Potts, B. C., & Shuster, D. E. (2010). The odyssey of marine pharmaceuticals: a current pipeline perspective. *Trends in Pharmacological Sciences, 31*, 255–265.

Menaa, F., Wijesinghe, U., Thiripuranathar, G., Althobaiti, N. A., Albalawi, A. E., Khan, B. A., & Menaa, B. (2021). Marine algae-derived bioactive compounds: A new wave of nanodrugs? *Marine Drugs, 19*, 484.

Molinski, T. F. (2004). Antifungal compounds from marine organisms. *Anti-Infective Agents in Medicinal Chemistry, 3*, 197–220.

Nagaoka, S., Shimizu, K., Kaneko, H., Shibayama, F., Morikawa, K., Kanamaru, Y., Otsuka, A., Hirahashi, T., & Kato, T. (2005). A novel protein C-phycocyanin plays a crucial role in the hypocholesterolemic action of Spirulina platensis concentrate in rats. *Journal of Nutrition, 135*, 2425–2430.

Nicholas, G. M., Li, R., MacMillan, J. B., & Molinski, T. F. (2002). Antifungal activity of bifunctional sphingolipids: intramolecular synergism within long-chain α, ω-bis-aminoalcohols. *Bioorganic & Medicinal Chemistry Letters, 12*, 2159–2162.

Nigam, M., Suleria, H. A., Farzaei, M. H., & Mishra, A. P. (2019). Marine anticancer drugs and their relevant targets: a treasure from the ocean. *DARU Journal of Pharmaceutical Sciences, 27*, 491–515.

Noguchi, N., Konishi, F., Kumamoto, S., Maruyama, I., Ando, Y., & Yanagita, T. (2013). Beneficial effects of Chlorella on glucose and lipid metabolism in obese rodents on a high-fat diet. *Obesity Research & Clinical Practice, 7*, 95–105.

Norris, R., Harnedy, P. A., & FitzGerald, R. J. (2013). Antihypertensive peptides from marine sources. In *Bioactive Compounds from Marine Foods: Plant and Animal Sources* (pp. 25–56). John Wiley & Sons.

Oh, K. B., Mar, W., Kim, S., Kim, J. Y., Oh, M. N., Kim, J. G., Shin, D., Sim, C. J., & Shin, J. (2005). Bis (indole) alkaloids as sortase A inhibitors from the sponge Spongosorites sp. *Bioorganic & Medicinal Chemistry, 15*, 4927–4931.

Olano, C., Méndez, C., & Salas, J. A. (2009). Antitumor compounds from marine actinomycetes. *Marine Drugs, 7*, 210–248.

Pandey, S., Sree, A., Dash, S. S., Sethi, D. P., & Chowdhury, L. (2013). Diversity of marine bacteria producing beta-glucosidase inhibitors. *Microbial Cell Factories, 12*, 1–7.

Peng, J., Yuan, J. P., Wu, C. F., & Wang, J. H. (2011). Fucoxanthin, a marine carotenoid present in brown seaweeds and diatoms: metabolism and bioactivities relevant to human health. *Marine Drugs, 9*, 1806–1828.

Perdicaris, S., Vlachogianni, T., & Valavanidis, A. (2013). Bioactive natural substances from marine sponges: new developments and prospects for future pharmaceuticals. *Natural Product Chemistry & Research, 1*, 1–8.

Pettit, G. R., Knight, J. C., Collins, J. C., Herald, D. L., Pettit, R. K., Boyd, M. R., & Young, V. G. (2000). Antineoplastic agents 430. Isolation and structure of cribrostatins 3, 4, and 5 from the Republic of Maldives Cribrochalina species. *Journal of Natural Products, 63*, 793–798.

Purushottama, G. B., Venkateshvaran, K., Pani Prasad, K., & Nalini, P. (2009). Bioactivities of extracts from the marine sponge *Halichondria panicea*. *Journal of Venomous Animals and Toxins Including Tropical Diseases, 15*, 444–459.

Rudi, A., Evan, T., Aknin, M., & Kashman, Y. (2000). Polycitone B and prepolycitrin A: two novel alkaloids from the marine ascidian *Polycitor africanus*. *Journal of Natural Products, 63*, 832–833.

Samarakoon, K., & Jeon, Y. J. (2012). Bio-functionalities of proteins derived from marine algae—A review. *Food Research International, 48*, 948–960.

Samuels, R., Mani, U. V., Iyer, U. M., & Nayak, U. S. (2002). Hypocholesterolemic effect of Spirulina in patients with hyperlipidemic nephrotic syndrome. *Journal of Medicinal Food, 5*, 91–96.

Sansone, C., & Brunet, C. (2019). Promises and challenges of microalgal antioxidant production. *Antioxidants, 8*, 199.

Sansone, C., Brunet, C., Noonan, D. M., & Albini, A. (2020). Marine algal antioxidants as potential vectors for controlling viral diseases. *Antioxidants, 9*, 392.

Senthilkumar, T., & Ashokkumar, N. (2012). Impact of *Chlorella pyrenoidosa* on the attenuation of hyperglycemia-mediated oxidative stress and protection of kidney tissue in streptozotocin-cadmium induced diabetic nephropathic rats. *Biomedical and Preventive Nutrition, 1*, 125–131.

Seo, C., Sohn, J. H., Oh, H., Kim, B. Y., & Ahn, J. S. (2009). Isolation of the protein tyrosine phosphatase 1B inhibitory metabolite from the marine-derived fungus *Cosmospora* sp. SF-5060. *Bioorganic & Medicinal Chemistry Letters, 19*, 6095–6097.

Sheih, I. C., Fang, T. J., & Wu, T. K. (2009). Isolation and characterization of a novel angiotensin I-converting enzyme (ACE) inhibitory peptide from the algae protein waste. *Food Chemistry, 115*, 279–284.

Shibata, S., Oda, K., Onodera-Masuoka, N., Matsubara, S., Kikuchi-Hayakawa, H., Ishikawa, F., Iwabuchi, A., & Sansawa, H. (2001). Hypocholesterolemic effect of indigestible fraction of *Chlorella regularis* in cholesterol-fed rats. *Journal of Nutritional Science and Vitaminology, 47*, 373–377.

Sima, P., & Vetvicka, V. (2011). Bioactive substances with anti-neoplastic efficacy from marine invertebrates: Porifera and Coelenterata. *World Journal of Clinical Oncology, 2*, 355.

Singh, H., Parida, A., Debbarma, K., Ray, D. P., & Banerjee, P. (2020). Common marine organisms: A novel source of medicinal compounds. *International Journal of Bioresource Science, 7*, 39–49.

Sperstad, S. V., Haug, T., Blencke, H. M., Styrvold, O. B., Li, C., & Stensvåg, K. (2011). Antimicrobial peptides from marine invertebrates: challenges and perspectives in marine antimicrobial peptide discovery. *Biotechnology Advances, 29*, 519–530.

Stevenson, C. S., Capper, E. A., Roshak, A. K., Marquez, B., Eichman, C., Jackson, J. R., Mattern, M., Gerwick, W. H., Jacobs, R. S., & Marshall, L. A. (2002). The identification and characterization of the marine natural product scytonemin as a novel antiproliferative pharmacophore. *Journal of Pharmacology and Experimental Therapeutics, 303*, 858–866.

Stevenson, C. S., Capper, E. A., Roshak, A. K., Marquez, B., Grace, K., Gerwick, W. H., Jacobs, R. S., & Marshall, L. A. (2002). Scytonemin—A marine natural product inhibitor of kinases key in hyperproliferative inflammatory diseases. *Inflammation Research, 51*, 112.

Stirk, W. A., Ördög, V., Van Staden, J., & Jäger, K. (2002). Cytokinin-and auxin-like activity in Cyanophyta and microalgae. *Journal of Applied Phycology, 14*, 215–221.

Sun, Z., Peng, X., Liu, J., Fan, K. W., Wang, M., & Chen, F. (2010). Inhibitory effects of microalgal extracts on the formation of advanced glycation end products (AGEs). *Food Chemistry, 120*, 261–267.

Tonks, N. K. (2006). Protein tyrosine phosphatases: From genes, to function, to disease. *Nature Reviews Molecular Cell Biology, 7*, 833–846.

Uranga, R. M., & Keller, J. N. (2010). Diet and age interactions with regards to cholesterol regulation and brain pathogenesis. *Current Gerontology and Geriatrics Research, 2010*, 1–11.

Varijakzhan, D., Loh, J. Y., Yap, W. S., Yusoff, K., Seboussi, R., Lim, S. H., Lai, K. S., & Chong, C. M. (2021). Bioactive compounds from marine sponges: Fundamentals and applications. *Marine Drugs, 19*, 246.

Vecina, J. F., Oliveira, A. G., Araujo, T. G., Baggio, S. R., Torello, C. O., Saad, M. J., & de Souza Queiroz, M. L. (2014). Chlorella modulates insulin signaling pathway and prevents high-fat diet-induced insulin resistance in mice. *Life Sciences, 95*, 45–52.

Wang, W., Wang, S. X., & Guan, H. S. (2012). The antiviral activities and mechanisms of marine polysaccharides: an overview. *Marine Drugs, 10*, 2795–2816.

Wijesekara, I., Yoon, N. Y., & Kim, S. K. (2010). Phlorotannins from *Ecklonia cava* (Phaeophyceae): Biological activities and potential health benefits. *Biofactors, 36*, 408–414.

Yang, S. K., Yusoff, K., Ajat, M., Thomas, W., Abushelaibi, A., Akseer, R., Lim, S. H., & Lai, K. S. (2019). Disruption of KPC-producing *Klebsiella pneumoniae* membrane via induction of oxidative stress by cinnamon bark (*Cinnamomum verum* J. Presl) essential oil. *PLoS One, 14*, e0214326.

Zhang, L., Xue, F., & Gai, R. (2001). Screening for β-glucosidase inhibitors from marine invertebrates extracts. *Chinese Journal of Marine Drugs, 27*, 11–14.

Zhao, C., Wu, Y., Yang, C., Liu, B., & Huang, Y. (2015). Hypotensive, hypoglycaemic and hypolipidaemic effects of bioactive compounds from microalgae and marine micro-organisms. *International Journal of Food Science & Technology, 50*, 1705–1717.

Zhao, K., Ding, M., Cao, H., & Cao, Z. X. (2012). In-vitro metabolism of glycyrrhetinic acid by human and rat liver microsomes and its interactions with six CYP substrates. *Journal of Pharmacy and Pharmacology, 64*, 1445–1451.

Index

B

C

N

O

P

Q

R

S

T

U

V